Y0-BRD-755

# THE PESTICIDE MANUAL

---

A World Compendium

# THE PESTICIDE MANUAL

## A World Compendium

SEVENTH EDITION

Editor CHARLES R. WORTHING, B.Sc., M.A., D.Phil.

Assistant Editor S. BARRIE WALKER, L.R.S.C., M.I.Inf.Sc.

Published by The British Crop Protection Council

| | |
|---|---|
| *First published* | *1968* |
| *Second edition* | *1971* |
| *Third edition* | *1972* |
| *Fourth edition* | *1974* |
| *Fifth edition* | *1977* |
| *Sixth edition* | *1979* |
| *Reprint* | *1980* |
| *Seventh edition* | *1983* |

British Library Cataloguing in Publication Data

The Pesticide Manual — 7th ed.
1. Pesticides—Handbooks, manuals, etc.
I Worthing, Charles R. II Walker, S. Barrie
III British Crop Protection Council
632′-95′0212 58951
ISBN 0-901436-77-1

*Printed in Great Britain by The Lavenham Press Limited*
*Lavenham, Suffolk*

# CONTENTS

# PREFACE

The continuing introduction of pesticidal chemicals and microbial agents, withdrawal of some earlier compounds, the adoption of many additional official common names and of some changes in the rules of chemical nomenclature recommended by the International Union of Pure and Applied Chemistry make this a suitable time to revise the *Pesticide Manual.*

The 6th edition published in 1979 and its reprint in 1980 have sold exceptionally well. Since then the Advisory Editorial Board has given considerable attention both to improving the information given in the text and to its presentation. Some of the changes made are obvious, others are less conspicuous.

The most significant change does not immediately affect the reader. The entire contents of the *Pesticide Manual* have been entered on a computer at the Commonwealth Agricultural Bureaux, Farnham Royal. The printed text and indexes have then been generated from the computer records. When necessary, individual entries can easily be retrieved and revised so that the preparation of future editions will be a much simpler process.

The intention has been to include all chemical and microbial agents used as active components of products to control crop pests and diseases, animal ectoparasites, and pests in public health. Details of herbicides and substances making them better able to be tolerated by crop plants are also included, together with plant growth regulators, pest repellents and synergists. Manufacturers were asked to update entries from previous editions, to add new compounds and to provide more precise values for physical properties and toxicological data. Thus phrases such as 'sparingly soluble in water' have often been replaced by expressions of the type 'solubility 1 mg/l water' or 'solubility $<1$ mg/l water'. Further progress can still be made, but in this regard this edition is a marked improvement over previous ones. Compounds of historic interest only, or those superseded by others of greater potential have been placed in a separate section but with fewer details.

Common names approved by the British Standards Insitution are correct up to May 1983; the French in addition to the English spellings of the common names approved by the International Organization for Standardization are included, as are modifications in the definition of some common names. Interpretations of the latest rules of chemical nomenclature have been incorporated. In general, methods of manufacture have been omitted. Methods are given in some patents and in *Pesticide Manufacturing and Toxic Materials Control Encyclopaedia.* The Advisory Editorial Board has evidence that, in some cases, methods other than those stated in previous editions are being used; the subject is also a sensitive one commercially. Fewer details of analytical methods are included than in previous editions, instead emphasis has been given to references on collaborative or other well-tried methods—for instance those recommended by the Collaborative International Pesticides Council Ltd or those listed in *Recommendations for Methods of Analysis for Pesticide Residues.* Chemical structures have been redrawn and presented more uniformly. No attempt has been made to include specific details, residue tolerances for individual crops, minimum intervals between treatment and harvest, or allowed uses. These details vary from country to country.

The collection of basic technical information on the chemicals included here is a task requiring the collaboration of the industrial laboratories in which they were developed or are being manufactured. In general, the response from manufacturers for the required information has been excellent though a prompter reply would have been appreciated in some cases. It is gratifying to add entries from additional manufacturers and we hope this trend will continue.

It is impossible to list here the many friends, spread over several continents, who have helped in supplying details or in placing us in contact with firms that have changed their address or ownership. To them and to the readers who notified us of errors in previous editions, we say a very sincere 'Thank you'.

In particular we wish to thank The Advisory Editorial Board for guidance and help, Dr E. K. Woodford, Managing Editor of The British Crop Protection Council (BCPC), for his patience and understanding, and Dr D. Rudd-Jones, Chairman of the Publications Committee of the BCPC, for his sympathetic help and support. We greatly appreciate the collaboration of the Systems Group of the Commonwealth Agricultural Bureaux, especially that of P. G. Beckingsale for his invaluable advice on entering data on the computer. We also thank M. J. Bone for expert presentation of chemical structures, Dr J. N. Davies for checking the original computer printouts of the entries, D. G. Sweeney for advice on the preparation of a computerised list of actual and potential entries and C. Waterhouse, of the Laboratory of the Government Chemist, for expert interpretation of the complicated rules of chemical nomenclature; and last, but far from least, the skill of Mrs W. E. Cossins, Mrs D. J. Crane, Mrs J. van de Poll and Mrs P. Uprichard for typing the entries and entering the records on the computer.

As in the 6th edition, four indexes are included (a) for the Wiswesser Line-Formulae, (b) for Molecular Formulae, (c) for code numbers given to the compounds by the manufacturers, licensees or official bodies such as WHO or USDA and (d) a name index covering chemical names, recognised common names, trade marks, and trivial names. Note that all these indexes refer to the entry numbers, *not* to the page on which the entry is sited.

We must stress that any factual errors, arising during transcription of the information supplied by the manufacturers are our responsibility. We should be grateful if readers would draw our attention to any errors or omissions so they can be corrected in the next edition.

Charles R. Worthing *Editor*
S. Barrie Walker *Assistant Editor*

Every effort has been made to ensure that the statements made in this Manual are correct but neither The British Crop Protection Council nor the Editors accept responsibility for any loss, damage or other accident arising from any error in the Manual.

## ABBREVIATIONS

The following abbreviations have been used, some being SI units.

ACS ..... American Chemical Society

ADI ..... acceptable daily intake

a.e. ...... acid equivalent — active ingredient expressed in terms of parent acid

AG ...... Aktiengesellschaft (Company)

a.i. ...... active ingredient

ANSI .... American National Standards Institute

AOAC ... Association of Official Analytical Chemists or, before 1966, Association of Official Agricultural Chemists

*AOAC Methods*.. Official Methods of Analysis of The Association of Official Analytical Chemists

ATP ..... Austrian Patent

AUP .... Australian Patent

BCPC ... British Crop Protection Council

BEP ..... Belgian Patent

BIOS .... British Intelligence Objective Sub-Committee

b.p. ..... boiling point at stated pressure

BPC ..... British Pharmocopoeia Commission

BSI ...... British Standards Institution

B.V. ..... Beperkt Vennootschap (Limited)

c. ....... *circa* (about)

*C.A.* ..... *Chemical Abstracts*

CAP ..... Canadian Patent

*cf* ........ compare

CHP ..... Swiss Patent

CIPAC .. Collaborative International Pesticides Analytical Committee Limited

Co. ...... Company

COLUMA . Comité de Lutte Contre les Mauvaises Herbes

Corp. .... Corporation

d ........ day(s)

$d^t_x$ ....... specific gravity (density of compound at t °C compared to that of water at x °C)

DDRP ... Democratic Republic of Germany Patent

DEAS ... Deutsche Auslegeschrift (an intermediate stage towards DEP which retains the same number)

decomp. .. with decomposition

DEOS ... Deutsche Offenlegungsschrift (the first stage towards DEP which retains the same number)

DEP ..... Federal Republic of Germany Patent

d.p. ..... dispersible powder

DRP ..... German Patent (before 1945)

e.c. ...... emulsifiable concentrate

| | |
|---|---|
| ECD ..... | electron-capture detection |
| ed. ...... | editor |
| Ed. ...... | Edition |
| *e.g.* ...... | for example |
| E-ISO .... | ISO name (English spelling) |
| EPA ..... | Environmental Protection Agency (of USA) |
| EPPO .... | European and Mediterranean Plant Protection Organisation |
| ESA ..... | Entomological Society of America |
| *et al.* ..... | and others (authors) |
| EUP ..... | European Patent (retains same number) |
| EUPA .... | European Patent Application |
| EWRC ... | European Weed Research Council |
| FAO ..... | Food and Agriculture Organization (of the United Nations) |
| FID ..... | flame-ionisation detection |
| F-ISO .... | ISO name (French spelling) |
| FP ...... | French patent |
| f.p. ...... | freezing point |
| FPD ..... | flame-photometric detection |
| FTP ..... | flame thermionic detection |
| g ........ | gram(s) |
| GBP ..... | British patent |
| gc-ms .... | combined gas chromatography-mass spectrometry |
| GIFAP ... | Groupement International des Associations Nationales de Fabricants de Produits Agrochimiques |
| glc ....... | gas-liquid chromatography |
| h ........ | hour(s) |
| ha ....... | hectare(s) ($10^4 m^2$) |
| HMSO ... | Her Majesty's Stationery Office |
| hplc ..... | high performance liquid chromatography (also known as high pressure liquid chromatography) |
| *ibid.* ..... | in the journal last mentioned |
| *idem*....... | by the author(s) last mentioned |
| *i.e.* ...... | that is |
| Inc. ...... | Incorporated |
| i.r. ...... | infrared |
| ISO ...... | International Standardization Organization |
| ITP ...... | Italian Patent |
| ITPA .... | Italian Patent Application |
| i.u. ...... | international unit (measure of activity of micro-organisms) |
| IUPAC ... | International Union of Pure and Applied Chemistry |
| JMAF ... | Japanese Ministry for Agriculture, Forestry and Fisheries |
| JMPR ... | Joint meeting of the FAO Panel of Experts on Pesticide Residues and the Environment and the WHO Expert Group on Pesticide Residues |

JPP ...... Japanese Patent

JPPA .... Japanese Patent Application (Kokai)

k ........ kilo, multiplier $10^3$ (1000) for SI units

kg ....... kilogram(s)

kPa ...... 1000 Pa

l ........ litre(s)

$LC_{50}$ ..... concentration required to kill 50% of the test organism

$LD_{50}$ ..... dose required to kill 50% of test organism

Ltd ...... Limited

m ....... metre; milli, multiplier $10^{-3}$ (0.001) for SI units

M ....... mega, multiplier $10^6$ (1000000) for SI units

M ........ molar

MAFF ... Ministry of Agriculture Fisheries and Food (England and Wales)

MCD .... microcoulometric detection

mg ...... milligram(s), $10^{-3}$ g, 0.001g

mm ...... millimetre(s), $10^{-3}$m, 0.001m

*m/m* .... proportion by mass

mmHg ... pressure equivalent to 1 mm of mercury (133.3 Pa)

m.p. ..... melting point

mPa ..... millipascal, $10^{-3}$ Pa (0.001 Pa)

MPa ..... megapascal, $10^6$ Pa (1000000 Pa)

n ........ nano, multiplier $10^{-9}$ for SI units

$n^t_D$ ....... refractive index for the sodium D lines at a temperature of t °C

NEL ..... no-effect level

NLP ..... Netherlands Patent

nm ...... nanometre(s), $10^{-9}$ m

NMR .... nuclear magnetic resonance

nPa ...... nanopascal, $10^{-9}$ Pa

NRDC ... National Research and Development Corporation

N.V. ..... Naamloze Vennotschap (Limited)

OCLALAV Organisation Commune de Lutte Antiacridienne et de Lutte Antiaviaire

o.m. ..... organic matter

OMS ..... Organisation Mondiale de la Santé = WHO

*op. cit.* ... in the book cited previously

p ........ pico, multiplier $10^{-12}$ for SI units

Pa ....... pascal

pH ...... $-\log_{10}$ hydrogen ion concentration

$pK_a$ ...... $-\log_{10}$ acid dissociation constant

*PMn* .... *Pesticide Manual* nth edition

post-em . .. after emergence

pPa ...... picopascal, $10^{-12}$ Pa

pre-em . ... before emergence

Reg. No. *Chemical Abstracts* Registry Number

r.h. ...... relative humidity

s ........ second(s)

S.A. ..... Société Anonyme (Company)

s.c. ....... suspension concentrate ('flowable')

s.p. ....... soluble powder

sp. ....... species (singular)

S.p.A. .... Société par Actions (Company)

spp. ...... species (plural)

t ........ tonne, 1000 kg

TD ...... toxic dose (lowest observed dose producing toxicity in stated species)

tech. ..... technical grade

TID ..... thermionic detection

tlc ....... thin-layer chromatography

UK ...... United Kingdom

ULV ..... ultra-low volume

USA ..... United States of America

USAID .. United States Agency for International Development

USDA ... United States Department of Agriculture

USP ..... United States of America Patent

u.v. ...... ultraviolet

v.p. ...... vapour pressure

WHO .... World Health Organisation (of the United Nations) = OMS

w.p. ..... wettable powder

w.s.c. .... water-soluble concentrate

WSSA ... Weed Science Society of America

y ........ year(s)

$[\alpha]_D^t$ ...... specification rotation (degrees) for sodium D lines at temperature t °C

$\mu$ ........ micro, multiplier ($10^{-6}$) for SI units

$\mu$g ....... microgram ($10^{-6}$ gram)

$\mu$l ....... microlitre ($10^{-6}$ litre)

$\mu$Pa ...... micropascal ($10^{-6}$ pascal)

$\varrho$ ........ density

t °C ...... temperature of t degrees Celsius (formerly Centigrade)

$>$ ....... greater than

$\geqslant$ ....... greater than or equal to

$<$ ....... less than

$\leqslant$ ....... less than or equal to

## THE WISWESSER LINE-FORMULA NOTATION SYMBOLS

All the international atomic symbols are used except K, U, V, W, Y, Cl and Br. Two-letter atomic symbols in organic notations are enclosed between hyphens. Single letters preceded by a blank space indicate ring positions.

Numerals preceded by a space are multipliers of preceding notation suffixes or within ring signs L...J and T...J show the number of multicyclic points in the ring structure.

Numerals not preceded by a space show ring sizes if within the ring signs—elsewhere numerals show the length of internally saturated, unbranched alkyl chains and segments.

Single letters not preceded by a blank space have the following meanings:

A Generic alkyl.
B Boron atom.
C Unbranched carbon atom multiply bonded to an atom other than carbon, or doubly bonded to two other carbon atoms.
D Proposed symbol for a chelate bond and initial symbol of a chelate notation.
E Bromine atom.
F Fluorine atom.
G Chlorine atom.
H When preceded by a locant within ring signs, shows the position of a carbon atom bonded to four other atoms—elsewhere H means hydrogen atom.
I Iodine atom.
J Sign for the end of a ring description.
K Nitrogen atom bonded to more than three other atoms.
L First symbol of a carbocyclin ring notation.
M Imino or imido -NH- group.
N Nitrogen atom, hydrogen free, attached to no more than three other atoms.
O Oxygen atom, hydrogen free; note that Ø represents the numeral zero.
P Phosphorus atom.
Q Hydroxyl group, -OH.
R Benzene ring.
S Sulphur atom.
T First symbol of a heterocyclic ring notation—or within ring signs indicates a ring containing two or more carbon atoms each bonded to four other atoms.
U Double bond; UU shows an acetylenic triple bond.
V Carbonyl connective, -CO- (carbon attached to three other atoms).
W Nonlinear (branching) dioxo group (as in $-NO_2$ or $-SO_2-$).
X Carbon atom attached to four atoms other than hydrogen.
Y Carbon atom attached to three atoms other than hydrogen or doubly bonded oxygen.
Z Amino or amido $-NH_2$ group.
& Punctuation mark showing the end of a side chain—or preceded by a space, sign of ionic salt, addition compound or suffixed information—or within ring signs indicates a ring not containing two or more carbon atoms that are bonded to four other atoms —or following a hyphen, shows certain spiro ring connections.
- Separator or connective or other special uses.
/ Encloses polymer notations; precedes each non-consecutive locant pair.
* (1) Points of attachment in polymer repeat units; (2) coincident atoms in polymer notations; (3) a multiplier symbol in inorganic notations.
. Space-filling symbol for inorganic notations.
Ø Zero.

## NOTES ON THE WISWESSER LINE-FORMULA NOTATION (WLN)

For most of the chemical compounds in this manual, the structures and molecular formulae are shown, together with the Wiswesser Line-Formula Notations. These Notations are strings of symbols constructed by strict rules to provide a compact, unique and unambiguous description of the molecular structure in linear form. The notations can be used in manual and computer-based indexing and retrieval systems.

Although the principles of encoding structures into WLNs require some weeks of training, any chemist can quickly learn to decode most notations. The list of notation symbols opposite will aide this process and the reader may like to practise on the notations in this manual. Salts have been coded using space && after the main structure notation. Thus the sodium salt of an organic acid is coded with the **-Na** atom replaced by **-H**.

For example sodium trifluoroacetate has been coded as QVXFFF &&Na SALT.

This technique helps to bring similar molecules closer together in Index I.

As a first example of a WLN code, consider monuron whose WLN is GR DMVN1&1. From the list opposite, it is seen that **G** is chlorine, **R** a benzene ring, **M** an NH group, **V** a carbonyl, **N** a branched nitrogen, 1 and **1** carbon alkyl chain. The **D** with a space in front indicates a ring position, and the **&** shows the end of the first alkyl chain. When these fragments are put together in the order shown by the notation, the structure is shown to be:

WLN characters

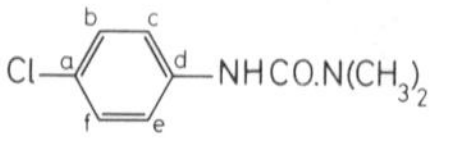

Cl–(a b c d e f benzene ring)–NHCO.N$(CH_3)_2$

G R MVN:

The notations for more complex cyclic compounds begin with a description of the ring system. As a second example take captan, T56 BVNV GUTJ CSXGGG

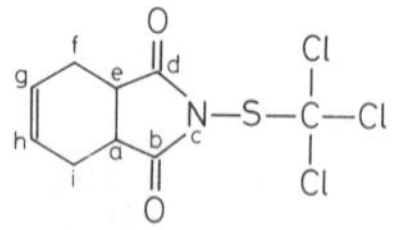

T56 BVNV GUT J CSXGGG

The opening **T** indicates a heterocyclic ring system. The two numerals following show a 5-membered and 6-membered ring fused together. In such bicyclic systems, the ring positions are lettered in order from the fusion point round the smaller and then the larger ring. The letter **B** with a space in front indicates the ring position of the following **V** or carbonyl group; immediately adjacent in the ring is a nitrogen **N** and then another carbonyl **V**. The position of the unsaturation **U** is shown preceded by its locant **(g)** and the second **T** means that the rings are otherwise saturated. The **J** closes the ring description. Finally the position and nature of the substituent group is shown: **S** stands for sulphur, **X** for a four-branched carbon, and the three **G** symbols for the three chlorines attached to it.

As a further example take the compound fluazifop-butyl.

The notation starts by describing the heterocyclic ring (T...J). In this case a pyridine ring (**T6NJ**). Locants are given to the substituents a (**b**) locant to the phenoxy group and an (**e**) locant to the trifluoromethyl group.

As the notation first proceeds into the phenoxy group, a new set of locants are ascribed to the atoms of the benzene ring, starting with an implied **a** locant where the ring is entered and leaving with a (**d**) locant into the **O** atom of the oxypropionate group.

Thus  T6NJ BOR DOY1&VO4

The **Y** atom shows a branching aliphatic chain and the **1&** a terminal methyl, the **&** returning the path to the **Y** atom and proceeding into the carboxylic ester group **VO4.** A further **&** is needed to return the locant path to the original pyridine ring, then completing the notation with the trifluoromethyl group at the (**e**) locant (**EXFFF**).

Thus the full notation T6NJ BOR DOY1&VO4& EXFFF.

In notations where salts or stereochemistry is involved a suffix can be added after a breaking sequence to show extra, non-WLN characteristics of the molecule. Thus in this case &&(RS) FORM

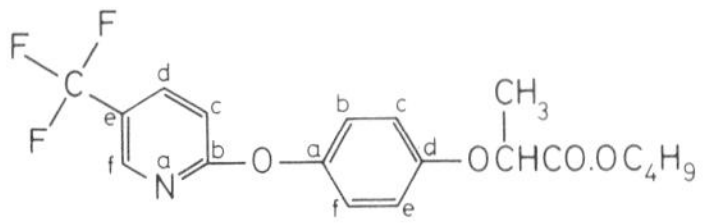

giving a final notation of

T6NJ BOR DOY1&VO4& EXFFF &&(RS) FORM

As a final example consider the compound oxadiazon.

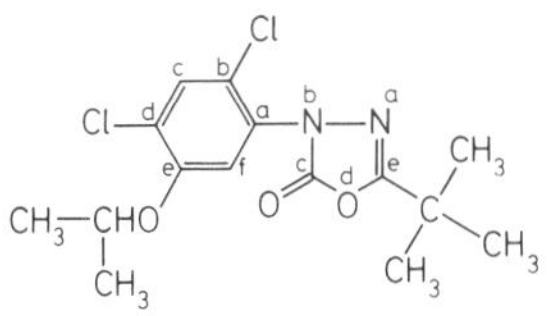

Again the notation starts by describing the ring system. In this example the hetero atoms (including the carbonyl) are adjacent in the ring and are thus strung together (**T5NNVOJ**). Note that bonding cannot be completed until the substituents are shown. Substituents are shown with the appropriate locants, the benzene ring being cited first. The simpler substituents, the chlorines are cited first along with their appropriate locants (**b** and **d**) in the benzene ring. The more complex isopropoxy group is cited last followed by an **&** to return the locant path to the original heteterocyclic ring. The notation is completed by citing the notation characters that form the *tert*-butyl group at the (**e**) locant (**EX1&1&1**).

Thus the final notation is

T5NNVOJ BR BG DG EOY1&1& EX18181

## GUIDE TO THE USE OF PART I OF THE MANUAL

**Entries**

Each compound or biological agent is described on a separate page or pages. Entries are arranged alphabetically in both the Main and Superseded Compounds Sections. The required compound is best located from the entry numbers which are in non-consecutive numerical order in both sections. These numbers are easily obtained from the following indexes:- **Index 1** Wiswesser Line-Formula Notation (explained on pp. xiii-xv), **Index 2** molecular formulae, **Index 3** official or manufacturers' code numbers, **Index 4** common names, trade marks or chemical names.

For ease of reference the information about each entry is grouped under the following sub-heads (see example on the facing page, in which the numbers refer to the paragraph numbers below).

**Heading**

**1** Entry number.

**2** Entry name, preferably the BSI common name. If the compound has no BSI common name then the names given by ISO (English spelling), ANSI, WSSA, BPC or ESA are used in that order of priority. Otherwise a well-known trival name (*e.g.* tar oils), chemical name (*e.g.* copper oxychloride) or IUPAC name (*e.g.* 1,3-dichloropropene) is used.

**3** Chemical structure.

**4** Molecular formula and relative molecular mass.

**5** Wiswesser Line-Formula Notation. A hyphen (-) is part of the notation and must be included but an equals sign ( = ) means run on to the next line *without* a space but omit the sign.

**6** Important derivatives of the title compound.

**Nomenclature and development**

**7** Common names recommended by BSI, E-ISO, F-ISO, ANSI, WSSA, BPC, ESA and JMAF are stated; national name or major spelling variations (excluding the addition or omission of a terminal 'e', accents or the use of 't' for 'th) are listed. The phrases 'draft E-ISO' and 'draft F-ISO' refer to names that, having passed the preliminary enquiry stage, are likely to be adopted but have not been ballotted by member bodies of ISO/TC 81. BSI and ISO names are correct up to at least May 1983.

**8** Preferred and alternative IUPAC names. A hyphen (-) at the end of a line is an essential part of the name and must be included; it, an opening parenthesis ( or bracket [ mean run straight on to the next line *without* a space. An equals sign ( = ) also means run on to the next line *without* a space but omit the = sign. In other cases a word ends on a line.

**9** **Chemical Abstracts** names under rules used for the 9th and 10th Collective Index periods, using the same conventions about continuation at the end of lines as in paragraph 8. *C.A.* Registry Number(s).

**10** Trivial names. Code numbers used by WHO (prefix OMS) and by USDA (prefix ENT, AI3 or AN4).

**11** Type of biological activity with first scientific reference using *Chemical Abstracts* style for journal abbreviations—see *Chemical Abstracts Service Source Index.*

**12** Discovering organisation or person (with protecting patents—which may have expired); code number(s) and principal trade mark(s) of manufacturing and marketing companies. Well-known trade marks of withdrawn formulations are shown in *italics*.

**Properties**

**13** Physical and chemical properties of the active ingredients and/or the technical product; also those of derivatives (salts and esters) used commercially.

**Uses**

**14** Principal uses.

**Toxicology**

**15** Oral and dermal $LD_{50}$ values. Chronic toxicity. Toxicity to wildlife.

**Formulations**

**16** Principal formulations, including mixtures with other active ingredients.

**Analysis**

**17** Methods, generally with relevant references, for product and residue analysis.

**18** Addresses of the leading chemical manufacturers or suppliers are given on pp.

**19** Page number.

**Note.** This specimen entry has been deliberately shortened and modified so that it illustrates the principles without unnecessary duplication. It is *not* the complete entry on picloram.

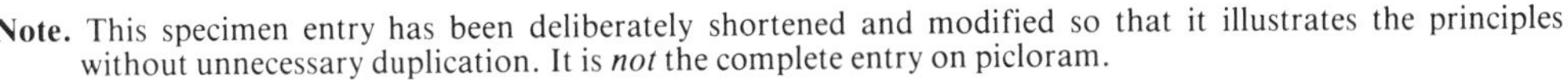

(1)→ 9750 Picloram ←(2)

$C_6H_3Cl_3N_2O_2$ (241.5) ←(4)
T6NJ BVQ CG DZ EG FG ←(5)

(3)→

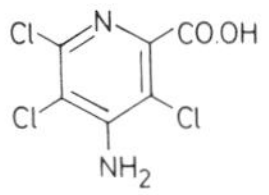

*picloram-potassium* ←(6)
$C_6H_2Cl_3KN_2O_2$ (280.6) ←(4)
T6NJ BVQ CG DZ EG FG &&K SALT ←(5)

**Nomenclature and development**

(7)→ Common name picloram (BSI, E-ISO, ANSI); piclorame (F-ISO).
(8)→ Chemical name (IUPAC) 4-amino-3,5,6-trichloropyridine-2-carboxylic acid; 4-amino-3,5,6-trichloropicolinic acid (I). (*C.A.*) 4-amino-3,5,6-tri= ←(9)
chloro-2-pyridinecarboxylic acid (9CI); (I) (8CI); Reg. No.
(10)→ *[1918-02-1]*. ENT 12 345. Its herbicidal properties were described
(11)→ by J. W. Hamaker *et al.* (*Science*, 1963, **141**, 363). Introduced by ←(12)
Dow Chemical Co. (USP 3 285 925) as code no. 'D-000'; trade mark 'Tordon'.

**Properties**

(13)→ Picloram is a colourless powder, decomposing at *c.*215 °C without melting; v.p. 82 $\mu$Pa at 35 °C. Solubility at 25 °C: 430 mg/l water; 19.8 g/l acetone. It is acidic, p$K_a$ 3.6, and forms water-soluble alkali metal salts, *e.g.* picloram-potassium, Reg. No. *[2545-60-0]*, solubility at 25 °C 400 g/l water.

**Uses**

(14)→ Picloram and its salts are herbicides which are rapidly absorbed by leaves and roots and translocated, accumulating in new growth. Most broad-leaved crops, except crucifers, are sensitive; most grasses are resistant. It is used at 2.2-3.3 kg a.e./ha alone or at 0.3-1.8 kg/ha in combination with 2,4-D against deep-rooted perennials. At the higher doses 50% loss from soil occurs in 30-330 d, depending upon soil conditions.

**Toxicology**

(15)→ Acute oral $LD_{50}$: for rats 8200 mg/kg; for mice 2000-4000 mg/kg. Acute percutaneous $LD_{50}$ for rabbits >4000 mg/kg. In 2-y feeding trials NEL for rats was 150 mg/kg daily. $LC_{50}$ (96-h) for rainbow trout is 19.3 mg/l. $LC_{50}$ for honeybees >1000 mg/kg.

**Formulations**

(16)→ These include: pellets (20 or 100 g a.i./kg); aqueous concentrates (240 g a.e. picloram-potassium/l). Mixtures include aqueous concentrates of salts: picloram + 2,4-D + dichlorprop; picloram + 2,4-D + MCPA + mecoprop.

**Analysis**

(17)→ Product analysis is by hplc (*CIPAC Handbook*, 1983, **1B**, in press). Residues may be determined by glc of derivatives (*AOAC Methods*, 1980, 26.000). Details of methods are available from Dow Chemical Co. ←(18)

(19)→ *page* 441 **9750** ←(1)

# PART I

# Compounds in Use

## Main Entries

This, the main part of *The Pesticide Manual*, lists pesticides in current use.

It includes all chemicals and microbial agents used as active ingredients of products for the control of crop pests and diseases, animal ectoparasites and pests in public health. It also contains plant growth regulators, pest repellants, synergists and substances (crop safeners) that reduce the phytotoxicity of herbicides to crop plants.

$C_4H_{10}NO_3PS$ (183.2)
1VMPO&S1&O1

```
      O
      ||
CH3SPNHCO.CH3
      |
      OCH3
```

**Nomenclature and development.**

Common name acephate (BSI, E-ISO, F-ISO, ANSI, JMAF). Chemical name (IUPAC) *O,S*-dimethyl acetylphosphoramidothioate (I). (*C.A.*) (I) (8 & 9CI); Reg. No. *[30560-19-1]*. ENT 27 822. Its insecticidal properties were described by J. M. Grayson (*Pest Control*, 1972, **40**, 30). Chemical structure-biological activity relationships of analogues were summarised by P. S. Magee (*Residue Rev.*, 1974, **53**, 3). Introduced by Chevron Chemical Co. (USP 3 716 600; 3 845 172) as code no. 'Ortho 12 420'; trade mark 'Orthene'.

**Properties.**

Technical grade acephate (purity 80-90%) is a colourless solid; m.p. 82-89 °C; v.p. 226 μPa at 24 °C; *d* 1.35. Solubility at room temperature: *c.* 650 g/l water; >100 g/l acetone, ethanol; <50 g/l aromatic solvents.

**Uses.**

It is a systemic insecticide of moderate persistence with residual activity lasting *c.* 10-15 d. It is effective against a wide range of aphids, leaf miners, lepidopterous larvae, sawflies and thrips at *c.* 50-100 g a.i./100 l, and is non-phytotoxic on many crop plants.

**Toxicology.**

Acute oral $LD_{50}$: for female rats 866 mg tech./kg, for males 945 mg/kg; for mice 361 mg/kg; for mallard ducks 350 mg/kg; for chickens 852 mg/kg; for ringneck pheasants 140 mg/kg. Acute percutaneous $LD_{50}$ for rabbits >2000 mg/kg; no irritation or sensitisation was observed in skin tests on guinea-pigs. In 2-y feeding trials: dogs showed depression of cholinesterase at 100 mg/kg diet (maximum dose level) but no other significant effect; rats showed depression of cholinesterase but no effect on weight gain or pathological effect at 30 mg/kg diet. No teratogenic, mutagenic or carcinogenic effect was observed. $LC_{50}$ (96-h) is: for rainbow trout >1000 mg/l; for bluegill 2050 mg/l; for largemouth black bass 1725 mg/l; for channel catfish 2230 mg/l; for goldfish 9550 mg/l.

**Formulations.**

These include: s.p. (250, 500 or 750 g a.i./kg); pressurised sprays (2.5 or 10 g/l); granules.

**Analysis.**

Product analysis is by glc (J. B. Leary, *Anal. Methods Pestic. Plant Growth Regul.*, 1973, **7**, 363). Residues may be determined by glc (*idem, ibid.*; *Pestic. Anal. Man.*, 1979, **I**, 201-H, 201-I). Particulars are available from Chevron Chemical Co.

$C_{14}H_7ClF_3NO_5$ (361.7)
WNR BVQ DOR BG DXFFF
*acifluorfen-sodium*
$C_{14}H_6ClF_3NNaO_3$ (383.7)
WNR BVQ DOR BG DXFFF &&Na
SALT

CO.OH
$CF_3$ —O— $NO_2$
Cl

**Nomenclature and development.**

Common name acifluorfen (BSI, E-ISO, ANSI, WSSA); acifluofène (F-ISO). Chemical name (IUPAC) 5-(2-chloro-α,α,α-trifluoro-*p*-tolyloxy)-2-nitrobenzoic acid. (*C.A.*) 5-[2-chloro-4-(trifluoromethyl)phenoxy]-2-nitrobenzoic acid (9CI); Reg. No. *[50594-66-6]* (acid), *[62476-59-9]* (sodium salt). Introduced as a herbicide by Mobil Chemical Co. who no longer manufacture or market it, and developed under licence by Rhône-Poulenc Phytosanitaire and by Rohm & Haas Co. (USP 3 979 437 to Mobil) as code no. 'MC 10 109' (Mobil), 'RH-6201' (Rohm & Haas); trade marks 'Tackle 2S' (Mobil), 'Blazer' (Rohm & Haas).

**Properties.**

Acifluorfen is a light tan to brown solid; m.p. 142-160 °C; v.p. 133 μPa at 20 °C; ρ 1.546 $g/cm^3$. Solubility at 23-25 °C: 120 mg tech./kg water; 500-600 g/kg acetone; 50 g/kg dichloromethane; 400-500 g/kg ethanol; <10 g/kg kerosene, xylene. It decomposes at 235 °C without boiling. No hydrolysis was observed at 40 °C at pH3 to pH9.

**Uses.**

It is being evaluated as a selective post-em. herbicide against broad-leaved weeds in groundnuts, rice, soyabeans and wheat. Promising results have been obtained with post-em. overall treatment in soyabeans at 280-840 g a.i./ha. It is effective against cocklebur, morning glory, ragweed and velvetleaf; also some grass species if sprayed at 1-3 leaf stage.

**Toxicology.**

Acute oral $LD_{50}$: for male rats 2025 mg/kg, for females 1370 mg/kg; for bobwhite quail 325 mg/kg; for mallard duck 2821 mg/kg. Acute percutaneous $LD_{50}$ for male and female rabbits >2000 mg/kg; moderate irritant to their skin, strong irritant to their eyes. Inhalation $LD_{50}$ (4-h) for male and female rats >6.9 mg/l air. Non-mutagenic in Ames and mouse lymphoma assays. $LC_{50}$ (96-h) is: for trout 17 mg/l; for bluegill 62 mg/l.

**Formulations.**

'Tackle 2S', 'Blazer', aqueous solutions of acifluorfen-sodium

$C_{14}H_{20}ClNO_2$ (269.8)
G1VN1O1&R B2 F2

**Nomenclature and development.**

Common name alachlor (BSI, E-ISO, ANSI, WSSA, JMAF); alachlore (F-ISO). Chemical name (IUPAC) 2-chloro-2',6'-diethyl-*N*-methoxymethylacetanilide; α-chloro-2',6'-diethyl-*N*-methoxymethylacetanilide. (*C.A.*) 2-chloro-*N*-(2,6-diethylphenyl)-*N*-(methoxymethyl)acetamide (9CI); 2-chloro-2',6'-diethyl-*N*-(methoxymethyl)acetanilide (8CI); Reg. No. *[15972-60-8]*. Its herbicidal properties were described by R. F. Husted *et al.* (*Proc. North Cent. Weed Control Conf.*, 1966, **21**, 44). Introduced by Monsanto Co. (USP 3 442 945; 3 547 620) as code no. 'CP 50 144'; trade mark 'Lasso'.

**Properties.**

Pure alachlor is a cream-coloured solid; m.p. 39.5-41.5 °C; b.p. 100 °C/0.02 mmHg; v.p. 2.9 mPa at 25 °C; $d^{25}_{15.6}$ 1.133. Solubility at 25 °C: 242 mg/l water; soluble in acetone, benzene, ethanol, ethyl acetate, sparingly soluble in heptane. Decomposes at 105 °C; stable to u.v. radiation; hydrolysed under strongly acid or alkaline conditions.

**Uses.**

It is a selective pre-em. or early post-em. herbicide used at 1.68-4.48 kg/ha to control annual grasses and many broad-leaved weeds in cotton, brassicas, maize, oilseed rape, peanuts, radish, soyabeans and sugarcane. It is mainly absorbed by germinating shoots rather than by roots and is translocated throughout the plant mainly to vegetative rather than to reproductive parts. It is rapidly metabolized in plants. It persists in soil 42-70 d, depending on conditions, loss being by microbial metabolism.

**Toxicology.**

Acute oral $LD_{50}$ for rats 930 mg tech./kg. Acute percutaneous $LD_{50}$: for rabbits 13 300 mg/kg; slight irritant to skin, non-irritant to eyes of rabbits. In 90-d feeding trials no treatment-related effects were observed in rats or dogs $\leqslant$200 mg/kg diet. $LC_{50}$ (8-d) for (10-15-d-old) mallard ducklings and bobwhite quail >5000 mg/kg diet. $LC_{50}$ (96-h) is: for rainbow trout 1.8 mg/l; for bluegill 2.8 mg/l.

**Formulations.**

'Lasso' (480 g a.i./l); granules (150 g/kg). Mixture, s.p. (alachlor + atrazine).

**Analysis.**

Product analysis is by glc with TCD (R. A. Conkin, *Anal. Methods Pestic. Plant Growth Regul.*, 1978, **10**, 255). Residues may be determined by glc with FID (*idem, ibid.*). Details are available from Monsanto Co.

$C_7H_{14}N_2O_2S$ (190.3)
1SX1&1&1UNOVM1

$$CH_3S.\overset{\displaystyle CH_3}{\underset{\displaystyle CH_3}{C}}.CH{=}NO.CO.NHCH_3$$

**Nomenclature and development.**

Common name aldicarb (BSI, E-ISO, ANSI); aldicarbe (F-ISO); *exception* Federal Republic of Germany. Chemical name (IUPAC) 2-methyl-2-(methylthio)propionaldehyde *O*-methylcarbamoyloxime (I). (*C.A.*) 2-methyl-2-(methylthio)propanal *O*-[(methylamino)carbonyl]oxime (9CI); (I) (8CI); Reg. No. *[116-06-3]*. OMS 771, ENT 27 093, AI3-27 093. Its insecticidal properties were described by M. H. J. Weiden *et al.* (*J. Econ. Entomol.*, 1965, **58**, 154). Introduced by the Union Carbide Corp. (USP 3 217 037) as code no. 'UC 21 149'; trade mark 'Temik'.

**Properties.**

Aldicarb forms colourless crystals; m.p. 98-100 °C; v.p. 13 mPa at 25 °C. Solubility at room temperature: 6 g/l water; it is practically insoluble in heptane; but soluble in most organic solvents. It is stable, except to concentrated alkali, is non-corrosive to metal containers and equipment and is non-flammable.

It is a cholinesterase inhibitor and was specifically designed to resemble *O*-acetylcholine structurally (L. K. Payne & M. H. J. Weiden, *J. Agric. Food Chem.*, 1966, **14**, 356). The sulphur atom of the molecule is readily oxidised to form the sulphoxide which is more slowly oxidised to the sulphone (aldoxycarb). The sulphoxide has solubility $>330$ g/l water (D. L. Bull *et al., ibid.*, 1967, **15**, 610).

**Uses.**

Aldicarb is a soil-applied systemic pesticide used against certain mites, nematodes and insects (especially aphids, whiteflies, leaf miners). Seed furrow, band or overall treatments (either pre-plant or at planting) as well as post-em. sidedress treatments at 0.56-11.25 kg a.i./ha are used. Soil moisture is required to release the active chemical from the granules, so irrigation or rainfall should follow application. Uptake by the roots is rapid and the plants are protected for up to 84 d. The persistence of aldicarb and its oxidation products in soil has been studied (J. H. Smelt *et al., Pestic. Sci.*, 1978, **9**, 279, 286, 293).

**Toxicology.**

Acute oral $LD_{50}$ for male rats 0.93 mg tech./kg. Acute percutaneous $LD_{50}$ for male rabbits 5.0 mg/kg. Rats were killed within 5 min by a dust concentration of 200 mg/m$^3$ air. In 2-y feeding trials rats receiving 0.3 mg/kg daily were unaffected. $LC_{50}$ (7-d) for 56-d old bobwhite quail was 2400 mg/kg diet. $LC_{50}$ (96-h) for rainbow trout is 8.8 mg/l.

**Formulations.**

Because of handling hazards only granular formulations in the 0.42-1.41 mm mesh size are available: on corn cob grits (50, 100 or 150 g a.i./kg); in the UK only, on coal (100 g/kg). Formulations based on gypsum have also been developed (50 or 100 g/kg). The formulation '*Ambush*' (20 g/kg corn cob) has been withdrawn.

**Analysis.**

Product analysis is by i.r. spectrometry (*CIPAC Handbook*, 1981, **1A**). Residues in plants, animal tissues, soil and water are determined by glc with FPD after oxidation (aldoxycarb) (R. R. Romine, *Anal. Methods Pestic. Plant Growth Regul.*, 1973, **7**, 147; J. C. Maitlan *et al., J. Agric. Food Chem.*, 1968, **16**, 549; J. H. Smelt *et al., loc. cit.*, pp. 279, 293). Methods based on colorimetry of a derivative were also used (D. P. Johnson *et al., J. Assoc. Off. Anal. Chem.*, 1966, **49**, 399; D. F. Lee & J. A. Rougham, *Analyst (London)*, 1971, **96**, 798).

$C_7H_{14}N_2O_4S$ (222.3)
WS1&X1&1&1UNOVM1

$$CH_3SO_2.\overset{\overset{CH_3}{|}}{\underset{\underset{CH_3}{|}}{C}}CH{=}NO.CO.NHCH_3$$

**Nomenclature and development.**

Common name aldoxycarb (BSI, E-ISO, ANSI); aldoxycarbe (F-ISO). Chemical name (IUPAC) 2-mesyl-2-methylpropionaldehyde *O*-methylcarbamoyloxime; 2-methyl-2-methylsulphonylpropionaldehyde *O*-methylcarbamoyloxime. (*C.A.*) 2-methyl-2-(methylsulfonyl)propanal *O*-[(methylamino)carbonyl]oxime (9CI); 2-methyl-2-methylsulfonylpropionaldehyde *O*-methylcarbamoyloxime; Reg. No. *[1646-88-4]*. ENT 29 261, AI3-29 261, AN4-9. Its insecticidal and nematicidal properties were described by M. H. J. Weiden *et al.* (*J. Econ. Entomol.*, 1965, **58**, 154). Introduced by Union Carbide Corp. (USP 3 217 037) as code no. 'UC 21 865'; trade mark 'Standak'.

**Properties.**

Aldoxycarb is a colourless crystalline solid; m.p. 140-142 °C; v.p. 12 mPa at 25 °C. Solubility at 25 °C: *c.* 9 g/l; 50 g/l acetone; 75 g/l acetonitrile; 41 g/l dichloromethane. Stable to light and heat, decomposed by concentrated alkali. It is corrosive to alloys.

**Uses.**

It is a systemic insecticide and nematicide and a potent inhibitor of cholinesterase. Applied to the soil, sprayed overall or as a band, followed by incorporation before sowing or planting it is effective at 0.5-11.25 kg a.i./ha, depending on pest species. Uptake through the root system protects plants against aphids, thrips, plant bugs and similar pests. For 25-56 d applied directly at 2-3 kg/ha in the water at transplant, it controls *Meloidogyne, Pratylenchus* and *Trichodorus* spp. on, for example, tobacco. Nematicidal and systemic insecticidal activity has been observed with cereals and cotton when used as a dressing at 0.5-2.0 kg/100 kg seed. It can be applied as a foliar spray at 3-4 kg/ha. Its persistence in soil has been studied (J. H. Smelt *et al., Pestic. Sci.*, 1978, **9**, 279).

**Toxicology.**

Acute oral $LD_{50}$: for male rats 26.8 mg tech./kg; for mallard duck 33.5 mg/kg. Acute percutaneous $LD_{50}$ (4 h) on intact skin for male rats 700-1400 mg/kg; there was no skin irritation or eye injury in the standard tests. Inhalation $LC_{50}$ (4 h dust exposure) for rats 120 mg/$m^3$ air. In life-span feeding trials NEL was: for mice 9.6 mg/kg; for rats 2.4 mg/kg. Teratology, reproduction and mutagenic studies were negative. $LC_{50}$ (8-d) was: for mallard duck $>$10 000 mg/kg diet; for bobwhite quail 5706 mg/kg diet. $LC_{50}$ (96-h) is: for trout 40 mg/l; for bluegill 55.5 mg/l.

**Formulations.**

These include: w.p., granules.

**Analysis.**

Product analysis is by i.r. spectrometry (*CIPAC Handbook*, 1980, **1A**,). Residues may be determined by glc with FPD (*AOAC Methods*, 1980, 29.076-29.080; R. T. Krause, *J. Assoc. Off. Anal. Chem.*, 1980, **63**, 1114; R. R. Romine, *Anal. Methods Pestic. Plant Growth Regul.*, 1973, **7**, 147; J. H. Smelt *et al., loc. cit.*).

$C_{12}H_8Cl_6$ (364.9)
L D5 C555 A D- EU JUTJ AG AG BG
IG JG KG

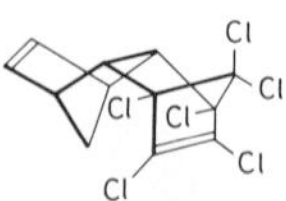


**Nomenclature and development.**

Common name for the pure compound HHDN (BSI, E-ISO, F-ISO, JMAF); *exception* USA; aldrin (for pure compound) (Denmark, USSR). Common name aldrin for a product containing ≥95% HHDN (BSI, E-ISO, ESA, JMAF); aldrine (F-ISO); *exception* (Denmark, USSR see above). Chemical name (IUPAC) (1*R*,4*S*,4a*S*,5*S*,8*R*,8*R*a*R*)-1,2,3,4,10,10-hexachloro-1,4,4a,5,8,8a-hexahydro-1,4:5,8-dimethanonaphthalene; 1,2,3,4,10,10-hexachloro-1,4,4a,5,8,8a-hexahydro-*exo*-1,4-*endo*-5,8-dimethanonaphthalene. (*C.A.*) (1α,4α,4aβ,5α,8α,8aβ)-1,2,3,4,10,10-hexachloro-1,4,4a,5,8,8a-hexahydro-1,4:5,8-dimethanonaphthalene (9CI); *endo,exo*-1,2,3,4,10,10-hexachloro-1,4,4a,5,8,8a-hexahydro-1,4:5,8-dimethanonaphthalene (8CI); Reg. No. *[309-00-2]*. OMS 194, ENT 15 949. Its insecticidal activity was described by C. W. Kearns *et al.* (*J. Econ. Entomol.*, 1949, **42**, 27). Introduced by J. Hyman & Co. and by Shell International Co. Ltd (USP 2 635 977 to Hyman) as code no. 'Compound 118' (to Hyman); trade mark 'Octalene'.

**Properties.**

Pure HHDN is a colourless crystalline solid; m.p. 104-104.5 °C; v.p. 10 μPa at 20 °C. Aldrin (95% HHDN) is a tan to dark brown solid; m.p. 49-60 °C; v.p. 700 μPa at 20 °C. Solubility at 27 °C: 27 μg/l water; >600 g/l acetone, benzene, xylene. Aldrin is stable >200 °C and at 4<pH<8, but oxidising agents and concentrated acids attack the unchlorinated ring. Compatible with most pesticides but can be corrosive because of the slow formation of hydrogen chloride on storage.

**Uses.**

Aldrin is an insecticide highly effective against a range of soil-dwelling pests, its principal use is in agriculture at 0.5-5.0 kg/ha. It acts by contact and ingestion. Also effective against termites being used for wood preservation and to combat ant infestations. Aldrin is readily epoxidised to dieldrin in living organisms.

**Toxicology.**

Acute oral $LD_{50}$: for rats 38-60 mg/kg; for birds 10-500 mg/kg. Acute percutaneous $LD_{50}$ for rats 98 mg/kg. Because aldrin is rapidly epoxidised in living tissue, the long term toxicology of dieldrin to laboratory animals (see 4580) can be regarded as representative of aldrin. $LC_{50}$ (24-h) to fish 0.018-0.089 mg/l. Aldrin/dieldrin were last reviewed at the FAO/WHO JMPR in 1977 when 0-0.0001 mg/kg was confirmed as the estimate of the ADI for man.

**Formulations.**

These include: e.c. (140-480 g/l); w.p. (400-700 g/kg); dust concentrates (750 g/kg). Urea may be added to prevent dehydrochlorination by certain carriers.

**Analysis.**

Technical product and formulation analysis is by i.r. spectrometry (*CIPAC Handbook*, 1970, **1**, 428; 1983, **1B**, in press; *AOAC Methods*, 1980, 6.182-6.187) or by glc (details available from the Shell International Co. Ltd). Residues may be determined by glc with ECD (*Anal. Methods Pestic. Plant Growth Regul.*, 1972, **6**, 268; *AOAC Methods*, 1980, 29.001-29.018; *Pestic. Anal. Man.*, 1979, **1**, 201-A, 201-G, 201-I; *Analyst (London)*, 1979, **104**, 425; P.A.Greve & W. B. F. Grevenstuk, *Meded. Fac. Landbouwwet. Rijksuniv. Gent*, 1975, **40**, 1115; G. M. Telling, *J. Chromatogr.*, 1977, **137**, 405; M. A. Luke *et al.*, *J. Assoc. Off. Anal. Chem.*, 1981, **64**, 1187).

$C_{19}H_{26}O_3$ (302.4)
L5V BUTJ B2U1 C1 DOV- BL3TJ A1 A1
C1UY1&1

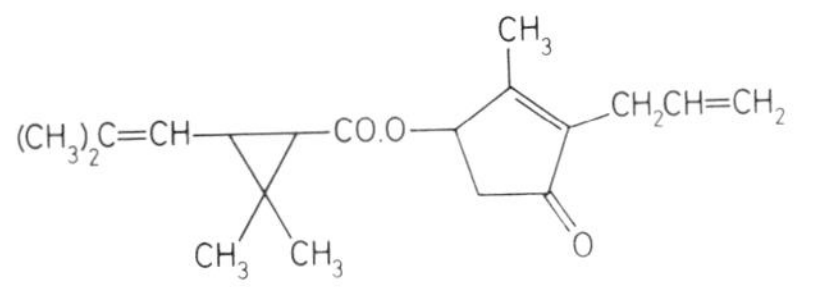

**Nomenclature and development.**

Common name allethrin (BSI, E-ISO, ESA, JMAF); alléthrine (F-ISO); palléthrine (France); *exception* (Federal Republic of Germany). Chemical name (IUPAC) (1*RS*)-3-allyl-2-methyl-4-oxocyclopent-2-enyl (1*RS*,3*RS*; 1*RS*,3*SR*)-2,2-dimethyl-3-(2-methylprop-1-enyl)cyclopropanecarboxylate; (1*RS*)-3-allyl-2-methyl-4-oxocyclopent-2-enyl (1*RS*)-*cis-trans*-2,2-dimethyl-3-(2-methylprop-1-enyl)cyclopropanecarboxylate; (1*RS*)-3-allyl-2-methyl-4-oxocyclopent-2-enyl (1*RS*)-*cis-trans*-chrysanthemate. *(C.A.)* 2-methyl-4-oxo-3-(2-propenyl)-2-cyclopenten-1-yl 2,2-dimethyl-3-(2-methyl-1-propenyl)cyclopropanecarboxylate (9CI); (±)-2,2-dimethyl-3-(2-methylpropenyl)cyclopropanecarboxylic acid ester with 2-allyl-4-hydroxy-3-methyl-2-cyclopenten-1-one (8CI); Reg. No. *[584-79-2]*. OMS 468; ENT 17 510. Its insecticidal properties were described by M. S. Schechter *et al.* (*J. Am. Chem. Soc.*, 1949, **71**, 3165). Its development and properties were reviewed by W. Barthel (*Wld. Rev. Pest. Control*, 1967, **6**, 59). Introduced by Sumitomo Chemical Co. as trade mark 'Pynamin'.

**Properties.**

Technical grade allethrin is a pale yellow to yellow-brown liquid; b.p. 140 °C/0.1 mmHg; $d_4^{25}$ 1.005-1.015; $n_D^{21}$ 1.5070; v.p. 16 mPa at 30 °C. Solubility at room temperature: >1 kg/kg hexane, methanol and xylene.

**Uses.**

Allethrin is a contact insecticide effective against household insect pests, but formulations show higher efficacy than those of natural pyrethrins. In the case of liquid formulations, insecticidal activities of allethrin are enhanced by pyrethrin synergists such as piperonyl butoxide or bis(2,3,3,3-tetrachloropropyl) ether, or *N*-(2-ethylhexyl)-8,9,10,-trinorborn-5-ene-2,3-dicarboximide.

**Toxicology.**

Acute oral $LD_{50}$: for male rats 1100 mg/kg, for females 685 mg/kg; for bobwhite quail 2030 mg/kg. Acute percutaneous $LD_{50}$ for rats >2500 mg/kg. $LC_{50}$ (96-h) is: for steelhead trout 17.5 mg/l; for channel catfish 30.1 mg/l.

**Formulations.**

These include: aerosol concentrates (1-6 g allethrin/l); powders; e.c. (810 g/kg) for mosquito coil; and premixtures (400 and 800 g/kg) for mosquito and housefly mat. Synergists [bis(2,3,3,3-tetrachloropropyl) ether, *N*-(2-ethylhexyl)-8,9,10-trinorborn-5-ene-2,3-dicarboximide, piperonyl butoxide] are included in aerosol concentrates and e.c. formulations. Mixtures include residual insecticides (chlorpyrifos).

**Analysis.**

Product analysis is by glc (*CIPAC Handbook*, 1980, **1A**, 1097). The proportions of stereosomers may be determined by glc of suitable derivatives (A. Murano, *Agric. Biol. Chem.*, 1972, **36**, 2203; F. E. Rickett, *Analyst (London)*, 1973, **98**, 687; A. Horiba *et al.*, *Agric. Biol. Chem.*, 1977, **41**, 2003; 1978, **42**, 671) or by nmr (F. E. Rickett & P. B. Henry, *Analyst (London)*, 1974, **99**, 330). Residues may be determined by glc (D. B. McClellan, *Anal. Methods Pestic., Plant Growth Regul. Food Addit.*, 1964, **2**, 25; *Anal. Methods Pestic. Plant Growth Regul.*, 1972, **6**, 283; J. Sherma, *ibid.*, 1976, **8**, 117). Details from McLaughlin Gormley King Co.

$C_8H_{12}ClNO$ (173.6)
G1VN2U1&2U1

$(CH_2{=}CHCH_2)_2NCO.CH_2Cl$

**Nomenclature and development.**

Common names allidochlor (BSI, E-ISO); CDAA (WSSA, JMAF); *exception* Austria. Chemical name (IUPAC) *N*,*N*-diallyl-2-chloroacetamide (I). (*C.A.*) 2-chloro-*N*,*N*-di-2-propenylacetamide (9CI); (I) (8CI); Reg. No. *[93-71-0]*. Developed from a study of herbicidal substituted 2-chloroacetamides (P. C. Hamm & A. J. Speziale, *J. Agric. Food Chem.*, 1956, **4**, 518) and introduced by Monsanto Co. (USP 2 864 683) as code no. 'CP 6343'; trade mark 'Randox'.

**Properties.**

Pure allidochlor is an amber-coloured oil; b.p. 74 °C/0.3 mmHg; v.p. 1.25 Pa at 20°C; $d^{25}$ 1.088. Solubility at 25 °C: 20 g/kg water; >500 g/kg chlorobenzene, chloroform, cyclohexanone, ethanol, xylene; at 36 °C >200 g/kg hexane. Stable to u.v. light; decomposes at 125 °C. Corrosive to steel.

**Uses.**

It is a selective pre-em. herbicide used to control annual grasses and certain broadleaf weeds in beans (various types), cabbages, celery, certain fruits, maize, onions, certain ornamentrals, sugarcane and sweet potatoes. Incorporated into soil after pre-em. application; also used post-em. and absorbed by cotyledons of grass seedlings. In plants, it is readily metabolised to glycollic acid and diallylamine. Loss from soil is mainly by microbial activity, but soil incorporation is needed because of its volatility.

**Toxicology.**

Acute oral $LD_{50}$ for rats 700 mg/kg. Acute percutaneous $LD_{50}$ for rabbits 830 mg tech./kg; corrosive to their skin and eyes. In 90-d feeding trials no adverse effect was observed in rats and dogs at 200 mg/kg diet (highest rate tested). $LC_{50}$ (8-d) for bobwhite quail and mallard duck >10 000 mg/kg diet. $LC_{50}$ (96-h) is: for rainbow trout 2.0 mg a.i. (as e.c.)/l; for bluegill 4.1 mg (as e.c.)/l.

**Formulations.**

An e.c. (480 g a.i./l); granules (200 g/kg).

**Analysis.**

Product and residue analysis is by glc.

$C_{17}H_{25}NO_5$ (323.4)
L6V BUTJ BY3&UNO2U1 CQ E1 E1
FVO1
*alloxydim-sodium*
$C_{17}H_{24}NNaO_5$ (345.4)
L6V BUTJ BY3&UNO2U1 CQ E1 E1
FVO1 &&Na SALT

**Nomenclature and development.**

Common name alloxydim (BSI, draft E-ISO); alloxydim for sodium salt (JMAF). Chemical name (IUPAC) methyl 3-[1-(allyloxyimino)butyl]-4-hydroxy-6,6-dimethyl-2-oxocyclohex-3-enecarboxylate. (*C.A.*) methyl 2,2-dimethyl-4,6-dioxo-5-[1-[(2-propenyloxy)amino]butylidene]cyclohexanecarboxylate (9CI); Reg. No. *[55634-91-8]* (alloxydim), *[66003-55-2]* (sodium salt). Its herbicidal properties were described by Y. Horono *et al.* (*Meet. Pestic. Sci. Soc. Jpn., 1st*, 1976) and A. Formigoni & Y. Horono (*Meded. Fac. Landbouwwet. Rijksuniv. Gent.*, 1977, **42**, 1597). Its sodium salt [methyl 4-hydroxy-6,6-dimethyl-2-oxo-3-[1-[(2-propenyloxy)imino]butyl]-3-cyclohexene-1-carboxylate sodium salt (C.A. 9CI)] was introduced by Nippon Soda Co Ltd, and later by BASF AG and May & Baker Ltd (JPPA 77-95 636 to Nippon Soda) as code no. 'NP-48' (Nippon Soda for alloxydim), 'NP-48 Na' (Nippon Soda for alloxydim-sodium), 'BAS 90 210H' (BASF for alloxydim-sodium); trade marks 'Kusagard' (Nippon Soda), 'Clout' (May & Baker).

**Properties.**

Alloxydim-sodium is a colourless crystalline solid; m.p. 185.5 °C (decomp.); v.p. 133 μPa at 25 °C. Solubility at 30 °C: >2 kg/kg water; 14 g/kg acetone; 15 g/kg butanone; <4 g/kg cyclohexanone, ethyl acetate, xylene; 1 kg/kg dimethylformamide; 50 g/kg ethanol; 619 g/kg methanol. It is very hygroscopic.

**Uses.**

Alloxydim-sodium is a selective herbicide effective post-em. against grass weeds and volunteer cereals in sugar beet, vegetables and broad-leaved crops at 0.5-1.0 kg a.i./ha. Tank mixes with other herbicides or split applications with herbicides effective against broad-leaved weeds are recommended to increase the range of herbicidal activity.

**Toxicology.**

Acute oral $LD_{50}$: for male rats 2322 mg/kg, for females 2260 mg/kg. Acute percutaneous $LD_{50}$ for male and female rats >5000 mg/kg; for Japanese quail 2970 mg/kg. In 2-y feeding trials no ill-effect was observed in rats and mice. $LC_{50}$ (48-h) for carp is 3500 mg a.i./l.

**Formulations.**

'NP-48 Na 75% SP', s.p. (750 g alloxydim-sodium/kg).

**Analysis.**

Product analysis is by u.v. spectrometry. Residues may be determined by hplc.

$AlP$ (57.96)
.AL..P
*phosphine*
$H_3P$ (34.00)
PHHH

AlP

**Nomenclature and development.**

Chemical name (IUPAC) aluminium phosphide is accepted in lieu of a common name (E-ISO); aluminium phosphide (JMAF). (*C.A.*) aluminum phosphide (8 & 9CI); Reg. No. *[20859-73-8]*. Introduced as a pesticide by the Dr. Werner Freyberg Chemische Fabrik (GBP 461 997; USP 2 117 158).

**Properties.**

It forms dark grey or yellowish crystals; m.p. >1000 °C. Though stable when dry, it reacts with moist air, violently with acids, producing phosphine and is used to generate the latter toxicant.

Phosphine (Reg. No. *[7803-51-2]*) is a gas, with an objectionable ammoniacal odour; b.p. −87.4 °C; f.p. −132.5 °C. Sparingly soluble in water. It is spontaneously flammable in air (due to the presence of traces of other hydrides of phosphorus) with an explosion limit of 26.1-27.1 g/m³.

**Uses.**

Phosphine is highly insecticidal and is produced from aluminium phosphide for fumigation purposes: silos using 1-3 g/t grain; goods at 0.75-1.5 g/m³ sackpile. Dosages vary with the fumigation conditions: (*Trop. Stored Prod. Inf.*, 1972, No.23, p. 6). Fumigation takes 3-10 d and should be undertaken only by trained personnel. Adequate airing after this period is checked by a gas detector. The residue is non-poisonous and is removed in the subsequent screening of the grain.

**Toxicology.**

Phosphine is a potent mammalian poison.

**Formulations.**

Aluminium phosphide formulations are normally based on products containing *c.* 570 g/kg. they include: 'Detia Gas-Ex-B' (W. Freyberg Chemische Fabrik): crepe paper bags (34 g powder, to produce 11 g phosphine); 'Detia Gas-Ex-T', 'Delecia Gastoxin' (Ernst Freyberg), 'Phostoxin' (Degesch AG), 'Celphos' (Excel Ind.): tablets (3 g, producing 1 g phosphine); 'Phostoxin', pellets (0.6 g, containing 55% aluminium phosphide, 40% ammonium carbamate and 5% aluminium oxide). These formulations are inserted by probe into the material to be fumigated, *e.g.* grain, the moisture content of which should be >10%. 'Phostoxin' evolves a non-flammable mixture of phosphine, ammonia and carbon dioxide.

**Analysis.**

Product and residue analysis depends upon determining the phosphine liberated by acid treatment. Measurement is by glc (B. Berck *et al.*, *J. Agric. Food Chem.*, 1970, **18**, 143; T. Dumas, *J. Assoc. Off. Anal. Chem.*, 1978, **61**, 51).

$C_9H_{17}N_5S$ (227.3)
T6N CN ENJ BS1 DMY1&1 FM2

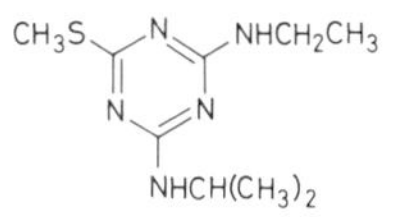

**Nomenclature and development.**

Common name ametryne (BSI, F-ISO); ametryn (E-ISO, ANSI, WSSA, JMAF). Chemical name (IUPAC) *N*-ethyl-*N*'-isopropyl-6-methylthio-1,3,5-triazine-2,4-diyldiamine; 2-ethylamino-4-isopropylamino-6-methylthio-1,3,5-triazine. (*C.A.*) *N*-ethyl-*N*'-(1-methylethyl)-6-(methylthio)-1,3,5-triazine-2,4-diamine (9CI); 2-(ethylamino)-4-( isopropylamino)-6-(methylthio)-*s*-triazine (8CI); Reg. No. *[834-12-8]*. Its herbicidal properties were reported by H. Gysin & E. Knüsli (*Adv. Pest Control Res.*, 1960, **3**, 289). Introduced by J. R. Geigy S.A. (now Ciba-Geigy AG), (GBP 814 948; CHP 337 019) as code no. 'G 34 162'; trade mark 'Gesapax', (in USA) 'Evik'.

**Properties.**

Ametryne is a colourless powder; m.p. 84-86 °C; v.p. 112 μPa at 20 °C; ρ 1.19 g/cm³ at 20 °C. Solubility at 20 °C: 185 g/l water; 500 g/l acetone; 600 g/l dichloromethane; 14 g/l hexane; 450 g/l methanol; 200 g/l octan-1-ol; 400 g/l toluene. Acidic (p$K_a$ 4.1); soluble in acid or alkali. On hydrolysis to the herbicidally inactive 6-hydroxy analogue 50% loss (calculated) occurs at 20 °C in 32 d at pH1, in >200 d at pH13.

**Uses.**

A selective herbicide used pre- and post-em. to control broad-leaved and grass weeds in bananas, citrus, cocoa, coffee, maize, oil palms, pineapples, sugarcane, tea, in potatoes (as a vine desiccant) and in non-crop areas. Review of degradation (H.O. Esser *et al.*, *Herbicides: Chemistry, Degradation and Mode of Action*, 1975, **1**, 129).

**Toxicology.**

Acute oral $LD_{50}$ for rats 1110 mg tech./kg. Acute percutaneous $LD_{50}$ for rats >3100 mg/kg; slight irritant to skin of rabbits, non-irritant to their eyes. In 2-y feeding studies NEL: for rats 1000 mg/kg diet (67 mg/kg daily); for dogs 1000 mg/kg diet (33 mg/kg daily).

**Formulations.**

'Gesapax' 50 & 80 WP, w.p. (500 or 800 g a.i./kg); 'Gesapax' 500 FW, s.c. (500 g/l). Mixtures include: 'Gardopax', ametryne + terbuthylazine (1:2 or 1:1); 'Gesapax combi', Reg. No. *[39324-65-7]*, ametryne + atrazine (1:1); 'Gesapax H', ametryne + 2,4-D (1:1.4); 'Gesapax plus', ametryne + sodium hydrogen methylarsonate (1:2); 'Gesatene', Reg. No. *[8076-68-4]*, ametryne + propetryne (1:1); 'Klinopalm', 2,4-D + sodium hydrogen methylarsonate (1:1.4:3); 'Topazol', ametryne + aminotriazole + ammonium thiocyanate (1.4:1.1:1).

**Analysis.**

Product analysis is by titration with acid or by glc (*CIPAC Handbook*, 1980, **1A**, 1102; FAO Specification CP/61; *AOAC Methods* 1980, 6.431-6.435). Residues are determined by glc (K. Ramsteiner *et al.*, *J. Assoc. Off. Anal. Chem.*, 1974, **57**, 192; E. Knüsli, *Anal. Methods Pestic., Plant Growth Regul. Food Addit.*, 1964, **4**, 13; B. G. Tweedy & R. A. Kahrs, *Anal. Methods Pestic. Plant Growth Regul.*, 1978, **10**, 493; T. H. Byast *et al.*, *Tech. Rep. ARC Weed Res. Organ.*, No. 15 (2nd Ed.), pp. 40, 68).

$C_{11}H_{16}N_2O_2$ (208.3)
1N1&R B1 DOVM1

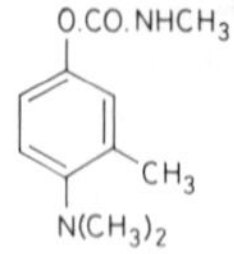

**Nomenclature and development.**

Common name aminocarb (BSI, E-ISO); aminocarbe (F-ISO). Chemical name (IUPAC) 4-dimethylamino-*m*-tolyl methylcarbamate. (*C.A.*) 4-(dimethylamino)-3-methylphenyl methylcarbamate (9CI); 4-(dimethylamino)-*m*-tolyl methylcarbamate (8CI); Reg. No. *[2032-59-9]*. ENT 25 784, OMS 170. Its insecticidal properties reported by G. Unterstenhöfer (*Meded. Landbouwhogesch. Opzoekingsstn. Staat Gent*, 1963, **28**, 758). Introduced by Bayer AG (DEP 1 145 162) as code no. 'Bayer 44 646', 'A 363'; trade mark 'Matacil'.

**Properties.**

A colourless crystalline solid; m.p. 93-94 °C. Slightly soluble in water; moderately soluble in aromatic solvents; soluble in polar organic solvents. Incompatible with highly alkaline pesticides.

**Uses.**

A non-systemic insecticide, mainly used against lepidopterous larvae and other chewing insects; generally recommended at 75 g a.i. / 100 l.

**Toxicology.**

Acute oral $LD_{50}$ for rats 30-50 mg/kg. Acute percutaneous $LD_{50}$ for rats 275 mg/kg. Acute intraperitoneal $LD_{50}$ for rats 21 mg/kg. Rats fed 2-y at 200 mg/kg diet showed no symptom of poisoning. Hazardous to honeybees.

**Formulations.**

A w.p.

**Analysis.**

Product analysis is by u.v. spectroscopy (H. Niessen & H. Frehse, *Pflanzenschutz-Nachr. (Engl. Ed.)*, 1963, **16**, 205). Details from Bayer AG. Residues may be determined by u.v. spectroscopy or by glc after conversion to a suitable derivative (R. J. Argauer, *J. Agric. Food Chem.*, 1969, **17**, 888; E. R. Holden *et al.*, *ibid.*, p.56; L. I. Butler & L. M. McDonough, *ibid.*, 1968, **16**, 403).

$C_2H_4N_4$ (84.08)
T5MN DNJ CZ

**Nomenclature and development.**

Common name amitrole (E-ISO, F-ISO, ANSI, WSSA); ATA (JMAF); the trivial name aminotriazole (BSI, France, USSR) is accepted in lieu of a common name. Chemical name (IUPAC) 1*H*-1,2,4-triazol-3-ylamine; 3-amino-1*H*-1,2,4-triazole. (*C.A.*) 1*H*-1,2,4-triazol-3-amine (9CI); 3-amino-*s*-triazole (8CI); Reg. No. *[61-82-5]*. ENT 25 445. Its herbicidal properties were reported by R. Behrens (*Proc. North Cent. Weed Control Conf.*, 1953, p. 61). Introduced by Amchem Products, Inc. (now Union Carbide Agricultural Products Co., Inc.) (USP 2 670 282) as trade mark 'Weedazol'.

**Properties.**

Aminotriazole is a crystalline powder; m.p. 157-159 °C. Solubility at 25 °C: 280 g/l water; insoluble in acetone, diethyl ether, non-polar solvents. It forms salts with most acids or bases and is a powerful chelating agent. It is corrosive to aluminium, copper and iron.

**Uses.**

It is a non-selective herbicide absorbed by roots and leaves, and translocated. It inhibits chlorophyll formation and regrowth from buds. It is used: around established apple and pear trees between harvest and the following summer; as a non-selective herbicide before planting kale, maize, oilseed rape, potatoes, wheat; on fallow land and in other non-crop situations. Its activity is enhanced by the addition of ammonium thiocyanate. General reviews have been published (E. Kröller, *Residue Rev.*, 1966, **12**, 163; M. C. Carter, *Herbicides: Chemistry, Degradation and Mode of Action*, 1975, **1**, p. 377).

**Toxicology.**

Acute oral $LD_{50}$ for rats 1100-24600 mg/kg. Acute percutaneous $LD_{50}$ for rats >10 000 mg/kg. In 476-d feeding trials rats receiving 50 mg/kg diet suffered no effect on growth or food intake, but the male rats developed an enlarged thyroid after 90 d. Rats fed 120 d at 500 mg/kg diet and returned to normal aminotriazole-free diet 14 d before sacrifice, appeared to have normal thyroids (T. H. Jukes & C. B. Schaeffer, *Science*, 1960, **132**, 296); it is doubtful whether this antithyroid action is carcinogenic (E. B. Ashwood, *J. Am. Med. Assoc.*, 1960, **172**, 1319; T. H. Jukes & C. B. Schaeffer, *loc. cit.*).

**Formulations.**

These include 'Weedazol', s.p. (500 g a.i./kg); 'Weedazol TL', a mixture of aminotriazole + ammonium thiocyanate. Mixtures include: aminotriazole + atrazine; aminotriazole + atrazine + 2,4-D; aminotriazole + atrazine + diuron; aminotriazole + diuron; aminotriazole + MCPA + simazine; aminotriazole + simazine.

**Analysis.**

Product analysis is by acid-base titration or by formation of a silver complex during titration with silver salts (*CIPAC Handbook*, 1983, **1B**, in press); other components of mixtures being determined by the appropriate methods (*loc. cit.*). Residues in soils may be determined by colorimetry (R. A. Herrett & A. J. Linck, *J. Agric. Food Chem.*, 1961, **9**, 466; J. Burke & R. W. Storherr, *J. Assoc. Off. Agric. Chem.*, 1961, **44**, 196; G. L. Sutherland, *Anal. Methods Pestic., Plant Growth Regul. Food Addit.*, 1964, **4**, 17; *Pestic. Anal. Man.*, 1979, **II**; H. Lokke, *J. Chromatogr.*, 1980, **200**, 234).

$C_{19}H_{23}N_3$ (293.4)
1R C1 DNU1N1&1UNR B1 D1

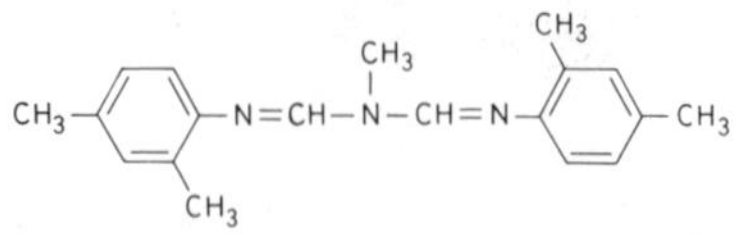

Nomenclature and development.

Common name amitraz (BSI, E-ISO, ANSI, BPC, JMAF); amitraze (F-ISO). Chemical name (IUPAC) *N*-methylbis(2,4-xylyliminomethyl)amine; *N*,*N*-bis(2,4-xylyliminomethyl)methylamine; incorrect name originally used 2-methyl-1,3-di(2,4-xylylimino)-2-azapropane. (*C.A.*) *N*'-(2,4-dimethylphenyl)-*N*-[[(2,4-dimethylphenyl)imino]methyl]-*N*-methylmethaniminamide (9CI); *N*-methyl-*N*'-2,4-xylyl-*N*-(*N*-2,4-xylylformimidoyl)formamidine (8CI); Reg. No. *[33089-61-1]*. OMS 1820, ENT 27 967. Its acaricidal properties were described by B. H. Palmer *et al.* (*Proc. Int. Congr. Acarol. 3rd*, 1971, p.687) for veterinary use and by D. M. Weighton *et al.* (*Meded. Fac. Landbouwwet. Rijksuniv. Gent*, 1972, **37**, 765) for crop use. Introduced by The Boots Co. Ltd (now FBC Limited) (GBP 1 327 935) as code no. 'BTS 27 419'; trade marks 'Mitac' for crop protection, 'Taktic' for animal health (FBC Limited), 'Triatox' (Wellcome).

**Properties.**

It forms colourless monoclinic needles; m.p. 86-86 °C (I. R. Harrison *et al.*, *Pestic. Sci.*, 1972, **3**, 679; 1973, **4**, 901); v.p. 51 μPa at 20 °C. Solubility at room temperature: *c.* 1 mg/l water; >300 g/l acetone, toluene. It is unstable at pH <7 and a slow deterioration of the *moist* compound occurs on prolonged standing.

**Uses.**

It is an acaricide and insectide effective against a wide range of phytopaghous mites and insects. All stages of mites are susceptible. It is effective against *Bemisia tabaci, Pyslla pyricola*, scale insects, mealy bugs and aphids. It also has activity against the eggs of various species of Lepidoptera such as *Heliothis zea*, *Spodoptera littoralis*, *Pectinophora gossypiella*, and *Earias* spp. It is relatively non-toxic to predatory insects.

Its main veterinary uses are against ticks, mites and lice of cattle, dogs, goats, pigs and sheep, including strains resistant to other chemical classes of ixodicide. The compound persists long enough on hair and wool to control all stages of the parasite. The unique repellent action of amitraz causes ticks to withdraw mouthparts rapidly and fall off the host animal.

**Toxicology.**

Acute oral $LD_{50}$: for rats 800 mg/kg; for mice >1600 mg/kg. Acute percutaneous $LD_{50}$ for rats >1600 mg/kg. In 2-y feeding trials no adverse effect was observed in rats receiving 50 mg/kg diet, or in dogs dosed 0.25 mg/kg daily. $LC_{50}$ (8-d) for mallard duck is 7000 mg/kg diet, for Japanese quail 1800 mg/kg diet. Low toxicity to honeybees: $LD_{50}$ by ingestion 12 μg a.i./bee, $LC_{50}$ by direct spray 3.6 g (20% e.c.)/l, residual contact/fumigant >10 g a.i. (as 20% e.c.)/l. $LC_{50}$ (48-h) is: for trout 2.7-4.0 mg/l; for Japanese carp 1.17 mg/l; $LC_{50}$ (96-h) is: for bluegill 1.3 mg/l, for harlequin 3.2-4.2 mg/l.

**Formulations.**

These include: 'Mitac', e.c. (200 g/l); w.p. (500 g/kg) for crop use. 'Tactic', e.c. (125 g/l) for farm animal use. 'Ectodex', e.c. (50 g/l) dog shampoo; d.p. (250 or 500 g/kg).

**Analysis.**

Product analysis is by glc. Residues may be determined by glc. Details of both methods are available from FBC Limited.

$H_6N_2O_3S$ (114.1)
ZSWQ &&$NH_4$ SALT

$NH_2SO_2.ONH_4$

**Nomenclature and development.**

Common name AMS (WSSA); traditional chemical name ammonium sulphamate (E-ISO, JMAF); sulfamate d'ammonium (F-ISO) is accepted in lieu of common name. Chemical name (IUPAC) ammonium sulphamidate. (*C.A.*) monoammonium sulfamate (8 & 9CI); Reg. No. *[773-06-0]*. Introduced as a herbicide by E. I. du Pont de Nemours and Co. (Inc.) and later by Albright and Wilson (Mfg) Ltd (USP 2 277 744 to du Pont) as trade marks 'Ammate' (du Pont), 'Amcide' (Albright & Wilson).

**Properties.**

Pure ammonium sulphamate forms colourless crystals; m.p. 131-132 °C. Solubility at 25 °C: 2.16 kg/kg water; soluble in formamide, glycerol, glycols. Decomposes at 160 °C, the gases acting as a fire retardant. Undergoes negligible hydrolysis at ordinary temperatures and pH, but the rate increases at higher temperatures or under more acid conditions. Corrosive to spray equipment.

**Uses.**

It is a non-selective herbicide applied as a foliar spray of aqueous solution or oil-water emulsion to control woody plants, certain annual and perennial herbaceous weeds. Application of crystals or concentrated solutions to frills, notches or cups cut in bark, controls undesirable trees and prevents resprouting of freshly-cut stumps. Poison ivy in orchards is controlled by localised application.

**Toxicology.**

Acute oral $LD_{50}$ for rats 3900 mg/kg. Repeated application of 50% aqueous solutions to the shaved skin of rats caused no irritation or systemic toxicity. In 105-d feeding trials in rats no adverse effect was observed at 10 000 mg/kg diet; growth inhibition was noted at 20 000 mg/kg diet.

**Formulations.**

'Ammate X'; 'Ammate X-NI' (950 g a.i./kg); 'Amcide', s.p. (970 g/kg).

**Analysis.**

Product analysis is by potentiometric titration with sodium nitrite (W. W. Bowler & E. A. Arnold, *Anal. Chem.*, 1947, **19**, 336). Residues may be determined by colorimetry (H. L. Pease *et al.*, *J. Agric. Food Chem.*, 1966, **14**, 140).

$C_{15}H_{16}N_2O_2$ (256.3)
T6N CNJ EXQR DO1&- AL3TJ

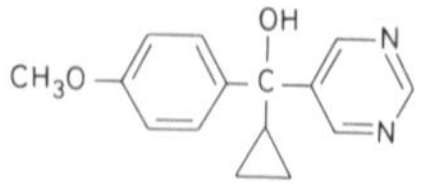

**Nomenclature and development.**

Common name ancymidol (BSI, E-ISO, ANSI); ancymidole (F-ISO). Chemical name (IUPAC) α-cyclopropyl-4-methoxy-α-(pyrimidin-5-yl)benzyl alcohol. (*C.A.*) α-cyclopropyl-α-(4-methoxyphenyl)-5-pyrimidinemethanol (9CI); α-cyclopropyl-α-(*p*-methoxyphenyl)-5-pyrimidinemethanol (8CI); Reg. No. *[12771-68-5]*. Its plant growth regulating properties were described by M. Snel & J. V. Gramlich (*Meded. Fac. Landbouwwet. Rijksuniv. Gent*, 1973, **38**, 1033). Introduced by Elanco Products, a division of Eli Lilly & Co. (GBP 1 218 623) as code no. 'EL-531'; trade marks 'A-Rest', 'Reducymol', '*Quel*' (discontinued).

**Properties.**

Pure ancymidol forms colourless crystals; m.p. 110-111 °C. Solubility at 25 °C: *c.* 650 mg/l water; >250 g/l acetone; 37 g/l hexane; >250 g/l methanol. The technical grade has v.p. 133 μPa at 50 °C.

**Uses.**

It is a plant growth regulator which reduces internode elongation in a wide range of species (E. E. Tschabold *et al., Plant Physiol.*, 1971, **48**, 519); the growth inhibition may be reversed by gibberellic acid (A. C. Leopold, *ibid.*, p. 537). Ancymidol has been evaluated on a range of commercial greenhouse plants including many cultivars of chrysanthemum, dahlia, Easter lily, poinsettia, tulip, and other ornamentals with excellent results. Delay of 1-3 d in flower maturation may occur at high application rates, but flower development is not otherwise affected.

**Toxicology.**

Acute oral $LD_{50}$: for rats 4500 mg/kg; for mice 5000 mg/kg; for chickens, dogs, monkeys >500 mg/kg. In rabbits there was a very slight skin irritation at 200 mg/kg and moderate eye irritation at a single dose of 56 mg. In 90-d feeding trials, rats and dogs receiving 8000 mg/kg diet suffered no ill-effect. $LC_{50}$ values were: for bluegill fingerlings 146 mg/l; rainbow trout fingerlings 55 mg/l; goldfish fingerlings >100 mg/l.

**Formulations.**

Aqueous solution.

**Analysis.**

Product and residue analysis is by glc with FID (R. Frank & E. Q. Day, *Anal. Methods Pestic. Plant Growth Regul.*, 1976, **8**, 475). Methods for the determination of ancymidol in formulations and in soil are available from Eli Lilly & Co.

$C_9H_5Cl_3N_4$ (275.5)
T6N CN ENJ BMR BG& DG FG

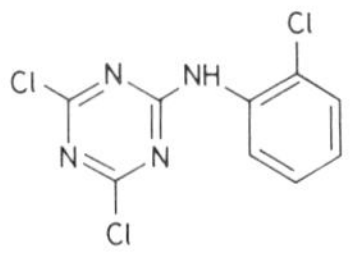

Cl, N, NH, N, N, Cl, Cl

**Nomenclature and development.**

Common name anilazine (BSI, E-ISO, F-ISO); triazine (JMAF). Chemical name (IUPAC) 2-chloro-*N*-(4,6-dichloro-1,3,5-triazin-2-yl)aniline; 2,4-dichloro-6-(2-chloroanilino)-1,3,5-triazine. (*C.A.*) 4,6-dichloro-*N*-(2-chlorophenyl)-1,3,5-triazin-2-amine (9CI); 2,4-dichloro-6-(*o*-chloroanilino)-*s*-triazine (8CI); Reg. No. *[101-05-3]*. ENT 26 058. Its fungicidal properties were described by C. N. Wolf *et al.* (*Science*, 1955, **121**, 61). Introduced by the Ethyl Corp. and later by Nippon Soda Co. Ltd and by Bayer AG (USP 2 720 480 to Ethyl Corp.) as code no. 'B-622' (Ethyl Corp.); trade marks 'Dyrene', 'Direz', 'Kemate' (all to Bayer AG).

**Properties.**

It forms tan crystals; m.p. 159-160 °C. Solubility at 30 °C: practically insoluble in water; 100 g/kg acetone; soluble hydrocarbons and chlorinated hydrocarbons. Stable in neutral and slightly acid media but hydrolysed on heating with alkali.

**Uses.**

It is a protectant fungicide effective against pathogens including *Botrytis, Septoria, Colletotrichum* spp. when applied, at intervals, at 1.7-2.5 g a.i./l. It is ineffective as a seed protectant.

**Toxicology.**

Acute oral $LD_{50}$: for rats 2710 mg/kg; for rabbits 460 mg/kg. Prolonged skin contact may cause irritation. In 2-y feeding trials rats receiving 5000 mg/kg diet suffered no ill-effect.

**Formulations.**

A w.p. (500 g a.i./kg).

**Analysis.**

Product analysis is by hydrolysis and measurement of the chloride liberated (P. F. Kane & K. G. Gillespie, *J. Agric. Food Chem.*, 1960, **8**, 29). Residues may be determined by glc (P. J. Wales & C. E. Mendoza, *J. Assoc. Off. Anal. Chem.*, 1970, **53**, 509; D. MacDougall *et al., Anal. Methods Pestic., Plant Growth Regul. Food Addit.*, 1964, **3**, 79; *Anal. Methods Pestic. Plant Growth Regul.*, 1972, **6**, 564).

$C_{13}H_{19}ClNO_3PS_2$ (367.8)
1Y1&NR DG&VISPS&O1&O1

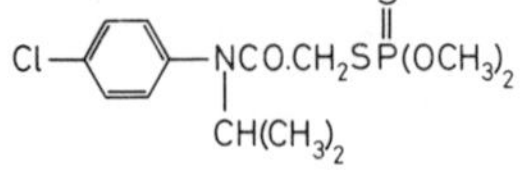

Cl–$C_6H_4$–NCO.$CH_2$SP(S)$(OCH_3)_2$
$CH(CH_3)_2$

**Nomenclature and development.**

Common name anilofos (BSI, draft E-ISO); *exception* (Japan). Chemical name (IUPAC) *S*-4-chloro-*N*-isopropylcarbaniloylmethyl *O,O*-dimethyl phosphorodithioate. (*C.A.*) *S*-[2-[(4-chlorophenyl)(1-methylethyl)amino]-2-oxoethyl] *O,O*-dimethyl phosphorodithioate. Reg. No. *[64249-01-0]*. Its herbicidal properties were described by P. Langeluddeke *et al.* (*Proc. Asian Pacific Weed Sci. Soc. Conf., 8th*, 1981, p. 449). Introduced by Hoechst AG as code no. 'Hoe 30 374'; trade marks 'Arozin', 'Rico'.

**Properties.**

Anilofos is a crystalline solid; m.p. 50.5-52.5 °C; v.p. 2.2 mPa at 60 °C. Solubility at 20 °C: 13.6 mg/l water; $>$1 kg/l acetone, chloroform, toluene; $>$200 g/l benzene, ethanol, ethyl acetate, dichloromethane; 12 g/l hexane.

**Uses.**

It is a herbicide effective against annual graminaceous weeds (*Echinochloa* spp., *Ischaemum rugosum, Leptochloa chiensis, Cyperus* spp., *Fimbristylis* spp.) and sedges (*Scirpus* spp., *Eleocharis acicularis*) in transplanted rice. The weeds are most susceptible in the seedling to 2.5 leaf stage, treatment being 4-12 d (according to climatic region) after the transplanting of paddy rice and using 300-400 g a.i. (as e.c.)/ha or 0.45 g a.i. (as granule)/ha.

**Toxicology.**

Acute oral $LD_{50}$: for male rats 830 mg tech./kg, for females 472 mg/kg; for male Japanese quail 3360 mg/kg, for females 2339 mg/kg. Acute percutaneous $LD_{50}$ for rats $>$2000 mg/kg; slightly irritant to skin and moderately to eyes of rabbits. $LC_{50}$ (96-h) is: for goldfish 4.6 mg/l; for rainbow trout 2.8 mg/l.

**Formulations.**

These include: e.c. (300 g a.i./l); granules (15 or 20 g a.i./kg). Mixtures of anilofos + 2,4-D-isobutyl are available for trial purposes.

$C_{14}H_8O_2$ (208.2)
L C666 BV IVJ

**Nomenclature and development.**

IUPAC name anthraquinone (I) (BSI, E-ISO, F-ISO) is accepted in lieu of a common name. (*C.A.*) 9,10-anthracenedione (9CI); (I) (8CI); Reg. No. *[84-65-1]*. Known for many years as a chemical, its use as a bird repellant was reported by F. Wenkel (*Hoefchen-Briefe (Engl. Ed.)*, 1951, **4**, 227), and it is marketed by Bayer AG as trade mark 'Morkit'.

**Properties.**

It forms yellow-green crystals; m.p. 285 °C, subliming; v.p. <10 Pa at 20 °C; $d_4^{40}$ 1.42-1.44. Solubility at 20 °C: 0.6 mg/l water; 1 g/kg ethanol; 2.6 g/kg benzene; 6.1 g/kg chloroform; 1 g/kg diethyl ether.

**Uses.**

Anthraquinone is used as a seed treatment to deter attack by birds, in particular rooks.

**Toxicology.**

It is non-poisonous, mice being unharmed by single oral doses of 5000 mg/kg.

**Formulations.**

These include: 'Morkit', powder for dry seed treatment; w.p. Combination products.

**Analysis.**

Residues are measured by glc (P. Maini, *J. Chromatogr.*, 1976, **128**, 174). Details are available from Bayer AG.

$C_{11}H_{10}N_2S$ (203.3)
L66J BMYZUS

NHCS.NH$_2$

**Nomenclature and development.**

Common name antu (BSI, E-ISO, F-ISO, JMAF). Chemical name (IUPAC) 1-(1-naphthyl)-2-thiourea (I). (*C.A.*) 1-naphthalenylthiourea (9CI); (I) (8CI); Reg. No. *[86-88-4]*. Trivial name α-naphthylthiourea. Its toxicity to rodents was described by C. F. Richter (*J. Am. Med. Assoc.*, 1945, **129**, 927; *Proc. Soc. Exp. Biol. Med.*, 1946, **63**, 364), a discovery protected by USP 2 390 848.

**Properties.**

Pure antu forms colourless crystals; m.p. 198 °C. The technical grade is a blue-grey powder. Solubility at room temperature: 600 mg/l water; 24.3 g/l acetone; 86 g/l triethyleneglycol. It is stable on exposure to air and to sunlight.

**Uses.**

It is a rodenticide specific for adult *Rattus norvegicus* for which the toxic dose is 6-8 mg/kg. It is less toxic to other *Rattus* spp. and tolerance is developed in rats by repeated administration of sub-lethal doses.

**Toxicology.**

It induces vomiting in dogs. In some countries it has been withdrawn from use because of the carcinogenicity of naphthylamines present as impurities.

**Formulations.**

These include: baits (10-30 g/kg) in suitable protein- or carbohydrate-rich materials; tracking powder (200 g/kg).

**Analysis.**

Product analysis is by reaction with silver nitrate and titration of the liberated nitric acid. (*CIPAC Handbook*, 1970, **1**, 16). Residues may be determined by colorimetry of a derivative (E. Bremais & K. G. Bergner, *Pharm. Zentralhalle*, 1950, **89**, 115).

$As_2O_3$ (197.8)
.AS2.O3

$As_2O_3$

**Nomenclature and development.**

Arsenious acid (JMAF); the traditional chemical name arsenous oxide (E-ISO); oxyde arsenieux (F-ISO) is accepted in lieu of a common name. Chemical name (IUPAC) diarsenic trioxide; arsenic trioxide. (*C.A.*) arsenic oxide ($As_2O_3$) (8 & 9CI); Reg. No. *[1327-53-3]*. Trivial names arsenious oxide,white arsenic. It has been used as a rodenticide since the 16th century.

**Properties.**

Arsenous oxide is a colourless solid which exists in 3 allotropic forms. The amorphous form is unstable, reverting to the octahedral form, m.p. 272 °C, subliming 125-150 °C; the rhombic form has m.p. 312 °C, v.p. 8.8 kPa at 312 °C. Solubility at 16 °C: 17 g/l water; practically insoluble in chloroform, diethyl ether, ethanol. It is stable in air but slowly is oxidised in acid media. It dissolves in alkali to form arsenites.

**Uses.**

Arsenous oxide is used as a rodenticide in baits, containing 100 g/kg soaked wheat or cereal offals, to control *Rattus norvegicus, R. rattus* and *Mus musculus*. It has also been used as a sheep dip to control ectoparasites.

**Toxicology.**

Acute oral $LD_{50}$: for rats 180-200 mg a.i. (in carbohydrate or protein)/kg, 300 mg (in bacon fat)/kg, 20 mg (in aqueous solution)/kg (E. W. Packman *et al., J. Agric. Food Chem.*, 1961, **9**, 271); for mice 34.4-63.5 mg/kg (J. W. E. Harrison, *AMA Arch. Ind. Health*, 1958, **17**, 118). It is extremely toxic to man, the minimum lethal dose is 2 mg/kg (H. O. Calvery, *J. Am. Med. Assoc.*, 1938, **111**, 1722). It is non-cumulative and is eliminated from the animal body in 7-42 d.

**Formulations.**

Paste (103 g/kg).

**Analysis.**

Product analysis is by titration with iodine (*WHO Specifications Insectic.* (2nd Ed.); *AOAC Methods*, 1980, 6.005-6.013). Residues may be determined by the Gutzeit method or by colorimetry (*ibid.*, 25.006-25.020).

$C_8H_{10}N_2O_4S$ (230.2)
ZR DSWMVO1
*asulam-sodium*
$C_8H_9N_2NaO_4S$ (252.2)
ZR DSWMVO1 &&Na SALT

$H_2N-C_6H_4-SO_2.NH.CO.OCH_3$

**Nomenclature and development.**

Common name asulam (BSI, E-ISO, ANSI, WSSA, JMAF); asulame (F-ISO); *exception* (Federal Republic of Germany). Chemical name (IUPAC) methyl sulphanilylcarbamate; methyl 4-aminophenylsulphonylcarbamate; methyl 4-aminobenzenesulphonylcarbamate. (*C.A.*) methyl [(4-aminophenyl)sulfonyl]carbamate (9CI); methyl sulfanilylcarbamate (8CI); Reg. No. *[3337-71-1]* (acid), *[2302-17-2]* (asulam-sodium). Its herbicidal properties were reported by H. J. Cottrell & B. J. Heywood (*Nature (London)*, 1965, **207**, 655). Introduced by May & Baker Ltd (GBP 1 040 541) as code no. 'M&B 9057'; trade mark 'Asulox'.

**Properties.**

Pure asulam forms colourless crystals; m.p. 142-144 °C (decomp.). Solubility at 20-25 °C: 4 g/l water; 300 g/l acetone; <20 g/l hydrocarbons, chlorinated hydrocarbons; 290 g/l methanol. The imido group is acidic (p$K_a$ 4.82), forming water-soluble salts. The calcium and magnesium salts are not precipitated by hard water.

**Uses.**

It is absorbed by leaves and roots causing a slow chlorosis in susceptible plants, and interferes with cell division and expansion. It is used post-em. to control: *Rumex* spp. in pasture and deciduous fruit orchards (1.0-1.7 kg a.i./ha), grasses in tropical tree crops (2.7-3.36 kg/ha), *Pteridium aquilinum* in pastures and forestry (4-6 kg/ha), *Avena fatua* in linseed (1 kg/ha); pre-em. and post-em. to control grasses in sugarcane (3.36-3.64 kg/ha), broad-leaved weeds and grasses in oilseed poppy (3-7 kg/ha).

**Toxicology.**

Acute oral $LD_{50}$ for mice, rabbits, rats >4000 mg asulam-sodium/kg; for chickens, pigeons, quail >2000 mg/kg. Acute percutaneous $LD_{50}$ for rats >1200 mg/kg. In 90-d feeding trials rats receiving 400 mg/kg diet suffered no significant effect. $LC_{50}$ (96-h) for channel catfish, goldfish, rainbow trout >5000 mg/l.

**Formulations.**

'Asulox', 'Asulox 40', asulam-sodium (400 g a.e./l). These include: 'Candex Liquid', asulam + atrazine (650 g total a.i./l). 'Graslam', asulam + MCPA + mecoprop (300 g total a.e./l). 'Talent', asulam + paraquat (368 g total a.i./l). 'Target', asulam + dalapon-sodium (635 g total a.i./kg).

**Analysis.**

Product analysis is by hydrolysis with colorimetry of a derivative (C. H. Brockelsby & D. F. Muggleton, *Anal. Methods Pestic. Plant Growth Regul.*, 1973, **7**, 497). Residues may be determined by the same method (*idem, ibid.*).

$C_8H_{14}ClN_5$ (215.7)
T6N CN ENJ BMY1&1 DM2 FG

**Nomenclature and development.**

Common name atrazine (BSI, E-ISO, F-ISO, ANSI, WSSA, JMAF). Chemical name (IUPAC) 6-chloro-*N*-ethyl-*N'*-isopropyl-1,3,5-triazinediyl-2,4-diamine; 2-chloro-4-ethylamino-6-isopropylamino-1,3,5-triazine. (*C.A.*) 6-chloro-*N*-ethyl-*N'*-(1-methylethyl)-1,3,5-triazine-2,4-diamine (9CI); 2-chloro-4-(ethylamino)-6-(isopropylamino)-*s*-triazine (8CI); Reg. No. *[1912-24-9]*. Its herbicidal properties were reported by H. Gysin & E. Knüsli (*Proc. Int. Congr. Crop Prot., 4th*, Hamburg, 1957). Introduced by J. R. Geigy S.A. (now Ciba-Geigy AG) (BEP 540 590; GBP 814 947) as code no. 'G 30 027'; trade mark 'Gesaprim'.

**Properties.**

Atrazine is a colourless powder; m.p. 175-177 °C; v.p. 40 µPa at 20 °C; ρ 1.187 g/cm³ at 20 °C. Solubility at 20 °C: 30 mg/l water; 52 g/kg chloroform; 12 g/kg diethyl ether; 28 g/kg ethyl acetate; 18 g/kg methanol; 10 g/kg octan-1-ol. It forms salts with acids and has $pK_a$ 1.7. Slow hydrolysis to the herbicidally-inactive 6-hydroxy analogue occurs at 70 °C under neutral conditions, and is more rapid in acid or alkali.

**Uses.**

Atrazine is a selective pre- and post-em. herbicide used in asparagus, forestry, grasslands, grass crops, maize, pineapple, sorghum, sugarcane and non-crop areas.

**Toxicology.**

Acute oral $LD_{50}$ for rats 1869-3080 mg tech./kg. Acute percutaneous $LD_{50}$ for rats >3100 mg/kg; slight irritant to skins of rabbits, non-irritant to their eyes. Inhalation $LC_{50}$ (1-h) for rats >710 mg/m³. In 2-y feeding trials NELs were: for rats 100 mg/kg diet (8 mg/kg daily); for dogs 150 mg/kg diet (5 mg/kg daily). $LC_{50}$ (96-h) was: for rainbow trout 4.5-8.8 mg/l; for carp 76->100 mg/l; for bluegill 16.0 mg/l.

**Formulations.**

'Gesaprim' 50 and 80 WP, w.p. (500 or 800 g a.i./kg); 'Gesaprim' 90 WEG, water-dispersible granules (900 g/kg); 'Gesaprim' 4L, liquid (40 g/l); 'Gesaprim' 500 FW, S.C. (500 g/l). 'Atratol', atrazine + prometon (15:1). 'Bicep', 'Primagram', 'Primextra', (Reg. No. *[59316-87-9]*) atrazine + metolachlor (1:1.3), (1:1) and (1:1.5, 1:2). 'Fanoprim', atrazine + bromofenoxim (1:1.3). 'Gesapax combi', (Reg. No. *[39324-65-7]*) atrazine + ametryne (1:1). 'Gesaprim combi', (Reg. No. *[8066-10-2]*) atrazine + terbutryne (1:1). 'Gesaprim D', atrazine + 2,4-D (2.5:1). 'Gesaprim S', (Reg. No. *[39331-45-8]*) atrazine + simazine (1:1). 'Vernam atrazine 10-5G' (100 g vernolate + 50 g atrazine/kg).

**Analysis.**

Product analysis is by titration (H. P. Bosshardt *et al.*, *J. Assoc. Off. Anal. Chem.*, 1971, **54**, 749) or by glc (R. T. Murphy *et al.*, *ibid.*, p. 697; *CIPAC Handbook*, 1980, **1A**, 1106; FAO Specification (CP/61); *AOAC Methods*, 1980 (13th Ed.), 6.431). Residues are determined by glc with ECD or FID; (K. Ramsteiner *et al.*, *J. Assoc. Off. Anal. Chem.*, 1974, **57**, 92; E. Knüsli, *Anal. Methods Pestic. Plant Growth Regul.*, 1972, **6**, 600; B. G. Tweedy & R. A. Kahrs, *ibid.*, 1978, **10**, 493; T. H. Byast *et al.*, *Tech. Rep. ARC Weed Res. Organ.*, No.15 (2nd Ed.), p.40).

$C_9H_{10}ClN_2O_5PS$ (324.7)
T56 BOVN FNJ D1SPO&O1&O1 HG

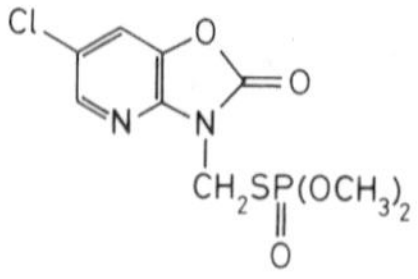

**Nomenclature and development.**

Common name azamethiphos (BSI, E-ISO, F-ISO). Chemical name (IUPAC) *S*-6-chloro-2,3-dihydro-2-oxo-oxazolo[4,5-*b*]pyridin-3-ylmethyl *O,O*-dimethyl phosphorothioate. (*C.A.*) *S*-[(6-chloro-2-oxooxazolo[4,5-*b*]pyridin-3(2*H*)-yl)methyl] *O,O*-dimethyl phosphorothioate (9CI); Reg. No. *[35575-96-3]*. OMS 1825. Its insecticidal and acaricidal properties were reported by R. Wyniger *et al.* (*Proc. 1977 Brit. Crop Prot. Conf. - Pests Dis.*, 1977, **2**, 1025). Introduced by Ciba-Geigy AG (BEP 769 051; GBP 1 347 373) as code no. 'CGA 18 809', 'GS 40 616'; trade mark 'Alfacron'.

**Properties.**

Azamethiphos forms colourless crystals; m.p. 89 °C; v.p. 4.9 μPa at 20 °C; ρ 1.60 g/cm$^3$ at 20 °C. Solubility at 20 °C: 1.1 g/l water; 13 g/kg benzene; 61 g/kg dichloromethane; 10 g/kg methanol; 5.8 g/kg octan-1-ol. On hydrolysis at 20 °C calculated times for 50% loss are: 800 h at pH 5, 260 h at pH 7, 4.3 h at pH9.

**Uses.**

Azamethiphos is a broad spectrum contact and stomach insecticide and acaricide with a quick knock-down effect and good residual activity. It is active against animal and public hygiene pests as well as against household insects.

**Toxicology.**

Acute oral $LD_{50}$ for rats 1180 mg/kg. Acute percutaneous $LD_{50}$ for rats >2150 mg/kg; non-irritant to skin of rabbits, slight irritant to their eyes. In 90-d feeding studies NEL was: for rats 20 mg/kg diet (2 mg/kg daily); for dogs 10 mg/kg diet (0.3 mg/kg daily). $LC_{50}$ (96-h) for rainbow trout 0.2 mg/1; for carp 6.0 mg/1; for bluegill 8.0 mg/1. It is toxic to honeybees, practically non-toxic to Japanese quail.

**Formulations.**

Formulations include: w.p. (10, 100 or 500 g a.i./kg) and aerosol generators.

**Analysis.**

Product analysis is by hplc. Residues are determined by hplc. Particulars are available from Ciba-Geigy AG.

$C_{12}H_{16}N_3O_3PS_2$ (345.4)
T66 BNNNVJ D1SPS&O2&O2

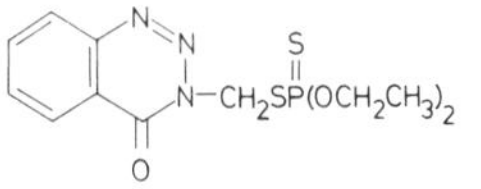

**Nomenclature and development.**

Common name azinphos-ethyl (BSI, E-ISO, F-ISO); triazotion (USSR). Chemical name (IUPAC) *S*-(3,4-dihydro-4-oxobenzo[*d*][1,2,3]triazin-3-ylmethyl) *O,O*-diethyl phosphorodithioate. (*C.A.*) *O,O*-diethyl *S*-[(4-oxo-1,2,3-benzotriazin-3(4*H*)-yl)methyl] phosphorodithioate (9CI); *O,O*-diethyl phosphorodithioate *S*-ester with 3-( mercaptomethyl)-1,2,3-benzotriazin-4(3*H*)-one (8CI); Reg. No. *[2642-71-9]*. ENT 22 014. Its herbicidal activity was described by E. E. Ivy *et al*., (*J. Econ. Entomol*., 1955, **48**, 293). It was developed by W. Lorenz and introduced by Bayer AG (USP 2 758 115; DEP 927 270) as code no. 'Bayer 16 259', 'R 1513'; trade mark 'Gusathion A' (in Federal Republic of Germany 'Gusathion K forte').

**Properties.**

Pure azinphos-ethyl forms colourless needles; m.p. 53 °C; b.p. 111 °C/1 x $10^{-3}$ mmHg; v.p. <29 μPa at 20 °C; $d_4^{20}$ 1.284; $n_D^{53}$ 1.5928. Solubility at 20 °C: 4-5 mg/l water; >1 kg/l dichloromethane, toluene. It is readily hydrolysed by alkali.

**Uses.**

It is a non-systemic insecticide and acaricide with good ovicidal properties and long persistence. It is used on a large range of crops, including beets, cereals, citrus, coffee, cotton, forage crops, grapes, hops, oilseed crops, some ornamentals, potatoes, rice, tobacco, top fruit, vegetables. Rates are generally 500-750 g a.i./ha. Metabolism in beans has been studied (W. Steffens & J. Wieneke, *Pflanzenschutz-Nachr. (Engl. Ed.)*, 1976, **29**, 35).

**Toxicology.**

Acute oral $LD_{50}$ for rats 12.5-17.5 mg/kg. Acute percutaneous $LD_{50}$ for rats 250 mg/kg (2-h). Intraperitoneal $LD_{50}$ for rats >7.5 mg/kg. In 90-d feeding trials rats receiving 2 mg/kg diet showed no sympton of poisoning. $LC_{50}$ (96-h): for goldfish 0.1 mg/l; for guppies 0.01-0.1 mg/l.

**Formulations.**

These include: e.c. (200-400 g a.i./l); w.p. (250-400 g/kg); ULV (500 g/l).

**Analysis.**

Product analysis is by colorimetric measurement of the phosphorodithioate moiety as a complex (*CIPAC Handbook*, 1970, **1**, 18; *FAO Specification* (CP/41)). Residues are measured by glc, details are available from Bayer AG.

$C_{10}H_{12}N_3O_3PS_2$ (317.1)
T66 BNNNVJ D1SPS&O1&O1

**Nomenclature and development.**

Common name azinphos-methyl (BSI, E-ISO, F-ISO); metiltriazotion (USSR). Chemical name (IUPAC) *S*-(3,4-dihydro-4-oxobenzo[*d*][1,2,3]triazin-3-ylmethyl) *O,O*-dimethyl phosphorodithioate. (*C.A.*) *O,O*-dimethyl *S*-[(4-oxo-1,2,3-benzotriazin-3(4*H*)-yl)methyl] phosphorodithioate (9CI); *O,O*-dimethyl phosphorodithioate *S* ester with 3-( mercaptomethyl)-1,2,3-benzotriazin-4(3*H*)-one(8CI); Reg. No. *[86-50-0]*. ENT 23 233, OMS 186; Its insecticidal/acaricidal properties were reported by E. E. Ivy *et al.* (*J. Econ. Entomol.*, 1955, **48**, 293). It was developed by W. Lorenz and introduced by Bayer AG (USP 2 758 115; DEP 927 270) as code no. 'Bayer 17 147', 'R 1582'; trade mark 'Guthion', 'Gusathion M'.

**Properties.**

Pure azinphos-methyl forms colourless crystals; m.p. 73-74 °C; v.p. $<1$ µPa at 20 °C; $d_4^{20}$ 1.44; $n_D^{76}$ 1.6115. Solubility at 20 °C: 33 mg/l water; $>1$ kg/l dichloromethane, toluene. Unstable $>200$ °C. Rapidly hydrolysed by cold alkali or acid.

**Uses.**

A non-systemic insecticide and acaricide of long persistence, chiefly effective against biting and sucking insect pests. Mainly used on citrus, cotton, grapes, maize, some ornamentals, top fruit and vegetables. Residues in plants, soil and water have been reviewed (C. A. Anderson *et al.*, *Residue Rev.*, 1974, **51**, 123; see also W. Steffens & J. Wieneke, *Pflanzenschutz-Nachr. (Engl. Ed.)*, 1976, **29**, 1; J. Wieneke & W. Steffens, *ibid.*, p.18).

**Toxicology.**

Acute oral $LD_{50}$: for female rats 16.4 mg/kg; for male guinea-pigs 80 mg/kg. Acute percutaneous $LD_{50}$ (2-h) for rats $>250$ mg/kg. In 2-y feeding trials no symptom of poisoning occurred in rats receiving 2.5 mg/kg diet. $LC_{50}$ (96-h) is: for goldfish $>1$ mg/l; for guppies 0.1 mg/l.

**Formulations.**

Formulations include: e.c. (200 g a.i./l); w.p. (200, 250, 400 or 500 g/kg); dust (50 g/kg). 'Gusathion MS', (Reg. No. *[8066-08-8]*), w.p. (azinphos-methyl + demeton-S-methyl sulphone).

**Analysis.**

Product analysis is by colorimetric measurement of the phosphorodithioate moiety as a complex (*CIPAC Handbook*, 1970, **1**, 25; FAO Specification (CP/4). Residue analysis is by glc *Analyst (London)*, 1977, **102**, 858; *Pestic. Anal. Manual*, 1979, **I**, 201-H,201-I; D. C. Abbott *et al.*, *Pestic. Sci.*, 1970, **1**, 10; R. Mestres *et al.*, *Trav. Soc. Pharm. Montpellier*, 1979, **39**, 323; D. H. MacDougall, *Anal. Methods Pestic. Plant Growth Regul.*, 1972, **6**, 397).

# Aziprotryne

$C_7H_{11}N_7S$ (225.3)
T6N CN ENJ BS1 DNNN FMY1&1

$CH_3S$, N, $NHCH(CH_3)_2$, N, N, $N_3$

**Nomenclature and development.**

Common name aziprotryne (BSI, E-ISO, F-ISO); aziprotryn (USA). Chemical name (IUPAC) 4-azido-*N*-isopropyl-6-methylthio-1,3,5-triazin-2-ylamine; 2-azido-4-isopropylamino-6-methylthio-1,3,5-triazine. (*C.A.*) 4-azido-*N*-(1-methylethyl)-6-(methylthio)-1,3,5-triazin-2-amine (9CI); 2-azido-4-(isopropylamino)-6-(methylthio)-*s*-triazine (8CI); Reg. No. *[4658-28-0]*. Its herbicidal activity was reported by D. H. Green *et al.* (*C. R. Journal. Etud. Herbic. Conf. COLUMA, 4th*, 1967, **1**, 1). Introduced by Ciba AG (now Ciba-Geigy AG). (BEP 656 233; GBP 1 093 376) as code no. 'C 7019' and trade mark 'Mesoranil'.

**Properties.**

Pure aziprotryne is a colourless crystalline powder; m.p. 94.5-95.5 °C; v.p. 267 μPa at 20 °C, ρ 1.40 g/cm³ at 20 °C. Solubility at 20 °C: 55 mg/l water; 27 g/kg acetone; 4 g/kg benzene; 37 g/kg dichloromethane; 12 g/kg ethyl acetate; 14 g/kg propan-2-ol. Stable in neutral and slightly acid media, slowly hydrolysed in slightly alkaline media.

**Uses.**

It is a pre-em. herbicide effective against a wide range of broad-leaved annual weeds and some grasses. It is used in transplanted brassica crops (excluding cauliflowers) after emergence of the weeds. Also applied to seeded fennel, onions, leeks and garlic.

**Toxicology.**

Acute oral $LD_{50}$ for rats 3600-5833 mg tech./kg. Acute percutaneous $LD_{50}$ for rats >3000 mg/kg; non-irritant to skin, slight irritant to eyes. Inhalation $LC_{50}$ (6-h) for rats >208 mg/m³. In 90-d feeding trials the NEL was: for rats 50 mg/kg daily; for dogs 50 mg/kg daily. $LC_{50}$ (96-h) for large-mouth bass and bluegill >1 mg/l. It is slightly toxic to birds and practically non-toxic to honeybees.

**Formulations.**

A w.p. (500 g a.i. /kg).

**Analysis.**

Product analysis is by glc. Residues are determined by glc using MCD. Particulars are available from Ciba-Geigy AG.

$C_{20}H_{35}N_3Sn$ (436.2)
T5NN DNJ A-SN- AL6TJ&- AL6TJ&-
AL6TJ

**Nomenclature and development.**

Common name azocyclotin (BSI, E-ISO, F-ISO). Chemical name (IUPAC) tri(cyclohexyl)-1*H*-1,2,4-triazol-1-yltin. (*C.A.*) 1-(tricyclohexylstannyl)-1*H*-1,2,4-triazole (9CI); Reg. No. *[41083-11-8]*. Its acaricidal activity was reported by I. Hammann *et al.* (*Pflanzenschutz-Nachr. (Engl. Ed.*), 1978, **31**, 61). Introduced by Bayer AG (DEPS 2 143 252) as code no. 'BAY BUE 1452'; trade mark 'Peropal'.

**Properties.**

It forms colourless crystals; m.p. 218.8 °C; v.p. $<5$ μPa at 25 °C. Solubility at 20 °C: $<1$ mg/kg water; $\leqslant 1$ g/kg dichloromethane, propan-2-ol. It is hydrolysed by aqueous acids with loss of the triazole group.

**Uses.**

Azocyclotin is a contact acaricide effective against all mobile stages, larvae and adult phytophagous mites.

**Toxicology.**

Acute oral $LD_{50}$ for rats 99 mg a.i./kg. Acute percutaneous $LD_{50}$ for male rats *c.* 1000 mg/kg. In 90-d feeding trials rats receiving 15 mg/kg diet showed no symptom of poisoning. $LC_{50}$ (96-h) for carp 0.012-0.025 mg a.i. (as 250 g/kg w.p.)/l; for rainbow trout 0.0012-0.0025 mg a.i. (as 250 g/kg w.p.)/l.

**Formulations.**

A w.p. (250 g a.i./kg).

**Analysis.**

Details of product analysis are available from Bayer AG. Residue analysis is by glc (E. Möllhoff, *ibid.*, 1977, **30**, 249).

**Nomenclature and development.**

Scientific name *Bacillus thuringiensis* Berliner. (*C.A.*) *Bacillus thuringiensis*. A trivial name Bt is sometimes used but, having no official standing, should be avoided.. This gram-positive bacterium was detected in 1902 in dying larvae of *Bombyx mori* Lin. by S. Ishiwata (cited by K. Ishikawa, *Pathology of the Silkworm*) and was characterised after isolation from larvae of *Ephestia kuehniella* Zell. by E. Berliner (*Z. Angew. Entomol.*, 1915, **2**, 29). It was first used as a microbial insecticide 'Sporeine' against lepidopterous larvae in 1938 (S. E. Jacobs, *Proc. Soc. Appl. Bacteriol.*, 1950, **13**, 83). Developed for the control of these pests by several companies, of which 3 currently market products under the trade marks 'Bactospeine' (distributor Biochem Products Ltd), 'Dipel' (Abbott Laboratories), 'Thuricide' (Sandoz AG). Also 'Bactimos' (Biochem Products Ltd), 'Technar' (Sandoz AG) and 'Vectobac' (Abbott Laboratories) obtained from a new variety *B. thuringiensis* var. *israelenis*.

**Properties.**

Produced by fermentation in well-controlled conditions; at sporulation each rod-shaped bacterium also produces a protein crystal. The dormant spores and crystals are harvested by centrifugation, low temperature evaporation or spray drying. The crystals are insoluble in water and are unstable in alkaline media or in the presence of certain enzymes.

The protein, of relatively high molecular mass, is broken down by the alkaline gut juices of susceptible insects forming smaller units that attack the lining of the larval gut, disrupting osmotic balance, paralysing mouthparts and gut, thus stopping feeding. In some host species the spores germinate in the gut and play an important role in pathogenicity, the larvae eventually dying of septicaemia.

An adenine nucleotide, the beta-exotoxin, is produced by some *B. thuringiensis* strains during the logarithmic phase of bacterial growth, but not by strains used in the commercial production of the spore-crystal complex. Separate minor products utilise the beta-exotoxin.

Formulated products are compatible with most insecticides, acaricides, fungicides and plant growth regulators but not with captafol, dinocap, alkaline sprays or, under some conditions, with foliar nutrients.

**Uses.**

*B. thuringiensis* not containing beta-toxin is a microbial insecticide effective only against larvae of many lepidopterous and dipterous species. Its specificity to pests, combined with safety to man and the natural enemies of many crop pests, make it ideal for use in pest management. Because the bacteria do not spread, it is important to treat the parts of the plant normally attacked by lepidoterous larvae, or zones of water in which dipterous larvae feed. It is non-phytotoxic and is used on cotton, ornamentals, pome fruits, soyabeans, tobacco and vegetables. *B. thuringiensis* var. *israelensis* is effective only against larvae of Diptera (mosquitoes and *Simulium* spp.) and is safe to the environment.

*continued overleaf*

**Toxicology.**

There is no evidence of acute or chronic toxicity of the spore-crystal complex to birds, dogs, fish, guinea-pigs, man, mice, rats or other test mammals (R. Fisher & L. Rosner *J. Agric. Food Chem.*, 1959, **7**, 687). Very slight inhalation and dermal irritation has been observed in test animals, probably due to the physical rather than the biological properties of the *B. thuringiensis* formulation.

**Formulations.**

The products are not markedly different. Typical agricultural formulations, containing spores and protein crystals, include: w.p. (16 000 i.u./mg) which are stable at 30% r.h. for ⩾3 y; liquid concentrates (4000 i.u./mg, with spray adjuvants); spray concentrates (9000 i.u./mg, without adjuvants), intended for aerial application and having a shelf-life of *c.* 0.5 y. Also dusts (160-800 i.u./mg), baits (320 i.u./mg, for high value crops where placement is critical). For larval control of mosquitoes and *Simulium* spp. products include: 'Bactimos', 'Technar' (1500 i.u./mg), 'Vectobac'.

**Analysis.**

The activity of *B. thuringiensis* is measured in i.u. relative to that of an appropriate standard product against *Trichoplusia ni* or *Aedes aegypti* in standardized bioassays. Assays based on the number of spores are not satisfactory.

# Barban

$C_{11}H_9Cl_2NO_2$ (258.1)
G2UU2OVMR CG

$NH.CO.OCH_2C{\equiv}CCH_2Cl$

Cl

**Nomenclature and development.**

Common name barban (BSI, E-ISO, ANSI, WSSA); barbane (F-ISO); barbanate (Republic of South Africa); chlorinat (USSR); CBN (JMAF); *exception* (Italy). Chemical name (IUPAC) 4-chlorobut-2-ynyl 3-chlorocarbanilate; 4-chlorobut-2-ynyl 3-chlorophenylcarbamate. (*C.A.*) 4-chloro-2-butynyl 3-chlorophenylcarbamate (9CI); 4-chloro-2-butynyl *m*-chlorophenylcarbamate (8CI); Reg. No. *[101-27-9]*. Its herbicidal properties were described by A. D. Brown (*Proc. North Cent. Weed Control Conf.*, 1958, **15**, 98). Introduced by the Spencer Chemical Co. (USP 2 906 614) as code no. 'CS-847'; trade mark 'Carbyne'.

**Properties.**

Pure barban is a crystalline solid; m.p. 75-76 °C; v.p. 50 µPa at 25 °C. Solubility at 25 °C: 11 mg/l water; 327 g/l benzene; 546 g/l 1,2-dichloroethane; 1.4 g/l hexane; 3.9 g/l kerosene; 279 g/l xylene. The technical grade is a tan solid, m.p. *c.* 60 °C, decomposing at 224 °C; ρ 1.403 $g/cm^3$ at 25 °C. It is stable under normal conditions of use. Hydrolysis is very rapid in alkali with displacement of the terminal chlorine substituent (50% loss in 58 s at 25 °C at pH13).

**Uses.**

It is a selective post-em. herbicide controlling wild oats in barley at the 1.5-2.5 leaf stage, and broad beans, field flax, lentils, lucerne, mustard, oilseed rape, safflower, soyabeans, sugar beet, sunflowers, wheat, Russian wild rye and other grasses.

**Toxicology.**

Acute oral $LD_{50}$ for rats 1376-1429 mg/kg. Acute percutaneous $LD_{50}$: for rats >1600 mg/kg; for rabbits >20 000 mg/kg. It is mildly irritating to the skin of rabbits and produces a delayed skin sensitisation in guinea-pigs; it has been reported as a skin sensitiser in man, skin contact should be avoided by using plastic protective clothing. Acute inhalation $LC_{50}$ (4-h) with dust for rats >27.4 mg/l (nominal concentration). In 2-y feeding trials NEL was: for rats 150 mg/kg diet; for dogs 5 mg/kg diet. It was not teratogenic in rats ≤87 mg/kg daily. In a 3-generation reproduction study in rats NEL was 957 mg/kg diet. Non-mutagenic to bacteria and mammals by in-vivo and in-vitro tests. $LC_{50}$ (8-d) for mallard duck and bobwhite quail >10 000 mg/kg diet. $LC_{50}$ (96-h) is: for goldfish and guppies 1.3 mg/l; for rainbow trout 0.6 mg/l; for bluegill 1.2 mg/l.

**Formulations.**

These include: 'Carbyne', 'B 25' (FBC Limited), e.c. (120-240 g a.i./l).

**Analysis.**

Product analysis is by u.v. spectrometry (K. J. Bombaugh & W. C. Bull, *J. Agric. Food Chem.*, 1961, **9**, 386). Residues may be determined by hplc (details from Velsicol Chemical Corp.) or by hydrolysis to 3-chloroaniline, derivatives of which are measured by glc (R. J. Harris & R. J. Whiteoak, *Analyst* (*London*), 1972, **97**, 294)

$C_{20}H_{23}NO_3$ (325.4)
1OVY1&NV1R&R B1 F1

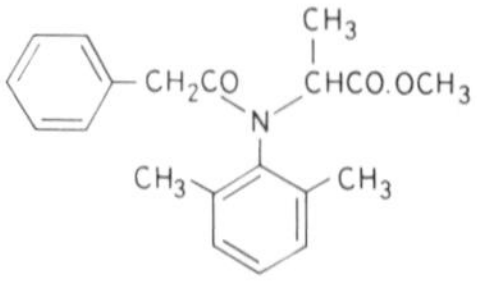

**Nomenclature and development.**

Common name benalaxyl (BSI, draft E-ISO). Chemical name (IUPAC) methyl *N*-phenylacetyl-*N*-2,6-xylyl-DL-alaninate. (*C.A.*) methyl *N*-(2,6-dimethylphenyl)-*N*-(phenylacetyl)-DL-alaninate (9CI); Reg. No. *[71626-11-4]*. Its fungicidal properties were described by Garavaglia *et al.* (*Att. del terzo Simposio Chimica degli Antiparassitari, Piacenza*, 1981). Introduced by Farmoplant S.p.A. (BEP 873 908; DEP 2 903 612; ITPA 19 896-1978) as code no. 'M 9834'; trade mark 'Galben'.

**Properties.**

Pure benalaxyl is a colourless solid; m.p. 78-80 °C; v.p. 670 μPa at 25 °C; $d_4^{25}$ 1.27. Solubility at 25 °C: 37 mg/l water; sparingly soluble in hexane and saturated hydrocarbons; readily soluble in most other organic solvents. It is stable at 25 °C in aqueous solution, buffered at pH4-9; but hydrolysed in concentrated alkaline media.

**Uses.**

It is a fungicide with systemic activity for the control of Oomycetes. It is recommended for use on grapes against *Plasmopara viticola* (12-15 g benalaxyl + 100-130 g mancozeb/100 l, or 12-15 g benalaxyl + 100-130 g zineb/100 l, or 16 g benalaxyl + 100 g folpet/100 l, or 15 g benalaxyl + 130 g copper (as copper oxychloride)/100 l); on hops against *Pseudoperonospora humuli* (20-25 g benalaxyl per rootstock) for primary infection, and (20 g benalaxyl + 70 g copper(as copper oxychloride)/100 l) for secondary infections; on potatoes and tomatoes against *Phytophthora infestans* (20-25 g benalaxyl + 160-195 g mancozeb/100 l); on tobacco against *Plasmopara tabacina* (20-25 g benalaxyl + 160-195 g mancozeb/100 l); and on turf against *Pythium* spp. (0.75-1.5 kg benalaxyl/ha).

**Toxicology.**

Acute oral $LD_{50}$: for rats 4200 mg tech./kg; for mice 680 mg/kg; for bobwhite quail >5000 mg/kg. Acute percutaneous $LD_{50}$ for rats >5000 mg/kg; no irritation nor sensitisation of skin has been observed. In a subchronic toxicity study NEL for rats was 1000 mg/kg diet. $LC_{50}$ (96-h) for trout is 3.75 mg/l.

**Formulations.**

These include: 'Galben' 2E, e.c. (240 g a.i./l); 'Galben' 25 WP, w.p. (250 g/kg); 'Galben' 5G, granules (50 g/kg). It is usually employed as w.p. of mixtures with contact fungicides: 'Galben M' 8-65 (80 g benalaxyl + 650 g mancozeb/kg); 'Galben Z' 8-65 (80 g benalaxyl + 650 g zineb/kg); 'Galben F' 8-50 (80 g benalaxyl + 500 g folpet/kg); 'Galben C' 4-33 (40 g benalaxyl + 330 g copper (as copper oxychloride)/kg); 'Galben C' 11-35 (110 g benalaxyl + 350 g copper (as copper oxychloride)/kg).

**Analysis.**

Details of analytical methods for product and residues are available from Montedison/Farmoplant S.p.A.

CH2CO.OH

$C_9H_6ClNO_3S$ (243.7)
T56 BNVSJ B1VQ IG
*benazolin-potassium*
$C_9H_5ClKNO_3S$ (281.8)
T56 BNVSJ B1VQ IG &&K SALT
*benazolin-ethyl*
$C_{11}H_{10}ClNO_3S$ (275.4)
T56 BNVSJ B1VO2 IG

**Nomenclature and development.**

Common name benazolin (BSI, E-ISO); bénazoline (F-ISO). Chemical name (IUPAC) 4-chloro-2,3-dihydro-2-oxobenzothiazol-3-ylacetic acid; 4-chloro-2-oxobenzothiazolin-3-ylacetic acid. (*C.A.*) 4-chloro-2-oxo-3(2*H*)-benzothiazoleacetic acid (9CI); 4-chloro-2-oxo-3-benzothiazolineacetic acid (8CI); Reg. No. *[3813-05-6]*. Its herbicidal properties were reported by E. L. Leafe (*Proc. Br. Weed Control Conf., 7th*, 1964, p.32). Introduced by The Boots Co. Ltd (now FBC Limited) (GBP 862 226; 1 243 006) as code no. 'RD 7693'; trade mark 'Cornox CWK'.

**Properties.**

Pure benazolin is a colourless crystalline solid: m.p. 193 °C; v.p. 396 nPa at 20 °C (calculated). Solubility at 20 °C: 600 mg benazolin/l water, but alkali metal and diethanolamine salts are readily soluble (600 g benazolin-potassium/l). Technical grade benazolin is *c.* 90% pure, m.p. 189 °C. Benazolin is stable except to concentrated alkali; its salts are compatible with similar salt formulations of the phenoxyalkanoic acids. Solutions of the alkali metal salts at pH 8.5-10.0 are mildly corrosive to aluminium, galvanised iron and tin-plate.

**Uses.**

Benazolin is a post-em. translocated growth regulator herbicide effective against *Galium aparine, Stellaria media* and, at higher dose rates, *Sinapis arvensis*. It is used for selective weed control in oilseed rape either alone or in mixtures with 3,6-dichloropicolinic acid. Synergism with dicamba has been shown, particularly against *Anthemis* and *Matricaria* spp. in cereals. In non-undersown cereals, it is combined with other herbicides such as dicamba + MCPA, dicamba + dichlorprop. In cereals undersown with grass/clover mixtures, it is combined with MCPA, MCPA + MCPB, MCPA + 2,4-DB. For grassland uses, where clover safety is not important, it is combined with dicamba + dichlorprop and, where clover safety is required, with MCPA + 2,4-DB.

**Toxicology.**

Acute oral $LD_{50}$: for mice 3200 mg a.i./kg; for rats >4800 mg/kg; for Japanese quail >10 204 mg tech./kg, 2856 mg (potassium salt)/kg. Acute percutaneous $LD_{50}$ for rats >5000 mg/kg. In 90-d feeding trials NEL: between 300 and 1000 mg/kg daily for rats; *c.* 300 mg/kg daily for dogs. A 30% solution of the potassium salt caused a mild irritation of the skin and eyes of rabbits, while a 1 in 24 dilution is non-irritant. $LC_{50}$ (24-h) for bluegill 204 mg/l.

**Formulations.**

Formulations: an aqueous solution of the potassium salt, (Reg. No. *[67338-65-2]*) (250 g a.i./l); as the ethyl ester, (Reg. No. *[25059-80-7]*) (280 g a.i./kg). 'Benazalox', (benazolin-ethyl + ethyl 3,6-dichloropicolinate) for use in oilseed rape. Esters, alkali metal or amine salts for use in cereals, 'Legumex extra', (benazolin + 2,4-DB + MCPA). 'Herbitox', (benazolin + dicamba + dichlorprop) as alkali metal salts for use in grassland.

**Analysis.**

Product and residue analysis is by glc after conversion to the methyl ester, details available from FBC Limited.

$C_{11}H_{13}NO_4$ (223.2)
T56 BO DO CHJ C1 C1 FOVM1

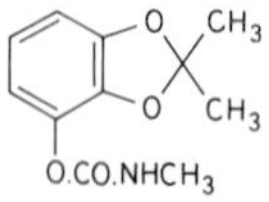

**Nomenclature and development.**

Common name bendiocarb (BSI, E-ISO, ANSI); bendiocarbe (F-ISO). Chemical name (IUPAC) 2,2-dimethyl-1,3-benzodioxol-4-yl methylcarbamate (I); 2,3-isopropylidenedioxyphenyl methylcarbamate (II). (*C.A.*) (I) (9CI); (II) (8CI); Reg. No. *[22781-23-3]*. OMS 1394. Its insecticidal properties were described by R. W. Lemon (*Proc. Br. Insectic. Fungic. Conf. 6th*, 1971, **2**, 570) and P. J. Brooker *et al.* (*Pestic. Sci.*, 1972, **3**, 735). Introduced by Fisons Ltd, Agrochemical Division (now FBC Limited) (GBP 1 220 056) as code no. 'NC 6897'; trade marks 'Ficam' (public health use), 'Garvox', 'Seedox', 'Tattoo' (all agricultural use).

**Properties.**

Pure bendiocarb is a colourless solid; m.p. 129-130 °C; v.p. 660 µPa at 25 °C. Solubility at 25 °C: 40 mg/l water; 200 g/kg acetone, chloroform, dioxane; 40 g/kg benzene, ethanol; 350 mg/kg hexane; <1 g/kg odourless kerosene. The partition ratio at 20 °C is 1:9 for water/hexane. It is stable to hydrolysis in water at pH 5; 50% decomposition (tested under EPA guidelines) occurs at 25 °C in 4 d at pH 7, the products being 2,3-isopropylidenedioxyphenol, methylamine and carbon dioxide.

**Uses.**

It is an insecticide acting by cholinesterase inhibition, effective as a contact and stomach poison and with systemic activity in some crop plants. It is active against many public health, industrial and storage pests, such as ants, cockroaches, fleas, flies, mosquitoes and moths. It is particularly useful inside buildings due to its low odour and lack of corrosive and staining properties. In agriculture it is used as a seed treatment and in granular formulations for the control of soil pests and some stem boring and leaf mining Diptera in maize and sugar beet, and in which residues have been below the limit of detection. (0.02 mg/kg). It is active against a broad range of foliar, stem boring and soil pests on other crops including brassicas, cereals, cocoa, mushrooms, oilseed rape, peas, potatoes, rice, rye grass, top fruit, turf, ornamentals, other vegetable crops and other tropical crops on which experimental work is in progress.

**Toxicology.**

Acute oral $LD_{50}$ for rats is in the range 40-156 mg/kg; other mammals respond similarly. Acute percutaneous $LD_{50}$ for rats 566-600 mg/kg. In 2-y feeding trials NEL for rats was 10 mg/kg diet; in 90-d trials rats receiving 250 mg/kg diet showed no visible toxic effects. $LC_{50}$ (96-h) for rainbow trout is 1.55 mg/l. Toxic to honeybees, $LD_{50}$ 0.1 µg/bee.

**Formulations.**

Formulations for public health include: residual spray, 'Ficam W' w.p. (800 g a.i./kg) (also 200 or 500 g/kg); 'Ficam Z', w.p. fly spray (184 g/kg). Also 'Ficam B' paint-on-bait (32 g/kg) for fly control; 'Ficam D' dust (10 g/kg); 'Ficam ULV', s.c. (250 g/l); mosquito coils; shelf paper and other consumer products. Agricultural formulations include 'Garvox 3G', granules (30 g/kg); 'Tattoo 3G' and 'Tattoo 5G', granules (30 or 50 g/kg); 'Garvox SC' and 'Seedox SC', s.c. (500 g/kg).

**Analysis.**

Product analysis is by hplc or glc. Analysis of residues in foodstuffs is based on hydrolysis to 2,3-isopropylidenedioxyphenol, a derivative of which is measured by glc with ECD (R. J. Whiteoak *et al.*, *Anal. Methods Pestic. Plant Growth Regul.*, 1978, **10**, 3).

# Benfluralin

$C_{13}H_{16}F_3N_3O_4$ (335.3)
FXFFR CNW ENW DN4&2

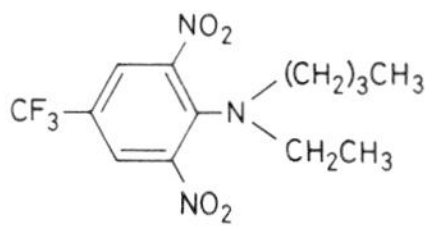

**Nomenclature and development.**

Common name benfluralin (BSI, E-ISO); benfluraline (F-ISO); benefin (WSSA); bethrodine (JMAF). Chemical name (IUPAC) *N*-butyl-*N*-ethyl-α,α,α-trifluoro-2,6-dinitro-*p*-toluidine (I); *N*-butyl-*N*-ethyl-2,6-dinitro-4-trifluoromethylaniline. (*C.A.*) *N*-butyl-*N*-ethyl-2,6-dinitro-4-(trifluoromethyl)benzenamine (9CI); (I) (8CI); Reg. No. *[1861-40-1]*. Its herbicidal properties were first described by J. F. Schwer (*Proc. North Cent. Weed Control Conf.*, 1965). Introduced by Eli Lilly & Co. (USP 3 257 190) as code no. 'EL-110'; trade marks 'Balan', 'Bonalan'.

**Properties.**

Pure benfluralin is a yellow-orange crystalline solid; m.p. 65-66.5 °C; v.p. 52 mPa at 30 °C. Solubility at 25 °C: $<1$ mg/l water; 650 g/l acetone; 24 g/l ethanol; 420 g/l xylene. It is decomposed by u.v. light but otherwise stable.

**Uses.**

It is a pre-em. herbicide incorporated into soil and effective at 1.0-1.35 kg a.i./ha for the control of annual grasses and broad-leaved weeds in chicory, cucumbers, endive, groundnuts, lettuce, lucerne, tobacco and other forage crops. At. 1.5 kg/ha it controls annual grass weeds in turf.

**Toxicology.**

Acute oral $LD_{50}$: for rats $>10\,000$ mg/kg; for mice $>5000$ mg/kg; for bobwhite quail, chickens, dogs, mallard duck, rabbits $>2000$ mg/kg. Skin application of 200 mg/kg and eye application to rabbits produced no irritation. In 2-y feeding trials NEL for rats was 1000 mg/kg diet. $LC_{50}$ for bluegill fingerlings is 0.37 mg/l.

**Formulations.**

These include: e.c. and granules.

**Analysis.**

Product analysis is by spectometry. Residues are analysed by glc with ECD (T. H. Byast *et al., Tech. Rep. ARC Weed Res. Organ.*, No. 15 (2nd Ed.), p. 11; W. S. Johnson & R. Frank, *Anal. Methods Pestic. Plant Growth Regul.*, 1976, **8**, 335). Details of analytical procedures and other information are available from Eli Lilly & Co.

$C_{13}H_{10}INO$ (323.1)
IR BVMR

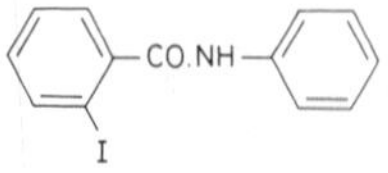

**Nomenclature and development.**

Common name benodanil (BSI, E-ISO, F-ISO). Chemical name (IUPAC) 2-iodobenzanilide (I). (*C.A.*) 2-iodo-*N*-phenylbenzamide (9CI); (I) (8CI); Reg. No. *[15310-01-7]*. Its fungicidal properties were reported by E. -H. Pommer *et al.*, at the 39th Deutsche Pflanzenschutztagung, Stuttgart, 1973 and by F. Löcher *et al.*, (*Meded. Fac. Landbouwwet. Rijksuniv. Gent*, 1974, **39**, 1079). Introduced by BASF AG as code number 'BAS 3170 F'; trade mark 'Calirus'.

**Properties.**

It is a colourless crystalline solid; m.p. 137 °C; v.p. $<$10 nPa at 20 °C. Solubility at 20 °C: 20 mg/kg water; 401 g/kg acetone; 77 g/kg chloroform; 93 g/kg ethanol; 120 g/kg ethyl acetate. Stable $\leqslant$50 °C; no hydrolysis was observed at 20 °C and pH1 or pH13.

**Uses.**

Benodanil is a systemic fungicide controlling *Puccinia striiformis* on wheat and barley and *P. hordei* on barley. It is effective against rust diseases of coffee, ornamentals and vegetables, and against *Marasmius oreades* in turf.

**Toxicology.**

Acute oral $LD_{50}$ for rats and guinea-pigs $>$6400 mg/kg. Acute percutaneous $LD_{50}$ for rats $>$2000 mg/kg; it is not irritating to rabbits' skin and eyes. In 90-d feeding trials NEL $\geqslant$100 mg/kg diet.

**Formulations.**

'Calirus', w.p. (500 g a.i./kg).

**Analysis.**

Product analysis is by hplc. Residues may be determined by glc with ECD.

CO.NH(CH2)3CH3
N
NHCO.OCH3
N

$C_{14}H_{18}N_4O_3$ (290.3)
T56 BN DNJ BVM4 CMVO1

**Nomenclature and development.**

Common name benomyl (BSI, E-ISO, F-ISO, ANSI, JMAF). Chemical name (IUPAC) methyl 1-(butylcarbamoyl)benzimidazol-2-ylcarbamate. (*C.A.*) methyl 1-[(butylamino)carbonyl]-1*H*-benzimidazol-2-ylcarbamate (9CI); methyl 1-(butylcarbamoyl)-2-benzimidazolecarbamate (8CI); Reg. No. *[17804-35-2]*. Its fungicidal activity was described by C. J. Delp & H. L. Klopping (*Plant Dis. Rep.*, 1968, **52**, 95). Introduced by E. I. du Pont de Nemours & Co. (Inc.) (NLP 6 706 331; USP 3 631 176) as code no. 'DuPont 1991'; trade mark 'Benlate'.

**Properties.**

Pure benomyl is a colourless crystalline solid; on heating it decomposes without melting; v.p. negligible at room temperature. Solubility at 25 °C: *c.* 2 mg/kg water at pH7; *c.* 18 g/kg acetone; *c.* 94 g/kg chloroform; *c.* 53 g/kg dimethylformamide; *c.* 4 g/kg ethanol; *c.* 400 g/kg heptane; *c.* 10 g/kg xylene. In some solvents dissociation occurs to form carbendazim (2000) and butyl isocyanate (M. Chiba & E. A. Cherniak, *J. Agric. Food Chem.*, 1978, **26**, 573). Decomposed on storage in contact with water and under moist conditions in soil. The mechanism of the acid-catalysed decomposition in aqueous media is known (J. P. Calmon & D. R. Sayag, *ibid.*, 1976, **24**, 314, 317).

**Uses.**

It is a protective and eradicant fungicide with systemic activity, effective against a wide range of fungi affecting field crops, fruits, nuts, ornamentals, and turf. It is also effective against mites, primarily as an ovicide. Its systemic properties are sometimes enhanced by surfactants. It is also used as pre- and post-harvest sprays or dips for the control of storage rots of fruits and vegetables. Typical rates are: on field and vegetable crops, 140-550 g a.i./ha; on tree crops 550-1100 kg/ha; for post-harvest uses 25-200 g/100 l. Benomyl *per se* is stable on the surface of banana skins (J. Cox *et al.*, *Pestic. Sci.*, 1974, **5**, 135; J. Cox & J. A. Pinegar, *ibid.*, 1976, **7**, 193). Carbendazim can be formed in soil, plants and animals under some conditions.

**Toxicology.**

Acute oral $LD_{50}$ for rats >10 000 mg a.i./kg. Acute percutaneous $LD_{50}$ for rabbits >10 000 mg/kg; negligible irritant to skin of guinea-pigs, temporary eye irritant to rabbits. In 2-y feeding trials NEL was: in rats >2500 mg/kg diet (top rate), no evidence of histopathological changes; in dogs 500 mg/kg diet. $LC_{50}$ (8-d) for mallard and bobwhite quail >500 mg/kg diet. $LC_{50}$ (96-h) is: for goldfish 4.2 mg/1; for rainbow trout 0.17 mg/1.

**Formulations.**

'Benlate', w.p. (500 g a.i./kg); 'Benlate' OD fungicide, oil dispersible (500 g/kg); 'Tersan' 1991 turf fungicide (500 g/kg). 'Benlate T' 20 fungicide, (Reg. No. *[8068-35-7]*), w.p. (200 g benomyl + 200 g thiram/kg).

**Analysis.**

Product analysis is by i.r. spectrometry (F. J. Baude *et al.*, *Anal. Methods Pestic. Plant Growth Regul.*, 1978, **10**, 157). Preferred method for determination of benomyl-derived residues in soils and plant tissue is cation-exchange hplc (J. J. Kirkland *et al.*, *J. Agric. Food Chem.*, 1973, **21**, 368; *Pestic. Anal. Man.*, 1979, **I**; N. Aharonson & A. Ben-Aziz, *J. Assoc. Off. Anal. Chem.*, 1973, **56**, 481; J. E. Farrow *et al.*, *Analyst (London)*, 1977, **102**, 752) and in cows' milk, excreta and tissues (J. J. Kirkland, *J. Agric. Food Chem.*, 1973, **21**, 171). Residues in soils and plants may be determined by fluorimetry or by colorimetry after conversion to benzimidazol-2-ylamine (or a derivative) (H. L. Pease & J. A. Gardiner, *ibid.*, 1969, **17**, 267; H. L. Pease & R. F. Holt, *J. Assoc. Off. Anal. Chem.*, 1971, **54**, 1399; J. L. Pease *et al.*, *Anal. Methods Pestic. Plant Growth Regul.*, 1973, **7**, 674).

$C_{14}H_{24}NO_4PS_3$ (397.5)
WSR&M2SPS&OY1&1&OY1&1

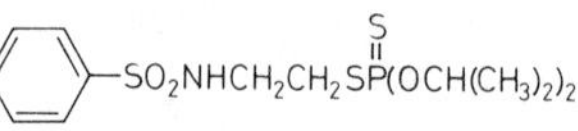

**Nomenclature and development.**

Common name bensulide (BSI, E-ISO, F-ISO, WSSA); SAP (JMAF). Chemical names (IUPAC) *S*-2-benzenesulphonamidoethyl *O,O*-di-isopropyl phosphorodithioate; *O,O*-di-isopropyl *S*-2-phenylsulphonylaminoethyl phosphorodithioate. (*C.A.*) *O,O*-bis(1-methylethyl) *S*-[2-[(phenylsulfonyl)amino]ethyl] phosphorodithioate (9CI); *O,O*-diisopropyl phosphorodithioate *S*-ester with *N*-(2-mercaptoethyl)benzenesulfonamide (8CI); Reg. No. *[741-58-2]*. Its herbicidal properties were described by D. D. Hemphill (*Res. Rep. North Cent. Weed Control Conf.*, 1962, pp. 104, 111) and it was introduced by the Stauffer Chemical Co. (USP 3 205 253) as code number 'R-4461'; trade marks 'Betasan' (for turf use), 'Prefar' (for crop use).

**Properties.**

Bensulide (99% pure) is a colourless solid; m.p. 34.4 °C, $d^{22}$ 1.25. Technical grade (95-98% pure) is an amber solid or supercooled liquid; $n_D^{30}$ 1.5438; v.p. $<$133 μPa at 20 °C. Solubility at 20 °C: 25 mg/l water; 300 mg/l kerosene; miscible with acetone, ethanol, 4-methylpentan-2-one, xylene.

**Uses.**

A pre-em. herbicide suitable for pre-plant use at 2.3-5.0 kg a.i./ha, on crops such as brassicas, cotton, cucurbits, and lettuces; or for use on established turf at 11-22 kg/ha per season. Its effects persist for 0.33-1.0 y according to rate of application. When applied to the roots bensulide is not translocated to the leaves but metabolites are.

**Toxicology.**

Acute oral $LD_{50}$; for male albino rats 360 mg/kg, for females 270 mg/kg. Acute percutaneous $LD_{50}$ for albino rats 3950 mg/kg; for rabbits $>$5000 mg/kg. In 90-d feeding trials it was well tolerated by rats $\leqslant$250 mg/kg diet and by dogs $\leqslant$625 mg/kg diet.

**Formulations.**

'Betasan E' and 'Prefar E', e.c. (348 or 480 g a.i./l). 'Betasan G', granules (36, 70 or 125 g/kg).

**Analysis.**

Product analysis is by hplc. Residues in crops and soils are determined by glc. Analytical methods are available from the Stauffer Chemical Co. See also (G. G. Patchett *et al.*, *Anal. Methods Pestic., Plant Growth Regul. Food Addit.*, 1967, **5**, 483; R. W. Buxton *et al.*, *Anal. Methods Pestic. Plant Growth Regul.*, 1972, **6**, 672).

$C_{10}H_{12}N_2O_3S$ (240.3)
T66 BMSWNVJ DY1&1

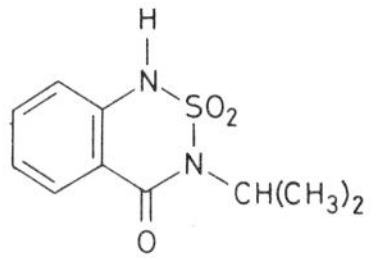

**Nomenclature and development.**

Common name bentazone (BSI, E-ISO, F-ISO, JMAF); bentazon (ANSI, Canada, WSSA); bendioxide (Republic of South Africa). Chemical name (IUPAC) 3-isopropyl-(1*H*)-2,1,3-benzothiadiazin-4(3*H*)-one 2,2-dioxide (I); 3-isopropyl-(1*H*)-benzo-2,1,3-thiadiazin-4-one 2,2-dioxide. (*C.A.*) 3-(1-methylethyl)-(1*H*)-2,1,3-benzothiadiazin-4(3*H*)-one 2,2-dioxide (9CI); (I) (8CI); Reg. No. *[25057-89-0]*. Its herbicidal properties were first described by A. Fischer (*Proc. Br. Weed Control Conf., 9th*, 1968, p.1042). Introduced by BASF AG as code no. 'BAS 351H'.

**Properties.**

It is a colourless crystalline powder; m.p. 137-139 °C; v.p. <10 nPa at 20 °C. Solubility at 20 °C: 500 mg/kg water; 1.51 kg/kg acetone; 33 g/kg benzene; 180 g/kg chloroform; 861 g/kg ethanol.

**Uses.**

It is a contact herbicide controlling *Anthemis* and *Matricaria* spp., *Chrysanthemum segetum, Galium aparine, Lapsana communis* and *Stellaria media* in winter and spring cereals when applied at 1.1-2.2 kg a.i. in 225-450 l water/ha after the cereal has reached the 3-4 leaf stage. Other crops are, *e.g.* maize, peas, rice, soyabeans. It is absorbed by the leaves and, having little effect on germinating seeds, is not used pre-em. Herbicidal activity is brief, giving no residual problems.

**Toxicology.**

Acute oral $LD_{50}$ for rats *c.* 1100 mg/kg. Acute percutaneous $LD_{50}$ for rats >2500 mg/kg; applications to the shaved skin of rabbits caused no irritation, but severe eye irritation healed after a week.

**Formulations.**

'Basagran' ('BAS 35 100H'; w.p. 500 g a.i./kg). 'Basagran liquid' ('BAS 35 107H'; 480 g/l). 'Basagran DP', (Reg. No. *[70281-42-4]*), ('BAS 35 801H'; solution, 250 g bentazone + 350 g dichlorprop-ethylammonium/l). 'Basagran MCPB', solution (bentazone + MCPB). 'Basagran M', (Reg. No. *[59472-65-0]*), ('BAS 43 300H'; 250 g bentazone + 125 g MCPA/l).

**Analysis.**

Product analysis is by hplc or by glc of the 1-methyl derivative. Residues are analysed by glc with ECD after conversion to the 1-methyl derivative (T. H. Byast *et al., Tech. Rep. ARC Weed Res. Organ.*, No. 15 (2nd Ed.), p.59).

$C_{18}H_{24}N_2O_4$ (332.4)
T5NOJ CMVR BO1 FO1& EX2&2&1

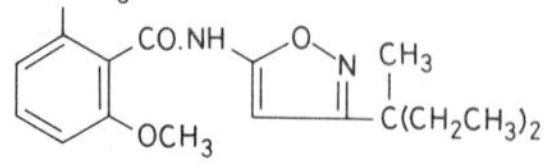

**Nomenclature and development.**

Common name benzamizole (BSI, draft E-ISO); *exception* (USA). Chemical name (IUPAC) *N*-[3-(1-ethyl-3-methylpropyl)isoxazol-5-yl]-2,6-dimethoxybenzamide. (*C.A.*) *N*-[3-(1-ethyl-1-methylpropyl)-5-isoxazoyl]-2,6-dimethoxybenzamide (9CI); Reg. No. *[82558-53-7]*. Its herbicidal properties were described by F. Huggenberger *et al.*(*Proc. Br. Crop Prot. Conf.-Weeds,* 1982, **1**, 47). Introduced by Eli Lilly & Co. (GBP 2 084 140) as code no. 'EL-107'.

**Properties.**

Pure benzamizole is a white crystalline solid; m.p. 176-179 °C. Solubility at 25 °C: 1-2 mg/l water; 50-100 g/l ethyl acetate, methanol; 70-80 mg/l hexane. Stable in water at pH5-9.

**Uses.**

A herbicide for the pre-em. control of broad-leaved weeds in cereals at 50-200 g a.i./ha. Weeds controlled include *Matricaria* spp., *Stellaria media, Polygonum, Veronica,* and *Viola* spp.

**Toxicology.**

Acute oral $LD_{50}$: for rats and mice >10 000 mg/kg; for dogs >5000 mg/kg. Dermal applications of 200 mg/kg to rabbits produced no irritation or toxicity and eye application produced only slight conjunctivitis.

**Formulations.**

These include s.c. (500 g a.i./l).

**Analysis.**

Product and residue analysis is by hplc, details are available from Eli Lilly & Co.

$C_{18}H_{18}ClNO_5$ (363.8)
2ONUYOVR&R CG BO1 FO1

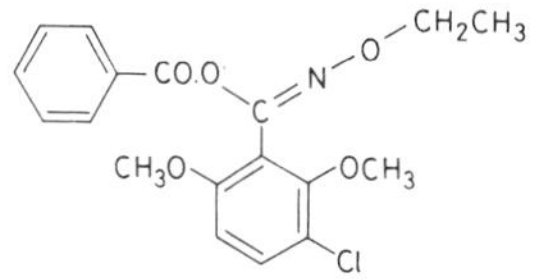

**Nomenclature and development.**

Common name benzoximate (BSI, E-ISO, F-ISO); benzomate (JMAF). Chemical name (IUPAC) 3-chloro-α-ethoxyimino-2,6-dimethoxybenzyl benzoate; ethyl *O*-benzoyl 3-chloro-2,6-dimethoxybenzohydroximate. (*C.A.*) Benzoic acid anhydride with 3-chloro-*N*-ethoxy-2,6-dimethoxybenzenecarboximidic acid; benzoic acid anhydride with 3-chloro-*N*-ethoxy-2,6-dimethoxybenzimidic acid (8CI); Reg. No. *[29104-30-1]*. Its acaricidal properties were reported (*Jpn. Pestic. Inf.*, 1972, (9), 41). Introduced by Nippon Soda Ltd (GBP 1 247 817) as code no. 'NA-53M'; trade marks 'Citrazon', in some European countries 'Aazomate'.

**Properties.**

It is a colourless crystalline solid, m.p. 73 °C. Solubility at 20 °C: practically insoluble in water; 650 g/l benzene; 1.46 kg/l dimethylformamide; 80 g/l hexane; 710 g/l xylene. Stable in acid media but decomposed by concentrated alkali.

**Uses.**

It is a non-systemic acaricide, effective against *Panonychus ulmi* and *P. citri* on apples and citrus at 10-13 g a.i./100 l. See also: A. Formigoni *et al.* (*Atti Giornate Fitopatol.*, 1973, p. 95), K. K. Siccama *et al.* (*Meded. Fac. Landbouwwet. Rijksuniv. Gent*, 1977, **42**, 1479).

**Toxicology.**

Acute oral $LD_{50}$: for Wistar rats >5000 mg/kg; for mice 12 000-14 500 mg/kg. Acute percutaneous $LD_{50}$ for rats and mice >15 000 mg/kg. In 2-y feeding trials rats receiving 400 mg/kg diet suffered no ill-effect. $LC_{50}$ (48-h) for carp is 1.75 mg tech./l.

**Formulations.**

'Citrazon 20 EC', e.c. (200 g a.i./l).

**Analysis.**

Product analysis is by u.v. spectrometry. Residues may be determined by colorimetry after conversion to a derivative.

$C_{18}H_{17}Cl_2NO_3$ (366.2)
GR BG DNVR&Y1&VO2

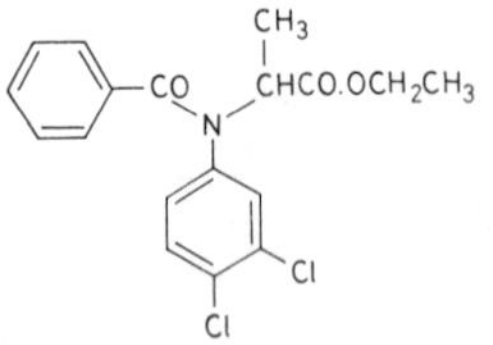

**Nomenclature and development.**

Common name benzoylprop-ethyl (BSI, E-ISO, F-ISO); benzoylprop (WSSA, for parent acid). Chemical name (IUPAC) ethyl *N*-benzoyl-*N*-(3,4-dichlorophenyl)-DL-alaninate (I); ethyl (±)-2-[*N*-(3,4-dichlorophenyl)benzamido]propionate; ethyl (±)-2-(*N*-benzoyl-3,4-dichloroanilino)propionate. (*C.A.*) (I) (8 & 9CI); Reg. No. *[22212-55-1]* (benzoylprop-ethyl), *[22212-56-2]* (parent acid). Its herbicidal properties were described by T. Chapman *et al.*, (*Symp. New. Herbic., 3rd*, 1969, p. 40). Introduced by Shell Research Ltd (GBP 1 164 160) as code no. 'WL 17 731'; trade mark 'Suffix'.

**Properties.**

Technical grade benzoylprop-ethyl (≥93% pure) is an off-white crystalline powder; m.p. 70-71 °C; v.p. 4.7 μPa at 20 °C. Solubility at 25 °C: *c.* 20 mg/l water; 700-750 g/l acetone; 250-300 g/l 'Shellsol A'. It is photochemically stable, and hydrolytically stable $3 \leqslant pH \leqslant 6$.

**Uses.**

Applied at 1.0-1.5 kg a.i./ha to wheat between the end of tillering and fourth node stage of the crop, it gives an 85-95% control of wild oat (*Avena barbata, A. fatua, A. ludoviciana, A. sterilis*). Wheat tolerates applications of 4 kg/ha. Field beans, oil seed rape and culinary mustard may also be safely treated. Crop competition ensures the effective control of wild oat, the stem elongation of which is inhibited (B. Jeffcoat & W. N. Harries, *Pestic. Sci.*, 1973, **4**, 891). The fate of the compound in plants, soils and animals has been reported (K. I. Beynon *et al.*, *ibid.*, 1974, **5**, 429, 443, 451; J. V. Crayford *et al.*, *ibid.*, 1976, **7**, 559).

**Toxicology.**

Acute oral $LD_{50}$: for rats 1555 mg/kg; for mice 716 mg/kg; for domestic fowl and several bird species >1000 mg/kg. Acute percutaneous $LD_{50}$ for rats >1000 mg/kg. In 90-d feeding trials no change of toxicological significance was seen in rats receiving 1000 mg/kg diet or dogs 300 mg/kg diet. $LC_{50}$ (96-h) for rainbow trout is 2.2 mg/l.

**Formulations.**

An e.c. (200 or 250 g a.i./l).

**Analysis.**

Product analysis is by glc (*CIPAC Handbook*, 1983, **1B**, in press) or by i.r. spectroscopy. Residues may be determined by glc with ECD (A. N. Wright & B. L. Mathews, *Pestic. Sci.*, 1976, **7**, 339). Details of methods can be obtained from Shell International Chemical Co. Ltd.

$C_9H_9N_3OS$ (207.2)
T56 BN DSJ CMVM1

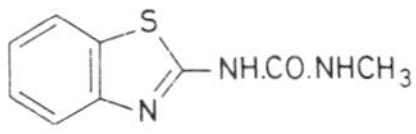

Nomenclature and development.

Common name benzthiazuron (BSI, E-ISO, F-ISO); *exception* Canada. Chemical name (IUPAC) 1-benzothiazol-2-yl-3-methylurea. (*C.A.*) *N*-2-benzothiazoly-*N*'-methylurea (9CI); 1-(2-benzothiazolyl)-3-methylurea (8CI); Reg. No. *[1929-88-0]*. Its herbicidal properties were reported by H. Hack (International Meeting on Selective Weed Control in Sugar Beet Crops, Marly-le-Roi, 9-10, March 1967). Introduced by Bayer AG (BEP 647 740; BP 1 004 469) as code no. 'Bayer 60 618'; trade mark 'Gatnon'.

**Properties.**

Pure benzthiazuron is a colourless powder, which decomposes, with sublimation, at 287 °C; v.p. 1.3 mPa at 90 °C. Solubility at 20 °C: 12 mg/l water; 5-10 mg/l acetone, chlorobenzene, xylene.

**Uses.**

Benzthiazuron is used as a pre-em. herbicide on sugar beet at rates of 3.2-6.4 kg a.i./ha.

**Toxicology.**

Acute oral $LD_{50}$ for rats 1280 mg/kg; percutaneous applications of 500 mg/kg to rats gave no symptom. In 60-d feeding trials all rats receiving 130 mg/kg daily survived.

**Formulations.**

A w.p. (800 g/kg). 'Merpelan', (Reg. No. *[12738-05-5]*), w.p. (600 g benzthiazuron + 125 g lenacil/kg).

**Analysis.**

Product analysis is by i.r. spectroscopy, details are available from Bayer AG. Residue analysis is by u.v. spectroscopy (H. J. Jarczyk & K. Vogeler, *Pflanzenschutz-Nachr. (Engl. Ed.)*,1967, **20**, 575).

# S-Benzyl O,O-di-isopropyl phosphorothioate

$C_{13}H_{21}O_3PS$ (288.3)
1Y1&OPO&S1R&OY1&1

$C_6H_5$—$CH_2SP(O)[OCH(CH_3)_2]_2$

**Nomenclature and development.**

Common name IBP (JMAF). Chemical name (IUPAC) *S*-benzyl *O,O*-di-isopropyl phosphorothioate (I). (*C.A.*) *O,O*-bis(1-methylethyl) *S*-phenylmethyl phosphorothioate (9CI); (I) (8CI); Reg. No. *[26087-47-8]*. Its fungicidal properties were reviewed by Y. Vesugi (*Jpn. Pestic. Inf.*, 1970, No. 2, p. 11). Introduced by Kumiai Chemical Industry Co. Ltd. as trade mark 'Kitazin P'.

**Properties.**

The pure compound is a yellow oil; b.p. 126 °C/0.04 mmHg; $n_D$ 1.5106. Solubility at 18 °C: 1 g/l water; soluble in most organic solvents. The technical grade is *c.* 94% pure.

**Uses.**

It is a systemic fungicide used to control *Piricularia oryzae* in rice. The e.c. is applied at 400-600 g a.i./ha in 1000 l/ha as soon as blast lesions appear; 1 or 2 sprays may be needed during the head-sprouting season. Its translocation and metabolism in the rice plant have been reported (H. Yamamoto *et al., Agric. Biol Chem.*, 1973, **37**, 1553).

**Toxicology.**

Acute oral $LD_{50}$ for mice 600 mg/kg. Acute percutaneous $LD_{50}$ for mice 4000 mg/kg. $LC_{50}$ for carp is 5.1 mg/l.

**Formulations.**

These include: e.c. (480 g a.i./l); dust (20 g/kg); granules (170 g/kg).

**Analysis.**

Product analysis is by tlc with colorimetry. Residues may be determined by glc with FPD.

# 5-Benzyl-3-furylmethyl (*E*)-(1*R*)-*cis*-2,2-dimethyl-3-(2-oxothiolan-3-ylidenemethyl)cyclopropanecarboxylate

$C_{23}H_{24}O_4S$ (396.5)
T5SVYTJ CU1- BL3TJ A1 A1 CVO1-
DT5OJ B1R &&(1R)-CIS-(E) FORM

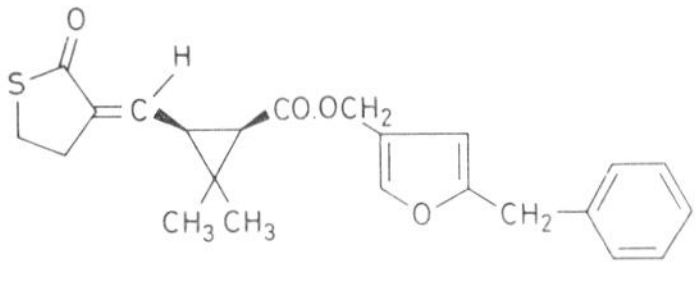

**Nomenclature and development.**

Chemical name (IUPAC) 5-benzyl-3-furylmethyl (*E*)-(1*R*,3*S*)-2,2-dimethyl-3-(2-oxothiolan-3-ylidenemethyl)cyclopropanecarboxylate; 5-benzyl-3-furylmethyl (*E*)-(1*R*)-*cis*-2,2-dimethyl-3-(2-oxothiolan-3-ylidenemethyl)cyclopropanecarboxylate. (*C.A.*) [1*R*-[1α,3α(*E*)]]-[5-(phenylmethyl)-3-furanyl]methyl 3-[(dihydro-2-oxo-3(2*H*)-thienylidene)methyl]-2,2-dimethylcyclopropanecarboxylate (9CI); Reg. No. *[58769-20-3]*. ENT 29 117, AI3-29117. Its insecticidal activity was described by J. Martel & J. Buendia (*Int. Congr. Pestic. Chem. IUPAC, 3rd*, 1974) and by J. Lhoste & F. Rauch (*Pestic. Sci.*, 1976, **7**, 247). Introduced by Roussel Uclaf (FP 2 097 244; USP 3 842 177) as code no. 'RU 15 525'; trade mark 'Kadethrin'.

**Properties.**

The technical grade (⩾93% pure *m*/*m* of stated stereoisomer) is a yellowish-brown fibrous mass; $[\alpha]_D$ 16-19°. Solubility 10 g/kg ethanol; soluble in acetone, benzene, dichloromethane, piperonyl butoxide, xylene; slightly soluble in deodorised kerosene. Rapidly decomposed in light; unstable to heat.

**Uses.**

It shows rapid knock-down action in addition to a lethal effect against many classes of insect, especially cockroaches, houseflies and mosquitoes. It is used in aerosols and sprays with bioresmethrin and piperonyl butoxide.

**Toxicology.**

Acute oral $LD_{50}$: for male rats 1324 mg/kg, for females 650 mg/kg; for dogs 1000 mg/kg. Acute percutaneous $LD_{50}$ for female rats >3200 mg/kg. Inhalation tests (10-d) with rats and guinea-pigs at 200 × normal dose in aerosols was tolerated perfectly. In 90-d feeding trials NEL was: for rats 25 mg/kg daily; for dogs 15 mg/kg daily. No teratogenic effect was observed in mice, rats or rabbits. It is rapidly metabolised and excreted by rats (K. Ohsawa & J. E. Casida, *J. Agric. Food Chem.*, 1980, **28**, 250).

**Formulations.**

Technical grade. Aerosol concentrates (with bioresmethrin and piperonyl butoxide).

**Analysis.**

Product analysis is by glc or hplc, details are available from Roussel Uclaf.

$C_{14}H_9Cl_2NO_5$ (342.1)
GR CG DOR DNW CVO1

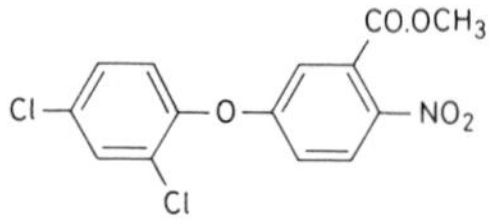

Nomenclature and development.

Common name bifenox (BSI, ANSI, WSSA, draft E-ISO). Chemical name (IUPAC) methyl 5-(2,4-dichlorophenoxy)-2-nitrobenzoate (I). (*C.A.*) (I) (8 & 9CI); Reg. No. *[42576-02-3]* formerly *[12680-11-4]*. Its insecticidal properties were described by W. M. Dest *et al.* (*Proc. Northeast. Weed Sci. Conf.*, 1973, **27**, 31). Introduced by Mobil Chemical Co. (who no longer manufacture or market it) and later by Eli Lilly & Co. and Rhône-Poulenc (GBP 1 232 368; USP 3 652 645; 3 776 715 all to Mobil Chemical Co.) as code no. 'MC-4379' (Mobil Chemical Co.); trade mark 'Modown' (Mobil and Rhône-Poulenc).

**Properties.**

Bifenox forms yellow crystals; m.p. 84-86 °C; v.p. 320 μPa at 30 °C. Solubility: 0.35 mg/1 water; 400 g/kg acetone, chlorobenzene; <50 g/kg ethanol; <10 g/kg aliphatic hydrocarbons; 300 g/kg xylene. Stable ⩽175 °C. At 22 °C stable in aqueous solution at pH5.0-7.3, rapidly hydrolysed at pH9.0. Stable to u.v. light between 290-400 nm, <5% degradation in 48 h. In a saturated aqueous solution, 50% loss occurs in 24 min at 250-400 nm and in *c.* 5 h for a thin film on soil.

**Uses.**

It is used for pre-em. and directed post-em. treatment to control broad-leaved weeds and some grasses in cereals, maize, rice, sorghum, soyabeans. Residual activity persists for 35-70 d depending upon rainfall, soil type and plant species.

**Toxicology.**

Acute oral $LD_{50}$ for rats >6400 mg tech./kg. Acute percutaneous $LD_{50}$ for rabbits >20 000 mg/kg; there is no serious eye irritation or inhalation toxicity. In 2-y feeding trials rats and dogs receiving 600 mg/kg diet showed no ill-effect. $LC_{50}$ (96-h) is: for rainbow trout 27 mg/1; for bluegill 0.47 mg/1.

**Formulations.**

'Alibi', bifenox + linuron. 'Ceridor', bifenox + mecoprop-(dimethylammonium).

**Analysis.**

Product analysis is by glc. Residues may be determined by glc with MCD.

$C_{15}H_{18}N_2O_6$ (322.3)
2Y1&R CNW ENW BOV1UY1&1

$(CH_3)_2C=CHCO.O$ $CH_3$
$O_2N$ $CHCH_2CH_3$
$NO_2$

**Nomenclature and development.**

Common name binapacryl (BSI, E-ISO, F-ISO, ANSI, JMAF). Chemical name (IUPAC) 2-*sec*-butyl-4,6-dinitrophenyl 3-methylbut-2-enoate; 2-*sec*-butyl-4,6-dinitrophenyl 3-methylcrotonate (I). (*C.A.*) 2-(1-methylpropyl)-4,6-dinitrophenyl 3-methyl-2-butenoate (9CI); (I) (8CI); Reg. No. *[485-31-4]*. ENT 25 793, OMS 571. Its biological properties were described by L. Emmel & M. Czech (*Anz. Schaedlingskd.*, 1960, **33**, 145). Introduced by Hoechst AG (GBP 855 736; DEP 1 099 787) as code no. 'Hoe 02 784'; trade marks 'Acricid', 'Endosan', 'Morocide', 'Morrocid'.

**Properties.**

Pure binapacryl is a colourless crystalline powder; m.p. 66-67 °C; v.p. 13 mPa at 60 °C; $\varrho$ 1.2 g/cm$^3$ at 20 °C. Solubility at 20 °C: *c.* 1 mg/l water at pH 5; *c.* 400 mg/l hexane; >500 g/l dichloromethane, ethyl acetate, toluene; *c.* 21 g/l methanol. Technical grade (purity ≥97%), m.p. 65-69 °C; $\rho$ 1.25-1.28 g/cm$^3$ at 20 °C. It is unstable in alkali and concentrated acids, suffers slight hydrolysis on long contact with water and is slowly decomposed by u.v. light. Incompatible with alkalies. It may be phytotoxic when mixed with some organophosphorus compounds.

**Uses.**

Binapacryl is a non-systemic acaricide which is effective against powdery mildews, used mainly against all stages of spider mites and powdery mildews of apples, citrus, cotton, pears and top fruit at 25-50 g a.i./100 l.

**Toxicology.**

Acute oral $LD_{50}$: for rats 150-225 mg/kg; for male mice 1600-3200 mg/kg; for female guinea-pigs 300 mg/kg; for dogs 450-640 mg/kg. Acute percutaneous $LD_{50}$ (acetone solution) for rabbits and mice 750 mg/kg; slightly irritating to the eyes. In 2-y feeding studies rats tolerated without symptom 200-500 mg/kg diet and dogs 20-50 mg/kg diet. Maximum tolerated dose is: for guppies 0.5 mg/l; for carp 1 mg/l; for trout 2 mg/l.

**Formulations.**

Formulations include: e.c. (384 g a.i./l); dispersion (500 g/l).

**Analysis.**

Product analysis is by hydrolysis to the parent phenol (dinoseb) measured by colorimetry (D. S. Farrington *et al., Analyst (London)*, 1983, **108**, 353) or by titration with titanium trichloride. Residues may be measured by the former method (*Pestic. Anal. Manual*, 1979, **I**; R. W. Buxton & T. A. Mohr, *Anal. Methods Pestic., Plant Growth Regul. Food. Addit.*, 1967, **5**, 235; *Anal. Methods Pestic. Plant Growth Regul.*, 1972, **6**, 314). Particulars of these methods may be obtained from Hoechst AG.

$C_{19}H_{26}O_3$ (302.4)
L5V BUTJ B2U1 C1 DOV- BL3TJ A1 A1
C1UY1&1 &&(RS)-CYCLOPENTYL
FORM

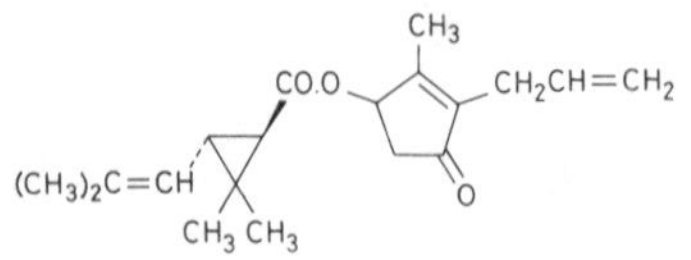

**Nomenclature and development.**

Common name bioallethrin (BSI, New Zealand); depalléthrin (France); the name *d-trans*-allethrin is used in USA. Chemical name (IUPAC) (1*RS*)-3-allyl-2-methyl-4-oxocylopent-2-enyl (1*R*,3*R*)-2,2-dimethyl-3-(2-methylprop-1-enyl)cyclopropane= carboxylate; (1*RS*)-3-allyl-2-methyl-4-oxocyclopent-2-enyl (1*R*)-*trans*-2,2-dimethyl-3-(2-methylprop-1-enyl)cyclopropanecarboxylate; (1*RS*)-3-allyl-2-methyl-4-oxocyclopent-2-enyl (1*R*)-*trans*-chrysanthemate. (*C.A.*) 2-methyl-4-oxo-3-(2-propenyl)-2-cyclopenten-1-yl 2,2-dimethyl-3-(2-methyl-1-propenyl)cyclopropanecarboxylate (9CI); *dl*-2-allyl-4-hydroxy-3-methyl-2-cyclopenten-1-one ester of *d-trans*-2,2-dimethyl-3-(2-methylpropenyl)cyclopropanecarboxylic acid; Reg. No. *[584-79-2]* formerly *[22431-63-6]*. ENT 16 275. The improved insecticidal activity of this pair of esters of the (1*R*)-*trans* acid was described by J. Lhoste *et al.* (*C. R. Seances Acad. Agric. Fr.*, 1967, **53**, 686). Introduced by Roussel Uclaf and by McLaughlin Gormley King Co. (USP 3 159 535 to Roussel Uclaf) as trade marks 'Bioallethrine' (to Roussel Uclaf in some countries), 'D-Trans' (to McLaughlin Gormley King Co.); 'Pynamin Forte' (Sumitomo Chemical Co.) comprises an 80:20 ratio of bioallethrin with the analogous esters of the (1*R*)-*cis*-acid; this mixture of stereoisomers has been referred to as *d*-allethrin.

**Properties.**

The material in 'D-Trans' is an amber viscous liquid; *d* 0.997; $n_D^{25}$ 1.500; flash point 65.6 °C (Tag open cup) and contains 90% *m*/*m* bioallethrin. That in 'Pynamin Forte' has b.p. 153 °C/0.4 mmHg; $d_4^{26.5}$ 1.000-1.020; $n_D^{20}$ 1.503-1.506; v.p. 16 mPa at 30 °C. Solubility: virtually insoluble in water; >1 kg/kg hexane, methanol, xylene.

**Uses.**

Bioallethrin is a potent contact, non-systemic insecticide which produces a rapid 'knockdown' and is used against household insect pests. Piperonyl butoxide and other synergists are included to inhibit metabolic detoxication.

**Toxicology.**

Acute oral $LD_{50}$: for male rats 425-575 mg ('D-Trans')/kg, for females 845-875 mg/kg, for males 1320 mg (as 'Pynamin Forte')/kg, for females 310 mg/kg. Acute percutaneous $LD_{50}$ for rats >2500 mg 'Pynamin Forte'/kg. In 90-d feeding trials NEL for rats was 750 mg/kg diet. $LC_{50}$ (8-d) for mallard duck and bobwhite quail was 5620 mg 'Pynamin Forte'/kg diet.

**Formulations.**

These include: aerosol concentrates, sprays in kerosene, dusts. Synergists are frequently included, especially piperoxyl butoxide, bis(2,3,3,3-tetrachloropropyl) ether, *N*-(2-ethylhexyl)-8,9,10-trinorborn-5-ene-2,3-dicarboximide. Mixtures include bioallethrin + bioresmethrin.

**Analysis.**

Product analysis is by glc (*CIPAC Handbook*, 1980, **1A**, 1097); proportions of stereoisomers may be determined by glc of suitable derivatives (A. Murano, *Agric. Biol. Chem.*, 1972, **36**, 2203; 1977, **41**, 2003; 1978, **42**, 671; F. E. Rickett, *Analyst (London)*, 1973, **98**, 687) or by nmr (F. E. Rickett & P. B. Henry, *ibid.*, 1974, **99**, 330). Residues may be determined by glc (D. B. McClellan, *Anal. Methods Pestic., Plant Growth Regul. Food Addit.*, 1964, **2**, 25; *Anal. Methods Pestic. Plant Growth Regul.*, 1972, **6**, 283; J. Sherma, *ibid.*, 1976, **8**, 117).

$C_{19}H_{26}O_3$ (302.4)
L5V BUTJ B2U1 C1 DOV- BL3TJ A1 A1
C1UY1&1 &&(RS)-CYCLOPENTENYL
FORM

**Nomenclature and development.**

Common name esdepalléthrine (France); the name *S*-bioallethrin (based on the BSI name for the racemic cyclopentenyl analogue) has no official standing. Chemical name (IUPAC) (1*S*)-allyl-2-methyl-4-oxocyclopent-2-enyl (1*R*,3*R*)-2,2-dimethyl-3-(2-methylprop-1-enyl)cyclopropanecarboxylate; (1*S*)-3-allyl-3-methyl-4-oxocylcopent-2-enyl (1*R*)-*trans*-2,2-dimethyl-3-(2-methylprop-1-enyl)cyclopropanecarboxylate; (1*S*)-3-allyl-3-methyl-4-oxocyclopent-2-enyl (1*R*)-*trans*-chrysanthemate. (*C.A.*) [1*R*-[1α(*S**),3β]]-2-methyl-4-oxo-3-(2-propenyl)-2-cyclopenten-1-yl 2,2-dimethyl-3-(2-methyl-1-propenyl)cyclopropanecarboxylate (9CI); Reg. No. *[28434-00-6]*. AI3-29 024. Its insecticidal properties were reported by F. Rauch *et al.* (*Meded. Fac. Landbouwwet. Rijksuniv. Gent.*, 1972, **37**, 755). Introduced by Roussel Uclaf as code no. 'RU 3054'; trade mark 'Esbiol'; 'Esbiothrin' for a product with ⩾72% esdepalléthrine and ⩾21% *m/m* other allethrin isomers.

**Properties.**

'Esbiol' contains ⩾90% esdepalléthrine. It is a yellow viscous liquid; $d^{20}$ 1.000-1.020; $[\alpha]_D^{20}$ ⩾−47° (5% in toluene). Solubility: sparingly soluble in water; 100 g/kg hexane; soluble in most organic solvents. Flash point *c.* 120 °C (open cup). 'Esbiothrin' has similar physical properties; $[\alpha]_D^{20}$ ⩾−37° (5% in toluene).

**Uses.**

Esdepalléthrine is by far the most potent constituent of allethrin (see 130), used in aerosols, coils or sprays indoors against flying and crawling insects, mosquitoes.

**Toxicology.**

Acute oral $LD_{50}$: for male rats 784 mg esdepalléthrine/kg, for females 1545 mg/kg; for chickens >5000 mg/kg. Acute percutaneous $LD_{50}$ for rabbits 1545 mg/kg. Acute inhalation $LC_{50}$: for mice *c.* 1500 mg/m³ air; for rats *c.* 1650 mg/m³ air. In 0.5-y feeding trials NEL for rats was 1000 mg/kg diet. $LC_{50}$ (8-d) for ducks and quail was >5000 mg/kg diet. $LC_{50}$ (96-h): for rainbow trout 10.5 μg/l; for bluegill 33.2 μg/l.

Acute oral $LD_{50}$: for male rats 670 mg 'Esbiothrin'/kg, for females 405 mg/kg. Other toxicological data are similar to those for esdepalléthrine.

**Formulations.**

Usually esdepalléthrine is combined with synergists and other insecticides as: aerosol concentrates, dusts, sprays in kerosene. 'Esbiothrin' is available as the technical grade.

**Analysis.**

Product analysis is by glc (*CIPAC Handbook*, 1980, **1A**, 1097); properties of stereoisomers may be determined by glc of suitable derivatives (A. Murano, *Agric. Biol. Chem.*, 1972, **36**, 2203; A. Horiba *et al., ibid.*, 1977, **41**, 2003; 1978, **42**, 671; F. E. Rickett, *Analyst (London)*, 1973, **98**, 687) by nmr (F. E. Rickett & P. B. Henry, *ibid.*, 1974, **99**, 330). Residues may be determined by glc (D. B. McClellan, *Anal. Methods Pestic., Plant Growth Regul. Food Addit.*, 1964, **2**, 25; *Anal. Methods Pestic. Plant Growth Regul.*, 1972, **6**, 283; J. Sherma, *ibid.*, 1976, **8**, 117).

$C_{22}H_{26}O_3$ (338.4)
T5OJ B1R& D1OV- BL3TJ A1 A1
C1UY1&1 &&(1R)-TRANS FORM

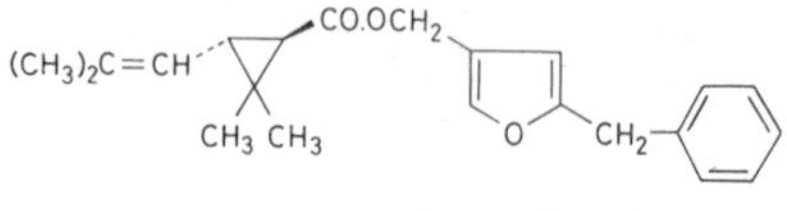

**Nomenclature and development.**

Common name bioresmethrin (BSI, E-ISO);bioresmethrine (F-ISO) Chemical name (IUPAC) 5-benzyl-3-furylmethyl (1*R*,3*R*)-2,2-dimethyl-3-(2-methylprop-1-enyl)cyclopropanecarboxylate; 5-benzyl-3-furylmethyl (1*R*)-*trans*-2,2-dimethyl-3-(2-methylprop-1-enyl)cyclopropanecarboxylate; 5-benzyl-3-furylmethyl (1*R*)-*trans*-chrysanthemate. (*C.A.*) [5-(phenylmethyl)-3-furanyl]methyl (1*R*)-*trans*-2,2-dimethyl-3-(2-methyl-1-propenyl)cyclopropanecarboxylate (9CI); 5-benzyl-3-furylmethyl (1*R*)-*trans*-chrysanthemate; Reg. No. *[28434-01-7]* formerly *[10453-54-0]*. *d-trans*-Resmethrin. ENT 27 622, AI3-27 622. Its insecticidal properties were described by M. Elliott *et al.* (*Nature (London)*, 1967, **213**, 493) and developed by Fisons Ltd., the FMC Corp., Roussel Uclaf, Sumitomo Chemical Co., the Wellcome Foundation. (GBP 1 168 797; 1 168 798; 1 168 799 all to NRDC) code no. 'NRDC 107','FMC 18 739','RU 11 484'; trade marks 'Combat White Fly Insecticide' (to Fisons Ltd), 'Resbuthrin' (to Wellcome), 'Biobenzyfuroline'; 'Chryson Forte' (to Sumitomo Chemical Co.) which contains 80:20 *m/m* bioresmethrin to cismethrin. The common name cismethrin (BSI and E-ISO); cismethrine (F-ISO); OMS 1800; is the corresponding (1*R*,3*S*)- or (1*R*)-*cis*-stereoisomer; Reg. No. *[35764-59-1]* formerly *[31182-61-3]*. This ratio of isomers has been referred to as *d*-resmethrin.

**Properties.**

Technical bioresmethrin is a viscous yellow to reddish brown liquid which partially solidifies on standing; m.p. 30-35 °C; b.p. 181 °C/0.01mmHg; $d^{20}$ 1.050 $[\alpha]_D^{20}$ −5 to −8° (ethanol). Films exposed to air or sunlight are decomposed (M. Elliott *et al., Proc. Insectic. Fungic. Conf., 7th*, 1973, **2**, 721). It is readily hydrolysed in alkaline media.

The technical 80:20 bioresmethrin-cismethrin is a colourless to yellowish liquid, sometimes partially crystalline at room temperature; $d_{25}^{25}$ 1.045; $n_D^{20}$ 1.527-1.530; v.p. 452 μPa at 20 °C. Solubility at 30 °C 1.2 mg/l water; at 25 °C >1 kg/kg hexane, methanol, xylene.

**Uses.**

Bioresmethrin is a potent contact insecticide effective against a wide range of insects including cockroaches, flies, mosquitoes and plant pests (I. C. Brookes *et al., Soap Chem. Spec.*, 1969, **45**(3), 62); and as a grain protectant. The toxicity to *Musca domestica* is 50× that of pyrethrum (M. Elliott *et al., Nature (London), loc. cit.*) but synergistic factors are lower than those of pyrethrum with the methylenedioxyphenyl synergists. Toxicity to plants is low.

**Toxicology.**

Acute oral $LD_{50}$: for rats 7070-8000 mg bioresmethrin/kg; for chicks >10 000 mg/kg. Acute percutaneous $LD_{50}$ for female rats >10 000 mg/kg. In 90-d feeding trials NEL was: for rats 1200 mg/kg diet; for dogs >500 mg/kg diet. Rats tolerated 4000 mg/kg diet for 60 d. In rats dosed daily with 80 mg/kg from d 8-16 of pregnancy no teratogenic effect was observed; there were foetal deaths at this dose but not at 40 mg/kg. $LC_{50}$ (96-h) for harlequin fish is 14 μg/l; $LC_{50}$ (48-h) for harlequins 18 μg/l, for guppies 0.5-1.0 mg/l. It is very toxic to honeybees (oral $LD_{50}$ 3 ng/bee; topical $LD_{50}$ 6.2 ng/bee).

**Formulations.**

These include: aerosol concentrates (1 g bioresmethrin/l); ready-to-use liquids (2.5 g/l) for oil- or water-based sprays. Combined products include 'Combat Vegetable Insecticide' (bioresmethrin + malathion); 'Chryson Forte' (bioresmethrin + cismethrin); bioresmethrin + bioallethrin.

**Analysis.**

Product analysis is by glc with FID (B. B. Brown, *Anal. Methods Pestic. Plant Growth Regul.*, 1973, **7**, 441) or by hplc (D. S. Gunew, *ibid.*, 1978, **10**, 19).

$C_{12}H_{10}$ (154.2)
RR

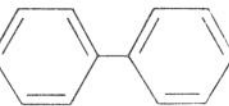

**Nomenclature and development.**

Chemical name (IUPAC) biphenyl (I) accepted (BSI, ISO) in lieu of a common name. (*C.A.*) 1,1'-biphenyl (9CI); (I) (8CI); Reg. No. *[92-52-4]*. Trivial name diphenyl. Its fungicidal activity against citrus rot fungi was described by G. B. Ramsey *et al.* (*Bot. Gaz.*, 1944, **106**, 74).

**Properties.**

It forms colourless crystals; m.p. 70.5 °C; b.p. 256 °C; *d* 1.041. Solubility: very low in water; soluble in most organic solvents.

**Uses.**

It is used to impregnate citrus fruit wraps to prevent attack by rot fungi.

**Toxicology.**

Acute oral $LD_{50}$ for rats 3280 mg/kg. Prolonged exposure of humans to vapour concentrations >5 mg/$m^3$ air is considered dangerous (J. Deichmann *et al., J. Ind. Hyg. Toxicol.*, 1947, **29**, 1).

**Analysis.**

Analysis is by u.v. spectrometry (F. A. Gunther *et al., Analyst (London)*, 1963, **88**, 35; 1966, **91**, 475; S. Norman *et al., J. Assoc. Off. Anal. Chem.*, 1966, **49**, 590; A. Rajzman, *Residue Rev.*, 1964, **8**, 1: S. W. Souci & G. Maier-Haarländer, *ibid.*, 1966, **16**, 103; *AOAC Methods*, 1980, 29.067-29.074).

$C_{14}H_8Cl_2N_4$ (303.1)
T6NN DNNJ CR BG& FR BG

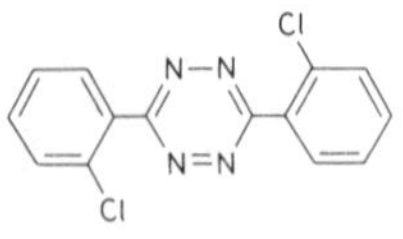

**Nomenclature and development.**

Chemical name (IUPAC) 3,6-bis(2-chlorophenyl)-1,2,4,5-tetrazine (I). (*C.A.*) (I) (9CI); Reg. No. *[74115-24-5]*. Trivial name *bisclofentezin* (rejected proposed common name). Its acaricidal properties were reported by K. M. G. Bryan *et al.* (*Proc. Br. Crop Prot. Conf. - Pests Dis.*, 1981, **1**, 67). Introduced by FBC Limited (EP 5912) as code no. 'NC 21 314'; trade mark 'Apollo'.

**Properties.**

The pure a.i. forms magenta crystals; m.p. 182-186 °C; v.p. 13 fPa at 25 °C. Solubility at 25 °C; <1 mg/kg water; 5 g/l acetone; 2.5 g/l benzene; 50 g/l chloroform; 1.7 g/kg cyclohexane; 1 g/l ethanol, hexane. Partition coefficient octan-1-ol/water 150. The a.i. and formulated products are stable to light, air and heat; of low flammability.

**Uses.**

This acaricide is a contact ovicide with some effect on young mobile stages. It shows some translaminar effect and has good persistence. It is particularly effective against *Panonychus ulmi* and has no effect on beneficial insect species. Used on apples, ornamentals and peas. It is applied to top fruit pre-blossom with a possible second application post-blossom to provide season-long control. It is also being used experimentally on citrus and cotton.

**Toxicology.**

Acute oral $LD_{50}$: for rats 3200 mg a.i./kg; for mallard duck >3000 mg/kg; for bobwhite quail >7500 mg/kg. Acute percutaneous $LD_{50}$ for rats >1332 mg/kg. Non-toxic to fish.

**Formulations.**

'Apollo', w.p. (500 g a.i./kg); s.c. (500 g/l).

**Analysis.**

Product analysis is by hplc. Residues in plants and soil may be determined by hplc with u.v. detection. Details are available from FBC Limited.

*Common name (BSI, draft E-ISO).

# Bis(2,4-dichlorophenyl) ethyl phosphate

$C_{14}H_{11}Cl_4O_4P$ (416.0)
GR CG DOPO&O2&OR BG DG

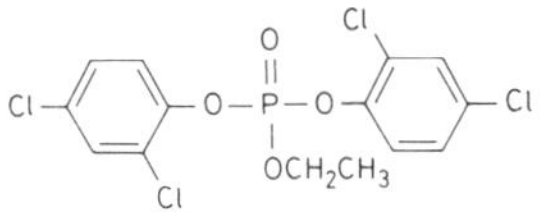

**Nomenclature and development.**

Common name phosdiphen (JMAF). Chemical name (IUPAC) bis(2,4-dichlorophenyl) ethyl phospate (I). (*C.A.*) (I) (8 & 9CI); Reg. No. *[36519-00-3]*. Its fungicidal properties were described by M. Hamada *et al.* (*Ann. Phytopathol. Soc. Japan*, 1971, **37**, 365). Introduced by Hokko Chemical Industry Co. Ltd. and Mitsui Toatsu Chemicals Inc. (JPP 634 549 to Hokko Chemical Industry Co. Ltd.) as code no. 'MTO-460' (Mitsui Toatsu Chemicals Inc.).

**Properties.**

The technical grade (95% pure) is a pale yellow liquid; b.p. 175 °C/0.2 mmHg; $d_4^{25}$ 1.405; $n_D^{20}$ 1.5628; v.p. 66 mPa at 20 C. Solubility at 20 °C: 0.7 mg/l water; freely soluble in acetone, hexane, xylene. At 25 °C 50% loss occurs in 7 d at pH 7, in 4 d at pH 9. It is stable $<$200 °C.

**Uses.**

A fungicide with specific action against *Pyricularia oryzae* on rice at 400-500 g a.i./ha.

**Toxicology.**

Acute oral $LD_{50}$: for male rats 6.2 g a.i./kg; for mice 5.3 g/kg. Acute percutaneous $LD_{50}$: for rats $>$5 g/kg; for mice 9.0-9.6 g/kg. Applications (24-h) of tech. to skin and eyes of rabbits produced no irritation. In 90-d feeding trials NEL for rats and mice was 200 mg/kg diet. No effect was observed on reproduction in, or teratogenicity to rats. Not mutagenic. $LC_{50}$ (48-h) for carp is 1.7 mg/l.

**Formulations.**

These include; e.c. (400 g a.i./l). 'Kasumiron', 'Hokumiron', 'Kasubaron', dust (20 g phosphate ester + 1 g kasugamycin/kg); w.p. (250 g phosphate ester + 10 g kasugamycin/kg).

**Analysis.**

Product and residue analysis is by glc. Details of methods are available from Hokko Chemical Industry Co. Ltd.

$C_{20}H_{23}N_3O$ (337.4)
T5NN DNJ AYOR DR&&YQX1&1&1

**Nomenclature and development.**

Common name bitertanol (BSI, draft E-ISO). Chemical name (IUPAC) *all-rac*-1-(biphenyl-4-yloxy)-3,3-dimethyl-1-(1*H*-1,2,4-triazol-1-yl)butan-2-ol [ratio of racemates (1*RS*,2*RS*) and (1*RS*,2*SR*) is *c.* 20:80]. (*C.A.*) β-([1,1-biphenyl]-4-yloxy)-α-(1,1-dimethylethyl)-1*H*-1,2,4-triazole-1-ethanol (9CI); Reg. No. *[55179-31-2]*. Fungicidal properties were described by W. Brandes *et al.* (*Pflanzenschutz-Nachr. (Engl. Ed.)*, 1979, **32**, 1). Introduced by Bayer AG (DEPS 2 324 010) as code no. 'BAY KWG 0599'; trade mark 'Baycor'.

**Properties.**

It forms colourless crystals: v.p. 1 μPa at 20 °C. Solubility at 20 °C: 5 mg/l water; 100-200 g/l dichloromethane; $\leqslant$ 10 g/kg light petroleum (b.p. 80-100 °C); 30-100 g/l propan-2-ol; 10-50 g/kg toluene.

**Uses.**

Effective fungicide used mainly for the control of diseases of fruit caused by *Venturia* and *Sclerotinia* spp. and of leaf spot diseases of groundnuts and bananas.

**Toxicology.**

Acute oral $LD_{50}$ for rats $>$5000 mg a.i./kg. Acute percutaneous $LD_{50}$ (24-h) for rats $>$5000 mg/kg. In 90-d feeding trials rats receiving 100 mg/kg diet showed no symptom of poisoning. $LC_{50}$ (48-h) for carp 2.5 mg/l; $LC_{50}$ (96-h) for rainbow trout 2.2-2.7 mg/l.

**Formulations.**

Formulations include: w.p., e.c., aerosol generators and dry seed treatments.

**Analysis.**

Details of product analysis are available from Bayer AG. Residue analysis is by glc (W. Specht & M. Tillkes, *ibid.*, 1980, **33**, 61).

$C_{17}H_{26}N_8O_5$ (422.4)
T6O CUTJ EMV1YZ2N1&YZUM FVQ B-AT6NVNJ DZ

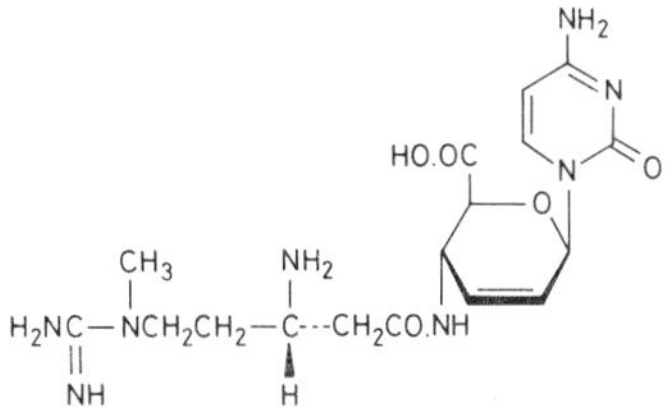

$H_2NC(=NH)-N(CH_3)CH_2CH_2-C(NH_2)(H)-CH_2CO.NH-$ [structure: cytosine-substituted dihydropyran with HO.OC]

**Nomenclature and development.**

Common name blasticidin-S (JMAF). Chemical name (IUPAC) 1-(4-amino-1,2-dihydro-2-oxopyrimidin-1-yl)-4-[(*S*)-3-amino-5-(1-methylguanidino)valeramido]-1,2,3,4-tetradeoxy-β-D-*erythro*-hex-2-enopyranuronic acid. (*C.A.*) (*S*)-4-[[3-amino-5-[(aminoiminomethyl)methylamino]-1-oxopentyl]amino]-1-[4-amino-2-oxo-1(2*H*)-pyrimidinyl]-1,2,3,4-tetradeoxy-β-D-erythrohex-2-enopyranuronic acid (9CI); 4-[3-amino-5-(1-methylguanidino)valeramido-1-[4-amino-2-oxo-1(2*H*)-pyrimidinyl]-1,2,3,4-tetradeoxy-β-D-erythro]hex-2-enopyranuronic acid (8CI); Reg. No. *[2079-00-7]* formerly *[11002-92-9]*, *[12767-55-4]*. This antibiotic was discovered by K. Fakunaga *et al.* (*Bull. Agric. Chem. Soc. Jpn.*, 1955, **19**, 18).

**Properties.**

Produced by the fermentation of *Streptomyces griseochromogenes*, blasticidin-S forms colourless needles; m.p. 253-255 °C; $[\alpha]_D$ +108.4°C. Solubility: soluble in water, acetic acid; insoluble in most common organic solvents.

**Uses.**

It is a contact fungicide used mainly to control *Piricularia oryzae* on rice at *c.* 10 g a.i. in 100 l water/ha. Potential phytoxicity limits its range of use.

**Toxicology.**

Acute oral $LD_{50}$ for mice 39.5 mg/kg. Acute percutaneous $LD_{50}$: for rats 3100 mg/kg; for mice 220 mg/kg. It can produce conjunctivitis. Carp are killed in water containing 8700 mg/l.

**Formulations.**

These include: e.c. (20 g a.i./l); w.p. (40 g/kg); dust (1.6 g/kg). The benzylaminobenzenesulphonate salt is normally used in these because it is less phytotoxic.

$B_4H_{20}Na_2O_{17}$ (381.4)
.NA..B4-O7 &&10$H_2O$
*disodium octaborate*
$B_8H_8Na_2O_{17}$ (412.5)
.NA2.B8-O13 &&4$H_2O$
*sodium metaborate*
$(BNaO_2)_x$
.NA..B-O2

$Na_2B_4O_7 \cdot 10\,H_2O$

**Nomenclature and development.**

The trivial name borax (i) (E-ISO, F-ISO) is accepted in lieu of a common name; the trivial names disodium octaborate (ii) and sodium metaborate (iii), respectively are used. Chemical names (IUPAC) disodium tetraborate (i); disodium octaborate; sodium metaborate. (*C.A.*) borax (i); disodium borate ($Na_2B_8O_{13}$) (ii); sodium borate ($NaBO_2$) (iii) (all 8 & 9CI); Reg. No. *[1303-96-4]* (formerly *[1344-90-7]*) (i), *[12008-41-2]* anhydrous salt, *[7775-19-1]* (iii). Borax has been used as a mild antiseptic fungicide and herbicide. Disodium borate and sodium metaborate were introduced as herbicides by the US Borax & Chemical Corp. (USP 2 998 310 (ii); 3 032 405 (iii)) trade marks 'Polybor' for (ii); 'Hybor' for mixture of (iii) with bromacil.

**Properties.**

Borax is a colourless crystalline solid, efflorescent at low r.h. and loses 5 ml of water of crystallisation at 100 °C; on rapid heating it 'melts' in its water of crystallisation at 75 °C. Solubility at 20 °C: 51.4 g/l water; soluble in ethylene glycol, glycerol; insoluble in ethanol. The aqueous solution is alkaline and so incompatible with some herbicides. It is a fire retardant.

Disodium octaborate is an amorphous powder, its composition being *c.* $Na_2B_8O_{13}.4H_2O$; m.p. *c.* 195 °C. Solubility at 20 °C 95 g/l water.

Sodium metaborate forms colourless crystals which 'melt' in their water of crystallisation at 55 °C. Solubility at 23 °C 480 g/l water.

**Uses.**

Borax and these other borates are non-selective herbicides used to control weeds on non-crop sites. They are frequently used in combination with sodium chlorate to reduce the fire hazard of the latter. Combinations with other herbicides are also used. Persistence of borax in soils is $\leqslant$2 y, but the duration of the protection given by these compounds is markedly dependent on rainfall and soil structure. Borax is also used in insect baits to control ants in food stores.

**Toxicology.**

Acute oral $LD_{50}$: for rats 4500-6000 mg borax/kg, 2330 mg/kg sodium metaborate/kg; for guinea pigs 5300 mg disodium octaborate/kg. Sodium metaborate is a moderate skin irritant.

**Formulations.**

Technical borax is herbicide; 'Nippon' insecticide. Mixtures include: borax + bromacil; borax + sodium chlorate; disodium octaborate + sodium chlorate; 'Hibor', sodium metaborate + bromacil.

**Analysis.**

Product analysis is by acid-base titration (*AOAC Methods*, 1980, 20.036). Residues may be determined by colorimetry (*ibid.*, 1980, 20.037-20.041).

**Nomenclature and development.**

The trivial name Bordeaux mixture (draft E-ISO) is accepted in lieu of a common name. Chemical name (IUPAC) a mixture, with or without stabilising agents, of calcium hydroxide and copper sulphate. (*C.A.*) Bordeaux mixture; Reg. No. *[8011-63-0]*. Trivial name (in USA) tribasic copper sulfate. The mixture was introduced by A. Millardet (*J. Agric. Prat. (Paris)*, 1885, **49**, 513). It may be used as a tank-mix or as pre-formed formulations.

**Properties.**

Prepared from solutions of copper sulphate and suspensions of calcium hydroxide, a blue precipitate is formed. Solubility: insoluble in water or common organic solvents; decomposed by acids; it dissolves in ammonium hydroxide forming a cuprammonium complex. It is incompatible with alkali-sensitive pesticides such as organophosphorus compounds or carbamates. It must not be prepared in metal vessels.

**Uses.**

Bordeaux mixture is used as a protective fungicide for foliage applications, the freshly prepared precipitate having a high tenacity. its use is limited to crop plants at stages of growth when its phytotoxicity is low. Major uses include the control of *Phytophthora infestans* on potatoes, *Venturia inaequalis* on apples and *Pseudoperonospora humuli* on hops.

**Formulations.**

These include: w.p.; flowable suspensions. It can be prepared as a tank-mix using copper sulphate pentahydrate (1.0 kg) with calcium hydroxide (1.25 kg) in water (100 l) for HV application, or (4 kg + 2 kg + 100 l) for LV application (R. Gair & D. J. Yarham, *Insecticide and Fungicide Handbook*, (5th Ed.), p. 128).

**Analysis.**

Product analysis is by digestion with sulphuric acid and determination of the copper content by electrolytic, volumetric, colorimetric or atomic absorption spectroscopic methods (*AOAC Methods*, 1980, 6.067-6.071; *CIPAC Handbook*, 1970, **1**, 226; 1983, **1B**, in press). Residues are determined by digestion with sulphuric acid and perchloric acid, and colorimetry of a complex (*AOAC Methods*, 1980, 25.038-25.043) or by atomic absorption spectroscopy (*ibid*, 25.044-25.048).

$C_{31}H_{23}BrO_3$ (523.4)
T66 BOVJ EQ D- GL66&TJ IR DR DE

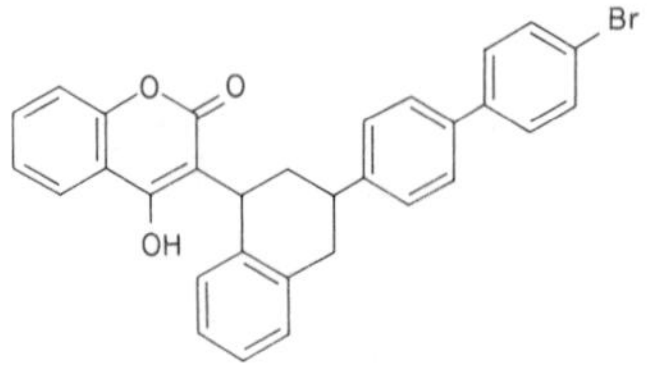

**Nomenclature and development.**

Common name brodifacoum (BSI, E-ISO, F-ISO). Chemical name (IUPAC) 3-[3-(4'-bromobiphenyl-4-yl)-1,2,3,4-tetrahydro-1-naphthyl]-4-hydroxycoumarin. (*C.A.*) 3-[3-(4'-bromo-[1,1'-biphenyl]-4-yl)-1,2,3,4-tetrahydro-1-naphthalenyl]-4-hydroxy-2*H*-1-benzopyran-2-one (9CI); Reg. No. *[56073-10-0]*. Its rodenticidal properties were described by R. Redfern *et al.* (*J. Hyg.*, 1976, **77**, 419). Introduced by Sorex (London) Ltd. and developed by ICI Plant Protection Division. (GBP 1 458 670 to Sorex) as code no. 'WBA 8119' (Sorex),'PP 581' (ICI); trade marks 'Talon', 'Ratak +' (to ICI).

**Properties.**

Brodifacoum is an off-white to fawn powder; m.p. 228-232 °C; v.p. <133μPa at 25 °C. Solubility at 20 °C: <10 mg/l water at pH 7; 6-20 g/l acetone; <0.6-6.0 mg/l benzene; 3 g/l chloroform. It is a weak acid which does not readily form water-soluble salts. There is no loss in 30 d in direct sunlight.

**Uses.**

It is an indirect anticoagulant active against rats and mice including strains resistant to other anticoagulants and against rodent species, such as hamsters, that are difficult to control with other anticoagulants. Its potency is such that, unlike other anticoagulants, a rodent may absorb a lethal dose by taking a 50 mg/kg bait as part of its food intake on only one occasion.

**Toxicology.**

Acute oral $LD_{50}$: for male rats 0.27 mg a.i./kg; for male rabbits 0.3 mg/kg; for male mice 0.4 mg/kg; for female guinea pigs 2.8 mg/kg; for cats *c.* 25 mg/ kg; for dogs 0.25-1.0 mg/kg; for chickens 4.5 mg/kg; for mallard duck 2.0 mg/kg. Acute percutaneous $LD_{50}$ for rats 50 mg tech./kg, 200 mg a.i. (as dust)/kg for 6-h exposure. In 42-d feeding trials rats receiving 0.1 mg/kg diet suffered no ill-effect. $LC_{50}$ (96-h) is: for bluegill 0.165 mg/l; for rainbow trout 0.051 mg/l.

**Formulations.**

Sorex brodifacoum Bait (20 and 50 mg a.i./kg); 'Talon', bait (50 mg/kg).

**Analysis.**

Product analysis is by u.v. spectroscopy (particulars from Sorex (London) Ltd) or by hplc (details from ICI Plant Protection Division).

$C_9H_{13}BrN_2O_2$ (261.1)
T6MVNVJ CY2&1 EE F1

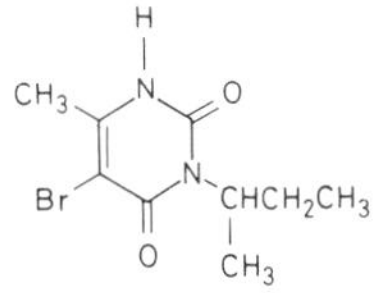

Nomenclature and development.

Common name bromacil (BSI, E-ISO, F-ISO, ANSI, WSSA, JMAF). Chemical name (IUPAC) 5-bromo-3-*sec*-butyl-6-methyluracil (I). (*C.A.*) 5-bromo-6-methyl-3-(1-methylpropyl)-2,4(1*H*,3*H*)-pyrimidinedione (9CI); (I) (8CI); Reg. No. *[314-40-9]*. The herbicidal activity of certain substituted uracils was reported by H. C. Bucha *et al.*, (*Science*, 1962, **137**, 537). Introduced by E. I. du Pont de Nemours & Co. (Inc.) (USP 3 325 357; 3 352 862; BEP 625 897) as code no. 'Du Pont Herbicide 976'; trade mark 'Hyvar' X.

**Properties.**

Pure bromacil is a colourless crystalline solid; m.p. 158-159 °C; v.p. 33 μPa at 25 °C. Solubility at 25 °C: 815 mg/l water; 201 g/kg acetone; 77 g/kg acetonitrile; 155 g/kg ethanol; 33 g/kg xylene. It is stable up to its m.p. It is decomposed slowly by concentrated acids; soluble in alkali (96 g/kg 0.9M sodium hydroxide). The water-soluble formulation is incompatible with ammonium sulphamate, No.2 diesel oil, xylene and liquid formulations of aminotriazole. Calcium salts may cause precipitation, but there is no problem with hard water.

**Uses.**

Bromacil is a non-selective inhibitor of photosynthesis absorbed mainly through the roots and recommended for general weed control on uncropped land at 5-15 kg a.i./ha; for subsequent annual maintenance 2-4 kg/ha is enough. It may also be used for annual weed control in established citrus plantations at 1.6-3.2 kg/ha; for perennial grasses and annual weeds at 3.2-8.0 kg/ha; in pineapple plantations at 1.8-5.5 kg/ha. At the higher rates its effect usually persists for more than one season.

**Toxicology.**

Acute oral $LD_{50}$ for rats 5200 mg/kg. Acute percutaneous $LD_{50}$ for rabbits >5000 mg/kg; mildly irritating to skin of young guinea-pigs, but not to older ones. Inhalation $LC_{50}$ for rats >4.8 mg/l. In 2-y feeding trials NEL for rats and dogs was 250 mg/kg diet. $LC_{50}$ (96-h) for mallard ducklings and bobwhite quail 10 000 mg/l. $LC_{50}$ (48-h) is: for bluegill 71 mg/l; for rainbow trout 75 mg/l; for carp 164 mg/l.

**Formulations.**

'Hyvar' X, w.p. (800 g a.i./kg). 'Hyvar' X-L, s.c. (240 g a.e./l; 219 g bromacil-lithium/kg). 'Krovar' I, (Reg. No. *[8071-35-0]*), w.p. (400 g bromacil + 400 g diuron/kg); 'Krovar' II, w.p. (530 g bromacil + 270 g diuron/kg).

**Analysis.**

Residue analysis is by glc (H. L. Pease, *J. Agric. Food Chem.*, 1966, **14**, 94; 1968, **16**, 54; H. Jarczyk, *Pflanzenschutz-Nachr. (Engl. Ed.)*, 1975, **28**, 319; H. L. Pease & J. F. Deye, *Anal. Methods Pestic., Plant Growth Regul. Food Addit.*, 1967, **5**, 335; *Anal. Methods Pestic. Plant Growth Regul.*, 1972, **6**, 663) and for residues in soil (T. H. Byast *et al., Tech. Rep. ARC Weed Res. Organ.*, No. 15 (2nd Ed.), p.47).

$C_{30}H_{23}BrO_4$ (527.4)
T66 BOVJ DYR&1YQR DR DE&& EQ

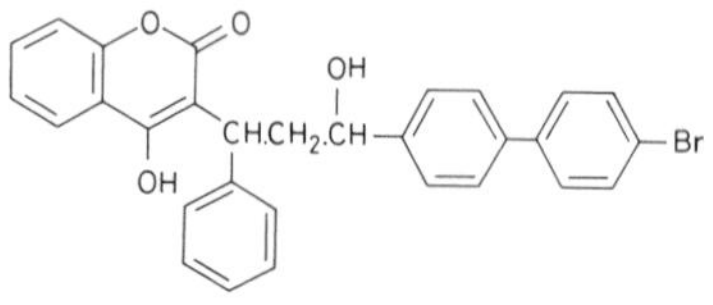

**Nomenclature and development.**

Common name bromadiolone (BSI, E-ISO, F-ISO); broprodifacoum (Republic of South Africa). Chemical name (IUPAC) 3-[3-(4'-bromobiphenyl-4-yl)-3-hydroxy-1-phenylpropyl]-4-hydroxycoumarin. (*C.A.*) 3-[3-[4'-bromo(1,1'-biphenyl)-4-yl]-3-hydroxy-1-phenylpropyl]-4-hydroxy-2*H*-1-benzopyran-2-one (9CI); 3-[α-[*p*-(*p*-bromophenyl)-β-hydroxyphenethyl]benzyl-4-hydroxycoumarin] (8CI); Reg. No. *[28772-56-7]*. Its rodenticidal properties were described by M. Grand (*Phytiatr. Phytopharm.*, 1976, **25**, 69). Introduced by Lipha S.A. (FP 96 651; USP 3 764 693; GBP 1 252 088) as code no. 'LM 637'; trade mark 'Maki'.

**Properties.**

Technical grade bromadiolone (97% pure) is a yellowish powder; m.p. 200-210 °C (mixture of two diastereoisomers). Solubility at 20 °C: 19 mg/l water; 730 g/l dimethylformamide; 8.2 g/l ethanol; 25 g/l ethyl acetate. It is stable $<200$ °C.

**Uses.**

It is an anticoagulant rodenticide, a single dose of a 50 mg/kg bait killing *Rattus norvegicus* and *R. rattus* from the 5th day. It is particularly palatable to rodents.

**Toxicology.**

Acute oral $LD_{50}$: for rats 1.125 mg/kg; for mice 1.75 mg/kg; for rabbits 1 mg/kg. In 90-d feeding trials the only effect observed in rats and dogs was reduction of prothrombin rating.

**Formulations.**

Baits (50 mg/kg) prepared from solutions (0.25 g a.i./l) or dry powder.

**Analysis.**

Details of product analysis are available from Lipha S.A.

$C_8H_5BrCl_6$ (393.8)
L55 A CUTJ AG AG BG CG DG EG
F1E

**Nomenclature and development.**

Common name bromocyclen (BSI, E-ISO, BPC);bromocyclène (F-ISO). Chemical name (IUPAC) 5-bromomethyl-1,2,3,4,7,7-hexachlorobicyclo[2.2.1]hept-2-ene (I); 5-bromomethyl-1,2,3,4,7,7-hexachloro-8,9,10-trinorborn-2-ene. (*C.A.*) (I) (9CI); 5-(bromomethyl)-1,2,3,4,7,7-hexachloro-2-norbornene (8CI); Reg. No. *[5895-04-5]*. Introduced as a veterinary insecticide and acaricide by Hoechst AG as trade mark 'Alugan', 'Bromodan' (in USA).

**Properties.**

The technical grade is *c.* 90% pure.

**Uses.**

It is an insecticide and acaricide effective against ectoparasites such as *Chorioptes, Cnemidocoptes, Notoedres, Otodectes, Psoroptes, Sarcoptes* spp. and *Cheyletiella parasitivorax* on cats, cattle, dogs, horses, pigs, sheep and some birds. The host may be dusted, bathed or treated with an aerosol of bromocyclen.

**Toxicology.**

Acute oral $LD_{50}$ for rats 12 500 mg/kg.

**Formulations.**

'Alugan' concentrate powder (425 g a.i./kg); 'Alugan' dusting powder (42.5 g/kg); 'Alugan' aerosol pressurised pre-pack.

$C_{13}H_7Br_2N_3O_6$ (461.0)
WNR CNW DONU1R DQ CE EE

**Nomenclature and development.**

Common name bromofenoxim (BSI, E-ISO, WSSA);bromophénoxime (F-ISO). Chemical name (IUPAC) 3,5-dibromo-4-hydroxybenzaldehyde 2,4-dinitrophenyloxime. (*C.A.*) 3,5-dibromo-4-hydroxybenzaldehyde *O*-(2,4-dinitrophenyl)oxime (8 & 9 CI); Reg. No. *[13181-17-4]* formerly *[37273-85-1]*. Its herbicidal properties were reported by D. H. Green *et al.*, (*Symp. New Herbic., 3rd*, 1969, p.77). Introduced by Ciba AG (now Ciba-Geigy AG) (BEP 675 444; GBP 1 096 037) as code no. 'C 9122'; trade mark 'Faneron'.

**Properties.**

Pure bromofenoxim forms cream-coloured crystals; m.p. 196-197 °C; v.p. 1.3 μPa at 20 °C; ρ 2.15 g/cm³ at 20 °C. Solubility at 20 °C: 0.1 mg/l water; 500 mg/l benzene; 200 mg/l hexane; 400 mg/l propan-2-ol. On hydrolysis 50% loss occurs at 70 °C in 41.4 h at pH 1, in 9.6 h at pH 5, in 0.76 h at pH 9.

**Uses.**

Bromofenoxim is a foliar-acting herbicide with a strong contact activity on annual dicotyledons. It has negligible activity through the soil. Suggested rates of application are: on winter-sown cereals, 1.5-2.5 kg a.i./ha when the weeds have emerged; on spring-sown cereals, 1.0-2.0 kg/ha to emerged broad-leaved weeds (up to 6 leaves). It is also used in combination with terbuthylazine to broaden its spectrum of weed control and extend its period of activity at rates of 0.8-1.25 kg mixed herbicide/ha.

**Toxicology.**

Acute oral $LD_{50}$ for rats 1217 mg tech./kg. Acute percutaneous $LD_{50}$ for rats >3000 mg/kg; slight irritant to skin and eyes of rabbits. Inhalation $LC_{50}$ (6-h) for rats >242 mg/m³ air. In 90-d feeding trials NEL was: for rats 300 mg/kg diet (20 mg/kg daily); for dogs 100 mg/kg diet (3 mg/kg daily). It is moderately to highly toxic to fish depending on the species; and practically non-toxic to birds and honeybees.

**Formulations.**

A s.c., 500 g a.i./l. Mixtures, w.p. (500 g total a.i./kg), s.c. (500 g/l) include: 'Faneron combi', Reg. No. *[39380-46-6]*, bromofenoxim + terbuthylazine (2:1); 'Faneron multi', (3:1); 'Mofix', (1.6:1). 'Faneron plus', bromofenoxim + terbuthylazine + mecoprop (3:1:6). 'Fanoprim', bromofenoxim + atrazine (1.3:1).

**Analysis.**

Product analysis is by hplc. Residues are determined by hplc with ECD. Particulars are available from Ciba-Geigy AG.

# Bromophos

$C_8H_8BrCl_2O_3PS$ (366.0)
GR DG BE EOPS&O1&O1

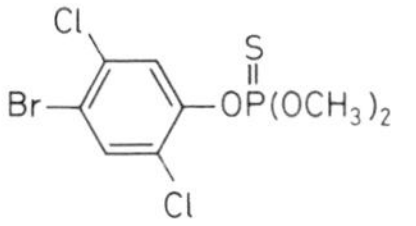

**Nomenclature and development.**

Common name bromophos (BSI, E-ISO, F-ISO). Chemical name (IUPAC) *O*-4-bromo-2,5-dichlorophenyl *O,O*-dimethyl phosphorothioate. (*C.A.*) *O*-(4-bromo-2,5-dichlorophenyl) *O,O*-dimethyl phosphorothioate (8 & 9CI); Reg. No. *[2104-96-3]*. OMS 658, ENT 27 162. Its insecticidal properties were described by R. Immel & G. Geisthardt (*Meded. Landbouwhogesch. Opzoekingsstn. Staat Gent*, 1964, **29**, 1242), and reviewed by D. Eichler (*Residue Rev.*, 1972, **41**, 65). Introduced by C. H. Boehringer Sohn/Cela GmbH (now Celamerck GmbH & Co. KG) (DEP 1 174 104; GBP 956 343) as code no. 'S-1942'; trade mark 'Nexion'.

**Properties.**

Pure bromophos forms yellowish crystals; m.p. 53-54 °C; b.p. 140-142 °C/0.01 mmHg; v.p. 17 mPa at 20 °C. Solubility at 20 °C: 0.7 mg/l water; 1.12 kg/l dichloromethane; 100 g/l methanol; 900 g/l xylene. It is stable in aqueous suspension, but is hydrolysed in distinctly alkaline media. It is compatible with all pesticides except sulphur and organometal fungicides.

**Uses.**

It is a non-systemic contact and stomach insecticide, active against Diptera, Hemiptera, certain Lepidoptera, Coleoptera and other insects. Recommended for crop protection at 25-75 g a.i./100 l; and for the control of flies and mosquitoes at 0.5 $g/m^2$. It is non-phytotoxic at insecticidal concentrations though suspect of damage under glass. It persists on sprayed foliage for 7-10 d.

**Toxicology.**

Acute oral $LD_{50}$: for rats 3750-8000 mg a.i./kg; for mice 3311-5900 mg/kg; for rabbits 720 mg/kg; for hens 9700 mg/kg; for honeybees 18.8-19.6 mg/kg. Acute percutaneous $LD_{50}$ for rabbits 2188 mg/kg. In 2-y feeding trials NEL was: for rats 0.63 mg/kg daily; for dogs 1.5 mg/kg daily. $LC_{50}$ for guppies is 0.5 mg/l; concentrations of 0.5-1.0 mg/l were non-fatal to mosquito fish in natural surroundings.

**Formulations.**

These include: e.c. (250 or 400 g a.i./l); w.p. (250 g/kg); dusts (20-50 g/kg); granules (50-100 g/kg); atomising concentrate (400 g/l); animal dip (200 g/l); coarse powder (30 g/kg).

**Analysis.**

Product analysis is by i.r. spectrometry (FAO Specification CP/70) or by titrimetric methods (*CIPAC Handbook*, 1983. **1B**, in press; R. D. Weeren & D. Eichler, *Anal. Methods Pestic. Plant Growth Regul.*, 1978, **10**, 31). Residues may be determined by glc (*idem, ibid.*; *Pestic. Anal. Man.*, 1979, **I**, 201-A, 201-G, 201-I; D. C. Abbott *et al., Pestic. Sci.*, 1970, **1**, 10; M. A. Luke *et al., J. Assoc. Off. Anal. Chem.*, 1981, **64**, 1187; *Analyst (London)*, 1980, **105**, 515). Determination of the halogenated phenol produced by hydrolysis can be used for both product and residue analysis.

$C_{10}H_{12}BrCl_2O_3PS$ (394.0)
GR DG BE EOPS&O2&O2

**Nomenclature and development.**

Common name bromophos-ethyl (BSI, E-ISO, F-ISO). Chemical name (IUPAC) *O*-4-bromo-2,5-dichlorophenyl *O,O*-diethyl phosphorothioate. (*C.A.*) *O*-(4-bromo-2,5-dichlorophenyl) *O,O*-diethyl phosphorothioate (8 & 9CI); Reg. No. *[4824-78-6]*. OMS 659,ENT 27 258. Its insecticidal properties were described by M. S. Mulla *et al.*, (*Mosq. News*, 1964, **24**, 312), and reviewed by D. Eichler (*Residue Rev.*, 1972, **41**, 65). Introduced by C. H. Boehringer Sohn/Cela GmbH (now Celamerck GmbH & Co. KG) (DEP 1 174 104; GBP 956 343) as code no. 'S-2225'; trade mark 'Nexagan'.

**Properties.**

Pure bromophos-ethyl is a colourless to pale yellow liquid; b.p. 122-133 °C/0.001 mmHg; v.p. 6.1 mPa at 30 °C; $d^{20}$ 1.52-1.55. Solubility at 20 °C: 0.14 mg/l water; completely soluble in most common solvents. At room temperature it is stable in aqueous suspension at $pH<9$, hydrolysed at $pH>9$, particularly at higher temperatures. It is compatible with other pesticides except sulphur and organometal fungicides.

**Uses.**

It is a non-systemic contact and stomach insecticide, active against Diptera, Hemiptera, certain Lepidoptera, Coleoptera and other insects, and with some acaricidal activity. Recommended for crop protection at 40-60 g a.i./100 l; against mites at 40-80 g/100 l; for treating cattle against ticks at 50-100 g/100 l; against flies and mosquitoes at 0.6-1.2 $g/m^2$. No phytotoxicity has been reported.

**Toxicology.**

Acute oral $LD_{50}$: for rats 48 mg/kg; for mice 210-550 mg/kg; for quail 200 mg/kg. Single oral doses of 100 mg/kg were lethal to guinea-pigs, of 125 mg/kg to sheep. Acute percutaneous $LD_{50}$ for rabbits 100-600 mg/kg. Acute inhalation $LC_{50}$ (1.7-h) for rats 2.5 mg/l. In 2-y feeding trials NEL was: for rats 78 mg/kg daily; for dogs 0.26 mg/kg daily. The $LC_{50}$ (96-h) is: for guppies 0.14-0.24 mg/l; for rainbow trout >0.4 mg/l.

**Formulations.**

These include: e.c. (400 or 800 g a.i./l); w.p. (250 g/kg); granules (50 g/kg); atomising concentrate (800-900 g/l); animal dips (400 g/l).

**Analysis.**

Product analysis is by i.r. spectrometry (FAO Specification CP/70) or by titrimetric methods (R. D. Weeren & D. Eichler, *Anal. Methods Pestic. Plant Growth Regul.*, 1978, **10**, 31). Residues may be determined by glc (*idem, ibid.*; *Pestic. Anal. Man.*, 1979, I, 201-A, 201-G, 201-I; D. C. Abbott *et al., Pestic. Sci.*, 1970, **1**, 10). Determination of the halogenated phenol produced by hydrolysis can be used for both product and residue analysis.

# Bromopropylate

$C_{17}H_{16}Br_2O_3$ (428.1)
ER DXQR DE&VOY1&1

**Nomenclature and development.**

Common name bromopropylate (BSI, E-ISO, F-ISO, ANSI);phenisobromolate (JMAF). Chemical name (IUPAC) isopropyl 4,4'-dibromobenzilate (I). (*C.A.*) 1-methylethyl 4-bromo-α-(4-bromophenyl)-α-hydroxybenzeneacetate (9CI); (I) (8CI); Reg. No. *[18181-80-1]*. ENT 27 552. Its acaricidal properties were reported by H. Grob *et al.*, (*Abstr. Int. Plant Prot. Congr., 6th*, 1967, p. 198). Introduced by J. R. Geigy S.A. (now Ciba-Geigy AG) (GBP 1 178 850; BEP 691 105; CHP 471 065) as code no. 'GS 19 851'; trade marks 'Neoron'; 'Acarol'.

**Properties.**

Pure bromopropylate forms colourless crystals; m.p. 77 °C; v.p. 11 μPa at 20 °C; ρ 1.59 g/cm$^3$ at 20 °C. Solubility at 20 °C: <0.5 mg/l water; 850 g/kg acetone; 750 g/kg benzene; 970 g/kg dichloromethane; 870 g/kg dioxane; 280 g/kg methanol; 90 g/kg propan-2-ol; 530 g/kg xylene. It is fairly stable to hydrolysis in neutral or slightly acid media.

**Uses.**

It is a contact acaricide with residual activity, recommended for use on citrus, cotton, grapes, ornamentals, pome and stone fruit, soyabeans, strawberries and vegetables at 37.5-60 g a.i./100 l; on field crops at 0.5-1.0 kg/ha. At somewhat higher rates it is also effective against mites resistant to organophosphorus compounds.

**Toxicology.**

Acute oral $LD_{50}$ for rats >5000 mg tech./kg. Acute percutaneous $LD_{50}$ for rats >4000 mg/kg; slightly irritant to skin and non-irritant to eyes of rabbits. In 2-y feeding trials NEL for rats was 500 mg/kg diet (*c.* 25 mg/kg daily); in 1-y trials for mice 1000 mg/kg diet (*c.* 143 mg/kg daily). $LC_{50}$ (96-h): for rainbow trout 0.35 mg/l; for bluegill 0.50 mg/l; for carp 2.4 mg/l. Toxicity to honeybees by direct application was negligible.

**Formulations.**

'Neoron 250'/'Acarol 250' and 'Neoron 500'/'Acarol 500', e.c. (250 or 500 g a.i./l, respectively).

**Analysis.**

Product analysis is by glc. Residues are determined by glc with ECD (*Pestic. Anal. Man.*, 1979, **1**, 201-A, 201-G; M. A. Luke *et al.*, *J. Assoc. Off. Anal. Chem.*, 1981, **64**, 1187).

$C_7H_3Br_2NO$ (276.9)
QR BE FE DCN
*bromoxynil-potassium*
$C_7H_2Br_2KNO$ (315.0)
QR BE FE DCN &&K SALT
*bromoxynil octanoate*
$C_{15}H_{17}Br_2NO_2$ (403.0)
NCR CE EE DOV7

**Nomenclature and development.**

Common name bromoxynil (BSI, E-ISO, F-ISO, ANSI, WSSA). Chemical name (IUPAC) 3,5-dibromo-4-hydroxybenzonitrile (I); 3,5-dibromo-4-hydroxyphenyl cyanide. (*C.A.*) (I) (8 & 9CI); Reg. No. *[1089-84-5]* (bromoxynil), *[2961-68-4]* (bromoxynil-potassium), *[1689-99-2]* (bromoxynil octanoate). ENT 20 852 (bromoxynil). The herbicidal properties of bromoxynil were described independently by R. L. Wain (*Nature (London)*, 1963, **200**, 28), by K. Carpenter & B. J. Heywood (*ibid.*, p. 28), and by Amchem Products Inc. (now Union Carbide Agricultural Products Co. Inc.). Its development was reviewed by B. J. Heywood (*Chem. Ind. (London)*, 1966, p. 1946). Introduced by May & Baker Ltd and by Amchem Products Inc. (GBP 1 067 033 to May & Baker Ltd; USP 3 397 054; 4 332 613 both to Union Carbide) as code no. 'M&B 10 064' (May & Baker for bromoxynil),'M&B 10 731','16 272 RP' (May & Baker and Rhône-Poulenc for the octanoate); trade marks 'Brominal' formerly '*Brominil*' (both to Amchem), 'Buctril' (May & Baker).

**Properties.**

Pure bromoxynil is a colourless solid; m.p. 194-195 °C. Solubility at 25 °C: 130 mg/l water; 170 g/l acetone; 90 g/l methanol; <20 g/l petroleum oils; 410 g/l tetrahydrofuran. The technical grade (*c.* 95% pure) has m.p. 188-192 °C. It is stable in sunlight and below m.p.

It is acidic, p$K_a$ 4.06, forming salts (m.p. *c.* 360 °C): bromoxynil-sodium (solubility at 20-25 °C: 42 g a.e./l water, 80 g/l acetone, 150 g/l 20% *m/V* aqueous acetone, 310 g/l 2-methoxyethanol, 430 g/l tetrahydrofurfuryl alcohol); bromoxynil-potassium (solubility 61 g a.e./l water, 70 g/l acetone, 240 g/l 20% *m/V* aqueous acetone, 130 g/l 2-methoxyethanol, 260 g/l tetrahydrofurfuryl alcohol).

Bromoxynil forms stable esters of which the octanoate [2,6-dibromo-4-cyanophenyl octanoate (IUPAC & *C.A.* 9CI)], and, occasionally, the heptanoate are used. Bromoxynil octanoate is a cream, waxy solid; m.p. 45-46 °C. Solubility at 20-25 °C: virtually insoluble in water; 100 g/l acetone, ethanol; 700 g/l benzene, xylene; 800 g/l chloroform, dichloromethane; 550 g/l cyclohexanone. It is stable to sunlight and at its m.p.; but is readily hydrolysed to bromoxynil at pH >9.

**Uses.**

Bromoxynil, its salts and esters are contact herbicides with some systemic activity; bromoxynil itself being the toxicant inhibiting photosynthesis and uncoupling oxidative phosphorylation. They are mainly used post-em. to control seedling broad-leaved weeds in cereal crops, flax, garlic, maize, onions, sorghum and in newly-sown turf. Frequently mixtures with other herbicides are used to extend the range of weeds controlled. They are decomposed in soils (50% loss in *c.* 10 d) (R. F. Collins, *Pestic. Sci.*, 1973, **4**, 181; G. H. Ingram & E. M. Pullin, *ibid.*, 1974, **5**, 287). Metabolism in plants and animals has been studied, the ester and nitrile groups being hydrolyzed with some debromination occurring (J. H. Buckland, *ibid.*, 1973, **4**, 149, 689).

$C_7H_3Br_2NO$ (276.9)
QR BE FE DCN
*bromoxynil-potassium*
$C_7H_2Br_2KNO$ (315.0)
QR BE FE DCN &&K SALT
*bromoxynil octanoate*
$C_{15}H_{17}Br_2NO_2$ (403.0)
NCR CE EE DOV7

**Toxicology.**

Acute oral $LD_{50}$: for rats 190 mg a.i./kg, 130 mg formulated salt/kg, 365 mg formulated octanoate/kg; for mice 110 mg a.i./kg, 100 mg formulated salt/kg, 306 mg formulated octanoate/kg; for pheasants 50 mg a.i./kg, 50 mg formulated salt/kg, 175 mg formulated octanoate/kg; for hens 240 mg a.i./kg, 120 mg salt/kg. Acute percutaneous $LD_{50}$ for rats >2000 mg a.i./kg, >2000 mg octanoate /kg; for rabbits, 3660 mg a.i./kg, 1675 mg tech. octanoate/kg, transient irritation of their eyes was observed with the a.i. In 90-d feeding trials NEL was: for rats 16.6 mg bromoxynil-potassium/kg daily, 15.6 mg bromoxynil-octanoate/kg daily; for dogs 5 mg bromoxynil octanoate/kg daily. $LC_{50}$ (48-h) is: for harlequin fish 5.0 mg bromoxynil-potassium/l; for rainbow trout 0.15 mg bromoxynil octanoate/l; for goldfish 0.46 mg/l; for catfish 0.063 mg/l. Sprays of 2.2 g bromoxynil-potassium/l or 3.4 g bromoxynil octanoate/l showed no contact toxicity to honeybees.

**Formulations.**

These include bromoxynil octanoate: 'Buctril 20' (200 g/l); 'Buctril 21', 'Pardner', e.c. (225 g/l); 'ME 4 Brominal' (butyrate + octanoate esters 480 g a.e./l). Mixtures include w.s.c. of sodium/potassium salts: 'Actril 4', 'Actril S', 'Dantril', 'Oxitril 4', 'Tetroxone' (bromoxynil + dichlorprop + ioxynil + MCPA); '3 + 3 Brominal' (bromoxynil butyrate + bromoxynil octanoate + MCPA-(2-butoxyethyl)), 'Doublet', 'Twin-Tak' (bromoxynil + ioxynil + isoproturon); Also esters: 'Brittox' (bromoxynil octanoate + ioxynil octanoate + mecoprop-isooctyl); 'Harness' (bromoxynil octanoate + MCPA-isooctyl + mecoprop-isooctyl); 'Oxytril P' (bromoxynil + dichlorprop-isooctyl + ioxynil).

**Analysis.**

Product analysis is by glc or by determination of bromine. Residues may be determined by glc of a derivative (H. S. Segal & M. L. Sutherland, *Anal. Methods Pestic., Plant Growth Regul. Food Addit.*, 1967, **5**, 347; *Anal. Methods Pestic. Plant Growth Regul.*, 1972, **6**, 605; T. H. Byast *et al., Tech. Rep. ARC Weed Res. Organ.*, No. 15 (2nd Ed.), p. 13) or by i.r. spectrometry.

$C_3H_6BrNO_4$ (200.0)
WNXE1Q1Q

Br
$HOCH_2C\,CH_2OH$
$NO_2$

**Nomenclature and development.**

Common name bronopol (BPC). Chemical name (IUPAC) 2-bromo-2-nitropropane-1,3-diol. (*C.A.*) 2-bromo-2-nitro-1,3-propanediol (8 & 9CI); Reg. No. *[52-51-7]*. Trivial name bronocot. Introduced in 1964 as a preservative for cosmetic and pharmaceutical preparations, and later as an agricultural bacteriostat (D. F. Spooner & S. B. Wakerley, *Proc. Br. Insectic. Fungic. Conf., 6th*, 1971, **1**, 201) by the Boots Co. Ltd (now FBC Limited) (GBP 1 193 954).

**Properties.**

Bronopol is a colourless to pale brown-yellow solid; m.p. 130 °C; v.p. 1.7 mPa at 20 °C. Solubility at 22 °C: 250 g/l water; very soluble in acetone, ethanol, 2-methoxyethanol. Technical product $>90\%$ pure. Slightly hygroscopic; stable under normal storage conditions, but unstable in aluminium containers. Mixtures with aldrin, captan, dieldrin, heptachlor and thiram have been prepared.

**Uses.**

Bronopol is a bacteriostat active against a wide range of plant pathogenic bacteria, especially *Xanthomonas malvacearum* causing Blackarm disease of cotton, against which it is applied as a seed treatment. It is non-phytotoxic on a wide range of crops including cotton.

**Toxicology.**

Acute oral $LD_{50}$: for rats 180-400 mg/kg; for mice 270-400 mg/kg; for dogs 250 mg/kg. Acute percutaneous $LD_{50}$ for rats $>1600$ mg/kg; a 0.5% aqueous solution is moderately irritant to rabbit skin on repeated application, and a single application of 1% to rabbits' eyes is slightly irritant. Acute inhalation $LC_{50}$ (6-h) for rats $>5$ g/m$^3$ air. In 72-d feeding trials rats receiving 1000 mg/kg diet showed no clinical or pathological sign of toxicity.

**Formulations.**

Seed treatment (120 g a.i./kg). Dust with captan, for treating cotton seed.

**Analysis.**

Product analysis is by glc; particulars from FBC Limited.

$C_{13}H_{19}NO_2$ (221.3)

(i) 3Y1&R COVM1

(ii) 2Y2&R COVM1

OCO.NHCH$_3$ CH$_3$ CH(CH$_2$)$_2$CH$_3$ (i)

OCO.NHCH$_3$ CH(CH$_2$CH$_3$)$_2$ (ii)

**Nomenclature and development.**

Common name bufencarb (BSI, E-ISO, ANSI); bufencarbe (F-ISO); *exception* Eire. The chemical name (IUPAC) of the main components derived from a reaction are: (i) 3-(1-methylbutyl)phenyl methylcarbamate (I) and (ii) 3-(1-ethylpropyl)phenyl methylcarbamate (II). (*C.A.*) (I) and (II) (9CI); *m*-(1-methylbutyl)phenyl methylcarbamate and *m*-(1-ethylpropyl)phenyl methylcarbamate (8CI); Reg. No. *[8065-36-9]* (bufencarb), *[2282-34-0]* (i), *[672-04-8]* (ii). OMS 227, ENT 27 127. Introduced as an insecticide by Chevron Chemical Co. (USP 3 062 864; 3 062 867) as code no. 'Ortho 5353'; trade mark 'Bux'.

**Properties.**

Produced by the action of methyl isocyanate on a reaction mixture of alkylated phenols, technical grade bufencarb comprises *c.* 65% of a 3:1 mixture of (i) and (ii) and 35% of insecticidally-inactive related isomers (mainly the 2- and 4-alkyl analogues). This product is a yellow to amber coloured solid; m.p. 26-39 °C; b.p. *c.* 125 °C/0.04 mm Hg; v.p. 4.0 mPa at 30 °C; $d^{26}$ 1.024. Solubility at room temperature: <50 mg/l water; very soluble in methanol, xylene; less so in hexane. It is stable in acid or neutral solutions, but the rate of hydrolysis increases with rise of pH or temperature. It is degraded in soil.

**Uses.**

It is a contact and stomach insecticide effective at 0.5-2.0 kg a.i./ha against a range of soil- and foliage-inhabiting insects, particularly *Chilo plejadellus, Lissorhoptrus oryzophilus*, corn rootworm larvae, rice green leafhoppers, rice planthoppers, pineapple root mealybugs.

**Toxicology.**

Acute oral $LD_{50}$ for rats 87 mg tech./kg. Acute percutaneous $LD_{50}$ for rabbits 680 mg/kg; irritation to their eyes is minimal. In 90-d feeding trials no effect was noted with beagle dogs and albino rats receiving $\leqslant$500 mg/kg diet. $LC_{50}$ (8-d) for pheasants and ducks was >27 000 mg/kg diet. $LC_{50}$ (96-h) is: for goldfish 0.56 mg/l; catfish 1.95 mg/l; for trout 0.064 mg/l.

**Formulations.**

'Bux Ten Granular', granules (100 mg a.i./kg); 'Bux 2 Emulsive', e.c. (240 g/l); 'Bux 360 Emulsive', e.c. (360 g/l); dusts (20 or 40 g/kg).

**Analysis.**

Product analysis is by glc (B. Tucker, *Anal. Methods Pestic. Plant Growth Regul.*, 1973, **7**, 179). Residues may be determined by glc (*idem, ibid.*). Details are available from Chevron Chemical Co.

$C_{13}H_{24}N_4O_3S$ (316.4)
T6N CNJ BM2 DOSWN1&1 E4 F1

CH3 N NHCH2CH3
N
CH3(CH2)3
O.SO2.N(CH3)2

**Nomenclature and development.**

Common name bupirimate (BSI, E-ISO, F-ISO, ANSI). Chemical name (IUPAC) 5-butyl-2-ethylamino-6-methylpyrimidin-4-yl dimethysulphamate. (*C.A.*) 5-butyl-2-(ethylamino)-6-methyl-4-pyrimidinyl dimethylsulfamate (9CI); Reg. No. *[41483-43-6]*. Its fungicidal properties were described by J. R. Finney *et al.* (*Proc. Br. Insectic. Fungic. Conf., 8th,* 1975, **2**, 667). Introduced by ICI Plant Protection Division (GBP 1 400 710) as code no. 'PP 588'; trade marks 'Nimrod', and 'Nimrod T' (mixture with triforine).

**Properties.**

A pale tan waxy solid; m.p. 50-51 °C; v.p. 67 μPa at 20 °C. Solubility at room temperature: 22 mg/l water; soluble in most organic solvents except paraffins. Easily hydrolysed by dilute acids. Unstable on prolonged storage >37 °C. Technical grade has m.p. *c.* 40-45 °C.

**Uses.**

A systemic fungicide specifically effective against powdery mildews, especially those of apple and glasshouse rose, at 5-15 g a.i./100 l HV. This specificity enhances its usefulness in integrated control programmes.

**Toxicology.**

Acute oral $LD_{50}$; for female rats, mice, rabbits and male guinea-pigs *c.* 4000 mg/kg; for pigeons >2700 mg/kg; for quail >5200 mg/kg. No clinical sign was noted in rats after 10 daily percutaneous treatments at 500 mg/kg; mild irritation was caused to the eyes of rabbits. In 2-y feeding trials NEL for rats was 100 mg/kg diet; in 90-d trials: for rats 1000 mg/kg diet, for dogs 15 mg/kg daily. $LC_{50}$ (96-h) for rainbow trout is 1.7 mg/l.

**Formulations.**

'Nimrod', e.c. (250 g a.i./l); w.p. (250 g/kg). 'Nimrod T', e.c. (62.5 g bupirimate + 62.5 g triforine/l).

**Analysis.**

Product analysis is by tlc and u.v. spectrometry. Residues may be determined by glc of a derivative. Details are available from ICI Plant Protection Division.

$C_{16}H_{23}N_3OS$ (305.4)
T6NVNYS FHJ AR& CY1&1
DUNX1&1&1

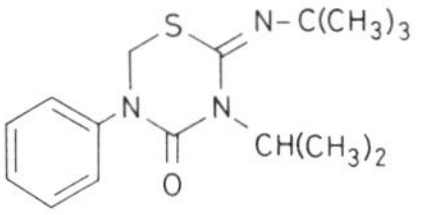

**Nomenclature and development.**

Common name buprofezin (BSI, draft E-ISO). Chemical name (IUPAC) 2-*tert*-butylimino-3-isopropyl-5-phenyl-1,3,5-thiadiazinan-4-one. (*C.A.*) 2-[(1,1-dimethylethyl)imino]tetrahydro-3-(1-methylethyl)-5-phenyl-4*H*-1,3,5-thiadiazin-4-one (9CI); Reg. No. *[69327-76-0]*. Its insecticidal properties were described by H. Kanno *et al.* (*Proc. Br. Crop Prot. Conf. Pests Dis.*, 1981, **1**, 59). Introduced by Nihon Nohyaku Co., Ltd (JPP 1 048 643) as code no. 'NNI-750'; trade mark 'Applaud'.

**Properties.**

It is a colourless crystalline solid; m.p. 104.5-105.5 °C; v.p. 1.25 mPa at 25 °C. Solubility at 25 °C: 0.9 mg/l water; 240 g/l acetone; 520 g/l chloroform; 80 g/l ethanol; 20 g/l hexane; 320 g/l toluene.

**Uses.**

It is an insecticide with persistent larvicidal action against some Coleoptera and Hemiptera, also Acarina.

**Toxicology.**

Acute oral $LD_{50}$: for male rats 2198 mg/kg, for females 2355 mg/kg; for male and female mice >10 000 mg/kg. $LC_{50}$ (48-h) for carp is 2.7 mg/l.

# Butachlor



$C_{17}H_{26}ClNO_2$ (311.9)
G1VN1O4&R B2 F2

$CH_2CH_3$
$CO.CH_3Cl$
N
$CH_2O(CH_2)_3CH_3$
$CH_2CH_3$

**Nomenclature and development.**

Common name butachlor (BSI, ANSI, WSSA, JMAF, draft E-ISO);butachlore (draft F-ISO). Chemical name (IUPAC) *N*-butoxymethyl-2-chloro-2',6'-diethylacetanilide; *N*-butoxymethyl-α-chloro-2',6'-diethylacetanilide. (*C.A.*) *N*-(butoxymethyl)-2-chloro-*N*-( 2,6-diethylphenyl)acetamide (9CI); *N*-(butoxymethyl)-2-chloro-2',6'-diethylacetanilide (8CI); Reg. No. *[23184-66-9]*. Its herbicidal properties were described by D. D. Baird & R. P. Upchurch (*Proc. South. Weed Control Conf., 23rd*, 1970, p.101). Introduced by Monsanto Co. (USP 3 442 945; 3 547 620) as code no. 'CP 53 619'; trade mark 'Machete'.

**Properties.**

Pure butachlor is an amber-coloured liquid; m.p. <-5 °C; b.p. 156 °C/0.5 mmHg; v.p. 600 μPa at 25 °C; ρ 1.070 g/cm³ at 25 °C. Solubility at 24 °C: 23 mg/l water; miscible with acetone, benzene, ethanol, ethyl acetate, hexane. Decomposes at 165 °C; stable to u.v. light. Corrosive to steel and black iron.

**Uses.**

Butachlor is a pre-em. herbicide for the control of annual grasses and certain broad-leaved weeds in rice, both seeded and transplanted. It shows selectivity in barley, cotton, peanuts, sugar beet, wheat and several brassica crops. Effective rates range from 1.0-4.5 kg a.i./ha. Activity is dependent on water availability such as rainfall following treatment, overhead irrigation or applications to standing water as in rice culture. It is absorbed mainly by germinating plant shoots, less so by roots; it is translocated throughout the plant, being concentrated in vegetative as opposed to reproductive organs, and is rapidly metabolised. It persists in soil 42-70 d, loss being mainly by microbial action.

**Toxicology.**

Acute oral $LD_{50}$: for rats 2000 mg tech./kg; for bobwhite quail and mallard ducks >10 000 mg/kg. Acute percutaneous $LD_{50}$: for rabbits >13 000 mg/kg; slightly irritating to eyes and moderately to skin of rabbits. In 2-y feeding trials no evidence of toxicity was observed in rats or dogs at 1000 mg/kg diet. $LC_{50}$ (8-d) is: for bobwhite quail 6597 mg/kg diet; for ducks >10 000 mg/kg diet. $LC_{50}$ (96-h) is: for carp 0.32 mg/l; for bluegill 0.44 mg/l.

**Formulations.**

An e.c. (600 g a.i./l); granules (50 g/kg).

**Analysis.**

Product analysis is by i.r. spectrometry. Residues may be determined by glc. Details are available from Monsanto Co.

$C_{13}H_{21}N_2O_4PS$ (332.4)
WNR D1 BOPS&O2&MY2&1

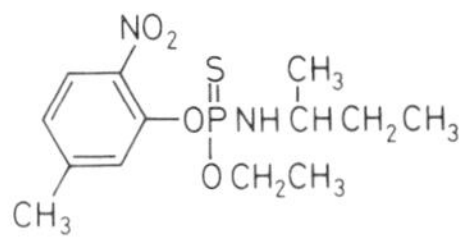

**Nomenclature and development.**

Common name butamifos (BSI, E-ISO, F-ISO, JMAF). Chemical name (IUPAC) *O*-ethyl *O*-6-nitro-*m*-tolyl *sec*-butylphosphoramidothioate. (*C.A.*) *O*-ethyl *O*-(5-methyl-2-nitrophenyl) (1-methylpropyl)phosphoramidothioate (9CI); Reg. No. *[36335-67-8]*. Its herbicidal properties were described by M. Ueda (*Jpn. Pestic. Inf.*, 1975, No. 23, p. 23). Introduced by Sumitomo Chemical Co. (GBP 1 359 727; USP 3 936 433) as code no. 'S-2846'; trade mark 'Cremart'.

**Properties.**

The technical grade is a yellow-brown liquid; v.p. 84 mPa at 27 °C; $d_{25}^{25}$ 1.188; $d_{\mathrm{D}}^{25}$ 1.5373. Solubility at 20 °C: 5.1 mg/l water; at room temperature >1 kg/kg acetone, methanol, xylene.

**Uses.**

It is a contact herbicide used pre-em. Effective against annual and especially graminaceous weeds, in beans, lawns and vegetables.

**Toxicology.**

Acute oral $LD_{50}$; for male rats 1070 mg/kg, for females 845 mg/kg. Acute percutaneous $LD_{50}$ for rats >5000 mg/kg; no skin or eye irritation was observed in rabbits. $LC_{50}$ (48-h) for carp is 2.4 mg/l.

**Formulations.**

These include: e.c. (500 g a.i./l); w.p. (500 g/kg).

**Analysis.**

Product analysis is by glc or by colorimetry (M. Horiba *et al.*, *Nippon Nogei Kagaku Kaishi*, 1979, **53** 111). Details of the methods for residue analysis are available from Sumitomo Chemical Co.

$C_{10}H_{16}N_4O_2S$ (256.3)
T5NVNTJ A1 DQ C- CT5NN DSJ
EX1&1&1

**Nomenclature and development.**

Common name buthidazole (BSI, E-ISO, F-ISO, ANSI, WSSA). Chemical name (IUPAC) 3-(5-*tert*-butyl-1,3,4-thiadiazol-2-yl)-4-hydroxy-1-methyl-2-imidazolidone. (*C.A.*) 3-[5-(1,1-dimethylethyl)-1,3,4-thiadiazol-2-yl]-4-hydroxy-1-methyl-2-imidazolidinone (9CI); Reg. No. *[55511-98-3]*. Its herbicidal properties were reported by R. F. Anderson (*Proc. Int. Velsicol Symp., 8th,* 1974).Introduced by Velsicol Chemical Corp. (USP 3 904 640) as code no. 'Vel-5026'; trade mark 'Ravage'.

**Properties.**

Pure buthidazole forms colourless crystals; m.p. 133-134 °C; $d^{25}$ 1.28; v.p. 5.3 μPa at 25 °C; decomposes at 237 °C; relatively resistant to u.v. degradation. Solubility at 25° C: 6.5 g/kg water; 150 g/kg acetone; 208 g/kg dimethylformamide; 45 g/kg toluene; 10 g/kg xylene. Technical grade (⩾86% pure) is a tan powder.

**Uses.**

Residual pre- and post-emergent herbicide in industrial and non-crop areas at 4.5-6.5 kg a.i./ha depending on rainfall. Used at lower rates against broad-leaved weeds and annual grasses in maize, pineapple and sugarcane. 50% loss usually takes >50 d.

**Toxicology.**

Acute oral $LD_{50}$: for rats 1575-2430 mg/kg; for mallard ducks 3060 mg/kg. Acute percutaneous $LD_{50}$ for rabbits >20 000 mg/kg; non-irritant to their skin or eyes. $LC_{50}$ (4-h) for rats to dust >20.8 mg/l. In 2-y feeding trials: NEL for rats was 300 mg/kg diet; liver tumours reported in mice ⩾1000 mg/kg diet. Not teratogenic in rabbits at 90 mg/kg daily, NEL for rats (3 generations) 1500 mg/kg diet. Non-mutagenic to mammalian and bacterial cells. Acute $LC_{50}$ (96-h) in fresh water was: for bluegill 126 mg/l; for rainbow trout 75 mg/l, for channel catfish 239 mg/l. $LC_{50}$ (8-d) for mallard ducks and bobwhite quail 10 000 mg/kg diet.

**Formulations.**

These include: 'Ravage', w.p. (500 or 750 g a.i./kg); pellets (50 or 100 g/kg).

**Analysis.**

Product analysis and soil residue determinations are by hplc. Details of methods available from Velsicol Chemical Corp.

$C_{21}H_{28}N_2S_2$ (372.6)
T6NJ CNUYS4&S1R DX1&1&1

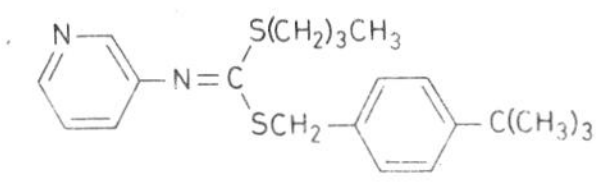

**Nomenclature and development.**

Common name buthiobate (BSI, E-ISO, F-ISO). Chemical name (IUPAC) butyl 4-*tert*-butylbenzyl *N*-(3-pyridyl)dithiocarbonimidate. (*C.A.*) butyl [4-(1,1-dimethylethyl)phenyl]methyl 3-pyridinylcarbonimidodithioate (9CI); Reg. No. *[51308-54-4]*. Its fungicidal properties were described by T. Kato *et al.* (*Agric. Biol. Chem.*, 1975, **39**, 169). Introduced by Sumitomo Chemical Co. (GBP 1 335 617; USP 3 832 351) as code no. 'S-1358'; trade mark 'Denmert'.

**Properties.**

Technical grade buthiobate is a red-brown viscous liquid; m.p. 31-33 °C; v.p. 60 μPa at 20 °C; $d_{25}^{25}$ 1.0865; $n_D^{26.5}$ 1.596. Solubility at 25 °C: 1 mg/l water; at 23 °C: >1 kg/kg methanol and xylene.

**Uses.**

It is a preventive, curative and persistent fungicide, having rain fastness. It is effective against powdery mildew on beans, fruit, ornamentals, vegetables and other crops. It is usually applied at 15-250 mg/l.

**Toxicology.**

Acute oral $LD_{50}$: for male rats 4400 mg/kg, for females 3200 mg/kg; for mallard duck >10 000 mg/kg; for bobwhite quail 21 804 mg/kg. Acute percutaneous $LD_{50}$ for rats >5000 mg/kg; no skin or eye irritation was observed in rabbits. $LC_{50}$ (48-h) for carp is 6.4 mg/l.

**Formulations.**

These include: e.c. (100 g a.i./l); w.p. (200 g/kg).

**Analysis.**

Product analysis is by glc (S. Irie *et al., Nippon Nogei Kagaku Kaishi*, 1977, **51**, 331). Details of the methods for residue analysis are available from Sumitomo Chemical Co.

$C_7H_{14}N_2O_2S$ (190.3)
1SY1&Y1&UNOVM1

$CH_3$ / $CH_3SCHCCH_3$ / $NOCO.NHCH_3$ (*E*)

$CH_3$ / $CH_3CCHSCH_3$ / $NOCO.NHCH_3$ (*Z*)

**Nomenclature and development.**

Common name butocarboxim (BSI, E-ISO);butocarboxime (F-ISO). Chemical name (IUPAC) 3-(methylthio)butanone *O*-methylcarbamoyloxime. (*C.A.*) 3-(methylthio)-2-butanone *O*-[(methylamino)carbonyl]oxime (9CI); 3-(methylthio)butanone *O*-( methylcarbamoyl)oxime (8CI); Reg. No. *[34681-10-2]*. Its insecticidal properties were described by M. Vulić *et al.* (*Meded. Fac. Landbouwwet. Rijksuniv. Gent*, 1973, **38**, 1175). Introduced by Wacker Chemie GmbH (DERP 2 036 491; USP 3 816 532; GBP 1 353 202) as code no. 'Co 755'; trade mark 'Drawin 755'.

**Properties.**

The (*E*)-isomer has m.p. 37 °C. The technical grade is liquid and obtained as an 85% solution in xylene containing the (*E*)- and (*Z*)-isomers in a ratio of 85:15. The isomeric mixture has v.p. *c.* 10.6 mPa at 20 °C. Completely miscible with aromatic hydrocarbons, esters, ketones; poorly soluble in aliphatic hydrocarbons, carbon tetrachloride. It is stable in sunlight and at temperatures $\leqslant$100 °C.

**Uses.**

It is a systemic insecticide effective against sucking insects. It has an excellent activity at 75 g a.i./100 l against *Aleurothrixus floccosus* in citrus cultures (M. Vulić & J. L. Beltran, *Z. Pflanzenkr. Pflanzenschutz*, 1977, **84**, 202) and at 125 kg/ha, *Bemesia tabaci* in cotton; good activity at 50-75 g/100 l against aphids on fruit, ornamental crops and vegetables and moderate activity against mites. Structure/activity of analogues was reported (T. A. Magee & L. E. Limpel, *J. Agric. Food Chem.*, 1977, **25**, 1376).

**Toxicology.**

Acute oral $LD_{50}$ for rats 158-215 mg/kg. Acute percutaneous $LD_{50}$ for albino rabbits 360 mg/kg. Subcutaneous $LD_{50}$ for rats 188 mg/kg. In 2-y feeding trials NEL for rats was 100 mg/kg diet; in 90-d trials with dogs 100 mg/kg diet. $LC_{50}$ (48-h) for Japanese quail was 1180 mg/kg diet. There was no carcinogenic or mutagenic activity in routine tests at the highest dose (300 mg/kg diet) in rats during 2 y, nor any effect on fertility, growth rate or mortality. No mutagenicity was observed with *Salmonella typhimurium*. $LC_{50}$ (48-h) is: for rainbow trout 35 mg/l; for golden ide 55 mg/l.

**Formulations.**

'Drawin 755', e.c. (500 g a.i./l) for agricultural and horticultural use; 'Drawin 75-5' solution (50 g/l) and spray (0.9 g/l) for garden and domestic use.

**Analysis.**

Product analysis is by i.r. spectrometry or by hplc. Residues of butocarboxim, its sulphoxide and sulphone (butoxycarboxim, 1590) may be determined by glc with TID. Details of these methods are available from Wacker Chemie GmbH.

$C_7H_{14}N_2O_4S$ (222.3)
WS1&Y1&Y1&UNOVM1

```
      CH3                      CH2
      |                        |
CH3SO2.CHCCH3           CH3CCHSO2.CH3
         ||                ||
         NOCO.NHCH3        NOCO.NHCH3
    (E)                      (Z)
```

**Nomenclature and development.**

Common name butoxycarboxim (BSI, E-ISO, JMAF); butoxycarboxime (F-ISO). Chemical name (IUPAC) 3-methylsulphonylbutanone *O*-methylcarbamoyloxime. (*C.A.*) 3-(methylsulfonyl)-2-butanone *O*-[(methylamino)carbonyl]oxime (9CI); 3-(methylsulfonyl)-2-butanone *O*-(methylcarbamoyl)oxime (8CI); Reg. No. *[34681-23-7]*. Its insecticidal properties and the special application method were described by M. Vulić & H. Bräunling (*Meded. Fac. Landbouwwet. Rijksuniv. Gent*, 1974, **39**, 847). Introduced by Wacker Chemie GmbH. (DERP 2 036 491; USP 3 816 532; GBP 2 353 202) as code no. 'Co 859'; trade mark 'Plant Pin'.

**Properties.**

Technical butoxycarboxim contains the (*E*)- and (*Z*)-isomers in a ratio of 85:15. The isomeric mixture has v.p. 266 μPa. Solubility: 208 g/l water; 172 g/l acetone; 5.3 g/l carbon tetrachloride; 186 g/l chloroform; 0.9 g/l cyclohexane; 100 mg/l heptane; 101 g/l propan-2-ol; 29 g/l toluene. It is stable to sunlight and to temperatures $\leqslant$100 °C. The pure (*E*)-isomer has m.p. 83 °C.

**Uses.**

It is a systemic insecticide effective against aphids and mites with special formulation and activity properties for soil application to potted plants. Structure/activity of analogues is reported (T. A. Magee & L. E. Limpel, *J. Agric. Food Chem.*, 1977, **25**, 1376).

**Toxicology.**

Acute oral $LD_{50}$: for rats 458 mg/kg; for rabbits 275 mg/kg; for hens 367 mg/kg. Acute percutaneous $LD_{50}$ for rats >11 000 mg/kg. Acute subcutaneous $LD_{50}$ for female rats 288 mg/kg. In 90-d feeding trials NEL for rats was 300 mg/kg diet, whilst at 1000 mg/kg diet only slight inhibition of erythrocyte and plasma cholinesterase was noted. Oral $LD_{50}$ for rats of the pasteboard stick formulation is >5000 mg/kg. $LC_{50}$ for carp is 1.75 g/l. But oxycarboxim is a metabolite of butocarboxim (1560) in plant and animal tissue, therefore the toxicological tests on the latter partly include butoxycarboxim.

**Formulations.**

Butoxycarboxim is formulated for the special use of soil application on potted house-plants. It is incorporated in 'Plant Pin' (pasteboard-pins, 40 mm × 8 mm each containing 50 mg a.i.).

**Analysis.**

Product analysis is by i.r. spectrometry or by hplc. Residues may be determined by glc of a derivative with TID. Details of both methods are available from Wacker Chemie GmbH.

# Butralin



$C_{14}H_{21}N_3O_4$ (295.3)
2Y1&MR BNW FNW DX1&1&1

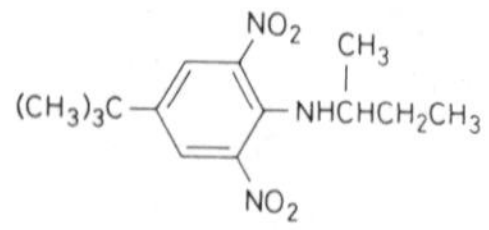

**Nomenclature and development.**

Common name butralin (BSI, E-ISO, ANSI, WSSA); butraline (F-ISO); *exception* (Eire, Japan). Chemical name (IUPAC) *N-sec*-butyl-4-*tert*-butyl-2,6-dinitroaniline (I). (*C.A.*) 4-(1,1-dimethylethyl)-*N*-(1-methylpropyl)-2,6-dinitrobenzenamine (9CI); (I) (8CI); Reg. No. *[33629-47-9]*. Its herbicidal properties were described by S. R. McLane *et al.* (*Proc. South. Weed Sci. Soc.*, 1971, **24**, 58). Introduced by Amchem Products Inc. (now Union Carbide Agricultural Products Co., Inc.) (USP 3 672 866) as code no. 'Amchem 70-25','Amchem A-280'; trade marks 'Amex', 'Tamex'.

**Properties.**

Butralin forms yellow-orange crystals, with a slight aromatic odour; m.p. 60-61 °C; b.p. 134-136 °C/0.5 mmHg; v.p. 1.7 mPa at 25 °C. Solubility at 24-26 °C: 1 mg/l water; 4.48 kg/kg acetone; 2.7 kg/kg benzene; 9.55 kg/kg butanone; 1.46 kg/kg carbon tetrachloride; 3.88 kg/kg xylene. Its flash point (Tag open cup) is 36 °C and it decomposes at 265 °C. Concentrates are stable on storage >3 y but should not be stored <−5 °C nor allowed to freeze. It is not corrosive to metals, but will permeate certain plastics and soften or swell certain types of rubber. It is stable to u.v. light.

**Uses.**

Butralin is a pre-em. herbicide, which should be incorporated with the soil soon after application, used at 1.12-3.4 kg a.i./ha (depending on soil type) for weed control in cotton and soyabeans. It is also used to control suckers on tobacco, being applied to stems at 125 mg/plant. It is not persistent in soil.

**Toxicology.**

Acute oral $LD_{50}$ for albino rats 12 600 mg tech./kg. Acute percutaneous $LD_{50}$ for albino rabbits 10 200 mg/kg. $LC_{50}$ (8-d) for bobwhite quail and mallard duck was 10 000 mg/kg diet. $LC_{50}$ (48-h) is: for bluegill 4.2 mg/l; for rainbow trout 3.4 mg/l.

**Formulations.**

'Amex', e.c. (480 g a.i./l) as herbicide. 'Tamex', e.c. (240 g/l) as plant growth regulator. Herbicidal mixtures include: 'Amexine' (butralin + atrazine); 'Linamex' (butralin + linuron); 'Monamex' (butralin + monolinuron).

**Analysis.**

Product analysis is by glc.

$C_{12}H_{13}ClN_2O$ (236.7)
GR DMVN1&Y1&1UU1

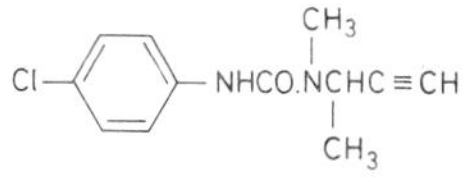

$Cl-C_6H_4-NHCO.N(CH_3)CH(CH_3)C\equiv CH$

**Nomenclature and development.**

Common name buturon (BSI, E-ISO, F-ISO);*exception* Portugal. Chemical name (IUPAC) 3-(4-chlorophenyl)-1-methyl-1-(1-methylprop-2-ynyl)urea. (*C.A.*) *N*'-(4-chlorophenyl)-*N*-methyl-*N*-(1-methyl-2-propynyl)urea (9CI); 3-(*p*-chlorophenyl)-1-methyl-1-(1-methyl-2-propynyl)urea (8CI); Reg. No. *[3766-60-7]*. Its herbicidal properties were described by A. Fischer (*Meded. Landbouwhogesch, Opzoekingsstn. Staat Gent*, 1964, **29**, 719). Introduced by BASF AG (DEP 1 108 977) as code no. 'H 95'; trade mark 'Eptapur'.

**Properties.**

It is a colourless solid; m.p. 145-146 °C; v.p. 10 nPa at 20 °C. Solubility at 20 °C: 30 mg/l water; 279 g/kg acetone; 9.8 g/kg benzene; 128 g/kg methanol. Technical grade has m.p. 132-142 °C. Stable under normal conditions ($1<pH<13$), but slowly decomposes in boiling water. Compatible with other herbicides.

**Uses.**

Buturon is a pre- and post-em. herbicide, absorbed mainly by the roots and recommended for use at 0.5-1.5 kg a.i./ha in cereals and maize for the control of shallow-germinating grasses and broad-leaved weeds. It is rapidly degraded in plants.

**Toxicology.**

Acute oral $LD_{50}$ for rats 3000 mg/kg. Applications (20-h) to the backs of rabbits produced a slight erythema but the ears were unaffected. In 120-d feeding trials rats receiving 500 mg/kg diet showed no ill-effect.

**Formulations.**

'Eptapur', w.p. (500 g a.i./kg).

**Analysis.**

Product analysis is by hydrolysis and potentiometric titration of the 4-chloroaniline produced. Residues may be determined by hydrolysis and colorimetric estimation of the 4-chloroaniline produced. Particulars of both methods are available from BASF AG.

$C_4H_{11}N$ (73.14)
ZY2&1

$$\begin{array}{c} \quad CH_3 \\ \quad | \\ CH_3CH_2CHNH_2 \end{array}$$

**Nomenclature and development.**

Butylamine (E-ISO) is proposed in lieu of a common name. Chemical name (IUPAC) (*RS*)-*sec*-butylamine (I); (*RS*)-2-aminobutane. (*C.A.*) 2-butanamine (9CI); (I) (8CI); Reg. No. *[13952-84-6]*. The fungicidal properties of this fumigant were described by J. W. Eckert & M. J. Kolbezen (*Nature (London)*, 1962, **194**, 188). Introduced by BASF AG as trade marks 'Butafume', '*Tutane*' (Eli Lilly & Co) is no longer marketed.

**Properties.**

It is a colourless liquid, with an ammoniacal odour; b.p. 63 °C; v.p. 18 kPa at 20 °C; $n_D^{20}$ 1.394; $d_4^{20}$ 0.724. Miscible with water and most organic solvents. It is a base, forming water-soluble salts with acids. It is stable but corrosive to aluminium, tin and some steels.

**Uses.**

It is a fungicide used to control many fruit-rotting fungi; aqueous solutions of its salts, containing 5-20 g amine/l, are used as dips or sprays on harvested fruit to prevent decay in transport or storage. Harvested fruit may be fumigated at 327 mg/m$^3$ for 4 h, or its equivalent; potatoes (seed or ware) at 280 ml/t for 2.5 h to control *Phoma exigua* var *foveata* and *Polyscytalum pustulans*. In neutral aqueous solution the control of decay of oranges due to *Penicillium digitatum* is largely due to the (*R*)-(-)-isomer (Reg. No. *[13250-12-9]*) (*idem, Phytopathology*, 1967, **57**, 98).

**Toxicology.**

Acute oral $LD_{50}$: for rats 380 mg amine/kg; for dogs 225 mg/kg; for hens 250 mg/kg. It is strongly irritant, but the percutaneous toxicity for rabbits is >2500 mg/kg. In 2-y feeding trials rats and dogs receiving 2500 mg/kg diet suffered no ill-effect. In teratology studies NEL was: for rats 2500 mg/kg diet; for rabbits 5000 mg/kg diet. In reproduction studies in rats NEL was 2500 mg/kg diet. $LC_{50}$ for bluegill fingerlings is >50 mg/l.

**Formulations.**

These include: 'Butafume', free amine (990 g/kg); concentrated aqueous solutions of appropriate salts.

**Analysis.**

Product analysis is by acid-base titration. Residues may be determined by glc of a derivative (*Anal. Methods Pestic. Plant Growth Regul.*, 1976, **8**, 251; *Pestic. Anal. Man.*, 1979, **II**; E. W. Day *et al.*, *J. Assoc. Off. Anal. Chem*, 1968, **51**, 39).

$C_{11}H_{23}NOS$ (217.4)
1Y1&1NVS2&1Y1&1

$[(CH_3)_2CHCH_2]_2NCO.SCH_2CH_3$

**Nomenclature and development.**

Common name butylate (BSI, E-ISO, WSSA); butilate (F-ISO); *exception* (Federal Republic of Germany). Chemical name (IUPAC) *S*-ethyl di-isobutyl(thiocarbamate) (I). (*C.A.*) *S*-ethyl bis(2-methylpropyl)carbamothioate (9CI); *S*-ethyl diisobutylthiocarbamate (8CI); Reg. No. *[2008-41-5]*. Its herbicidal properties were reported by R. A. Gray *et al.* (*Proc. North Cent. Weed Control Conf.*, 1962, **19**, 19). Introduced by Stauffer Chemical Co. (USP 2 913 327) as code no. 'R-1910'; trade mark 'Sutan'.

**Properties.**

Butylate is a clear liquid, with an aromatic odour; b.p. 130 °C/10 mmHg; v.p. 170 mPa at 25 °C, $d^{25}$ 0.9402, $n_D^{30}$ 1.4702. Solubility at 20°C: 46 mg/l water; miscible with acetone, ethanol, kerosene, 4-methylpentan-2-one, xylene. Stable <200 °C.

**Uses.**

It is toxic to germinating weed seeds and is incorporated into the soil immediately prior to sowing for the control of broad-leaved and grass weeds in maize. The rates recommended are: for the control of annual grasses and nut grass 3 kg a.i./ha; for the control of broad-leaved weeds in addition to grasses 4 kg/ha, or 3 kg butylate + either atrazine or cyanazine. It is non-persistent.

**Toxicology.**

Acure oral $LD_{50}$: for male albino rats 3500 mg tech./kg, for females 3970 mg/kg. Acute percutaneous $LD_{50}$ for rabbits >5000 mg/kg; a mild irritant to their skin and non-irritating to their eyes. In 90-d feeding trials, 40 mg/kg daily was well tolerated by rats and dogs.

**Formulations.**

'Sutan 6E', e.c. (720 g a.i./l). Mixtures (Reg. No. *[55947-96-1]*) with *N,N*-diallyl-2,2-dichloroacetamide, 'Sutan +' E, e.c. (720 or 800 g butylate/l), 'Sutan +' 10G, granules (100 g butylate/kg). 'Sutan +' atrazine 18-6G (Reg. No. *[8070-81-3]*, granules (180 g butylate + 60 g atrazine/kg).

**Analysis.**

Product analysis is by glc. Residues in crops or soils are determined by glc or colorimetry, after conversion to a derivative (J. E. Barney, *Anal. Methods Pestic. Plant Growth Regul.*, 1973, **7**, 641). Analytical methods are available from the Stauffer Chemical Co.

# 2-*sec*-Butylphenyl methylcarbamate 1780

$C_{12}H_{17}NO_2$ (207.3)
2Y1&R BOVM1

$CH_3NHCO.O$ $CH_3$
$CHCH_2CH_3$

**Nomenclature and development.**

Common name BPMC (JMAF). Chemical name (IUPAC) 2-*sec*-butlyphenyl methylcarbamate. (*C.A.*) 2-(1-methylpropyl)phenyl methylcarbamate (9CI); *o-sec*-butylphenyl methylcarbamate (8CI); Reg. No. *[3766-81-2]*. Its insecticidal properties were described by R. L. Metcalf *et al.* (*J. Econ. Entomol.*, 1962, **55**, 889). Introduced by Sumitomo Chemical Co., by Kumiai Chemical Industry Co. Ltd, by Mitsubishi Chemical Industries and by Bayer AG as code no. 'Bayer 41 367c'; trade marks 'Osbac' (to Sumitomo), 'Bassa' (to Kumiai), 'Baycarb' (to Bayer).

**Properties.**

The pure compound has m.p. 31-32 °C. The technical grade is a pale yellow or pale reddish liquid; $n_D^{20}$ 1.514; v.p. 48 mPa at 20 °C, $d_4^{30}$ 1.035. Solubility at 30 °C: 610 mg/l water; at room temperature: >1 kg/kg acetone, benzene, toluene, xylene. It is unstable to alkali and to concentrated acids.

**Uses.**

It is an insecticide controlling sucking insects (leaf and plant hoppers), bugs and weevils on rice at 500 g a.i./ha. A combination with fenitrothion is applied for the simultaneous control of rice borers, rice leaf and plant hoppers, at 2-*sec*-butylphenyl methylcarbamate + fenitrothion (300 + 450 g/ha).

**Toxicology.**

Acute oral $LD_{50}$: for male rats 623.4 mg/kg, for females 657.0 mg/kg. Acute percutaneous $LD_{50}$ for rats >5000 mg/kg. $LC_{50}$ (48-h) for carp is 12.6 mg/l.

**Formulations.**

These include: e.c. (500 g a.i./l); dust (20 g/kg); microgranule (30 g/kg). 'Sumibas' (Reg. No. *[65252-61-9]*) (Sumitomo Chemical Co.), e.c., ULV (300 g 2-*sec*-butylphenyl methylcarbamate + 450 g fenitrothion/l); 'Bay-Bassa', e.c. (2-*sec*-butylphenyl methylcarbamate + fenthion).

**Analysis.**

Product analysis is by u.v. spectrometry or by hplc (S. Sakaue *et al., Nippon Nogei Kagaku Kaishi*, 1981, **55**, 1237). The analytical method for 'Sumibas' is available from Sumitomo Chemical Co. Ltd.

$C_{28}H_{44}O$ (396.7)
L56 FYTJ A1 BY1&1U1Y1&Y1&1 FU2U-
BL6YYTJ AU1 DQ

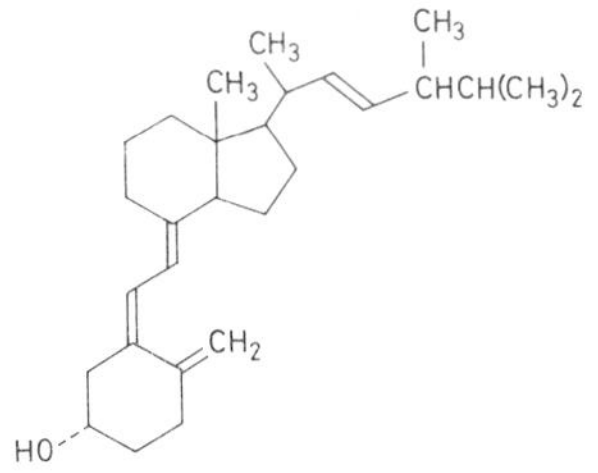

**Nomenclature and development.**

Common name calciferol (BPC); ergocalciferol (US Pharmacopoeia); also known as Vitamin $D_2$. Chemical name (IUPAC) (3β,5*Z*,7*E*,22*E*)-9,10-secoergosta-5,7,10(19),22-tetraen-3-ol (I). (*C.A.*) (I) (9CI); ergocalciferol (8CI); Reg. No. *[50-14-6]*. Its properties were reviewed by H. H. Inhoffen (*Angew. Chem.*, 1960, **72**, 875). Its rodenticidal properties were described by M. Hadler (*Proc. Br. Pest Control Conf., 4th*). Introduced by Sorex Ltd (GBP 1 371 135) as 'Sorexa C.R.' for a combination of calciferol and warfarin.

**Properties.**

Calciferol forms colourless crystals; m.p. 115-118 °C. Solubility: 50 mg/l water; 69.5 g/l acetone; 10 g/l benzene; 1 g/l hexane.

**Uses.**

Calciferol is an essential natural vitamin, but in high doses produces a lethal hypervitaminosis characterised by hypercalcaemia and increase in serum cholesterol. Evidence (Hadler, *loc. cit.*) suggests that admixture with warfarin increases the rodenticidal efficacy in many species. The mixture is semi-acute in action, 1-2 feeds being sufficient to cause death. This advantage, combined with excellent acceptability, is responsible for its success in the control of mice.

**Toxicology.**

Acute oral $LD_{50}$ for rats 56 mg/kg; for mice 23.7 mg/kg. Sub-acute oral $LD_{50}$ (5-d) for rats 7 mg/kg daily.

**Formulations.**

Ready-to-use bait on canary seed (1 g calciferol + 250 mg warfarin/kg); oil concentrate (20 g/l).

**Analysis.**

Product analysis is by colorimetry or by u.v. spectrometry: details are available from Sorex Ltd.

$CaS_x$
.CA..S-X

$CaS_x$

**Nomenclature and development.**

Common name lime sulfur (ESA, JMAF), which applies to an aqueous solution of calcium polysulphides. Chemical name (IUPAC) calcium polysulphide (E-ISO); polysulfure de calcium (F-ISO) used in lieu of a common name. (*C.A.*) calcium sulfide ($CaS_x$) (8 & 9CI); Reg. No. *[1344-81-6]*. Eau Grison (after Grison who used it as a fungicide in 1852). Introduced in the 19th Century as a fungicide and for the control of scale insects.

**Properties.**

Lime sulphur, produced by dissolving sulphur in aqueous suspensions of calcium hydroxide, is a deep orange malodorous liquid $d^{15.6} \geqslant 1.28$. It is an aqueous solution of calcium polysulphides, with traces of calcium thiosulphate. It is slightly alkaline. It is decomposed by carbon dioxide, by acids and by the soluble salts of metals that form insoluble sulphides; sulphur, hydrogen sulphide and the metal sulphide being formed. It is of limited compatibility with other pesticides.

**Uses.**

Lime sulphur is a fungicide itself, effective against powdery mildews. The residue of sulphur left by decomposition acts as a protectant fungicide. Lime sulphur is also effective against *Comstockaspis perniciosus* and other scales, an activity attributed to its softening action on their wax (G. D. Schafer, *Mich. Agric. Exp. Stn. Tech. Bull.*, 1911, No. 11; 1915, No. 21). It is phytotoxic, especially to sulphur-shy varieties. It has also been used in sheep dips against mites.

**Formulations.**

Lime sulphur is used without formulation, usually at dilutions of *c.* 10 ml/l.

**Analysis.**

Product analysis for content of total polysulphide-sulphur ($S_x$), sulphide sulphur and thiosulphate (*AOAC Methods*, 1980 6.107-6.108; *MAFF Ref. Bk. (Tech. Bull.)* 1958, No. 1; *CIPAC Handbook*, 1980, **1A**, 1279; J. R. Gray, *J. Sci. Food Agric.*, 1956, **7**, 3; FAO Specification (CP/58)).

$C_{10}H_{10}Cl_8$ (approx.) (431.8 approx.)

**Nomenclature and development.**

Common name camphechlor (BSI, E-ISO);camphéchlore (F-ISO);toxaphene (Belgium, Canada, USA);polychlorcamphene (USSR);*exception* (France). Chemical name (IUPAC) a reaction mixture of chlorinated camphenes containing 67-69% chlorine. (*C.A.*) toxaphene (8 & 9CI); Reg. No. *[8001-35-2]*. ENT 9735. Its insecticidal properties were described by W. LeRoy Parker & J. R. Beacher (*Del. Univ. Agric. Exp. Stn. Bull.*, 1947, No. 264). Introduced by Hercules Inc. (now Boots-Hercules Agrochemical Co.). (USP 2 565 471; 2 657 164) as code no. 'Hercules 3956'; trade mark 'Toxaphene', which in some countries has been dedicated to the public and is there used as a common name.

**Properties.**

It is a yellow wax of mild terpene-like odour, softening in the range 70-95 °C; v.p. 27-53 Pa at 25 °C; $d^{25}$ 1.65. Solubility at room temperature *c.* 3 mg/l water; readily soluble in organic solvents including petroleum oils. It is dehydrochlorinated by heat, by strong sunlight and by certain catalysts such as iron. It is incompatible with strongly alkaline pesticides and is non-corrosive in the absence of moisture.

The components present have been examined (J. E. Casida *et al., J. Agric. Food Chem.*, 1974, **22**, 653; 1975, **23**, 991); 26 components account for about 40% of the product; a heptachlorobornane and a mixture of 2 octachlorobornanes are more toxic to insects and mice than other components and are similarly biodegradable.

**Uses.**

It is a non-systemic contact and stomach insecticide with some acaricidal action. It is non-phytotoxic, except to cucurbits, and is used in the control of many insects on corn, cotton, fruit, small grains, vegetables and the control of *Cassia obtusifolia* in soyabeans. It is also used for the control of animal ectoparasites.

**Toxicology.**

Acute oral $LD_{50}$ for rats 80-90 mg/kg. Acute percutaneous $LD_{50}$ for rats 780-1075 mg/kg. In 2-y feeding trials NEL for rats was 25 mg/kg diet. There was an accumulation in the body fat proportional to the dose fed but elimination was rapid when the intake was stopped.

**Formulations.**

These include: e.c. (480, 720, 900 or 960 g a.i./l).

**Analysis.**

Product analysis is by total chlorine (*CIPAC Handbook*, 1970, **1**, 132; FAO Specifications (CP/35; CP/68)). Residues may be determined by glc with EDC (*Anal. Methods Pestic. Plant Growth Regul.*, 1972, **6**, 514) or by spectrometry of a derivative (C. L. Dunn, *Anal. Methods Pestic., Plant Growth Regul. Food Addit.*, 1964, **2**, 523).

$C_{10}H_9Cl_4NO_2S$ (349.1)
T56 BVNV GUTJ CSXGGYGG

O
N–SCCl₂.CHCl₂
O

**Nomenclature and development.**

Common name captafol (BSI, E-ISO, F-ISO, ANSI);difolatan (JMAF). Chemical name (IUPAC) *N*-(1,1,2,2-tetrachloroethylthio)cyclohex-4-ene-1,2-dicarboximide; 1,2,3,6-tetrahydro-*N*-(1,1,2,2-tetrachloroethylthio)phthalimide; 3a,4,7,7a-tetrahydro-*N*-(1,1,2,2-tetrachloroethanesulphenyl)phthalimide. (*C.A.*) 3a,4,7,7a-tetrahydro-2-[(1,1,2,2-tetrachloroethyl)thio]-1*H*-isoindole-1,3(2*H*)-dione (9CI); *N*-[(1,1,2,2-tetrachloroethyl)thio]-4-cyclohexene-1,2-dicarboximide (8CI); Reg. No. *[2425-06-1]*. Its fungicidal properties were described by W. D. Thomas *et al.* (*Phytopathology*, 1962, **52**, 754). Introduced by Chevron Chemical Co. as code no. 'Ortho-5865'; trade mark 'Difolatan'.

**Properties.**

Pure captafol is a colourless crystalline solid; m.p. 160-161 °C; v.p. negligible at room temperature. Solubility: 1.4 mg/l water; slightly soluble in most organic solvents. The technical grade is a light tan powder, with a characteristic odour. It is unstable under strongly alkaline conditions, and slowly decomposes at its m.p.

**Uses.**

It is a protectant non-systemic fungicide applied to foliage at *c.* 200 g a.i./100 l. The stability of deposits which retain their fungicidal activity has led to spray programmes requiring fewer applications/season when applied at higher rates during the dormant season on pome fruits. It is widely used to control foliage and fruit diseases of tomatoes, coffee berry disease, potato blight, tapping panel disease of rubber and many other diseases.

It is also used in the lumber and timber industries to reduce losses from wood rot fungi in logs and wood products.

**Toxicology.**

Acute oral $LD_{50}$ for rats: 5000-6200 mg a.i./kg; 2500 mg w.p. (administered as an aqueous suspension)/kg. Acute percutaneous $LD_{50}$ for rabbits $>$15 400 mg/kg; some people develop an allergy to captafol. In 2-y feeding trials no ill-effect was observed in rats receiving 500 mg/kg diet; in dogs 10 mg/kg daily. $LC_{50}$ (10-d) was: for pheasants $>$23 070 mg/kg diet. $LC_{50}$ (96-h) is: for rainbow trout 0.5 mg/l; for goldfish 3.0 mg/l; for bluegill 0.15 mg/l.

**Formulations.**

These include 'Ortho Difolatan 80W', w.p. (800 g a.i./kg); 'Ortho Difolatan 4 Flowable', s.c. (480 g/l).

**Analysis.**

Product analysis is by glc with FID or by hplc (A. A. Carlstrom & J. B. Leary, *Anal. Methods Pestic. Plant Growth Regul.*, 1978, **10**, 173); details available from the Chevron Chemical Co. Residues may be determined by glc with FPD or ECD (*idem, ibid.*; *Pestic. Anal. Man.*, 1979, **I**, 201-I; R. Mestres *et al., Trav. Soc. Pharm. Montpellier*, 1979, **39**, 323).

$C_9H_8Cl_3NO_2S$ (300.6)
T56 BVNV GUTJ CSXGGG

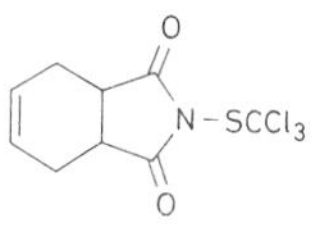

**Nomenclature and development.**

Common name captan (BSI, E-ISO, JMAF);captane (F-ISO);*exception* (Republic of South Africa). Chemical name (IUPAC) *N*-(trichloromethylthio)cyclohex-4-ene-1,2-dicarboximide; 1,2,3,6-tetrahydro-*N*-(trichloromethylthio)phthalimide; 3a,4,7,7a-tetrahydro-*N*-(trichloromethanesulphenyl)phthalimide. (*C.A.*) 3a,4,7,7a-tetrahydro-2-[(trichloromethyl)thio]-1*H*-isoindole-1,3(2*H*)-dione (9CI); *N*-[(trichloromethyl)thio]-4-cyclohexene-1,2-dicarboximide (8CI); Reg. No. *[133-06-2]*. ENT 26 538. Its fungicidal properties were described by A. R. Kittleston (*Science*, 1952, **115**, 84). Introduced by the Standard Oil Development Co., later by the Chevron Chemical Co. (USP 2 553 770; 2 553 771; 2 553 776) as code no. 'SR 406'; trade marks 'Orthocide 406', 'Orthocide'.

**Properties.**

Captan forms colourless crystals; m.p. 178 °C; v.p. <1.3 mPa at 25 °C. Solubility at 25 °C: 3.3 mg/l water; 21 g/kg acetone; 70 g/kg chloroform; 23 g/kg cyclohexanone; insoluble in petroleum oils; 1.7 g/kg propan-2-ol; 20 g/kg xylene. The technical grade (90-95% pure) is a colourless to beige amorphous solid, with a pungent odour; m.p. 160-170 °C. It is unstable under alkaline conditions and decomposes at or near its m.p. It is itself non-corrosive though its decomposition products are corrosive.

**Uses.**

It is a fungicide used, generally at 120 g a.i./100 l, to control diseases of many fruit, ornamental and vegetable crops, including *Venturia inaequalis* of apple and *V. pirina* of pear. It should not be mixed with oil sprays. It is also used as a spray, root dip or seed treatment, to protect young plants against rots and damping-off.

**Toxicology.**

Acute oral $LD_{50}$ for rats 9000 mg/kg. It may cause skin irritation. In 2-y feeding trials NEL for rats was 1000 mg/kg diet. No teratogenic or mutagenic effects have been observed.

**Formulations.**

These include: 'Orthocide 50WP', w.p. (500 g a.i./kg; also 800 or 830 g/kg); dusts (50 or 100 g/kg); also dusts (600-750 g/kg) and w.p. for seed treatment. Seed treatment, w.p. combined with other fungicides, or with insecticides.

**Analysis.**

Product analysis is by i.r. spectrometry or by total chlorine content after alkaline hydrolysis (*CIPAC Handbook*, 1970, **1**, 171; FAO Specification (CP/57)). Residues may be determined by glc (*AOAC Methods*, 1980, 29.001-29.018; *Pestic. Anal. Man.*, 1979, **I**, 201-A, 201-G, 201-I; R. Mestres *et al., Trav. Soc. Pharm. Montpellier*, 1979, **39**, 323; J. N. Ospenson, *Anal. Methods Pestic., Plant Growth Regul. Food Addit.*, 1964, **3**, 7; *Anal. Methods Pestic. Plant Growth Regul.*, 1972, **6**, 546).

$C_{12}H_{11}NO_2$ (201.2)
L66J BOVM1

OCO.NHCH$_3$

**Nomenclature and development.**

Common name carbaryl (BSI, E-ISO, F-ISO, ANSI, BPC);sevin (USSR);NAC (JMAF);*exception* Sweden. Chemical name (IUPAC) 1-naphthyl methylcarbamate (I). (*C.A.*) 1-naphthalenyl methylcarbamate (9CI); (I) (8CI); Reg. No. *[63-25-2]*. OMS 29, OMS 629, ENT 23 969. Its insecticidal properties were described by H. L. Haynes *et al.* (*Contrib. Boyce Thompson Inst.*, 1957, **18**, 507). Introduced by Union Carbide Corp. (USP 2 903 478) as 'Experimental Insecticide 7744'; trade mark 'Sevin'.

**Properties.**

A colourless crystalling solid; m.p. 142 °C; v.p. <665 mPa at 26 °C; $d_{20}^{20}$ 1.232. Solubility: at 30 °C 120 mg/l water; at 25 °C 400-450 g/kg dimethylformamide, dimethyl sulphoxide. Technical grade ≥99% pure. Stable <70 °C, and to light. Hydrolysis to 1-naphthol is rapid at pH >9. Incompatible with strongly alkaline pesticides.

**Uses.**

Carbaryl is a contact and stomach insecticide with slight systemic properties recommended for use at 0.25-2.0 kg a.i./ha against many insect pests of cotton, fruit, vegetables and other crops. There is no evidence of phytotoxicity at these rates. It is also used to reduce the number of fruits on heavily laden apple trees.

**Toxicology.**

Acute oral $LD_{50}$ for male rats 850 mg/kg. Acute percutaneous $LD_{50}$: for rats >4000 mg/kg; for rabbits >2000 mg/kg. In 2-y feeding trials rats receiving 200 mg/kg diet suffered no ill-effect. $LC_{50}$ to fish 5-13 mg/l. Toxic to honeybees.

**Formulations.**

These include: w.p. (500, 800 or 850 g a.i./kg); granules (50 g/kg); dusts (50 or 100 g/kg); bait pellets (50 g/kg); micronised suspensions in molasses, in non-phytotoxic oil or in aqueous media (220, 300, 400, 440 or 480 g/l) and as true solutions in organic solvents.

**Analysis.**

Product analysis is by i.r. spectroscopy (*AOAC Methods*, 1980, 6.320-6.323; *CIPAC Handbook*, 1970, **1**, 185; 1980, **1A**, 1113; FAO Specification (CP/55)); hydrolysis and the determination of the methylamine so produced has also been used. Residues may be determined by glc (*AOAC Methods*, 1980, 29.058-29.063; *Pestic. Anal. Man.*, 1979, **I**, 201-I; R. T. Krause, *J. Assoc. Off. Anal. Chem.*, 1980, **63**, 1114; *Anal. Methods Pestic. Plant Growth Regul.*, 1972, **6**, 478); or by hydrolysis to 1-naphthol, a derivative of which is measured colorimetrically (H. A. Stansbury & R. Miskus, *Anal. Methods Pestic., Plant Growth Regul. Food Addit.*, 1964, **2**, 437; *AOAC Methods*, 1980, 29.082-29.090).

# Carbendazim

$C_9H_9N_3O_2$ (191.2)
T56 BM DNJ CMVO1
*carbendazim phosphate*
$C_9H_{12}N_3O_6P$ (289.2)
T56 BM DNJ CMVO1 &&$H_3PO_4$ SALT

**Nomenclature and development.**

Common name carbendazim (BSI, E-ISO);carbendazime (F-ISO);carbendazol (JMAF). Chemical name (IUPAC) methyl benzimidazol-2-ylcarbamate. (*C.A.*) methyl 1*H*-benzimidazol-2-ylcarbamate (9CI); methyl 2-benzimidazolecarbamate (8CI); Reg. No. *[10605-21-7]*. Trivial names: MBC, BMC. Its fungicidal properties were described by H. Hampel & F. Löcher (*Proc. Br. Insectic. Fungic. Conf.*, 1973, **1**, pp. 127, 301). Introduced by BASF AG, Hoechst AG and E. I. du Pont de Nemours & Co. (Inc.) (USP 3 657 443; GBP 1 190 614 du Pont) as code no. 'BAS 346F' (BASF), 'Hoe 17 411' (Hoechst); trade marks 'Bavistin' (BASF), 'Derosal' (Hoechst), 'Delsene' (du Pont).

**Properties.**

Pure carbendazim is a colourless solid; m.p. 310 °C (decomp.); v.p. <100 nPa at 20 °C. Solubility at 20 °C: 28 mg/l water at pH 4, 8 mg/l at pH 7, 7 mg/l at pH 8; 300 mg/l acetone; 100 mg/l chloroform; 68 mg/l dichloromethane; 5 g/l dimethylformamide; 300 g/l ethanol. It is stable in acid, forming water-soluble salts, e.g. carbendazim phosphate (Reg. No. *[52316-55-9]*), 9 g/l at pH2. Subject to microbial degradation in soils. The technical grade is ⩾99% pure.

**Uses.**

Carbendazim is a systemic fungicide controlling a wide range of pathogens of cereals, fruit, grapes, ornamentals and vegetables. It is absorbed by the roots and green tissues of plants. Injections of solutions of salts, especially the hydrochloride, hypophosphite and phosphate, into the trunks have given some control of Dutch elm disease (D. J. Clifford *et al., Pestic. Sci.*, 1976, **7**, 91).

**Toxicology.**

Acute oral $LD_{50}$: for rats >15 000 mg a.i./kg; for dogs >2500 mg/kg; for quail >10 000 mg/kg. Acute percutaneous $LD_{50}$: for rats >2000 mg/kg; for rabbits >10 000 mg/kg; non-irritant to eyes of rabbits and skin of guinea-pigs. In 2-y feeding trials NEL for rats and dogs was 300 mg/kg diet. $LC_{50}$ (96-h) for rainbow trout is 0.36 mg/l.

**Formulations.**

These include: 'Derosal', w.p. (594 g a.i./kg), dispersion (188 g/l); 'Bavistin', w.p. (500 g/kg); 'Fungicide BLP', water-soluble liquid (7 g carbendazim phosphate/l). 'Cosmic', w.p. (carbendazim + maneb + tridemorph). 'Delsene M', (Reg. No. *[52080-81-6]*), w.p. (100 g carbendazim + 640 g maneb/kg). 'Delsene MX', (Reg. No. *[62713-28-4]*), w.p. (62 g carbendazim + 738 g mancozeb/kg). 'Granosan', seed treatment, powder (150 g carbendazim + 600 maneb/kg). 'Mastiff', liquid (carbendazim + chlormequat chloride).

**Analysis.**

Product analysis is by titration against perchloric acid in acetic acid. Residues in crops may be determined using methods for benomyl, hplc, (J. J. Kirkland *et al., J. Agric. Food Chem.*, 1973, **21**, 368; *Pestic. Anal. Man.*, 1979, **I**; J. E. Farrow *et al., Analyst (London)*, 1977, **105**, 1185) or fluorimetry or colorimetry of derivatives (H. L. Pease & J. A. Gardiner *J. Agric. Food Chem.*, 1969, **17**, 267; N. Aharonson & A. Ben-Aziz, *J. Assoc. Off. Anal. Chem.*, 1973, **56**, 1330).

$C_{12}H_{16}N_2O_3$ (236.3)
2MVY1&OVMR

**Nomenclature and development.**

Common name carbetamide (BSI, E-ISO, F-ISO, ANSI, WSSA);*exception* (Federal Republic of Germany). Chemical name (IUPAC) (*R*)-1-(ethylcarbamoyl)ethyl carbanilate; (*R*)-1-(ethylcarbamoyl)ethyl phenylcarbamate; D-(−)-*N*-ethyl-2-(phenylcarbamoyloxy)propionamide. (*C.A.*) (*R*)-*N*-ethyl-2-[[(phenylamino)carbonyl]oxy]propanamide (9CI); D-*N*-ethyllactamide carbanilate ester (8CI); Reg. No. *[16118-45-0]*. Its herbicidal properties were described by J. Desmoras *et al.* (*C. R. Journ. Etud. Herbic. Conf. COLUMA, 2nd*, 1963, p. 14). Introduced by Rhône-Poulenc Phytosanitaire (GBP 959 204; BEP 597 035; USP 3 177 061) as code no. '11 561 RP'; trade mark 'Legurame'.

**Properties.**

Pure carbetamide forms colourless crystals; m.p. 119 °C; v.p. negligible at 20 °C. The technical grade has m.p. >110 °C. Solubility at 20 °C: *c.* 3.5 g/l water; soluble in acetone, dichloromethane, dimethylformamide, ethanol, methanol. It is stable under normal storage conditions.

**Uses.**

It is a selective herbicide effective post-em. against grasses and some broad-leaved weeds. It is used at 2 kg a.i./ha to control weeds in brassicas, chicory, red clover, endive, lucerne and oilseed rape. it persists in soil *c.* 60 d under normal field conditions.

**Toxicology.**

Acute oral $LD_{50}$: for rats 11 000 mg/kg; for mice 1250 mg/kg; for dogs 1000 mg/kg. Percutaneous applications of 500 mg/kg are non-toxic to rabbits. In 90-d feeding trials no effect was observed with: rats receiving 3200 mg/kg diet; dogs 12 800 mg/kg diet.

**Formulations.**

'Legurame liquide', e.c. (300 g a.i./l); 'Legurame P.M.', 'Carbetamex', w.p. (700 g/kg).

**Analysis.**

Product analysis is by titration of the ethylamine liberated on hydrolysis (J. Desmoras *et al., Anal. Methods Pestic. Plant Growth Regul.*, 1973, **7**, 509). Residues may be determined by hydrolysis to aniline which is measured by colorimetry of a derivative (*idem, ibid.*).

$C_{12}H_{15}NO_3$ (221.3)
T56 BOT&J C1 C1 IOVM1

OCO.NHCH3 / O / CH3 / CH3

**Nomenclature and development.**

Common name carbofuran (BSI, E-ISO, F-ISO, ANSI). Chemical name (IUPAC) 2,3-dihydro-2,2-dimethylbenzofuran-7-yl methylcarbamate. (*C.A.*) 2,3-dihydro-2,2-dimethyl-7-benzofuranyl methylcarbamate (8 & 9CI); Reg. No. *[1563-66-2]*. OMS 864, ENT 27 164. Its insecticidal properties were described by F. L. McEwen & A. C. Davis (*J. Econ. Entomol.*, 1965, **58**, 369) and E. J. Armburst & G. C. Gyrisco (*ibid.*, p. 940). Introduced by the Agricultural Chemical Div. of the FMC Corp. and by Bayer AG (USP 3 474 170; 3 474 171 (both to FMC); DEPS 1 493 646 to Bayer) as code no. 'FMC 10 242','Bay 70 143'; trade marks 'Furadan' (FMC), 'Curaterr', 'Yaltox' (both to Bayer).

**Properties.**

Carbofuran is a crystalline solid; m.p. 150-152 °C; v.p. 2.7 mPa at 33 °C; $d_{20}^{20}$ 1.180. Solubility at 25 °C: 700 mg/l water; 150 g/kg acetone; 140 k/kg acetonitrile; 40 g/kg benzene; 90 g/kg cyclohexanone; 270 g/kg dimethylformamide; 250 g/kg dimethyl sulphoxide; 300 g/kg 1-methyl-2-pyrrolidone; it is essentially insoluble in conventional formulation solvents used in agriculture. It is unstable in alkaline media.

**Uses.**

It is a systemic acaricide, insecticide and nematicide, applied to foliage at 0.25-1.0 kg a.i./ha for the control of insects and mites, or applied to the seed furrow at 0.5-4.0 kg/ha for the control of soil-dwelling and foliar-feeding insects, or broadcast at 6-10 kg/ha for the control of nematodes. Its activity against these pests has been summarised (B. Homeyer, *Pflanzenschutz-Nachr. (Engl. Ed.)*, 1975, **28**, 3) and those of maize (P. Villeroy & P. Pourcharesse, *ibid.*, p. 55; K. Küthe, *ibid.*, p. 144), brassicas (T. J. Martin & D. B. Morris, *ibid.*, p. 92) and rice (K. Iwaya & G. Kollmer, *ibid.*, p. 137).

**Toxicology.**

Acute oral $LD_{50}$: for rats 8-14 mg a.i. (in corn oil)/kg; for dogs 19 mg a.i. (dry powder)/kg. Acute percutaneous $LD_{50}$ for rabbits 2550 mg a.i. (as w.p.)/kg. In 2-y feeding trials no effect was observed on rats receiving 25 mg/kg diet, on dogs receiving 20 mg/kg diet, nor on rats receiving 10 mg/kg diet for 3 generations nor on dogs receiving 50 mg/kg diet for one generation. $LD_{50}$ (10-d) for pheasants was 960 mg a.i. (as 10% granule)/kg diet. $LC_{50}$ (96-h) for trout is 0.28 mg/l. It is metabolised in the liver and excreted in the urine of animals, 50% being lost in 6-12 h; in soils 50% is lost in 30-60 d. 2,3-Dihydro-3-hydroxy-2,2-dimethylbenzofuran-7-yl methylcarbamate, which is of low toxicity to insects and mammals, is one of the products formed.

**Formulations.**

These include: w.p. (750 g a.i./kg); flowable paste (480 g/l); granules (20, 30, 50 or 100 g/kg).

**Analysis.**

Product analysis is by i.r. spectrometry. Residues may be determined by glc (E. Möllhoff, *ibid.*, p. 370; *AOAC Methods*, 1980, 29.058-29.063; R. F. Cooke *et al.*, *J. Agric. Food Chem.*, 1969, **17**, 277; R. F. Cooke, *Anal. Methods Pestic. Plant Growth Regul.*, 1973, **7**, 187).

$CS_2$ (76.13)
SCS

$CS_2$

**Nomenclature and development.**

The chemical name (IUPAC) carbon disulphide is accepted in lieu of a common name (BSI, E-ISO); sulfure de carbone (F-ISO). (*C.A.*) carbon disulfide (8 & 9CI); Reg. No. *[75-15-0]*. Trivial name carbon bisulphide Used as an insecticide in 1854 by Garreau (see *Science*, 1926, **64**, 326).

**Properties.**

It is a colourless mobile liquid; b.p. 46.3 °C; m.p. −108.6 °C; v.p. 47 kPa at 25 °C; $d_4^{20}$ 1.2628; $n_D^{18}$ 1.6295. Solubility at 32 °C: 2.2 g/l water; miscible with chloroform, diethyl ether, ethanol. Its vapour is 2.63 times as dense as air, and is extremely flammable with flash point *c.* 20 °C and it ignites spontaneously *c.* 125-135 °C. Its impurities have an unpleasant odour.

**Uses.**

It is an insecticide used for fumigation of nursery stock and for soil treatment against insects and nematodes. It is also used in some countries as an insecticide in mixtures with carbon tetrachloride (to reduce fire hazard) for fumigating stored grain.

**Toxicology.**

The vapour is highly poisonous, producing giddiness and vomiting in 30 min at 6.8 g/m$^3$ air; repeated daily exposures to 227 mg/m$^3$ air caused ill health.

**Formulations.**

Soil applications are made using the compound alone, as emulsions or solutions in alkali (thiocarbonates).

**Analysis.**

Residues in grain may be determined by glc (S. G. Heuser & K. A. Scudamore, *J. Sci. Food Agric.*, 1969, **20**, 566) or by colorimetry of a derivative (C. L. Dunning, *J. Assoc. Off. Agric. Chem.*, 1957, **40**, 168). The concentration in air may be measured by drawing the air through an ethanolic solution of diethylamine and copper(II) acetate and colorimetry of the complex produced.

# Carbon tetrachloride

$CCl_4$ (153.8)
GXGGG

$CCl_4$

**Nomenclature and development.**

The traditional name carbon tetrachloride (I) (BSI, E-ISO); tétrachlorure de carbon (F-ISO) is accepted in lieu of a common name. Chemical name (IUPAC) (I); tetrachloromethane (II). (*C.A.*) (II) (9CI); (I) (8CI); Reg. No. *[56-23-5]*. ENT 27 164. Its use as an insecticidal fumigant was described by W. E. Britton (*Conn. Agric. Exp. Stn. Rep.*, 1908, No. 31).

**Properties.**

Carbon tetrachloride is a colourless liquid; b.p. 76 °C; m.p. −23 °C; v.p. 15 kPa at 25 °C; $d_{25}^{25}$ 1.588; $n_D^{20}$ 1.4607. Solubility at 25 °C: 280 mg/kg water; miscible with most organic solvents. The vapour is dense, 5.32 times that of air. It is non-flammable and non-explosive; though generally inert, it is decomposed by water at high temperatures.

**Uses.**

It is of low insecticidal activity, but is used for grain disinfestation when long exposures are possible, its main advantages being low absorption by treated grain. It is often used in mixtures with more potent fumigants, such as ethylene dichloride, to reduce the fire hazard of the latter. It has also been used as an anthelmintic in veterinary practice.

**Toxicology.**

Acute oral $LD_{50}$: for rats 5730-9770 mg/kg; for mice 12 800 mg/kg; for rabbits 6380-9975 mg/kg. It is a general anaesthetic, prolonged exposure causing irritation of the mucous membranes, headache and nausea; repeated exposure to high concentrations causes liver damage.

**Formulations.**

Carbon tetrachloride + carbon disulphide; carbon tetrachloride + ethylene dibromide + ethylene dichloride; carbon tetrachloride + ethylene dichloride.

**Analysis.**

Residues in cereals are determined by glc (S. G. Heuser & K. A. Scudamore, *J. Sci. Food Agric.*, 1969, **20**, 566; *Analyst (London)*, 1974, **99**, 570; *Pestic. Sci.*, 1973, **4**, 1).

$C_{11}H_{16}ClO_2PS_3$ (342.9)
GR DS1SPS&O2&O2

$Cl-C_6H_4-SCH_2SP(=S)(OCH_2CH_3)_2$

**Nomenclature and development.**

Common name carbophenothion (BSI, E-ISO, F-ISO, ANSI). Chemical name (IUPAC) *S*-4-chlorophenylthiomethyl *O*,*O*-diethyl phosphorodithioate. (*C.A.*) *S*-[[(4-chlorophenyl)thio]methyl] *O*,*O*-diethyl phosphorodithioate (9CI); *S*-[[(*p*-chlorophenyl)thio]methyl] *O*,*O*-diethyl phosphorodithioate (8CI); Reg. No. *[786-19-6]*. OMS 244, ENT 23 708. Insecticide and acaricide (*Agric. Chem.*, 1956, **11**(11), 91). Introduced by Stauffer Chemical Co. (USP 2 793 224) as code no. 'R-1303'; trade mark Trithion .

**Properties.**

Pure carbophenothion is a colourless liquid with a mercaptan-like odour; b.p. 82 °C/0.01 mmHg; v.p. 1.07 mPa at 25 °C; $n_D^{25}$ 1.597. Solubility: <1 mg/l water; miscible with acetone, ethanol, kerosene, 4-methylpentan-2-one, xylene. Technical grade (95% pure) is a light amber liquid; $d_{20}^{20}$ 1.285. It is relatively stable to hydrolysis and heat (<80 °C); oxidation occurs to the phosphorothioate on leaf surfaces (B. J. Luberoff *et al.*, *ibid.*, 1958, **13**(3), 83).

**Uses.**

Carbophenothion is a non-systemic acaricide and insecticide with a long residual action. It is used, in combination with petroleum oil, as a spray to control overwintering aphids, mites and scale insects on dormant deciduous fruit trees; and as an acaricide on citrus trees and cotton; in combination with parathion-methyl as a spray to control mites on grapevines, also with sulphur against mildew *Oidium* spp.; in combination with mevinphos to control mites on top fruit; as a seed treatment against *Delia coarctata* on wheat.

**Toxicology.**

Acute oral $LD_{50}$: for male albino rats 79.4 mg/kg, for females 20.0 mg/kg. Acute percutaneous $LD_{50}$ for rabbits 1850 mg/kg.

**Formulations.**

'Trithion E', e.c. (240, 480 or 960 g a.i./l); 'Trithion 25WP', w.p. (250 g/kg); 'Trithion 3 Dust' (30 g/kg); 'Trithion 10G', granules (100 g/kg). Mixtures: 200 g carbophenothion + 200 g parathion-methyl/l; 150 g carbophenothion + 100 g mevinphos/l; 12 g carbophenothion + 12 g parathion-methyl + 700 g sulphur/kg.

**Analysis.**

Product analysis is by glc. Residues may be determined by glc (*AOAC Methods*, 1980, 29.039-29.044; *Pestic. Anal. Man.*, 1979, **1**, 201-H, 201-I; D. C. Abbott *et al.*, *Pestic. Sci.*, 1970, **1**, 10). Analytical methods are available from the Stauffer Chemical Co.

$C_{12}H_{13}NO_2S$ (235.3)
T6O DS BUTJ B1 CVMR

**Nomenclature and development.**

Common name carboxin (BSI, E-ISO, ANSI);carboxine (F-ISO);carbathiin (Canada);*exceptions* (Denmark, Federal Republic of Germany). Chemical name (IUPAC) 5,6-dihydro-2-methyl-1,4-oxathi-ine-3-carboxanilide; 5,6-dihydro-2-methyl-1,4-oxathi-in-3-carboxanilide; 2,3-dihydro-6-methyl-5-phenylcarbamoyl-1,4-oxathiin. (*C.A.*) 5,6-dihydro-2-methyl-*N*-phenyl-1,4-oxathiin-3-carboxamide (9CI); 5,6-dihydro-2-methyl-1,4-oxathiin-3-carboxanilide (8CI); Reg. No. *[5234-68-5]*. Its fungicidal properties were described by B. von Schmeling & M. Kulka (*Science*, 1966, **152**, 659). Introduced by Uniroyal Inc. (USP 3 249 499; 3 393 202; 3 454 391) as code no. 'D 735'; trade mark 'Vitavax'.

**Properties.**

The technical grade (>97% pure) is a colourless solid; m.p. 91.5-92.5 °C, a dimorphic form has m.p. 98-100 °C; v.p. <133 Pa at 20 °C. Solubility at 25 °C: 170 mg/l water; 600 g/kg acetone; 1500 g/kg dimethyl sulphoxide; 110 g/kg ethanol; 210 g/kg methanol. It is compatible with all except highly alkaline or acidic pesticides.

**Uses.**

It is a systemic fungicide used for seed treatments of cereals against bunts and smuts and with other fungicides for the control of most other seed-borne and soil-borne seedling diseases; also against *Rhizoctonia* spp. on cotton, groundnuts and vegetables. The dimorphic forms do not differ in fungicidal activity.

**Toxicology.**

Acute oral $LD_{50}$ for rats 3820 mg/kg. Acute percutaneous $LD_{50}$ for rabbits >8000 mg/kg. Acute inhalation $LC_{50}$ for rats >20 mg/l air. In 2-y feeding trials albino rats receiving 600 mg/kg diet suffered no detectable symptom. $LC_{50}$ (96-h) is: for bluegill 1.2 mg/l; for rainbow trout 2.0 mg/l. $LC_{50}$ (48-h) for water flea is 84.4 mg/l. $LC_{50}$ (8-d) is: for mallard duck >4640 mg/kg diet; for bobwhite quail >10 000 mg/kg diet.

**Formulations.**

Carboxin is formulated alone or in combination with other fungicides, w.p., s.c. or liquid seed treatments are available for application to a wide range of crops at 2-4 g/kg seed. Mixtures with captan, gamma-HCH, maneb, phenylmercury acetate, thiram are available.

**Analysis.**

Product analysis is by hplc or i.r. spectroscopy. Details of methods are available from Uniroyal Inc. Residue analysis is by hydrolysis to aniline and determination by colorimetry of a derivative (green tissue) (J. R. Lane, *J. Agric. Food Chem.*, 1970, **18**, 409) or (for seed) by glc (H. R. Siskin & J. E. Newell, *ibid.*, 1971, **19**, 738; G. M. Stone, *Anal. Methods Pestic. Plant Growth Regul.*, 1976, **8**, 319).

$C_7H_{15}N_3O_2S_2$ (237.3)
ZVS1Y1SVZN1&1
*cartap hydrochloride*
$C_7H_{16}ClN_3O_2S_2$ (273.8)
ZVS1Y1SVZN1&1 &&HCl SALT

$(CH_3)_2NCH(CH_2SCONH_2)_2$

**Nomenclature and development.**

Common name cartap (BSI, E-ISO, F-ISO, JMAF). Chemical name (IUPAC) *S,S'*-2-dimethylaminotrimethylene bis(thiocarbamate) (I); 1,3-di(carbamoylthio)-2-dimethylaminopropane. (*C.A.*) *S,S'*-[2-(dimethylamino)-1,3-propanediyl] dicarbamothioate (9CI); (I) (8CI); Reg. No. *[15263-53-3]* (free base), *[22042-59-7]* (hydrochloride), *[15263-52-2]* (monohydrochloride). Its insecticidal properties were described by M. Sakai *et al.* (*Jpn. J. Appl. Entomol. Zool.*, 1967, **11**, 125) its action and structure-activity relationships reviewed (K. Konishi, *Pestic. Chem. (Congr. Pestic. Chem., 2nd*, 1971), 1972, **1**, 179; M. Sakai & Y. Sato, *ibid.*, p. 445; M. Sakai, *Jpn. Pestic. Inf.*, 1971, No. 6, p. 15; 1978, No. 34, p. 22). The hydrochloride was introduced by Takeda Chemical Industries Ltd. (GBP 1 126 204; USP 3 332 943; FP 1 452 338) as code No. 'TI-1258'; trade marks 'Padan', 'Cadan', 'Patap', 'Sanvex', 'Thiobel', 'Vegetox'.

**Properties.**

Cartap hydrochloride is a colourless crystalline solid; m.p. 179-181 °C. Solubility at 25 °C: *c.* 200 g/l water; slightly soluble in ethanol, methanol. The technical grade is 97% pure. It is stable under acidic conditions, but hydrolysed in neutral or alkaline media. It is virtually non-corrosive though susceptible to moisture.

**Uses.**

Cartap hydrochloride paralyses the insect by ganglionic blocking action on the central nervous system. It is used against boll weevils at 560 and 1120 g/ha; cabbage caterpillars at 550 g/ha; *Chilo plejadellus* on rice at 600 g a.i./ha; *Epilachna varivestis* at 560 g/ha; *Leptinotarsa decemlineata* on potatoes at 375-550 g/ha; *Plutella xylostella* at 500 g/ha. It is tolerated by a wide range of crops, and is non-persistent.

**Toxicology.**

Acute oral $LD_{50}$ for rats 325-345 mg/kg. Acute percutaneous $LD_{50}$ for mice >1000 mg/kg. $LC_{50}$ (48-h) for carp is 1.3 mg/l. It is moderately toxic to honeybees.

**Formulations.**

These include: s.p. (250 or 500 g a.i./kg); dust (20 g/kg); granules (40 or 100 g/kg).

**Analysis.**

Product and residue analysis is by glc or by oscillopolarography (K. Nishi *et al., Anal. Methods Pestic. Plant Growth Regul.*, 1973, **7**, 371).

$C_8H_{11}Cl_3O_6$ (309.5)
T55 BO DO FOTJ CXGGG GYQ1Q HQ

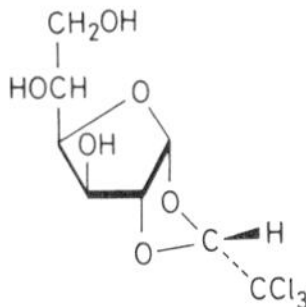


**Nomenclature and development.**

The trivial names chloralose (BSI, E-ISO, F-ISO); glucochloralose (BSI, E-ISO); glucochloral (F-ISO) are accepted in lieu of a common name. Chemical name (IUPAC) (*R*)-1,2-*O*-(2,2,2-trichloroethylidene)-α-D-glucofuranose (I). (*C.A.*) (I) (9CI); chloralose (8CI); Reg. No. *[15879-93-3]* formerly, *[39598-39-5]*, previously *[14798-36-8]*. Trivial name alphachloralose. It has been used for many years on grain as a bird repellent.

**Properties.**

Alphachloralose is a crystalline powder, m.p. 187 °C; $[\alpha]_D^{22}$ +19°. Solubility at 15 °C: 4.44 g/l water; soluble in acetic acid, diethyl ether. Chloralose also exists in a beta-form, Registry No. *[16376-36-6]*, m.p. 227-230 °C, which is less soluble than the alpha-form in water, ethanol, diethyl ether. Alphachloralose will reduce Fehlings solution only after prolonged standing, and is hydrolysed to its components by acid.

**Uses.**

Chloralose is a narcotic rendering birds easier to kill by other means; it is also used as a rodenticide in baits against mice, usually only by trained personnel. It retards metabolism and lowers body temperature to a fatal extent in small mammals. It is rapidly metabolised and hence non-cumulative.

**Toxicology.**

Acute oral $LD_{50}$: for rats 400 mg/kg; for mice 32 mg/kg; for birds 32-178 mg/kg.

**Formulations.**

Bait (15 g a.i./kg grain) against birds; baits (≤40 g/kg) against mice.

**Analysis.**

Product analysis is by hydrolysis and estimation of the trichloroacetaldehyde produced (K. C. Barrons & R. W. Hummer, *Agric. Chem.*, 1951, **6**(6), 48; G. I. Mills, *Anal. Chim. Acta*, 1952, **7**, 70).

# Chloramben

$C_7H_5Cl_2NO_2$ (206.0)
ZR BG EG CVQ

CO.OH
Cl
Cl
$NH_2$

**Nomenclature and development.**

Common name chloramben (BSI, E-ISO, ANSI, WSSA); chlorambène (F-ISO); exception (India); *amben* (former WSSA name). Chemical name (IUPAC) 3-amino-2,5-dichlorobenzoic acid (I). (*C.A.*) (I) (8 & 9CI); Reg. No. *[133-90-4]*. It was introduced as a herbicide by Amchem Products, Inc. (now Union Carbide Agrochemical Products Co., Inc.) (USP 3 014 063; 3 174 842) as code no. 'ACP M-629'; trade marks 'Amiben' (formerly '*Amoben*').

**Properties.**

Chloramben is a colourless crystalline solid; m.p. 200-201 °C; v.p. 930 mPa at 100 °C. Solubility at 25 °C: 700 mg/l water; 172 g/l ethanol. There is no problem of precipitation with hard water.

**Uses.**

It is a selective pre-planting incorporated and pre-em. herbicide used at 2-4 g a.i./kg to control grasses and broad-leaved weeds in seedling asparagus, navy beans, groundnuts, maize, sweet potatoes, pumpkins, soyabeans, squash, sunflowers and certain ornamentals. It is rapidly leached in soil. The *N*-glycoside was isolated from soyabeans in amounts equivalent to the chloramben applied, but little was recovered from barley, a susceptible crop (S. R. Colby, *Science*, 1965, **150**, 619).

**Toxicology.**

Acute oral $LD_{50}$ for male albino rats 5620 mg/kg. Acute percutaneous $LD_{50}$ for albino rats >3160 mg/kg; a single application of 3 mg caused a mild irritation which subsided within 24 h. In 2-y feeding trials rats receiving 10 000 mg/kg diet suffered no ill-effect.

**Formulations.**

These include: 'Amiben', aqueous solution (240 g a.e./l); 'Amiben' Granular 10%', granules (100 g a.e./kg).

**Analysis.**

Product analysis is by u.v. spectrometry (*CIPAC Handbook*, 1983, **1B**, in press) or volumetric methods (AOAC Methods; details of methods are available from Union Carbide Agricultural Products Co., Inc.). See also: H. S. Segal & M. L. Sutherland, *Anal. Methods Pestic., Plant Growth Regul. Food Addit.*, 1967, **5**, 321; *Anal. Methods Plant Growth Regul.*, 1972, **6**, 588.

$C_9H_{10}BrClN_2O_2$ (293.5)
GR BE EMVN1&O1

Br-C6H3(Cl)-NH.CO.N(CH3)OCH3

**Nomenclature and development.**

Common name chlorbromuron (BSI, E-ISO, WSSA, ex-ANSI);chlorobromuron (F-ISO). Chemical name (IUPAC) 3-(4-bromo-3-chlorophenyl)-1-methoxy-1-methylurea (I). (*C.A.*) *N'*-(4-bromo-3-chlorophenyl)*N*-methoxy-*N*-methylurea (9CI); (I) (8CI); Reg. No. *[13360-45-7]*. Its herbicidal properties were described by D. H. Green *et al.* (*Proc. Br. Weed Control Conf., 8th*, 1966, **2**, 363). Introduced by Ciba AG (now Ciba-Geigy AG) (BEP 662 268; GBP 965 313) as code no. 'C 6313'; trade mark 'Maloran'.

**Properties.**

Pure chlorbromuron is a colourless powder; m.p. 95-97 °C; v.p. 53 μPa at 20 °C; ρ 1.69 g/cm$^3$ at 20 °C. Solubility at 20 °C: 35 mg/l water; 460 g/kg acetone; 72 g/kg benzene; 170 g/kg dichloromethane; 89 g/kg hexane; 12 g/kg propan-2-ol. Decomposition is slow in neutral, slightly acid or slightly alkaline media.

**Uses.**

It is a herbicide suitable for: pre-em. use on carrots, peas, potatoes, soyabeans, sunflowers at 1.0-2.5 kg/ha; post-em. use on carrots and transplanted celery at 0.75-1.5 kg/ha. At these rates it persists in soil >56 d.

**Toxicology.**

Acute oral $LD_{50}$ for rats >5000 mg tech./kg. Acute percutaneous $LD_{50}$ for rats >2000 mg/kg; slight irritant to skin and eyes of rabbits. Inhalation $LC_{50}$ (6-h) for rats >1054 mg/m$^3$ air. In 90-d feeding trials NEL was: for rats 316 mg/kg diet (21.0 mg/kg daily); for dogs >316 mg/kg diet (10.5 mg/kg daily). $LC_{50}$: for rainbow trout and bluegill 5 mg/l; for crucian carp 8 mg/l. Practically non-toxic to birds and honeybees.

**Formulations.**

'Maloran': w.p. (500 g a.i./kg).

**Analysis.**

Product analysis is by glc. Residues may be determined by hydrolysis to 4-bromo-3-chloroaniline a derivative of which is determined by colorimetry, or by glc with ECD (G. Voss, *Anal. Methods Pestic. Plant Growth Regul.*, 1973, **7**, 569). Direct glc with ECD may also be used (T. H. Byast *et al., Tech. Rep. ARC Weed Res. Organ.*, No. 15 (2nd Ed.), p. 49).

$C_{11}H_{10}ClNO_2$ (223.7)
GR CMVOY1&1UU1

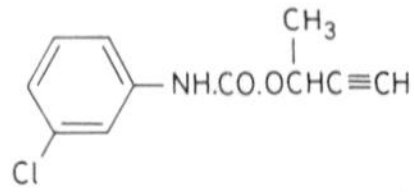

**Nomenclature and development.**

Common name chlorbufam (BSI, E-ISO);chlorbufame (F-ISO);BIPC (JMAF). Chemical name (IUPAC) 1-methylprop-2-ynyl 3-chlorocarbanilate; 1-methylprop-2-ynyl 3-chlorophenylcarbamate. (*C.A.*) 1-methyl-2-propynyl (3-chlorophenyl)carbamate (9CI); 1-methyl-2-propynyl *m*-chlorocarbanilate (8CI); Reg. No. *[1967-16-4]*. Its herbicidal properties were first described by A. Fischer (*Z. Pflanzenkr. Pflanzenpathol. Pflanzenschutz*, 1960, **67**, 577). Introduced by BASF AG (DEP 1 034 912; 1 062 482) as trade marks 'BiPC'; 'Alicep' (Reg. No. *[8065-18-7]*) a mixture with chloridazon, '*Alipur*' (Reg. No. *[8015-55-2]*) a mixture with cycluron is no longer available.

**Properties.**

Chlorbufam forms colourless crystals; m.p. 39-40 °C; v.p. 2.1 mPa at 20 °C. Solubility at 20 °C: 540 mg/l water; 280 g/kg acetone; 95 g/kg ethanol; 286 g/kg methanol. It is unstable at $1 > pH > 13$. Alcohols may cause transesterification.

**Uses.**

Chlorbufam is a pre-em. herbicide generally used as a mixture with chloridazon at 3-6 kg product/ha (for use in flower-bulb crops, leeks and onions). The main weeds controlled are: *Apera spica-venti, Raphanus raphanistrum, Sinapsis alba, Stellaria media, Urtica circus, Atriplex, Matricaria, Poa* and *Polygonum* spp. Chlorbufam persists in sandy loam for *c*. 56 d.

**Toxicology.**

Acute oral $LD_{50}$ for rats 2500 mg/kg. A slight temporary erythema was caused by 15-min application to the backs of white rabbits, significant after 20-h application.

**Formulations.**

'Alicep', w.p. (200 g chlorbufam + 250 g chloridazon/kg).

**Analysis.**

Product analysis is by acid hydrolysis to 3-chloroaniline, determined by titration. Residues may be determined by hydrolysis and colorimetry of the resulting 3-chloroaniline.

$C_{10}H_6Cl_8$ (409.8)
L C555 A IUTJ AG AG BG DG EG HG IG JG

**Nomenclature and development.**

Common name chlordane (BSI, E-ISO, F-ISO, JMAF). Chemical name (IUPAC) 1,2,4,5,6,7,8,8-octachloro-2,3,3a,4,7,7a-hexahydro-4,7-methanoindene (I); 1,2,4,5,6,7,8,8-octachloro-3a,4,7,7a-tetrahydro-4,7-methanoindene. (*C.A.*) 1,2,4,5,6,7,8,8-octachloro-2,3,3a,4,7,7a-hexahydro-4,7-methano-1*H*-indene (9CI); (I) (8CI); Reg. No. *[57-47-9]*. OMS 1437,ENT 9932. Its insecticidal properties, first described by C. W. Kearns *et al.* (*J. Econ. Entomol.*, 1945, **38**, 661), were discovered independently by R. Riemschneider (*Chim. Ind. (Paris)*, 1950, **64**, 695). Introduced by Velsicol Chemical Corp. (USP 2 598 561 (Velsicol Chemical Corp.)); GBP 618 432 (J. Hyman) as code no. 'Velsicol 1068','M 410'; trade mark 'Octachlor' (Velsicol Chemical Corp.).

**Properties.**

Produced by the Diels-Alder addition of cyclopentadiene to hexachlorocyclopentadiene and subsequent chlorination, technical chlordane (Reg. No. *[12789-03-6]*) is a viscous amber liquid; $d^{25}$ 1.59-1.63; $n_D^{25}$ 1.56-1.57; v.p. *c.* 61 mPa at 25 °C. The refined product has v.p. 1.3 mPa at 25 °C. Solubility at 25° C: 0.1 mg/l water; completely miscible with acetone, cyclohexanone, ethanol, deoderised kerosene, propan-2-ol, trichloroethylene.

Technical chlordane contains 60-75% of chlordane isomers, the major components being two stereoisomers whose nomenclature at C(1) and C(2) has been confused in the literature. The alpha or *cis*-isomer (1α,2α,3aα,4β,7β,7aα) Reg. No. *[5103-71-9]* (formerly *[22212-52-8]*) has m.p. 106-107 °C. The *trans*-isomer (1α,2β,3aα,4β,7β,7aα) *[5103-74-2]* usually known as gamma- occasionally as beta-chlordane has m.p. 104-105 °C. (The term gamma-chlordane has also been applied to the 2,2,4,5,6,7,8,8-octachloro-isomer *[5564-34-7]*). The remainder of the technical grade comprises other stereoisomers (each ⩽7%) and heptachlor.

**Uses.**

It is a persistent, non-systemic contact and stomach insecticide with some fungicidal action. It is used on land against ants, coleopterous pests, cutworms, grasshoppers, *Solenopsis geminata*, subterranean termites (including *Coptotermes* spp.) and many other insect pests. It also controls household insects, pests of man and domestic animals, is used as a wood preservative, a protective treatment for underground cables and to reduce earthworm populations in lawns. It may be applied to soil, directly to foliage or as a seed treatment. Insecticidal activities of its components have been compared (S. J. Cristol, *Adv. Chem. Ser.*, 1950, No. 1, p. 541).

**Toxicology.**

Acute oral $LD_{50}$ for rats 457-590 mg/kg; for bobwhite quail 83 mg/kg. Acute percutaneous $LD_{50}$ for rabbits >200 but <2000 mg/kg; extremely irritating to their eyes, but produces only mild irritation to their skin; it is not a dermal sensitiser to guinea-pigs. $LC_{50}$ (4-h) following exposure to an aerosol >200 mg/l (nominal concentration). In 2-y feeding trials NEL for dogs was 3 mg/kg diet, the target organ being the liver. $LC_{50}$ (8-d) was: for mallard duck 795 mg/kg diet; for bobwhite quail 421 mg/kg diet. It is not a teratogen in rabbits at 15 mg/kg daily. NEL in a 3-generation study in rats was 60 mg/kg diet. Results of in-vivo and in-vitro studies indicate it not mutagenic. The US National Academy of Sciences has ruled it is carcinogenic to certain strains of mice but there is evidence that it may act as a promoter rather than as an initiator of carcinogenesis. $LC_{50}$ (96-h) is: for rainbow trout 90 μg/l; for bluegill 70 μg/l; $LC_{50}$ (48-h) for *Daphnia* is 590 μg/l.

*continued overleaf*

$C_{10}H_6Cl_8$ (409.8)
L C555 A IUTJ AG AG BG DG EG
HG IG JG

**Formulations.**

These include: e.c. (480-960 g/l); solutions in oil (300 g/l); w.p. (250-400 g/kg); granules (50-330 g/kg); dusts (50-100 g/kg).

**Analysis.**

Product analysis is by total chlorine content (*AOAC Methods*, 1980, 6.220-6.222; *CIPAC Handbook*, 1970, **1**, 203; 1980, **1A**, 1119); or by i.r. spectrometry (M. Malina *et al., J. Assoc. Off. Anal. Chem.*, 1972, **55**, 972; M. Malina, *ibid.*, 1973, **56**, 591). Residues may be determined by glc with ECD (*Pestic. Anal. Man.*, 1979, **I**, 201-A, 201-G, 201-H; W. P. Cochrane *et al., J. Assoc. Off. Anal. Chem.*, 1975, **58**, 1051; *Anal. Methods Pestic. Plant Growth Regul.*, 1972, **6**, 315; H. K. Suzuki *et al., ibid.*, 1978, **10**, 45; *AOAC Methods*, 1980, 29.001-29.002, 29.008-29.019).

$C_{10}H_{13}ClN_2$ (196.7)
GR C1 DNU1N1&1
*chlordimeform hydrochloride*
$C_{10}H_{14}Cl_2N_2$ (233.2)
GR C1 DNU1N1&1 &&HCl SALT

Cl–C6H3(CH3)–N=CH.N(CH3)2

**Nomenclature and development.**

Common name chlordimeform (BSI, E-ISO, ANSO);chlordiméforme (F-ISO);chlorophenamidine (JMAF);chlorodimeform (New Zealand). Chemical name (IUPAC) $N^2$-(4-chloro-*o*-tolyl)-$N^1$,$N^1$-dimethylformamidine (I). (*C.A.*) $N'$-(4-chloro-2-methylphenyl)-*N*,*N*-dimethylmethanimidamide (9CI); (I) (8CI); Reg. No. *[6164-98-3]* (chlordimeform), *[19750-95-9]* (hydrochloride). OMS 1209 (chlordimeform), ENT 27 567 (hydrochloride). Its acaricidal properties were described by V. Dittrich (*J. Econ. Entomol.*, 1966, **59**, 889). Introduced by Schering AG and by Ciba AG (now Ciba-Geigy AG) (GBP 1 039 930; DEP 1 172 081; USP 3 378 437) as code no. 'Schering 36 268','C 8514' (Ciba-Geigy); trade marks 'Fundal', formerly '*Spanone*' (both Schering), 'Galecron' (Ciba-Geigy).

**Properties.**

Pure chlordimeform forms colourless crystals; m.p. 32 °C; b.p. 163-165 °C/14 mmHg; v.p. 48 mPa at 20 °C; $d^{30}$ 1.10. Solubility at 20 °C: 250 mg/l water; >200 g/l acetone, benzene, chloroform, ethyl acetate, hexane, methanol. It is hydrolysed in neutral and acidic media, first to 4-chloro-*o*-tolylformamide and then to 4-chloro-*o*-toluidine. The technical grade is >96% pure.

It forms salts with acids, *e.g.* the hydrochloride; m.p. 225-227 °C (decomp.) with solubility: >500 g/l water; 1 g/l benzene, hexane; 10-20 g/l chloroform; >300 g/l methanol. An aqueous solution of the hydrochloride (5 g/l; pH3-4) is stable for several days at 20 °C.

**Uses.**

Chlordimeform is an acaricide, active mainly against eggs and immature mite stages, and used as an ovicide at 50 g a.i./100 l. it is also effective against the eggs and early instars of Lepidoptera, *e.g. Chilo suppressalis, Heliothis* spp., *Laspeyresia pomonella, Spodoptera littoralis, Trichloplusia ni* at 0.2-1.0 kg/ha.

**Toxicology.**

Acute oral $LD_{50}$: for rats 340 mg chlordimeform/kg, 355 mg hydrochloride/kg; for rabbits 625 mg chlordimeform/kg. Acute percutaneous $LD_{50}$ for rabbits >4000 mg hydrochloride/kg; irritation to them was slight. In 2-y feeding trials NEL was: for rats 0.1 mg/kg daily; for dogs 6.25 mg/kg daily. $LC_{50}$ (24-h) is: for trout 11.7 mg/l; for bluegill 1 mg/l; $LC_{50}$ (48-h) for Japanese killifish 33 mg/l. It is non-toxic to honeybees.

**Formulations.**

These include: 'Fundal 800', 'Fundal SP' in USA, s.p. (800 g a.i./kg) as hydrochloride; 'Fundal 500 EC', 'Fundal 4 EC' in USA, e.c., 'Galecron 50 EC' (500 g a.i./l) as base.

**Analysis.**

Product analysis is by acid-base titration or by glc (G. Voss *et al., Anal. Methods Pestic. Plant Growth Regul.*, 1973, **7**, 211). Residues may be determined by glc of a derivative (*idem, ibid.*; *Pestic. Anal. Man.*,1979, **II**; H. Geissbühler *et al., J. Agric. Food Chem.*, 1971, **19**, 365).

$C_8H_5Cl_3O_2$ (239.5)
QV1R BG CG FG
*chlorfenac-sodium*
$C_8H_4Cl_3NaO_2$ (261.5)
QV1R BG CG FG &&Na SALT

CH2CO.OH
Cl Cl
Cl

**Nomenclature and development.**

Common name chlorfenac (BSI, E-ISO, F-ISO);fenac (WSSA);*exception* USA. Chemical name (IUPAC) (2,3,6-trichlorophenyl)acetic acid (I). (*C.A.*) 2,3,6-trichlorobenzeneacetic acid (9CI); (I) (8CI); Reg. No. *[85-34-7]*. Introduced as a herbicide by Amchem Products Inc. (now Union Carbide Agricultural Products Co., Inc.) and Hooker Chemical Corp. (GBP 860 310 to Hooker) as trade mark 'Fenac' (Amchem).

**Properties.**

Chlorfenac is a colourless solid; m.p. 156 °; v.p. 1.1 Pa at 100 °C. Solubility at 28 °C: 200 mg/l water; soluble in most organic solvents. It forms water-soluble salts with alkalis, *e.g.* chlorfenac-sodium (Reg. No. *[2439-00-1]*).

**Uses.**

Chlorfenac, usually applied as the sodium salt, is a herbicide absorbed primarily by roots; it resists leaching and is persistent in soil. It is used to control annual grasses, broad-leaved weeds, field bindweed, couch grass and other perennial weeds in industrial and other non-crop situations at rates up to 18 kg a.e./ha. It is also used pre-em. for weed control in sugarcane at 2.8-3.4 kg a.e./ha.

**Toxicology.**

Acute oral $LD_{50}$ for rats 576-1780 mg/kg. Acute percutaneous $LD_{50}$ for rabbits 1440-3160 mg/kg. In 2-y feeding trials rats receiving 2000 mg/kg diet showed no ill-effect.

**Formulations.**

'Fenatrol', aqueous concentrate of chlorfenac-sodium. Also mixtures: chlorfenac + aminotriazole + atrazine; chlorfenac + bromacil; chlorfenac + 2,4-D.

**Analysis.**

Product analysis is by glc. Residues may be determined by glc of suitable derivatives (G. Yip, *J. Assoc. Off. Anal. Chem.*, 1972, **45**, 367).

$C_{10}H_{10}Cl_2O_2$ (233.1)
GR D1YGVO1

Cl—C$_6$H$_4$—$CH_2CHClCO.OCH_3$

**Nomenclature and development.**

Common name chlorfenprop-methyl (BSI, E-ISO, F-ISO);*exception* (Canada, USA). Chemical name (IUPAC) methyl (±)-2-chloro-3-(4-chlorophenyl)propionate. (*C.A.*) methyl α,4-dichlorobenzenepropanoate (9CI); methyl *p*,α-dichlorohydrocinnamate (8CI); Reg. No. *[14437-17-3]*. Its herbicidal properties were described by L. Eue, *Z. Pflanzenkr. Pflanzenpathol. Pflanzenschutz*, 1968, Sonderheft IV, 211. Only the (-)-enantiomer (Reg. No. *[59404-06-7]*) is herbicidal (T. Schmidt *et al., Z. Naturforsch. C. Biosci.*, 1976, **31C**, 252). Chlorfenprop-methyl was introduced by Bayer AG (GBP 1 077 194; FP 1 476 247) as code no. 'Bayer 70 533'; trade mark 'Bidisin'.

**Properties.**

Pure chlorfenprop-methyl is a colourless liquid, with a fennel-like odour; m.p. >−20 °C, b.p. 110-113 °C/0.1 mmHg; v.p. 930 mPa at 50 °C; $n_D^{20}$ 1.532. Solubility at 20 °C: 40 mg/l water; soluble in acetone, aromatic hydrocarbons, diethyl ether and fatty oils. It is a light brown liquid, $d_4^{20}$ 1.30.

**Uses.**

It is a specific contact herbicide for the control of wild oats used only after their emergence. They are most susceptible between the one-leaf stage and tillering. A rate of 4 kg a.i. in 200-400 l water/ha is effective against *Avena fatua* but not against *A. sterilis*. Sprays are well-tolerated by cereals (other than oats), by fodder crops, peas and sugar beet (W. Kampe, *Pflanzenschutz-Nachr. (Engl. Ed.)*, 1973, **26**, 299; H. J. Hewston, *ibid.*, p. 317; H. Hack, *ibid.*, p. 353; P. Zonderwijk, *ibid.*, 1974, **27**, 3).

**Toxicology.**

Acute oral $LD_{50}$ for rats *c.* 1190 mg/kg; for guinea-pigs and rabbits 500-1000 mg/kg; for dogs >500 mg/kg; for chickens *c.* 1500 mg/kg. Acute percutaneous $LD_{50}$ for rats >2000 mg/kg. In 90-d feeding trials rats receiving 1000 mg/kg diet suffered no ill-effect (E. Loser & G. Kimmerle, *ibid.*, 1973, **26**, 368). $LC_{50}$ (96-h) for goldfish 1-10 mg/l.

**Formulations.**

An e.c. (500 or 800 g a.i./l).

**Analysis.**

Product analysis is by glc. Details of methods are available from Bayer AG. Residues are determined by glc (H. J. Jarczyk, *ibid.*, 1968, **21**, 360).

$C_{12}H_8Cl_2O_3S$ (303.2)
GR DSWOR DG

Cl–C6H4–SO2.O–C6H4–Cl

**Nomenclature and development.**

Common name chlorfenson (BSI, E-ISO, F-ISO);ovatran (Argentina);chlorofénizon (France);ovex (ANSI, Canada);ephirsulphonate (USSR);CPCBS (JMAF). Chemical name (IUPAC) 4-chlorophenyl 4-chlorobenzenesulphonate. (*C.A.*) 4-chlorophenyl 4-chlorobenzenesulfonate (9CI); *p*-chlorophenyl *p*-chlorobenzenesulfonate (8CI); Reg. No. *[80-33-1]*. ENT 16 358. Its acaricidal activity was described by E. E. Kenaga & R. W. Hummer (*J. Econ. Entomol.*, 1949, **42**, 996). Introduced by Dow Chemical Co. (USP 2 528 310) as code no. 'K 6451'; trade mark 'Ovotran'; though no longer marketed or manufactured by Dow Chemical Co. it is available from several companies, usually in mixtures with insecticides or other acaricides (see below).

**Properties.**

Pure chlorfenson is a colourless crystalline solid, with a characteristic odour; m.p. 86.5 °C; v.p. negligible at 25 °C. Solubility at 25 °C: practically insoluble in water; 1300 g/kg acetone; 10 g/kg ethanol; 780 g/kg xylene. It is hydrolysed by alkali. The technical grade is a colourless to tan flaky solid, m.p. *c.* 80 °C.

**Uses.**

It is a non-systemic acaricide with long residual ovicidal activity. It is effective against mites of citrus and other fruit, ornamentals and vegetable crops at 17-32 g a.i./100 l.

**Toxicology.**

Acute oral $LD_{50}$: for rats *c.* 2000 mg/kg; for Japanese quail 4600 mg/kg. Acute percutaneous $LD_{50}$ $>$10 000 mg/kg for rats. It may cause skin irritation. In 130-d feeding trials no apparent effect was observed on rats receiving 300 mg/kg diet, $LC_{50}$ (48-h) for carp is 3.2 mg/l.

**Formulations.**

These include: 'Ovitox' (Phyteurop), 'Sappiron' (Nippon Soda Co. Ltd.), 'Trichlorfenson' (Bourgeois), w.p. (500 g a.i./kg). 'Erysit Super' (Schering), 'Fac Super' (Sopra), chlorfenson + prothoate. 'Mitran WP' (Nippon Soda Co. Ltd.), (Reg. No. *[70161-99-8]*), w.p. (250 g chlorfenson + 250 g chlorfenethol/kg). 'Naftil acaricide' (Pépro), w.p. chlorfenson + carbaryl.

**Analysis.**

Product analysis is by alkaline hydrolysis and back titration (*CIPAC Handbook*, 1970, **1**, 213; *FAO Plant Prot. Bull.*, 1964, **12**, 17). Residues may be determined by glc or by colorimetry of a derivative produced after alkaline hydrolysis (A. H. Kutschinski & E. N. Luce, *Anal. Chem.*, 1952, **24**, 1188; F. A. Gunther & L. R. Jepson, *J. Econ. Entomol.*, 1954, **47**, 1027).

$C_{12}H_{14}Cl_3O_4P$ (359.6)
GR CG DYU1GOPO&O2&O2

**Nomenclature and development.**

Common name chlorfenvinphos (BSI, E-ISO, F-ISO, BPC);CVP (JMAF);*exception* USA. Chemical name (IUPAC) 2-chloro-1-(2,4-dichlorophenyl)vinyl diethyl phosphate (I). (*C.A.*) 2-chloro-1-(2,3-dichlorophenyl)ethenyl diethyl phosphate (9CI); (I) (8CI); Reg. No. *[470-90-6]* (formerly *[2701-86-2]*) (*Z* + *E*)-isomers, *[18708-87-7]* (*Z*)-isomer, *[18708-86-6]* (*E*)-isomer. OMS 1328, ENT 24 969. Its insecticidal properties were described by W. F. Chamberlain *et al.* (*J. Econ. Entomol.*, 1962, **55**, 86). Introduced by the Shell International Chemical Company Ltd, Ciba AG (now Ciba-Geigy AG) and by Allied Chemical Corp. (who no longer produce or market it) (USP 2 956 075; 3 116 201 Shell) as code no. 'SD 7859' (Shell),'C 8949' (Ciba-Geigy),'GC 4072' (Allied); trade mark 'Birlane' (Shell), 'Sapecron' (Ciba-Geigy).

**Properties.**

Produced by the reaction of triethyl phosphite with 2,2,2',4'-tetrachloroacetophenone, the technical material ( $\geqslant$90% (*Z*)- and (*E*)-isomers, typical ratio (*Z*:*E*) 8.5:1) is an amber-coloured liquid, with a mild odour; m.p. -19 to -23 °C; b.p. 167-170 °C/0.5 mmHg; v.p. (for pure compound) 530 μPa at (extrapolated) at 20 °C: $n_D^{20}$ 1.5281 (*Z*)-, 1.4210 (*E*)-isomer; ρ 1.36 g/cm³ at 20 °C. Solubility at 23 °C 145 mg/l water; fully miscible with acetone, dichloromethane, ethanol, hexane, kerosene, propylene glycol, xylene. Slowly hydrolysed by water or acid, 50% decomposition occurs at 38 °C in $>$700 h at pH 1.1, $>$400 h at pH 9.1, but unstable in alkali - at 20 °C 50% loss occurs in 1.28 h at pH 13. It may corrode brass, iron and steel on prolonged contact and the e.c. formulations are corrosive to tin plate.

**Uses.**

Chlorfenvinphos may be used either as a soil insecticide for the control of cutworms, root flies and root worms at 2-4 kg a.i./ha or as a foliar insecticide to control *Leptinotarsa decemlineata* on potato, scale insects on citrus at 200-400 g/ha where it also exhibits ovicidal activity against mite eggs, and of stem borers on maize, rice and sugarcane at 550-2200 g/ha. It controls whiteflies (*Bemisia* spp.) on cotton at 400-750 g/ha but whitefly parasites are not affected. It is non-phytotoxic at these rates. The metabolism and breakdown are reviewed by K. I. Beynon *et al.* (*Residue Rev.*, 1973, **47**, 55). Chlorfenvinphos may also be used in public health programmes especially against mosquito larvae.

**Toxicology.**

Acute oral $LD_{50}$: for rats 9.9-39 mg tech./kg; for dogs $>$12 000 mg/kg; for pigeons 13.8 mg/kg; for pheasants 100 mg/kg. Acute percutaneous $LD_{50}$ depends on carrier and conditions: for rats it ranges from 31-245 mg/kg; for rabbits 417-4700 mg/kg., non-irritant to their skin or eyes. Acute inhalation $LC_{50}$ (4-h) for rats *c.* 50 mg/m³ air. In 2-y feeding trials NEL for rats and dogs was 1 mg/kg diet (0.05 mg/kg daily). $LC_{50}$ (24-h): for harlequin fish 0.36 mg/l; for guppies 0.54 mg/l; $LC_{50}$ (96-h) for harlequin fish 0.3 mg/l. It is slightly toxic to honeybees. It is rapidly and completely metabolised in mammals and metabolites are eliminated within a few days. ADI for man 0.002 mg/kg (*JMPR* 1971).

**Formulations.**

These include: e.c. (240 g a.i./l); w.p. (250 g/kg); dust (50 g/kg); granules (100 g/kg); emulsifiable citrus spray oils (37.5 or 50 g/l) liquid seed treatments.

**Analysis.**

Product analysis is by i.r. spectrometry or glc (*CIPAC Handbook*, 1980, **1A**, 1131; FAO Specification (CP/66)). Particulars from the Shell International Chemical Co. Residues may be determined by glc (*Pestic. Anal. Man.*, 1979, **I**, 201-I; D. C. Abbott *et al., Pestic. Sci.*, 1970, **1**, 10; R. Mestres *et al., Trav. Soc. Pharm. Montpellier*, 1979, **39**, 323; K. I. Beynon *et al., J. Sci. Food Agric.*, 1968, **19**, 302).

$C_{14}H_9ClO_3$ (260.7)
L B656 HHJ EG HVQ HQ
*chlorflurecol-methyl*
$C_{15}H_{11}ClO_3$ (274.7)
L B656 HHJ EG HVO1 HQ

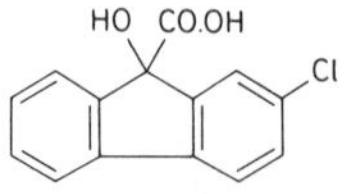

**Nomenclature and development.**

Common name chlorflurenol (E-ISO, F-ISO);chlorflurecol (BSI, Canada, Denmark);*exception* (Poland, USA). Chemical name (IUPAC) 2-chloro-9-hydroxyfluorene-9-carboxylic acid (I). (*C.A.*) 2-chloro-9-hydroxy-9*H*-fluorene-9-carboxylic acid (9CI); (I) (8CI); Reg. No. *[2464-37-1]*. The effects on plant growth of derivatives of fluorene-9-carboxylic acids were first described by G. Schneider (*Naturwissenschaften*, 1964, **51**, 416) who proposed they should be called morphactins (G. Schneider *et al., Nature (London)*, 1965, **208**, 1013). The methyl ester of chlorflurecol, chlorflurecol-methyl (chlorflurenol-methyl), (Reg. No. *[2536-31-4]*) was introduced by E. Merck (now Celamerck GmbH Co.). (GBP 1 051 652; 1 051 653; 1 051 654) as code no. 'IT 3456'.

**Properties.**

Chlorflurecol-methyl forms cream-coloured crystals; m.p. 136-142 °C; v.p. 6.6 mPa at 25 °C; $d^{20}$ *c.* 1.5. Solubility at 25 °C: *c.* 18 mg/l water; *c.* 260 g/l acetone; *c.* 70 g/l benzene; *c.* 80 g/l ethanol. It is stable under normal storage conditions, light excluded, compatible with other growth regulators and with maleic hydrazide formulated as 'MH 30'.

**Uses.**

Chlorflurecol-methyl is used as a plant growth retardant at 2-4 kg a.i./ha and for soil application for weed suppression at 0.5-1.5 kg/ha; also as tank mixes with other growth-active compounds.

**Toxicology.**

Acute oral $LD_{50}$: for rats >12 800 mg/kg; for quails >10 000 mg/kg. Acute percutaneous $LD_{50}$ for rats >10 00 mg/kg. Acute intraperitoneal $LD_{50}$: for rats 1417 mg/kg; for mice 1670 mg/kg. In 2-y feeding trials no adverse effect was observed; in rats receiving 3000 mg/kg diet; in dogs 300 mg/kg diet. $LC_{50}$ (96-h) is: for bluegill 7.2 mg/l; for carp *c.* 9 mg/l; for rainbow trout 0.015 mg/l.

**Formulations.**

An e.c. (125 g a.i./kg).

**Analysis.**

Product analysis is by u.v. spectroscopy or by chromatographic methods (W. P. Cochrane *et al., J. Assoc. Off. Anal. Chem.*, 1977, **60**, 728). Residues may be determined by colorimetry of a derivative or by glc of an ester (E. Amadori & W. Heupt, *Anal. Methods Pestic. Plant Growth Regul.*, 1978, **10**, 525).

$C_{10}H_8ClN_3O$ (221.6)
T6NNVJ BR& DG EZ

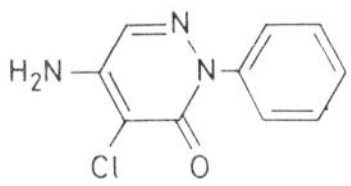

**Nomenclature and development.**

Common name chloridazon (BSI, E-ISO);chloridazone (F-ISO);pyrazon, a name formerly used by BSI and in many countries, is retained (ANSI, WSSA, Canada, Denmark, Poland);PAC (JMAF). Chemical name (IUPAC) 5-amino-4-chloro-2-phenylpyridazin-3(2*H*)-one. (*C.A.*) 5-amino-4-chloro-2-phenyl-3(2*H*)-pyridazinone (8 & 9CI); Reg. No. *[1698-60-8]*, was *[58858-18-7]*. Its herbicidal properties were reported by A. Fischer (*Weed Res.*, 1962, **2**, 177). Introduced by BASF AG (DEP 1 105 232) as code no. 'H 119'; trade mark 'Pyramin'.

**Properties.**

A pale yellow solid; m.p. 205-206 °C; v.p. <10 μPa at 20 °C. Solubility at 20 °C: 400 mg/l water; 28 g/kg acetone; 0.7 g/kg benzene; 34 g/kg methanol.

**Uses.**

Chloridazon is effective against broad-leaved weeds particularly for use on sugar beet and beet crops at 1.6-3.3 kg a.i./ha, applied pre-em. or after the late cotyledon stage. It persists in sandy loam for about 140 d and is decomposed to 5-amino-4-chloropyridazin-3-one which is non-phytotoxic. Also used in combination with other herbicides.

**Toxicology.**

Acute oral $LD_{50}$ for rats 2424 mg a.i. (as w.p.)/kg. Application to backs and ears of white rabbits (20-h) caused slight temporary erythema. In 2-y feeding trials rats receiving 300 mg/kg diet showed no detectable toxic effect.

**Formulations.**

'Pyramin', w.p. (800 g tech./kg). 'Pyramin FL', s.c. (430 g/l). 'Alicep', (Reg. No. *[8065-18-7]*) w.p. (250 g chloridazon + 200 g chlorbufam/kg). 'Herald' (May & Baker Ltd), (chloridazon + chlorpropham + fenuron + propham). 'Murbectol' (Murphy Chemical Ltd), (chloridazon + fenuron + propham). 'Pyradex' (Reg. No. *[58811-21-5]*), w.p. (330 g chloridazon + 200 g di-allate/kg). 'Spectron' (FBC Limited), (280 g chloridazon + 115 g ethofumesate/l).

**Analysis.**

Product analysis is by i.r. spectrometry. Residues in plants are determined by colorimetric measurement of the aniline produced on hydrolysis, particulars from BASF AG. Residues in soil are determined by hplc with u.v. detection (T. H. Byast *et al., Tech. Rep. ARC Weed Res. Organ.*, No. 15 (2nd Ed.), p. 65).

$C_5H_{12}ClO_2PS_2$ (234.7)
G1SPS&O2&O2

$ClCH_2SP(=S)(OCH_2CH_3)_2$

**Nomenclature and development.**

Common name chlormephos (BSI, E-ISO, F-ISO). Chemical name (IUPAC) *S*-chloromethyl *O,O*-diethyl phosphorodithioate. (*C.A.*) *S*-(chloromethyl) *O,O*-diethyl phosphorodithioate (8 & 9CI); Reg. No. *[24934-91-6]*. Its insecticidal properties were described by F. Colliot *et al.* (*Proc. Br. Insectic. Fungic. Conf., 7th*, 1973, **2**, 557). Introduced by Murphy Chemical Ltd. and developed under licence by Rhône-Poulenc Phytosanitaire (GBP 1 258 922; 817 360; 902 795 to Murphy) as code no. 'MC 2188' (Murphy); trade mark 'Dotan' (Rhône-Poulenc).

**Properties.**

Pure chlormephos is a colourless liquid; b.p. 81-85 °C/0.1 mmHg; v.p. 7.6 Pa at 30 °C; $n_D$ 1.5244; *d* 1.260. Solubility at 20 °C: 60 mg/l water; miscible with most organic solvents. The technical grade is *c.* 90-93% pure. It is stable to water but hydrolysed by dilute acid or alkali at 80 °C.

**Uses.**

It is a contact insecticide, effective when applied to soil for the control of millipedes, white grubs and wireworms in maize and sugar beet at 2-4 kg a.i./ha for overall application and 0.3-0.4 kg a.i./for band treatment.

**Toxicology.**

Acute oral $LD_{50}$ for rats 7 mg a.i./kg. Acute percutaneous $LD_{50}$ for rats 27 mg a.i./kg; >1600 mg (as 5% granules)/kg, value depending on the species tested. In 90-d feeding trials NEL was 0.39 mg/kg diet. $LC_{50}$ for fish is 1.5 mg/l.

**Formulations.**

Granules (50 g a.i./kg).

**Analysis.**

Product analysis is by glc with FID, (V. P. Lynch, *Anal. Methods Pestic. Plant Growth Regul.*, 1978, **10**, 49). Residues may be determined by glc with TID (*idem, ibid.*).

$ClCH_2CH_2\overset{+}{N}(CH_3)_3$

$C_5H_{13}ClN$ (122.6)
G2K1&1&1
*chlormequat chloride*
$C_5H_{13}Cl_2N$ (158.1)
G2K1&1&1 &&Cl SALT

**Nomenclature and development.**

Common name chlormequat (BSI, E-ISO, F-ISO). Chemical name (IUPAC) 2-chloroethyltrimethylammonium ion. (*C.A.*) 2-chloro-*N*,*N*,*N*-trimethylethanaminium (9CI); (2-chloroethyl)trimethylammonium (8CI); Reg. No. *[7003-89-6]* (for ion), *[991-81-5]* (for chloride salt). The chloride is sometimes known by the trivial names chlorocholine chloride,CCC. The plant growth-regulating properties of chlormequat chloride were described by N. E. Tolbert (*Plant Physiol.*, 1960, **35**, 380; *J. Biol. Chem.*, 1960, **235**, 475). Introduced in collaboration with Michigan State University by American Cyanamid Co. (GBP 944 807; FP 1 264 866; BEP 593 961) as code no. 'AC 38 555'; trade mark 'Cycocel'.

**Properties.**

Chlormequat chloride is a colourless crystalline solid, with a fish-like odour; it begins to decompose at 245 °C; v.p. 10 μPa at 20 °C. Solubility at 20 °C: >1 kg/kg water; <1 g/kg chloroform; insoluble in cyclohexane; 320 g/kg ethanol. The technical grade is 97-98% pure. The solid is extremely hygroscopic but its aqueous solutions are stable though corrosive to unprotected metals. It may be stored in containers of glass or high-density plastic, or of metal lined with rubber or epoxy resin.

**Uses.**

Chlormequat chloride is a plant growth regulator which influences the habit of certain plants by shortening and strengthening the stem, *e.g.* in wheat and poinsettias. It can also influence the developmental cycle resulting in increased flowering and harvest, *e.g.* in pears and tomatoes. An intensification of chlorophyll formation is often seen. The root system may also be increased, resulting in yield increases under dry conditions. Wheat is protected against damage by *Cercosporella herpotrichoides*. Chlormequat chloride is rapidly degraded in soil by enzyme activity and there is no influence on soil microflora or fauna.

**Toxicology.**

Acute oral $LD_{50}$: for male rats 670 mg tech./kg; for female mice 1020 mg/kg; for male guinea-pigs 620 mg/kg; for chickens 920 mg/kg. Acute percutaneous $LD_{50}$ for male rabbits 440 mg/kg. In 2-y feeding trials rats receiving 1000 mg/kg diet suffered no ill-effect. In rats 96% of the administered compound was excreted unchanged in the urine and faeces. The toxicity of chlormequat chloride is reduced by the addition of choline chloride (DEP 1 215 436) as several tests in laboratory animals showed.

**Formulations.**

Aqueous solutions (118, 400, 500 or 725 g chlormequat chloride/l); dust (650 g/kg). In addition, a mixture (460 g chlormequat chloride + 320 g choline chloride/l) is sold as 'WR62', 'Cycocel 460' (American Cyanamid Co. and BASF AG), 'BAS 06200W' and 'CCC Extra' (BASF AG).

**Analysis.**

Product analysis is by potentiometric titration with silver nitrate; chlormequat chloride and choline chloride can thus be determined separately. Residues may be determined by glc after reaction with sodium phenylsulphide (F. Tafuri *et al.*, *Analyst (London)*, 1970, **95**, 675). Details and procedures also available from American Cyanamid Co.

$C_2H_3ClO_2$ (94.50)
QV1G
*sodium chloracetate*
$C_2H_2ClNaO_2$ (116.5)
QV1G &&Na SALT

$ClCH_2CO.OH$

**Nomenclature and development.**

The chemical name chloroacetic acid (I) (BSI, E-ISO);and trivial name monochloroacetic acid (BSI, E-ISO);acid chloracetique (F-ISO) are accepted in lieu of a common name. Chemical name (IUPAC) (I). (*C.A.*) (I) (8 & 9CI); Reg. No. *[79-11-8]* (acid), *[3926-62-3]* (sodium salt). Trivial names for sodium salt: SMA, SMCA. Its herbicidal properties were described by A. E. Hitchcock *et al.* (*Proc. Northeast. Weed Control Conf.*, 1951, p. 105) and those of its sodium salt by T. C. Breese & A. F. J. Wheeler (*Proc. Br. Weed Control Conf., 3rd*, 1956, p. 759). Sodium chloroacetate was introduced by ICI Plant Protection Division as trade mark '*Monoxone*' (but is no longer manufactured or marketed by this firm).

**Properties.**

Chloroacetic acid forms a deliquescent solid and exists in 3 crystalline forms: alpha, m.p. 63 °C; beta, m.p. 55-56 °C; gamma, m.p. 50 °C. The melt has b.p. 189 °C. Solubility: very soluble in water; soluble in benzene, chloroform, diethyl ether, ethanol. The technical product grade has m.p. 61-63 °C. It is corrosive.

The sodium salt is a colourless crystalline solid, solubility at 20 °C 850 g/l water. The technical grade is *c.*90% pure.

**Uses.**

Sodium chloroacetate is a post-em. contact herbicide used to control a wide range of annual weeds at seedling stage in Brussels sprouts (20 kg a.i. in 225 l water/ha), kale (20-25 kg in 225-560 l/ha), leeks and onions (25 kg in 315-560 l/ha). It is also used in combination with atrazine for total weed control on industrial sites and other non-crop land.

**Toxicology.**

Acute oral $LD_{50}$: for rats 650 mg sodium chloroacetate/kg; for mice 165 mg/kg. it may cause irritation to the skin and eyes. The health of rats receiving 700 mg/kg diet for several months was not affected. $LC_{50}$ (48-h) for rainbow trout is 900 mg/l.

**Formulations.**

'Herbon Somon' (Cropsafe), s.p. (900 g sodium chloroacetate/kg). 'A-Plus Granules' (Diamond Shamrock Agrochemicals Ltd), s.p. (sodium chloroacetate + atrazine).

**Analysis.**

Product analysis is by chlorine content determined by alkaline hydrolysis and titration with silver nitrate, correcting for chloride ion and sodium dichloroacetate originally present.

$C_{16}H_{14}Cl_2O_3$ (325.2)
GR DXQR DG&VO2

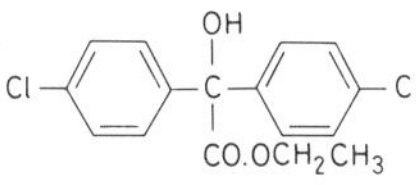

**Nomenclature and development.**

Common name chlorobenzilate (BSI, E-ISO, F-ISO, JMAF). Chemical name (IUPAC) ethyl 4,4'-dichlorobenzilate (I) (*C.A.*) ethyl 4-chloro-α-(4-chlorophenyl)-α-hydroxybenzeneacetate (9CI); (I) (8CI); Reg. No. *[510-15-6]*. ENT 18 596. Its acaricidal properties were described by R. Gasser (*Experientia*, 1952, **8**, 65). Introduced by J. R. Geigy S.A. (now Ciba-Geigy AG) (BEP 511 234; GBP 705 037) as code no. 'G 23 992'; trade marks 'Akar', 'Folbex', and in USA 'Acaraben'.

**Properties.**

Chlorobenzilate is a colourless solid; m.p. 36-37.5 °C; v.p. 120 μPa at 20 °C; b.p. 156-158 °C/0.07 mmHg; ρ 1.2816 g/cm$^3$ at 20 °C; $n_D^{20}$ 1.5740. Solubility at 20 °C: 10 mg/l water; 1 kg/kg acetone, dichloromethane, methanol, toluene; 600 g/kg hexane; 700 g/kg octan-1-ol.

**Uses.**

It is a non-systemic acaricide with little insecticidal action. It is recommended for use against phytophagous mites on citrus, cotton, grapes, soyabeans, tea and vegetables at 30-60 g a.i./100 l or 1.0-1.5 kg/ha.

**Toxicology.**

Acute oral $LD_{50}$ for rats 2784-3880 mg tech./kg. Acute percutaneous $LD_{50}$ for rats >10 000 mg/kg; non-irritant to skins of rabbits. In 2-y feeding trials NEL was: for rats 40 mg/kg diet (*c.* 2.7 mg/kg daily); for dogs 500 mg/kg diet (*c.* 16.0 mg/kg daily). $LC_{50}$ is: for rainbow trout 0.60 mg/l; for bluegill 1.80 mg/l. Practically non-toxic to birds and honeybees.

**Formulations.**

'Akar 338', e.c. (250 g a.i./l); w.p. (250 g/kg). 'Akar 50', e.c. (500 g/l).

**Analysis.**

Product analysis is by glc (E. Bartsch *et al.*, *Residue Rev.*, 1971, **39**, 1; A. Margot & K. Stammbach, *Anal. Methods Pestic., Plant Growth Regul. Food Addit.*, 1964, **2**, 65; *CIPAC Handbook*, 1980, **1A**, 1145; *AOAC Methods*, 1980, 6.431). Residues are determined by glc with MCD or ECD (*Anal. Methods Pestic., Plant Growth Regul. Food Addit.*, 1972, **6**, 319; G. Formica *et al.*, *Meded. Fac. Landbouwwet. Rijksuniv. Gent*, 1975, **40**, 1135; E. Bartsch *et al.*, *loc. cit.*; *Pestic. Anal. Manual* 1979, **I**, 201-A, 201-G; R. Mestres *et al.*, *Trav. Soc. Pharm. Montpellier*, 1979, **39**, 323).

$C_{10}H_{13}ClN_2S$ (228.7)
GR C1 DMYUS&N1&1

Cl—⟨ring⟩—NHCS.N(CH$_3$)$_2$, CH$_3$

**Nomenclature and development.**

Common name chloromethiuron (BSI, E-ISO, F-ISO). Chemical name (IUPAC) 3-(4-chloro-*o*-tolyl)-1,1-dimethylthiourea. (*C.A.*) *N'*-(4-chloro-2-methylphenyl)-*N,N*-dimethylthiourea (9CI); 3-(4-chloro-*o*-tolyl)-1,1-dimethyl-2-thiourea (8CI); Reg. No. *[28217-97-2]*. Its acaricidal properties were reported by M. von Orelli *et al.* (*Proc. World Vet. Congr., 20th*, 1975, p. 659). Introduced by Ciba-Geigy AG (BEP 678 543; GBP 1 138 714) as code no. 'CGA 13 444'; trade mark 'Dipofene'.

**Properties.**

Pure chloromethiuron forms colourless crystals, m.p. 175 °C; v.p. 1.06 μPa at 20 °C; ρ 1.34 g/cm$^3$ at 20 °C. Solubility at 20 °C: 50 mg/l water; 37 g/kg acetone; 40 g/kg dichloromethane; 50 mg/l hexane; 5 g/kg propan-2-ol. On hydrolysis 50% loss occurs in *c.* 1 y at $5<pH<9$.

**Uses.**

It controls all tick species including strains resistant to other ixodicides. It can be used on cattle, sheep, horses and dogs in plunge dips at 1.8 g a.i./l.

**Toxicology.**

Acute oral $LD_{50}$ for rats 2500 mg tech./kg. Acute percutaneous $LD_{50}$ for rats >2150 mg/kg; non-irritant to rabbits' skin, slightly irritant to their eyes. In 90-d feeding trials NEL was: for rats 10 mg/kg diet (1 mg/kg daily); for dogs 50 mg/kg diet (2 mg/kg daily). $LC_{50}$ (96-h) for rainbow trout, bluegill, carp >49 mg/l.

**Formulations.**

'Dipofene 600 FW', s.c. (600 g tech./kg).

**Analysis.**

Product analysis is by acidimetric titration. Residues are determined by glc with ECD. Details are available from Ciba-Geigy AG.

$C_8H_8Cl_2O_2$ (207.1)
1OR BG EG DO1

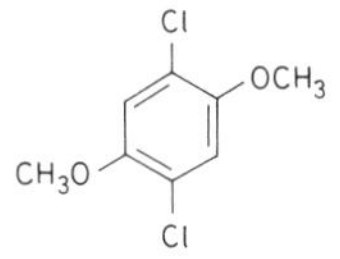

## Nomenclature and development.

Common name chloroneb (BSI, E-ISO, ANSI);chloronébe (F-ISO). Chemical name (IUPAC) 1,4-dichloro-2,5-dimethoxybenzene (I). (*C.A.*) (I) (8 & 9CI); Reg. No. *[2675-77-6]*. Its fungicidal activity was described by M. J. Fielding & R. C. Rhodes (*Proc. Cotton Dis. Counc.*, 1967, **27**, 56). Introduced by E. I. du Pont de Nemours & Co. (Inc.) (USP 3 265 564) as code no. 'Soil Fungicide 1823'; trade mark 'Demosan'.

## Properties.

Pure chloroneb is a colourless crystalline solid with a musty odour; m.p. 133-135 °C; b.p. 268 °C; v.p. 400 mPa at 25 °C. Its solubility at 25 °C: 8 mg/l water; 115 g/kg acetone; 133 g/kg dichloromethane; 118 g/kg dimethylformamide; 89 g/kg xylene. It is stable at least up to 268 °C and at ambient temperatures in the presence of alkali or acid, but is subject to microbial decomposition in soil under moist conditions.

## Uses.

It is a systemic fungicide which is taken up by roots, concentrated in the roots and in lower stem portions, rendering them fungistatic. It is highly fungistatic to *Rhizoctonia* spp., moderately so to *Pythium* spp., poorly to *Fusarium* spp. It is used as a supplemental seed treatment for beans and soyabeans (1.63 g a.i./kg), for cotton (2.44 g/kg); or as an in-furrow treatment for beans and soyabeans (68 g/ha), for cotton (90-135 g/ha). It is also used on turf grass to control snow mould (*Typhula* spp.) (1.3-2.0 kg/ha) or *Pythium* blight (880 g/ha).

## Toxicology.

Acute oral $LD_{50}$ for rats >11 000 mg/kg; for mallard duck and bobwhite quail >5000 mg/kg. In acute percutaneous trials lowest dose causing fatality for rabbits >5000 mg/kg; a 50% aqueous suspension of the w.p. caused no irritation to skin of guinea-pigs, repeated application did not result in skin sensitisation. In 2-y feeding trials in rats growth and food consumption were reduced at 2500 mg/kg diet; no other compound-related effects were observed at this or lower doses upon microscopic or gross examination; there were no changes in organ body weight or ratios.

## Formulations.

'Demosan' 65W, w.p. (650 g a.i./kg); 'Tersan SP' Turf fungicide (650 g/kg).

## Analysis.

Product analysis is by glc (H. L. Pease & R. W. Reiser, *Anal. Methods Pestic. Plant Growth Regul.*, 1973, **7**, 657). Residues may be determined by glc (*idem, ibid.*,; H. L. Pease, *J. Agric. Food Chem.*, 1967, **15**, 917).

$C_{23}H_{15}ClO_3$ (374.8)
L56 BV DV CHJ CVYR&R DG

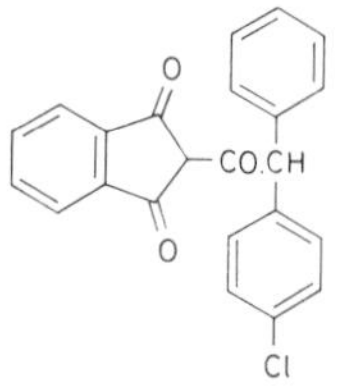

**Nomenclature and development.**

Common name chlorophacinone (BSI, E-ISO, F-ISO, JMAF). Chemical name (IUPAC) 2-[2-(4-chlorophenyl)-2-phenylacetyl]indan-1,3-dione; 2-[2-(4-chlorophenyl)-2-phenylacetyl]indane-1,3-dione. (*C.A.*) 2-[(4-chlorophenyl)phenylacetyl]-1*H*-indene-1,3(2*H*)-dione (9CI); 2-[(*p*-chlorophenyl)phenylacetyl]-1,3-indandione (8CI); Reg. No. *[3691-35-8]*. Introduced as a rodenticide by Lipha S.A. (USP 3 153 612; FP 1 269 638) as code no. 'LM 91'; trade marks 'Caid', 'Liphadione', 'Raviac' (all Lipha), 'Drat' (May & Baker), 'Quick' (Rhône-Poulenc), 'Saviac' (Aulagne-Chimiotechnic).

**Properties.**

Chlorophacinone is a yellow crystalline solid; m.p. 140 °C; v.p. negligible at 20 °C. Solubility: sparingly soluble in water; soluble in acetone, ethanol, ethyl acetate. It is stable, resistant to weathering and is non-corrosive; compatible with cereals, fruits, roots and other potential bait substrates.

**Uses.**

It is an anticoagulant rodenticide, a single dose of a 50 mg/kg bait killing *Rattus norvegicus* from the 5th d. It is normally incorporated as 50-250 mg/kg bait. It does not induce 'bait-shyness'. In mammals it uncouples oxidative phosphorylation in addition to its anticoagulant action.

**Toxicology.**

Human volunteers tolerated a single dose of 20 mg a.i. with an uneventful recovery without treatment. A solution of 5 mg in 2 ml liquid paraffin applied to 100 $cm^2$ of a rabbit's shaved skin caused only a slight reduction of prothrombin rating. Administration of 15 daily doses of 2.25 mg to grey partridges produced no ill-effect.

**Formulations.**

As 'Caid', 'Liphadione', solution (2.5 g a.i./l oil); 'Raviac', 'Quick', 'CX 14', prepared bait (50 mg/kg) on whole, cracked or milled grain.

**Analysis.**

It may be extracted from baits then oxidised to 4-chlorobenzophenone, which is estimated by glc.

# 1-(4-Chlorophenoxy)-1-(imidazol-1-yl)-3,3-dimethylbutanone

$C_{15}H_{17}ClN_2O_2$ (292.5)
T5N CNJ AYOR DG&VX1&1&1

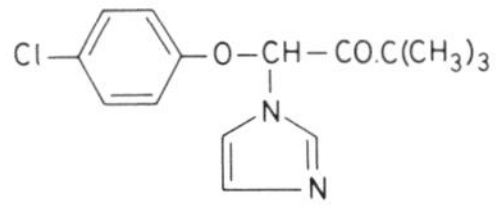

Cl–⟨C6H4⟩–O–CH–CO.C$(CH_3)_3$
N (imidazol-1-yl)

**Nomenclature and development.**

Chemical name (IUPAC) 1-(4-chlorophenoxy)-1-(imidazol-1-yl)-3,3-dimethylbutanone. (*C.A.*) 1-(4-chlorophenoxy)-1-(1*H*-imidazol-1-yl)-3,3-dimethyl-2-butanone (9CI); Reg. No. *[38083-17-9]*. Introduced by Bayer AG (DEOS 2 600 800) as code no. 'BAY MEB 6401'; trade mark 'Baysan'.

**Properties.**

It is a colourless crystalline solid; m.p. 95.5 °C; v.p. 1.0 mPa (extrapolated) at 50 °C. Solubility at 20 °C: 5.5 mg/l water; 400-600 g/kg cyclohexanone; 100-200 g/kg propan-2-ol.

**Uses.**

It is a fungicide, effective against *Aspergillus, Candida, Paecilomyces* and *Penicillium* spp. on various household materials, utensils and parts of buildings.

**Toxicology.**

Acute oral $LD_{50}$: for male rats 400 mg/kg.

**Formulations.**

An aerosol concentrate (5 g/l) combined with benzalkonium chloride (alkyl(benzyl)dimethylammonium chloride, 'Dimanin A').

**Analysis.**

Product analysis is by glc. Details of methods are available from Bayer AG.

$CCl_3NO_2$ (164.4)
WNXGGG

$Cl_3CNO_2$

**Nomenclature and development.**

The trivial name chloropicrin (BSI, E-ISO);chloropicrine (F-ISO) is accepted in lieu of a common name. Chemical name (IUPAC) trichloronitromethane (I). (*C.A.*) (I) (8 & 9CI); Reg. No. *[76-06-2]*. It has been used as an insecticide since 1908 (GBP 2387).

**Properties.**

Chloropicrin is a colourless liquid, m.p. −64 °C; b.p. 112.4 °C; v.p. 3.2 kPa at 25 °C; $n_D^{20}$ 1.595; $d_4^{20}$ 1.656. Solubility at 0 °C: 2.27 g/l water; miscible with acetone, benzene, carbon tetrachloride, diethyl ether, methanol. It is non-flammable and chemically rather inert. It attacks iron, zinc and other light metals; the formation of a protective coating permits storage in iron or galvanised containers.

**Uses.**

It is an insecticide used for the fumigation of stored grain. Also of soil to control nematodes and other soil-dwelling pests; it is effective against soil fungi, except those forming sclerotia. It is highly phytotoxic. Traces are also added to methyl bromide and other fumigants to warn operators of the presence of the major components.

**Toxicology.**

It is lachrymatory and highly toxic. The intense irritation of the mucous membranes serves to warn of hazards to humans. Exposure of cats, guinea-pigs and rabbits to concentrations of 800 mg/$cm^3$ was lethal in 20 min.

**Formulations.**

It is used without formulation. 'Di-Trapex' (chloropicrin + 1,2-dichloropropane + 1,3-dichloropropene) is used to fumigate soil.

**Analysis.**

Chloropicrin is determined by glc. For determination in air collect in propan-2-ol and convert to a derivative which is measured colorimetrically (L. Feinsilver & F. W. Oberst, *Anal. Chem.*, 1953, **25**, 820).

$C_{17}H_{16}Cl_2O_3$ (339.2)
GR DXQR DG&VOY1&1

Cl–C₆H₄–C(OH)(CO.OCH(CH₃)₂)–C₆H₄–Cl

**Nomenclature and development.**

Common name chloropropylate (BSI, E-ISO, F-ISO, ANSI, JMAF). Chemical name (IUPAC) isopropyl 4,4'-dichlorobenzilate (I). (*C.A.*) 1-methylethyl 4-chloro-α-(4-chlorophenyl)-α-hydroxybenzeneacetate (9CI); (I) (8CI); Reg. No. *[5836-10-2]*. ENT 26 999. Its acaricidal properties were described by F. Chabousson (*Phytiatr.-Phytopharm.*, 1956, **5**, 203). Introduced by J. R. Geigy S.A. (now Ciba-Geigy AG) (BEP 511 234; GBP 705 037) as code no. 'G 24 163'; trade mark 'Rospin'.

**Properties.**

Pure chloropropylate forms colourless crystals; m.p. 73 °C; b.p. 148-150 °C/0.5 mmHg; v.p. 44 μPa at 20 °C; ρ 1.35 g/cm$^3$ at 20 °C. Solubility at 20 °C: 1.5 mg/l water; 700 g/l acetone, dichloromethane; 50 g/l hexane; 300 g/l methanol; 130 g/l octan-1-ol; 500 g/l toluene.

**Uses.**

A non-systemic contact acaricide suitable for use at 30-60 g a.i./100 l on cotton, fruit, nuts, ornamentals, sugar beet, tea and vegetables, applied to obtain full coverage of foliage. It is non-phytotoxic at these rates.

**Toxicology.**

Acute oral $LD_{50}$ to rats >5000 mg tech/kg. Non-irritant to skin and eyes of rabbits. In 2-y feeding trials NEL was: for rats 40 mg/kg diet; for dogs 500 mg/kg diet. $LC_{50}$ (96-h): for rainbow trout 0.45 mg/l; for bluegill 0.66 mg/l; for goldfish 0.6 mg/l. It is practically non-toxic to birds, slightly toxic to honeybees.

**Formulations.**

'Rospin 25', e.c. (250 g/l).

**Analysis.**

Product analysis is by glc (E. Bartsch *et al.*, *Residue Rev.*, 1971, **39**, 1; *CIPAC Handbook*, 1980, **1A**, 1149; *AOAC Methods*, 1980, 6.431-6.435). Residues may be determined by glc with MCD. Particulars are available from Ciba-Geigy AG.

$C_8Cl_4N_2$ (265.9)
NCR BG CG DG FG ECN

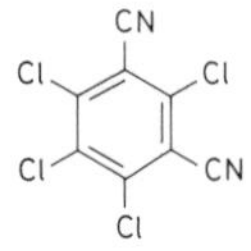

**Nomenclature and development.**

Common name chlorothalonil (BSI, E-ISO, F-ISO, ANSI);TPN (JMAF). Chemical name (IUPAC) tetrachloroisophthalonitrile (I). (*C.A.*) 2,4,5,6-tetrachloro-1,3-benzenedicarbonitrile (9CI); (I) (8CI); Reg. No. *[1897-45-6]*. Its fungicidal properties were described by N. J. Turner *et al.* (*Contrib. Boyce Thompson Inst.*, 1964, **22**, 303). Introduced by the Diamond Alkali Co. (now Diamond Shamrock Corp.) (USP 3 290 353; 3 331 735) as trade marks 'Bravo', 'Daconil 2787', 'Exotherm Termil'.

**Properties.**

Pure chlorothalonil forms colourless, odourless crystals; m.p. 250-251 °C; b.p. 350 °C; v.p. 1.3 Pa at 40 °C. Solubility at 25 °C: 0.6 mg/kg water; 20 g/kg acetone, butanone, dimethyl sulphoxide; 30 g/kg cyclohexanone, dimethylformamide; $<$10 g/kg kerosene; 80 g/kg xylene. The technical grade (*c.* 98% pure) has a slightly pungent odour. It is thermally stable under normal storage conditions, is stable to alkaline and acid aqueous solutions and to u.v. light.

**Uses.**

It is effective against a broad range of plant pathogens attacking many agronomic and vegetable crops at 0.63-2.52 kg a.i. ('Bravo 500 W-75')/ha; on turf at 0.46-1.8 g a.i. ('Daconil 2787 W-75')/$m^2$, on several ornamentals for control of *Botrytis* spp. and certain other diseases at 222-333 g a.i. ('Daconil 2787 W-75')/100 l; for sublimation in glasshouses at 2 g a.i./28 $m^3$ using 'Exotherm Termil' to control *Botrytis* spp. on many ornamentals.

Chlorothalonil ('Nopocide') is also used as a preservative in paints and adhesives.

**Toxicology.**

Acute $LD_{50}$ for albino rats $>$10 000 mg/kg. Acute percutaneous $LD_{50}$ for albino rabbits $>$10 000 mg/kg; a single application of 100 mg to the eyes of rabbits caused marked irritation and corneal opacities. In 2-y feeding trials no ill-effect was observed at the highest doses, in rats at 60 mg/kg diet, in dogs at 120 mg/kg diet (*WHO Pestic. Residues Ser.*, 1973, **4**, 109). $LC_{50}$ (8-d) was: for mallard ducklings $>$21 500 mg/kg diet; for bobwhite quail 5200 mg/kg diet. $LC_{50}$ (estimated) is: for rainbow trout 0.25 mg/l; for bluegill 0.39 mg/l; for channel catfish 0.43 mg/l.

**Formulations.**

These include: 'Bravo W-75' and 'Daconil 2787 W-75 Fungicide', w.p. (750 g a.i./kg); 'Bravo 500' and 'Daconil Flowable', s.c. (500 g/l); 'Exotherm Termil', for sublimation in glasshouses.

**Analysis.**

Product analysis is by glc (D. L. Ballee *et al., Anal. Methods Pestic. Plant Growth Regul.*, 1976, **8**, 263). Residues may be determined by glc (*idem, ibid.*; *Pestic. Anal. Man.*, 1979, **I**, 201-A, 201-G, 201-I). Details of methods are available from Diamond Shamrock Corp.

$C_{15}H_{15}ClN_2O_2$ (290.7)
GR DOR DMVN1&1

Cl—⟨benzene⟩—O—⟨benzene⟩—NH.CO.N$(CH_3)_2$

**Nomenclature and development.**

Common name chloroxuron (BSI, E-ISO, F-ISO, ANSI, WSSA, JMAF);chloroxifenidim (USSR). Chemical name (IUPAC) 3-[4-(4-chlorophenoxy)phenyl]-1,1-dimethylurea. (*C.A.*) *N*'-[4-(4-chlorophenoxy)phenyl]-*N*,*N*-dimethylurea (9CI); 3-[*p*-(*p*-chlorophenoxy)phenyl]-1,1-dimethylurea (8CI); Reg. No. *[1982-47-4]*. Its herbicidal properties were reported (*Symp. New Herbic., 3rd*, 1961, p. 88). Introduced by Ciba AG (now Ciba-Geigy AG) (BEP 593 743; GBP 913 383) as code no. 'C 1983'; trade mark 'Tenoran'.

**Properties.**

Pure chloroxuron is a colourless powder; m.p. 151-152 °C; v.p. 239 nPa at 20 °C; ρ 1.34 g/cm$^3$ at 20 °C. Solubility at 20 °C: 4 mg/l water; 44 g/kg acetone; 106 g/kg dichloromethane; 35 g/kg methanol; 4 g/kg toluene. No significant hydrolysis occurred at 30 °C at pH1.

**Uses.**

It is absorbed by roots and leaves and is recommended for the control of weeds in carrots, celery, conifer seeds, lawns and sports fields, leeks, onions, ornamentals (incl. seed production), strawberries.

**Toxicology.**

Acute oral $LD_{50}$ for rats 3000 mg tech./kg. Acute percutaneous $LD_{50}$ for rats >3000 mg/kg; slightly irritant to skin and eyes of rabbits. Inhalation $LC_{50}$ (6-h) for rats >1350 mg/m$^3$ air. In 120-d feeding trials NEL for rats was 30 mg/kg daily; in 90-d trials in dogs 400 mg/kg diet (16.7 mg/kg daily). $LC_{50}$ is: for rainbow trout >100 mg/l; for crucian carp >150 mg/l; for bluegill 28 mg/l. Slightly toxic to birds, practically non-toxic to honeybees.

**Formulations.**

'Tenoran' WP, w.p. 500 g a.i./kg.

**Analysis.**

Product analysis is by determination of the dimethylamine produced by hydrolysis (G. Voss *et al., Anal. Methods Pestic. Plant Growth Regul.*, 1973, **7**, 569; *AOAC Methods*, 1980, 6.245-6.249). Residues may be determined by hydrolysis to 4-(4-chlorophenoxy)aniline, derivatives of which are measured colorimetrically or by glc with ECD (G. Voss *et al., loc. cit.*). Alternatively, glc with ECD may be used direct (T. H. Byast *et al., Tech. Rep. ARC Weed Res. Organ.*, No. 15 (2nd Ed.), p. 49).

$C_{19}H_{32}Cl_2P$ (362.3)
GR CG D1P4&4&4
*chlorphonium chloride*
$C_{19}H_{32}Cl_3P$ (397.8)
GR CG D1P4&4&4 &&Cl SALT

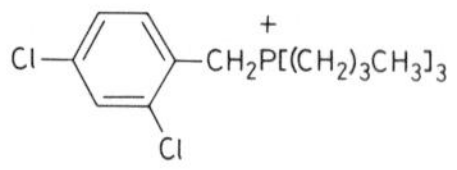

**Nomenclature and development.**

Common name chlorphonium (BSI, E-ISO);chlorfonium (F-ISO). Chemical name (IUPAC) tributyl(2,4-dichlorobenzyl)phosphonium ion. (*C.A.*) tributyl[(2,4-dichlorophenyl)methyl]phosphonium (9CI); tributyl(2,4-dichlorobenzyl)phosphonium (8CI); Reg. No. *[115-78-6]* for chloride. The chloride was introduced as a plant growth regulator by the Mobil Chemical Co. (who no longer manufacture or market it) (USP 3 268 323) as trade marks 'Phosfon' (Mobil Chemical Co.), 'Phosfleur' (Perifleur Products Ltd).

**Properties.**

Chlorphonium chloride is a colourless crystalline solid; m.p. 114-120 °C; vapour pressure negligible. Solubility: soluble in water, acetone, alcohols; insoluble in diethyl ether, hexane.

**Uses.**

Chlorphonium chloride is used to reduce internodal length in ornamentals (chrysanthemums and petunias) to increase the breaking of lateral shoots (geraniums and petunias) and to increase the number of buds and flowers (azaleas, geraniums and rhododendrons). It is most effective as a soil treatment for potted plants. In addition, treatment of stock plants enhances the uniformity of cuttings taken from them.

**Toxicology.**

Acute oral $LD_{50}$ for rats 210 mg/kg. Acute percutaneous $LD_{50}$ for rabbits 750 mg/kg. The technical grade and formulations are irritating to skin and eyes. $LC_{50}$ (96-h) for rainbow trout is 115 mg/l.

**Formulations.**

'Phosfleur', granule (15 g chlorphonium chloride/kg); 'Phosfleur' liquid (100 g/l).

$C_{12}H_{14}ClN_2O_3PS$ (332.7)
NCYR BG&UNOPS&O2&O2

CN S
C=NOP(OCH$_2$CH$_3$)$_2$
Cl

**Nomenclature and development.**

Common name chlorphoxim (BSI, E-ISO);chlorphoxime (F-ISO). Chemical name (IUPAC) 2-(2-chlorophenyl)-2-(diethoxyphosphinothioyloxyimino)acetonitrile; *O,O*-diethyl 2-chloro-α-cyanobenzylideneamino-oxyphosphonothioate. (*C.A.*) 7-(2-chlorophenyl)-4-ethoxy-3,5-dioxa-6-aza-4-phosphaoct-6-ene-8-nitrile 4-sulfide (9CI); formerly 2-chloro-α-[(diethoxyphosphinothioyloxy)imino]benzene acetonitrile (9CI); *o*-chlorophenylglyoxylonitrile oxime *O,O*-diethyl phosphorothioate (8CI); Reg. No. *[14816-20-7]*. OMS 1197. Its insecticidal properties were described by J. E. Hudson & W. O. Obudho (*Mosq. News*, 1972, **32**, 37). Introduced by Bayer AG (DEP 1 238 902) as code no. 'BAY SRA 7747'; trade mark 'Baythion C'.

**Properties.**

A colourless solid; m.p. 66.5 °C; not distillable; v.p. $<1$ mPa at 20 °C. Solubility at 20 °C: 1.7 mg/kg water; 400-600 g/kg cyclohexane, toluene.

**Uses.**

An insecticide used to control mosquitoes and simulium flies.

**Toxicology.**

Acute oral $LD_{50}$ for rats $>2500$ mg/kg; acute percutaneous $LD_{50}$ for rats $>500$ mg/kg. In 90-d feeding trials mice showed no effect at 50 mg/kg diet. $LC_{50}$ (96-h) for rainbow trout 0.1-1.0 mg/l; $LC_{50}$ (48-h) for carp 8.5 mg/l.

**Formulations.**

A w.p. (500 g a.i./kg) ULV (200 g/l).

**Analysis.**

Product analysis is by i.r. spectrometry. Residues are determined by glc (D. B. Leuck & M. C. Bowman, *J. Econ. Entomol.*, 1973, **66**, 798).

$C_{10}H_{12}ClNO_2$ (213.7)
GR CMVOY1&1

NH.CO.OCH(CH$_3$)$_2$
Cl

**Nomenclature and development.**

Common name chlorpropham (BSI, E-ISO, WSSA);chlorprophame (F-ISO);IPC (JMAF);chlor-IFC (USSR). Chemical name (IUPAC) isopropyl 3-chlorocarbanilate; isopropyl 3-chlorophenylcarbamate. (*C.A.*) 1-methylethyl 3-chlorophenylcarbamate (9CI); isopropyl *m*-chlorocarbanilate (8CI); Reg. No. *[101-21-3]*. Trivial names CIPC, chloro-IPC. ENT 18 060. Its herbicidal properties were described by E. D. Witman & W. F. Newton (*Proc. Northeast. Weed Control Conf.*, 1951, p. 45). Introduced in 1951 (USP 2 695 225).

**Properties.**

Pure chlorpropham is a solid, m.p. 41.4 °C; $d^{30}$ 1.180; $n_D^{20}$ (supercooled) 1.5395. Solubility at 25 °C: 89 mg/l water; 100 g/kg kerosene; miscible with alcohols, aromatic hydrocarbons, most organic solvents. The technical grade (98.5% pure) has m.p. 38.5-40.0 °C. Stable $<$100 °C, but slowly hydrolysed in acidic or alkaline media.

**Uses.**

It is a pre-em. herbicide used alone or, more frequently, with other herbicides to increase the range of weeds controlled. it controls many germinating weeds and established chickweed in carrots, bulbous ornamentals, leeks, lettuce, and onions. It is also used to inhibit the sprouting of ware potatoes (P. C. Martin & E. S. Schultz, *Am. Potato J.*, 1952, **29**, 268).

**Toxicology.**

Acute oral $LD_{50}$ for rats 5000-7000 mg/kg. In 2-y feeding trials rats receiving 2000 mg/kg diet showed no ill-effect. Fish were not affected by concentrations of 5 mg/l.

**Formulations.**

These include e.c. of various concentrations and for specific crops. These include: chlorpropham + diuron; chlorpropham + fenuron; chlorpropham + linuron; chlorpropham + pentanochlor; chlorpropham + propham.

**Analysis.**

Product analysis is by glc or by i.r. spectroscopy (*Anal. Methods Pestic. Plant Growth Regul.*, 1972, **6**, 612), or by titration of the carbon dioxide or 3-chloroaniline formed on hydrolysis (*CIPAC Handbook*, 1970, **1**, 223; FAO Specification CP/73). Residues may be determined by glc or hplc of derivatives (*Anal. Methods Pestic. Plant Growth Regul., loc. cit.*; T. H. Byast *et al., Tech. Rep. ARC Weed Res. Organ.*, No. 15, (2nd Ed.), p. 17).

$C_9H_{11}Cl_3NO_3PS$ (350.6)
T6NJ BOPS&O2&O2 CG EG FG

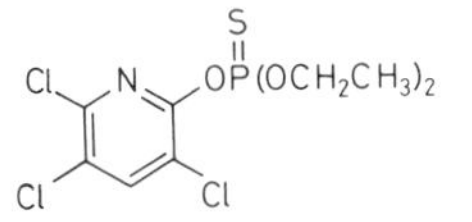

**Nomenclature and development.**

Common name chlorpyrifos (BSI, E-ISO, ANSI, BPC); chlorpyriphos (F-ISO, JMAF);chlorpyriphos-éthyl (France ). Chemical name (IUPAC) *O,O*-diethyl *O*-3,5,6-trichloro-2-pyridyl phosphorothioate. (*C.A.*) *O,O*-diethyl *O*-(3,5,6-trichloro-2-pyridinyl) phosphorothioate (9CI); *O,O*-diethyl *O*-(3,5,6-trichloro-2-pyridyl) phosphorothioate (8CI); Reg. No. *[2921-88-2]*. OMS 971, ENT 27 311. Its insecticidal properties were described by E. E. Kenaga *et al.* (*J. Econ. Entomol.*, 1965, **58**, 1043). Introduced by Dow Chemical Co. (USP 3 244 586) as code no. 'Dowco 179'; trade marks 'Dursban', 'Lorsban'.

**Properties.**

It forms colourless crystals with a mild mercaptan odour; m.p. 42-43.5 °C; v.p. 2.5 mPa at 25 °C. Solubility at 25 °C: 2 mg/l water; 6.5 kg/kg acetone; 7.9 kg/kg benzene; 6.3 kg/kg chloroform; 450 g/kg methanol. The rate of hydrolysis in water increases with pH and temperature, the presence of copper and possibly of other metals that can form chelates. Under laboratory conditions, 50% hydrolysis takes from 1.5 d (water at pH8 and 25 °C) to 100 d (phosphate buffer at pH7 and 15 °C). It is compatible with non-alkaline pesticides but is corrosive to copper and brass.

**Uses.**

It has a broad range of insecticidal activity and is effective by contact, ingestion and vapour action, but is not systemic. Used for the control of flies, household pests, mosquitoes (larvae and adults) and of various crop pests in soil and on foliage; also used for control of ectoparasites on cattle and sheep. Its volatility is great enough to form insecticidal deposits on nearby untreated surfaces. It is non-phytotoxic at insecticidal concentrations. It is degraded in soil, initially to 3,5,6-trichloropyridin-2-ol which is subsequently degraded to organochlorine compounds and carbon dioxide. It persists in soil for 60-120 d.

**Toxicology.**

Acute oral $LD_{50}$: for rats 135-163 mg/kg; for guinea-pigs 500 mg/kg; for chickens 32 mg/kg; for rabbits 1000-2000 mg/kg. Acute percutaneous $LD_{50}$ (in solutions) for rabbits is *c.* 2000 mg/kg. In 2-y feeding trials NEL, based on blood plasma cholinesterase activity, was: for rats 0.03 mg/kg daily; for dogs 0.01 mg/kg daily. It is rapidly detoxified in rats, dogs and other animal species. $LC_{50}$ (96-h) for rainbow trout is 0.003 mg/l.

**Formulations.**

These include: w.p. (250 g a.i./kg); e.c. (240 or 480 g/l); ULV (240 g/l); granules (50, 75 or 100 g/kg).

**Analysis.**

Product analysis is by hplc with uv detection (*J. Assoc. Off. Anal. Chem.*, 1981, **64**, 503). Residues may be determined by glc (*Pestic. Anal. Man.*, 1979, 201-A, 201-G, 201-H, 201-I; R. Mestres *et al., Trav. Soc. Pharm. Montpellier*, 1977, **39**, 323; *Ann. Falsif. Expert. Chim.*, 1979, **72**, 577). Details of both methods are available from Dow Chemical Co.

# Chlorpyrifos-methyl



$C_7H_7Cl_3NO_3PS$ (322.5)
T6NJ BOPS&O1&O1 CG EG FG

**Nomenclature and development.**

Common name chlorpyrifos-methyl (BSI, E-ISO, ANSI);chlorpyriphos-methyl (F-ISO, JMAF). Chemical name (IUPAC) *O,O*-dimethyl *O*-3,5,6-trichloro-2-pyridyl phosphorothioate (I). (*C.A.*) *O,O*-dimethyl *O*-(3,5,6-trichloro-2-pyridinyl) phosphorothioate (9CI); (I) (8CI); Reg. No. *[5598-13-0]*. OMS 1155, ENT 27 520. Its insecticidal properties were described by R. H. Rigterink & E. E. Kenaga (*J. Agric. Food Chem.*, 1966, **14**, 304). Introduced by Dow Chemical Company (USP 3 244 586) as code no. 'Dowco 214'; trade mark 'Reldan'.

**Properties.**

It forms colourless crystals, with a slight mercaptan odour; m.p. 45.5-46.5 °C; v.p. 5.6 mPa at 25 °C. Solubility at 24 °C: 4 mg/l water; 6.4 kg/kg acetone; 5.2 kg/kg benzene; 3.5 kg/kg chloroform; 230 g/kg hexane; 300 g/kg methanol. It is stable under normal storage conditions. It is relatively stable in neutral media, but it is hydrolysed under both acidic (pH4-6) and more readily under alkaline (pH8-10) conditions.

**Uses.**

It is an insecticide with a broad range of activity; effective by contact, ingestion and by vapour action. It is used to control aquatic larvae, flies, household pests, mosquitoes (adult), pests of stored grain, and various foliar crop pests. It is not persistent in soil. It is degraded initially to 3,5,6-trichloropyridin-2-ol which is subsequently degraded to organic chlorine compounds and carbon dioxide.

**Toxicology.**

Acute oral $LD_{50}$: for rats 1630-2140 mg/kg; for guinea-pigs 2250 mg/kg; for rabbits 2000 mg/kg; for chickens (capsule application) >7950 mg/kg. Acute percutaneous $LD_{50}$ for rabbits >2000 mg/kg. In 2-y feeding trials NEL, based on plasma cholinesterase levels, was 0.1 mg/kg daily for dogs and rats. It is readily metabolised in rats and other animals. $LC_{50}$ (96-h) for rainbow trout is 0.3 mg/l; but it is toxic to crustaceans, $LC_{50}$ (36-h) for crayfish 0.004 mg/l.

**Formulations.**

'Reldan EC', e.c. (240 or 500 g a.i./l); ULV (25 g a.i./l) for the treatment of stored grain.

**Analysis.**

Product analysis is by hplc with u.v. detection. Residues may be determined by glc (*Pestic. Anal. Man.*, 1979, **I**, 201-H, 201-I; J. Desmarchelier *et al., Pestic. Sci.*, 1977, **8**, 473). Details of these methods are available from Dow Chemical Co.

$C_{12}H_{12}ClN_5O_4S$ (357.8)
T6N CN ENJ BO1 DMVMSWR BG& F1

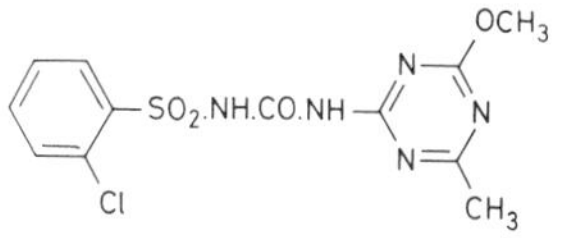

**Nomenclature and development.**

Common name chlorsulfuron (BSI, E-ISO, ANSI, WSSA). Chemical name (IUPAC) 1-( 2-chlorophenylsulphonyl)-3-(4-methoxy-6-methyl-1,3,5-triazin-2-yl)urea. (*C.A.*) 2-chloro-*N*-[[(4-methoxy-6-methyl-1,3,5-triazin-2-yl)amino]carbonyl]benzenesulfonamide (9CI); Reg. No. *[64902-72-3]*. Its herbicidal activity was described by P. G. Jensen (*Weeds Weed Control*, 1980, 21st, 24). Introduced by E. I. du Pont de Nemours & Co. (Inc.) (USP 4 124 405) as code no. 'DPX 4189'; trade mark 'Glean'.

**Properties.**

Pure chlorsulfuron forms colourless crystals; m.p. 174-178 °C; v.p. 612 μPa at 25 °C. Solubility at 25 °C: 100-125 mg/l water (solution has pH 4.1), 300 mg/l (at pH 5), 27.9 g/l (at pH 7); at 22 °C: 57 g/l acetone; 102 g/l dichloromethane; 10 mg/l hexane; 14 g/l methanol; 3 g/l toluene. Stable to light when dry. It decomposes at 192 °C. Aqueous solutions and deposits on plants or soil are decomposed. At 20 °C and pH 5.7-7.0 50% loss by hydrolysis occurs in 4-8 weeks. Loss from soil is by microbial action.

**Uses.**

It is a broad range selective herbicide applied pre-em., early post-em., pre-plant or early post-plant incorporated to control broadleaf weeds in cereals. It is absorbed by foliage and roots of both resistant and susceptible species, with acropetal and basipetal translocation. Selectivity is due to detoxication in resistant species; action is by inhibition of cell division in growing tips of roots and shoots of susceptible species.

**Toxicology.**

Acute oral $LD_{50}$: for male rats 5545 mg/kg, for females 6293 mg/kg. Acute percutaneous $LD_{50}$ for rabbits >3400 mg/kg; temporary slight irritant to eyes and non-irritant to the skin of rabbits. In 2-y feeding trials there was no ill-effect; in rats at 100 mg/kg diet; in mice at 500 mg/kg diet; in 120-d trials in dogs no ill-effects was observed at 2500 mg/kg diet. $LC_{50}$ (8-d) for mallard duck and bobwhite quail was >5000 mg/kg diet. $LC_{50}$ (96-h) for bluegill and rainbow trout is >250 mg/l. Tests for oncogenicity, mutagenicity and teratogenicity were negative.

**Formulations.**

Dispersible granules (750 g a.i./kg). 'Glean C', w.p. (chlorsulfuron + methabenzthiazuron); 'Glean TP', (chlorsulfuron + bromoxynil octanoate + ioxynil octanoate).

$C_{10}H_6Cl_4O_4$ (332.0)
1OVR BG CG EG FG DVO1

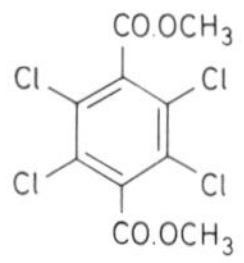

**Nomenclature and development.**

Common name chlorthal-dimethyl (BSI, E-ISO, F-ISO);DCPA (WSSA);TCTP (JMAF). Chemical name (IUPAC) dimethyl tetrachloroterephthalate (I). (*C.A.*) dimethyl 2,3,5,6-tetrachlorobenzene-1,4-dicarboxylate (9CI); (I) (8CI); Reg. No. *[1861-32-1]*. Its herbicidal properties were described by P. H. Schuldt *et al.* (*Proc. Northeast. Weed Control Conf.*, 1960, p. 42). Introduced by Diamond Alkali Co. (now Diamond Shamrock Corp.) (USP 2 923 634) as code no. 'DAC 893'; trade mark 'Dacthal'.

**Properties.**

Chlorthal-dimethyl forms colourless crystals, m.p. 156 °C, v.p. <67 Pa at 40 °C. Solubility at 25 °C: <0.5 mg/l water; 100 g/kg acetone; 250 g/kg benzene; 120 g/kg dioxane; 170 g/kg toluene; 140 g/kg xylene. It is stable in the pure state and in w.p. formulations.

**Uses.**

It is a pre-em. herbicide recommended at 6-14 kg/ha for the control of annual weeds in many crops. In most soils it is hydrolysed with a 50% loss in 100 d.

**Toxicology.**

Acute oral $LD_{50}$ for rats >3000 mg/kg. Acute percutaneous $LD_{50}$ for albino rabbits >10 000 mg/kg; a single application of 3 mg to their eyes produced a mild irritation which subsided within 24 h. It is metabolised in animals to the monomethyl ester and to the parent acid (chlorthal) which are eliminated in the urine.

**Formulations.**

'Dacthal W-75', w.p. (750 g a.i./kg); 'Dimethyl-T', w.p. (750 g/kg); 'Dacthal G-5', granules (50 g/kg). Chlorthal-dimethyl + methazole.

**Analysis.**

Product analysis is by i.r. spectrometry (*AOAC Methods*, 1980, 6.259-6.261) or by glc (*ibid.*, 6.262-6.264). Residues may be determined by glc (H. P. Burchfield & E. E. Storris, *Anal. Methods Pestic., Plant Growth Regul. Food Addit.*, 1964, **4**, 67; *Anal. Methods Pestic. Plant Growth Regul.*, 1972, **6**, 616). Details of methods are available from Diamond Shamrock Corp.

$C_7H_5Cl_2NS$ (206.1)
SUYZR BG FG

CS.NH2
Cl Cl

**Nomenclature and development.**

Common name chlorthiamid (BSI, E-ISO);chlortiamide (F-ISO);DCBN (JMAF). Chemical name (IUPAC) 2,6-dichloro(thiobenzamide) (I). (*C.A.*) 2,6-dichlorobenzenecarbothiamide (9CI); (I) (8CI); Reg. No. *[1918-13-4]*. Its herbicidal properties were described by H. Stanford (*Proc. Br. Weed Control Conf., 7th*, 1964, p.208). Introduced by Shell Research Ltd (GBP 987 253) as code no. 'WL 5792'; trade mark 'Prefix'.

**Properties.**

It is an almost colourless solid; m.p. 151-152 °C; v.p. 130 μPa at 20 °C. Solubility at 21 °C: 950 mg/l water; 50-100 g/kg aromatic and chlorinated hydrocarbons. Stable <90 °C, and in acid solution but is converted to dichlobenil (see 4130) in alkaline solution.

**Uses.**

Chlorthiamid is toxic to germinating seeds. It is absorbed by the roots and, to some extent, by the leaves though there is little downward translocation. For total weed control on non-crop areas it is recommended at 17-28 kg a.i./ha. The 7.5% granular formulation is preferred for selective weed control and is recommended at 9.2 kg a.i./ha in apples, 6.75-9.2 kg/ha in black currants and gooseberries, 9.0-13.2 kg/ha in vines, 2.5-4.6 kg/ha in forest plantations. It is also useful as a 'spot' treatment for the control of weeds such as *Cardus* and *Rumex* spp. It should be applied in early spring before any vegetative growth takes place. Chlorthiamid is converted to dichlobenil in soil and the initial time for 50% loss from soil (chlorthiamid + dichlobenil) is *c.* 35 d under dry and 14 d under wet conditions. The fate in crops, soils and animals has been reviewed (K. I. Beynon & A. N. Wright, *Residue Rev.*, 1972, **43**, 23).

**Toxicology.**

Acute oral $LD_{50}$: for rats 757 mg/kg; for mice and chickens 500 mg/kg. Acute percutaneous $LD_{50}$ for rats >1000 mg/kg. In 90-d feeding trials NEL for rats was 100 mg/kg diet. $LC_{50}$ (24-h) for harlequin fish 41 mg/l.

**Formulations.**

These are primarily granules (75, 100 or 150 g a.i./kg).

**Analysis.**

Product analysis is by titration with silver nitrate (*CIPAC Handbook*, 980, **1A**, 1158). Residues may be determined by glc (K. I. Beynon *et al.*, *J. Sci. Food Agric.*, 1966, **17**, 151; T. H. Byast *et al.*, *Tech. Rep., ARC Weed Res. Organ.*, No.15 (2nd Ed.), p.13). Details available from Shell International Chemical Co. Ltd.

$C_{11}H_{15}Cl_2O_3PS_2$ (361.2)

(i) 2OPS&O2&OR BG EG DS1

(ii) 2OPS&O2&OR CG DG FS1

(iii) 2OPS&O2&OR BG DG ES1

**Nomenclature and development.**

Common name chlorthiophos (BSI, E-ISO, F-ISO, ANSI) for a reaction mixture. Chemical name (IUPAC) *O*-2,5-dichloro-4-(methylthio)phenyl *O,O*-diethyl phosphorothioate (i, main component); *O*-4,5-dichloro-2-(methylthio)phenyl *O,O*-diethyl phosphorothioate (ii, small quantities); *O*-2,4-dichloro-5-(methylthio)phenyl *O,O*-diethyl phosphorothioate (iii, minor component). (*C.A.*) *O*-[2,5-dichloro-4-(methylthio)phenyl] *O,O*-diethyl phosphorothioate (i) (8 & 9CI); *O*-[4,5-dichloro-2-(methylthio)phenyl] *O,O*-diethyl phosphorothioate (ii) (8 & 9CI); *O*-[2,4-dichloro-5-(methylthio)phenyl] *O,O*-diethyl phosphorothioate (iii) (8 & 9CI); Reg. No. *[60238-56-4]* for mixed isomers, *[21923-23-9]* for isomer (i). OMS 1342, ENT 27 635. Its insecticidal properties were described by H. Holtmann & E. Raddatz (*Proc. Br. Insectic. Fungic. Conf., 6th*, 1971, **2**, 485). Introduced by C. H. Boehringer Sohn/Cela GmbH (now Celamerck GmbH & Co. KG) (DEP 1 298 990; GBP 1 210 826) as code no. 'S 2957'; trade mark 'Celathion'.

**Properties.**

Chlorthiophos is a yellow-brown liquid which tends to crystallise <25 °C; b.p. 153-158 °C/0.013 mmHg; v.p. 530 μPa at 25 °C; $d^{20}$ 1.345. Solubility at 20 °C: *c.* 0.3 mg/l water; miscible with all common organic solvents. On hydrolysis 50% loss occurs in 42 d at pH <9.

**Uses.**

It is a non-systemic contact and stomach insecticide, effective against Diptera, Hemiptera, Lepidoptera and with some acaricidal activity. Rates 25-75 g tech./100 l (300-1500 g/ha) depend on pests and the crop.

**Toxicology.**

Acute oral $LD_{50}$: for male rats 10.7 mg/kg, for females 7.8 mg/kg; for mice 91.4 mg/kg. Acute percutaneous $LD_{50}$: for male rats 153 mg/kg, for females 121 mg/kg; for rabbits 50-58 mg/kg. In 2-y feeding trials NEL, based on erythrocyte cholinesterase inhibition, for rats was 1.6 mg/kg diet; in 1-y trials for dogs 1 mg/kg diet. $LC_{50}$ (8-d) is: for bobwhite quail 213 mg/kg diet; for mallard duck 0.198 mg/kg diet. $LC_{50}$ (96-h) is: for bluegill 1.3 mg/l; for rainbow trout 0.019 mg/l.

**Formulations.**

These include: e.c. (500 g tech./l); w.p. (400 g/kg); granules (50 g/kg); dust (50 g/kg); ULV (500 g/kg).

**Analysis.**

Product analysis is by bromometric titration; residue analysis by glc. Details of methods are available from Celamerck GmbH & Co. KG.

# Chlortoluron

$C_{10}H_{13}ClN_2O$ (212.7)
GR B1 EMVN1&1

$CH_3$—C₆H₃(Cl)—NH.CO.N$(CH_3)_2$

**Nomenclature and development.**

Common name chlortoluron (BSI, France, New Zealand);chlorotoluron (E-ISO, F-ISO). Chemical name (IUPAC) 3-(3-chloro-*p*-tolyl)-1,1-dimethylurea (I). (*C.A.*) *N*'-(3-chloro-4-methylphenyl)-*N*,*N*-dimethylurea (9CI); (I) (8CI); Reg. No. *[15545-48-9]*. Its herbicidal properties were described by Y. L'Hermite *et al*., (*C. R. Journ. Etud. Herbic. Conf., COLUMA*, *5th*, 1969, **II**, 349). Introduced by Ciba AG (now Ciba-Geigy AG) (BEP 728 267; GBP 1 255 258) as code no. 'C 2242'; trade mark 'Dicuran'.

**Properties.**

Pure chlortoluron is a colourless powder; m.p. 147-148 °C; v.p. 17.3 μPa at 20 °C; ρ 1.40 g/cm$^3$ at 20 °C. Solubility at 20 °C: 70 mg/l water; 50 g/kg acetone; 24 g/kg benzene; 43 g/kg dichloromethane; 15 g/kg propan-2-ol. On hydrolysis at 30 °C 50% loss (calculated) occurs in 1.48-3.81 y at pH 1-13.

**Uses.**

It is effective both as a residual soil-acting herbicide and as a contact foliar-spray against grass weeds and many broad-leaved weeds of cereal crops, especially against *Alopecurus myosuroides*. It is used at 1.5-3.0 kg a.i./ha on winter cereals immediately after sowing; for post-em., at the same rates when the cereal is at the 3-leaf stage to the end of tillering. It is combined with mecoprop ('Lumeton forte') to improve the control of *Galium, Papaver* and *Veronica* spp.

**Toxicology.**

Acute oral $LD_{50}$ for rats >10 000 mg tech./kg. Acute percutaneous $LD_{50}$ for rats >2000 mg/kg; non-irritant to skin and eyes of rabbits. In 90-d feeding trials the NEL was: for rats 800 mg/kg diet (52 mg/kg daily); for dogs 600 mg/kg diet (23 mg/kg daily). $LC_{50}$ (96-h): for rainbow trout 20-35 mg/l; for bluegill 40-50 mg/l; for crucian carp >100 mg/l. It is practically non-toxic to birds.

**Formulations.**

'Dicurane' 80 WP, w.p. (800 g/kg); 500 FW, s.c. (500 g/l). 'Dicuran extra', chlortoluron + terbutryne (1:6); 'Lumeton forte', (Reg. No. *[53028-44-7]*), chlortoluron + mecoprop (1:1.25); 'Piotal', chlortoluron + terbuthylazine + terbutryne (5:4:1).

**Analysis.**

Product analysis is by acidimetric determination of dimethylamine liberated on hydrolysis (*AOAC Methods*, 1980, 6.244; *CIPAC Handbook*, 1980, **1A**, 1151). Residues may be determined by hplc (T. H. Byast *et al*., *Tech. Rep. ARC Weed Res. Organ*., No. 15 (2nd Ed.), p.49) or by alkaline hydrolysis to 3-chloro-*p*-toluidine a derivative of which is determined by glc with TID, MCD, or ECD. Particulars are available from Ciba-Geigy AG.

$C_{13}H_{11}Cl_2NO_5$ (332.1)
T5OVNV EHJ CR CG EG& EVO2 E1

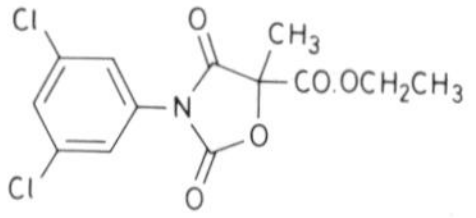

**Nomenclature and development.**

Common name chlozolinate (BSI, draft E-ISO). Chemical name (IUPAC) ethyl (±)-3-(3,5-dichlorophenyl)-5-methyl-2,4-dioxo-oxazolidine-5-carboxylate. (*C.A.*) (±)-ethyl 3-(3,5-dichlorophenyl)-5-methyl-2,4-dioxo-5-oxazolidinecarboxylate (9CI); Reg. No. *[72391-46-9]*. Its fungicidal properties were described by Di Toro *et al.* (*Atti Giornate Fitopatol.*, Siusi, 1980). Introduced by Farmoplant S.p.A. (BRP 874 406; DEP 2 906 574; ITPA 20 579-1978) as code no. 'M 8164'; trade mark 'Serinal'.

**Properties.**

Pure chlozolinate is a colourless solid; m.p. 113-114 °C; v.p. 13 μPa at 25 °C; $d_4^{25}$ 1.42. Solubility at 25 °C: ≤2 mg/l water; soluble in most organic solvents. When dissolved, it is hydrolysed 5<pH>9.

**Uses.**

It is a fungicide for the control of *Botrytis cinerea, Monilia* and *Sclerotinia* spp. and some diseases of ornamental crops. It is recommended for use on grapes and strawberries against *B. cinerea*, on stone and pome fruits against *M. laxa* and *M. fructigena*, and on vegetables against *Botrytis* and *Sclerotinia* spp. using 0.75-1.0 kg a.i./ha.

**Toxicology.**

Acute oral $LD_{50}$: for rats >4500 mg tech./kg; for mice >10 000 mg/kg; for bobwhite quail >9000 mg/kg. Acute percutaneous $LD_{50}$ for rats >5000 mg/kg; neither irritation nor sensitisation of skin has been observed. In a subchronic toxicity trial NEL for rats was 20 mg/kg daily. $LC_{50}$ (96-h) for trout is 27.5 mg/l.

**Formulations.**

These include: 'Serinal' 20 WP, 'Serinal' 50 WP (w.p. 200 or 500 g a.i./kg).

**Analysis.**

Details of analytical methods for product and residues are available from Montedison/Farmoplant S.p.A.

$C_{11}H_{14}ClNO_4$ (259.7)
G1YO1&OR BOVM1

O.CO.NHCH$_3$ / OCHCH$_2$Cl / OCH$_3$

**Nomenclature and development.**

Common name cloethocarb (BSI, draft E-ISO);cloethocarbe (draft F-ISO). Chemical name (IUPAC) 2-(2-chloro-1-methoxyethoxy)phenyl methylcarbamate (I). (*C.A.*) (I) (9CI); Reg. No. *[51487-69-5]*. Its insecticidal properties were described by V. Harries *et al.* (*Meded. Fac. Landbouwwet. Rijksuniv. Gent*, 1978, **45**, 739). Introduced by BASF AG. (DEP 2 231 249; GBP 1 426 233; USP 3 962 316) as code no. 'BAS 263I'; trade mark 'Lance'.

**Properties.**

Clotheocarb is a colourless crystalline solid; m.p. 80 °C. Solubility at 20 °C: 1.3 g/kg water; > 1 kg/kg acetone, chloroform; 153 g/kg ethanol. It is hydrolysed by concentrated alkali or acid.

**Uses.**

It is an insecticide with contact, systemic and stomach-poison activity. It is being tested experimentally on coffee, lucerne, maize, oilseed rape, peanuts, potatoes, rice, sorghum, soyabeans, sugarcane, tobacco.

**Toxicology.**

Acute oral $LD_{50}$ for rats 35.4 mg/kg. Acute percutaneous $LD_{50}$ for rats 4000 mg/kg.

**Formulations.**

Granules (50 or 200 g a.i./kg).

**Analysis.**

Product analysis is by hplc with u.v. detection. Residues may be determined by alkaline hydrolysis with hplc-determination of a derivative.

(approx.) $ClCu_2H_3O_3$ (213.6)
.CU2.Q3.G

$Cu_2Cl(OH)_3$

**Nomenclature and development.**

The trivial name copper oxychloride (E-ISO);oxychlorure de cuivre (F-ISO) is accepted in lieu of a common name for basic copper oxychloride. Chemical name (IUPAC) dicopper chloride trihydroxide (approximate composition). (*C.A.*) copper(II) chloride hydroxide (8 & 9CI); Reg. No. *[1332-65-6]*. Introduced as a fungicide in the early 1900s as trade marks 'Recop' (Sandoz), 'Cupravit' (Bayer), '*Fernacot*', '*Pere-col*' (ICI Plant Protection Division).

**Properties.**

It is a green to blue-green powder, the composition varies with the conditions of manufacture. Solubility: practically insoluble in water; dissolves in dilute acids forming Cu(II) salts; soluble in ammonium hydroxide, forming a complex ion; insoluble in organic solvents. Corrosive to metal containers.

**Uses.**

A protectant fungicide, uses including the control of *Phytophthora infestans* on potatoes and *Pseudoperonospora humuli* on hops.

**Toxicology.**

Acute oral $LD_{50}$ for male rats 1440 mg/kg.

**Formulations.**

These include w.p., water-dispersible granules, or paste.

**Analysis.**

The copper content and formulation characteristics may be determined (*CIPAC Handbook*, 1970, **1**, 226).

$C_{19}H_{15}ClO_4$ (342.8)
T66 BOVJ DYR DG&1V1 EQ

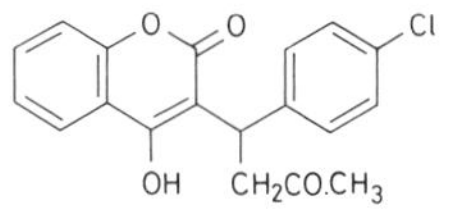

**Nomenclature and development.**

Common name coumachlor (BSI, E-ISO);coumachlore (F-ISO). Chemical name (IUPAC) 3-[1-(4-chlorophenyl)-3-oxobutyl]-4-hydroxycoumarin; 3-(α-acetonyl-4-chlorobenzyl)-4-hydroxycoumarin. (*C.A.*) 3-[1-(4-chlorophenyl)-3-oxobutyl]-4-hydroxy-2*H*-1-benzopyran-2-one (9CI); 3-(α-acetonyl-*p*-chlorobenzyl)-4-hydroxycoumarin (8CI); Reg. No. *[81-82-3]*. Its rodenticidal activity was described by M. Reiff & R. Wiesmann (*Acta Trop.*, 1951, **8**, 97). Introduced by J. R. Geigy S.A. (now Ciba-Geigy AG) (BEP 500 937; GBP 701 111) as code no. 'G 23 133'; trade marks 'Tomorin', 'Ratilan'.

**Properties.**

Pure coumachlor forms colourless crystals; m.p. 169 °C; v.p. <10 mPa at 20 °C; ρ 1.40 g/cm$^3$. Solubility at 20 °C: 0.5 mg/l water (pH 4.5); 100 g/kg acetone; >500 g/kg dimethylformamide; 30 g/kg methanol; 10 g/kg octan-1-ol.

**Uses.**

Coumachlor is an anticoagulant rodenticide which does not induce bait shyness. The $LD_{50}$ with repeated administration to rats is 0.1-1.0 mg/kg daily.

**Toxicology.**

Acute oral $LD_{50}$ for rats 187 mg/kg. Acute percutaneous $LD_{50}$ for rats 33 mg/kg; non-irritant to skin and eyes of rabbits.

**Formulations.**

These include: ready to use bait (300 mg a.i./kg); paraffin bait blocks (400 mg/kg); tracking powder (10 g/kg); bait additive (10 g/kg); concentrate for tracking powder (100 g/kg).

**Analysis.**

Product analysis is by glc or hplc. Particulars are available from Ciba-Geigy AG.

$C_{19}H_{16}O_3$ (292.6)
T66 BOVJ EQ D- GL66&TJ

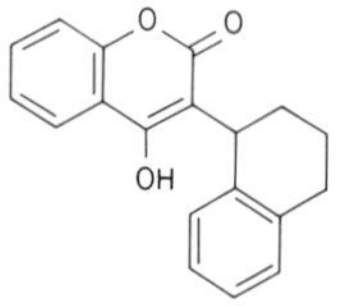

**Nomenclature and development.**

Common name coumatetralyl (BSI, E-ISO, F-ISO);coumarins (JMAF) - *name also applies to* warfarin. Chemical name (IUPAC) 4-hydroxy-3-(1,2,3,4-tetrahydro-1-naphthyl)coumarin (I). (*C.A.*) 4-hydroxy-3-(1,2,3,4-tetrahydro-1-naphthalenyl)-2*H*-1-benzopyran-2-one (9CI); (I) (8CI); Reg. No. *[5836-29-3]*. Introduced as a rodenticide by Bayer AG (DEP 1 079 382; USP 2 952 689) as trade mark 'Racumin'.

**Properties.**

Pure coumatetralyl is a colourless powder; m.p. 172-176 °C. Solubility at 20 °C: 4 mg/l water; 50-100 g/l dichloromethane; 20-50 g/l propan-2-ol. Stable $\leqslant$150 °C.

**Uses.**

It is an anticoagulant rodenticide which does not induce bait shyness. The sub-chronic $LD_{50}$ (5-d) for rats is 0.3 mg/kg daily. Its development was reviewed (G. Hermann & S. Hombrecher, *Pflanzenschutz-Nachr. (Engl. Ed.)*, 1962, **15**, 89).

**Toxicology.**

Sub-acute oral $LD_{50}$ (8-d) for hens >50 mg/kg daily. $LC_{50}$ (96-h) for fish *c.* 1000 mg/l.

**Formulations.**

'Racumin 57' (7.5 g a.i./kg), used as a tracking powder or for bait (1 part 'Racumin 57' + 19 parts bait material). Ready-prepared baits (375 mg/kg).

**Analysis.**

Analysis is by u.v. spectroscopy, particulars from Bayer AG.

$C_7H_{10}ClN_3$ (171.6)
T6N CNJ BG DN1&1 F1

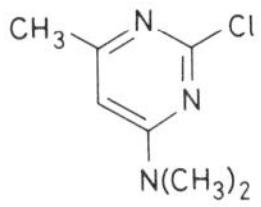

CH3 N Cl
N
N(CH3)2

**Nomenclature and development.**

Common name crimidine (BSI, E-ISO, F-ISO). Chemical name (IUPAC) 2-chloro-*N*,*N*,6-trimethylpyrimidin-4-ylamine; 2-chloro-4-dimethylamino-6-methylpyrimidine. (*C.A.*) 2-chloro-*N*,*N*,6-trimethyl-4-pyrimidinamine (9CI); 2-chloro-4-(dimethylamino)-6-methylpyrimidine (8CI); Reg. No. *[535-89-7]*. Introduced as a rodenticide in the early 1940s by Bayer AG (USP 2 219 858 to Winthrop) as trade mark 'Castrix'.

**Properties.**

The technical product (containing *c*. 20% of the non-rodenticidal 4-chloro-*N*,*N*,6-trimethylpyrimidin-2-ylamine) is a brownish wax. Pure crimidine has m.p. 87 °C; b.p. 140-147 °C/4 mmHg. Solubility at 20 °C: 9.4g/l water; 600-1200 g/l dichloromethane; 100-200 g/l propan-2-ol.

**Uses.**

Crimidine is used as a rodenticide; the poisoned rats are non-toxic to predators.

**Toxicology.**

Acute oral $LD_{50}$: for rats 1.25 mg/kg; for rabbits 5 mg/kg. It is not a cumulative poison, birds are not harmed by the ingestion of a few grains of the formulation.

**Formulations.**

'Castrix-Giftkörner', impregnated wheat grains (1g a.i./kg).

**Analysis.**

Product and residue analysis is by hplc (E. Möllhoff, *Pflanzenschutz-Nachr. (Engl. Ed.)*, 1980, **33**, 86). Details available from Bayer AG.

$C_{14}H_{19}O_6P$ (314.3)
1YR&OV1UY1&OPO&O1&O1

**Nomenclature and development.**

Common name crotoxyphos (BSI, E-ISO, F-ISO, ESA);crotoxyfos (BPC). Chemical name (IUPAC) dimethyl (*E*)-1-methyl-2-(1-phenylethoxycarbonyl)vinyl phosphate; 1-phenylethyl 3-(dimethoxyphosphinyloxy)isocrotonate. (*C.A.*) 1-phenylethyl (*E*)-3-[(dimethoxyphosphinyl)oxy]-2-butenoate (9CI); α-methylbenzyl (*E*)-3-hydroxycrotonate ester with dimethyl phosphate (8CI); Reg. No. *[7700-17-6]*. OMS 239, ENT 24 717. Its insecticidal properties were described by C. P. Weidenback & R. L. Younger (*J. Econ. Entomol.*, 1962, **55**, 793). Introduced by the Shell Development Co. (USP 3 068 268; 3 116 201) as code no. 'SD 4294'; trade mark 'Ciodrin'.

**Properties.**

Produced by the reaction of trimethyl phosphite with 1-phenylethyl 2-chloro-3-oxobutyrate, the technical grade (80% pure) is a light straw-coloured liquid; b.p. 135 °C/0.03 mmHg; v.p. 1.9 mPa at 20 °C; $n_D^{25}$ 1.5505; $d_{20}^{20}$ 1.2. Solubility at room temperature: 1 g/l water; soluble in acetone, ethanol, propan-2-ol, xylene. In aqueous solution at 38 °C, 50% is hydrolysed in 87 h at pH1, 35 h at pH9. It is slightly corrosive to copper, lead, mild steel, tin and zinc, but will not attack fibreglass, reinforced polyester, rigid PVC, or the usual lacquers used for lining drums. It is incompatible with most mineral carriers, except synthetic silicas such as 'Colloidal Silica K320' and 'Hisil 233'.

**Uses.**

It is an insecticide with a rapid action and moderately persistent residual effect on a wide range of external insect pests of livestock and is recommended for the control of flies, mites and ticks on cattle and pigs. Its fate in animals, soils and water has been reviewed (K. I. Beynon *et al.*, *Residue Rev.*, 1973, **47**, 55); residues in milk and meat are negligible.

**Toxicology.**

Acute oral $LD_{50}$: for rats 52.8 mg/kg; for mice 90 mg/kg. Acute percutaneous $LD_{50}$ for rabbits 384 mg/kg. In 90-d feeding trials no effect on growth nor histopathological change was observed: for male rats receiving 900 mg/kg diet, for females 300 mg/kg diet. $LC_{50}$ (48-h) for sheepshead minnow is $>1$ mg/l.

**Formulations.**

An e.c. (240 g a.i./l).

**Analysis.**

Product analysis is by i.r. spectroscopy or by glc. Residues may be determined by glc with FPD (P. E. Porter, *Anal. Methods Pestic., Plant Growth Regul. Food Addit.*, 1967, **5**, 243; *Anal. Methods Pestic. Plant Growth Regul.*, 1972, **6**, 325).

$Cu_2O$ (143.1)
.CU2.O

$Cu_2O$

**Nomenclature and development.**

The traditional chemical name cuprous oxide (E-ISO);oxyde cuivreux (F-ISO) is accepted in lieu of a common name. Chemical name (IUPAC) copper(I) oxide; dicopper oxide. (*C.A.*) copper oxide ($Cu_2O$) (8 & 9CI); Reg. No. *[1317-39-1]*. Its fungicidal properties as a seed protectant were described by J. G. Horsfall (*N. Y. St. Agric. Exp. Stn. Bull.*, 1932, No. 615). It was subsequently used for foliage protection as trade marks 'Caocobre', 'Copper-Sandoz' (both Sandoz), 'Perenox' (ICI Plant Protection Division), 'Yellow Cuprocide' (Rohm & Haas Co.).

**Properties.**

Cuprous oxide is an amorphous powder ranging in colour from yellow to red. Solubility: practically insoluble in water and organic solvents; soluble in dilute mineral acid, to form copper(I) salt (or copper(II) salt plus metallic copper); and in aqueous solutions of ammonia and its salts. It is liable to oxidation to copper(II) oxide and to conversion to a carbonate on exposure to moist air. It is corrosive to aluminium.

**Uses.**

Cuprous oxide is a protective fungicide, used mainly for seed treatment and for foliage application against blight, downy mildews and rusts; it is non-phytotoxic except to brassicas and 'copper-shy' varieties particularly under adverse weather conditions.

**Toxicology.**

Acute oral $LD_{50}$ for rats 470 mg/kg. There have been instances of fatal copper poisoning in sheep grazing in orchards sprayed with cuprous oxide, but there is no significant history of ill-effect on birds or honeybees. Under conditions of moderate use and cultivation the hazard to earthworms, and hence to soil structure, is not considered significant.

**Formulations.**

A w.p. or water dispersible micro-granule (500 g copper/kg); a stabiliser is necessary to delay oxidation and formation of carbonate.

**Analysis.**

Total copper is determined by electrolytic or iodometric methods (*AOAC Methods*, 1980, 6.015-6.016; *MAFF Ref. Book*, 1958, No. 1, p. 16). Metallic copper in cuprous oxide may be determined (L. C. Hurd & A. R. Clark, *Ind. Eng. Chem. Anal. Ed.*, 1936, **8**, 380). Methods are available for total copper content of w.p. and dusts (*CIPAC Handbook*, 1970, **1**, 226; 1983, **1B**, in press). Residues may be determined by reaction with concentrated sulphuric acid, and colorimetric estimation of derivatives or by atomic absorption spectroscopy (*AOAC Methods*, 1980, 25.038-25.048).

$C_9H_{13}ClN_6$ (240.7)
T6N CN ENJ BMX1&1&CN DM2 FG

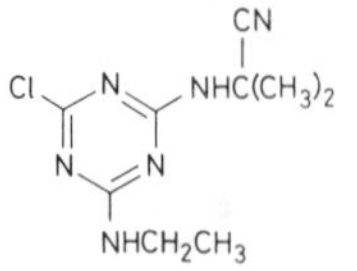

**Nomenclature and development.**

Common name cyanazine (BSI, E-ISO, F-ISO, WSSA). Chemical name (IUPAC) 2-(4-chloro-6-ethylamino-1,3,5-triazin-2-ylamino)-2-methylpropionitrile; 2-chloro-4-(1-cyano-1-methylethylamino)-6-ethylamino-1,3,5-triazine. (*C.A.*) 2-[[4-chloro-6-(ethylamino)-1,3,5-triazin-2-yl]amino]-2-methylpropanenitrile (9CI); 2-[[4-chloro-6-(ethylamino)-*s*-triazin-2-yl]amino]-2-methylpropionitrile (8CI); Reg. No. *[21725-46-2]*. Its herbicidal properties were described by W. J. Hughes *et al.*, (*Proc. North Cent. Weed Control Conf.*, 1966, p.27). Introduced by Shell Research Ltd (GBP 1 132 306) as code no. 'WL 19 805','SD 15 418'; trade marks 'Bladex' (a w.p. for use on maize), 'Fortrol' (a liquid suspension for use on peas).

**Properties.**

Technical cyanazine (≥95% pure) is a colourless crystalline solid; m.p. 166.5-167.0 °C; v.p. 200 nPa at 20 °C. Solubility at 25 °C: 171 mg/l water; 15 g/l benzene; 210 g/l chloroform; 45 g/l ethanol; 15 g/l hexane. Stable to heat (1.8% decomposition after 100 h at 75 °C), light, and to hydrolysis ($5 \leqslant pH \leqslant 9$).

**Uses.**

Cyanazine is a pre- and post-em. herbicide of short persistence. It is valuable for general weed control (a) applied pre-em. to the crop at 1-3 kg a.i./ha for broad beans, maize and peas; (b) applied post-em. in barley and wheat during the early tillering stage at 0.26-0.33 kg/ha in combination with hormone weedkillers for the control of hard-to-kill broad-leaved weeds. Other crops for which it is used include: cotton, flower bulbs, forestry, potatoes, soyabeans, sugarcane. Laboratory tests show that the nitrile group is hydrolysed in plants and soil to the corresponding carboxylic acid. The chlorine atom may also be replaced by a hydroxy group, conjugates involving this group are sometimes formed. These degradation products have not been detected in crops following field use (K. I. Beynon *et al.*, *Pestic. Sci.*, 1972, **3**, 293, 379, 389, 401).

**Toxicology.**

Acute oral $LD_{50}$: for rats the value is from 182-380 mg a.i./kg, depending on the concentration of cyanazine and the carrier used; for mice 380 mg/kg; for quail 400 mg/kg; for mallard duck >2000 mg/kg. Acute percutaneous $LD_{50}$ for rats is >1200 mg/kg. In 2-y feeding trials NEL was: for rats 12 mg/kg diet; for dogs 25 mg/kg diet. It is rapidly metabolised and eliminated from the body by rats and dogs within *c.* 4 d. $LC_{50}$ (48-h): for harlequin fish 10 mg/l; for sheepshead minnow 18 mg/l.

**Formulations.**

These include: w.p. (500 g a.i./kg); s.c. (500 g/l); dry flowable (900 g/kg). 'Cleaval', (Reg. No. *[59915-10-5]*), cyanazine + mecoprop. 'Envoy', cyanazine + MCPA. 'Stay-Kleen', cyanazine + linuron. Mixture, cyanazine + atrazine. Mixture, cyanazine + dichlorprop. Mixture, cyanazine + ioxynil octanoate. Some are also available as water dispersible granules in addition to w.p. or s.c. Details may be obtained from Shell International Chemical Co. Ltd.

**Analysis.**

Product analysis is by glc or hplc (*Anal. Methods Pestic. Plant Growth Regul.*, 1978, **10**, 275) or i.r. spectrometry. Residues in plants may be determined by glc with ECD or FID (*ibid.*) and in soils by glc or hplc (*ibid.*; T. H. Byast *et al.*, *Tech. Rep. ARC Weed Res. Organ.*, No. 15 (2nd Ed.) pp. 40, 74).

$C_{15}H_{14}NO_2PS$ (303.3)
NCR DOPS&R&O2

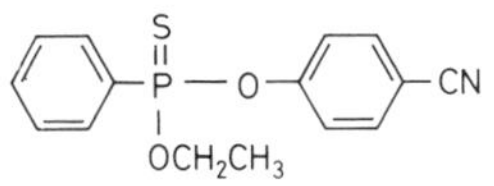

**Nomenclature and development.**

Common name cyanofenphos (BSI, E-ISO);cyanophenphos (F-ISO);CYP (JMAF). Chemical name (IUPAC) *O*-4-cyanophenyl *O*-ethyl phenylphosphonothioate. (*C.A.*) *O*-( 4-cyanophenyl) *O*-ethyl phenylphosphonothioate (9CI); *O*-ethyl phenylphosphonothioate acid *O*-ester with *p*-hydroxybenzonitrile (8CI); Reg. No. *[13067-93-1]*. ENT 25 832a. Its insecticidal properties were described by Y. Nishizawa (*Bull. Agric. Chem. Soc. Japan*, 1960, **24**, 744). Introduced by Sumitomo Chemical Co. (JPP 410 930; 410 925; GBP 929 738) as code no. 'S-4087'; trade mark 'Surecide'.

**Properties.**

Technical grade cyanofenphos is a white or slightly yellowish solid; m.p. 71-78 °C; v.p. 1.37 μPa at 20 °C. Solubility at 30 °C: 0.6 mg/l water; 89 g/kg methanol; at 27 °C: 515 g/kg xylene.

**Uses.**

It is an insecticide effective against penetrating and chewing insects, cotton bollworms, gall midges, rice borers and other insects on fruits and vegetables.

**Toxicology.**

Acute oral $LD_{50}$; for rats 89 mg/kg. Acute percutaneous $LD_{50}$; for male rats >1000 mg/kg, for females 640 mg/kg. $LC_{50}$ (48-h) for carp is 1.45 mg/l.

**Formulations.**

These include: e.c. (250 g a.i./l); dusts (15 or 30 g/kg).

**Analysis.**

Product analysis is by glc. Residues in milk and meat may be determined by glc using FPD (J. Miyamoto *et al., Nihon Noyaku Gakkaishi*, 1977, **2**, 1).

$C_9H_{10}NO_3PS$ (243.2)
NCR DOPS&O1&O1

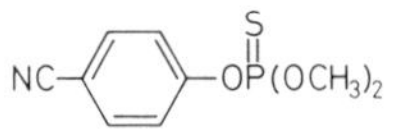

$NC-C_6H_4-OP(S)(OCH_3)_2$

**Nomenclature and development.**

Common name cyanophos (BSI, E-ISO, F-ISO);CYAP (JMAF). Chemical name (IUPAC) *O*-4-cyanophenyl *O,O*-dimethyl phosphorothioate. (*C.A.*) *O*-(4-cyanophenyl) *O,O*-dimethyl phosphorothioate (9CI); *O,O*-dimethyl phosphorothioate *O*-ester with *p*-hydroxybenzonitrile (8CI); Reg. No. *[2636-26-2]*. OMS 226, OMS 869. Its insecticidal properties were described by Y. Nishizawa (*Agric. Biol. Chem.*, 1961, **25**, 597). Introduced by Sumitomo Chemical Co. (JPP 405 852; 415 199) as code no. 'S-4084'; trade mark 'Cyanox'.

**Properties.**

Pure cyanophos has m.p. 14-15 °C. The technical grade is an amber liquid; v.p. 105 mPa at 20 °C; $n_D^{25}$ 1.535-1.545; $d_4^{25}$ 1.255-1.265. Solubility at 30 °C: 46 mg/l water; at 20 °C: 27 g/kg hexane; >1 kg/kg methanol, xylene. Stable to storage >2 y under normal conditions. It is incompatible with alkaline materials.

**Uses.**

It is an insecticide used at 25-50 g a.i./100 l to control lepidopterous pests and sucking insects on fruit, ornamentals and vegetables. It is approved as a locust control insecticide by FAO and OCLALAV and can also be used for the control of household pests such as cockroaches, houseflies and mosquitoes.

**Toxicology.**

Acute oral $LD_{50}$: for male rats 580 mg/kg, for females 610 mg/kg. Acute percutaneous $LD_{50}$ for mice >2500 mg/kg. $LC_{50}$ (48-h) for carp is 5 mg/l.

**Formulations.**

These include: for agricultural use e.c. (500 g a.i./l), w.p. (400 g/kg), ULV (300 g/l), dust (30 g/kg); for public health use liquid (10 g/l), e.c. (50 g/l).

**Analysis.**

Product analysis is by glc or by u.v. spectrometry. Residues may be determined by glc with FTD. Details of methods are available from Sumitomo Chemical Co.

$C_{11}H_{21}NOS$ (215.4)
L6TJ AN2&VS2

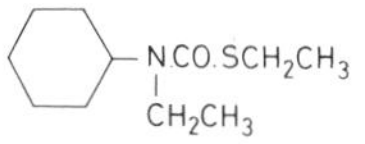

## Nomenclature and development.

Common name cycloate (BSI, E-ISO, F-ISO, WSSA); hexylthiocarbam (JMAF). Chemical name (IUPAC) *S*-ethyl *N*-cyclohexyl-*N*-ethyl(thiocarbamate) (I). (*C.A.*). *S*-ethyl cyclohexylethylcarbamothioate (9CI); (I) (8CI); Reg. No. *[1134-23-2]*. Its herbicidal properties were reported (*Nat. Weed Comm. Can. Dep. Agric. Res. Rep.*, 1963, p.51). Introduced by Stauffer Chemical Co. (USP 3 175 897) as code no. 'R-2063'; trade mark 'Ro-Neet'.

## Properties.

Cycloate (99% pure) is a clear liquid, with an aromatic odour; m.p. 11.5 °C (supercools readily); b.p. 145 °C/10 mmHg; v.p. 830 mPa at 25 °C; $d_4^{30}$ 1.016; $n_D^{30}$ 1.5054. Solubility at 20 °C: 75 mg/l water; miscible with acetone, benzene, ethanol, kerosene, 4-methylpentan-2-one, xylene. Stable, 50% unchanged >10 y at 70 °C.

## Uses.

Cycloate is toxic to germinating weed seeds and is used to control annual broad-leaved weeds, grasses and *Cyperus* spp. in sugar beet and spinach by pre-plant soil incorporation at 3-4 kg a.i./ha. It is non-persistent.

## Toxicology.

Acute oral $LD_{50}$ for male and female rats 2710 mg/kg. Acute percutaneous $LD_{50}$ for rabbits >4640 mg/kg; non-irritating to their eyes. In 90-d feeding studies, no toxic symptom was noted in dogs receiving ≤240 mg/kg daily. $LC_{50}$ (96-h) for rainbow trout is 4.5 mg/l.

## Formulations.

'Ro-Neet E', e.c. (720 g a.i./l); 'Ro-Neet 10G', granules (100 g/kg). Mixture (Reg. No. *[37341-18-7]*), w.p. (375 g cycloate + 63 g lenacil/kg).

## Analysis.

Product analysis is by glc (*CIPAC Handbook*, 1980, **1A**, 1190; *AOAC Methods*, 1980, 6.426-6.430). Residues in crops and soil are determined by glc or colorimetry after conversion to a derivative (J. R. Lane, *Anal. Methods Pestic., Plant Growth Regul. Food Addit.*, 1967, **5**, 491; W. A. Ja *et al.*, *Anal. Methods Pestic. Plant Growth Regul.*, 1972, **6**, 686). Analytical methods are available from the Stauffer Chemical Co.

It is the aim of the editors of *The Pesticide Manual* to present accurate and up-to-date information and its contents have been stored in a computer. This will simplify modification of the text for future editions. The contents and periodic updates of the *Manual* will be available on-line as the Pesticide Databank from 1984.

The easiest way to find a given compound is by its ***Entry Number***. Because the majority of pesticides are known by various synonyms it is recommended that the reader ascertains the required ***Entry Number*** from one or more of the Indexes.

**Index 1** Wiswesser Line-Formula Notation (for description of this see p. 613).

**Index 2** Molecular formulae.

**Index 3** Code numbers used by official bodies, the manufacturers and suppliers. This index is arranged numerically and whether the qualifying letters are, for example, BAY or 'Bayer' is immaterial.

**Index 4** Chemical, common, and trivial names and trade marks.

Each of these give you the ***Entry Number***. A simple number (e.g. 1230) indicates that the entry is in the ***Main Section***, while a preceding asterisk (*2400) shows it is in the ***Superseded Compounds Section*** printed on coloured paper so that it is easy to find.

**Read the Section on How to Use Part I of the Manual,** page xvi.

Remember: an equals sign ( = ) at the end of a line means run on to the next line (omitting = ); chemical names ending in -, ( or [ run straight on. Otherwise the word finishes on a given line.

$C_{15}H_{23}NO_4$ (281.4)
T6VMVTJ E1YQ- BL6VTJ D1 F1

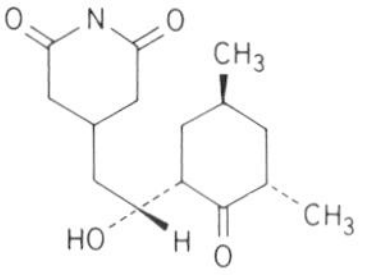

**Nomenclature and development.**

Common name cycloheximide (BSI, E-ISO, F-ISO). Chemical name (IUPAC) 4-(2*R*)-2-[(1*S*,3*S*,5*S*)-(3,5-dimethyl-2-oxocyclohexyl)]-2-hydroxyethylpiperidine-2,6-dione; 3-(2*R*)-2-[(1*S*,3*S*,5*S*)-3,5-dimethyl-2-oxocyclohexyl]-2-hydroxyethylglutarimide. (*C.A.*) [1*S*-[1α(*S*★),3α,5β]]-4-[2-(3,5-dimethyl-2-oxocyclohexyl)-2-hydroxyethyl]-2,6-piperidinedione (9CI); 3-[2-(3,5-dimethyl-2-oxocyclohexyl)-2-hydroxyethyl]glutarimide (8CI); Reg. No. *[66-81-9]*. It was isolated by A.Whiffen *et al.* (*J. Bacteriol.*, 1946, **52**, 610) as an antibiotic effective against certain fungi pathogenic to man. Its structure and stereochemistry have been established, and its agricultural uses reviewed (J. H. Ford *et al., Plant Dis. Rep.*, 1958, **42**, 680). Introduced for crop protection by the Upjohn Co. as trade mark 'Acti-dione'.

**Properties.**

Produced in culture by *Streptomyces griseus* and recovered as a by-product of streptomycin manufacture, cycloheximide forms colourless crystals; m.p. 115.5-117 °C; $[\alpha]^{25}_{D}$ −6.8° (20 g/l water). Solubility at 2 °C: 21 g/l water; soluble in chloroform, propan-2-ol, other common organic solvents; sparingly soluble in saturated hydrocarbons. Though stable in neutral or acid solution, it is rapidly decomposed by alkali at room temperature to form 2,4-dimethylcyclohexanone. Chlordane is reported to cause a rapid loss of activity.

**Uses.**

It inhibits the growth, in culture, of many plant pathogenic fungi; it is used to control powdery mildews on ornamentals, rusts and leafspot on grasses. Its plant growth-regulating properties are used for promoting the abscission of fruit, *e.g.* olives and oranges.

**Toxicology.**

Acute oral $LD_{50}$: for rats 2 mg/kg; for mice 133 mg/kg; for guinea-pigs 65 mg/kg; for monkeys 60 mg/kg.

**Formulations.**

These include: 'Acti-dione PM' (270 mg a.i./kg); 'Acti-dione TGF' (21 g/kg); 'Acti-Aid' (42 g/l); 'Acti-dione', w.p. (1 g/kg). 'Acti-dione' thiram (Reg. No. *[8069-74-7]*) (7.5 g cycloheximide + 750 g thiram/kg); 'Acti-dione RZ' (13 g cycloheximide + 750 g quintozene/kg).

**Analysis.**

Product analysis is by bioassay (G. C. Prescott *et al., J. Agric. Food Chem.*, 1956, **4**, 343; W. W. Kilgore, *Anal. Methods Pestic., Plant Growth Regul. Food Addit.*, 1964, **3**, 1).

$C_{23}H_{19}ClF_3NO_3$ (449.9)
L3TJ A1 A1 BVOYCN&R COR&&
C1UYGXFFF &&(1RS)-CIS-(Z) FORM

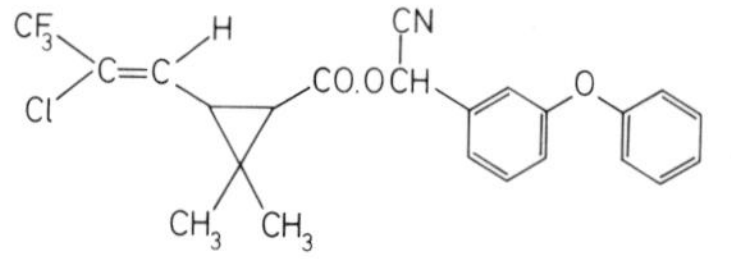

**Nomenclature and development.**

Common name cyhalothrin (BSI, BPC, draft E-ISO). Chemical name (IUPAC) (*RS*)-α-cyano-3-phenoxybenzyl (*Z*)-(1*RS*,3*RS*)-(2-chloro-3,3,3-trifluoropropenyl)-2,2-dimethylcyclopropanecarboxylate; (*RS*)-α-cyano-3-phenoxybenzyl (*Z*)-(1*RS*)-*cis*-3-(2-chloro-3,3,3-trifluoropropenyl)-2,2-dimethylcyclopropanecarboxylate. (*C.A.*) cyano(3-phenoxyphenyl)methyl 3-(2-chloro-3,3,3-trifluoro-1-propenyl)-2,2-dimethylcyclopropanecarboxylate (9CI); Reg. No. *[68085-85-8]*. Its insecticidal properties were described by P. D. Bentley *et al.* (*Pestic. Sci.*, 1980, **11**, 156) and its acaricidal properties by V. K. Stubbs (*Austral. Vet. J.*, 1982, **59**, 152). Introduced by ICI Australia (AUP 521 136; USP 4 183 948; GBP 2 000 764) as code no. 'PP 563'; trade mark 'Grenade'.

**Properties.**

Technical grade cyhalothrin (purity ⩾90% *m*/*m*, of which ⩾95% are *cis*-isomers) is a yellow to brown viscous oil; b.p. 187-190 °C/0.2 mmHg; $n_D^{24}$ 1.534. Solubility: <1 mg/l water; partitioning occurs with aqueous solutions of alcohols or ketones; miscible in all proportions with alcohols, aliphatic, aromatic and halogenated hydrocarbons, esters, ethers and ketones. No decomposition or change in *cis*/*trans* ratio occurred in 90 d at 50 °C. Hydrolysis by water is slow at pH7-9, more rapid at pH>9.

**Uses.**

It is used to control animal ectoparasites, especially *Boophilus microplus* or *Haematobia irritans* on cattle, and lice and ked on sheep. It is applied as a cattle dip every 21 d at 70 mg/l and as a spray around livestock houses.

**Toxicology.**

Acute oral $LD_{50}$: for male rats 243 mg tech./kg, for females 144 mg/kg; for guinea-pigs >5000 mg/kg; for rabbits >1000 mg/kg. Moderately irritant to eyes and mildly to skin of rabbits. Orally-administered cyhalothrin is rapidly eliminated from rats in their urine and faeces; this ester grouping is extensively hydrolysed, both moieties forming polar conjugates.

**Formulations.**

An e.c. (200 g tech./l).

**Analysis.**

Details of analytical methods are available from ICI Australia.

$C_{18}H_{34}OSn$ (385.2)
L6TJ A-SN-Q- AL6TJ&- AL6TJ

Sn
OH

**Nomenclature and development.**

Common name cyhexatin (BSI, E-ISO, F-ISO, ANSI); tricyclohexyltin hydroxide (JMAF). Chemical name (IUPAC) tricyclohexyltin hydroxide. (*C.A.*) tricyclohexylhydroxystannane (8 & 9CI); Reg. No. *[13121-70-5]*. ENT 27 395-X. Its acaricidal properties were described by W. E. Allison *et al.* (*J. Econ. Entomol.*, 1968, **61**, 1254). It was developed from a joint project of Dow Chemical Co. and M & T Chemicals Inc. and introduced by Dow Chemical Co. (USP 3 264 177; 3 389 048) as code no. 'Dowco 213'; trade mark 'Plictran'.

**Properties.**

It is a colourless crystalline powder; apparent m.p. 245 °C; v.p. negligible at 25 °C. Solubility at 25 °C: <1 mg/l water; 1.3 g/kg acetone; 216 g/kg chloroform; 37 g/kg methanol; 10 g/kg toluene. It is stable to 100 °C in aqueous suspensions from slightly acid (pH6) to alkaline. It is degraded when exposed to u.v. light.

**Uses.**

It is an acaricide effective by contact against the motile stages of a wide range of phytophagous mites, the usual dosage being 20-30 g a.i./100 l. It is non-phytotoxic to deciduous fruit, most ornamentals grown in the open, wind-break tree species, vegetables and vines. Citrus (immature foliage and fruit in the early stages of development), as well as the seedlings and immature foliage of some glasshouse-grown ornamentals and vegetables, are susceptible to possible injury, usually in the form of localised spotting.

**Toxicology.**

Acute oral $LD_{50}$: for rats 540 mg tech./kg; for guinea-pigs 780 mg/kg; for young chickens 650 mg/kg. Acute percutaneous $LD_{50}$ for rabbits >2000 mg/kg. In 2-y feeding trials NEL was: for dogs 0.75 mg/kg daily; for mice 3 mg/kg daily; for rats 1 mg/kg daily. $LC_{50}$ (8-d) is: for bobwhite quail 520 mg/kg diet; for mallard ducklings 3189 mg/kg diet. $LC_{50}$ (24-h) is: for large-mouth bass 0.06 mg/l; for goldfish 0.55 mg/l. Virtually non-hazardous to honeybees (dermal $LD_{50}$ 32 μg/bee), most predacious mites and insects at the recommended rates of use.

**Formulations.**

'Plictran' 50W, w.p. (500 g a.i./kg). In Europe, Middle East and Africa: 'Plictran' 25W, w.p. (250 g/kg); 'Plictran' 600F, s.c. (600 g/l). 'Dorvert', s.c. (150 g cyhexatin + 50 g tetradifon/l).

**Analysis.**

Residues may be determined by glc (*Pestic. Anal. Man.*, 1979, **II**; W. O. Gauer, *J. Agric. Food Chem.*, 1974, **22**, 252; E. Möllhoff *Pflanzenschutz-Nachr (Engl. Ed.)*, 1974, **30**, 249) or by atomic absorption spectroscopy (J. L. Love & J. E. Patterson, *J. Assoc. Off. Anal. Chem.*, 1978, **61**, 627). Analytical methods for the determination of cyhexatin and the principal metabolite, dicyclohexyltin oxide are available from Dow Chemical Co.

$C_7H_{10}N_4O_3$ (198.2)
2MVMVYCN&UNO1

$$CH_3CH_2NHCO.NHCO.\underset{}{\overset{CN}{\overset{|}{C}}}=NOCH_3$$

**Nomenclature and development.**

Common name cymoxanil (BSI, ANSI, draft E-ISO). Chemical name (IUPAC) 1-(2-cyano-2-methoxyiminoacetyl)-3-ethylurea. (*C.A.*) 2-cyano-*N*-[(ethylamino)carbonyl]-2-(methoxyimino)acetamide (9CI); Reg. No. *[57966-95-7]*. Its fungicidal properties were described by J. M. Serres & G. A. Carraro (*Meded. Fac. Landbouwwet. Rijksuniv. Gent*, 1976. **41**, 645). Introduced by E.I. du Pont de Nemours & Co. Inc. (USP 3 957 847) as code no. 'DPX-3217'; trade mark 'Curzate'.

**Properties.**

Pure cymoxanil is a colourless crystalline solid; m.p. 160-161 °C; $d^{25}$ 1.31; v.p. 80 µPa (extrapolated) at 25 °C. Solubility at 25 °C: 1 g/kg water; 105 g/kg acetone; 2 g/kg benzene; 103 g/kg chloroform; 185 g/kg dimethylformamide; <1 g/kg hexane. It is stable at 2<pH<7.3. Loss from soil is 50% in <7 d. The technical grade is 97-99% pure.

**Uses.**

It is a fungicide with local systemic action and can control certain diseases during the incubation period. It is primarily effective against the Peronosporales: *Peronospora, Phytophthora* and *Plasmopara* spp. It is used in combination with preventative fungicides to improve residual activity. Typical rates in mixtures for control of *Plasmopara viticola* and *Phytophthora infestans* are 100-120 g a.i./ha. For example *Plasmopara viticola* is usually controlled by treatment 3-6 d after infection and *Phytophthora infestans* 1-3 d after inoculation.

**Toxicology.**

Acute oral $LD_{50}$: for male rats 1196 mg tech./kg, for females 1390 mg/kg; for guinea-pigs 1096 mg/kg. Acute percutaneous $LD_{50}$ for rabbits >3000 mg/kg; not a skin irritant or sensitiser; very slight irritant to eyes. In 2-y feeding trials NEL for rats was 100 mg/kg diet; body weight effects were observed at the highest dose (2500 mg/kg diet). $LC_{50}$ (8-d) was: for quail 2847 mg/kg diet; for mallard duck >10 000 mg/kg diet. $LC_{50}$ (96-h) is: for rainbow trout 18.7 mg/l; for bluegill 13.5 mg/kg.

**Formulations.**

Mixtures include: cymoxanil + Bordeaux mixture + folpet + zineb; cymoxanil + Bordeaux mixture + zineb; cymoxanil + captafol + copper(II) oxychloride; cymoxanil + captafol + folpet; cymoxanil + copper(II) oxychloride + copper(II) sulphate + mancozeb; cymoxanil + copper(II) sulphate + folpet; cymoxanil + copper(II) sulphate + mancozeb; cymoxanil + folpet; cymoxanil + folpet + mancozeb; cymoxanil + folpet + zineb; cymoxanil + mancozeb ('Fytospore', '*Curzate M*'); cymoxanil + propineb; cymoxanil + zinc ammoniate ethylenebis(dithiocarbamate) poly(ethylenebisthiuram disulphide).

**Analysis.**

Product analysis is by hplc with u.v. detection. Residues may be determined by glc. Details are available from E.I. du Pont de Nemours & Co. Inc.

$C_{10}H_7N_3O$ (185.2)
NCYR&UNO1CN &&(Z) FORM

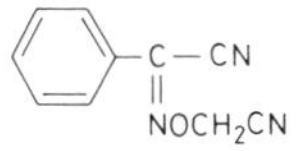

**Nomenclature and development.**

Common name cyometrinil (BSI, draft E-ISO). Chemical name (IUPAC) (*Z*)-cyanomethoxyimino(phenyl)acetonitrile. (*C.A.*) (*Z*)-α-[cyanomethoxyimino]benzeneacetonitrile (9CI); Reg. No. *[63278-33-1]* (unstated stereochemistry). Introduced by Ciba-Geigy AG (BEP 845 827; GBP 1 524 596) as code no. 'CGA 43 089'; trade mark 'Concep'.

**Properties.**

Pure cyometrinil forms colourless crystals; m.p. 55-56 °C; ρ 1.260 g/cm$^3$; v.p. 46.5 μPa at 20 °C. Solubility at 20 °C: 95 mg/l water; 550 g/kg benzene; 700 g/kg dichloromethane; 230 g/kg methanol; 74 g/kg propan-2-ol. At 20 °C, 50% hydrolysis (calculated) occurs in >200 d at pH 5-7, 19.4 d at pH9. Exothermic decomposition >300 °C.

**Uses.**

Increases the tolerance of some crop plants to certain chloroacetanilide herbicides.

**Toxicology.**

Acute oral $LD_{50}$ for rats 2277 mg. tech./kg. Acute percutaneous $LD_{50}$ for rats >3100 mg/kg; non-irritant to skin and eyes of rabbits. In 90-d feeding trials NEL for dogs was >100 mg/kg diet (3.1 mg/kg daily). $LC_{50}$ (96-h) is: for rainbow trout 5.6 mg/l; for carp 11.7 mg/l; for bluegill 10.9 mg/l. In laboratory trials it is slightly toxic to birds.

**Analysis.**

Product analysis is by glc. Residues are determined by glc using TID. Particulars are available from Ciba-Geigy AG.

$C_{22}H_{19}Cl_2NO_3$ (416.3)
L3TJ A1 A1 BVOYCN&R COR&&
C1UYGG

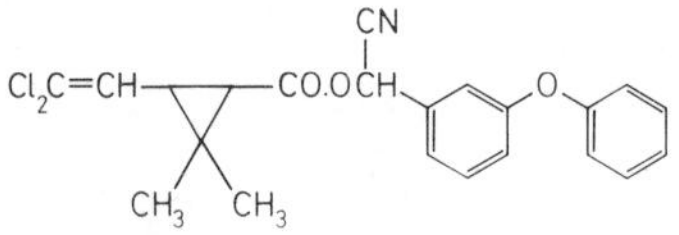

**Nomenclature and development.**

Common name cypermethrin (BSI, E-ISO, ANSI, BPC);cyperméthrine (F-ISO). Chemical name (IUPAC) (*RS*)-α-cyano-3-phenoxybenyzyl (1*RS*,3*RS*; 1*RS*,3*SR*)-3-(2,2-dichlorovinyl)-2,2-dimethylcyclopropanecarboxylate; (*RS*)-α-cyano-3-phenoxybenzyl (1*RS*)-*cis-trans*-3-(2,2-dichlorovinyl)-1,1-dimethylcyclopropanecarboxylate. The ratio of *cis*/*trans* isomers, which may vary with the manufacturing process, should be stated. (*C.A.*) (*RS*)-cyano(3-phenoxyphenyl)methyl (1*RS*)-*cis-trans*-3-(2,2-dichloroethenyl)-2,2-dimethylcyclopropanecarboxylate (9CI); Reg. No. *[52315-07-8]* formerly *[69865-47-0]*. OMS 2002. It was discovered by M. Elliott *et al.* (*Pestic. Sci.*, 1975, **6**, 537) and its performance in field trials reported by M. H. Breese (*ibid.*, 1977, **8**, 264) and by M. H. Breese & D.P. Highwood (*Proc. Br. Crop Prot. Conf. - Pests Dis.*, 1977, **2**, 641). It has been developed by Ciba-Geigy AG, ICI Plant Protection Division, Shell International Chemical Co. Ltd. (GBP 1 413 491) as code no. 'NRDC 149','PP 383' (ICI),'WL 43 467' (Shell); trade marks 'Polytrin' (to Ciba-Geigy AG), 'Ambush C', 'Cymbush', 'Imperator', 'Kafil Super', 'CCN52' (all to ICI), 'Ripcord', 'Stockade' (for agronomic use), 'Barricade' (for veterinary use), 'Folcord' (for public health use) (all to Shell).

**Properties.**

Technical grade cypermethrin is a viscous yellowish-brown semi-solid mass, which is liquid at 60 °C. The pure compound has v.p. 190 nPa at 20 °C (extrapolated) (B. T. Grayson *et al., Pestic. Sci.*, 1982, **13**, 552). Solubility at 21 °C: 0.01-0.2 mg tech./l water; at 20 °C: >450 g/l acetone, chloroform, cyclohexane, ethanol, xylene; 103 g/l hexane. It is stable <220 °C; photochemical decomposition has been observed in laboratory tests, but field data indicate this does not occur in practice sufficiently to affect biological performance adversely. It is more stable in acid than alkaline media, with optimum stability at pH4. Its degradation in soil has been reported (T. R. Roberts & M. E. Standen, *ibid.*, 1977, **8**, 305).

**Uses.**

Cypermethrin is a stomach and contact insecticide effective against a wide range of insect pests particularly Lepidoptera, in cereals, citrus, cotton, forestry, fruit, rape, soyabeans, tobacco, tomatoes, vegetables, vines and other crops at 20-75 g a.i./ha. Control of coffee leaf miner and leaf-mining Lepidoptera on top fruit is achieved at 10-20 g/ha. If applied before infestations become well established, it will also give protection against plant sucking Hemiptera in most crops. Soil surface sprays (≤75 g/ha) give good control of *Euxoa* spp. It has good residual activity on treated plants and no case of phytotoxicity has been reported even on sensitive ornamentals.

It also controls ectoparasites on cattle (*Boophilus microplus* at 150 mg/l bath, including strains resistant to organophosphorus compounds). Sheep scab (*Psoroptes* spp.), lice and ked are controlled by a single treatment at 10 mg/l. Good knockdown and residual control of biting flies in and around animal housing have been obtained following direct spray application to animals or structural surfaces. Control of mosquitoes and other nuisance or disease-carrying insects in dwelling houses is excellent at 50-75 mg/$m^2$ with good persistence (42-72 d) on most surfaces.

$C_{22}H_{19}Cl_2NO_3$ (416.3)
L3TJ A1 A1 BVOYCN&R COR&&
C1UYGG

**Toxicology.**

Oral $LD_{50}$ values for cypermethrin depend on such factors as: carrier, *cis/trans* ratio of the sample, species, sex, age and degree of fasting. Values reported sometimes differ markedly. Typical values for acute oral $LD_{50}$ are: for rats 251-4123 mg/kg; for mice 138 mg/kg; for chickens >2000 mg/kg. Acute percutaneous $LD_{50}$ for rabbits >2400 mg/kg; it is a slight skin irritant, a mild eye irritant and may have a weak skin sensitising potential. In 2-y feeding trials no compound-related toxicological effects were observed in rats receiving 100 mg/kg diet and dogs 300 mg/kg diet. $LD_{50}$ (96-h) for brown trout under laboratory conditions is 2.0-2.8 μg/l; however, under field conditions, fish are not at risk from normal agricultural usage. It is highly toxic to honeybees in laboratory tests, but field applications at recommended dosages to not put hives at risk.

Cypermethrin was last reviewed by the FAO/WHO JMPR in 1981 where ⩽0.05 mg/kg was proposed as an ADI for man.

**Formulations.**

These include: e.c. (25-400 g a.i./l) for agronomic use; ULV (10-50 g/l); 'Barricade' and 'Stockade', e.c. (25-200 g/l) for veterinary use and 'Folcord' for public health use. Most commercial formulations contain a *cis/trans* isomer ratio of 45:55 ± 10% *m/m*. 'Polytrin C', e.c. (40 g cypermethrin + 400 g profenofos/l). Some e.c. formulations ('Ripcord') contain additional insecticides.

**Analysis.**

Product analysis is by hplc or by glc with FID. Residues may be determined by glc with ECD. Details of methods are available from Shell International Chemical Co. Ltd. or ICI Plant Protection Division.

$C_{14}H_{14}ClNO_3$ (279.7)
T5OVTJ CNR CG&V- AL3TJ &&(±)
FORM

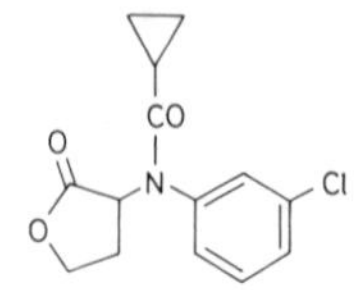

**Nomenclature and development.**

Common name cyprofuram (BSI, draft E-ISO). Chemical name (IUPAC) (±)-α-[*N*-(3-chlorophenyl)cyclopropanecarboxamido]-γ-butyrolactone. (*C.A.*) *N*-(3-chlorophenyl)-*N*-(tetrahydro-2-oxo-3-furanyl)cyclopropanecarboxamide (9CI); Reg. No. *[69581-33-5]* (no stereochemistry specified). Its fungicidal properties were described by D. Baumert & H. Buschhaus (*Meded. Fac. Landbouwwet. Rijksuniv. Gent*, 1982, **47**, 979). Introduced by Schering AG (BP 1 603 730) as code no. 'SN 78 314'; trade mark 'Vinicur'.

**Properties.**

It is a colourless crystalline solid; m.p. 95-96 °C; v.p. 6.6 μPa at 25 °C. Solubility at room temperature: 574 mg/l water; 500 g/kg acetone, dichloromethane; 330 g/kg cyclohexanone; 50 g/kg ethanol; 17 g/kg octan-1-ol; 60 g/kg xylene. Stable in acid media, but hydrolysed under alkaline conditions; at 25 °C 50% loss occurs in 37 d at pH7, 12 h at pH9.

**Uses.**

Cyprofuram is a systemic fungicide active against Oomycetes. It is recommended against *Plasmopara viticola* and *Phytophthora infestans* but only in combination with contact fungicides.

**Toxicology.**

Acute oral $LD_{50}$: for rats 174 mg/kg; for mice 296-314 mg/kg; for ducks, partridge and quail >2000 mg/kg. Acute percutaneous $LD_{50}$ for rabbits >1000 mg/kg. $LC_{50}$ (96-h) is: for trout 25 mg/l; for carp 33 mg/l.

**Formulations.**

'Vinicur F 50-SC', s.c. (71 g cyprofuram + 460 g folpet/l). 'Vinicur M-SC', s.c. (46 g cyprofuram + 325 g mancozeb/l).

**Analysis.**

Details of analytical methods are available from Schering AG.

$C_8H_6Cl_2O_3$ (221.0)
QV1OR BG DG
*2,4-D-(2-butoxyethyl)*
$C_{14}H_{18}Cl_2O_4$ (311.2)
GR CG DO1VO2O4

Cl—C6H3(Cl)—OCH2CO.OH

**Nomenclature and development.**

Common name 2,4-D (BSI, E-ISO, WSSA); 2,4-PA (JMAF). Chemical name (IUPAC) (2,4-dichlorophenoxy)acetic acid (I). (*C.A.*) (I) (8 & 9CI); Reg. No. *[94-75-7]*. The potent effects of its salts on plant growth were first described by P. W. Zimmerman & A. E. Hitchcock (*Contrib. Boyce Thompson Inst.*, 1942, **12**, 321) and its history is covered in *The Hormone Weedkillers*. Various esters, salts and mixtures with other herbicides have been marketed by many companies over the past 40 years; those mentioned below are currently the more important commercially.

**Properties.**

2,4-D is a colourless powder; m.p. 140.5 °C; v.p. 53 Pa at 160 °C. Solubility at 25 °C: 620 mg/l water; soluble in aqueous alkali, alcohols, diethyl ether; insoluble in petroleum oils. It is corrosive.

It is a strong acid, p$K_a$ 2.64, and forms water-soluble salts with alkali metals or amines. A sequestering agent is included in commercial formulations to prevent precipitation of the calcium or magnesium salts by hard water. The more important salts and their *C.A.* Registry Nos. are: sodium *[2702-72-9]* (solubility at room temperature 45 g/l water); bis(2-hydroxyethyl)ammonium *[5742-19-8]*; diethylammonium *[20940-37-8]*; dimethylammonium *[2008-39-1]*; (2-hydroxyethyl)ammonium *[3599-58-4]*; methylammonium *[51173-63-8]*; tris(2-hydroxyethyl)ammonium *[2569-10-9]* (solubility at 30-32 °C 4.4 kg/kg water).

Esters frequently encountered are: 2,4-D-isopropyl, Reg. No. *[94-11-1]*, colourless liquid, b.p. 130 °C/1 mmHg, crystallising in 2 forms, m.p. 5-10 °C and 20-25 °C, v.p. 1.4 Pa at 25 °C, practically insoluble in water soluble in alcohols, most oils; 2,4-D-(2-butoxyethyl), Reg. No. *[1929-73-3]*; 2,4-D-butyl, Reg. No. *[94-80-4]*; 2,4-D-iso-octyl, Reg. No. *[25168-26-7]* (formerly *[1280-20-2]*).

**Uses.**

2,4-D, its salts and esters are systemic herbicides, widely used for weed control in cereals and other crops at 0.28-2.3 kg/ha, the highest rate persisting in soil *c.* 30 d.

**Toxicology.**

Acute oral $LD_{50}$ for rats: 375 mg 2,4-D/kg, 666-805 mg 2,4-D-sodium/kg, 700 mg 2,4-D-isopropyl/kg. The toxicity and hazards to man, domestic animals and wildlife have been reviewed (J. M. Way, *Residue Rev.*, 1969, **26**, 37).

**Formulations.**

Amine salts are usually marketed as solutions of declared a.e. content, methanol being included in the formulation to delay crystallisation on storage. Esters such as the 2-butoxyethyl and iso-octyl are formulated as e.c. Numerous mixtures with various other herbicides including dichlorprop, MCPA, mecoprop, 2,4,5-T are available.

**Analysis.**

Product analysis of 2,4-D, salts, esters and mixed combination products are by acid-base titration or by glc (*CIPAC Handbook*, 1970, **1**, 241; 1980, **1A**, 1194; *Herbicides 1977*, pp. 6-21). Residues may be determined by glc of derivatives (*Pestic Anal. Man.*, 1979, **I**, 201-D; *Anal. Methods Pestic. Plant Growth Regul.*, 1972, **6**, 630; T. H. Byast *et al., Tech. Rep. ARC Weed Res. Organ.*, No. 15 (2nd Ed.), p. 31).

$C_3H_4Cl_2O_2$ (143.0)
QVXGG1
*dalapon-sodium*
$C_3H_3Cl_2NaO_2$ (165.0)
QVXGG1 &&Na SALT

$CH_3CCl_2CO.OH$

**Nomenclature and development.**

Common name dalapon (BSI, ANSI, WSSA, Canada, Federal Republic of Germany, France);proprop (Republic of South Africa);DPA (JMAF). In most other countries the chemical name is used. Chemical name (IUPAC) 2,2-dichloropropionic acid (I). (*C.A.*) 2,2-dichloropropanoic acid (9CI); (I) (8CI); Reg. No. *[75-99-0]* (dalapon), *[127-20-8]* (dalapon-sodium). The sodium salt (dalapon-sodium) was introduced as a herbicide by Dow Chemical Co. (USP 2 642 354) as trade marks 'Dowpon', 'Radapon'.

**Properties.**

Dalapon is a colourless liquid; b.p. 185-190 °C. Dalapon-sodium is a hygroscopic powder, decomposing at 166.5 °C without melting. Solubility at 25 °C: 900 g/kg water; 1.4 g/kg acetone; 20 mg/kg benzene; 160 mg/kg diethyl ether; <185 g/kg ethanol; 179 g/kg methanol. Dalapon has p$K_a$ 1.74-1.84, depending on the ionic strength of the solution; the calcium and magnesium salts are very soluble in water. It is subject to hydrolysis, slight at 25 °C but comparatively rapid ≥50 °C; so aqueous solutions should not be kept for any length of time. Alkali causes dehydrochlorination above 120 °C. Solutions of dalapon-sodium are corrosive to iron.

**Uses.**

Dalapon-sodium is a selective herbicide, translocated from roots and foliage. it is used to control annual and perennial grasses at ≤37 kg a.e./ha on non-crop areas, and at >1.7 kg/ha on many crops. It is readily broken down by soil micro-organisms (B. E. Day *et al.*, *Soil Sci.*, 1963, **95**, 326).

**Toxicology.**

Acute oral $LD_{50}$ for rats 7570-9300 mg dalapon-sodium/kg. The solid and concentrated solutions produced a transient eye irritation; although some skin irritation may occur there is no evidence to suggest percutaneous toxicity. In 2-y feeding trials no effect was observed in rats receiving 15 mg/kg daily, there was slight increase in kidney weight at 50 mg/kg daily. $LC_{50}$ (5-d) for pheasants, Japanese quail,mallard duck was >500 mg dalapon-sodium/kg diet. $LC_{50}$ (96-h) for rainbow trout, channel catfish and goldfish >100 mg/l. The toxic effects of dalapon have been reviewed (E. E. Kenaga, *Residue Rev.*, 1974, **53**, 104).

**Formulations.**

These include: 'Dowpon', s.p. (850 g dalapon-sodium/kg; 740 g a.e./kg) and other formulations of dalapon-sodium (usually of this strength). Mixtures include: dalapon-sodium + aminotriazole; dalapon-sodium + aminotriazole + simazine; dalapon-sodium + bromacil; dalapon-sodium + diuron; dalapon-sodium + diuron + MCPA; dalapon-sodium + MCPA.

**Analysis.**

Technical dalapon and dalapon-sodium are analysed by decomposition of a complex formed with mercury(II) nitrate and copper(II) nitrate (*CIPAC Handbook*, 1970, **1**, 274; 1980, **1A**, 1197); glc of a suitable ester may also be used (*Herbicides 1977*, p. 45). Residues are determined by the glc of a suitable ester (T. H. Byast *et al.*, *Tech. Rep. ARC Weed Res. Organ.*, No. 15 (2nd Ed.), p. 29) details are also available from Dow Chemical Co. See also: G. N. Smith & E. H. Yonkers, *Anal. Methods Pestic., Plant Growth Regul. Food Addit.*, 1964, **4**, 79; *Anal. Methods Pestic. Plant Growth Regul.*, 1972, **6**, 621.

It is the aim of the editors of *The Pesticide Manual* to present accurate and up-to-date information and its contents have been stored in a computer. This will simplify modification of the text for future editions. The contents and periodic updates of the *Manual* will be available on-line as the Pesticide Databank from 1984.

The easiest way to find a given compound is by its ***Entry Number***. Because the majority of pesticides are known by various synonyms it is recommended that the reader ascertains the required ***Entry Number*** from one or more of the Indexes.

**Index 1** Wiswesser Line-Formula Notation (for description of this see pp. 613).

**Index 2** Molecular formulae.

**Index 3** Code numbers used by official bodies, the manufacturers and suppliers. This index is arranged numerically and whether the qualifying letters are, for example, BAY or 'Bayer' is immaterial.

**Index 4** Chemical, common, and trivial names and trade marks.

Each of these give you the ***Entry Number***. A simple number (e.g. 1230) indicates that the entry is in the ***Main Section***, while a preceding asterisk (*2400) shows it is in the ***Superseded Compounds Section*** printed on coloured paper so that it is easy to find.

**Read the Section on How to Use Part I of the Manual,** page xvi.

Remember: an equals sign ( = ) at the end of a line means run on to the next line (omitting = ); chemical names ending in -, ( or [ run straight on. Otherwise the word finishes on a given line.

$C_6H_{12}N_2O_3$ (160.2)
QV2VMN1&1

$(CH_3)_2NNHCO.CH_2CH_2CO.OH$

**Nomenclature and development.**

Common name daminozide (BSI, E-ISO, F-ISO, ANSI). Chemical name (IUPAC) *N*-dimethylaminosuccinamic acid. (*C.A.*) butanedioic acid mono(2,2-dimethylhydrazide) (9CI); succinic acid mono(2,2-dimethylhydrazide) (8CI); Reg. No. *[1596-84-5]*. Trivial name SADH. Its effects on plant growth were described by J. A. Riddell *et al.* (*Science*, 1962, **136**, 391). Introduced by Uniroyal Inc. (USP 3 240 799; 3 334 991) as code no. 'B-995'; trade marks 'Alar', 'B-Nine'.

**Properties.**

The technical grade (>99% pure) forms colourless crystals; m.p. 154-156 °C; v.p. <133 Pa at 20 °C. Solubility at 25 °C: 100 g/kg water; 25 g/kg acetone; 50 g/kg methanol.

**Uses.**

It is a plant growth regulator used on certain fruit to improve the balance between vegetative growth and fruit production and to improve fruit quality and synchronise maturity. It is also used to modify the stem length of ornamental plants and reduce vegetative growth of groundnuts.

**Toxicology.**

Acute oral $LD_{50}$ for rats 8400 mg/kg. Acute percutaneous $LD_{50}$ for rabbits >16 000 mg/kg. Acute inhalation $LC_{50}$ for rats >147 mg/l air. In 2-y feeding trials no ill-effect was observed on rats and dogs at 3000 mg tech./kg diet (highest dose tested). $LC_{50}$ (8-d) for mallard duck and bobwhite quail >10 000 mg/kg diet. Lactation and 3-generation study in rats at 300 mg/kg diet showed no significant effect on fertility or reproductive capacity; it was not teratogenic at 500 mg/kg daily (highest dose tested). It shows no mutagenic potential in in-vivo and in-vitro tests. $LC_{50}$ (96-h) is: for bluegill 423 mg/l; for rainbow trout 149 mg/l. $LC_{50}$ (48-h) for water flea is 98.5 mg/l.

**Formulations.**

These include: 'Alar 85', s.p. (850 g a.i./kg); 'B-Nine S.P.', s.p. (850 g/kg) for use on ornamentals; 'Kylar', s.p. (850 g/kg) for use on groundnuts.

**Analysis.**

Product analysis is by bromide/bromate titration. Method available from Uniroyal Chemical Inc. Residues may be determined by glc of a silylated derivative (J. R. Lane, *Anal. Methods Pestic., Plant Growth Regul. Food Addit.*, 1967, **5**, 499; *Pestic. Anal. Man.*, 1979, **II**) or by alkaline hydrolsis to liberate 1,1-dimethylhydrazine which is determined colorimetrically (V. P. Lynch, *J. Sci. Food Agric.*, 1969, **20**, 13; J. W. Dicks, *Pestic. Sci.*, 1971, **2**, 176. *Anal. Methods Pestic. Plant Growth Regul.*, 1972, **6**, 697; V. P. Lynch, *ibid.*, 1976, **8**, 491).

$C_5H_{10}N_2S_2$ (162.3)
T6NYS ENTJ A1 BUS E1

**Nomenclature and development.**

Common name dazomet (BSI, E-ISO, F-ISO, WSSA, JMAF); tiazon (USSR). Chemical name (IUPAC) 3,5-dimethyl-1,3,5-thiadiazinane-2-thione; tetrahydro-3,5-dimethyl-1,3,5-thiadiazine-2-thione. (*C.A.*) tetrahydro-3,5-dimethyl-2*H*-1,3,5-thiadiazine-2-thione (8 & 9CI); Reg. No. *[533-74-4]*. *DMTT* (former WSSA name). Originally prepared by M. Delépine (*Bull. Soc. Chim. Fr.*, 1897, **15**, 891), it was introduced as a soil fumigant as code no. 'N-521' (Stauffer Chemical Co.),'Crag fungicide 974' (Union Carbide Corp.); trade marks 'Salvo' (Stauffer Chemical Co.), 'Mylone' (Union Carbide Corp.); 'Basamid' (BASF AG).

**Properties.**

Pure dazomet forms colourless crystals; m.p. 104-105 °C (decomp.); v.p. 370 μPa at 20 °C. Solubility at 20 °C: 3 g/kg water; 173 g/kg acetone; 51 g/kg benzene; 391 g/kg chloroform; 400 g/kg cyclohexane; 15 g/kg ethanol; 6 g/kg diethyl ether. The technical grade is ⩾98% pure. Dazomet is moderately stable but is sensitive to heat above 35 °C, and to moisture. Acid hydrolysis yields carbon disulphide but, in soil, dazomet breaks down to methyl(methylaminomethyl)dithiocarbamic acid which in turn yields methyl isothiocyanate. Its behaviour in soil is summarised by N. Drescher & S. Otto (*Residue Rev.*, 1968, **23**, 49).

**Uses.**

Dazomet, by virtue of the release of methyl isothiocyanate (8390), is a soil fumigant effective for the control of nematodes, germinating weeds and soil fungi, such as: *Fusarium, Pythium, Rhizoctonia*, and *Verticillium* spp. and *Colletotrichum atramentarium*, when incorporated into glasshouse soils at 400-600 kg a.i./ha. At these rates it is also effective against millipedes, soil insects and wireworms. It is strongly phytotoxic and treated soils should not be planted until shown to be free of the compound and its decomposition products, generally within 8-24 d (depending on soil temperature) by the normal germination of cress seed sown on a sample of the treated soil (*Pest and Disease Control Handbook*, 1979, p. 7.4).The granular formulation is also recommended for outdoor use at 200-400 kg/ha depending on soil type; soil temperature should be at least 7 °C at a depth of 15 cm, the granules incorporated evenly to a depth of 20-22 cm and the surface of the treated soil sealed with polythene, by flooding with water or by compaction with a heavy roller to compact the surface. Cultivation is needed to aid release of the methyl isothiocyanate and the cress test should be applied as above.

**Toxicology.**

Acute oral $LD_{50}$: for rats *c.* 640 mg/kg; for mice *c.* 280 mg/kg. The dust is an irritant to the skin and eyes of rabbits.

**Formulations.**

Dust (850 g a.i./kg); 'Basamid', granular (980-1000 g/kg).

**Analysis.**

Product analysis is by acid hydrolysis, absorption of the carbon disulphide produced and iodometric titration (H. A. Stansbury, *Anal. Methods Pestic., Plant Growth Regul. Food Addit.*, 1964, **3**, 119).

$C_{10}H_{10}Cl_2O_3$ (249.1)
QV3OR BG DG

$Cl-C_6H_3(Cl)-O(CH_2)_3CO.OH$

**Nomenclature and development.**

Common name 2,4-DB (BSI, Canada, France, New Zealand). Chemical name (IUPAC) 4-(2,4-dichlorophenoxy)butyric acid (I). (*C.A.*) 4-(2,4-dichlorophenoxy)butanoic acid (9CI); (I) (8CI); Reg. No.*[94-82-6]* (acid), *[10433-59-7]* (sodium salt), *[19480-40-1]* (potassium salt), *[2758-42-1]* (methylammonium salt). Its plant growth regulating properties were described by M. E. Synerholm & P. W. Zimmerman (*Contrib. Boyce Thompson Inst.*, 1947, **14**, 369). Introduced as a herbicide by May & Baker Ltd (CAP 570 065) as code no. 'M&B 2878'; trade mark 'Embutox'.

**Properties.**

Pure 2,4-DB forms colourless crystals; m.p. 119-119.5 °C. Solubility at 25 °C: 46 mg/l water; soluble in acetone, benzene, ethanol, diethyl ether. It has p$K_a$ 4.8 and forms water-soluble alkali metal and amine salts but hard water will precipitate its calcium and magnesium salts. The acid, salts and esters are stable. Technical 2,4-DB has m.p. 115-119 °C.

**Uses.**

2,4-DB is a translocatable herbicide effective against broad-leaved weeds but is more selective than 2,4-D because its activity is dependent on beta-oxidation to the latter within the plant. It may be used on lucerne, undersown cereals and grassland at 1.5-3.0 kg a.i./ha alone or in mixture with MCPA.

**Toxicology.**

Acute oral $LD_{50}$ for rats 700 mg 2,4-DB/kg, 1500 mg 2,4-DB-sodium/kg; for mice *c.* 400 mg 2,4-DB-sodium/kg.

**Formulations.**

These include: 'Embutox', solution of mixed potassium and sodium salts of 2,4-DB (300 g a.e./l); 'Embutox E', e.c. (400 g 2,4-DB-(iso-octyl)/l). Mixtures include: 'Embutox Plus', solution of potassium and sodium salts (2,4-DB + MCPA); 'Butoxone', amine salts (2,4-DB + 2,4-D + MCPA); 'Legumex Extra', solution of potassium and sodium salts (2,4-DB + benazolin + MCPA).

**Analysis.**

Product and residue analysis is by glc after conversion to suitable esters (W. H. Gutenmann, *Anal. Methods Pestic., Plant Growth Regul. Food Addit.*, 1967, **5**, 369) *Anal. Methods Pestic. Plant Growth Regul.*, 1972, **6**, 630, 636; (T. H. Byast *et al., Tech. Rep. ARC Weed Res. Organ.*, No. 15 (2nd Ed.), p. 31).

$C_{14}H_9Cl_5$ (354.5)
GXGGYR DG&R DG

Cl–C6H4–CH(CCl3)–C6H4–Cl

**Nomenclature and development.**

Common name DDT (BSI, draft E-ISO, ESA, JMAF) dicophane (BPC); chlorophenothane (US Pharmacopoeia) for a mixture of isomers; the major component is known as *pp*'-DDT (BSI);para,para'-DDT (Canada). Chemical name (IUPAC) of major component 1,1,1-trichloro-2,2-bis(4-chlorophenyl)ethane; 1,1,1-trichloro-di-(4-chlorophenyl)ethane. (*C.A.*) 1,1'-(2,2,2-trichloroethylidene)bis[4-chlorobenzene] (9CI); 1,1,1-trichloro-2,2-bis(*p*-chloroethane) (8CI); Reg. No. *[50-29-3]*. Trivial name dichlorodiphenyltrichloroethane. OMS 16, ENT 1506. Its insecticidal properties were discovered by P. Müller (P. Langer *et al., Helv. Chim. Acta*, 1944, **27**, 892). Introduced by J. R. Geigy S.A. (now Ciba-Geigy AG) who no longer manufacture or market it. (CHP 226 180; GBP 547 871) trade marks '*Gesarol*', '*Guesarol*', '*Neocid*'.

**Properties.**

Produced by condensation of chlorobenzene with trichloroacetaldehyde (a reaction first investigated in 1874) the main component is *pp*'-DDT but the technical product, a waxy solid of indefinite m.p., contains ⩽ 30% *op*'-DDT [1,1,1-trichloro-2-(2-chlorophenyl)-2-(4-chlorophenyl)ethane] which, being of insecticidal value, is not usually removed. *pp*'-DDT forms colourless crystals; m.p. 108.5 °C; v.p. 25 µPa at 20 °C. Solubility: practically insoluble in water; moderately soluble in hydroxylic or polar organic solvents, petroleum oils; readily soluble in most aromatic or chlorinated solvents. Technical DDT has similar solubility characteristics to *pp*'-DDT.

*pp*'-DDT is dehydrochlorinated at temperatures above its m.p. to the non-insecticidal 1,1-dichloro-2,2-bis(4-chlorophenyl)ethylene (sometimes known as DDE), a reaction catalysed by iron(III) or aluminium chlorides, by u.v. light and, in solution, by alkali. It is generally stable to oxidation but should not be stored in iron containers. Dehydrochlorination of DDT may occur >50 °C. Transformation of residues during cooking and food processing has been discussed (T. E. Archer, *Residue Rev.*, 1976, **61**, 29).

**Uses.**

DDT is a potent non-systemic stomach and contact insecticide which is persistent on solid surfaces and readily partitions into animal fats where it may accumulate. It has little activity against phytophagous mites and is phytotoxic to cucurbits and some varieties of barley. Its metabolism by microbial systems has been reviewed (R. E. Johnson, *ibid.*, p. 1).

**Toxicology.**

Acute oral $LD_{50}$ for rats 113-118 mg/kg. Acute percutaneous $LD_{50}$ for female rats 2510 mg/kg. in 160-d feeding trials NEL in rats was 1 mg/kg diet. Though stored in body fat and excreted in milk, 17 humans who ate 35 mg/man daily (*c.* 0.5 mg/kg daily) for 1.75 y suffered no ill-effect. The bioconcentration of DDT in the environment (A. Bevenue, *ibid.*, p. 37) and effects on reproduction of higher animals (G. W. Ware, *ibid.*, 1975, **59**, 119) have been discussed.

*Continued overleaf*

$C_{14}H_9Cl_5$ (354.5)
GXGGYR DG&R DG

**Formulations.**

These include: e.c., w.p., and dusts.

**Analysis.**

Product analysis is by glc or i.r. spectrometry (*CIPAC Handbook*, 1983, **1B**, in press) which have largely replaced classical methods based on chlorine content and, for *pp*'-DDT, setting point, (*ibid.*, 1970, **1**, 280; FAO Specification (CP/37); *Organochlorine Insecticides 1973*).

Conversion to TDE [1,1-dichloro-2,2-bis(4-chlorophenyl)ethane] which is no longer available commercially, also occurs *in vivo* and in the environment and must be considered when DDT-derived residues are considered. Residues are normally determined by glc (*AOAC Methods*, 1980, 29.001-29.018; *Pestic. Anal. Man.*, 1979, **I**, 201-A, 201-G, 201-I; P. A. Greve & W. B. F. Grevenstuk, *Meded. Fac. Landbouwwet. Rijksuniv., Gent*, 1975, **40**, 1115; D. C. Holmes & N. F. Wood, *J. Chromatogr.*, 1972, **67**, 173; G. M. Telling *et al., ibid.*, 1977, **137**, 405; R. Mestres *et al., Trav. Soc. Pharm. Montpellier*, 1976, **36**, 43; 1979, **39**, 323; *Analyst (London)*, 1979, **104**, 425; M. K. Baldwin *et al., Pestic. Sci.*, 1977, **8**, 110). Various colorimetric methods are now only of historical interest (R. Miskus, *Anal. Methods Pestic., Plant Growth Regul. Food Addit.*, 1964, **2**, 97; *Anal. Methods Pestic. Plant Growth Regul.*, 1972, **6**, 340).

$C_{22}H_{19}Br_2NO_3$ (505.2)
L3TJ A1 A1 BVOYCN&R COR&&
C1UYEE &&(1R)-CIS-(S) FORM

**Nomenclature and development.**

Common name deltamethrin (BSI, draft E-ISO);deltamethrine (France). Chemical name (IUPAC) (*S*)-α-cyano-3-phenoxybenzyl (1*R*,3*R*)-3-(2,2-dibromovinyl)-2,2-dimethylcyclopropanecarboxylate; (*S*)-α-cyano-3-phenoxybenzyl (1*R*)-*cis*-3-(2,2-dibromovinyl)-2,2-dimethylcyclopropanecarboxylate. (*C.A.*) (1*R*-[1α(*S**),3*α*])-cyano(3-phenoxyphenyl)methyl 3-(2,2-dibromoethenyl)-2,2-dimethylcyclopanecarboxylate (9CI); Reg. No. *[52918-63-5]*. Trivial name *decamethrin* (rejected common name proposal). OMS 1998. This single isomer, first described by M. Elliott *et al.* (*Nature (London)*, 1974, **248**, 710), and reviewed in *Deltamethrin Monograph* was introduced by Roussel Uclaf (GBP 1 413 491 to NRDC) as code no. 'NRDC 161','RU 22 974' (Roussel Uclaf); trade mark 'Decis' for agronomic uses, 'K-Othrine' for domestic industrial public health and stored products uses, 'Butox', 'Butoflin' both for veterinary uses.

**Properties.**

Esterification of the parent acid with the racemic mandelonitrile yielded the (*RS*)-α-cyano-3-phenoxybenzyl (1*R*)-*cis*-isomeric pair, known by the code no. 'NRDC 156', (Reg. No. *[52820-00-5]*), (M. Elliott *et al.*, *Pestic. Sci.*, 1978, **9**, 105). A solution of this in hexane deposited crystals of one isomer, deltamethrin, on cooling. The technical material produced industrially by Roussel Uclaf contains ≥98% deltamethrin *m*/*m* and is a colourless crystalline powder; m.p. 98-101 °C; [α]D + 61° (40 g/l benzene); v.p 2μPa at 25 °C. Solubility at 20 °C: <2 μg/l water; 500 g/l acetone; 450 g/l benzene, dimethyl sulphoxide; 750 g/l cyclohexanone; 900 g/l dioxane; 15 g/l ethanol; 250 g/l xylene. It is stable on exposure to air and sunlight, and more stable in acid than in alkaline media.

**Uses.**

It is a potent insecticide, effective as a contact and stomach poison against a wide range of insects. It controls numerous insect pests of field crops at 11 g a.i./ha (J. Martel & R. Colas, *Beltwide Cotton Insect Conf.*, 1976; J. J. Hervé *et al.*, *Proc. 1977 Br. Crop Prot. Conf. - Pests Dis.*, 1977, **2**, 613). It has a very good residual activity also for outdoor uses (cattle dip, field crops, mosquito control) and for indoor uses (crawling and flying insects, stable flies and stored products insects).

**Toxicology.**

Acute oral $LD_{50}$: for rats ranges from 135- >5000 mg/kg depending upon carrier and conditions of the study; for dogs >300 mg/kg; for ducks >4640 mg/kg. Acute percutaneous $LD_{50}$ for rats >2000 mg/kg. Acute inhalation $LD_{50}$ for rats 600 mg/$m^3$. Acute oral $LD_{50}$ for formulations in rats was: 1080 mg (of 150 g/l e.c.)/kg; 535 mg (of 250 g/l e.c.)/kg; >5000 mg (of 5 g/l ULV)/kg; no $LD_{50}$ could be attained at rates of 16 000 mg (of 25 g/kg w.p.)/kg or 40 000 mg (of 25 g/l s.c.)/kg. In 2-y feeding trials NEL was: for mice 12 mg/kg diet; for rats 2.1 mg/kg diet. $LC_{50}$ (8-d) is: for ducks >4640 mg/kg diet; for quail >10 000 mg/kg diet. In laboratory trials $LC_{50}$ for fish is *c.* 1-10 μg/l; for honeybees $LD_{50}$ 0.05 μg/bee, however in normal conditions outdoors it is harmless to honeybees. Aquatic fauna, particularly crustacea, may be affected, but fishes are not harmed under normal conditions of use.

**Formulations.**

'Decis': e.c. (25 g a.i./l); ULV concentrate (3-10 g/l); s.c. (25 g/l); granules 0.5 g/kg. 'K-Othrin': concentrate for reformulation (50 or 300 g/l); e.c. (15 or 25 g/l); ULV (100 g/l); w.p. (25 g/l); s.c. (7.5 or 25 g/l); dusts (0.1-0.5 g/kg). 'Butox': e.c. (50 g/l). 'Decis B' (Procida), (25 g deltamethrin + 400 g heptenophos/l). 'Thripstick' (Aquaspersions Ltd). deltamethrin + polybutenes.

**Analysis.**

Product analysis is by hplc (C. Meinard *et al.*, *J. Chromatog.*, 1979, **176**, 140; D. Mourot *et al.*, *ibid.*, 1979, **173**, 412); i.r. spectroscopy of bromine determination may also be used (Roussel Uclaf - unpublished methods). Residues in plant tissues may be determined by glc with ECD (R. Mestres *et al.*, *Trav. Soc. Pharm. Montpellier*, 1978, **38**, 183; 1979, **39**, 329). Residues may be determined in soils using glc with ECD (B. D. Hill, *J. Agric. Food Chem.*, in press) and in milk and animal tissues (M. H. Akhtar, *J. Chromatog.*, in press).

$C_8H_{19}O_3PS_2$ (258.3)

*demeton-O* (i)
2S2OPS&O2&O2

*demeton-S* (ii)
2S2SPO&O2&O2

$CH_3CH_2SCH_2CH_2OP(S)(OCH_2CH_3)_2$

(i)

$CH_3CH_2SCH_2CH_2SP(O)(OCH_2CH_3)_2$

(ii)

**Nomenclature and development.**

Common name demeton (BSI, ESA) for a reaction product comprising demeton-O and demeton-S (BSI, E-ISO, F-ISO);mercaptofos and mercaptofos teolevy (USSR). Chemical names (IUPAC) (i) *O,O*-diethyl *O*-2-ethylthioethyl phosphorothioate and (ii) *O,O*-diethyl *S*-2-ethylthioethyl phosphororthioate. (*C.A.*) (i) *O,O*-diethyl *O*-[2-(ethylthio)ethyl] phosphorothioate and (ii) *O,O*-diethyl *S*-[2-(ethylthio)ethyl] phosphorothioate (8 & 9CI); Reg. No. *[298-03-3]* (i), *[126-75-0]* (ii), *[8065-48-3]* formerly *[8000-97-3]* (i + ii). Originally known as diethyl 2-ethylthioethyl phosphorothionate (i) and diethyl 2-ethylthioethyl phosphorothiolate (ii). ENT 17 295 (demeton). Its insecticidal properties were described by G. Unterstenhöfer (*Meded. Landbouwhogesch. Opzoekingsstn. Staat Gent*, 1952, **17**, 75). Introduced by Farbenfabriken Bayer AG (now Bayer AG) (DEP 836 349; USP 2 571 989) as code no. 'Bayer 10 756','E-1059'; trade mark 'Systox'.

**Properties.**

Reaction of 2-ethylthioethanol with *O,O*-diethyl phosphorochloridothioate gives demeton as *c.* 65:35 mixture of (i) and (ii). The technical product is a light yellow oil with a pronounced mercaptan odour. It is hydrolysed by concentrated alkali, but is compatible with most non-alkaline pesticides except water-soluble mercury compounds. The mechanism of isomerisation of (i) to (ii) was examined by T. R. Fukuto & R. L. Metcalf (*J. Am. Chem. Soc.*, 1954, **76**, 5103) and the self-alkylation of demeton described by D. F. Heath, (*Nature (London)*, 1957, **179**, 377). Pure demeton-O (i) is a colourless oil; b.p. 123 °C/1mmHg; v.p. 38 mPa at 20 °C; $d_4^{21}$ 1.119; $n_D^{18}$ 1.4900. Solubility at room temperature 60 mg/l water; soluble in most organic solvents. Demeton-S (ii) is a colourless oil; b.p. 128 °C/1mmHg; v.p. 35 mPa at 20 °C; $d_4^{21}$ 1.132; $n_D^{18}$ 1.5000. Solubility at room temperature 2 g/l water; soluble in most organic solvents.

**Uses.**

Demeton is a systemic insecticide and acaricide with some fumigant action effective against sap-feeding insects and mites. Demeton-S rapidly penetrates foliage. The thioether sulphur of both isomers is oxidised, metabolically, to the sulphoxide and sulphone (R. B. March *et al.*, *J. Econ. Entomol.*, 1955, **48**, 355). No marked phytotoxicity has been observed at field concentrations.

**Toxicology.**

Acute oral $LD_{50}$: for male rats 6-12 mg demeton/kg; for female rats 2.5-4.0 mg demeton/kg; for male rats 30 mg demeton-O/kg, 1.5 mg demeton-S/kg, 2.3 mg (ii-sulphoxide)/kg, 1.9 mg (ii-sulphone)/kg. Acute percutaneous $LD_{50}$ for male rats 14 mg demeton/kg. In 1.83-y feeding trials NEL for rats was 2 mg/kg diet (0.1 mg/kg daily) (*WHO Pestic. Residue Ser.*, 1974, **3**, 217). $LC_{50}$ (24-h) for rainbow trout 1-10 mg demeton/l; $LC_{50}$ (48-h) for carp 15.2 mg/l.

**Formulations.**

These include e.c. of various a.i. content.

**Analysis.**

Product analysis is by glc (J. G. F. Ernst (*J. Chromatog.*, 1977, **133**, 245) or by alkaline hydrolysis and titration of the excess of alkali (*CIPAC Handbook*, 1970, **1**, 302) or of the thiol with iodine (*ibid*, 1980, **1A**, 1198). Residues are determined by glc (*Pestic. Anal. Man.*, 1979, **I**, 201-H, 201-I); D. C. Abbott *et al.*, *Pestic. Sci.*, 1970, **1**, 10) or by oxidation to phosphoric acid which is measured by standard colorimetric procedures (D. MacDougall, *Anal. Methods Pestic., Plant Growth Regul. Food Addit.*, 1964, **2**, 451; *Anal. Methods Pestic. Plant Growth Regul.*, 1972, **6**, 483).

$C_6H_{15}O_3PS_2$ (230.3)
2S2SPO&O1&O1

$CH_3CH_2SCH_2CH_2SP(O)(OCH_3)_2$

**Nomenclature and development.**

Common name demeton-S-methyl (BSI, E-ISO, F-ISO);methyl demeton (JMAF);methyl-mercaptofos teolovy (USSR);*exception* USA. Chemical name (IUPAC) *S*-2-ethylthioethyl *O,O*-dimethyl phosphorothioate. (*C.A.*) *S*-[2-(ethylthio)ethyl] *O,O*-dimethyl phosphorothioate (8 & 9CI); Reg. No. *[919-86-8]*. Trivial name 2-ethylthioethyl dimethyl phosphorothiolate. In 1954 Farbenfabriken Bayer AG (now Bayer AG) introduced demeton-methyl (BSI) (Reg. No. *[8022-00-2]*) which contained demeton-S-methyl and demeton-O-methyl, (*O*-2-ethylthioethyl *O,O*-dimethyl phosphorothioate, Reg. No. *[867-27-6]*). An improved manufacturing process gave demeton-S-methyl, introduced by Bayer AG in 1957 (see G. Schrader *Die Entwicklung neuer insektizider Phosphorsäure-Ester*) (DEP 836 349; USP 2 571 989) as code no. 'Bayer 18 436','Bayer 25/154'; trade mark 'Metasystox i'.

**Properties.**

Technical demeton-S-methyl is a pale yellow oil; b.p. 89 °C/0.15 mmHg; v.p. 48 mPa at 20 °C; $n_D^{20}$ 1.5065; $d_4^{20}$ 1.207. Solubility at 20 °C: 3.3 g/kg water; 600 g/kg dichloromethane, propan-2-ol.

**Uses.**

Demeton-S-methyl is a systemic and contact insecticide used on most agricultural and horticultural crops at 7.5-22 g a.i./100 l HV or 245 g/ha LV. Sprays are phytotoxic to some ornamentals especially certain chrysanthemum cultivars. It is metabolised in plants to the sulphoxide (oxydemeton-methyl) (9240) and sulphone (3930) (T. R. Fukuto *et al.*, *J. Econ. Entomol.*, 1955, **48**, 348; 1956, **49**, 147; 1957, **50**, 399).

**Toxicology.**

Acute oral $LD_{50}$: for rats 57-106 mg/kg; for male guinea-pigs 110 mg/kg. Acute percutaneous $LD_{50}$ for male rats 302 mg/kg. $LC_{50}$ (48-h) for carp and Japanese killifish 10-40 mg/l.

**Formulations.**

These include e.c. of various a.i. content: 'Metasystox 55' (Bayer), 'Campbell's DSM', 'Demetox' (ICI Plant Protection Division), 'Duratox' (Shell).

**Analysis.**

Product analysis is based on alkaline hydrolysis determining the acid released (*CIPAC Handbook*, 1970, **1**, 312) or the thiol by titration with iodine (*ibid*). Residues may be determined by glc (J. H. van der Merwe & W. B. Taylor, *Pflanzenschutz-Nachr. (Engl. Ed.)*, 1971, **24**, 259; *Pestic. Anal. Man.*, 1979, **I**, 201-H, 201-I; D. C. Abbott *et al.*, *Pestic. Sci.*, 1970, **1**, 10) or by oxidation to phosphoric acid measured by standard colorimetric procedures (E. Q. Laws & D. J. Webley, *Analyst (London)*, 1959, **84**, 28; H. Tietz & H. Frehse, *Hoefchen-Briefe (Engl. Ed.)*, 1960, **13**, 212; D. MacDougall, *Anal. Methods Pestic., Plant Growth Regul. Food Addit.*, 1964, **2**, 295).

$C_6H_{15}O_5PS_2$ (262.3)
WS2&2SPO&O1&O1

$CH_3CH_2SO_2CH_2CH_2SP(O)(OCH_3)_2$

**Nomenclature and development.**

Common name demeton-S-methylsulphon (E-ISO); déméton-S-méthylsulfone (F-ISO); demeton-S-methyl sulphone (BSI); *exception* (USA). Chemical name (IUPAC) *S*-2-ethylsulphonylethyl *O,O*-dimethyl phosphorothioate. (*C.A.*) *S*-[2-(ethylsulfonyl)ethyl] *O,O*-dimethyl phosphorothioate (8 & 9CI); Reg. No. *[17040-19-6]*. Introduced by Bayer AG (DEP 948 241) as code no. 'Bayer 20 315', 'E 158', 'M3/158'; trade mark 'Metaisosystoxsulfon'.

**Properties.**

Pure demeton-S-methyl sulphone is a colourless to yellow crystalline solid; m.p. 60 °C; b.p. 120 °C/0.03 mm Hg; v.p. 660 µPa at 20 °C. Solubility at 20 °C: /l water; >600g/kg dichloromethane; <10g/kg toluene.

**Uses.**

Demeton-S-methyl sulphone is a systemic insecticide effective against sap-feeding insects, sawflies and mites. Its range of action coincides with that of demeton-S-methyl of which it is a metabolic product in plants and animals. It is normally used in combination with other insecticides, *e.g.* with azinphos-methyl to control a wide range of pests on pome fruit (25 g azinphos-methyl + 7.5 g demeton-S-methyl sulphone/100 l).

**Toxicology.**

Acute oral $LD_{50}$ for rats *c.* 37.5 mg a.i./kg. Acute percutaneous $LD_{50}$ for rats *c.* 500 g/kg. Acute intraperitoneal $LD_{50}$ for rats *c.* 20.8 mg/kg. $LC_{50}$ for golden orfe 102 mg/l.

**Formulations.**

'Gusathion MS' (Reg. No. *[8066-08-8]*), w.p. (75 g demeton-S-methyl sulphone + 250 g azinphos-methyl/kg).

**Analysis.**

Product analysis is by hydrolysis with alkali, the excess of which is determined by titration (*CIPAC Handbook*, 1970, **1**, 312). Residues are determined by glc (M. C. Bowman *et al.*, *J. Assoc. Off. Anal. Chem.*, 1969, **52**, 157; J. H. van der Merwe & W. B. Taylor, *Pflanzenschutz-Nachr. (Engl. Ed.)*, 1971, **24**, 259) or by oxidation to phosphoric acid which is measured by standard colorimetric methods. Details of methods are available from Bayer AG.

$C_8H_8Cl_2O_5S$ (303.1)
WSQO2OR BG DG
*2,4-DES-sodium*
$C_8H_7Cl_2NaO_5S$ (325.1)
WSQO2OR BG DG &&Na SALT

Cl–$C_6H_3$(Cl)–$OCH_2CH_2O.SO_2.OH$

**Nomenclature and development.**

Common name disul (E-ISO, F-ISO);2,4-DES (BSI, Australia, USSR);sesone (WSSA). Chemical name (IUPAC) 2-(2,4-dichlorophenoxy)ethyl hydrogen sulphate. (*C.A.*) 2-(2,4-dichlorophenoxy)ethyl hydrogen sulfate (8 & 9CI); Reg. No. *[149-26-8]*. SES. The herbicidal properties of sodium 2-(2,4-dichlorophenoxy)ethyl sulphate (2,4-DES-sodium, Reg. No. *[136-78-7]*), were reported by L. J. King *et al.* (*Contrib. Boyce Thompson Inst.*, 1950, **16**, 191). This was introduced by Union Carbide Corp. (who no longer manufacture or market it). (USP 2 573 769) as trade marks '*Crag Herbicide I*', 'Crag Sesone'; available in England and Wales as 'Herbon Blue' (Cropsafe Ltd) as a mixture with simazine.

**Properties.**

2,4-DES-sodium forms colourless crystals; m.p. 170 °C; v.p. negligible at room temperature. Solubility 250 g/kg water; insoluble in most organic solvents except methanol. The calcium salt is sufficiently soluble to avoid precipitation by hard water. 2,4-DES-sodium is hydrolysed by alkali to form 2-(2,4-dichlorophenoxy)ethanol and sodium sulphate.

**Uses.**

2,4-DES-sodium is not itself phytotoxic, but is converted in moist soil to 2-(2,4-dichlorophenoxy)ethanol which is oxidised to 2,4-D. It is used as a mixture with simazine to control annual weeds pre-em. in maize, soft fruits, roses, certain established perennial crops and certain ornamental trees and shrubs.

**Toxicology.**

Acute oral $LD_{50}$ for rats 730 mg/kg. in 2-y trials rats receiving 2000 mg/kg diet suffered no ill-effect. It is dangerous to fish.

**Formulations.**

'Herbon Blue', w.p. (2,4-DES-sodium + simazine).

**Analysis.**

Analysis is by colorimetry of a derivative (J. N. Hogsett & G. L. Funk, *Anal. Chem.*, 1954, **26**, 849).

$C_{16}H_{16}N_2O_4$ (300.3)
2OVMR COVMR

NHCO.OCH$_2$CH$_3$
NHCO.O

**Nomenclature and development.**

Common name desmedipham (BSI, E-ISO, ANSI, WSSA);desmédiphame (F-ISO). Chemical name (IUPAC) ethyl 3-phenylcarbamoyloxycarbanilate; ethyl 3-phenylcarbamoyloxyphenylcarbamate; 3-ethoxycarbonylaminophenyl phenylcarbamate. (*C.A.*) ethyl [3-[[(phenylamino)carbonyl]oxy]phenyl]carbamate (9CI); ethyl *m*-hydroxycarbanilate carbanilate (ester) (8CI); Reg. No. *[13684-56-5]*. Its herbicidal properties were described by F. Arndt & G. Boroschewski (*Symp. New Herbic., 3rd*, 1969, p.141). Introduced by Schering AG (GBP 1 127 050) as code no. 'Schering 38 107'; trade mark 'Betanal AM'.

**Properties.**

Pure desmedipham forms colourless crystals; m.p. 120 °C; v.p. 400 nPa at 25 °C. Solubility at room temperature: 9 mg/l water; 400 g/l acetone; 1.6 g/l benzene; 80 g/l chloroform; 180 g/l methanol. The technical grade is *c.* 96% pure. On hydrolysis 50% loss occurs in *c.* 70 d at pH 5, *c.* 20 h at pH 7, 10 min at pH 9, with the formation of ethyl 3-hydroxycarbanilate and aniline.

**Uses.**

It is a post-em. herbicide used to control broad-leaved weeds, including *Amaranthus retroflexus*, in beet crops, in particular sugar beet. It is frequently sprayed in combination with phenmedipham. Application rates are 800-1000 g a.i./ha in 200-300 l water broadcast, or one-third of this amount for band sprays. It acts through the leaves only, and does not depend on soil type and humidity under normal growing conditions. Due to its wide safety margin to the crop, spraying is merely timed according to the development stage of the weeds, with an optimum of weed control, when they have 2-4 true leaves.

**Toxicology.**

Acute oral $LD_{50}$: for rats 10 250 mg a.i./kg; for quail 390 mg a.i. (as e.c./kg). Acute percutaneous $LD_{50}$ for rabbits >318 mg a.i. (as e.c.)/kg. $LC_{50}$ (96-h) is: for rainbow trout 0.6 mg a.i. (as e.c.)/l; for bluegill 2.1 mg a.i. (as e.c.)/l.

**Formulations.**

'Betanal AM' ('Betanex' in Canada and USA), e.c. (157 g a.i./l). 'Betanal AM 11', e.c. (80 g desmedipham + 80 g phenmedipham/l).

**Analysis.**

Product analysis is by quantitative tlc, hplc or colorimetry of a derivative. Residues may be determined by hydrolysis to aniline using derivatives suitable for glc (C. -H. Roeder *et al., Anal. Methods Pestic. Plant Growth Regul.*, 1978, **10**, 293) or colorimetry.

$C_8H_{15}N_5S$
T6N CN ENJ BS1 DMY1&1 FM1

$CH_3S$ N $NHCH(CH_3)_2$
N N
$NHCH_3$

**Nomenclature and development.**

Common name desmetryne (BSI, F-ISO, JMAF);desmetryn (E-ISO, WSSA);*exception* (Portugal). Chemical name (IUPAC) *N*-isopropyl-*N'*-methyl-6-methylthio-1,3,5-triazine-2,4-diyldiamine; 2-isopropylamino-4-methylamino-6-methylthio-1,3,5-triazine. (*C.A.*) *N*-methyl-*N'*-(1-methylethyl)-6-(methylthio)-1,3,5-triazine-2,4-diamine (9CI); 2-( isopropylamino)-4-(methylamino)-6-(methylthio)-*s*-triazine (8CI); Reg. No. *[1014-69-3]*. Its herbicidal properties were described by J. G. Elliott & T. I. Cox (*Proc. Br. Weed Control Conf., 6th*, 1962, **2**, 759). Introduced by J. R. Geigy S.A. (now Ciba-Geigy AG) (CHP 337 019; GBP 814 948) as code no. 'G 34 360'; trade mark 'Semeron'.

**Properties.**

Pure desmetryne is a colourless powder; m.p. 84-86 °C; v.p. 133 μPa at 20 °C; ρ 1.172 $g/cm^3$ at 20 °C: Solubility at 20 °C 580 mg/l water; 230 g/kg acetone; 200 g/kg dichloromethane; 2.6 g/kg hexane; 300 g/kg methanol; 100 g/kg octan-1-ol; 200 g/kg toluene. No significant hydrolysis can be detected at 70 °C at $5<pH<13$.

**Uses.**

Desmetryne is a selective post-em. herbicide, translocated from leaves and roots and of brief persistence in soil. It is effective for the control of *Chenopodium album* and other broad-leaved and grassy weeds in brassica crops, except broccoli and cauliflower. The recommended rate is *c.* 500 g a.i./ha.

**Toxicology.**

Acute oral $LD_{50}$ for rats 1390 mg tech./kg. In 90-d feeding trials NEL was: for rats 200 mg/kg diet (13 mg/kg daily); for dogs 200 mg/kg (6.6 mg/kg daily). It has negligible toxicity to wild life.

**Formulations.**

'Semeron' 25WP, w.p. (250 g a.i./kg).

**Analysis.**

Product analysis is by glc (*CIPAC Handbook*, 1983, **1B**, in press). Residues may be determined by glc (K. Ramsteiner *et al.*, *J. Assoc. Off. Anal. Chem.*, 1974, **57**, 192).

# Dialifos



$C_{14}H_{17}ClNO_4PS_2$ (393.8)
T56 BVNVJ CY1GSPS&O2&O2

**Nomenclature and development.**

Common name dialifos (BSI, E-ISO);dialiphos (F-ISO);dialifor (ANSI, JMAF). Chemical name (IUPAC) *S*-2-chloro-1-phthalimidoethyl *O,O*-diethyl phosphorodithioate. (*C.A.*) *S*-[2-chloro-1-(1,3-dihydro-1,3-dioxo-2*H*-isoindol-2-yl)ethyl] *O,O*-diethyl phosphorodithioate (9CI); *O,O*-diethyl phosphorodithioate *S*-ester with *N*-(2-chloro-1-mercaptoethyl)phthalimide (8CI); Reg. No. *[10311-84-9]*. ENT 27 320. Its insecticidal properties were described by W. R. Cothran *et al.* (*J. Econ. Entomol.*, 1967, **60**, 1151). Introduced by Hercules Inc. (now Boots-Hercules Agrochemical Co.) (GBP 1 091 738; USP 3 355 353) as code no. 'Hercules 14 503'; trade mark 'Torak'.

**Properties.**

Pure dialifos forms colourless crystals; m.p. 67-69 °C; v.p. 133 mPa at 35 °C. Insoluble in water; slightly soluble in aliphatic hydrocarbons, alcohols; very soluble in acetone, cyclohexanone, 3,5,5-trimethylcyclohex-2-enone, xylene. The technical product is hydrolysed by alkali, at room temperature 50% loss at pH 8 in 2.5 h.

**Uses.**

It is a non-systemic insecticide and acaricide effective in controlling many insects and mites on apples, citrus, grapes, nut trees, oilseed rape, potatoes and vegetables.

**Toxicology.**

Acute oral $LD_{50}$: for female rats 5 mg/kg, for males 43-53 mg/kg; for male mice 39 mg/kg, for females 65 mg/kg; for male dogs 97 mg/kg; for mallard ducks 940 mg/kg. Acute percutaneous $LD_{50}$ for rabbits 145 mg/kg. $LC_{50}$ (24-h) for rainbow trout is 0.55-1.08 mg/l. It is of low toxicity to honeybees, $LD_{50}$ 34-38 μg/bee.

**Formulations.**

An e.c. (240 or 480 g tech./l).

**Analysis.**

Product analysis is by u.v. spectrometry. Residues may be determined by glc with TID (*Pestic. Anal. Man.*, 1979, **I**, 201-A, 201-G, 201-I). Details of methods are available from Boots-Hercules Agrochemical Co.

$C_{10}H_{17}Cl_2NOS$ (270.2)
G1UYG1SVNY1&1&Y1&1

$[(CH_3)_2CH]_2NCO.SCH_2(Cl)C=C(Cl)H$ (E)

$[(CH_3)_2CH]_2NCO.SCH_2(Cl)C=C(H)Cl$ (Z)

**Nomenclature and development.**

Common name di-allate (BSI, E-ISO);diallate (WSSA, F-ISO). Chemical name *S*-2,3-dichloroallyl di-isopropyl(thiocarbamate). (*C.A.*) *S*-(2,3-dichloro-2-propenyl) bis(1-methylethyl)carbamothioate (9CI); *S*-(2,3-dichloroallyl) diisopropylthiocarbamate (8CI); Reg. No. *[2303-16-4]*. Its herbicidal properties were described by L. H. Hannah (*Proc. North Cent. Weed Control Conf.*, 1959, p. 50). Introduced by Monsanto Co. (USP 3 330 643; 3 330 821) as code no. 'CP 15 336'; trade mark 'Avadex'.

**Properties.**

Pure di-allate is an amber-coloured liquid; b.p. 97 °C/0.15 mmHg; m.p. 25-30 °C; v.p. 20 mPa at 25 °C; $d^{25}_{15.6}$ 1.188. Solubility at 25 °C: 14 mg/l water; miscible with acetone, ethanol, ethyl acetate, kerosene, propan-2-ol, xylene. It is stable to u.v. light, decomposes >200 °C.

It exists as geometric isomers which have been separated (F. H. A. Rummens, *Weed Sci.*, 1975, **23**, 7). The (*Z*)-isomer has m.p. <0 °C, $n^{40}_{D}$ 1.5124; the (*E*)-isomer, m.p. 36 °C, $n^{40}_{D}$ 1.5097. The (*E*)- is 65% less effective than the (*Z*)-isomer in reducing height of wild oats by 50%; a mixture (*Z*-, 42%) + (*E*-, 58%) is 2.6 × less effective than the (*Z*)-isomer (F. H. A. Rummens *et al., ibid.*, p. 11).

**Uses.**

It is a pre- or post-planting incorporated herbicide which controls wild oats in barley, flax, forage legumes, lentils, maize, peas, potatoes, soyabeans and sugar beet. it is absorbed by the emerging wild oat coleoptile, toxicity being mainly at cell division. it competes with moisture for adsorption on soil. Loss is mainly by microbial breakdown (50% loss in 30 d) provided it is fully incorporated in the soil.

**Toxicology.**

Acute oral $LD_{50}$: for rats 395 mg/kg; for dogs 510 mg/kg. Acute percutaneous $LD_{50}$ for rabbits 2000-2500 mg/kg; a moderate irritant to their skin and eyes. In 90-d feeding trials: rats receiving 400 mg/kg diet suffered weight loss, irritability, hyperactivity and mild cardiac changes but no deaths occurred at 1200 mg/kg diet (the highest dose tested). In beagle dogs adverse effects were observed at 600, but not at 125 mg/day. $LC_{50}$ for 10- to 15-day-old mallard ducklings and bobwhite quail chicks was >5000 mg/kg diet. $LC_{50}$ (96-h) is: for rainbow trout 7.9 mg/l; for bluegill 5.9 mg/l.

**Formulations.**

'Avadex', granules (100 g/kg); e.c. 480 g/l.

**Analysis.**

Product analysis is by glc. Residues may be determined by glc. Details of methods are available from Monsanto Co.

$C_8H_{11}Cl_2NO$ (208.1)
GYGVN2U1&2U1

$Cl_2CHCO.N(CH_2CH{=}CH_2)_2$

**Nomenclature and development.**

Chemical name (IUPAC) *N*,*N*-diallyl-2,2-dichloroacetamide. (*C.A.*) 2,2-dichloro-*N*,*N*-di-2-propenylacetamide (9CI); Reg. No. *[37764-25-3]*. Its use to enhance herbicidal selectivity was reported by F. Y. Chang *et al.* (*Can. J. Plant Sci.*, 1972, **52**, 707). Introduced by the Stauffer Chemical Co. (USP 4 137 070) as code no. 'R-25 788'.

**Properties.**

The compound (>99% pure) is a clear viscous liquid: v.p. 800 mPa at 25 °C; $d_{20}^{20}$ 1.202: $n_D^{30}$ 1.499. Technical grade (*c.* 95% pure) is amber to brown; m.p. 5.0-6.5 °C; $d_{20}^{20}$ 1.192-1.204. Solubility *c.* 5 g/l water; *c.* 15 g/l kerosene; miscible with acetone, ethanol, 4-methylpentan-2-one, xylene. Corrosive to mild steel. Unstable >100 °C; violent decomposition when heated with iron.

**Uses.**

*N*,*N*-Diallyl-2,2-dichloroacetamide increases the tolerance of maize to thiocarbamate herbicides, including butylate, EPTC and vernolate, at rates of 140-700 g a.i./ha. The mixture with herbicide is applied when the soil is dry enough to permit incorporation in the top 7.5-10 cm and the maize is then planted without delay. Metabolism in maize, and rats and degradation in soil was reported (J. B. Miallis *et al.*, *Chemistry and Mode of Action of Herbicide Antidotes*, p. 109), the uptake by plant roots studied (R. A. Gray & G. K. Joo, *ibid.*, p. 67) and the mode of action investigated (R. E. Wilkinson, *ibid.*, p. 85). G. R. Stephenson (*J. Agric. Food Chem.*, 1978, **26**, 137) compared the chemical structure/biological activity of analogues.

**Toxicology.**

Acute oral $LD_{50}$: for male albino rats 2080 mg/kg; for females 2030 mg/kg. Acute percutaneous $LD_{50}$ for rabbits >5000 mg/kg; a mild irritant to skin and non-irritating to eyes of rabbits.

**Formulations.**

Mixtures (Reg. No. *[55947-96-1]*): 'Sutan +E', e.c. (with 720 or 800 g butylate/l); 'Sutan + 10G', granules (with 100 g butylate /kg). Mixtures (Reg. No. *[51990-04-6]*): 'Eradicane E', e.c. (with 720 or 800 g EPTC/l); 'Eradicane 10G', granules (with 100 g EPTC/kg). Mixture (Reg. No. *[68924-90-3]*), 'Surpass E', e.c. (with 800 g vernolate /l).

$C_{12}H_{21}N_2O_3PS$ (304.3)
T6N CNJ BY1&1 DOPS&O2&O2 F1

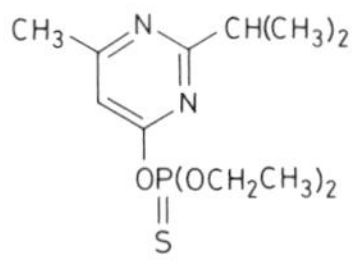

**Nomenclature and development.**

Common name diazinon (BSI, E-ISO, F-ISO, ANSI, BPC, JMAF);*dimpylate* (former BPC name). Chemical name (IUPAC) *O,O*-diethyl *O*-2-isopropyl-6-methylpyrimidin-4-yl phosphorothioate. (*C.A.*) *O,O*-diethyl *O*-[6-methyl-2-(1-methylethyl)-4-pyrimidinyl] phosphorothioate (9CI); *O,O*-diethyl *O*-(2-isopropyl-6-methyl-4-pyrimidinyl) phosphorothioate (8CI); Reg. No. *[333-41-5]*. OMS 469, ENT 19 507. Its insecticidal properties were described by R. Gasser (*Z. Naturforsch. Teil B*, 1953, **8**, 225). Introduced by J. R. Geigy S.A. (now Ciba-Geigy AG) (BEP 510 817; GBP 713 278) as code no. 'G 24 480'; trade marks 'Basudin', 'Diazitol', 'Neocidol', 'Nucidol'.

**Properties.**

Pure diazinon is a clear colourless liquid; b.p. 83-84 °C/0.0002 mmHg; v.p. 97 μPa at 20 °C; ρ 1.11 g/cm³ at 20 °C; $n_D^{20}$ 1.4978-1.4981. Solubility at 20 °C: 40 mg/l water; completely miscible in acetone, benzene, cyclohexane, dichloromethane, diethyl ether, ethanol, octan-1-ol, toluene. At 20 °C 50% hydrolysis occurs in 11.77 h at pH3.1, 185 d at pH7.4, 6.0 d at pH 10.4. Decomposes >120 °C.

**Uses.**

A non-systemic insecticide, its main applications are in fruit trees, horticultural crops, maize, potatoes, rice, sugarcane, tobacco and vineyards for a wide range of sucking and leaf-eating insects. Used also against flies in glasshouses, mushroom houses and against flies and ticks in veterinary practice. Decreases in residue levels in plants and animals have been discussed (E. Bartsch, *Residue Rev.*, 1974, **51**, 37).

**Toxicology.**

Acute oral $LD_{50}$ for rats 300-400 mg tech./kg. Acute percutaneous $LD_{50}$ for rats >2150 mg/kg; slightly irritant to skin and eyes of rabbits. Inhalation $LC_{50}$ (4-h) for rats 3500 mg/m³ air. In 90-d feeding trials NEL was: for rats 0.1 mg/kg daily; for dogs 0.02 mg/kg daily. $LC_{50}$ (96-h) for rainbow trout 2.6-3.2 mg/l; for bluegill 16 mg/l; for carp 7.6-23.4 mg/l. In laboratory trials diazinon is toxic to honeybees and birds.

**Formulations.**

Typical formulations for agricultural and horticultural use include: granules, 'Basudin 5' (50 g a.i./kg), 'Basudin 10' (100 g/kg); w.p. 'Basudin 40WP' (400 g/kg); seed treatment, 'Basudin 50SD', dust (500 g/kg); e.c. 'Basudin 60EC' (600 g/l); 'Diazitol Liquid'; 'Basudin Ulvair 500'; 'Basudin 20 Mushroom Aerosol'. For veterinary use: 'Neocidol 60'/'Nucidol 60', e.c. (600 g/l).

**Analysis.**

Product analysis is by glc (*CIPAC Handbook*, 1980, **1A**, 1199; *Anal. Methods Pestic. Plant Growth Regul.*, 1972, **6**, 345; *AOAC Methods*, 1980, 6.431-6.435; D. O. Eberle, *J. Assoc. Off. Anal. Chem.*, 1974, **57**, 48). Residues may be determined by glc with TID or MCD (D. O. Eberle & D. Novak, *ibid.*, 1969, **52**, 1067; *AOAC Methods*, 1980, 29.001-29.018, 29.044-29.049 29.050-29.054; *Pestic. Anal. Man.*, 1979, **1**, 201-A, 201-G, 201-H, 201-I; D. C. Abbott *et al.*, *Pestic. Sci.*, 1970, **1**, 10; *Analyst (London)*, 1980, **105**, 515). For review of analytical methods see D. O. Eberle (*Residue Rev.*, 1974, **51**, 1).

$C_8H_6Cl_2O_3$ (221.0)
QVR BG EG FO1
*dicamba-dimethylammonium*
$C_{10}H_{13}Cl_2NO_3$ (266.1)
QVR BG EG FO1 &&$(CH_3)_2$NH SALT
*dicamba-sodium*
$C_8H_5Cl_2NaO_3$ (243.0)
QVR BG EG FO1 &&Na SALT

**Nomenclature and development.**

Common name dicamba (BSI, E-ISO, F-ISO, ANSI, WSSA);dianat (USSR);MDBA (JMAF). Chemical name (IUPAC) 3,6-dichloro-*o*-anisic acid (I). (*C.A.*) 3,6-dichloro-2-methoxybenzoic acid (9CI); (I) (8CI); Reg. No. *[1918-00-9]* (acid), *[2300-66-5]* (dimethylammonium salt), *[10007-85-9]* (potassium salt), *[1982-69-0]* (sodium salt). Its herbicidal properties were described by R. A. Darrow & R. H. Haas (*Proc. South. Weed Conf., 14th*, 1961, p. 202). Introduced by Velsicol Chemical Corp. (USP 3 013 054) as code no. 'Velsicol 58-CS-11'; trade marks 'Banvel', 'Mediben'.

**Properties.**

The technical product is a pale buff crystalline solid, purity 80-90% *m*/*m*, the remainder mainly being 3,5-dichloro-*o*-anisic acid. Pure dicamba is a colourless solid; m.p. 114-116 °C; $d^{25}$ 1.57; v.p. 4.5 mPa at 25 °C. It decomposes *c.* 200 °C on attempted distillation. Solubility at 25 °C: 6.5 g/l water; 810 g/l acetone; 260 g/l dichloromethane; 1.18 kg/l dioxane; 922 g/l ethanol; 130 g/l toluene; 78 g/l xylene. It is resistant to oxidation and to hydrolysis under normal conditions.

Dicamba is acidic, forming salts that are appreciably soluble in water: sodium salt, 360 g a.e./l water; potassium salt, 480 g a.e./l water; dimethylammonium salt, 720 g a.e./l water.

**Uses.**

It is a foliar- or soil- applied herbicide which is readily absorbed by leaves and roots and is translocated throughout the plant. It effectively controls many annual and perennial broad-leaved weed species in asparagus, cereals, grain, maize, perennial seed grasses, sorghum, sugarcane, turf between cropping systems. It is also used for woody brush and vine control in pasture, rangeland and cropland. Dosage varies with specific use and ranges from 105 g to 11.2 kg a.i./ha. Dicamba is rapidly degraded in soil with 50% loss in $<$14 d.

**Toxicology.**

Acute oral $LD_{50}$ for rats 1700 mg/kg; for mallard duck 2000 mg/kg. Acute percutaneous $LD_{50}$ for rabbits $>$2000 mg/kg; it is extremely irritating and corrosive to their eyes and moderately irritating to their skin; in a 21-d dermal study in rabbits exposed to $\geqslant$100 mg/kg daily evidence of slight to moderate irritation was observed microscopically. It is a moderate dermal sensitiser to guinea-pigs.

In 2-y feeding trials no adverse effect was noted at the highest levels: for rats 500 mg/kg diet; for dogs 50 mg/kg diet. Enlargement of the hepatocytes of the liver was observed in mice in a dose-related incidence and degree at 100-10 000 mg/kg diet. $LC_{50}$ (8-d) for mallard duck and bobwhite quail were $>$10 000 mg/kg diet. $LC_{50}$ (96-h) for rainbow trout and bluegill is 135 mg/l. It was not teratogenic in rabbits at 10 mg/kg daily or in rats at 400 mg/kg daily. No effect was observed in a 3-generation reproduction study in rats at the top dose, 500 mg/kg diet. Bacterial and mammalian in-vitro and in-vivo tests showed it is not a mutagen; however, EPA has obtained weakly positive results from bacterial mutogenicity assays.

$C_8H_6Cl_2O_3$ (221.0)
QVR BG EG FO1
*dicamba-dimethylammonium*
$C_{10}H_{13}Cl_2NO_3$ (266.1)
QVR BG EG FO1 &&$(CH_3)_2$NH SALT
*dicamba-sodium*
$C_8H_5Cl_2NaO_3$ (243.0)
QVR BG EG FO1 &&Na SALT

**Formulations.**

'Banvel Herbicide' (480 g a.e./l) as dimethylammonium salt; 'Banvel II' (240 g a.e./l) as sodium salt; granules (50 or 100 g a.e./kg). 'Banvel K' (Reg. No. *[51602-02-9]*) (150 g dicamba + 300 g 2,4-D/l) as dimethylammonium salts. 'Banvel M' (150 g dicamba + 300 g MCPA/l) as dimethylammonium salts. Other mixtures include: dicamba + dichlorprop (potassium salts); dicamba + dichlorprop + MCPA (potassium and sodium salts); dicamba + MCPA (Reg. No. *[8003-31-4]*) (potassium and sodium salts); dicamba + MCPA + mecoprop (potassium and sodium salts); dicamba + mecoprop (potassium salts); dicamba + mecoprop + 2,4,5-T (amine salts).

**Analysis.**

Product analysis is by i.r. spectrometry (*AOAC Methods*, 1980, 6.265-6.269; *CIPAC Handbook*, 1980, **1A**, 1204; M. A. Malina, *Anal. Methods Pestic. Plant Growth Regul.*, 1973, **7**, 545). Residues in plants and soil may be determined by glc of a suitable ester (*idem, ibid.*,; H. K. Suzuki *et al., ibid.*, 1978, **10**, 305; T. H. Byast *et al.*, *Tech. Rep. ARC Weed Res. Organ.*, No. 15, (2nd Ed.), p. 70).

$C_7H_3Cl_2N$ (172.0)
NCR BG FG

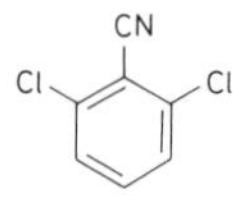

**Nomenclature and development.**

Common name dichlobenil (BSI, E-ISO, F-ISO, ANSI, WSSA); DBN (JMAF). Chemical name (IUPAC) 2,6-dichlorobenzonitrile (I). (*C.A.*) (I) (8 & 9CI); Reg. No. *[1194-65-6]*. Its herbicidal properties were described by H. Koopman & J. Daams (*Nature (London)*, 1960, **186**, 89). Introduced by Philips-Duphar B.V. (now Duphar B.V.). (NLP 572 662; USP 3 027 248) as code no. 'H 133'; trade mark 'Casoron'.

**Properties.**

Technical grade dichlobenil (≥94% pure) is a colourless to off-white crystalline solid with an aromatic odour; m.p. 139-145 °C (145-146 °C for the pure compound); b.p. 270 °C; v.p. 73 μPa at 20 °C; volatile in steam. Solubility at 20 °C: 18 mg/l water; 100 g/l dichloromethane; at 8 °C: *c.* 50 g/l in acetone, dioxane, xylene; < 10 g/l in non-polar solvents. Stable in sunlight and <270 °C. Stable to acids but rapidly hydrolysed by alkali to 2,6-dichlorobenzamide.

**Uses.**

It is a herbicide that inhibits actively-dividing meristems. It is used as a selective herbicide, pre-em. and post-em, controlling annual and perennial weeds at the seedling and later stages of growth. For control in fruit and other crops at 2.5-10 kg a.i./ha; for control of aquatic weeds at 4.5-12 kg/ha; for total weed control at ≤20 kg/ha. Its fate in crops, soils and animals has been reviewed (K. I. Beynon & A. N. Wright, *Residue Rev.*, 1972, **43**, 23; A. Verloop, *ibid.*, p. 55).

**Toxicology.**

Acute oral $LD_{50}$: for male and female rats ≥3160 mg/kg; for male mice 2126 mg/kg, for females 2056 mg/kg. Acute percutaneous $LD_{50}$ for albino rabbits 1350 mg/kg. In 2-y feeding trials and a 3-generation feeding study NEL for rats was 20 mg/kg diet. In 84-d feeding trials NEL for rats and rabbits was 50 mg/kg diet. $LC_{50}$ (8-d) was: for Japanese quail >5000 mg/kg diet; for pheasant *c.* 1500 mg/kg diet. $LC_{50}$ (48-h) is: for guppies > 18 mg/l; for water fleas 9.8 mg/l.

**Formulations.**

'Casoron 133', w.p. (450 g a.i./kg); 'Casoron G', granules (67.5 g/kg); 'Casoron G-SR', granules (200 g/kg). 'Fydulan G' (Reg. No. *[66278-58-8]*), (67.5 g dichlobenil + 100 g dalapon-sodium/kg); 'Fydulex G', granules (75 g dichlobenil + 150 g dalapon-sodium/kg); 'Fydumas G', granules (34 g dichlobenil + 10 g simazine/kg); 'Fydumas G Forte', granules (67.5 dichlobenil + 20 g simazine/kg); 'Fydusit G', granules (135 g dichlobenil + 15 g bromacil/kg).

**Analysis.**

Product analysis is by glc (*CIPAC Handbook*, 1983, **1B**, in press; A. van Rossum, *Anal. Methods Pestic. Plant Growth Regul.*, 1978, **10**, 311) or by spectrometry. Residues may be determined by glc (K. I. Beynon *et al., J. Sci. Food Agric.*, 1966, **17**, 151; T. H. Byast *et al., Tech. Rep. ARC Weed Res. Organ.*, No. 15 (2nd Ed.), p. 13).

$C_9H_{11}Cl_2FN_2O_2S_2$ (333.2)
GXGFSNR&SWN1&1

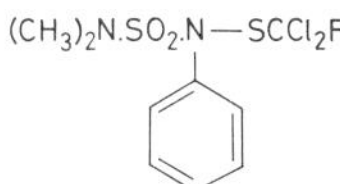

**Nomenclature and development.**

Common name dichlofluanid (BSI, E-ISO);dichlofluanide (F-ISO); dichlorfluanid (JMAF);*exception* USA. Chemical name (IUPAC) *N*-dichlorofluoromethylthio-*N*'*N*'-dimethyl-*N*-phenylsulphamide; *N*-dichlorofluoromethanesulphenyl-*N*',*N*'-dimethyl-*N*-phenylsulphamide. (*C.A.*) 1,1-dichloro-*N*-[(dimethylamino)sulfonyl]-1-fluoro-*N*-phenylmethanesulfenamide (9CI); *N*-[(dichlorofluoromethyl)thio]-*N*',*N*'-dimethyl-*N*-phenylsulfamide (8CI); Reg. No. *[1085-98-9]*. Its fungicidal properties were described by H. W. K. Müller (*Erwerbsobstbau*, 1964, **6**, 67) and it is a member of the aryl(dichlorofluoromethylthio) compounds described by E. Kühle (*Angew. Chem.*, 1964, **76**, 807). Introduced by Bayer AG (DEAS 1 193 498) as code no. 'Bayer 47 531', 'Kü 13-032-c'; trade marks 'Euparen', 'Elvaron'.

**Properties.**

It is a colourless powder; m.p. 105.0-105.6 °C; v.p. 133 μPa at 20 °C. It is practically insoluble in water; solubility 15 g/l methanol; 70 g/l xylene. It is sensitive to light (T. Clark & D. A. M. Watkins, *Pestic. Sci.*, 1978, **9**, 225) but discoloration does not affect its biological activity. It is decomposed by strongly alkaline media and by polysulphides.

**Uses.**

Dichlofluanid is a protective fungicide, with a broad range of activity, used for the control of *Venturia* spp. on apples and pears, *Botrytis* spp. and downy mildews. It has some effect on powdery mildews and against red spider mites, but does not affect *Phytoseiulus persimilis*, the predator used in biological control of *Tetranychus urticae* in glasshouses. It is applied at 75-100 g a.i./100 l HV.

**Toxicology.**

Acute oral $LD_{50}$: for male rats 500-2500 mg/kg, for females *c.* 525 mg/kg. Acute percutaneous $LD_{50}$ (4-h) for male rats 1000 mg/kg. In 2-y feeding trials on male and female rats, NEL was 1500 mg/kg diet. $LC_{50}$ (24-h) for goldfish and guppies 1-10 mg/l.

**Formulations.**

A w.p. (500 g a.i./kg); dust (75 g/kg).

**Analysis.**

Product analysis is by reaction with sodium methoxide and, ultimately, titration of the chloride (*CIPAC Handbook*, 1983, **1B**, in press). Residues may be determined by glc (K. Vogeler & H. Niessen, *Pflanzenschutz-Nachr. (Engl. Ed.)*, 1967, **20**, 534; *Methodensammlung Rückstandsanalytik Pflanzenschutzmitteln*, 1982, No. 6 (XII-6, 8, 12, 19)). Details of these methods may be obtained from Bayer AG.

$C_{10}H_4Cl_2FNO_2$ (260.1)
T5VNVJ BR DF& DG EG

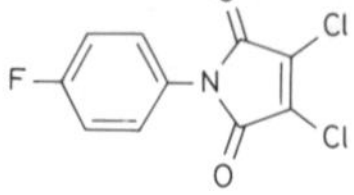

**Nomenclature and development.**

Common name fluoromide (JMAF). Chemical name (IUPAC) 2,3-dichloro-*N*-4-fluorophenylmaleimide. (*C.A.*) 3,4-dichloro-1-(4-fluorophenyl)-1*H*-pyrrole-2,5-dione (9CI); Reg. No. *[41205-21-4]*. Its fungicidal activity was described (*Jpn. Pestic. Inf.*, 1978, No. 34, p. 26). Introduced by Mitsubishi Chemical Industries Ltd and Kumiai Chemical Industry Co. Ltd (JPP 712 681) as code no. 'MK-23'; trade mark 'Sparticide'.

**Properties.**

It forms pale yellow crystals; m.p. 240.5-241.8 °C. Solubility at 20 °C: 5.9 mg/l water; 840 mg/kg methanol. It is stable in neutral or slightly acid media but hydrolysed by alkali.

**Uses.**

It is a fungicide used at 2-5 kg a.i./ha against apple fruit spot, melanose and scab of citrus, coffee berry disease, pink disease of rubber.

**Toxicology.**

Acute oral $LD_{50}$: for rats or mice >15 000 mg/kg. Acute percutaneous $LD_{50}$ for mice >5000 mg/kg. In 2-y feeding trials NEL was >600 but <2000 mg/kg diet. $LC_{50}$ (48-h) for carp is 5-6 mg/l.

**Formulations.**

A w.p. (500 or 750 g a.i./kg).

**Analysis.**

Product and residue analysis is by glc.

$C_{13}H_{10}Cl_2O_2$ (269.1)
QR DG B1R BQ EG

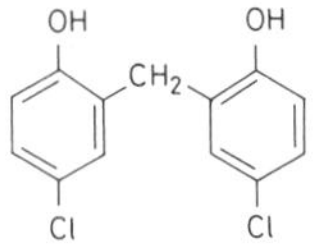

OH OH
$CH_2$
Cl Cl

**Nomenclature and development.**

Common name dichlorophen (BSI, E-ISO, BPC);dichlorophène (F-ISO). Chemical name (IUPAC) 4,4'-dichloro-2,2'-methylenediphenol; bis(5-chloro-2-hydroxyphenyl)methane, 5,5'-dichloro-2,2'-dihydroxydiphenylmethane. (*C.A.*) 2,2'-methylenebis(4-chlorophenol) (8 & 9CI); Reg. No. *[97-23-4]*. Its properties in controlling mildew on cotton were described by P B. Marsh & M. L. Butler (*Ind. Eng. Chem.*, 1946, **38**, 701). Introduced by Sindar Corp., by BDH Ltd and by May & Baker Ltd (USP 2 334 408) as code no. 'G4' (Sindar); trade mark 'Super Mosstox' (May & Baker).

**Properties.**

Pure dichlorophen forms colourless, odourless crystals; m.p. 177-178 °C; v.p. 13 nPa at 25 °C. Solubility at 25 °C: 30 mg/l water; 800 g/l acetone; 530 g/l ethanol. The technical grade is a light tan powder with a slight phenolic odour and m.p. ⩾164 °C. It is acidic (p$K_1$ 7.6; p$K_2$ 11.6) and readily soluble in aqueous alkali, forming salts.

**Uses.**

It is a fungicide and bactericide recommended for the protection of textiles and materials including horticultural benches and equipment against moulds and algae, and used to control moss in turf.

It is also used as 'Antiphen' in the treatment of tapeworm infestation in man and animals and is the basis of a preparation against athlete's foot.

**Toxicology.**

Acute oral $LD_{50}$: for guinea-pigs 1250 mg/kg; for dogs 2000 mg/kg. In 90-d feeding trials rats receiving 2000 mg/kg diet showed no evidence of toxicity. Fifty human patients treated with dichlorophen (6 g) on 2 successive days showed no ill-effect (H. Most, *J. Am. Med. Assoc.*, 1963, **874**, 185).

**Formulations.**

Dichlorophen is used as such; as 'Panacide', an aqueous solution of the monosodium salt (400 g/l); as 'Super Mosstox' e.c.

**Analysis.**

Product analysis is by u.v. spectrometry (J. R. Clements & S. H. Newburger, *J. Assoc. Off. Agric. Chem.*, 1954, **37**, 190).

$C_{13}H_9Cl_2NO_4$ (314.1)
WNR BO1 DOR BG DG

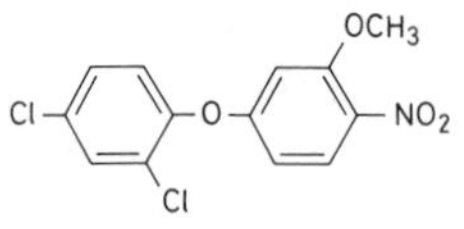

Nomenclature and development.

**Nomenclature and development.**

Common name* chlormethoxynil (JMAF). Chemical name (IUPAC) 5-(2,4-dichlorophenoxy)-2-nitroanisole; 2,4-dichlorophenyl 3-methoxy-4-nitrophenyl ether. (*C.A.*) 4-(2,4-dichlorophenoxy)-2-methoxy-1-nitrobenzene (9CI); Reg. No. *[32861-85-1]*. Introduced as a herbicide by Nihon Nohyaku Co. Ltd (JPP 600 441) as code no. 'X-52'; trade mark 'Ekkusagoni'.

**Properties.**

It forms yellow crystals; m.p. 113-114 °C. Solubility at 15 °C: 0.3 mg/l water; 200 g/kg acetone; 150 g/kg benzene; 100 g/kg dimethyl sulphoxide. Stable to acid, alkali and light.

**Uses.**

A contact herbicide used pre-em. at 1.5-2.5 kg a.i./ha in transplanted rice to control barnyard grass and other annual weeds.

**Toxicology.**

Acute oral $LD_{50}$: for rats > 10 000 mg/kg; for mice 33 000 mg/kg. Acute percutaneous $LD_{50}$ for rats > 5000 mg/kg. $LC_{50}$ (48-h) for carp 237 mg/l.

**Formulations.**

These include w.p. (700 g a.i./kg); granules (70 g/kg).

**Analysis.**

Product analysis is by glc with FID (F. Yamane & K. Tsuchiya, *Anal. Methods Pestic. Plant Growth Regul.*, 1978, **10**, 267). Residues may be determined by glc with ECD (*idem, ibid.*).

*Common name chlomethoxyfen (BSI, draft E-ISO)

$C_{14}H_{13}Cl_2O_2PS$ (347.2)
GR CG DOPS&R&O2

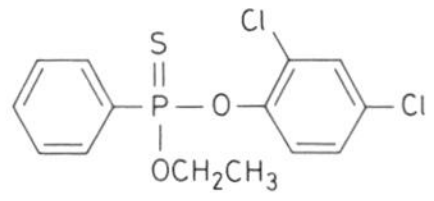

**Nomenclature and development.**

Common name EPBP (JMAF). Chemical name (IUPAC) *O*-2,4-dichlorophenyl *O*-ethyl phenylphosphonothioate. (*C.A.*) *O*-(2,4-dichlorophenyl) *O*-ethyl phenylphosphonothioate (8 & 9CI); Reg. No. *[3792-59-4]*. Introduced by Nissan Chemical Industries Ltd (USP 3 318 764) as code no. 'S-7'; trade mark 'S-Seven'.

**Properties.**

The pure compound is a light brown oil; b.p. 206 °C/5 mmHg; v.p. 509 Pa at 200 °C; $n_D^{20}$ 1.5956; $d_4^{24}$ 1.312. It is practically insoluble in water; soluble in most organic solvents. The technical grade is *c.* 90% pure.

**Uses.**

It is a non-systemic contact and stomach-acting insecticide used mainly to control soil-dwelling pests such as *Agrotis fucosa, Phyllotreta striolata, Rhizoglyphus echinopus* and *Hylemya* spp. etc., using the dust at 1.8-3.0 kg a.i./ha. Its properties are incompatible with propanil.

**Toxicology.**

Acute oral $LD_{50}$ for mice 275 mg/kg. Acute subcutaneous $LD_{50}$ for mice 784 mg/kg.

**Formulations.**

Dust (30 g a.i./kg).

**Analysis.**

Product analysis is by glc. Residues may be determined by glc.

$C_6H_3Cl_2NO_2$ (192.0)
T6NJ BVQ CG FG

Cl N CO.OH Cl

**Nomenclature and development.**

The names 3,6-dichloropicolinic acid (I) (BSI, E-ISO, WSSA);acide dichloro-3,6 picolinique (F-ISO) are accepted in lieu of a common name. Chemical name (IUPAC) 3,6-dichloropyridine-2-carboxylic acid; (I) (*C.A.*) 3,6-dichloro-2-pyridinecarboxylic acid (8 & 9CI); Reg. No. *[1702-17-6]*. Trivial name 3,6-DCP. Its herbicidal properties were described by T. Haagsma (*Down Earth*, 1975, **30**(4), 1). Introduced by Dow Chemical Co. (USP 3 317 549) as code no. 'Dowco 290'; trade mark 'Lontrel'.

**Properties.**

It forms colourless crystals; m.p. 151-152 °C, v.p. 1.6 mPa at 25 °C. Solubility at 20 °C: 9 g/kg water; 250 g/kg acetone, cyclohexanone; <11 g/kg xylene. It is acidic (p$K_a$ 2.33) and forms salts, for example 2-hydroxyethylammonium (Reg. No. *[57754-85-5]*) solubility at 25 °C 560 g/l and potassium (solubility, >300 g/l). Solutions of the salts are corrosive to aluminium, steel and tin plate.

**Uses.**

3,6-Dichloropicolinic acid is a systemic, post-em. herbicide, effective against many species of Compositae, Leguminosae, Solanaceae and Umbelliferae. It is selective in graminaceous crops, as well as in various broad-leaved crops including brassicas, sugar beet and other beet crops, flax, strawberries and onion-type crops. Applied in cereals (not undersown) at 50-80 g a.e./ha in combination with other herbicides. In non-graminaceous crops it is applied at 70-100 g a.e./ha, either in combination with other herbicides, against annual weeds such as *Matricaria* spp., or alone at 100-200 g a.e./ha for control of perennial weeds such as *Cirsium arvense*. It is not metabolised in plants. It undergoes microbial degradation in soils; when incorporated at levels ranging from 0.25-1.0 mg/kg, 50% is lost by this route in *c.* 49 d under conditions favourable to microbial activity.

**Toxicology.**

Acute oral $LD_{50}$: for rats >4300-5000 mg/kg; for mallard duck 1465 mg/kg. Acute percutaneous $LD_{50}$ for rabbits >2000 mg/kg; it is a severe eye irritant. In 2-y feeding trials NEL for rats was 50 mg/kg daily. $LC_{50}$ (8-d) for mallard duck and bobwhite quail was >4640 mg/kg diet. $LC_{50}$ (96-h) is: for rainbow trout 103.5 mg/l; for bluegill 125.4 mg/l. Non-toxic to honeybees; oral $LD_{50}$, (48-h) and contact $LD_{50}$ >100 μg/bee.

**Formulations.**

These include: aqueous concentrates containing 2-hydroxyethylammonium 3,6-dichloropicolinate, 'Lontrel SF 100', 'Matrigon'. Likewise mixtures as aqueous solutions, e.c., s.c. or w.p.: 'Benzalox', w.p. (50 g 3,6-dichloropicolinic acid + 300 g benazolin/kg); 'Sel-oxone' (15 g 3,6-dichloropicolinic acid + 510 g mecoprop/kg); 'Seppic MMD' (17.5 g 3,6-dichloropicolinic acid + 100 g MCPA + 450 g mecoprop/kg); with bifenox + mecoprop; with bromoxynil + MCPA + mecoprop; with ioxynil + MCPA + mecoprop; with propyzamide. Also e.c. with triclopyr-(2-butoxyethyl).

**Analysis.**

Product analysis is by glc. Residues may be determined by glc of a suitable dreivative (T. H. Byast *et al.*, *Tech. Rep. ARC Weed Res. Organ.*, No. 15 (2nd Ed.), p. 31; A. J. Pik & G. W. Hodgson, *J. Assoc. Off. Anal. Chem.*, 1976, **59** 264). Details of methods are available from Dow Chemical Co.

*Common name clopyralid (BSI, draft E-ISO)

$C_3H_6Cl_2$ (113.0)
GY1&1G

Cl
|
$ClCH_2CHCH_3$

**Nomenclature and development.**

Chemical name (IUPAC) 1,2-dichloropropane (I) (BSI, E-ISO); dichloro-1,2 propane (F-ISO) is accepted in lieu of a common name. (*C.A.*) (I) (8 & 9CI); Reg. No. *[78-87-5]*. Trivial name propylene dichloride. ENT 15 406. Its properties as an insecticidal fungicide were described by I. E. Neifert *et al.* (*U.S. Dep. Agric. Bull.*, 1925, No. 1313).

**Properties.**

It is a colourless liquid, b.p. 95.4 °C; f.p. −70 °C; v.p. 27.9 kPa at 19.6 °C; $d_{20}^{20}$ 1.1595; $n_D^{25}$ 1.437. Solubility at 20°C: 2.7 g/kg water; soluble in ethanol, diethyl ether. It is flammable, flash point (Cleveland open cup) 21 °C.

**Uses.**

1,2-Dichloropropane is an insecticidal fumigant. It is a component of 'D-D' mixture and 'Vidden D' (see 4410). It is far less toxic to nematodes (C. R. Youngston & C. I. A. Goring, *Plant Dis. Rep.*, 1970, **54**, 196; M. V. McKenry & I. J. Thomason, *Hilgardia*, 1974, **42**, 422) than (*Z*)- and (*E*)-1,3-dichloropropenes, but moves more freely in soil, from which the greatest loss is by volatilisation (T. R. Roberts & G. Stoydin, *Pestic. Sci.*, 1976, **7**, 325).

**Toxicology.**

It is strongly narcotic but low concentrations cause an irritation of the respiratory tract. Exposure of guinea-pigs, rabbits and rats to the vapour has been studied (L. A. Heppel *et al., J. Ind. Hyg. Toxicol.*, 1946, **28**, 1).

**Formulations.**

It is a component with (*Z*)- and (*E*)-1,3-dichloropropene of 'D-D' and 'Vidden D' (see following entry).

**Analysis.**

Product and residue analysis is by glc.

**Nomenclature and development.**

This reaction mixture is often referred to by a trade mark 'D-D' (Shell Chemical Co.). Chemical name (IUPAC) 1,2-dichloropropane (previous entry) + 1,3-dichloropropene (following entry). (*C.A.*) 1,2-dichloropropane + 1,3-dichloro-1-propene (8 & 9CI); Reg. No. *[8003-19-8]*. ENT 8420. Its properties as a soil fumigant were described by W. Carter (*Science*, 1943, **97**, 383). Introduced by Shell Chemical Co. and by Dow Chemical Co. as trade marks 'D-D' (Shell), 'Vidden D' (Dow).

**Properties.**

Produced by the high temperature chlorination of propene (H. P. A. Gross & G. Hearne, *Ind. Eng. Chem.*, 1939, **31**, 1530) it is a mixture of chlorinated hydrocarbons containing ≥50% *m/m* (*E*)- and (*Z*)-1,3-dichloropropenes (see 4420), Reg. No. *[542-75-6]*, the other main constituent being 1,2-dichloropropane (see 4400), Reg. No. *[78-87-5]*. It has an organic chlorine content of ≥55.0% *m/m*.

It is a clear amber liquid, with a pungent odour. It flash distils over the range 59-115 °C; v.p. 4.6 kPa at 20 °C; ρ 1.17-1.22 g/cm$^3$ at 20 °C. Solubility at room temperature is *c.* 2 g/kg water; fully miscible with esters, halogenated solvents, hydrocarbons, ketones. The mixture is stable at ≤500 °C but reacts with dilute organic bases, concentrated acids, halogens and some metal salts. It is corrosive to some metals - *e.g.*, aluminium, magnesium and their alloys.

**Uses.**

The mixture is a pre-plant nematicide effective against soil nematodes including root knot, meadow, sting and dagger, spiral and sugar beet nematodes. The mixture is usually applied by injection into the soil or through tractor-drawn hollow tines, to a depth of 15-20 cm at 150-400 kg/ha (occasionally to a maximum of 1000 l/ha) depending on soil type and the following crop. The soil surface is sealed by rolling. Because the components are highly phytotoxic, a 7-d pre-planting interval should be allowed for every 75 l applied/ha. In wet or cold conditions (soil temperature <15 °C) rather longer intervals may be required. It does not accumulate in the soil (M. V. McKenry & I. J. Thomason, *Hilgardia*, 1974, **42**, 393; T. R. Roberts & G. Stoydin, *Pestic. Sci.*, 1976, **7**, 325); and has no permanent harmful effect on soil micro-organisms. Tainting of potato tubers has been reported and attributed to impurities such as 2,2-dichloropropane and 1,2,3-trichloropropane (C. J. Shepherd, *Nature (London)*, 1952, **170**, 1073) but the product is widely used before planting potatoes.

**Toxicology.**

Acute oral $LD_{50}$: for rats 140 mg/kg; for mice 300 mg/kg. Acute percutaneous $LD_{50}$: for rats 779 mg/kg; for rabbits 2100 mg/kg. It is a severe irritant to skin and eyes and may cause sensitisation. In 2-y feeding trials no effects related to the product were observed in rats receiving ≤120 mg product/kg diet. $LC_{50}$ (96-h) for harlequin fish is 4-5 mg/l. Its toxicity to honeybees is very low. ($LD_{50}$ >60 µg/bee).

**Formulations.**

The mixture is used without formulation. 'Ditrapex' (mixture with chloropicrin).

**Analysis.**

Product and residue analysis is by glc (T. R. Roberts & G. Stoydin, *loc. cit.*; C. E. Castro, *Anal. Methods Pestic., Plant Growth Regul. Food Addit.*, 1964, **3**, 151; *Anal. Methods Pestic. Plant Growth Regul.*, 1972, **6**, 710). Details can be obtained from Shell International Chemical Co. Ltd and from Dow Chemical Co.

# 1,3-Dichloropropene

$C_3H_4Cl_2$ (111.0)
G2U1G

$ClCH_2(H)C{=}C(H)Cl$ (E) $ClCH_2(H)C{=}C(Cl)H$ (Z)

**Nomenclature and development.**

The chemical name (IUPAC) 1,3-dichloropropene (I) (BSI, E-ISO); dichloro-1,3 propene (F-ISO) is accepted in lieu of a common name. (*C.A.*) 1,3-dichloro-1-propene (9CI); (I) (8CI); Reg. No. *[542-75-6]*. Its properties as a soil fumigant were described by Dow Chemical Co. (*Down Earth*, 1956, **12**(2), 7), who introduced it as trade mark 'Telone'; it is also a component of 'D-D' (Shell Co.), 'Vidden D' (Dow).

**Properties.**

The technical product (92% pure) is a colourless to amber-coloured liquid, with a sweetish odour; f.p. $<-50$ °C; b.p. 104 °C; flash point 28 °C (Tag closed cup), $d_4^{25}$ 1.217; v.p. 3.7 kPa at 20 °C. Solubility at 20 °C: 1 g/kg water; miscible with acetone, benzene, carbon tetrachloride, heptane, methanol.

It is a mixture of (*E*)-isomer, Reg. No. *[10061-02-6]*, b.p. 104.2 °C, $d_4^{20}$ 1.224, $n_D^{20}$ 1.4682; and (*Z*)-isomer *[10061-01-5]*, b.p. 112 °C, $d_4^{20}$ 1.217, $n_D^{20}$ 1.4730.

**Uses.**

It is a soil fumigant and nematicide, controlling root-knot nematodes at 187 l/ha at soil temperatures of 5-27 °C; a double rate is needed on highly organic soils. It is non-persistent and is hydrolysed in soil to the corresponding 3-chloroallyl alcohols (C. E. Castro & N. O. Belser, *J. Agric. Food Chem.*, 1966, **14**, 69; T. R. Roberts & G. Stoydin, *Pestic. Sci.*, 1976, **7**, 325; M. V. McKenry & I. J. Thomason, *Hilgardia*, 1974, **42**, 393). The (*Z*)-isomer is more toxic than the (*E*)-isomer to nematodes (*idem, ibid.*; C. R. Youngson & C. A. I. Goring, *Plant Dis. Rep.*, 1970, **54**, 196).

**Toxicology.**

Acute oral $LD_{50}$ for rats 250-500 mg/kg. It is a skin vesicant and very irritating and damaging to the eyes. In 0.5-y inhalation studies NEL was: for rats 1 ppm; for dogs, guinea-pigs, rabbits 3 ppm. $LC_{50}$ (8-d) for mallard duck and bobwhite quail $>10\,000$ mg/kg diet. $LC_{50}$ (96-h) is: for bluegill 7.09 mg/l; for rainbow trout 3.94 mg/l. $LC_{50}$ (48-h) for *Daphnia* 6.2 mg/l.

**Formulations.**

'Telone II' [92% (*Z*)- + (*E*)-1,3-dichloropropene]. These include: 'Telone' [42% (*Z*)- + 36.5% (*E*)-1,3-dichloropropene with 20.5% 1,2-dichloropropane and related compounds]. 1-Chloro-2,3-epoxyethane (epichlorohydrin) (1%) is added as a stabiliser in countries where this is permitted. See also 1,2-dichloropropane + 1,3-dichloropropene, previous entry.

**Analysis.**

Product analysis is by glc: details available from Dow Chemical Co. Residues may be determined by glc (M. V. McKenry & I. J. Thomason, *loc. cit.*; T. R. Roberts & G. Stoydin, *loc. cit.*).

$C_9H_8Cl_2O_3$ (235.1)
QVY1&OR BG DG

**Nomenclature and development.**

Common name dichlorprop (BSI, E-ISO, F-ISO, WSSA);2,4-DP (USSR). Chemical name (IUPAC) (±)-2-(2,4-dichlorophenoxy)propionic acid (I). (*C.A.*) (±)-2-(2,4-dichlorophenoxy)propanoic acid (9CI); (I) (8CI); Reg. No. *[120-36-5]*. Trivial name 2,4-DP. Although its growth regulating properties were known earlier (P. W. Zimmerman & A. E. Hitchcock, *Proc. Am. Soc. Hortic. Sci.*, 1944, **45,** 353), it was not introduced as a herbicide commercially until 1961, by The Boots Co. Ltd (now FBC Limited) as code no. 'RD 406'; trade mark 'Cornox RK'.

**Properties.**

Commercial dichlorprop is a racemate, only the (+)-isomer being herbicidally active. Dichlorprop is a colourless crystalline solid; m.p. 117.5-118.1 °C; v.p. negligible at room temperature. Solubility at 20 °C: 350 mg/l water; 595 g/l acetone; 85 g/l benzene; 69 g/l toluene; 51 g/l xylene. Technical grade has m.p. 114 °C. It forms salts (solubility at 20 °C): potassium (900 g a.e./l water); sodium (660 g/l); bis(2-hydroxyethyl)ammonium (750 g/l). Dichlorprop is corrosive to metals in the presence of water; concentrated solutions (480 g a.e./l) do not corrode tin plate or iron if pH≥8.6 and temperature <70 °C. Esters of lower alcohols are liquids that are volatile; dichlorprop-isooctyl and dichlorprop-(2-butoxyethanol), less so.

**Uses.**

Dichlorprop is a post-em. translocated herbicide, particularly effective for the control of *Polygonum persicaria, P. lapathifolium* and *P. convolvulus*. It also controls *Galium aparine* and *Stellaria media* but is not consistently effective against *P. aviculare*. It is used at 2.7 kg a.e./ha and may be used alone or in combination with other herbicides (see formulations). It is also used, at much lower rates, to prevent premature drop of apples.

**Toxicology.**

Acute oral $LD_{50}$: for rats 800 mg/kg; for mice 400 mg/kg. Acute percutaneous $LD_{50}$ for mice 1400 mg/kg; it is not an irritant to eyes at 10 g/l or skin at 24 g/l. In 98-d feeding trials no toxic effect was observed in rats receiving 12.4 mg/kg daily, though slight liver hypertrophy occurred at 50 mg/kg daily. $LC_{50}$ (48-h) for bluegill was 165 mg dichlorprop-(dimethylammonium)/l; 16 mg dichlorprop-isooctyl/l; 1.1 mg dichlorprop-(2-butoxyethyl)/l.

**Formulations.**

These include: aqueous solutions of alkali metal or amine salts, 'Polyclene', dichlorprop-potassium; e.c. of esters; e.c. based on dried non-hygroscopic salts. Mixtures as salts or esters. 'Basagran DP' (Reg. No. *[70281-42-4]*), BAS 35 801H, solution (250 g bentazone + 350 g dichlorprop-ethylammonium/l). 'Certrol PA' (Reg. No. *[39278-61-0]*) dichlorprop + ioxynil + MCPA. 'Cornoxynil', dichlorprop-(iso-octyl) + bromoxynil octanoate. 'Herbitox', dichlorprop + benazolin + dicamba. 'Mayclene', dichlorprop + 3,6-dichloropicolinic acid + MCPA. 'Oxytril P', (Reg. No. *[55172-48-0]*) dichlorprop-(iso-octyl) + bromoxynil octanoate + ioxynil octanoate. 'Tetroxone M', dichlorprop + bromoxynil + ioxynil + MCPA. 'Ustilan NK25', dichlorprop + aminotriazole + ethidimuron. Dichlorprop + 2,4-D; dichlorprop + MCPA; dichlorprop + mecoprop.

**Analysis.**

Product analysis is by glc of the methyl ester or by titratable acid content (*CIPAC Handbook,* 1980, **1A**, 1211). Residue analysis is by glc (T. H. Byast *et al., Tech. Rep., ARC Weed Res. Organ.* No. 15 (2nd Ed.), p. 61).

$C_4H_7Cl_2O_4P$ (221.0)
GYGU1OPO&O1&O1

$Cl_2C{=}CHOP(O)(OCH_3)_2$

**Nomenclature and development.**
Common name dichlorvos (BSI, E-ISO, F-ISO, BPC, ESA); DDVP (JMAF); DDVF (USSR). Chemical name (IUPAC) 2,2-dichlorovinyl dimethyl phosphate (I). (*C.A.*) 2,2-dichloroethenyl dimethyl phosphate (9CI); (I) (8CI); Reg. No. *[62-73-7]*. Trivial name DDVP. OMS 14, ENT 20 738. Its insecticidal properties were described by Ciba AG (now Ciba-Geigy AG) (GBP 775 085) but was given an incorrect structure; it was later reported by A. M. Martson *et al.* (*J. Agric. Food Chem.*, 1955, **3**, 319) as an insecticidal impurity in trichlorphon. Introduced by Ciba-Geigy AG, Shell Chemical Co., and Bayer AG (GBP 775 085 to Ciba-Geigy; USP 2 956 073 to Shell) as code no. 'Bayer 19 149' trade marks 'Nogos', 'Nuvan' (both to Ciba-Geigy), 'Vapona' (to Shell), 'Dedevap' (to Bayer).

**Properties.**
Dichlorvos is a colourless to amber liquid, with an aromatic odour; b.p. 35 °C/0.05 mmHg; v.p. 1.6 Pa at 20 °C; $d_4^{25}$ 1.415, $n_D^{25}$ 1.4523. Solubility at 20 °C: *c.* 10 g/l water; 2-3 g/kg kerosene; miscible with most organic solvents and aerosol propellants. It is stable to heat but is hydrolysed by water, a saturated aqueous solution at room temperature is converted to dimethyl hydrogen phosphate and dichloroacetaldehyde at a rate of *c.* 3%/d, more rapidly in alkali. It is corrosive to iron and mild steel.

**Uses.**
It is a contact and stomach-acting insecticide with fumigant and penetrant action. It is used: as a household and public health fumigant, especially against Diptera and mosquitoes; for the protection of stored products at 0.5-1.0 g a.i./100 $m^3$; for crop protection against sucking and chewing insects at 300-1000 g/ha. It is non-phytotoxic (except to some chrysanthemum cultivars) and non-persistent. It is used as an anthelmintic by incorporation in animal feeds.

**Toxicology.**
Acute oral $LD_{50}$ for rats 56-108 mg/kg. Acute percutaneous $LD_{50}$ for rats 75-210 mg/kg. In 90-d feeding trials rats receiving 1000 mg/kg diet showed no intoxication. Inhalation $LC_{50}$ (4-h): for mice 13.2 mg/$m^3$; for rats 14.8 mg/$m^3$. $LC_{50}$ (24-h) for bluegill is 1 mg/l. It is highly toxic to honeybees and toxic to birds.

**Formulations.**
The water content and acidity of the technical product must be controlled to delay hydrolysis. Formulations include: 'Nogos 50EC', 'Nogos 100EC', 'Nuvan 50EC', 'Nuvan 100EC', e.c. (500 and 1000 g a.i./l); 'Nuvan 100SC', an oil soluble concentrate (1000 g/l); aerosol concentrates (4-10 g/l); granules (5 g/kg). The safety of resin strips is discussed by J. W. Gillet *et al.* (*Residue Rev.*, 1972, **44**, 115, 161).

**Analysis.**
Product analysis is by glc, by i.r. spectrometry or by reaction with excess of iodine which is estimated by titration (*CIPAC Handbook*, 1980, **1A**, 1214). Residues may be determined by glc (*Pestic. Anal. Man.*, 1979, **I**, 201-H, 201-I; D. C. Abbott *et al., Pestic. Sci.*, 1970, **1**, 10; *Analyst (London)*, 1973, **98**, 19; 1977, **102**, 858; R. Mestres *et al., Trav. Soc. Pharm. Montpellier*, 1979, **39**, 323; *Ann. Falsif. Expert. Chim*, 1979, **72**, 577. For the sampling of atmospheres see, *Anal. Methods Pestic. Plant Growth Regul.*, 1972, **6**, 529).

# Diclobutrazol



$C_{15}H_{19}Cl_{2}N_{3}O$ (328.2)
T5NN DNJ AY1R BG DG&YQX1&1&1
&&(2RS,3RS) FORM

**Nomenclature and development.**

Common name diclobutrazol (BSI, draft E-ISO, F-ISO). Chemical name (IUPAC) (2*RS*,3*RS*)-1-(2,4-dichlorophenyl)-4,4-dimethyl-2-(1*H*-1,2,4-triazol-1-yl)pentan-3-ol. (*C.A.*) (*R**,*R**)-(±)-β-[(2,4-dichlorophenyl)methyl]-α-(1,1-dimethylethyl)-1*H*-1,2,4-triazole-1-ethanol (9CI); Reg. No. *[75736-33-3]* (stated stereoisomers), *[66345-62-8]* (unstated stereochemistry). Its fungicidal properties were described by K. J. Bent & A. M. Skidmore (*Proc. 1979 Br. Crop Prot. Conf. - Pests Dis.*, 1977, **2**, 477). Introduced by ICI Plant Protection Division (GBP 1 596 698) as code no. 'PP296'; trade mark 'Vigil'.

**Properties.**

Diclobutrazol is an off-white crystalline solid; m.p. 147-149 °C; v.p. *c.* 2.7 μPa at 20 °C; ρ 1.25 g/cm$^3$. Solubility at room temperature: 9 mg/l water; ⩽50 g/l acetone; chloroform, ethanol, methanol; and is slightly in xylene; moderately in 2-butoxyethanol cyclohexanone, dimethylformamide and glycol ethers. Stable at 50 °C for 0.5 y, and down to −5 °C.

**Uses.**

It is a broad spectrum systemic fungicide giving excellent control of powdery mildews on apples, cereals, cucurbits, grapevines and of rusts on cereals and coffee.

**Toxicology.**

Acute oral $LD_{50}$: for rats 4000 mg a.i./kg; for mice >1000 mg/kg; for guinea-pigs, rabbits 4000 mg/kg; for mallard duck >9461 mg/kg. Acute percutaneous $LD_{50}$ for rats and rabbits >1000 mg/kg. It is non-irritant to skin of rats; slightly irritant to skin and mildly irritant to eyes of rabbits. In 90-d feeding trials NEL for rats was 2.5 mg/kg daily; in 0.5 y NEL for dogs was 15 mg/kg daily. $LC_{50}$ (96-h) at 15 °C for rainbow trout 9.6 mg/l. The oral and contact $LD_{50}$ values are 50 μg/honeybee. It has no adverse effect on earthworm populations, or on soil microarthropods 30-245 d after application of sprays at 0.2 and 2.0 kg/ha.

**Formulations.**

'Vigil', s.c. (125 g a.i./l). 'Vigil K', s.c. (100 g diclobutrazol + 200 g carbendazim/l); 'Vigil T', s.c. (50 g diclobutrazol + 520 g captafol/l).

**Analysis.**

Residues in crops are determined by glc with ECD.

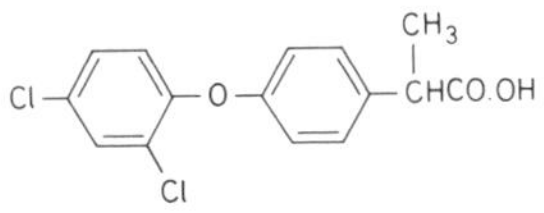

$C_{15}H_{12}Cl_2O_4$ (327.2)
QVY1&OR DOR BG DG
*diclofop-methyl*
$C_{16}H_{14}Cl_2O_4$ (341.2)
GR CG DOR DOY1&VO1

**Nomenclature and development.**

Common name diclofop (BSI, E-ISO, F-ISO, ANSI, WSSA); the ester present, *e.g.* diclofop-methyl should be stated. Chemical name (IUPAC) (*RS*)-2-[4-(2,4-dichlorophenoxy)phenoxy]propionic acid; methyl (*RS*)-2-[4-(2,4-dichlorophenoxy)phenoxy]propionate. (*C.A.*) (*RS*)-2-[4-(2,4-dichlorophenoxy)phenoxy]propanoic acid (9CI); methyl 2-[4-(2,4-dichlorophenyoxy)phenoxy]propanoate; Reg. No. *[40843-25-2]* (acid), *[51338-27-3]* (methyl ester). The herbicidal properties of diclofop-methyl were described by P. Langelüddeke *et al.* (*Mitt. Biol. Bundesanst. Land.-Forstwirtsch. Berlin-Dahlem*, 1975, **165**, 169). Introduced by Hoechst AG (DEP 2 136 828; 2 223 894) as code no. 'Hoe 23 408'; trade mark 'Hoe-Grass','Hoegrass', 'Hoelon', 'Illoxan', 'Iloxan'.

**Properties.**

Diclofop-methyl is a colourless crystalline solid; m.p. 39-41 °C; v.p. 34.4 μPa at 20 °C. Solubility at 22 °C 3 mg/kg water; at 20 °C: 2.49 kg/l acetone; 2.28 kg/l diethyl ether; 110 g/l ethanol; 60 g/l light petroleum (b.p. 60-95 °C); 25.3 kg/l xylene. Technical grade ⩾93% pure.

**Uses.**

Diclofop-methyl is a post-em. herbicide effective for the control of annual grasses such as *Avena fatua, A. ludoviciana, Echinochloa crus-galli, Eleusine indica, Setaria faberi, S. lutescens, S. viridis, Panicum dichotomiflorum, Lolium multiflorum, Leptochloa* spp. and volunteer maize. It is effective after uptake by foliage and roots and is selective to barley, carrots, celery, clover, cucumbers, field and dwarf beans, groundnuts, lettuces, lucerne, peas, potatoes, rape-seed, soyabeans, spinach, sugar beet, tomatoes, wheat. Dosage: 0.7-1.1 kg a.i./ha. Studies on degradation in plants and soil (A. E. Smith, *J. Agric. Food Chem.*, 1977, **25**, 893; R. Martens, *Pestic. Sci.*, 1978, **9**, 127) show a rapid formation of diclofop, which is decomposed further. Diclofop-methyl and diclofop are both herbicidal but appear to act at different sites in *Avena* spp. and cereals (M. A. Shimabukuro *et al.*, *Pestic. Biochem. Physic.*, 1978, **8**, 199). Diclofop-methyl (*R*)-(+)-enantiomer ( Reg. No. *[71283-65-3]*) shows significantly greater herbicidal activity by foliar application to 3 weed species than does the (*S*)-(-)-enantiomer( Reg. No. *[75021-72-6]*) but there was less selectivity by soil application (H. J. Nestler & H. Bieringer, *Z. Naturforsch. B. Anorg. Chem. Org. Chem.*, 1980, **35**B, 366).

**Toxicology.**

Acute oral $LD_{50}$: for rats 563-693 mg/kg; for Japanese quail >10 000 mg/kg. Acute percutaneous $LD_{50}$: for rats >5000 mg/kg. In 2-y feeding trials NEL for rats was 20 mg/kg diet; 1.25-y NEL for dogs 8 mg/kg diet. $LC_{50}$ (96-h) for rainbow trout 0.35 mg/l water.

**Formulations.**

Include e.c. (190, 284, 360 or 378 g diclofop-methyl/l).

**Analysis.**

Details of glc methods are available from Hoechst AG.

$C_6H_4Cl_2N_2O_2$ (207.0)
ZR BG FG DNW

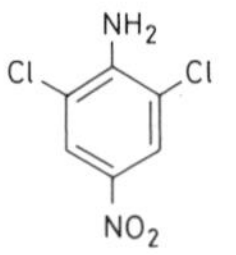

**Nomenclature and development.**

Common name dicloran (BSI, New Zealand);dichloran (Canada, Republic of South Africa);dichlorane (France);CNA (JMAF). Chemical name (IUPAC) 2,6-dichloro-4-nitroaniline (I). (*C.A.*) 2,6-dichloro-4-nitrobenzenamine (9CI); (I) (8CI); Reg. No. *[99-30-9]*. Its fungicidal properties were described by N. G. Clark *et al.* (*Chem. Ind. (London)*, 1960, p.572). Introduced by The Boots Co. Ltd. (now FBC Limited) (GBP 845 916) as code no. 'RD 6584'; trade mark 'Allisan' (Boots Co.- now FBC Limited), 'Botran' (Upjohn Co.).

**Properties.**

Pure dicloran is a yellow crystalline solid; m.p. 195 °C; v.p. 160 μPa at 20 °C. Solubility: 7 mg/l water; 34 g/l acetone; 4.6 g/l benzene; 12 g/l chloroform; 6 mg/l cyclohexane; 40 g/l dioxane. The technical grade ( ⩾90% pure) is brownish-yellow. Stable to hydrolysis and oxidation.

**Uses.**

It is a protectant fungicide which causes hyphal distortion but has little effect on spore germination. Effective against a wide range of fungal pathogens, particularly certain species of *Botrytis, Rhizopus* and *Sclerotinia*, it is used in a number of fruit, ornamental and vegetable crops and is basically non-phytotoxic but young lettuce plants, between stages of transplanting and the development of 2 true leaves, display some susceptibility to damage.

**Toxicology.**

Acute oral $LD_{50}$: for rats 4000 mg a.i./kg; for mice 1500-2500 mg/kg. Acute percutaneous $LD_{50}$ for mice >5000 mg/kg. In 2-y feeding trials rats receiving 1000 mg/kg diet suffered no ill-effect. It is rapidly metabolised in rats and excreted in the urine as the sulphate conjugate of 3,5-dichloro-4-aminophenol. Metabolised to polar compounds in plants.

**Formulations.**

These include: w.p. (500 or 750 g a.i./kg); dusts (40-80 g/kg); smoke generators. 'Turbair Botryticide', ULV (dicloran + thiram).

**Analysis.**

Product analysis is by i.r. spectrometry (P. G. Marshall, *Analyst (London)*, 1960, **85**, 681). Residues may be determined by glc (*Pestic. Anal. Man.*, 1979, **I**, 201-A, 201-G, 201-I; M. A. Luke *et al.*, *J. Assoc. Off. Anal. Chem.*, 1981, **64**, 1187; *Anal. Methods Pestic. Plant Growth Regul.*, 1972, **6**, 553).

$C_{14}H_9Cl_5O$ (370.5)
GXGGXQR DG&R DG

Cl–C₆H₄–C(OH)(CCl₃)–C₆H₄–Cl

**Nomenclature and development.**

Common name dicofol (BSI, E-ISO, F-ISO, ESA);kelthane (JMAF);*exception* (Austria, Federal Republic of Germany). Chemical name (IUPAC) 2,2,2-trichloro-1,1-bis(4-chlorophenyl)ethanol; 2,2,2-trichloro-1,1-di-(4-chlorophenyl)ethanol. (*C.A.*) 4-chloro-α-(4-chlorophenyl)-α-(trichloromethyl)benzenemethanol (9CI); 4,4'-dichloro-α-(trichloromethyl)benzhydrol (8CI); Reg. No. *[115-32-2]*. ENT 23 648. Its acaricidal properties were described by J. S. Barker & F. B. Maugham (*J. Econ. Entomol.*, 1956, **49**, 458). Introduced by Rohm & Haas Co. (GBP 778 209; USP 2 812 280; 2 812 362; 3 102 070; 3 194 730) as code no. 'FW-293'; trade mark 'Kelthane'.

**Properties.**

Pure dicofol is a colourless solid; m.p. 78.5-79.5 °C. The technical grade is a brown viscous oil (*c.* 80% pure) $d^{25}$ 1.45. It is practically insoluble in water; soluble in most aliphatic and aromatic solvents. It is hydrolysed by alkali to 4,4'-dichlorobenzophenone and chloroform, but is compatible with all but highly alkaline pesticides. The w.p. formulations are sensitive to solvents and surfactants and these may affect acaricidal activity and phytotoxicity.

**Uses.**

It is a non-systemic acaricide with little insecticidal activity, recommended for the control of mites on a wide range of crops at 0.56-4.5 kg a.i./ha. Although residues in soil decrease rapidly, traces may remain for $\geqslant$1 y.

**Toxicology.**

Acute oral $LD_{50}$ for rats 668-842 mg/kg. Acute percutaneous $LD_{50}$ for rabbits 1870 mg/kg. In 1-y feeding trials dogs receiving 300 mg/kg diet showed no evidence of toxicity (R. Blackwell Smith *et al., Toxicol. Appl. Pharmacol.*, 1959, **1**, 119).

**Formulations.**

These include: 'Kelthane AP', w.p. (185 g a.i./kg); 'Kelthane 35', w.p. (350 g/kg); 'Kelthane EC', e.c. (185 g/l); 'Kelthane MF', e.c. (420 g/l); dicofol 20 and dicofol emulsion (200 g/l); 'Kelthane Dust Base' (300 g/kg). 'Childion', e.c. (400 g dicofol + 125 g tetradifon/l).

**Analysis.**

Product analysis is by glc (*AOAC Methods*, 1980, 6.283-6.288). Residues may be determined by glc (*Pestic. Anal. Man.*, 1979, 201-A, 201-G, 201-I; M. A. Luke *et al., J. Assoc. Off. Anal. Chem.*, 1981, **64**, 1187; *Anal. Methods Pestic. Plant Growth Regul.*, 1972, **6**, 415).

$C_8H_{16}NO_5P$ (237.2)
1OPO&O1&OY1&U1VN1&1
&&(E) FORM

**Nomenclature and development.**

Common name dicrotophos (BSI, E-ISO, F-ISO, ESA). Chemical name (IUPAC) (*E*)-2-dimethylcarbamoyl-1-methylvinyl dimethyl phosphate; 3-dimethoxyphosphinyloxy-*N*,*N*-dimethylisocrotonamide. (*C.A.*) (*E*)-3-(dimethylamino)-1-methyl-3-oxo-1-propenyl dimethyl phosphate (9CI); dimethyl phosphate ester with (*E*)-3-hydroxy-*N*,*N*-dimethylcrotonamide; Reg. No. *[141-66-2]* (*E*)-isomer, *[18250-63-0]* (*Z*)-isomer, *[3735-78-3]* mixed isomers. OMS 253, ENT 24 482. Its insecticidal properties were described by R. A. Corey (*J. Econ. Entomol.*,1965, **58**, 112). Introduced by Ciba AG (now Ciba-Geigy AG) and later by Shell Chemical Co., USA, (BEP 552 284; GBP 829 576 both to Ciba-Geigy; USP 2 956 073;3 068 268 both to Shell) as code no. 'C-709' (Ciba-Geigy),'SD 3562' (Shell); trade marks 'Carbicron', 'Ektafos' (Ciba-Geigy), 'Bidrin' (Shell).

**Properties.**

Produced by reacting trimethyl phosphite with 2-chloro-*N*,*N*-dimethyl-3-oxobutyramide, the technical grade, containing about 85% (*E*)-isomer, is an amber liquid: b.p. 400 °C at 760 mmHg; v.p. 13 mPa at 20 °C; $d_{20}^{20}$ 1.21. Solubility: totally miscible with water and organic solvents such as acetone, ethanol, 4-hydroxy-4-methylpentan-2-one, propan-2-ol; <10 g/kg diesel oil, kerosene. Pure dicrotophos is a yellowish liquid; b.p. 130 °C/0.1 mmHg; v.p. 9.3 mPa at 20 °C; ρ 1.216 g/cm³ at 20 °C; $n_D^{23}$ 1.4680. It is stable when stored in glass or polythene containers ⩽40 °C, but is decomposed after prolonged storage at 55 °C. On hydrolysis, at 20 °C, 50% loss (calculated) occurs in 88 d at pH5, 23 d at pH9. It is corrosive to cast iron, mild steel, brass and some grades of stainless steel.

**Uses.**

Dicrotophos, the (*E*)-isomer (which is insecticidally more active than the (*Z*)-), is a systemic insecticide and acaricide of moderate persistence. It is effective against sucking, boring and chewing pests at dose rates of 300-600 g a.i./ha and is recommended for use on coffee, cotton, rice and other crops. It is non-phytotoxic except to certain varieties of fruit under some conditions. Its metabolism and breakdown have been reviewed (K. I. Beynon *et al.*, *Residue Rev.*, 1973, **47**, 55).

**Toxicology.**

Acute oral $LD_{50}$ for rats 17-22 mg/kg; for birds 1.2-12.5 mg/kg. Acute percutaneous $LD_{50}$ for rats ranges from 111-136 mg/kg to 148-181 mg/kg, depending on the carrier and the conditions of the test; for rabbits 224 mg/kg, slight irritant to skin and eyes of rabbits. Inhalation $LC_{50}$ (4-h) *c.* 90 mg/m³ air. In 2-y feeding trials NEL was: for rats 1.0 mg/kg diet (0.05 mg/kg daily); for dogs 1.6 mg/kg diet (0.04 mg/kg daily). NEL in a 3-generation reproduction study with rats was 2 mg/kg daily. It is not neurotoxic to hens. $LC_{50}$ (24-h) is: for mosquito fish 200 mg/l; for harlequin fish >1000 mg/l. It is very toxic to honeybees but because surface residues rapidly decline little effect is seen in practice.

**Formulations.**

These include: water-soluble concentrates 'Bidrin' (850-1030 g a.i./l), 240 g/l, 'Carbicron 50 SCW', 'Carbicron 100 SCW'; e.c. (400 or 500 g/l);'Carbicron 250 ULV' (for spraying undiluted), ULV solutions (240-500 g/l).

**Analysis.**

Product analysis is by i.r. spectroscopy or glc. Residues may be determined by glc (M.C. Bowman & M. Beroza, *J. Agric. Food Chem.*, 1967, **15**, 465; B. Y. Giang & H. F. Beckman, *ibid.*, 1968, **16**, 899) or by hydrolysis and measurement of the dimethylamine formed (R. T. Murphy, *ibid.*, 1965, **13**, 242). See also: P. E. Porter, *Anal. Methods Pestic. Plant Growth Regul.*, 1972, **6**, 287.

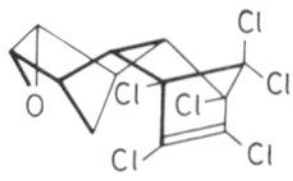


$C_{12}H_8Cl_6O$ (380.9)
T E3 D5 C555 A D- FO KUTJ AG AG BG JG KG LG

**Nomenclature and development.**

Common name (for pure compound) HEOD (BSI, E-ISO, F-ISO, JMAF);dieldrin (for pure compound) Denmark, USSR. The name dieldrin (for material containing >85% HEOD) (BSI, E-ISO, BPC, ESA, JMAF);dieldrine (F-ISO);*exceptions* Denmark, USSR (as above). Chemical name (IUPAC) (1*R*,4*S*,4a*S*,5*R*,6*R*,7*S*,8*S*,8a*R*)-1,2,3,4,10,10-hexachloro- 1,4,4a,5,6,7,8,8a-octahydro-6,7-epoxy-1,4:5,8-dimethanonaphthalene; 1,2,3,4,10,10-hexachloro-6,7-epoxy-1,4,4a,5,6,7,8,8a-octahydro-*endo*-1,4,-*exo*-5,8-dimethanonaphthalene. (*C.A.*) (1aα,2β,2aα,3β,6β,6aα,7β,7aα)-3,4,5,6,9,9-hexachloro-1a,2,2a,3,6,6a,7,7a-octahydro-2,7:3,6-dimethanonaphth[2,3-*b*]oxirene (9CI); *endo,exo*-1,2,3,4,10,10-hexachloro-6,7-epoxy-1,4,4a,5,6,7,8,8a-octahydro-1,4:5,8-dimethanonaphthalene (8CI); Reg. No. *[60-57-1]*. OMS 18, ENT 16 225. Its insecticidal properties were described by C. W. Kearns *et al.* (*J. Econ. Entomol.*, 1949, **42**, 127). Introduced by J. Hyman & Co. and later by the Shell International Chemical Co. Ltd (USP 2 676 547 to Hyman) as code no. 'Compound 497' (to Hyman); trade mark 'Octalox' (to Hyman).

**Properties.**

Technical dieldrin consists of buff to light tan flakes (setting point >95 °C) with a mild odour; m.p. 175-176 °C; v.p. 400 μPa at 20 °C; ρ 1.62 g/cm$^3$ at 20 °C. Solubility at 20 °C 0.186 mg/l water. Dieldrin is stable to alkali, mild acids and to light. It reacts with concentrated mineral acids, acid catalysts, acid oxidising agents and active metals (iron, copper).

**Uses.**

Dieldrin is a non-systemic and persistent insecticide of high contact and stomach activity to most insects. It is used extensively to control locusts and tropical disease vectors, such as *Glossina* spp. and mosquitoes. Dieldrin is not recommended as a foliar spray on edible crops because of residue difficulties. Industrial uses include timber preservation, termite-proofing of plastic and rubber coverings of electrical and telecommunication cables, of plywood and building boards and as a termite barrier in building construction.

**Toxicology.**

Acute oral $LD_{50}$: for rats 46 mg/kg; for birds 10-500 mg/kg. Acute percutaneous $LD_{50}$ for rats 50-120 mg/kg. In long-term exposure NEL for rats and dogs is 0.1 mg/kg diet (0.005 mg/kg daily). The data available indicate that dieldrin and aldrin do not produce malignant tumours in rats, dogs, monkeys at all tolerated dose levels in long-term feeding studies. The significance of the marginal increase in incidence of hepatic tumours in mice, following lifetime exposure to dieldrin, is doubtful. $LC_{50}$ (24-h) for fish 0.018-0.089 mg/kg. Aldrin/dieldrin were last reviewed at the FAO/WHO JMPR 1977 when 0-0.0001 mg/kg was confirmed as the estimate of the ADI for man.

**Formulations.**

These include: e.c. (180-200 g a.i./l); w.p. (500 or 750 g/kg); granules (20 or 50 g/kg), (the addition of urea may be necessary for stability with certain carriers); oil solutions for ULV application (180-200 g/l). Seed treatments, frequently incorporating fungicides.

**Analysis.**

Product analysis is by glc (details of methods are available from Shell International Chemical Co. Ltd.) or by i.r. spectrometry (*CIPAC Handbook*, 1970, **1**, 405; 1983, **1B**, in press; FAO Specification (CP/36), *Organochlorine Insecticides 1973*; *AOAC Methods* 1980, 6.182-6.187). Residues can be determined by glc with ECD (*Anal. Methods Pestic. Plant Growth Regul.*, 1972, **6**, 268; *AOAC Methods*, 1980, 29.001-29.018; *Pestic. Anal. Man.*, 1979, **1**, 201-A, 201-G, 201-I; *Analyst (London)*, 1979, **104**, 425; P. A. Greve & W. B. F. Grevenstuk, *Meded. Fac. Landbouwwet. Rijksuniv. Gent*, 1975, **40**, 1115; G. M. Telling *et al.*, *J. Chromatogr.*, 1977, **137**, 405; M. A. Luke *et al.*, *J. Assoc. Off. Anal. Chem.*, 1981, **64**, 1187).

$C_{10}Cl_{10}$ (474.6)
L5 AHJ AG BG CG DG EG A- AL5
AHJ AG BG CG DG EG

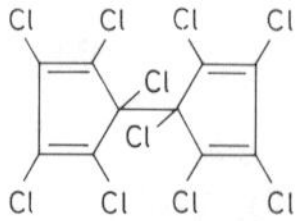

**Nomenclature and development.**

Common name dienochlor (BSI, E-ISO); diénochlore (F-ISO). Chemical name (IUPAC) perchloro-1,1'-bicyclopenta-2,4-dienyl. (*C.A.*) 1,1',2,2',3,3',4,4',5,5'-decachlorobi-2,4-cyclopentadien-1-yl (8 & 9CI); Reg. No. *[2227-17-0]*. ENT 25 718. Its acaricidal properties were described by W. W. Allen *et al.* (*J. Econ. Entomol.*, 1964, **57**, 187). Introduced by Hooker Chemical Corp. and later by Zoecon Corp. (owned by Sandoz AG) (USP 2 732 409; 2 934 470 to Hooker Chemical Corp.) as trade mark 'Pentac'.

**Properties.**

It is a tan crystalline solid; m.p. 122-123 °C; v.p. 1.3 mPa at 25 °C. Solubility: sparingly soluble in water; slightly soluble acetone, aliphatic hydrocarbons, ethanol; moderately soluble in aromatic hydrocarbons. It is stable in acid and alkaline media, but is decomposed by direct sunlight or >130 °C.

**Uses.**

It is a specific acaricide used to control mites on glasshouse roses at 250 mg a.i./l water using 2 applications with a 14-d interval; on other ornamentals 1 application is sufficient except for severe attacks. It acts by interference with oviposition.

**Toxicology.**

Acute oral $LD_{50}$ for male albino rats > 3160 mg tech./kg. Acute percutaneous $LD_{50}$ for albino rabbits > 3160 mg/kg; a single application of 3 mg to their eyes produced a slight irritation which subsided by the 6th day.

**Formulations.**

'Pentac', w.p. (500 g a.i./kg).

**Analysis.**

Product analysis is by i.r. spectroscopy.

$C_{14}H_{18}ClNO_3$ (283.8)
QV1NV1GR B2 F2
*diethatyl-ethyl*
$C_{16}H_{22}ClNO_3$ (311.8)
2OV1NV1GR B2 F2

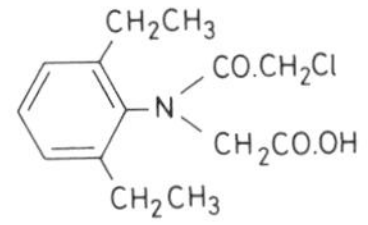

**Nomenclature and development.**

Common name diethatyl (BSI, E-ISO, F-ISO, ANSI, WSSA) for parent acid. Chemical name (IUPAC) *N*-chloroacetyl-*N*-(2,6-diethylphenyl)glycine (I). (*C.A.*) (I) (8 & 9CI); Reg. No. *[38725-95-0]* (acid), *[58727-55-8]* (ethyl ester). Herbicidal activity of its ethyl ester, diethatyl-ethyl, was reported by S. K. Lehman (*Proc. North Cent. Weed Control Conf., 29th*, 1972) and this was introduced by Hercules Inc. (now Boots-Hercules Agrochemical Co.) as code no. 'Hercules 22 234'; trade mark 'Antor'.

**Properties.**

Diethatyl-ethyl is crystalline; m.p. 49-50 °C. Solubility at 25 °C: 105 mg/l water; soluble in common organic solvents and chlorobenzene, cyclohexanone, 3,5,5-trimethylcyclohex-2-enone, xylene. Hydrolysed at 5>pH>12. Technical product and formulations are stable for >2 y under normal conditions.

**Uses.**

Diethatyl-ethyl (5-10 l/ha) is used in groundnuts, potatoes, red beet, soyabeans, spinach, sugar beet, sunflowers and winter wheat to control grasses (*Alopercurus, Echinochola, Digitaria, Panicum, Setaria* spp.) in mixtures with herbicides effective against broad-leaved weeds. It is also effective against broad-leaved weeds including *Solanum nigrum, Matricaria, Veronica* and *Amaranthus* spp. It is applied pre-em., either soil-incorporated or to the soil surface.

**Toxicology.**

Acute oral $LD_{50}$ for rats 2300-3700 mg/kg. Acute percutaneous $LD_{50}$: for rabbits 4000 mg/kg. In 24-h tests: bluegill had TL 7.14 mg/l, NEL 2.4 mg/l; rainbow trout TL 10.3 mg/l, NEL 1.0 mg/l. Non-toxic at 43.5 μg/bee.

**Formulations.**

An e.c. (480 g tech. diethatyl-ethyl/l), crystallisation can occur below 0 °C.

**Analysis.**

Product analysis is by glc or hplc. Residues are determined by glc. Details of methods available from Boots-Hercules Agrochemical Co.

$C_{12}H_{17}NO$ (191.3)
2N2&VR C1

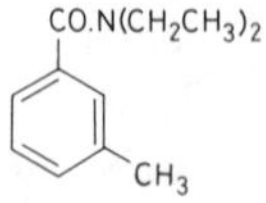

$CO.N(CH_2CH_3)_2$ ... $CH_3$

**Nomenclature and development.**

Common name diethyltoluamide (BPC) and accepted in lieu of a common name (BSI, E-ISO);deet (ESA). Chemical name (IUPAC) *N,N*-diethyl-*m*-toluamide (I). (*C.A.*) *N,N*-diethyl-3-methylbenzamide (9CI); (I) (8CI); Reg. No. *[134-62-3]*. ENT 20 218. Its repellent action against insects was reported by I. H. Gilbert *et al.* (*J. Econ. Entomol.*, 1955, **48**, 741). Introduced by Hercules Inc. (now Boots-Hercules Agrochemical Co., who no longer manufacture or market it) (USP 2 408 389) as trade mark 'Metadelphene'.

**Properties.**

Pure diethyltoluamide is a colourless to amber liquid, b.p. 111 °C/1 mmHg; $n_D^{25}$ 1.5206. Solubility: practically insoluble in water; miscible with ethanol, cottonseed oil, propan-2-ol, propylene glycol. The technical grade (85-95% *m*-isomer) has $d^{24}$ 0.996-0.998, viscosity 13.3 mPa s at 30 °C.

**Uses.**

It is an insect repellent especially effective against mosquitoes. The *o*- and *p*-isomers are less effective than the *m*-isomer.

**Toxicology.**

Acute oral $LD_{50}$ for male albino rats *c.* 2000 mg/kg. In 200-d feeding trials rats receiving 10 000 mg/kg diet suffered no ill-effect (A. M. Ambrose *et al., Toxicol. Appl. Pharmacol.*, 1959, **1**, 97). The undiluted compound may irritate mucous membranes, but daily applications to the face and arms caused only mild irritation.

**Formulations.**

'Off'.

**Analysis.**

Product analysis is by i.r. spectrometry or by glc.

$C_{31}H_{24}O_3$ (444.5)
T66 BOVJ EQ D- GL66&TJ IR DR

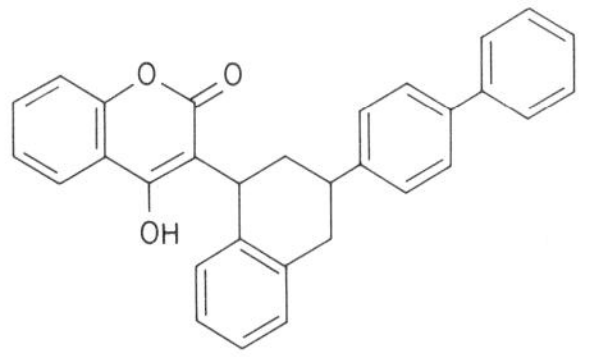

**Nomenclature and development.**

Common name difenacoum (BSI, E-ISO, F-ISO). Chemical name (IUPAC) 3-(3-biphenyl-4-yl-1,2,3,4-tetrahydro-1-naphthyl)-4-hydroxycoumarin. (*C.A.*) 3-[3-(1,1'-biphenyl)-4-yl-1,2,3,4-tetrahydro-1-naphthalenyl]-4-hydroxy-2*H*-1-benzopyran-2-one (9CI); Reg. No. *[56073-07-5]*. Its rodenticidal properties were reported by M. Hadler (*J. Hyg.*, 1975, **74**, 441). Introduced by Sorex (London) Ltd and later by ICI Plant Protection Division (GBP 1 458 670 to Sorex) as trade marks 'Neosorexa' (to Sorex), 'Ratak'(to ICI).

**Properties.**

Difenacoum is an off-white powder; m.p. 215-219 °C; v.p. 160 μPa at 45 °C. Solubility: $<$10 mg/l water at pH7; $>$ 50 g/l acetone, chloroform; 600 mg/l benzene. Stable $\leqslant$100 °C; no loss in sunlight at 30 °C for 30 d. It is a weak acid but its sodium and potassium salts are sparingly soluble in water.

**Uses.**

It is an indirect anticoagulant rodenticide which is very potent and is effective against rats and most mice resistant to other anticoagulants, using 180 g per baiting point for rats, 30 g per point for mice.

**Toxicology.**

Acute oral $LD_{50}$: for male rats 1.8 mg/kg; for male mice 0.8 mg/kg; for female guinea-pigs, 50 mg/kg; for pigs $>$50 mg/kg; for chickens and dogs ; for cats 100 mg/kg; Acute percutaneous $LD_{50}$: for rats $>$50 mg/kg; for rabbits 1000 mg tech./kg. Sub-acute oral $LD_{50}$ (5-d) for male rats 0.16 mg/kg daily.

**Formulations.**

'Neosorexa' concentrate (1 g a.i./kg); 'Neosorexa' ready-to-use bait (50 mg/kg); 'Ratak' (50 mg/kg).

**Analysis.**

Product analysis is by hplc (D. E. Munday & A. F. Machin, *J. Chromatogr.*, 1977, **139**, 321) or by u.v. spectrometry (details from Sorex (London) Ltd). Residues may be determined by hplc (*idem, ibid.*).

# Difenoxuron 4700

$C_{16}H_{18}N_2O_3$ (286.3)
1OR DOR DMVN1&1

$CH_3O$—$C_6H_4$—O—$C_6H_4$—$NH.CO.N(CH_3)_2$

**Nomenclature and development.**

Common name difenoxuron (BSI, E-ISO, F-ISO). Chemical name (IUPAC) 3-[4-(4-methoxyphenoxy)phenyl]-1,1-dimethylurea. (*C.A.*) *N*'-[4-(4-methoxyphenoxy)phenyl]-*N*,*N*-dimethylurea (9CI); 3-[*p*-(*p*-methoxyphenoxy)phenyl]-1,1-dimethylurea (8CI); Reg. No. *[14214-32-5]*. Its herbicidal properties were reported by L. Ebner & J. Schuler (*Proc. Br. Weed Control Conf., 7th*, 1964, **2**, 711). Introduced by Ciba-Geigy AG (BEP 593 743; GBP 913 383) as code no. 'C 3470'; trade mark 'Lironion'.

**Properties.**

Pure difenoxuron forms colourless crystals; m.p. 138-139 °C; v.p. 1.24 nPa at 20 °C; $\rho$ 1.30 g/cm$^3$ at 20 °C. Solubility at 20 °C: 20 mg/l water; 63 g/kg acetone; 8 g/kg benzene; 156 g/kg dichloromethane; 52 mg/kg hexane; 10 g/kg propan-2-ol. No significant hydrolysis observed at 30 °C at pH 1 or 13.

**Uses.**

It is used as a selective herbicide at 2.5 kg a.i./ha in onions at the 'whip' stage or later. Also in leeks and garlic.

**Toxicology.**

Acute oral $LD_{50}$ for rats $>$7750 mg tech./kg. Acute percutaneous $LD_{50}$ for rats $>$2150 mg/kg; non-irritant to skin and eyes of rabbits. Inhalation $LC_{50}$ (6-h) for rats $>$660 mg/m$^3$ air. In 90-d feeding trials NEL was: for rats 50 mg/kg daily; for dogs 200 mg/kg daily. $LC_{50}$ (48-h) for trout 5-10 mg/l.

**Formulations.**

A w.p. (500 g a.i./kg).

**Analysis.**

Product analysis is by acidimetric titration. Residues may be determined by glc with NSD; particulars are available from Ciba-Geigy AG.

$C_{17}H_{17}N_2$ (249.3)
T5KNJ A1 B1 CR& ER
*difenzoquat methyl sulphate*
$C_{18}H_{20}N_2O_4S$ (360.4)
T5KNJ A1 B1 CR& ER &&$CH_3OSO_3$ SALT

**Nomenclature and development.**

Common name difenzoquat (BSI, E-ISO, F-ISO, ANSI, WSSA). Chemical name (IUPAC) 1,2-dimethyl-3,5-diphenylpyrazolium ion. (*C.A.*) 1,2-dimethyl-3,5-diphenyl-1*H*-pyrazolium (9CI); Reg. No. *[49866-87-7]* ion, *[43222-48-6]* methyl sulphate. The herbicidal properties of the methyl sulphate (in *C.A.* methyl sulfate) were described by T. R. O'Hare & C. B. Wingfield (*Proc. North Cent. Weed Control Conf.*, 1973). This salt was introduced by American Cynamid Co. (BEP 792 801; USP 3 882 142) as code no. 'AC 84 777'; trade marks 'Avenge', 'Finaven'.

**Properties.**

The pure salt is a colourless hygroscopic solid. The technical grade (96% pure) has m.p. 150-160 °C. Solubility at 25 °C: 765 g salt/l; 20 g/l acetone; 71 g/l 1,2-dichloroethane; 23 g/l propan-2-ol; 1.1 g/l xylene. Stable to light, in water at low pH but alkaline conditions cause precipitation.

**Uses.**

Difenzoquat methyl sulphate is a post-em. herbicide for control of wild oats, *Avena fatua, A. sterilis* and *A. ludoviciana* in wheat and barley (R. J. Winfield & J. J. B. Caldicott, *Pestic. Sci.*, 1975, **6**, 297).

**Toxicology.**

Acute oral $LD_{50}$ for male albino rats 470 mg tech./kg. Acute percutaneous $LD_{50}$ for male white rabbits, 3540 mg/kg. In 2-y feeding trials rats receiving 500 mg/kg suffered no ill-effect. $LC_{50}$ (96-h): for bluegill 696 mg difenzoquat/l; for rainbow trout 694 mg/l. $LC_{50}$ (8-d) in feeding trials: for bobwhite quail $>4640$ mg/kg; for mallard duck 10 388 mg/kg.

**Formulations.**

Aqueous solutions and solid formulations.

**Analysis.**

Product analysis is by colorimetry (W. A. Steller, *Anal. Methods Pestic. Plant Growth Regul.*, 1980, **11**, 291). Residues are determined by glc of a derivative (W. A. Steller, *loc. cit.*).

$C_{14}H_9ClF_2N_2O_2$ (310.7)
GR DMVMVR BF FF

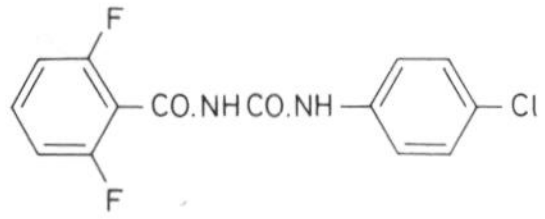

**Nomenclature and development.**

Common name diflubenzuron (BSI, E-ISO, F-ISO, ANSI). Chemical name (IUPAC) 1-(4-chlorophenyl)-3-(2,6-difluorobenzoyl)urea. (*C.A.*) *N*-[[(4-chlorophenyl)amino]carbonyl]-2,6-difluorobenzamide (9CI); Reg. No. *[35367-38-5]*. OMS 1804, ENT 29 054. Its insecticidal properties were described by J. J. Van Daalen *et al.* (*Naturwissenschaften* 1972, **59**, 312) and were reviewed by A. C. Grosscurt (*Pestic. Sci.*, 1978, **9**, 373) and W. Maas *et al.* (*Chemie Pflanzenschutz und Schadlbekampf. Mittel.*, 1980, **6**, 423). Introduced by Philips-Duphar B.V. (now Duphar B.V.) (GBP 1 324 293; USP 3 748 356; USP 3 989 842) as code no. 'PH 60-40'; 'PDD 6040-I' (both to Duphar B.V.),'TH 6040' (to T.H. Agriculture & Nutrition Co. Inc.); trade mark 'Dimilin'.

**Properties.**

Technical grade diflubenzuron (≥95% pure) is an off-white to yellow crystalline solid; m.p. 210-230 °C (230-232 °C for the pure compound); v.p. <13 μPa. Solubility at 20 °C: 0.1 mg/l water; 6.5 g/l acetone; at 25 °C: 104 g/l dimethylformamide; 24 g/l dioxane; <10 g/l apolar solvents. Decomposition: <0.5% after 1-d storage at 100 °C; <0.5% after 7-d at 50 °C. The solid is stable to sunlight. Decomposition at 20 °C in aqueous solution after 21 d in the dark is: 4% at pH5.8, 8% at pH7, 26% at pH9. It is rapidly degraded in soil, 50% degradation occurring in <7 d; its fate in soil, plants and animals has been reviewed (A. Verloop & C.D. Ferrell, *ACS Symp. Series*, 1977, No. 37, 237).

**Uses.**

It belongs to a new group of insecticides (K. Wellings *et al.*, *J. Agric. Food Chem.*, 1973, **21**, 348, 993), effective as a stomach and contact poison, acting by inhibition of chitin synthesis and so interfering with the formation of the cuticle. Hence all stages of insects that form new cuticles should be susceptible to diflubenzuron exposure. It has no systemic activity and does not penetrate plant tissue, hence sucking insects are in general unaffected, forming the basis of selectivity in favour of many insect predators and parasites. It is effective at 1.5-30 g a.i./100 l of water against: leaf-feeding larvae and leaf miners in forestry (*Lymantria dispar, Thaumetopoea pityocampa*), top fruit (*Cydia pomonella, Psylla* spp.), citrus (*Phyllocoptruta oleivora*), field crops including cotton and soyabeans (*Anthonomus grandis, Anticarsia gemmatalis*), horticultural crops (*Pieris brassicae*). Also the larvae of Sciaridae and Phoridae in mushrooms (1 g/$m^2$ casing at case mixing, or as a drench in 2.5 l of water to the finished casing); mosquito larvae (20-45 g/ha water surface); fly larvae (*Stomoxys calcitrans, Musca domestica*) as a surface application in animal housings (0.5-1.0 /$m^2$ surface).

**Toxicology.**

Acute oral $LD_{50}$ for male and female rats and mice >4640 mg/kg. Acute percutaneous $LD_{50}$ for rabbits >2000 mg/kg. Acute intraperitoneal $LD_{50}$ for mice >2150 mg/kg. In 2-y feeding trials NEL for rats was 40 mg/kg diet. No effect was observed in teratogenic, mutagenic and oncogenic studies. $LC_{50}$ (8-d) for mallard duck and bobwhite quail was 4640 mg/kg diet. $LC_{50}$ (96-h) is: for bluegill 135 mg/l; for rainbow trout 140 mg/l. Relatively non-hazardous to honeybees and predatory insects.

**Formulations.**

'Dimilin' WP-25, w.p. (250 g a.i./kg); 'Dimilin' G1 and G4 granules (10 or 40 g a.i./kg); 'Dimilin' ODC-45 (450 g/l) for ULV application after dilution with an appropriate low viscosity, low volatility, non-phytotoxic oil (for thermal fogging a suitable diluent should be used).

**Analysis.**

Product analysis is by hplc. Residues may be determined by hplc or by glc after hydrolysis to 4-chloroaniline which is converted to a derivative (B. Rabenort *et al, Anal. Methods Pestic. Plant Growth Regul.*, 1978, **10**, 57).

$C_{12}H_{14}N_2S$ (218.3)
T5NYSJ A1 BUNR B1 D1

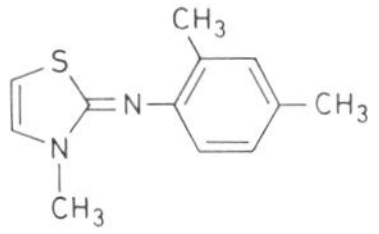

**Nomenclature and development.**

Chemical name (IUPAC) *N*-(2,3-dihydro-3-methyl-1,3-thiazol-2-ylidene)-2,4-xylidine; *N*-(3-methyl-4-thiazolin-2-ylidene)-2,4-xylidine. (*C.A.*) 2,4-dimethyl-*N*-(3-methyl-2(3*H*)-thiazolylidene)benzenamine (9CI); Reg. No. *[61676-87-7]*. Its ixodicidal activity was reported by R. M. Immler *et al.* (*Proc. Br. Crop Prot. Conf. - Pests Dis.*, 1977, **2**, 383). Introduced by Ciba-Geigy AG (BEP 841 504; GBP 1 527 807) as code no. 'CGA 50 439'; trade mark 'Tifatol'.

**Properties.**

The pure compound forms colourless crystals; m.p. 44 °C; v.p. 2.4 mPa at 20 °C; $\rho$ 1.19 g/cm$^3$ at 20 °C. Solubility at 20 °C: 150 mg/l water (pH 9); 800 g/kg benzene, dichloromethane, methanol; 110 g/kg hexane. No significant hydrolysis occurred during 28 d at each pH-value at temperatures $\leqslant$70 °C. At 21 °C p*K* is 5.2.

**Uses.**

Used in dips or as a spray it controls all tick species, including strains resistant to organochlorine, organophosphorus and carbamate ixodicides at concentrations $>$0.1 g a.i./l.

**Toxicology.**

Acute oral $LD_{50}$ for rats 725 mg/kg. Acute percutaneous $LD_{50}$ for rats $>$3100 mg/kg; slight irritant to skin and eyes of rabbits. $LC_{50}$ (96-h) is: for rainbow trout 12 mg/l; for carp 32 mg/l. It is practically non-toxic to birds.

**Formulations.**

'Tifatol', e.c. (300 g a.i./l).

**Analysis.**

Product analysis is by glc. Residues are determined by glc using FPD or PND. Particulars are available from Ciba-Geigy AG.

$C_{12}H_{18}O_7$ (274.3)
T B556 CO EO GO JO LOTJ D1 D1
FVQ K1 K1
*dikegulac-sodium*
$C_{12}H_{17}NaO_7$ (296.3)
T B556 CO EO GO JO LOTJ D1 D1
FVQ K1 K1 &&Na SALT

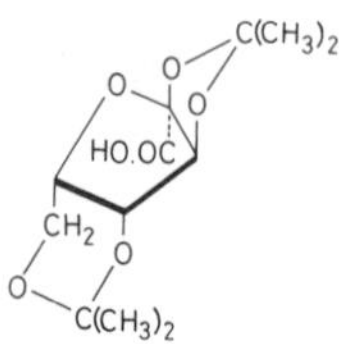


**Nomenclature and development.**

Common name dikegulac (BSI, E-ISO, F-ISO, ANSI). Chemical name (IUPAC) 2,3:4,6-di-*O*-isopropylidene-α-L-*xylo*-2-hexulofuranosonic acid (I). (*C.A.*) 2,3:4,6-bis-*O*-(1-methylethylidene)-α-L-*xylo*-2-hexulofuranosonic acid (9CI); (I) (8CI); Reg. No. *[18476-77-1]* (acid), *[52508-35-7]* (sodium salt). The plant growth regulating activity of the sodium salt was described by P. Bocion *et al.* (*Nature (London)*, 1975, **258**, 142). Dikegulac was introduced as a herbicide and plant growth regulator by F. Hoffman-La Roche & Co. and dikegulac-sodium by Dr R. Maag Ltd as code no. 'Ro 07-6145' (Hoffmann-La Roche for acid); trade marks for dikegulac-sodium 'Atrinal' (to Maag), 'Cutlass' (to ICI Plant Protection Division).

**Properties.**

Dikegulac-sodium is a colourless powder, m.p. >300 °C; v.p. <13 nPa at room temperature. Solubility at room temperature: 590 g/kg water; <10 g/kg acetone, cyclohexanone, hexane; 63 g/kg chloroform; 230 g/kg ethanol; 390 g/kg methanol. It is insensitive to light and is stable, stored at room temperature in the dry state, for ⩾3 y.

**Uses.**

Dikegulac-sodium is a systemic plant growth regulator which reduces apical dominance and increases side-branching and flower-bud formation on ornamental plants and temporarily retards longitudinal growth on hedges and woody ornamentals. An additional side-branching effect on the lower, older parts of hedges is initiated which improves the foliar coverage. The addition of a surfactant is necessary and hedges should not be treated until at least 4 y old. Rates of use range from 0.06-6.0 g a.i./l.

**Toxicology.**

Acute oral $LD_{50}$: for rats 18 000-31 000 mg dikegulac-sodium/kg; for mice 19 500 mg/kg. Acute percutaneous $LD_{50}$ for rats >2000 mg/kg; aqueous solutions (200 g/l) caused no skin irritation to guinea-pigs and (30 g/l) no eye irritation to rabbits. In 90-d feeding trials no significant effect was observed; in rats receiving 2000 mg/kg daily; in dogs 3000 mg/kg daily. $LC_{50}$ (5-d) for Japanese quail, mallard duck and chickens was >50 000 mg/kg diet. $LC_{50}$ (96-h) is: for bluegill >10 000 mg/l; for goldfish, harlequin fish, rainbow trout and Japanese black carp >5000 mg/l. $LD_{50}$ by oral and topical application to honeybees was >0.1 mg/bee.

**Formulations.**

'Atrinal', soluble concentrate (200 g dikegulac-sodium/l). A special surfactant (37.5 g 2-[nonylphenoxy(polyethoxy)]ethanol/l) is used in conjunction with it.

**Analysis.**

Product and residue analysis is by glc after conversion to an ester of dikegulac.

$C_{13}H_{18}ClNO_2$ (255.7)
G1VN2O1&R B1 F1

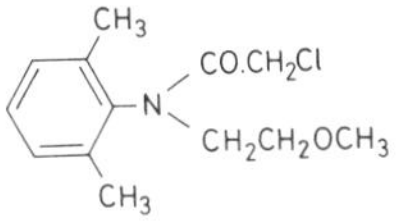

**Nomenclature and development.**

Common name dimethachlor (BSI, E-ISO, WSSA); diméthachlore (F-ISO). Chemical name (IUPAC) 2-chloro-*N*-(2-methoxyethyl)acet-2',6'-xylidide (*C.A.*) 2-chloro-*N*-(2,6-dimethylphenyl)-*N*-(2-methoxyethyl)acetamide (9CI); Reg. No. *[50563-36-5]*. Its herbicidal properties were described by J. Cortier *et al.* (*C. R. Journ. Étud. Herbic. Conf., COLUMA, 9th*, 1977, **1**, 99). Introduced by Ciba-Geigy AG (BEP 795 021; GBP 1 422 473) code no. CGA '17 020'; trade mark 'Teridox'.

**Properties.**

Pure dimethachlor forms colourless crystals; m.p. 47 °C; v.p. 2.1 mPa at 20 °C; ρ 1.21 g/cm$^3$ at 20 °C. Solubility at 20 °C: 2.1 g/l water; >800 g/kg benzene, dichloromethane, methanol; 340 g/kg octan-1-ol. On hydrolysis at 20 °C, 50% loss occurs (calculated) in >200 d at 1<pH<9, 9.3 d at pH13.

**Uses**

Its is a selective herbicide for use on rape, effective against annual broad-leaved weeds and grasses, *e.g. Alopercurus myosuroides*, *Apera spica-venti* and *Poa annua.* It is used immediately after sowing at 1.25-2.00 kg a.i./ha depending on the o.m. content of the soil.

**Toxicology.**

Acute oral $LD_{50}$ for rats 1600 mg tech./kg. Acute percutaneous $LD_{50}$ for rats >3170 mg/kg; slight irritant to skin and eyes of rabbits. In 90-d feeding trials NEL was: for rats 700 mg/kg diet (47 mg/kg daily); for dogs 350 mg/kg diet (10.2 mg/kg daily). $LC_{50}$ (96-h): for rainbow trout 3.9 mg/l; for crucian carp 8 mg/l; for bluegill 15 mg/l. It is slightly toxic to birds and honeybees.

**Formulations.**

'Teridox 500 EC', e.c. (500 g a.i./l).

**Analysis.**

Product analysis is by glc. Residues may be determined by glc with FID. Particulars are available from Ciba-Geigy AG.

$C_{11}H_{21}N_5S$ (255.4)
T6N CN ENJ BS1 DMY1&Y1&1 FM2

**Nomenclature and development.**

Common name dimethametryn (BSI, E-ISO, JMAF);diméthamétryne (F-ISO). Chemical name (IUPAC) *N*-(1,2-dimethylpropyl)-*N*'-ethyl-6-methylthio-1,3,5-triazine-2,4-diyldiamine; 2-(1,2-dimethylpropylamino)-4-ethylamino-6-methylthio-1,3,5-triazine. (*C.A.*) *N*-(1,2-dimethylpropyl)-*N*'-ethyl-6-(methylthio)-1,3,5-triazine-2,4-diamine (9CI); 2-(1,2-dimethylpropylamino)-4-(ethylamino)-6-(methylthio)-*s*-triazine (8CI); Reg. No. *[22936-75-0]*. Its herbicidal properties were described by D. H. Green & L. Ebner (*Proc. Br. Weed Control Conf., 11th*, 1972, p. 822). Introduced by Ciba-Geigy AG (BEP 714 992; GBP 1 191 585) as code no. 'C 18 898'; trade mark 'Avirosan'.

**Properties.**

Pure dimethametryn forms colourless crystals; m.p. 65 °C; b.p. 151-153 °C/0.05 mmHg; v.p. 186 μPa at 20 °C; ρ 1.098 g/cm$^3$ at 20 °C. Solubility at 20 °C: 50 mg/l water; 650 g/l acetone; 800 g/l dichloromethane; 60 g/l hexane; 700 g/l methanol; 350 g/l octan-1-ol; 600 g/l toluene. No significant hydrolysis occurred during 28 d at $5 \leqslant pH \leqslant 9$ at 70 °C.

**Uses.**

Dimethametryn is a selective herbicide active against broad-leaved weeds in rice. It is marketed in combination with piperophos to control both mono- and di-cotyledonous weeds. Rates of 1-2 kg total a.i./ha are used in transplanted rice and 2-3 kg/ha in seeded rice.

**Toxicology.**

Acute oral $LD_{50}$ for rats 3000 mg tech/kg. Acute percutaneous $LD_{50}$ for rats >2150 mg/kg; non-irritant to skin and slightly irritant to eyes of rabbits. In 90-d feeding trials NEL was: for rats 300 mg/kg diet (*c.* 27 mg/kg daily); for dogs >1000 mg/kg diet (*c.* 29 mg/kg daily). $LC_{50}$ (96-h): for rainbow trout 5 mg/l; for crucian carp 8 mg/l.

**Formulations.**

Mixture (Reg. No. *[12701-69-8]*), dimethametryn + piperophos (1:4): 'Avirosan' 3G, 'Avirosan' 5.5G granules (30 or 55 g total a.i./kg); 'Avirosan' 500EC, e.c. (500 g/l).

**Analysis.**

Product analysis is by glc. Residues may be determined by glc with TID. Particulars are available from Ciba-Geigy AG.

$C_6H_{10}O_4S_2$ (210.3)
T6SW DSW BUTJ E1 F1

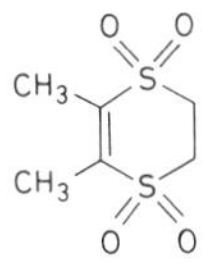

**Nomenclature and development.**

Common name dimethipin (BSI, ANSI, draft E-ISO). Chemical name (IUPAC) 2,3-dihydro-5,6-dimethyl-1,4-dithi-ine 1,1,4,4-tetraoxide. (*C.A.*) 2,3-dihydro-5,6-dimethyl-1,4-dithiin 1,1,4,4-tetraoxide (9CI); Reg. No. *[55290-64-7]*. Its plant growth regulating properties were described by R. B. Ames *et al.* (*Proc. Beltwide Cotton Produc. Res. Conf.*, 1974, p. 61). Introduced by Uniroyal Chemical Co. (USP 3 920 438) as code no. 'N 252'; trade mark 'Harvade'.

**Properties.**

Technical grade dimethipin is a colourless crystalline solid; m.p. 162-167 °C. Solubility at 25 °C: 3 g/kg water; 180 g/kg acetone; 10 g/kg xylene.

**Uses.**

It is an effective defoliant for cotton, nursery stock, rubber trees and vines. It enhances the maturation and subsequently reduces the seed moisture content at harvest of rice and sunflower.

**Toxicology.**

Acute oral $LD_{50}$ for rats 1180 mg/kg. Acute percutaneous $LD_{50}$ for rabbits $>$12 000 mg/kg. $LC_{50}$ (8-d) for mallard duck is $>$10 000 mg/kg diet; for bobwhite quail 4522 mg/kg diet. Long term studies on rats and mice have shown no oncogenic effect in either species when administered in the diet. $LC_{50}$ (96-h) is: for bluegill 28 mg/l; for rainbow trout 8 mg/l; $LC_{50}$ (48-h) for *Daphnia* is 20 mg/l.

**Analysis.**

Product analysis is by glc or i.r. spectrometry. Residues may be determined by glc using a sulphur-specific detector. Details of methods are available from Uniroyal Chemical Co.

$C_{11}H_{19}N_3O$ (209.3)
T6N CNJ BN1&1 DQ E4 F1

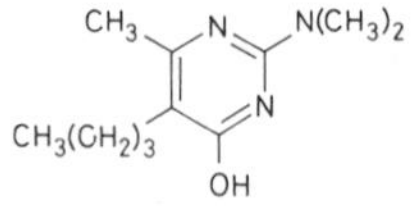

**Nomenclature and development.**

Common name dimethirimol (BSI, E-ISO, JMAF);diméthyrimol (F-ISO). Chemical name (IUPAC) 5-butyl-2-dimethylamino-6-methylpyrimidin-4-ol. (*C.A.*) 5-butyl-2-( dimethylamino)-6-methyl-4(1*H*)-pyrimidinone (9CI); 5-butyl-2-(dimethylamino)-6-methyl-4-pyrimidinol (8CI); Reg. No. *[5221-53-4]*. Its systemic fungicidal properties were described by R. S. Elias *et al.* (*Nature (London)*, 1968, **219**, 1160). Introduced by ICI Plant Protection Division (GBP 1 182 584) as code no 'PP 675'; trade mark 'Milcurb'.

**Properties.**

Dimethirimol forms colourless needles; m.p. 102 °C; v.p. 1.46 mPa at 30 °C. Solubility at 25 °C: 1.2 g/l water; 45 g/l acetone; 1.2 kg/l chloroform; 65 g/l ethanol; 360 g/l xylene. Stable to acidic and alkaline solutions, dissolving in aqueous solutions of strong acids to form salts. Acid formulations should not be stored in galvanised containers.

**Uses.**

A fungicide effective against powdery mildews of chrysanthemum, cineraria and cucurbits. When applied to foliage it enters the plant and moves in the transpiration stream. Relatively stable and loosely absorbed when applied to soil and, following root uptake, can protect the plant for $\geqslant$42 d. Soil applications of $\leqslant$1 g a.i./plant are safe on established plants but young plants with restricted root system should not be overdosed. No significant effect on insect pests, red spider mites or their predators.

**Toxicology.**

Acute oral $LD_{50}$: for rats 2350 mg/kg; for mice 800-1600 mg/kg; for guinea-pigs 500 mg/kg; for hens 4000 mg/kg; for honeybees 4000 mg/kg. Daily applications (14 d) of 500 mg/kg to the shaved skin of rabbits had no effect. In 2-y feeding trials NEL was: for dogs 25 mg/kg; for rats 300 mg/kg diet. $LC_{50}$ at 16 °C for brown trout fingerlings is: (96-h) 28 mg/l, (48-h) 33 mg/l, (24-h) 42 mg/l.

**Formulations.**

Aqueous solutions of the hydrochloride: 'Milcurb' (125 g dimethirimol/l); '*PP 675*' (12.5 g/l) discontinued. The solution is acidic and should not be used in galvanised steel.

**Analysis.**

Product analysis is by glc (J. E. Bagness & W. G. Sharples, *Analyst (London)*, 1974, **99**, 225) or by u.v. spectrometry. Residue analysis is by u.v. spectrometry. Details available from ICI Plant Protection Division.

$C_5H_{12}NO_3PS_2$ (229.2)
1OPS&O1&S1VM1

$CH_3NHCOCH_2SP(S)(OCH_3)_2$

**Nomenclature and development.**

Common name dimethoate (BSI, E-ISO, F-ISO, ANSI, JMAF);fosfamid (USSR). Chemical name *O,O*-dimethyl *S*-methylcarbamoylmethyl phosphorodithioate. (*C.A.*) *O,O*-dimethyl *S*-[2-(methylamino)-2-oxoethyl] phosphorodithioate (9CI); *O,O*-dimethyl phosphorodithioate *S*-ester with 2-mercapto-*N*-methylacetamide (8CI); Reg. No. *[60-51-5]*. OMS 94, OMS 111, ENT 24 650. Its insecticidal properties were described by E. I. Hoegberg & J. T. Cassaday (*J. Am. Chem. Soc.*, 1951, **73**, 557) (*Ital. Agric.*, 1955, **92**, 747). Introduced by American Cyanamid Co., by BASF AG, by Boehringer Sohn (now Celamerck GmbH), by Montecatini S.p.A. (now Montedison S.p.A.) (USP 2 494 283 to American Cyanamid; DEP 1 076 662 to Celamerck; GBP 791 824 to Montedison) as code no. 'E.I. 12 880' (American Cyanamid),'L 395' (Montedison S.p.A.); trade marks 'Cygon' (American Cyanamid), 'Perfekthion' (BASF AG), 'Roxion' (Celamerck GmbH), 'Fostion MM', 'Rogor' (Montedison S.p.A.).

**Properties.**

Pure dimethoate forms colourless crystals; m.p. 51-52 °C; v.p. 1.1 mPa at 25 °C; $d_4^{50}$ 1.281; $n_D^{65}$ 1.5334. The technical grade (96% pure) forms white to greyish crystals; m.p. 45-47 °C. Solubility at 21 °C: 25 g/l water; at 20 °C: >300 g/kg alcohols, benzene, chloroform, dichloromethane, ketones, toluene; >50 g/kg carbon tetrachloride, saturated hydrocarbons; >50 g/kg octan-1-ol. It is relatively stable in aqueous media at 2<pH<7; 50% loss occurs in 12 d at pH9, and is incompatible with alkaline pesticides. It decomposes on heating, initially forming the *O,S*-dimethyl analogue.

**Uses.**

It is a contact and systemic acaricide and insecticide, effective at 300-700 g a.i./ha against a broad range of insects and mites on a wide range of crops. It is effective against houseflies and Diptera of medical importance. It is non-phytotoxic at recommended rates, except to a few citrus, fig, nut and olive varieties.

**Toxicology.**

Acute oral $LD_{50}$: for male rats 500-600 mg a.i./kg, for females 570-680 mg/kg; for males 180-325 mg tech./kg, for females 240-336 mg/kg; for male pheasants 15 mg/kg; for female ducks 40 mg/kg. Acute percutaneous $LD_{50}$ for rats >800 mg/kg; no irritation was observed after application of 130 mg a.i. (as e.c.)/20 $cm^2$ shaved skin of rabbits. In feeding trials (90-d to 1-y) NEL in rats for cholinesterase inhibition was 1.0-32 mg/kg diet. Human volunteers in trials ≤57 d showed NEL for cholinesterase inhibition at 15 mg daily. $LC_{50}$ (96-h) for mosquito fish is 40-60 mg/l. $LD_{50}$ for honeybees is 0.9 μg/bee.

**Formulations.**

These include: e.c. (200, 400 or 600 g tech./l); 'Rogor AS' 300 g/l for ULV application; w.p. (200 g/kg); granules (50 g/kg).

**Analysis.**

Product analysis is by glc (*CIPAC Handbook*, 1980, **1A**, 1225) or by tlc and estimation of phosphorus in the appropriate spots (*ibid.*, 1983, **1B**, in press), Residues may be determined by glc (*Pestic. Anal. Man.*, 1979, **I**, 201-H, 201-I; D. C. Abbott *et al., Pestic. Sci.*, 1970, **1**, 10; *Analyst (London)*, 1977, **102**, 858; 1980, **105**, 515; W. Wagner & H. Frehse, *Pflanzenschutz-Nachr. (Engl. Ed.)*, 1976, **29**, 54; J. E. Boyd, *Anal. Methods Pestic. Plant Growth Regul.*, 1972, **6**, 357; P. de Pietri-Tonelli *et al., Residue Rev.*, 1965, **11**, 60).

$C_2H_7AsO_2$ (138.0)
Q-AS-O&1&1
*sodium dimethylarsinate*
$C_2H_6AsNaO_2$ (159.9)
Q-AS-O&1&1 &&Na SALT

$(CH_3)_2As(=O).OH$

**Nomenclature and development.**

Common name dimethylarsinic acid (I) (draft BSI, draft E-ISO) as alternative to trivial name cacodylic acid (WSSA, draft E-ISO) is under consideration in lieu of a common name. Chemical name (IUPAC) (I). (*C.A.*) (I) (9CI); hydroxydimethylarsine oxide (8CI); Reg. No. *[75-60-5]*. Its herbicidal properties were described by R. G. Mowev & J. F. Cornman (*Proc. Northeast. Weed Control Conf.*, 1959, **13**, 62). Introduced by Ansul Chemical Co. (which no longer exists) and later by Crystal Chemical Co. (USP 3 056 668) as trade marks '*Ansar*', 'Phytar'.

**Properties.**

Pure dimethylarsinic acid forms colourless crystals; m.p. 192-198 °C. Solubility at 25 °C: 2 kg/kg water; soluble in short-chain alcohols; insoluble in diethyl ether. The technical grade is 65% pure, sodium chloride being one impurity. It is decomposed by powerful oxidising or reducing agents. Its aqueous solution is mildly corrosive. The acid has $pK_a$ 6.29 (D. Wauchope, *J. Agric. Food Chem.*, 1976, **24**, 717) and at *c*. pH7 forms a water-soluble sodium salt (Reg. No. *[124-65-2]*), which is deliquescent.

**Uses.**

Dimethylarsinic acid is used as a non-selective post-em. herbicide to control weeds in non-crop situations, for lawn renovation at 11-17 kg/ha, as a desiccant and defoliant for cotton at 1.1-1.7 kg/ha and for killing unwanted trees by injection. It is virtually inactivated on contact with soil.

**Toxicology.**

Acute oral $LD_{50}$ for rats 1350 mg tech./kg; it is non-irritating to the skin and eyes of rabbits.

**Formulations.**

These include: 'Phytar 560', w.s.c. sodium dimethylarsinate (300 g a.e. + surfactant/l).

**Analysis.**

Product analysis is by acid/base titration of the oxidised product (E. A. Dietz & L. O. Moore, *Anal. Methods Pestic. Plant Growth Regul.*, 1978, **10**, 385). Residues may be determined by oxidation followed by reduction to arsine which is determined by colorimetry (*idem, ibid.*).

# Dimethyl phthalate

$C_{10}H_{10}O_4$ (194.2)
1OVR BVO1

$CO.OCH_3$
$CO.OCH_3$

**Nomenclature and development.**

The chemical name (IUPAC) dimethyl phthalate (I) (BSI, E-ISO); phtalate de diméthyle (F-ISO) is accepted in lieu of a common name. (*C.A.*) dimethyl 1,2-benzenedicarboxylate (9CI); (I) (8CI); Reg. No. *[131-11-3]*. Trivial name DMP. ENT 262. It was introduced as an insect repellent during the 1939-45 war, having long been used as a plasticiser.

**Properties.**

Dimethyl phthalate is a colourless viscous liquid; b.p. 282-285 °C, 147.6 °C/10 mmHg; v.p. 1.3Pa at 20 °C; $d_{20}^{20}$ 1.194, $n_D^{20}$ 1.5168. Solubility at room temperature 4.3 g/kg water; soluble in diethyl ether, ethanol, petroleum oils and most organic solvents. It is hydrolysed by alkali.

**Uses.**

It is an insect repellent used for personal protection from biting insects.

**Toxicology.**

Acute oral $LD_{50}$ for rats 8200 mg/kg. Acute percutaneous $LD_{50}$ (9 d) for rats >4800 mg/kg. It may cause smarting if applied to the eye or mucous membranes. In 2-y feeding trials there was no effect on the growth rate of rats receiving 20 000 mg/kg diet.

**Formulations.**

Applied alone or incorporated into creams.

**Analysis.**

Product analysis is by glc or by alkaline hydrolysis and estimation of the resulting phthalic acid by standard methods, or by estimation of methoxy groups by standard methods.

# Dimethyl(4-piperidinocarbonyloxy-2,5-xylyl)sulphonium toluene-4-sulphonate

$C_{23}H_{31}NO_5S_2$ (465.6)
T6NTJ AVOR B1 E1 DS1&1
&&4-$CH_3C_6H_4SO_3H$ SALT

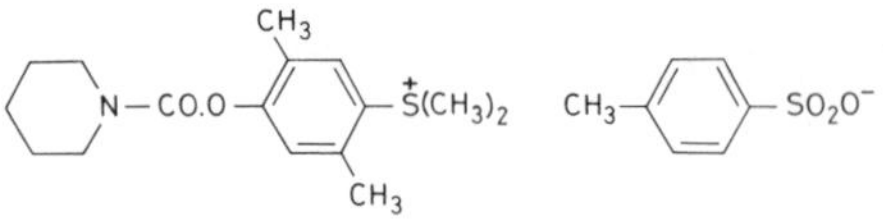

**Nomenclature and development.**

Dimethyl(4-piperidinocarbonyloxy-2,5-xylyl)sulphonium toluene-4-sulphonate (IUPAC). (*C.A.*) [2,5-dimethyl-4-[(1-piperidinylcarbonyl)oxy]phenyl]dimethylsulfonium 4-methylbenzenesulfonate Reg. No. *[68721-60-8]*. Its plant growth-regulating properties were described by J. F. Garrod *et al.* (*Monograph British Plant Growth Regulator Group*, 1980, No. 4, p. 67). Investigated by Boots Co. Ltd (now FBC Limited). (GBP 1 593 541) as code no. 'BTS 44 584'.

**Uses.**

It retards the growth of various crops, including beans, chrysanthemums, cotton, soyabeans and sunflowers.

**Toxicology.**

Acute oral $LD_{50}$ for rats >400 mg/kg.

**Formulations.**

Water-soluble powder (900 g a.i./kg).

$C_8H_{12}NO_5PS_2$ (297.3)
ZSWR DOPS&O1&O1

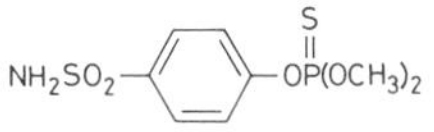

$NH_2SO_2$–C₆H₄–$OP(S)(OCH_3)_2$

**Nomenclature and development.**

Chemical name (IUPAC) *O,O*-dimethyl *O*-4-sulphamoylphenyl phosphorothioate. (*C.A.*) *O,O*-dimethyl *O*-(4-aminosulfonylphenyl)phosphorothioate (9CI); *O,O*-dimethyl phosphorothioate *O*-ester with *p*-hydroxybenzenesulfonamide (8CI); Reg. No. *[115-93-5]*. Trivial name cythioate. Introduced for the control of animal ectoparasites by American Cyanamid Co. (USP 3 435 100) as code no. 'CL 26 691'; trade mark 'Cyflee'.

**Properties.**

It is a colourless solid; m.p. 73-74 °C. Insoluble in water; soluble in acetone, benzene, diethyl ether, ethanol.

**Uses.**

It is used for repeated oral administration in dogs to control fleas (Ctenocephalides), mange mites (Demodex), and ticks (Ixodes); and in cats to control fleas (Ctenocephalides).

**Toxicology.**

Acute oral $LD_{50}$ for rats 160 mg tech./kg. Acute percutaneous $LD_{50}$ for rabbits >2500 mg/kg.

**Formulations.**

These include: tablets (30 mg); liquid (15 mg/ml) for oral administration.

$C_{11}H_{13}F_3N_4O_4$ (322.2)
FXFFR BZ CNW ENW DN2&2

$NO_2$ / $CF_3$ / $N(CH_2CH_3)_2$ / $NH_2$ $NO_2$

**Nomenclature and development.**

Common name dinitramine (BSI, E-ISO, F-ISO, ANSI, WSSA). Chemical name (IUPAC) $N^1,N^1$-diethyl-2,6-dinitro-4-trifluoromethyl-*m*-phenylenediamine. (*C.A.*) $N^3,N^3$-diethyl-2,4-dinitro-6-(trifluoromethyl)-1,3-benzenediamine (9CI); $N^1,N^1$-diethyl-2,6-dinitro-4-(trifluoromethyl)-1,3-benzenediamine (9CI); $N^4,N^4$-diethyl-α,α,α-trifluoro-3,5-dinitrotoluene-2,4-diamine (8CI); Reg. No. *[2091-05-2]*. Introduced by US Borax & Chemical Corp. (who no longer manufacture or market it) and later under licence by Wacker GmbH (USP 3 617 252 to US Borax) as code no. 'USB 3584'; trade mark 'Cobex'.

**Properties.**

Pure dinitramine is a yellow crystalline solid; m.p. 98-99 °C; v.p. 479 μPa at 25 °C, and decomposing at ⩾70 °C. Solubility at 25 °C: 1.1 mg/l water; 640 g/kg acetone; 120 g/kg ethanol. The technical grade is >83% pure *m/m*. Both pure and technical dinitramine showed no significant decompostion after 2-y storage at ambient temperatures. It is subject to photo-degradation.

**Uses.**

It is a pre-em. herbicide which is incorporated in the soil against many annual grasses and broad-leaved weeds in beans, carrots, cotton, groundnuts, soyabeans, sunflowers, swedes, turnips, transplanted brassicas, transplanted peppers and transplanted tomatoes at 0.4-0.8 kg a.i./ha. It has little post-em. activity.

**Toxicology.**

Acute oral $LD_{50}$ for rats 3000 mg a.i./kg. Acute percutaneous $LD_{50}$ for rabbits >6800 mg/kg. In 90-d feeding trials no ill-effect was observed for rats and beagle dogs receiving 2000 mg/kg diet. In 2-y trials no carcinogenic response was observed in rats receiving 100 or 300 mg/kg diet. $LC_{50}$ (96-h) is: for trout 6.6 mg/l; for bluegill 11 mg/l; for catfish 3.7 mg/l.

**Formulations.**

An e.c. (250 g/kg; 240 g/l).

**Analysis.**

Product analysis is by glc, details from Wacker Chemie GmbH. Residues may be determined by glc with ECD (H. C. Newsom & E. M. Mitchell, *J. Agric. Food Chem.*, 1972, **20**, 1222; H. C. Newsom, *Anal. Methods Pestic. Plant Growth Regul.*, 1976, **8**, 359).

$C_{14}H_{18}N_2O_7$ (326.3)
2Y1&R CNW ENW BOVOY1&1

$(CH_3)_2CHO.CO.O$ $CH_3$
$O_2N$ $CHCH_2CH_3$
$NO_2$

**Nomenclature and development.**

Common name dinobuton (BSI, E-ISO, F-ISO). Chemical name (IUPAC) 2-*sec*-butyl-4,6-dinitrophenyl isopropyl carbonate (I). (*C.A.*) 1-methylethyl 2-(1-methylpropyl)-4,6-dinitrophenyl carbonate (9CI); (I) (8CI); Reg. No. *[973-21-7]*. OMS 1056, ENT 27 244. Its acaricidal and fungicidal properties were described by M. Pianka & C. B. F. Smith (*Chem. Ind. (London)*, 1956, p. 1216). Introduced by Murphy Chemical Ltd (now owned by Dow Chemical Co.) and currently manufactured by KenoGard AB (GBP 1 019 451) as code no. 'MC 1053' (Murphy); trade marks 'Acrex' (KenoGard and formerly Murphy), '*Sytasol*' (formerly Murphy).

**Properties.**

Pure dinobuton forms pale yellow crystals; m.p. 61-62 °C; v.p. negligible at room temperature. Solubility: practically insoluble in water; soluble in ethanol, aliphatic hydrocarbons, fatty oils; highly soluble in aromatic hydrocarbons, lower aliphatic ketones. The technical grade (97% pure) has m.p. 58-60 °C. It is non-corrosive but, to prevent hydrolysis, alkaline components must be avoided in its formulation. It is deactivated by carbaryl.

**Uses.**

It is a non-systemic acaricide and a fungicide active against powdery mildews; it is recommended for glasshouse and field use against red spider mites and powdery mildews of apples, cotton and vegetables at 50 g a.i./100 l.

**Toxicology.**

Acute oral $LD_{50}$: for mice 2540 mg/kg; for rats 140 mg/kg; for hens 150 mg/kg. Acute percutaneous $LD_{50}$ for rats >5000 mg/kg. NEL was for dogs 4.5 mg/kg daily; for rats 3-6 mg/kg daily. It acts as a metabolic stimulant, high doses causing loss of body weight. Its persistence in soils is brief.

**Formulations.**

An e.c. (300 g a.i./l).

**Analysis.**

Product and residue analysis is by hydrolysis to dinoseb (5190) which is measured colorimetrically (D. S. Farrington *et al.*, *Analyst (London)*, 1983, **103**, 353; *CIPAC Handbook*, in press; V. P. Lynch, *Anal. Methods Pestic. Plant Growth Regul.*, 1976, **8**, 275). Details are available from KenoGard AB.

$C_{18}H_{24}N_2O_6$ (364.3)

(i, n = 0) 6Y1&R CNW ENW BOV1U2
(i, n = 1) 5Y2&R CNW ENW BOV1U2
(i, n = 2) 4Y3&R CNW ENW BOV1U2

(ii, n = 0) 6Y1&R CNW ENW DOV1U2
(ii, n = 1) 5Y2&R CNW ENW DOV1U2
(ii, n = 2) 4Y3&R CNW ENW DOV1U2

**Nomenclature and development.**

Common name dinocap (BSI, E-ISO, F-ISO);DPC (JMAF);*exception* (Republic of South Africa) for an isomeric reaction mixture. It was originally thought to be the chemical (IUPAC) 2-(1-methylheptyl)-4,6-dinitrophenyl crotonate **(I, i, n = 0)** but is actually a mixture of isomers (see below). (*C.A.*) 2-(1-methylheptyl)-4,6-dinitrophenyl (*E*)-2-butenoate (9CI); (I) (8CI); Reg. No. *[131-72-6]* (single compound), *[34300-45-3]* (actual dinocap). ENT 24 727. Introduced by Rohm & Haas Co. (USP 2 526 660; 2 810 767) as code no. 'CR-1693'; trade marks 'Karathane', formerly '*Arathane*' (both to Rohm & Haas)

**Properties.**

Alkylation of phenol by octan-1-ol, and nitration of the product gives a mixture containing 2,4-dinitro-6-octylphenols corresponding to (i, n = 0, 1, 2) and 2,6-dinitro-4-octylphenols (ii, n = 0, 1, 2) in which octyl is a mixture of 1-methylheptyl, 1-ethylhexyl and 1-propylpentyl substituents. Esterification with crotonyl chloride provides dinocap, a dark brown liquid; b.p. 138-140 °C/0.05 mmHg; *d* 1.10. Solubility: sparingly soluble in water; soluble in most organic solvents.

The commercial material comprises 2.0-2.5 parts of (i)-isomers to 1.0 of (ii). Various workers have shown that compounds of type (i) are more effective as acaricides, and those of type (ii) as fungicides (A. H. M. Kirby & L. D. Hunter, *Nature (London)*, 1965, **208**, 189; A. H. M. Kirby *et al., Ann. Appl. Biol.*, 1966, **57**, 21; R. J. W. Byrde *et al., ibid.*, p. 223).

**Uses.**

Dinocap is a non-systemic acaricide and contact fungicide, recommended to control powdery mildews on various fruits, grape vines and ornamentals. it also reduces the population of *Panonychus ulmi* on top fruit. It is also used in seed treatments.

**Toxicology.**

Acute oral $LD_{50}$ for rats 980-1190 mg/kg (F. S. Larson *et al., Arch. Int. Pharmacodyn. Ther.*, 1959, **119**, 31). In 1-y feeding trials dogs receiving 50 mg/kg diet suffered no loss in weight; cataracts were produced in white Pekin ducks at this dose level.

**Formulations.**

These include: w.p. (250, 500 or 830 g a.i./kg). Mixtures with gamma-HCH and other fungicides for seed treatments. Mixtures with various other fungicides, especially systemic ones, to broaden and prolong the range of activity.

**Analysis.**

Residues may be determined by glc (*Pestic. Anal. Man.*, 1979, **I**, 201-A, 201-G, 201-I; A. Ambrus *et al., J. Assoc. Off. Anal. Chem.*, 1981, **64**, 733; C.E. Johansson, *Pestic. Sci.*, 1975, **6**, 97; W. Specht & M. Tillkes, *Fresenius' Z. Anal. Chem.*, 1980, **301**, 300; *Anal. Methods Pestic. Plant Growth Regul.*, 1972, **6**, 568).

$C_{10}H_{12}N_2O_5$ (240.2)
WNR BQ ENW CY2&1

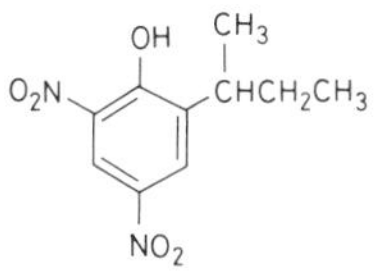

**Nomenclature and development.**

Common name dinoseb (BSI, E-ISO, ANSI, WSSA); dinosèbe (F-ISO); DNBP (JMAF). Chemical name (IUPAC) 2-*sec*-butyl-4,6-dinitrophenol (I). (*C.A.*) 2-(1-methylpropyl)-4,6-dinitrophenol (9CI); (I) (8CI); Reg. No. *[88-85-7]*. Trivial name DNBP. ENT 1122. Its herbicidal properties were described by A. S. Crafts (*Science*, 1945, **101**, 417). Developed by Dow Chemical Co. (who no longer manufacture or market it) (USP 2 192 197) as code no. 'DN 289'; trade mark '*Premerge*' (Dow).

**Properties.**

Pure dinoseb is an orange solid; m.p. 38-42 °C. The technical grade (95-98% pure) is an orange-brown solid; m.p. 30-40 °C. Solubility at room temperature: *c.* 100 mg/l water; soluble in petroleum oils and most organic solvents. It is acidic (p$K_a$ 4.62) forming salts, some of which are water-soluble, with inorganic and organic bases. In the presence of water it is corrosive to mild steel.

**Uses.**

It is a contact herbicide used as the ammonium or an amine salt for post-em. weed control in cereals, undersown cereals, seedling lucerne and peas at $\leqslant$2 kg a.i./ha. Solutions in oil are used for pre-em. control of annual weeds in beans, peas and potatoes, for pre-harvest desiccation of hops, leguminous seed crops and potatoes and for control of runners and suckers in raspberries and strawberries.

**Toxicology.**

Acute oral $LD_{50}$ for rats 58 mg/kg. Acute percutaneous $LD_{50}$ for rabbits 80-200 mg/kg. In 180-d feeding trials rats receiving 100 mg/kg diet suffered no ill-effect.

**Formulations.**

Aqueous concentrates of amine or ammonium salts, solutions in oil.

**Analysis.**

Product analysis is by colorimetry of solutions in alkali (*CIPAC Handbook*, 1970, **1**, 337; D. S. Farrington *et al.*, *Analyst (London)*, 1983, **108**, 353). Residues are determined by glc of a derivative (*Anal. Methods Pestic. Plant Growth Regul.*, 1972, **6**, 639).

$C_{12}H_{14}N_2O_6$ (282.2)
2Y1&R CNW ENW BOV1

$CH_3CO.O$ $CH_3$
$O_2N$ $CHCH_2CH_3$
$NO_2$

**Nomenclature and development.**

The common name dinoseb is approved (BSI, E-ISO, ANSI, WSSA); dinosèbe (F-ISO); DNBP (JMAF). Its acetate ester has common name dinoseb acetate (BSI, E-ISO); dinosèbe-acétate (F-ISO); DNBPA (JMAF). Chemical name (IUPAC) 2-*sec*-butyl-4,6-dinitrophenyl acetate (I). (*C.A.*) 2-(1-methylpropyl)-4,6-dinitrophenyl acetate (9CI); (I) (8CI); Reg. No. *[2813-95-8]*. Its herbicidal properties were described by K. Härtel (*Meded. Landbouwhogesch. Opzoekingsstn. Staat Gent*, 1960, **25**, 1422). Introduced by Hoechst AG (DEP 1 088 757; GBP 909 372) as code no. 'Hoe 02904'; trade mark 'Aretit', 'Ivosit'.

**Properties.**

Pure dinoseb acetate is a brown oil with an aromatic vinegar-like odour; m.p. 26-27 °C; v.p. 183 mPa at 60 °C. The technical grade (*c.* 94% pure) is a viscous brown oil. Solubility at 20 °C: 1.6 g tech./l water; soluble in aromatic solvents. The ester is slowly hydrolysed in the presence of water and is sensitive to acid or alkali. The technical product is slightly corrosive.

**Uses.**

Dinoseb acetate is a post-em. herbicide recommended against annual broad-leaved weeds in beans, cereals, lucerne, maize, peas and potatoes at 1.5-2.5 kg a.i./ha. It is also used in combination with monolinuron pre-em. in dwarf beans and potatoes.

**Toxicology.**

Acute oral $LD_{50}$ for rats 60-65 mg/kg. The application of 200 mg w.p./kg 5 times to the shaved skin of rabbits caused no irritation. In 2-y feeding trials NEL was: for rats 100 mg/kg diet; for dogs 8 mg/kg diet.

**Formulations.**

These include: w.p. (376 g a.i./kg); e.c. (492 g/l). 'Aresin combi' (Reg. No. *[8070-87-9]*), w.p. (dinoseb acetate + monolinuron).

**Analysis.**

Product analysis is by colorimetry after hydrolysis to dinoseb (D. S. Farrington *et al.*, *Analyst (London)*, 1983, **108**, 353). Residues in soil may be determined by glc after conversion to the corresponding methyl ether (T. H. Byast *et al.*, *Tech. Rep. ARC Weed Res. Organ.*, No. 15, (2nd Ed.), p. 82).

$C_{10}H_{12}N_2O_5$ (240.2)
WNR BQ ENW CX1&1&1

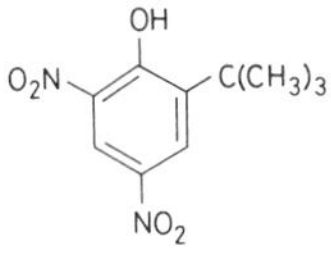

**Nomenclature and development.**

Common name dinoterb (BSI, E-ISO);dinoterbe (F-ISO). Chemical name (IUPAC) 2-*tert*-butyl-4,6-dinitrophenol (I). (*C.A.*) 2-(1,1-dimethylethyl)-4,6-dinitrophenol (9CI); (I) (8CI); Reg. No. *[1420-07-1]*. Its herbicidal properties were described by P. Poignant & P. Crisinel (*C. R. Journ. Etud. Herbic. Conf. COLUMNA, 4th*, 1967, p. 196). Introduced by Pépro (now a subsidiary of Rhône-Poulenc Phytosanitaire) and by Murphy Chemical Ltd (who no longer manufacture or market it) and currently available as amine salts or as mixtures with other herbicides. (FP 1 475 686; 1 532 332; GBP 1 126 658; USP 3 565 601 all to Pépro).

**Properties.**

Dinoterb is a yellow solid; m.p. 125.5-126.5 °C. Solubility: practically insoluble in water; *c.* 100 g/kg alcohols, glycols, aliphatic hydrocarbons; *c.* 200 g/kg cyclohexanone, dimethyl sulphoxide, ethyl acetate. It is acidic, forming water-soluble salts (dinoterb-ammonium, Reg. No. *[6365-83-9]*, dinoterb-bis(2-hydroxyethyl)ammonium).

**Uses.**

It is a contact herbicide used: post-em. to control annual weeds in cereals, lucerne and maize; pre-em. in beans and peas.

**Toxicology.**

Acute oral $LD_{50}$ for mice *c.* 25 mg/kg. Acute percutaneous $LD_{50}$ for guinea-pigs 150 mg/kg.

**Formulations.**

These include 'Herbogil Liquid D', 'Nixone', e.c. dinoterb-bis(2-hydroxyethyl)ammonium (250 g a.e./l). Mixtures include: 'DM 68', dinoterb + mecoprop; 'Tolkan S', dinoterb + isoproturon; dinoterb + nitrofen.

**Analysis.**

Product analysis is by glc of a derivative (*CIPAC Handbook*, 1983, **1B**, in press).

$C_{11}H_{13}NO_4$ (223.2)
T5O COTJ BR BOVM1

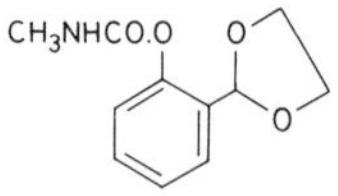

**Nomenclature and development.**

Common name dioxacarb (BSI, E-ISO, ANSI, JMAF);dioxacarbe (F-ISO). Chemical name (IUPAC) 2-(1,3-dioxolan-2-yl)phenyl methylcarbamate (I). (*C.A.*) (I) (9CI); *o*-1,3-dioxolan-2-ylphenyl methylcarbamate (8CI); Reg. No. *[6988-21-2]*. OMS 1102, ENT 27 389. Its insecticidal properties were described by F. Bachmann & J. B. Legge (*J. Sci. Food Agric., Suppl.*, 1968, p. 39). Introduced by Ciba AG (now Ciba-Geigy AG) (BEP 670 630; GBP 1 122 633) as code no. 'C 8353'; trade marks 'Elocron', 'Famid'.

**Properties.**

Pure dioxacarb forms colourless crystals; m.p. 114-115 °C; v.p. 40 μPa at 20 °C; ρ 1.46 g/cm³ at 20 °C. Solubility at 20 °C: 6 g/l water; 280 g/l acetone; 235 g/l cyclohexanone; 550 g/l dimethylformamide: 80 g/l ethanol; <1 g/l kerosene; 9 g/l xylene. In aqueous media at 20 °C, 50% loss (calculated) occurs in: 85 h at pH5; 24.6d at pH7; 15 h at pH9. It is rapidly decomposed in soil and is unsuitable for use against soil pests.

**Uses.**

It is a contact and stomach insecticide used against cockroaches (including those resistant to organochlorine and organophosphorus insecticides) and against a wide range of household and stored product pests, for wall application at 0.5-2.0 g a.i./m². It is effective against a wide range of sucking and chewing foliage pests including aphids resistant to organophosphorus compounds, *Leptinotarsa decemlineata*, plant and leaf hoppers of rice, capsids on cocoa; recommended against sucking pests at 250-500 g/ha, against chewing insects 500-750 g/ha. It has a rapid 'knockdown' action and persists on foliage 5-7d, on wall surfaces *c.* 0.5 y depending on porosity.

**Toxicology.**

Acute oral $LD_{50}$ for rats 72 mg tech./kg; Acute percutaneous $LD_{50}$ for rats *c.* 3000 mg/kg; slightly irritant to skin and eyes of rabbits. Inhalation $LC_{50}$ for rats *c.* 160 mg/m³ air. In 90-d feeding trials NEL was: for rats 10 mg/kg daily; for dogs 2 mg/kg daily (marginal). $LC_{50}$ (96-h): for rainbow trout 29 mg/l; for carp 32 mg/l. In laboratory trials it is toxic to honeybees and slightly toxic to birds.

**Formulations.**

These include: 'Elecron 50 WP', w.p. (500 g/kg); 'Elecron 40 SCW', a.c. (400 g/l).

**Analysis.**

Product analysis is by acidimetric titration (A. Becker & W. Zeizler, *Fresenius' Z. Anal. Chem.*, 1971, **257**, 125). Residues may be determined by glc with ECD of a derivative. Particulars are available from Ciba-Geigy AG.

$C_{12}H_{26}O_6P_2S_4$ (456.5)
T6O DOTJ BSPS&O2&O2 CSPS&O2&O2

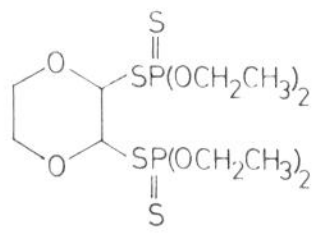

**Nomenclature and development.**

Common name dioxathion (BSI, E-ISO, F-ISO, ANSI);delnav (Turkey, USSR);dioxane phosphate (JMAF);*exception* Italy. Chemical name (IUPAC) *S,S'*-(1,4-dioxane-2,3-diyl) *O,O,O',O'*-tetraethyl bis(phosphorodithioate); *S,S'*-(1,4-dioxane-2,3-diyl) *O,O,O',O'*-tetraethyl di(phosphorodithioate) (I); 1,4-dioxan-2,3-diyl *S,S*-di(*O,O*-diethyl phosphorodithioate). (*C.A.*) (I) (9CI); *S,S-p*-dioxane-2,3-diyl bis(*O,O*-diethyl phosphorodithioate) (8CI); Reg. No. *[78-34-2]*. ENT 22 879. Its insecticidal properties were described by W. R. Diveley *et al.* (*J. Am. Chem. Soc.*, 1959, **81**, 139). Introduced by Hercules Inc. (now Boots-Hercules Agrochemical Co.) (USP 2 725 328; 2 815 350) as code no. 'Hercules AC258'; trade mark 'Delnav'.

**Properties.**

Technical grade dioxathion (68-75% pure, 24% *cis*-, 48% *trans*-isomers, *c.* 30% related compounds) is a brown liquid; $d_4^{26}$ 1.257; $n_D^{20}$ 1.5409. Solubility: insoluble in water; 10 g/kg hexane, kerosene; soluble in most organic solvents. It is stable in water at neutral pH, but hydrolysed by alkali or on heating. Unstable to iron or tin surfaces or when mixed with certain carriers.

**Uses.**

It is a non-systemic insecticide and acaricide used for treatment of livestock to control external pests, including ticks. Recommended against phytophagous mites on fruit (including citrus), nuts and ornamentals. The *cis*- is more toxic to flies and rats than the *trans*-isomer (B. W. Arthur & J. E. Casida, *J. Econ. Entomol.*, 1959, **52**, 20).

**Toxicology.**

Acute oral $LD_{50}$: for male albino rats 43 mg/kg, for females 23 mg/kg. Acute percutaneous $LD_{50}$: for male albino rats 235 mg/kg, for female 63 mg/kg. Cholinesterase inhibition was observed in female rats receiving 10 mg/kg diet. Relatively harmless to honeybees.

**Formulations.**

These include: e.c. of various strengths.

**Analysis.**

Product analysis is by i.r. spectroscopy, details from Boots-Hercules Agrochemical Co. Residues may be determined by glc (*Pestic. Anal. Man.*, 1979, **I**, 201-H, 201-I; D. C. Abbott *et al.*, *Pestic. Sci.*, 1970, **1**, 10).

$C_{12}H_{12}N_2O_3$ (232.2)
T5O COTJ B1ONUYR&CN &&(Z) FORM

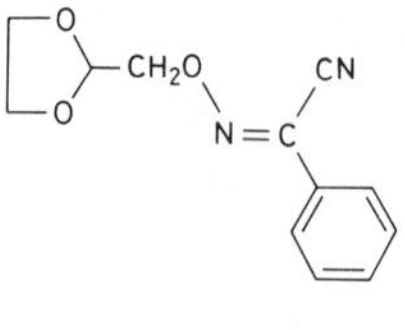

**Nomenclature and development.**

Chemical name (IUPAC) (*Z*)-1,3-dioxolan-2-ylmethoxyimino(phenyl)acetonitrile. (*C.A.*) α-[(1,3-dioxolan-2-yl)methoxyimino]benzeneacetonitrile (9CI); Reg. No. *[74782-23-3]*. Its activity as a safener to metolachlor injury was described by T. R. Dill *et al.* (*Abstr. Ann. Meeting Weed Sci. Soc. Am.*, 1982, 20). Introduced by Ciba-Geigy AG (EUPA 0 011 047) as code no. 'CGA 92 194'; trade mark 'Concep II'.

**Properties.**

The pure compound forms colourless crystals; m.p. 77.7 °C; v.p. 520 μPa at 20 °C; ρ 1.33 $g/cm^3$ at 20 °C. Solubility at 20 °C: 20 mg/l water; 250 g/kg acetone; 300 g/kg cyclohexanone; 450 g/kg dichloromethane; 220 g/kg toluene; 150 g/kg xylene. Stable ⩽240 °C.

**Uses.**

It is a safener which protects sorghum, hybrids as well as various yellow endosperm, sweet sorghum and Sudan grass varieties, from injury by metolachlor. It is applied as a seed treatment at 1-2 g a.i./kg seed, allowing the safe use of metolachlor for the control of a wide range of grasses. Sorghum tolerance is maintained when metolachlor is applied in combination with triazines (atrazine, propazine, terbuthylazine, terbutryne) for additional activity on broad-leaved weeds.

**Toxicology.**

Acute oral $LD_{50}$ for rats >5000 mg tech./kg. Acute percutaneous $LD_{50}$ for rats >5000 mg/kg. Inhalation $LC_{50}$ (4-h) for rats *c.* 1495 $mg/m^3$ air. In laboratory trials it is slightly toxic to birds.

**Formulations.**

'Concep II', w.p. (700 g a.i./kg).

**Analysis.**

Analytical methods not yet specified. Residues are determined by glc using FID. Particulars are available from Ciba-Geigy AG.

*Common name oxabetrinil (BSI, draft E-ISO)

# Diphacinone

$C_{23}H_{16}O_3$ (340.4)
L56 BV DV CHJ CVYR&R

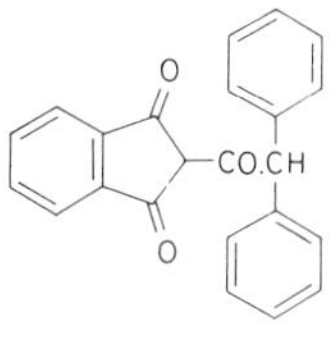

**Nomenclature and development.**

Common name diphacinone (BSI, E-ISO, F-ISO, ANSI);diphacin (Turkey);diphacins (JMAF);diphenadione (BPC);*exception* (Italy). Chemical name (IUPAC) 2-( diphenylacetyl)indan-1,3-dione (I). (*C.A.*) 2-(diphenylacetyl)-1*H*-indene-1,3(2*H*)-dione (9CI); 2-(diphenylacetyl)-1,3-indandione (8CI); Reg. No. *[82-66-6]*. Its rodenticidal activity was described by J. T. Correll *et al.* (*Proc. Soc. Exp. Biol. Med.*, 1952, **80**, 139). Introduced by Velsicol Chemical Corp. and Upjohn Co. (who no longer manufacture or market it) (USP 2 672 483) as trade marks 'Diphacin', 'Ramik' (to Velsicol).

**Properties.**

Technical grade diphacinone is a yellow powder (95% pure); m.p. 145 °C; $d^{25}$ 1.281; v.p. 13.7 nPa at 25 °C; decomposes at 338 °C without boiling. Solubility: 0.3 mg/kg water; 29 g/kg acetone; 204 g/kg chloroform; 2.1 g/kg ethanol; 1.8 g/kg heptane; 73 g/kg toluene; 50 g/kg xylene. It is acidic, forming water-soluble alkali metal salts, diphacinone-sodium solubility 6.5 g/kg. Stable at pH 6-9 for 14 d, it is partly converted to a tautomer at pH 4 and hydrolysed in <24 h. Diphacinone is rapidly decomposed in water by sunlight.

**Uses.**

It is an anticoagulant rodenticide, used to control mice, rats, prairie dogs (*Cynomys* spp.), ground squirrels, voles, and other rodents.

**Toxicology.**

Acute oral $LD_{50}$: for rats 2.3 mg/kg; for dogs *c.* 3 mg/kg; for mallard duck 3158 mg/kg. Acute percutaneous $LD_{50}$ for rats <200 mg/kg, not a primary irritant to their skin or eyes. In a 21-d percutaneous toxicity study in rabbits NEL was 0.1 mg/kg daily. a delayed skin sensitization study in guinea-pigs showed it is neither a skin irritant nor a sensitizer but very toxic, one guinea-pig dying at 2.5 mg. Acute inhalation $LC_{50}$ (4-h) for rats <2 mg/l following exposure to dust. Chronic $LD_{50}$ for albino rats is 0.1 mg/kg daily. It was not mutagenic in the Ames test. $LC_{50}$ (96-h) is: for bluegill 7.6 mg/l; for rainbow trout 2.8 mg/l; for channel catfish 2.1 mg/l. A 56-d secondary poisoning trial with bait (50 mg a.i./kg) revealed no hazard to sparrow hawks under conditions likely to be encountered in nature.

**Formulations.**

'Diphacin' dry concentrates (1 g a.i./kg) for mixing with cereal bait. 'Gold Crest', 'Promar', 'Ramik' weather resistant pellets or meal (50 mg/kg). All products are for professional applications only.

**Analysis.**

Analytical methods are available from Velsicol Chemical Corp.

$C_{16}H_{17}NO$ (239.3)
1N1&VYR&R

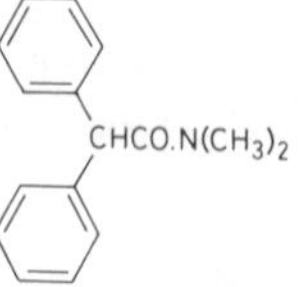


**Nomenclature and development.**

Common name diphenamid (BSI, E-ISO, ANSI, WSSA, JMAF); difénamid (F-ISO); *exception* (Federal Republic of Germany). Chemical name (IUPAC) *N*,*N*-dimethyldiphenylacetamide. (*C.A.*) *N*,*N*-dimethyl-α-phenylbenzeneacetamide (9CI); *N*,*N*-dimethyl-2,2-diphenylacetamide (8CI); Reg. No. *[957-51-7]*. Its herbicidal properties were described by E. F. Alder *et al.* (*Proc. North Cent. Weed Control Conf.*, 1960, p. 55). Introduced by Eli Lilly & Co. and by The Upjohn Co. (USP 3 120 434) as code no. 'L-34 314' (to Lilly); trade marks 'Dymid' (Lilly), 'Enide' (Upjohn).

**Properties.**

Pure diphenamid forms colourless crystals, m.p. 134.5-135.5 °C; v.p. negligible at room temperature. Solublity at 27 °C: 260 mg/l water; moderately soluble in polar organic solvents. The technical grade is an off-white solid, m.p. 132-134 °C, and is stable at room temperature for 7 d at pH3 and pH9.

**Uses.**

It is a selective pre-em. herbicide used to control most grass weeds and many broad-leaved weeds in a wide range of crops including cotton, soyabeans, potatoes, brassicas, fruit trees, soft fruit, ornamentals, tobacco and vegetables. it is generally used at 4-6 kg a.i./ha. It is metabolised in plants and animals by *N*-demethylation.

**Toxicology.**

Acute oral $LD_{50}$ for rats 1050 mg/kg. Acute percutaneous $LD_{50}$ for rats >225 mg/kg. In 2-y feeding trials dogs and rats receiving 2000 mg/kg diet suffered no unusual effect on their physiology or fertility.

**Formulations.**

These include: 'Enide 50W', 'Enide 90W', w.p. (500 or 900 g a.i./kg).

**Analysis.**

Residues may be determined by glc with FID (G. A. Boyack *et al.*, *J. Agric. Food Chem.*, 1966, **14**, 312; J. B. Tepe *et al.*, *Anal. methods Pestic., Plant Growth Regul. Food Addit.*, 1967, **5**, 375; *Anal. Methods Pestic. Plant Growth Regul.*, 1972, **6**, 637).

$C_{11}H_{21}N_5S$ (255.4)
T6N CN ENJ BS2 DMY1&1 FMY1&1

$CH_3CH_2S$ — triazine ring — $NHCH(CH_3)_2$; $NHCH(CH_3)_2$

**Nomenclature and development.**
Common name dipropetryn (BSI, E-ISO, ANSI, WSSA); dipropétryne (F-ISO). Chemical name (IUPAC) 6-ethylthio-*N*,*N*'-di-isopropyl-1,3,5-triazine-2,4-diyldiamine; 2-ethylthio-4,6-bis(isopropylamino)-1,3,5-triazine. (*C.A.*) 6-ethylthio-*N*,*N*'-bis(1-methylethyl)-1,3,5-triazine-2,4-diamine (9CI); 2-(ethylthio)-4,6-bis(isopropylamino)-*s*-triazine (8CI); Reg. No. *[4147-51-7]*. Its herbicidal properties were described by G. A. Buchanan & D. L. Thurlow (*Ala. Agric. Exp. Stn. Highlights*, 1968, **15**(2), 7). Introduced by J. R. Geigy S.A. (now Ciba-Geigy AG) (CHP 337 019; GBP 814 948) as code no. 'GS 16 068'; trade marks 'Sancap' (in USA), 'Cotofor'.

**Properties.**
Pure dipropetryn is a colourless powder; m.p. 104-106 °C; v.p. 97 μPa at 20 °C; ρ 1.120 $g/cm^3$ at 20 °C. Solubility at 20 °C: 16 mg/l water; 270 g/l acetone; 300 g/l dichloromethane; 9 g/l hexane; 190 g/l methanol; 130 g/l octan-1-ol; 220 g/l toluene. On hydrolysis at 25 °C 50% loss occurs in: 26 ± 1.9 d at pH1; >2.5 y at 7<pH<13.

**Uses.**
It is a pre-em. herbicide used for weed control in cotton and water-melons at 1.25-3.5 kg a.i./ha.

**Toxicology.**
Acute oral $LD_{50}$ for rats 3900-4200 mg/kg. Acute percutaneous $LD_{50}$ for rabbits >10 000 mg/kg; mild irritant to skin, non-irritant to eyes of rabbits. In 133-d feeding trials NEL in 98-d trials for rats was 400 mg/kg diet (27 mg/kg daily); for dogs was 400 mg/kg diet (13 mg/kg daily). $LC_{50}$ (96-h): for rainbow trout 2.7 mg/l; for bluegill 1.6 mg/l. In laboratory trials dipropetryn is practically non-toxic to birds and honeybees.

**Formulations.**
Include 80 WP, w.p. (800 g/kg); 500 FW, s.c. (500 g/l). 'Cotodon', dipropetryn + metolachlor (1.5:1).

**Analysis.**
Product analysis is by glc or by titration with perchloric acid in acetic acid. Residues may be determined by glc with FPD (B. G. Tweedy & R. A. Kahrs, *Anal. Methods Pestic. Plant Growth Regul.*, 1978, **10**, 493; K. Ramsteiner *et al.*, *J. Assoc. Off. Anal. Chem.*, 1974, **57**, 192). Particulars available from Ciba-Geigy AG.

# Dipropyl pyridine-2,5-dicarboxylate 5320

$C_{13}H_{17}NO_4$ (251.3)
T6NJ BVO3 EVO3

**Nomenclature and development.**

Chemical name (IUPAC) dipropyl pyridine-2,5-dicarboxylate; dipropyl isocinchomeronate. (*C.A.*) dipropyl 2,5-pyridinedicarboxylate (8 & 9CI); Reg. No. *[136-45-8]*. ENT 17 595. Its insect repellent properties were described by L. D. Goodhue & R. E. Stansbury (*J. Econ. Entomol.*, 1953, **46**, 982). Introduced by McLaughlin Gormley Co. (USP 2 757 120; 3 067 091; 2 824 824) as trade mark 'MGK Repellent 326'.

**Properties.**

Pure dipropyl pyridine-2,5-dicarboxylate is an amber liquid, with a mild aromatic odour; b.p. 150 °C/1mmHg; $n_D^{25}$ 1.4979; $d^{20}$ 1.082. The technical grade has $d^{20}$ 1.106-1.130. Solubility: practically insoluble in water; miscible with ethanol, kerosene, methanol, propan-2-ol. It is decomposed by sunlight, and hydrolysed by alkali. It is incompatible with alkaline pesticides or with high concentrations of dichlorvos.

**Uses.**

It is a fly repellent, effective against *Musca domestica, Chrysops* and *Tabatius* spp., especially *M. vetustissima* of Australia. it lacks persistence at high humidity, a defect overcome by the addition of 7-hydroxy-4-methylcoumarin.

**Toxicology.**

Acute oral $LD_{50}$ for rats 5230-7230 mg/kg. Acute percutaneous $LD_{50}$ for rats 9400 mg/kg. In 90-d feeding trials with rats no ill-effect was observed $\leq$20 000 mg/kg diet. $LC_{50}$ (8-d) for mallard duck and bobwhite quail was $>$5000 mg/kg diet. $LC_{50}$ (96-h) is: for rainbow trout 1.59 mg/l; for bluegill 1.77 mg/l.

**Formulations.**

Sprays in oil or other carriers for personal use (2-50 g a.i./l). Mixtures with diethyltoluamide or 1,4,4a,5a,6,9,9a,9b-octahydrodibenzofuran-4a-carbaldehyde.

**Analysis.**

Product analysis is by glc or by hydrolysis to pyridine-2,5-dicarboxylic acid which is measured by u.v. spectrometry, details are available from McLaughlin Gormley King Co. Residues may be determined by the same method.

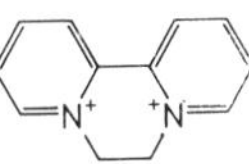

$C_{12}H_{12}N_2$ (184.2)
T B666 GK JK&T&J
*diquat dibromide*
$C_{12}H_{12}Br_2N_2$ (344.0)
T B666 GK JK&T&J &&2Br SALT

**Nomenclature and development.**

Common name diquat (BSI, E-ISO, F-ISO, ANSI, WSSA, JMAF);deiquat (Federal Republic of Germany);region (USSR). Chemical name (IUPAC) 9,10-dihydro-8a,10a-diazoniaphenanthrene ion; 1,1'-ethylene-2,2'-bipyridyldiylium ion. (*C.A.*) 6,7-dihydrodipyrido[1,2-*a*:2',1'-*c*]pyrazinediium (8 & 9CI); Reg. No. *[2764-72-9]* free ion, *[85-00-7]* dibromide, *[6385-62-2]* dibromide monohydrate. Its herbicidal properties were described by R. C. Brian *et al.*, (*Nature (London)*, 1958, **181**, 446). Introduced by ICI Plant Protection Division (GBP 785 732) as code no. 'FB/2'; trade marks 'Reglone' (ICI), Aquacide (Chipman Chemical Co. Ltd); 'Pathclear', 'Weedol' (ICI),'Cleansweep'(Shell Chemical Co. Ltd).

**Properties.**

Normally, the solid dibromide is not isolated from the technical product (>95% pure). The dibromide monohydrate forms colourless to yellow crystals, decomposing >300 °C; v.p. <13 μPa. Solubility at 20 °C 700 g/l water; slightly soluble in alcohols and hydroxylic solvents; practically insoluble in non-polar organic solvents; unstable at 9<pH<12, oxidation and opening of the pyridine rings occurring. It may be inactivated by inert clays and by anionic surfactants. The unformulated compound is corrosive to most common metals, especially aluminium and zinc, but the formulated product contains corrosion inhibitors enabling aqueous solutions to be applied through spray machinery.

**Uses.**

Diquat is a contact desiccant and herbicide with some systemic properties. It is rapidly absorbed by green plant tissues which are killed on exposure to light. It is inactivated upon contact with soil. Uses include potato haulm destruction (420-840 g diquat/ha); seed crop desiccation (392-784 g/ha); aquatic weed control (588-2950 g/ha). Where grasses predominate, a diquat + paraquat formulation is used.

**Toxicology.**

Acute oral $LD_{50}$: for rats 231 mg/kg; for mice 125 mg/kg; for dogs 100-200 mg/kg; for cows 30 mg/kg; for hens 200-400 mg/kg. The only pathological symptom associated with long-term feeding trials is the occurrence of bilateral cataracts. In 2-y feeding trials NEL in rats was 25 mg/kg diet; cataracts appeared after 124 d at 35 mg/kg diet. In 4-y studies NEL in dogs was 50 mg/kg diet. Diquat is absorbed through human skin only after prolonged exposure; shorter exposure can cause irritation and a delay in the healing of cuts and wounds. It is an irritant to eyes and can cause temporary damage to nails and nose bleeding if inhaled.

**Formulations.**

'Reglone', aqueous concentrate (140 or 200 g diquat/l) as dibromide. 'Cleansweep', aqueous concentrate (100 g diquat + 100 g paraquat/l) (Shell Chemical Co. Ltd). 'Pathclear', w.p. (diquat + paraquat + simazine). 'Weedol', water-soluble granules (25 g diquat + 25 g paraquat/kg) as dibromide and dichloride respectively.

**Analysis.**

Product analysis is by u.v. spectroscopy (*AOAC Methods*, 1980, 6.332-6.333; *CIPAC Handbook*, 1970, **1**, 342); impurities may be measured by glc (*ibid.*, 1980, **1A**, 1245; *Herbicides 1977*, p. 48). Residues may be determined by colorimetry (M. G. Ashley, *Pestic. Sci.*, 1970, **1**, 101; A. Calderbank *et al., Analyst (London)*, 1961, **86**, 569; A. Calderbank & S. H. Yuen, *ibid.*, 1965, **90**, 95; 1966, **91**, 625; *J. E. Pack, Anal. Methods Pestic., Plant Growth Regul. Food Addit.*, 1967, **5**, 397; J. B. Leary, *Anal. Methods Pestic. Plant Growth Regul.*, 1978, **10**, 321; *Pestic. Anal. Man.*, 1979, **II**; T. H. Byast *et al., Tech. Rep. ARC Weed Res. Organ.*, No. 15 (2nd Ed.), p. 35) or by polarography (*idem., loc. cit.*). Details are available from ICI Plant Protection Division.

$C_8H_{19}O_2PS_3$ (274.4)
2S2SPS&O2&O2

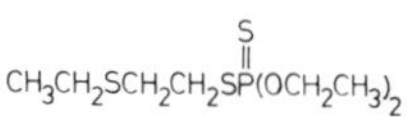

**Nomenclature and development.**

Common name disulfoton (BSI, E-ISO, F-ISO); M-74 (USSR); ethylthiodemeton (JMAF). Chemical name (IUPAC) *O,O*-diethyl *S*-2-ethylthioethyl phosphorodithioate. (*C.A.*) *O,O*-diethyl *S*-[2-(ethylthio)ethyl] phosphorodithioate (8 & 9CI); Reg. No. *[298-04-4]*. Trivial name thiodemeton (rejected common name proposal). ENT 23 347. Described by G. Schrader (*Die Entwicklung neuer insektizider Phosphorsäure*). Introduced by Bayer AG and later by Sandoz AG (DEP 917 668; 947 369; USP 2 759 010 all to Bayer AG) as code no. 'Bayer 19 639','S 276' (Bayer); trade marks 'Disyston' ('Di-Syston' in USA), 'Dithiosystox' (all to Bayer AG), 'Frumin AL', 'Solvirex' (both to Sandoz AG).

**Properties.**

Pure disulfoton is a colourless oil, with a characteristic odour; b.p. 62 °C/0.01 mmHg; v.p. 24 mPa at 20 °C; $d_4^{20}$ 1.144, $n_D^{20}$ 1.5496. Solubility at 22 °C: 25 mg/l water; readily soluble in most organic solvents. At 20 °C, 50% loss by hydrolysis occurs in 3.04 y at pH1-5; at 70 °C: in 1.2 d at pH7, 7.2 h at pH9. It is stable upon storage (shelf life at 20 °C ⩾2 y).

**Uses.**

It is a systemic insecticide and acaricide used mainly for treatment of seed and application as granules to soil or plants. It is metabolised in plants, animals and soil to the sulphoxide and sulphone, to the corresponding phosphorothioate analogues (of demeton-S), and finally, to derivatives of *O,O*-diethyl hydrogen phosphate and 2-ethylthioethyl mercaptan (R. L. Metcalf *et al., J. Econ. Entomol.*, 1957, **50**, 338; I. Takase, *Pflanzenschutz-Nachr. (Engl. Ed.),* 1972, **25**, 43).

**Toxicology.**

Acute oral $LD_{50}$ for rats 2.6-12.0 mg/kg. Acute percutaneous $LD_{50}$ for male rats *c.* 20 mg/kg. In 90-d feeding trials NEL for rats was 1 mg/kg diet.

**Formulations.**

These include: 'Disyston', 'Frumin AL', seed treatment powder (500 g a.i./kg); 'Disyston', 'Solvirex', granules (50-100 g/kg) on various mineral carriers or carbon, also 'Solvigran P' (75 g/kg), 'Ekatin TD' (50 g/kg). 'Doubledown', granules (60 g disulfoton + 40 g fonofos/kg). 'Ethimeton 6' (30 g disulfoton + 30 g diazinon/kg).

**Analysis.**

Product analysis is by glc or by paper chromatography, followed by combustion and subsequent determination of phosphate by standard methods. Residues are determined by glc (*Pestic. Anal. Man.*, 1979, **I**, 201-A, 201-G, 201-H, 201-I; D.C. Abbott *et al., Pestic. Sci.*, 1970, **1**, 10; *Analyst (London)*, 1980, **105**, 515). Details available from Sandoz AG.

# Dithianon

$C_{14}H_4N_2O_2S_2$ (296.3)
T C666 BV DS GS IVJ ECN FCN

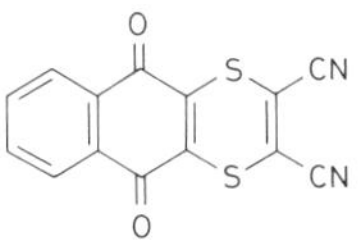

**Nomenclature and development.**

Common name dithianon (BSI, E-ISO, F-ISO, JMAF);*exception* (Italy). Chemical name (IUPAC) 5,10-dihydro-5,10-dioxonaphtho[2,3-*b*]-1,4-dithi-in-2,3-dicarbonitrile (I); 2,3-dicyano-1,4-dithia-anthraquinone. (*C.A.*) (I) (9CI); 5,10-dihydro-5,10-dioxonaphtho[2,3-*b*]-*p*-dithiin-2,3-dicarbonitrile (8CI); Reg. No. *[3347-22-6]*. Its fungicidal properties were described by J. Berker *et al.* (*Proc. Br. Insectic. Fungic. Conf., 2nd*, 1963, p. 351). Introduced by E. Merck (now Celamerck GmbH Co. KG) (GBP 857 383) as code no. 'IT-931','MV 119A'; trade mark 'Delan'.

**Properties.**

Pure dithianon forms brown crystals; m.p. 225 °C; v.p. 66 μPa at 25 °C. Solubility at 20 °C: 0.5 mg g/l water; 10 g/l acetone; 8 g/l benzene; 12g/l chloroform. It is decomposed under alkaline conditions (pH >7) and is incompatible with petroleum oil sprays.

**Uses.**

It is a protectant fungicide effective against many foliar diseases of pome and stone fruit but not against powdery mildews. Pathogens controlled include *Venturia* spp. (apple, cherry and pear), *Microthyriella rubi* (apple), *Coccomyces hiemalis* and *Stigmina carpophila* (cherry), *Monilla* spp., *Taphrina deformans* and *Tranzschelia discolor* (apricot and peach), *Plasmopara viticola* (grape), *Pseudoperonospora humuli* (hop), *Elsinoe fawcetti* and *Phomopsis citri* (citrus), *Mycosphaerella fragrariae* and *Diplocarpon earliana* (strawberry), *Colletotrichum coffeanum* (coffee), *Cronartium ribicola, Drepanopeziza ribis* and *Plasmopara ribicola* (currant). The average rate is 50 g a.i./100 l. It is non-phytotoxic at fungicidal concentrations.

**Toxicology.**

Acute oral $LD_{50}$: for rats 638 mg/kg; for guinea-pigs 110 mg/kg; for male quails 280 mg/kg, for females 430 mg/kg. In 2-y feeding trials no toxic effect was observed in rats at 20 mg/kg diet and dogs at 40 mg/kg diet. Contact toxicity for honeybees is >100 μg/bee.

**Formulations.**

A w.p. (750 g a.i./kg); dispersion in water (250 g/l).

**Analysis.**

Product analysis is by colorimetry (E. Amodori & W. Heupt, *Anal. Methods Pestic. Plant Growth Regul.*, 1978, **10**, 181). Details are available from Celamerck GmbH & Co. KG residues may be determined by colorimetry (*idem, ibid.*; H. Sieper & H. Pies, *Fresenius' Z. Anal. Chem.*, 1968, **242**, 234).

$C_9H_{10}Cl_2N_2O$ (233.1)
GR BG DMVN1&1

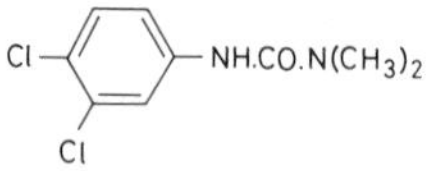

Cl—C6H3(Cl)—NH.CO.N(CH3)2

**Nomenclature and development.**

Common names diuron (BSI, E-ISO, F-ISO, ANSI, WSSA);dichlorfenidim (USSR);DCMU (JMAF) *exception* Sweden. Chemical name (IUPAC) 3-(3,4-dichlorophenyl)-1,1-dimethylurea (I). (*C.A.*) *N*'-(3,4-dichlorophenyl)-*N*,*N*-dimethylurea (9CI); (I) (8CI); Reg. No. *[330-54-1]*. Its herbicidal properties were described by H. C. Bucha & C. W. Todd (*Science*, 1951, **114**, 493). Introduced by E. I. du Pont de Nemours & Co. (Inc.) (USP 2 655 445) as trade mark 'Karmex'.

**Properties.**

Pure diuron is a colourless crystalline solid; m.p. 158-159 °C; v.p. 412 μPa at 50 °C. Solubility at 25 °C: *c*. 42 mg/l water; at 27 °C 53 g/kg acetone; sparingly soluble in hydrocarbons. Its rate of hydrolysis is negligible at ordinary temperatures and in neutral pH range but is greater under acid or alkaline conditions or at elevated temperatures. Decomposes at 180-190 °C. It is degraded in soil by demethylation, 50% loss occurs in 90-180 d (G. D. Hill *et al.*, *Agron. J.*, 1955, **47**, 93; T. J. Sheets, *J. Agric. Food Chem.*, 1964, **12**, 30).

**Uses.**

It is a herbicide that inhibits photosynthesis and is used for general weed control on non-crop areas at 10-30 kg a.i./ha; for subsequent annual maintenance 5-10 kg/ha will prevent re-infestation by seedlings. It is also used selectively pre-em. on crops such as asparagus, citrus, cotton, pineapple, sugarcane, temperate tree and bush fruits at 0.6-4.8 kg/ha. Phytotoxic residues in soil disappear within 1 season at these lower rates.

**Toxicology.**

Acute oral $LD_{50}$ for rats 3400 mg/kg. The w.p. caused irritation to the eyes and mucous membranes of rabbits, but a 50% water paste was not irritating to the intact skin of guinea-pigs. In 2-y feeding trials NEL was: for rats 250 mg/kg diet; for dogs 125 mg/kg diet (H. C. Hodge *et al.*, *Food Cosmet. Toxicol.*, 1967, **5**, 513). $LC_{50}$ (8-d) was: for bobwhite quail 1730 mg/kg diet; for Japanese quail, mallard ducklings, ring-necked pheasant chicks >5000 mg/kg. $LC_{50}$ for fish ranges from 3-60 mg/l depending on exposure, size and species.

**Formulations.**

'Karmex', w.p. (800 g a.i./kg). 'Krovar I', (Reg. No. *[8071-35-0]*), w.p. (400 g diuron + 400 g bromacil); 'Krovar II', w.p. (270 g diuron + 530 g bromacil). 'Velpar K-4', (Reg. No. *[73347-92-9]*), w.p. (468 g diuron + 132 g hexazinone). Other mixtures include: diuron + aminotriazole; diuron + aminotriazole + atrazine; diuron + chlorpropham; diuron + dalapon + MCPA; diuron + linuron + terbacil; diuron + paraquat; diuron + propham

**Analysis.**

Product analysis is by hydrolysis and titration of the amine liberated (*CIPAC Handbook*, 1980, **1A**, 1251), details available from E. I. du Pont de Nemours & Co. (Inc.). Residues may be determined by glc (T. H. Byast *et al.*, *Tech. Rep. ARC Weed Res. Organ.*, No.15 (2nd Ed.), p.49; *Anal. Methods Pestic. Plant Growth Regul.*, 1972, **6**, 664; W. E. Bleidner, *J. Agric. Food Chem.*, 1954, **2**, 682). Hydrolysis to 3,4-dichloroaniline, a derivative of which is measured colorimetrically, may also be used (H. L. Pease, *ibid.*, 1962, **10**, 279; R. L. Dalton & H. L. Pease, *J. Assoc. Off. Anal. Chem.*, 1962, **45**, 377).

$C_7H_6N_2O_5$ (198.1)
WNR BQ C1 ENW

**Nomenclature and development.**

Common name DNOC (BSI, E-ISO, F-ISO, WSSA, JMAF). Chemical name (IUPAC) 4,6-dinitro-*o*-cresol (I). (*C.A.*) 2-methyl-4,6-dinitrophenol (9CI); (I) (8CI); Reg. No. *[534-52-1]*. Trivial name DNC. ENT 154. Introduced as an insecticide by Fr Bayer & Co. (now Bayer AG) in 1892, and as a herbicide by G. Truffaut et Cie in 1932 (GBP 3301 - Bayer; GBP 425 295 - Truffaut) as trade marks 'Antinonnin' (Bayer), 'Sinox' (Truffaut).

**Properties.**

It forms yellowish crystals; m.p. 86 °C; v.p. 14 mPa at 25 °C. Solubility at 15 °C 130 mg/l water; soluble in most organic solvents. The ammonium, calcium, potassium and sodium salts are appreciably soluble in water. Technical grade (95-98% pure), m.p. 83-85 °C. It is usually moistened with up to 10% water to reduce the risk of explosion though it is corrosive to mild steel in the presence of moisture.

**Uses.**

DNOC is a non-systemic stomach poison and contact insecticide, ovicidal to the eggs of certain insects and spider mites. It is strongly phytotoxic and its use as an insecticide is limited to dormant sprays especially for fruit trees or on waste ground - *e.g.* against locusts. It is used as a contact herbicide for the control of broad-leaved weeds in cereals at $\leqslant$10 kg/ha and, in e.c. formulations, for the pre-harvest desiccation of potatoes and leguminous seed crops.

**Toxicology.**

Acute oral $LD_{50}$: for rats 25-40 mg/kg; for sheep 200 mg DNOC-sodium/kg. In 0.5-y feeding trials rats tolerated 100 mg/kg diet. DNOC acts as a cumulative poison in man though there is little evidence of accumulation in laboratory animals (D. G. Harvey *et al.*, *Br. Med. J.*, 1951, **2**, 13; E. King & D. G. Harvey, *Biochem. J.*, 1953, **53**, 185, 196).

**Formulations.**

Pastes of the sodium or amine salts, ammonium sulphate being added as 'activator'; solutions in petroleum oil.

**Analysis.**

Product analysis is by conversion to the ammonium salt which is measured by colorimetry (D. S. Farrington, *Analyst (London)*, 1983, **108**, 353) or by titration with titanium trichloride in an inert atmosphere (W. Fischer, *Z. Anal. Chem.*, 1938, **112**, 91; *MAFF Ref. Book.* No.1, 1958; *CIPAC Handbook*, 1970, **1**, 348). Residues may be determined by colorimetry in alkaline solution (A. W. Avens *et al.*, *J. Econ. Entomol.*, 1948, **41**, 432; W. H. Parker, *Analyst (London)*, 1949, **74**, 646).

$C_{18}H_{35}NO$ (281.5)
T6N DOTJ C1 E1 A- AL-12-TJ
*dodemorph acetate*
$C_{20}H_{39}NO_3$ (341.5)
T6N DOTJ C1 E1 A- AL-12-TJ
&&$CH_3COOH$ SALT

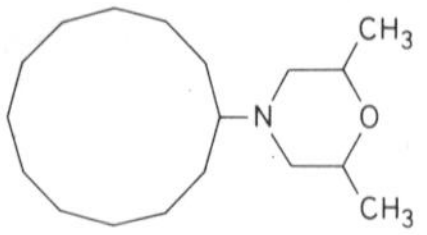

**Nomenclature and development.**

Common name dodemorph (BSI, E-ISO); dodémorphe (F-ISO). Chemical name (IUPAC) 4-cyclododecyl-2,6-dimethylmorpholine (I). (*C.A.*) (I) (8 & 9CI); Reg. No. *[1593-77-7]*, dodemorph was originally reported in the literature with an incorrect structure *[35842-13-2]*. The fungicidal properties of morpholines with large groups attached to the nitrogen were described by K. H. König *et al.* (*Angew. Chem.*, 1965, 77, 327); those of dodemorph by J. Kradel & E. -H. Pommer (*Proc. Br. Insectic. Fungic. Conf., 4th*, 1967, **1**, 170). BASF AG introduced the acetate salt (dodemorph acetate, Reg. No. *[31717-87-0]*) (DEP 1 198 125) as code no. 'BAS 238F'; trade mark 'Meltatox'.

**Properties.**

Dodemorph contains *cis*-2,6-dimethylmorpholine- (*c.* 60% *m/m*) and *trans*-2,6-dimethylmorpholine-moieties (*c.* 40%). The *trans*-isomer is a colourless oil. The predominating *cis*-isomer is a colourless solid with characteristic odour; m.p. 72-73 °C, v.p. 480 μPa at 20 °C; solubility: <100 mg/kg water; >1 kg/kg chloroform; 66 g/kg ethanol.

Dodemorph acetate is a colourless solid; m.p. 63-64 °C; *d c.* 0.93; v.p. 2.5 mPa at 20 °C; solubility: <100 mg/kg water; >1 kg/kg chloroform; 170 g/kg ethanol. It is stable in unopened containers for > 1 y. The products of thermal decomposition are flammable.

**Uses.**

Dodemorph is an eradicant and systemic fungicide recommended against powdery mildews on roses outdoors and underglass and on other ornamentals.

**Toxicology.**

The acute oral $LD_{50}$ for rats is 1800 mg a.i. (as 4.5 ml 40% formulation)/kg; it causes severe skin and eye irritation to rabbits. $LC_{50}$ for guppies is *c.* 40 mg/l. It is harmless to honeybees.

**Formulations.**

'Meltatox' ('BAS F238'), e.c. (400 g a.i./l). 'Badilin Rosenfluid' ('BAS 31 002F', Reg. No. *[61144-31-8]*) (280 g dodemorph + 58 g dodine/l).

**Analysis.**

Product analysis is by potentiometric titration with perchloric acid in glacial acetic acid. Residues in crops may be analysed by colorimetry of a complex formed with methyl orange.

$C_{15}H_{33}N_3O_2$ (287.4)
MUYZM12 &&$CH_3COOH$ SALT
*free base*
$C_{13}H_{29}N_3$ (227.4)
MUYZM12

$$\overset{+}{NH_2} \quad \text{(=)} \quad CH_3(CH_2)_{11}NHCNH_2 \;\; CH_3CO.O^-$$

**Nomenclature and development.**

Common name (for the acetate salt) dodine (BSI, E-ISO, F-ISO, ANSI);doguadine (France);tsitrex (USSR);*dodine acetate* (BSI, prior to 1969);(for the free base) guanidine (JMAF), *dodine* (BSI, prior to 1969). Chemical name (IUPAC) 1-dodecylguanidinium acetate. (*C.A.*) dodecylguanidine monoacetate (8 & 9CI); Reg. No. *[2439-10-3]* (acetate), *[112-65-2]* (free base). Its fungicidal properties were described by B. Cation (*Plant Dis. Rep.*, 1957, **41**, 1029). Introduced by American Cyanamid Co. (USP 2 867 562) as code no. 'CL 7521','AC 5223'; trade marks 'Cyprex', 'Melprex'.

**Properties.**

Dodine forms colourless crystals; m.p. 136 °C. Solubility at 25 °C: 63 mg/l water; 7-23 g/l low molecular weight alcohols; soluble in acids; insoluble in most organic solvents. Stable under moderately alkaline or acid conditions, but the free base is liberated by concentrated alkali. Dodine is compatible with anionic surfactants and with hard water.

**Uses.**

Dodine is a protective fungicide recommended for the control of a number of major fungal diseases of fruit, nut and vegetable crops and on certain ornamentals and shade trees. Foliar application is recommended, especially against *Venturia* spp. (on apple and pear) and cherry leaf spot, against which dodine has some eradicant action. Usual rate is 30-80 g a.i./100 l.

**Toxicology.**

Acute oral $LD_{50}$: for male rats *c.* 1000 mg tech./kg; for mallard ducks 1142 mg/kg. Acute percutaneous $LD_{50}$ for rabbits >1500 mg/kg. In 2-y feeding trials, rats receiving 800 mg/kg diet suffered a slight retardation of growth with no effect on reproduction or lactation. $LC_{50}$ (48-h) for harlequin fish 0.53 mg a.i. (as 650 g/kg formulation)/l (T. E. Tooby *et al.*, *Chem. Ind. (London)*, 1975, p. 523). $LD_{50}$ for honeybees >11 μg/bee for topical application.

**Formulations.**

These include: w.p. (650 or 800 g a.i./kg); liquid (200 or 250 g/l); dust (750 g/kg). 'Badilin Rosenfluid' (BAS 31 002 F), Reg. No. *[61144-31-8]*, (280 g dodemorph + 58 g dodine/l).

**Analysis.**

Product analysis is by titration in non-aqueous media (*CIPAC Handbook*, 1983, **1B**, in press; *AOAC Methods*, 1980, 6.346-6.347). Residues may be determined by colorimetry of a complex formed with bromocresol purple (*ibid.*, 29.108-29.111; *Pestic. Anal. Man.*, 1979, **II**; W. H. Newsome, *J. Agric. Food Chem.*, 1976, **24**, 997; G. L. Sutherland, *Anal. Methods Pestic., Plant Growth Regul. Food Addit.*, 1964, **3**, 41).

$C_{10}H_8ClN_3O_2$ (237.6)
T5NOVYJ DUNMR BG& E1

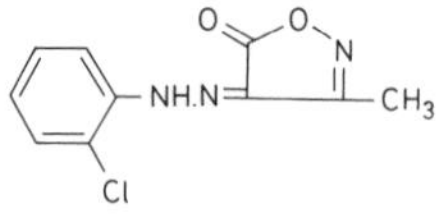

**Nomenclature and development.**

Common name drazoxolon (BSI, E-ISO, F-ISO). Chemical name (IUPAC) 4-(2-chlorophenylhydrazono)-3-methyl-5-isoxazolone. (*C.A.*) 3-methyl-4,5-isoxazoledione 4-[(2-chlorophenyl)hydrazone] (9CI); 3-methyl-4,5-isoxazoledione 4-[(*o*-chlorophenyl)hydrazone] (8CI); Reg. No. *[5707-69-7]*. Its fungicidal properties were described by M. J. Geoghegan (*Proc. Br. Insectic. Fungic. Conf., 4th*, 1967, **1**, 451). Introduced by ICI, Plant Protection Division, (GBP 999 097) as code number 'PP 781'.

**Properties.**

Drazoxolon is a yellow crystalline solid; m.p. 167 °C; v.p. 533 μPa at 30 °C. Solubility: almost insoluble in water, aliphatic hydrocarbons; 40 g/kg aromatic hydrocarbons; *c.* 100 g/kg chloroform; 10 g/kg ethanol; 50 g/kg ketones. Insoluble in and stable to acids; dissolving in aqueous alkali to form salts. Non-corrosive to materials usually used for packaging and spray appliances but, if long storage is anticipated, polythene liners should be used in spray drums. Aqueous formulations should not be stored in metal containers. It is incompatible with lime sulphur; mixtures with dodine are phytotoxic to apples.

**Uses.**

It is effective in controlling powdery mildews on blackcurrants, roses and other crops and for the control of certain other foliar diseases; also as a seed treatment to control *Pythium* and *Fusarium* spp. on such crops as beans, maize, grass and peas. Also for the control of *Ganoderma* spp. on rubber trees.

**Toxicology.**

Acute oral $LD_{50}$: for rats 126 mg/kg; for mice 129 mg/kg; for rabbits 100-200 mg/kg; for guinea-pigs 12.5-25 mg/kg; for dogs 17 mg/kg; for hens *c.* 100 mg/kg. It has no marked irritant effect on the skin and eyes, but some sensitisation may occur and prolonged contact should be avoided. It causes some lung irritation and vapour concentrations $>0.5$ mg/m$^3$ air must be avoided. In 90-d feeding trials there was no significant abnormality in rats receiving 30 mg/kg diet nor in dogs receiving 2 mg/kg daily. $LC_{50}$ (96-h) for brown trout 0.55 mg/l.

**Formulations.**

These include: 'Mil-Col' and 'SAIsan', aqueous suspension (300 g a.i./l). 'Ganocide', a grease formulation (100 g/kg) for use on rubber trees.

**Analysis.**

Product and residue analysis is by colorimetry (S. H. Yuen, *Anal. Methods Pestic. Plant Growth Regul.*, 1973, **7**, 665). Details of methods are available from ICI PLC, Plant Protection Division.

$C_{14}H_{15}O_2PS_2$ (310.4)
2OPO&SR&SR

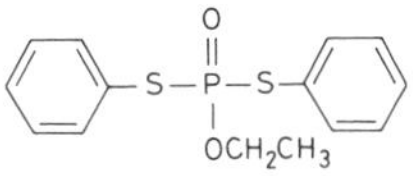

**Nomenclature and development.**

Common name edifenphos (BSI, E-ISO, F-ISO);EDDP (JMAF). Chemical name (IUPAC) *O*-ethyl *S*,*S*-diphenyl phosphorodithioate (I). (*C.A.*) (I) (8 & 9CI); Reg. No. *[17109-49-8]*. Its fungicidal properties were described by H. Scheinpflug & H. F. Jung (*Pflanzenschutz-Nachr. (Engl. Ed.)*, 1968, **21**, 79). Introduced by Bayer AG (BEP 686 048; DEAS 1 493 736) as code no. 'Bayer 78 418'; trade mark 'Hinosan'.

**Properties.**

A clear yellow to light-brown liquid, with a thiophenol-like odour: b.p. 154 °C/0.01 mm Hg; $d_4^{20}$ c. 1.23; $n_D^{22}$ 1.61. Practically insoluble in water; soluble in acetone, xylene. At 25 °C 50% loss occurs in 19 d at pH7, in 2 d at pH9.

**Uses.**

This fungicide has a specific action against *Piricularia oryzae* on rice at 30-50 g a.i./l water using 800-1200 l/ha; 1 or 2 applications to wet paddy in the nursery, 2 or 3 applications after transplanting or in fields of broadcast rice before tillering has ceased. It is also effective against *Pellicularia sasakii* and *Cochiobolus miyabeanus* and is well tolerated by rice varieties at effective fungicidal rates. It should not be used within 10 d before or after an application of propanil.

**Toxicology.**

Acute oral $LD_{50}$: for male rats 340 mg tech. (in ethanol + propylene glycol)/kg, for females 150 mg/kg; for mice 218 mg/kg; for guinea pigs and rabbits 350-400 mg/kg; for hens *c.* 750 mg tech. (in 'Lutrol')/kg. Acute percutaneous $LD_{50}$ (4-h) for rats >1230 mg/kg; a 24-h application of the technical product to a rabbit's ear caused no injury. In 2-y feeding trials NEL was: for male rats 5 mg/kg diet, for females 15 mg/kg diet. $LC_{50}$ for mirror carp: 1.3 mg a.i. (using e.c. of 500 g a.i./l) /l, and 0.9 mg a.i. (using dust of 20 g a.i./kg)/l.

**Formulations.**

These include: e.c. (300, 400 or 500 g a.i./l); dusts (15, 20 or 25 g/kg).

**Analysis.**

Details of methods for product and residue analysis are available from Bayer AG. Residues in rice may be determined by glc with TID (K. Vogeler, *ibid.*, p. 317).

$C_7H_{10}ClN_5O_2$ (231.6)
T6N CN ENJ BM2 DM1VQ FG
*eglinazine-ethyl*
$C_9H_{14}ClN_5O_2$ (259.7)
T6N CN ENJ BM2 DM1VO2 FG

Cl, NHCH$_2$CO.OH, N, N, N, CH$_3$CH$_2$NH

**Nomenclature and development.**

Common name eglinazine (BSI, E-ISO, F-ISO, WSSA). Chemical name (IUPAC) *N*-(4-chloro-6-ethylamino-1,3,5-triazin-2-yl)glycine (I). (*C.A.*) (I) (9CI); *N*-[4-chloro-6-(ethylamino)-*s*-triazin-2-yl]glycine (8CI); Reg. No. *[68228-19-3]* (eglinazine), *[6616-80-4]* (eglinazine-ethyl). The ethyl ester, eglinazine-ethyl, was introduced as a herbicide by Nitrokémia Ipartelepek as code no. 'MG-06'.

**Properties.**

Pure eglinazine-ethyl forms colourless crystals, m.p. 228-230 °C; v.p. 27 μPa at 20 °C. Solubility at 25 °C: 300 mg/l water; 200 g/l acetone; 20 g/l hexane; 40 g/l xylene. It is stable at room temperature at pH5-8, but on heating it is hydrolysed by acid or alkali to give the corresponding herbicidally-inactive hydroxy triazine. It is stable to light. It decomposes at 250 °C. It is rapidly decomposed in soil, 50% loss occurs in 12-18 d. The technical grade is *c.* 96% pure.

**Uses.**

Eglinazine-ethyl is a selective pre-em. herbicide used in cereals at 3 kg a.i./ha, and is especially effective against *Apera spica-venti* and *Matricaria inodora* though its herbicidal range is narrow.

**Toxicology.**

Acute oral $LD_{50}$: for rats and mice >10 000 mg/kg; for guinea pigs >3375 mg/kg. Acute percutaneous $LD_{50}$ for rats and rabbits is >10 000 mg/kg; it is not a skin or eye irritant. Acute intraperitoneal $LD_{50}$ for rats and mice >7100 mg/kg. There is no evidence of cumulative toxicity.

**Formulations.**

A w.p. (500 g a.i./kg).

**Analysis.**

Product analysis is by glc. Residues may be determined by glc with TID.

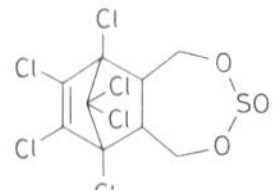


$C_9H_6Cl_6O_3S$ (406.9)
T C755 A EOSO KUTJ AG AG BG FO JG KG LG

**Nomenclature and development.**

Common name endosulfan (BSO, E-ISO, F-SIO, ANSI);thiodan (Iran, USSR);benzoepin (JMAF);*exception* (Italy). Chemical name (IUPAC) *C,C'*-(1,4,5,6,7,7-hexachloro-8,9,10-trinorborn-5-en-2,3-ylene)(dimethyl sulphite) 6,7,8,9,10,10-hexachloro-1,5,5a,6,9,9a-hexahydro-6,9-methano-2,4,3-benzodioxathiepin 3-oxide (I). (*C.A.*) (I) (9CI); 1,4,5,6,7,7-hexachloro-5-norbornene-2,3-dimethanol cyclic sulfite (8CI). Reg. No. *[115-29-7]*. OMS 570, ENT 23 979. Its insecticidal properties were described by W. Finkenbrink (*Nachrichtenbl. Dtsch. Pflanzenschutzdienstes* (*Braunschweig*), 1956, **8**, 183). Introduced by Hoechst AG and, in the USA, by FMC Corp. (DEP 1 015 797; USP 2 799 685; GBP 810 602 all to Hoechst AG) as code no. 'Hoe 02671','FMC 5462'; trade marks 'Thiodan' (to Hoechst AG and FMC Corp.), 'Cyclodan', 'Beosit', 'Malix', 'Thimul', 'Thifor'.

**Properties.**

Technical endosulfan ($\geqslant$94% pure) is a brown crystalline solid with an odour of sulphur dioxide; m.p. 70-100 °C; v.p. 1.2 Pa at 80 °C. Solubility at 22 °C: alpha-endosulfan 0.32 mg/l water, beta-endosulfan 0.33 mg/l; at 20 °C: 200 g a.i./l dichloromethane, ethyl acetate, toluene; *c.* 65 g/l ethanol; *c.* 24 g/l hexane. It is stable to sunlight, unstable in alkaline media and subject to slow hydrolysis to the diol and sulphur dioxide.

Endosulfan is a mixture of two stereoisomers: alpha-endosulfan, endosulfan (I), (Reg. No. *[959-98-8]* formerly *[33213-66-0]*), stereochemistry 3α,5aβ,6α,9α,9aβ, m.p. 109 °C, is 64-67% of the technical grade; beta-endosulfan, endosulfan (II) (Reg. No. *[33213-65-9]* formerly *[891-86-1]* and *[19670-15-6]*), stereochemistry 3α,5aα,6β,9β,9aβ, m.p. 213.3 °C, is 29-32%. Earlier reports on the stereochemistry of these isomers gave conflicting reports (W. Reimschneider, *World Rev. Pest Control*, 1963, **2**(4), 29).

**Uses.**

It is a wide range non-systemic contact and stomach insecticide effective against numerous insects and certain mites on cereals, coffee, cotton, fruit, oilseeds, potatoes, tea, vegetables and numerous other crops. The two stereoisomers have comparable $LD_{50}$ values for *Musca domestica* (D. A. Lindquist & P. A. Dahm, *J. Econ. Entomol.*, 1957, **50**, 483).

**Toxicology.**

Acute oral $LD_{50}$: for rats 80-110 mg tech. (in oil)/kg, 76 mg alpha-isomer/kg, 240 g beta-isomer/kg; for dogs 76.7 mg tech./kg; for mallard duck 205-245 mg/kg; for ring-necked pheasants 620-1000 mg/kg. Acute percutaneous $LD_{50}$ for rabbits 359 mg (in oil)/ kg. In 2-y feeding trials rats receiving 30 mg/kg diet showed no ill-effect; in 1-y feeding trials NEL for dogs was 3 mg/kg diet. It is highly toxic to fish ($LC_{50}$ (96-h) for golden ide 2 μg/l water) but, in practical use, should be harmless to wildlife and to honeybees.

It is metabolised, in plants and mammals, to the corresponding sulphate which is toxicologically similar to endosulfan (the sulphite). For most fruits and vegetables 50% of the residue is lost in 3-7 d. There is no accumulation in milk, fat or muscle and it is excreted as conjugates of the diol and as other highly polar metabolites depending on the species (H. Maier-Bode, *Residue Rev.*, 1968, **22**, 1).

**Formulations.**

These include: e.c. (161, 357 and 480 g a.i./l); w.p. (164, 329, 470 g/kg); dusts (30-47 g/kg); granules (10, 30, 40 or 50 g/kg); ULV (242, 497 or 604 g/l). Combinations include: endosulfan + dimethoate; endosulfan + parathion-methyl.

**Analysis.**

Product analysis is by i.r. spectrometry (*CIPAC Handbook*, 1970, **1**, 360) or by glc (G. Zweig & T. E. Archer, *J. Agric. Food Chem.*, 1960, **8**, 190, 403; FAO Specification (CP/49); *Organochlorine Insecticides 1973*). Residues may be determined by glc with MCD (*AOAC Methods*, 1980, 29.029-29.034; *Pestic. Anal. Man.*, 1979, 201-A, 201-G, 201-I; G. M. Telling *et al.*, *J. Chromatog.*, 1977, **137**, 405; M. A. Luke *et al.*, *J. Assoc. Off. Anal. Chem.*, 1981, **64**, 1187; *Anal. Methods Pestic. Plant Growth Regul.*, 1972, **6**, 488, 511).

$C_8H_{10}O_5$ (186.2)
T55 A AOTJ CVQ DVQ
*endothal-sodium*
$C_8H_8Na_2O_5$ (230.2)
T55 A AOTJ CVQ DVQ &&2Na SALT

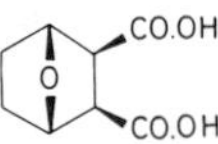


**Nomenclature and development.**

Common name endothal (BSI, France, New Zealand);endothall (ANSI, Canada, WSSA) salts being indicated in the normal manner;endothal-sodium (defined as such) (E-ISO, F-ISO);*exception* (Italy). Chemical name (IUPAC) 7-oxabicyclo[2.2.1]heptane-2,3-dicarboxylic acid (I); 3,6-epoxycyclohexane-1,2-dicarboxylic acid. (*C.A.*) (I) (8 & 9CI); Reg. No. *[145-73-3]* (acid). Its herbicidal properties were described by N. Tischler *et al.* (*Proc. Northeast. Weed Control Conf.*, 1951, p. 51). Introduced by Sharples Chemical Corp. (now merged with Pennwalt Corp.) (USP 2 550 494; 2 576 080; 2 576 081)

**Properties.**

Produced by reduction of the Diels-Alder condensation product of furan and maleic anhydride, the *rel*-(1*R*,2*S*,3*R*,4*S*)-isomer (Reg. No. *[28874-46-6]*), formerly *exo-cis* isomer (in *C.A.* usage *endo-cis* isomer) forms a hydrate which is a colourless solid, m.p. 144 °C. Solubility at 20 °C: 100 g/kg water; 70 g/kg acetone; 100 mg/kg benzene; 75 g/kg dioxane; 280 g/kg methanol. It is stable to light but forms an anhydride at 90 °C. It is a dibasic acid (p$K_1$ 3.4, p$K_2$ 6.7) forming water-soluble amine and alkali metal salts. It is non-corrosive to metals and non-flammable. Of the 3 stereoisomers of endothal, the one named above is the most effective herbicide (USP 2 550 494).

**Uses.**

Amine, ammonium and alkali metal salts of endothal are recommended for the pre-em. and post-em. control of weeds in red beet, spinach and sugar beet at 2-6 kg a.i./ha. It is used as a desiccant on lucerne and on potato, for the defoliation of cotton and to control algae and aquatic weeds.

**Toxicology.**

Acute oral $LD_{50}$ for rats: 51 mg endothal/kg, 182-197 mg endothal-sodium (for 19.2% formulation)/kg. The latter causes light to moderate skin irritation. In 2-y feeding trials rats receiving 1000 mg/kg diet suffered no ill-effect.

**Formulations.**

These include: disodium, diammonium and mono(ethyldimethylammonium) salts.

**Analysis.**

Product analysis is by glc with FLD (R. Carlson *et al., Anal. Methods Pestic. Plant Growth Regul.*, 1978, **10**, 327). Residues may be determined by glc of a derivative, which is measured by MCD (*idem, ibid.*). Details of methods are available from Pennwalt Corp.

$C_{12}H_8Cl_6O$ (380.9)
T E3 D5 C555 A D- FO KUTJ AG AG
BG JG KG LG

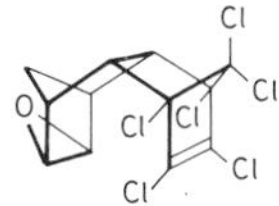


**Nomenclature and development.**

Common name endrin (BSI, E-ISO, ESA, JMAF);endrine (F-ISO);nendrin (Republic of South Africa);*exception* (India). Chemical name (IUPAC) (1*R*,4*S*,4a*S*,5*S*,7*R*,8*R*,8a*R*)-1,2,3,4,10,10-hexachloro-1,4,4a,5,6,7,8,8a-octahydro-6,7-epoxy-1,4:5,8-dimethanonaphthalene; 1,2,3,4,10,10-hexachloro-6,7-epoxy-1,4,4a,5,6,7,8,8a-octahydro-*exo*-1,4-*exo*-5,8-dimethanonaphthalene. (*C.A.*) (1aα,2β,2aβ,3α,6α,6aβ,7β,7aα)-3,4,5,6,9,9-hexachloro-1a,2,2a,3,6,6a,7,7a-octahydro-2,7:3,6-dimethanonaphth[2,3-*b*]oxirene (9CI); *endo,endo*-1,2,3,4,10,10-hexachloro-6,7-epoxy-1,4,4a,5,6,7,8,8a-octahydro-1,4:5,8-dimethanonaphthalene (8CI); Reg. No. *[72-20-8]*. OMS 197, ENT 17 251. It has been developed by Shell International Chemical Co. after being introduced by J. Hyman & Co. (USP 2 676 132 to Hyman) as code no. 'Experimental Insecticide 269' (Hyman & Co.).

**Properties.**

Pure endrin is a colourless crystalline solid; m.p. 226-230 °C (decomp.). The technical product ( $\geqslant$92% pure) is a light tan solid decomposing >200 °C; v.p. 20 nPa at 20 °C; ρ 1.64 g/cm$^3$ at 20 °C. Solubility of endrin; negligible in water; moderately in acetone, benzene; sparingly in alcohols, petroleum hydrocarbons.

**Uses.**

Endrin is a foliar insecticide which acts against a wide range of pests particularly Lepidoptera. It can be used at 0.2-0.5 kg a.i./ha on cotton, maize, sugar cane, upland rice and many other crops.

**Toxicology.**

Acute oral $LD_{50}$: for rats 7.5-17.5 mg a.i./kg; for birds 0.75-5.64 mg/kg. Acute percutaneous $LD_{50}$ for female rats 15 mg/kg. In 2-y feeding trials NEL for rats and dogs was 1 mg/kg diet. In 2-y feeding trials rats and mice showed no carcinogenic response. It is rapidly metabolised in animals to hydrophilic metabolites which are excreted (M. R. Baldwin *et al.*, *Pestic. Sci.*, 1976, **7**, 575). Continuous exposure to endrin in diets results in only a limited accumulation in body fat. $LC_{50}$ (24-h) for fish 0.75-5.6 mg/l.

**Formulations.**

These include: e.c. (192 or 200 g a.i./l); w.p. (500 g/kg); granules (10-50 g/kg).

**Analysis.**

Product analysis is by glc, (details available from Shell International Chemical Co.) or by i.r. spectroscopy (*CIPAC Handbook*, 1970, **1**, 378; 1983, **1B**, in press; FAO Specification (CP/46); *Organochlorine Insecticides* 1973). Residues may be determined by glc with ECD (*AOAC Methods* 1980, 29.001-29.018; *Pestic. Anal. Man.*, 1979, **I**, 201-A, 201-G, 201-I: *Analyst (London)*, 1979, **104**, 425; G. M. Telling *et al.*, *J. Chromatog.*, 1977, **137**, 405; M. A. Luke *et al.*, *J. Assoc. Off. Anal. Chem.*, 1981, **64**, 1187; *Anal. Methods Pestic. Plant Growth Regul.*, 1972, **6**, 393).

$C_{14}H_{14}NO_4PS$ (323.3)
WNR DOPS&R&O2

S
||
P—O—C6H4—NO2 (phenyl–P; OCH2CH3)

**Nomenclature and development.**

Common name EPN (ESA, JMAF). Chemical name (IUPAC) *O*-ethyl *O*-4-nitrophenyl phenylphosphonothioate. (*C.A.*) *O*-ethyl *O*-(4-nitrophenyl) phenylphosphonothioate (9CI); *O*-ethyl *O*-(*p*-nitrophenyl) phenylphosphonothioate (8CI); Reg. No. *[2104-64-5]*. OMS 219, ENT 17 298. Introduced by E. I. du Pont de Nemours & Co. (Inc.) and subsequently by Nissan Chemical Industries Ltd. and Velsicol Corp. (USP 2 503 390 to du Pont).

**Properties.**

Pure EPN is a light yellow crystalline powder; m.p. 36 °C; v.p. 126 μPa at 25 °C. Practically insoluble in water; soluble in most organic solvents. The technical grade is a dark amber-coloured liquid; $d^{25}$ 1.27; $n_D^{30}$ 1.5978. Stable in neutral and acidic media, but hydrolsed by alkali and so incompatible with alkaline pesticides. On heating in a sealed tube it is converted to the *S*-ethyl isomer (R. L. Metcalf & R. B. March, *J. Econ. Entomol.*, 1953, **46**, 288).

**Uses.**

It is a non-systemic insecticide and acaricide with contact and stomach action. It is effective at 0.5-1.0 kg a.i./ha against a wide range of lepidopterous larvae, especially bollworms (*Heliothis* spp. and *Pectinophora gossypiella*) and *Alabama argillacea* on cotton, rice stem borers and other leaf-eating larvae on fruit and vegetables. It is non-phytotoxic except to some apple varieties.

**Toxicology.**

Acute oral $LD_{50}$ is: for male rats 33-42 mg/kg, for females 14 mg/kg; for mice 50-100 mg/kg; for bobwhite quail 220 mg/kg; for ring-necked pheasants >165 mg/kg. All dogs tested survived at 20 mg/kg, all were killed by 45 mg/kg. Acute percutaneous $LD_{50}$ for male white rats 230 mg/kg, for females 25 mg/kg. In 2-y feeding trials NEL was for male rats 150 mg/kg diet, for females 75 mg/kg diet; only effect observed at 3 × these doses (highest tested) was retarded growth. In 1-y trial with male and female dogs NEL was 2 mg/kg daily (H. C. Hodge *et al.*, *J. Pharmacol, Exp. Therap.*, 1954, **112**, 29). $LC_{50}$ is: for rainbow trout 0.21 mg/l; for bluegill 0.37 mg/l.

**Formulations.**

These include e.c. (450 or 480 g a.i./l), granules (40 g/kg). Mixture, e.c. (EPN + parathion-methyl).

**Analysis.**

Product analysis is by glc, details from E. I. du Pont de Nemours and Co. (Inc.). Residues may be determined by glc (*AOAC Methods*, 1980, 29.039-29.043; H. L. Pease & J. J. Kirkland, *J. Agric. Food Chem.*, 1967, **15**, 187).

$C_9H_{19}NOS$ (189.3)
3N3&VS2

$[CH_3(CH_2)_2]_2NCO.SCH_2CH_3$

**Nomenclature and development.**

Common name EPTC (BSI, E-ISO, F-ISO, WSSA, JMAF). Chemical name (IUPAC) *S*-ethyl dipropyl(thiocarbamate) (I). (*C.A.*) *S*-ethyl dipropylcarbamothioate (9CI); (I) (8CI); Reg. No. *[759-94-4]*. Its herbicidal properties were described by J. Antognini *et al.* *(Proc. Northeast. Weed Control Conf.*, 1957, p.2). Introduced by Stauffer Chemical Co. (USP 2913 327) as code no. 'R-1608'; trade mark 'Eptam'; mixtures with *N,N*-diallyl-2,2-dichloroacetamide are sold as 'Eradicane'.

**Properties.**

EPTC (99.5% pure) is a clear colourless liquid, with an aromatic odour; b.p. 127 °C/20 mmHg; v.p. 4.5 Pa at 25 °C; $d_4^{30}$ 0.9546; $n_D^{30}$ 1.475. Solubility at 24 °C: 375 mg/l water; miscible with acetone, ethanol, kerosene, 4-methylpentan-2-one, xylene. Stable $<$200 °C.

**Uses.**

It kills germinating weed seeds and inhibits bud development from underground portions of some perennial weeds. Recommended for pre-plant use (3-6 kg a.i./ha) on a wide range of crops. Incorporation into the soil mechanically or by irrigation is necessary to avoid loss by volatilisation. It is particularly useful for the control of *Agropyron repens* and perennial *Cyperus* spp.

**Toxicology.**

Acute oral $LD_{50}$: for male albino rats 2550 mg/kg, for females 2525 mg/kg. Acute percutaneous $LD_{50}$ for rabbits $>$5000 mg/kg; a mild irritant to their skin and eyes. Rats fed 326 mg/kg daily for 21 d showed no symptom other than excitability and weight loss. $LC_{50}$ (7-d) for bobwhite quail was 20 000 mg/kg diet. $LC_{50}$ (96-h)was: for bluegill 27 mg/l; for rainbow trout 19 mg/l.

**Formulations.**

These include; 'Eptam E', e.c. (720, 800 or 840 g a.i./l); 'Eptam G', granules (23, 50, 100 or 200 g/kg). 'Eradicane', (Reg. No. *[51990-04-6]*), EPTC + *N,N*-diallyl-2,2-dichloroacetamide: 'Eradicane E', e.c. (720 or 800 g EPTC/l); 'Eradicane 10G', granules (100 g EPTC/kg).

**Analysis.**

Product analysis is by glc (*AOAC Methods*, 1980, 6.426-6.430; *CIPAC Handbook*, in press). Residues are determined by glc: for crops (G. G. Patchett *et al.*, *Anal. Methods Pestic. Plant Growth Regul. Food Addit.*, 1964, **4**, 117; W. Y. Ja *et al.*, *Anal. Methods Pestic. Plant Growth Regul.*, 1972, **6**, 644) and soils (T. H. Byast *et al.*, *Tech. Rep. ARC Weed Res. Organ.*, No. 15 (2nd Ed.), p.17). Analytical methods are available from the Stauffer Chemical Co.

$C_{11}H_{25}ClO_6Si$ (316.9)
1O2O-SI-2GO2O1&O2O1

$(CH_3OCH_2CH_2O)_3SiCH_2CH_2Cl$

**Nomenclature and development.**

Common name etacelasil (BSI, E-ISO, F-ISO). Chemical name (IUPAC) 2-chloroethyltris(2-methoxyethoxy)silane. (*C.A.*) 6-(2-chloroethyl)-6-(2-methoxyethoxy)-2,5,7,10-tetraoxa-6-silaundecane (9CI); Reg. No. *[37894-46-5]*. Its use as an abscission agent was described by J. Rufener & D. Pietà (*Riv. Ortoflorofruttic. Ital.*, 1974, **4**, 274). Introduced by Ciba-Geigy AG (BEP 773 498; GBP 1 371 804) code no. 'CGA 13 586'; trade mark 'Alsol'.

**Properties.**

Pure etacelasil is a colourless liquid; b.p. 85 °C/0.001 mmHg; v.p. 27 mPa at 20 °C; ρ 1.10 g/cm³ at 20 °C. Solubility at 20 °C 25 g/l water; completely miscible with benzene, dichloromethane, hexane, methanol, octan-1-ol. On hydrolysis at 20 °C 50% loss (calculated) occurs in 50 min at pH 5, 160 min at pH 6, 43 min at pH7, 23 min at pH 8.

**Uses.**

It is an abscission agent for olives acting by releasing ethylene. Depending on the olive variety, a spray (1-2 g a.i./l) is recommended with harvest 6-10 d after application.

**Toxicology.**

Acute oral $LD_{50}$ for rats 2066 mg tech./kg. Acute percutaneous $LD_{50}$ for rats >3100 mg/kg; slight irritant to skin, and non-irritant to eyes of rabbits. Inhalation $LC_{50}$ (4-h) for rats >3700 mg/m³ air. In 90-d feeding trials NEL was: for rats 20 mg/kg daily; for dogs 10 mg/kg daily. $LC_{50}$ (96-h) for rainbow trout, crucian carp, bluegill >100 mg/l. It is practically non-toxic to birds.

**Formulations.**

'Alsol' 800 SCW, soluble concentrate (800 g a.i./l).

**Analysis.**

Product analysis is by glc. Residues may be determined by glc with MCD. Particulars are available from Ciba-Geigy AG.

# Etaconazole

$C_{14}H_{15}Cl_2N_3O_2$ (328.2)
T5O COTJ BR BG DG& D2 B1- AT5NN DNJ

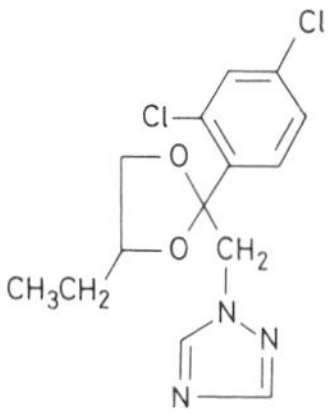

**Nomenclature and development.**

Common name etaconazole (BSI, draft E-ISO). Chemical name (IUPAC) (±)-1-[2-(2,4-dichlorophenyl)-4-ethyl-1,3-dioxolan-2-ylmethyl]-1*H*-1,2,4-triazole. (*C.A.*) 1-[[2-(2,4-dichlorophenyl)-4-ethyl-1,3-dioxolan-2-yl]methyl]-1*H*-1,2,4-triazole (9CI); Reg. No. *[60207-94-4]*, formerly *[71245-23-3]*. Its fungicidal properties were described by T. Staub *et al.* (*Abstr. Int. Congr. Plant Protect.*, 1979, 310). It was introduced by Ciba-Geigy AG as an agricultural fungicide, having been invented by Janssen Pharmaceutica (GBP 1 522 657; BEP 859 579) as code no. 'CGA 64 251'; trade mark 'Sonax' (Ciba-Geigy AG).

**Properties.**

Pure etaconazole forms colourless crystals; m.p. ranging over 75-93 °C (depending on ratio of stereoisomers); ρ 1.40 g/cm$^3$ at 20 °C; v.p. 30.6 μPa at 20 °C. Solubility at 20 °C: 80 mg/l water; 300 g/kg acetone; 700 g/kg dichloromethane; 7.5 g/kg hexane; 400 g/kg methanol; 100 g/kg propan-2-ol; 250 g/kg toluene. It is stable to hydrolysis, and to temperatures <350 °C.

**Uses.**

It is a broad spectrum systemic fungicide with protective and curative action against Ascomycetes, Basidiomycetes and Deuteromycetes; it has no activity on Oomycetes. It is especially active on powdery mildews, rusts, *Sclerotinia* and *Venturia* spp. on various crops. Post-harvest treatments are active against fungi including *Diplodia, Geotrichum* and *Pencillium* spp. and others. (J. D. Van Geluwe *et al.*, *ibid.*, p. 495; J. D. Gilpatrick, *ibid.*, p. 312).

**Toxicology.**

Acute oral $LD_{50}$ for rats 1343 mg/kg. Acute percutaneous $LD_{50}$ for rats >3100 mg/kg; slight irritant to skin and eyes of rabbits. $LC_{50}$ (96-h) in laboratory trials is: for rainbow trout 2.5-2.9 mg/l; for carp 4 mg/l.

**Formulations.**

These include 'Sonax 10', 'Vangard 10', w.p. (100 g a.i./kg). 'Sonax C 52', w.p. (20 g etaconazole + 500 g captan/kg).

**Analysis.**

Product analysis is by glc with TID. Residues may be determined by glc with FID. Particulars are available from Ciba-Geigy AG.

$C_{13}H_{14}F_3N_3O_4$ (333.3)
FXFFR CNW ENW DN2&1Y1&U1

**Nomenclature and development.**

Common name ethalfluralin (BSI, E-ISO, ANSI, WSSA);éthalfluraline (F-ISO). Chemical name (IUPAC) *N*-ethyl-α,α,α-trifluoro-*N*-(methylallyl)-2,6-dinitro-*p*-toluidine (I); *N*-ethyl-*N*-(methylallyl)-2,6-dinitro-4-trifluoromethylaniline. (*C.A.*) *N*-ethyl-*N*-(2-methyl-2-propenyl)-2,6-dinitro-4-(trifluoromethyl)benzenamine (9CI); Reg. No. *[55283-68-6]*. Its herbicidal properties were described by G. Skylakakis *et al.* (*Proc. Br. Weed Control Conf., 12th*, 1974, **2**, 795). Introduced by Eli Lilly & Co. (USP 3 257 190) code no. 'EL-161'; trade marks 'Sonalan', 'Sonalen'.

**Properties.**

Pure ethalfluralin is a yellow-orange crystalline solid; m.p. 55-56 °C; v.p. 110 μPa at 25 °C. Solubility at 25 °C: 0.2 mg/l water at pH7; >500 g/l acetone; >82 g/l methanol; >500 g/l xylene.

**Uses.**

It is a pre-planting herbicide which controls susceptible weed species as their seeds germinate; established weeds are tolerant. When incorporated in soil at 1.0-1.25 kg a.i./ha, it has a residual action on numerous broad-leaved and annual grass weeds in dry beans, cotton, soyabeans. A mixture with atrazine is recommended for pre-em. and post-em. surface application in maize.

**Toxicology.**

Acute oral $LD_{50}$: for rats and mice >10 000 g/kg; for cats, dogs, bobwhite quail and mallard duck >200 mg/kg. Acute percutaneous $LD_{50}$ for rabbits >2000 mg/kg; there were only slight irritant effects to skin and eyes. In 2-y feeding trials no ill-effect was observed in rats or mice receiving 100 mg/kg diet. $LC_{50}$ is: for bluegill 0.012 mg/l; for rainbow trout 0.0075 mg/l; for goldfish 0.1 mg/l.

**Formulations.**

These include: e.c. and w.p. 'Maizor' (Reg. No. *[64867-15-8]*), 'Grindor', w.p. (ethalfluralin + atrazine).

**Analysis.**

Product analysis is by glc with FID or by spectrophotometry (E. W. Day, *Anal. Methods Pestic. Plant Growth Regul.*, 1978, **10**, 341). Residues are determined by glc with ECD (*idem, ibid.*). Details available from Eli Lilly & Co.

$C_2H_6ClO_3P$ (144.5)
QPQO&2G

$ClCH_2CH_2P(=O)(OH)_2$

**Nomenclature and development.**

Common name ethephon (ANSI, Canada);chorethephon (New Zealand). Chemical name (IUPAC) 2-chloroethylphosphonic acid (I). (*C.A.*) (I) (8 & 9CI); Reg. No. *[16672-87-0]*. Introduced as a plant growth regulator by Amchem Products Inc. (now Union Carbide Agricultural Products Co. Inc.) and GAF Corp. (USP 3 879 188; 3 896 163; 3 897 486) as trade marks 'Ethrel', 'Florel', 'Cerone' (all to Amchem), '*Cepha*' (to GAF Corp.).

**Properties.**

It is a colourless waxy solid; m.p. 74-75 °C; *d* 1.58. Solubility: *c.* 1 kg/l water, short-chain alcohols, glycols; sparingly soluble in non-polar organic solvents; insoluble in kerosene, diesel oil. It is corrosive. It is stable at pH $\leqslant$ 3, decomposing to liberate ethylene at higher pH values.

**Uses.**

It is used to accelerate the pre-harvest ripening of fruit and vegetables including: apples, blackberries, black currants, blueberries, citrus and coffee; post-harvest ripening of fruit: bananas, citrus, mangoes. Also: yellowing of tobacco leaves and temporary growth inhibition of tobacco and tomato transplants; stimulation of latex flow in rubber trees and of resin flow in pine trees; acceleration of boll opening and of defoliation in cotton; prevention of lodging in cereals and maize; induction of hull split in walnuts, flower abscission in cloves, increases in fruit set and yield of cucumbers; induction of flowering in pineapples and ornamental Bromeliads; promotion of earlier defoliation of apple nursery stock, roses and tallhedge blackthorn; elimination of undesirable fruit on apples and crabapples; stimulation of lateral branching of azaleas, geraniums and roses; modification of sex expression in cucumbers and squash; shortening of stem length in forced daffodils. It acts by releasing ethylene within plant tissue. Plant response depends upon climatic conditions so the rates may vary with area and country.

**Toxicology.**

Acute oral $LD_{50}$: for rats 4229 mg/kg; for bobwhite quail 1000 mg/kg. Acute percutaneous $LD_{50}$ for rabbits 5730 mg/kg; it is an eye irritant. In 2-y feeding trials rats receiving $\leqslant$12 500 mg/kg diet showed no ill-effect except at top dose levels towards the end of the trial. $LC_{50}$ (8-d) for mallard duck was $>$10 000 mg/kg diet. $LC_{50}$ (96-h) is: for bluegill 300 mg/l; for rainbow trout 350 mg/l.

**Formulations.**

'Ethrel' liquid concentrate (240 or 480 g a.i./l); 'Ethrel C' (in the UK, ICI Plant Protection Division); 'Cerone', 'Prep', liquid concentrates (480 g/l). 'Florel', (40 g/l). 'Terpal' (BASF), liquid concentrate (mixture with mepiquat chloride).

**Analysis.**

Product analysis is by measuring the ethylene produced on treatment with concentrated alkali. Residues may be determined by conversion to the dimethyl ester, measured by glc with FID or FPD.

# Ethidimuron 5690

$C_7H_{12}N_4O_3S_2$ (264.2)
T5NN DSJ CSW2 EN1&VM1

$CH_3CH_2SO_2$ — S — $NCO.NHCH_3$ ($CH_3$); N—N

**Nomenclature and development.**

Common name ethidimuron (BSI, E-ISO, F-ISO). Chemical name (IUPAC) 1-(5-ethylsulphonyl-1,3,4-thiadiazol-2-yl)-1,3-dimethylurea. (*C.A.*) *N*-[5-(ethylsulfonyl)-1,3,4-thiadiazol-2-yl]-*N,N'*-dimethylurea (9CI); 1-(5-ethylsulfonyl-1,3,4-thiadiazol-2-yl)-1,3-dimethylurea (8CI); Reg. No. *[30043-49-3]*. Its herbicidal properties were described by L. Eue *et al*., (*C. R. Journ. Etud. Herbic. Conf. COLUMA, 4th*, 1973, **1**, 14). Introduced by Bayer AG (BEP 743 615) as code no. 'BAY MET 1486'; trade mark 'Ustilan'.

**Properties.**

It forms colourless crystals; m.p. 156 °C; v.p. <1 μPa at 20 °C. Solubility at 20 °C: 3 g/kg water; 106 g/l dichloromethane; 6 g/l propan-2-ol. Unstable to alkali. Decomposes at 217 °C.

**Uses.**

Ethidimuron is a herbicide for total weed destruction on non-crop areas at 7 kg a.i./ha and for use in sugarcane plantations at 0.7-1.4 kg/ha.

**Toxicology.**

Acute oral $LD_{50}$: for rats >5000 mg/kg; for mice >2500 mg/kg; for Japanese quail 300-400 mg/kg; for canaries 1000 mg/kg. Acute percutaneous $LD_{50}$ for rats >1000 mg/kg. In 90-d feeding trials NEL for rats was >1000 mg/kg diet. $LC_{50}$ for golden orfe >1000 mg/l. It is not harmful to honeybees.

**Formulations.**

'Ustilan', w.p. (700 g a.i./kg). 'Ustilan NK25', ethidimuron + aminotriazole + dichlorprop.

**Analysis.**

Product analysis is by i.r. spectrometry, details available from Bayer AG. Residues may be determined by glc with TID (H. J. Jarczyk, *Pflanzenschutz-Nachr. (Engl. Ed.)*, 1979, **32**, 186).

$C_{11}H_{15}NO_2S$ (225.3)
2S1R BOVM1

$OCO.NHCH_3$
$CH_2SCH_2CH_3$

**Nomenclature and development.**

Common name ethiofencarb (BSI, E-ISO);éthiophencarbe (F-ISO). Chemical name (IUPAC) α-ethylthio-*o*-tolyl methylcarbamate (I); 2-ethylthiomethylphenyl methylcarbamate. (*C.A.*) 2-[(ethylthio)methyl]phenyl methylcarbamate (9CI); (I) (8CI); Reg. No. *[29973-13-5]*. Its insecticidal properties were described by J. Hammann & H. Hoffmann (*Pflanzenschutz-Nachr. (Engl. Ed.)*, 1974, **27**, 267). Introduced by Bayer AG (DEOS 1 910 588; BEP 746 649) as code no. 'BAY HOX 1901'; trade mark 'Croneton'.

**Properties.**

It is a yellow oil; m.p. 33.4 °C; v.p. 13 mPa at 30 °C; $d_4^{20}$ 1.147. Solubility at 20 °C: 1.82 g/kg water; >600 g/kg dichloromethane; propan-1-ol, toluene.

**Uses.**

It is a soil- and foliar-applied systemic insecticide with a specific effect against aphids at *c.* 50 g a.i./100 l (R. Homeyer, *ibid.*, 1976, **29**, 254). Its breakdown in soil and water have been studied (G. Dräger, *ibid.*, 1977, **30**, 18).

**Toxicology.**

Acute oral $LD_{50}$: for male rats 411-499 mg/kg; for mice 224-256 mg/kg; for Japanese quail 155 mg/kg; for canaries *c.* 100 mg/kg; for hens *c.* 1000 mg/kg. Acute percutaneous $LD_{50}$ for male rats >1150 mg/kg. In 2-y feeding trials NEL for rats was 330 mg/kg diet. $LC_{50}$ (96-h) is: for carp 10-20 mg/l; for golden orfe 8-10 mg/l; for goldfish 20-40 mg/l.

**Formulations.**

'Croneton', e.c. (500 g a.i./l); granules (100 g/kg).

**Analysis.**

Details of the methods for product analysis may be obtained from Bayer AG. Residues may be determined by glc (*idem, ibid.*, 1974, **27**, 144).

$C_9H_{22}O_4P_2S_4$ (384.5)
2OPS&O2&S1SPS&O2&O2

$(CH_3CH_2O)_2P(=S)SCH_2SP(=S)(OCH_2CH_3)_2$

**Nomenclature and development.**

Common name ethion (BSI, E-ISO, F-ISO, ANSI, JMAF);diethion (France, India, Republic of South Africa);*exception* (Italy, Portugal, Turkey). Chemical name (IUPAC) *O,O,O',O'*-tetraethyl *S,S'*-methylene bis(phosphorodithioate); *S,S'*-methylene bis(*O,O*-diethyl phosphorodithioate). (*C.A.*) *O,O,O',O'*-tetraethyl *S,S'*-methylene di(phosphorodithioate) (9CI); *S,S'*-methylene *O,O,O',O'*-tetraethyl di(phosphorodithioate) (8CI); Reg. No. *[563-12-2]*. ENT 24 105. Its insecticidal properties were reported (*Chem. Eng. News*, 1957, **35**, 87). Introduced by FMC Corp. (GBP 872 221; USP 2 873 228) as code no. 'FMC 1240'.

**Properties.**

Pure ethion is a colourless to amber-coloured liquid; f.p. −12 ° to −15 °C; v.p. 200 μPa at 25 °C; $d_4^{20}$ 1.22; $n_D^{20}$ 1.5490. Solubility: sparingly soluble in water; soluble in most organic solvents including kerosene, petroleum oils. The technical grade has $d_4^{20}$ 1.215-1.230, $n_D^{20}$ 1.530-1.542. It is slowly oxidised in air and is hydrolysed by acids and alkalis.

**Uses.**

It is a non-systemic acaricide and insecticide used on apples, particularly with petroleum oils on dormant trees, to kill eggs and scales. It is non-phytotoxic except to apple varieties maturing at the same time as or before Early McIntosh. Other uses include the control of *Boophilus* spp. on cattle and of citrus mite.

**Toxicology.**

Acute oral $LD_{50}$: for rats 208 mg a.i./kg; for female rats 24.4 mg tech./kg. Acute percutaneous $LD_{50}$ for rabbits 915 mg/kg. In 2-y feeding trials NEL was: for rats 6 mg/kg diet (0.3 mg/kg daily); for dogs 2 mg/kg diet (0.05 mg/kg daily).

**Formulations.**

These include: w.p. (250 g a.i./kg); e.c. (480 or 960 g/l); dust (40 g/kg); granules (50, 80 or 100 g/kg). Mixtures include: ethion + oils; ethion + dicofol; ethion + dicofol + parathion-methyl.

**Analysis.**

Product analysis is by i.r. spectrometry (*CIPAC Handbook*, 1983, **1B**, in press; *AOAC Methods*, 1980, 6.352-6.355). Residues may be determined by glc (*AOAC Methods*, 1980, 29.001-29.018; *Pestic. Anal. Man.*, 1979, **I**, 201-A, 201-G, 201-H, 201-I; D. C. Abbott *et al., Pestic. Sci.*,, 1970, **1**, 10; *Anal. Methods Pestic. Plant Growth Regul.*, 1972, **6**, 396).

$C_{11}H_{19}N_3O$ (209.3)
T6N CNJ BM2 DQ E4 F1

$CH_3$ N $NHCH_2CH_3$
N
$CH_3(CH_2)_3$
OH

**Nomenclature and development.**

Common name ethirimol (BSI, E-ISO); éthyrimol (F-ISO). Chemical name (IUPAC) 5-butyl-2-ethylamino-6-methylpyrimidin-4-ol. (*C.A.*) 5-butyl-2-(ethylamino)-6-methyl-4(1*H*)-pyrimidinone (9CI); 5-butyl-2-(ethylamino)-6-methyl-4-pyrimidinol (8CI); Reg. No. *[23947-60-6]*. Its fungicidal properties were described by R. M. Bebbington *et al.* (*Chem. Ind. (London)*, 1969, p.1512). Introduced by ICI Plant Protection Division (GBP 1 182 584) as code no. 'PP 149'.

**Properties.**

A colourless crystalline solid; m.p. 159-160 °C (with a phase change at *c.* 140 °C); v.p. 267 μPa at 25 °C. Solubility at room temperature 200 mg/l water; sparingly soluble in acetone; slightly soluble in ethanol, 4-hydroxy-4-methylpentan-2-one; soluble in chloroform, trichloroethylene and aqueous solutions of strong acids and bases. Stable in alkaline and acid solutions. Non-corrosive to metals, but acidic solutions should not be stored in galvanised steel.

**Uses.**

A systemic fungicide effective against powdery mildews of cereals and other field crops. It is effective when applied as a seed dressing or to the soil in the root zone; the plant roots then continue to absorb it giving protection to foliage throughout critical growth stages. When used as a foliar spray it penetrates the leaves and moves towards the leaf margins. Ethirimol has no effect on soil fauna.

**Toxicology.**

Acute oral $LD_{50}$: for female rats 6340 mg/kg; for male rabbits 1000-2000 mg/kg; for female guinea-pigs 500-1000 mg/kg; for hens 4000 mg/kg; for honeybees >1.6 μg/bee (>10 000 mg/kg). Acute percutaneous $LD_{50}$ for rats >1000 mg/kg; application of 50 mg/l to the eyes of rabbits for 1 d caused only slight irritation; it has no effect on their skin. In 2-y feeding trials NEL was: for rats 200 mg/kg diet; for dogs 30 mg/kg daily. $LC_{50}$ (96 h) for brown trout fingerlings is 20 mg/l. A spray (20 g a.i./l) has no harmful effect when sprayed upon honeybees.

**Formulations.**

'Milstem', an aqueous suspension (col) (500 g a.i./l) used undiluted as a seed treatment. 'Milgo E', col (250 g/l) used as a spray. 'Milcurb Super', e.c. (250 g/l). 'Milcap' (Reg. No. *[55948-12-4]*), (70 g ethirimol + 360 g captafol/l).

**Analysis.**

Product analysis is by glc (J. E. Bagness & W. G. Sharples, *Analyst (London)*, 1974, **99**, 225), or by u.v. spectrometry. Residues are determined by u.v. spectrometry (M. J. Edwards, *Anal. Methods Pestic. Plant Growth Regul.*, 1976, **8**, 285). Details are available from ICI Plant Protection Division.

$C_{13}H_{18}O_5S$ (286.3)
T56 BOT&J CO2 D1 D1 GOSW1



**Nomenclature and development.**

Common name ethofumesate (BSI, E-ISO, F-ISO, ANSI, WSSA). Chemical name (IUPAC) (±)-2-ethoxy-2,3-dihydro-3,3-dimethylbenzofuran-5-yl methanesulphonate. (*C.A.*) (±)-2-ethoxy-2,3-dihydro-3,3-dimethyl-5-benzofuranyl methanesulfonate (8 & 9CI); Reg. No. *[26225-79-6]*. Its herbicidal properties were described by R. K. Pfeiffer (*Symp. New Herbic., 3rd*, 1969, p. 1). Introduced by Fisons Ltd Agrochemical Division (now FBC Limited) (GBP 1 271 659) as code no. 'NC 8438'; trade mark 'Nortron'.

**Properties.**

It is a colourless crystalline solid; m.p. 70-72 °C; v.p. 86 μPa at 25 °C. Solubility at 25 °C: 110 mg/l water; 400 g/kg acetone, benzene, chloroform, dioxane; 100 g/kg ethanol; 4 g/kg hexane. It is stable to hydrolysis in water at pH 7; 50% loss in 30 d at pH 1 gives 2,3-dihydro-2-hydroxy-3,3-dimethylbenzofuran-5-yl methanesulphonate and ethanol; and in 10 d at pH 13 2-ethoxy-2,3-dihydro-3,3-dimethylbenzofuran-5-ol.

**Uses.**

Ethofumesate is a pre- and/or post-em. herbicide particularly selective in sugar and other beet crops, ryegrass and the other pasture grasses at 1.0-2.0 kg a.i./ha. It is effective in controlling a wide range of important grasses and broad-leaved weeds, with a good persistence of activity in the soil. In beet crops 1.0-3.0 kg/ha can be used, but ethofumesate is normally recommended at 0.5-2.0 kg/ha in tank-mixtures or co-formulations with other residual or contact herbicides for use in beet. A high degree of tolerance is also shown by strawberries, sunflowers and tobacco depending on the time of application. Onions show intermediate tolerance. It is biologically degraded in soil, the rate depending on climate, soil type and microbial activity. The time for 50% loss ranges from <35 d under moist, warm conditions, to >98 d under dry, cold conditions.

**Toxicology.**

Acute oral $LD_{50}$ for rats >6400 mg/kg. Acute percutaneous $LD_{50}$ for rats >1440 mg/kg. In 2-y feeding trials NEL for rats was >1000 mg/kg diet. $LC_{50}$ (5-d) for ducks and quail >10 000 mg/kg diet. $LC_{50}$ (24-h) to guppies was 15 mg tech./l. Honeybees were unaffected by 16 μg/bee in both contact and oral tests.

**Formulations.**

An e.c. (200 g a.i./l); s.c. (500 g/l). 'Magnum' and 'Spectron', (Reg. No. *[50641-63-9]*), s.c. (ethofumesate + chloridazon). 'Morlex', w.p. (ethofumesate + chlorpropham + fenuron + propham).

**Analysis.**

Product analysis is by glc with FPD (R. J. Whiteoak *et al.*, *Anal. Methods Pestic. Plant Growth Regul.*, 1978, **10**, 353). Details are available from FBC Limited. Residues of ethofumesate and of conjugates (which are first hydrolysed to aglycones) are determined by glc of a derivative with FPD (R. J. Whiteoak *et al.*, *loc. cit.*). Residues in soil are determined similarly (T. H. Byast *et al.*, *Tech. Rep. ARC Weed Res. Organ.*, No.15, (2nd Ed.), p.27).

$C_8H_{19}O_2PS_2$ (242.3)
3SPO&S3&O2

$CH_3CH_2OP(=O)(SCH_2CH_2CH_3)_2$

**Nomenclature and development.**

Common name ethoprophos (BSI, E-ISO, F-ISO);ethoprop (ANSI, Society of Nematologists, (USA)). Chemical name (IUPAC) *O*-ethyl *S*,*S*-dipropyl phosphorodithioate (I). (*C.A.*) (I) (8 & 9CI); Reg. No. *[13194-48-4]*. Its nematicidal properties were reported by S. J. Locascio (*Proc. Fla. St. Hortic. Soc.*, 1966, **79**, 170). Introduced by Mobil Chemical Co. (who no longer manufacture or market it) and later by Rhône-Poulenc (USP 3 112 244; 3 268 393 to Mobil Chemical Co.) as code no. 'VC9-104' (Mobil Chemical Co.); trade marks 'Mocap', 'Prophos'.

**Properties.**

Ethoprophos is a clear pale yellow liquid; b.p. 86-91 °C/0.2 mmHg; v.p. 46.5 mPa at 26 °C; $d_4^{20}$ 1.094. Solubility: 750 mg/l water; >300 g/kg acetone, cyclohexane, 1,2-dichloroethane, diethyl ether, ethanol, ethyl acetate, petroleum spirit, xylene. It is stable in water up to 100 °C at pH7, but rapidly hydrolysed at 25 °C and pH9.

**Uses.**

It is a non-systemic, non-fumigant nematicide, which is also effective against soil-dwelling insects and used at 1.6-6.6 kg a.i./ha.

**Toxicology.**

Acute oral $LD_{50}$ for albino rats 62 mg/kg; for rabbits 55 mg/kg; for mallard duck 61 mg/kg; for hens 5.6 mg/kg. Acute percutaneous $LD_{50}$ for albino rabbits 26 mg/kg, the granular formation being much less toxic. In 90-d feeding trials rats and dogs receiving 100 mg/kg diet showed depression of cholinesterase levels but no other effect on pathology or histology. $LC_{50}$ (96-h) is: for goldfish 13.6 mg/l; for bluegill 2.07 mg/l; for rainbow trout 13.8 mg/l.

**Formulations.**

'Mocap', granules (100 g a.i./kg); e.c. (700 g/l).

**Analysis.**

Product analysis is by glc. Residues may be determined by glc with MCD or FPD.

$C_2H_4Br_2$ (187.9)
E2E

$BrCH_2CH_2Br$

**Nomenclature and development.**

Common name EDB (JMAF); the traditional name ethylene dibromide (BSI, E-ISO); dibromure d'ethylene (F-ISO) is accepted in lieu of a common name. Chemical name (IUPAC) 1,2-dibromoethane (I). (*C.A.*) (I) (8 & 9CI); Reg. No. *[106-93-4]*. Trivial name EDB. ENT 15 349. Its fumigant insecticidal properties were described by I. E. Neifert *et al.* (*U.S. Dep. Agric. Bull.*, 1925, No. 1313). Introduced by Dow Chemical Co. (USP 2 448 265; 2 473 984) as trade marks 'Bromofume', 'Dowfume 85'.

**Properties.**

It is a colourless liquid; b.p. 131.5 °C; m.p. 9.3 °C; v.p. 1.5 kPa at 25 °C; $d_{25}^{25}$ 2.172; $n_D^{20}$ 1.5379. Solubility at 30 °C: 4.3 g/kg water; soluble in diethyl ether, ethanol and most organic solvents. It is non-flammable.

**Uses.**

It is an insecticidal fumigant used against pests of stored products (J. Aman *et al., Ann. Appl. Biol.*, 1946, **33**, 389); for the treatment of fruit and vegetables (J. W. Balock, *Science*, 1951, **114**, 122); for the spot treatment of flour mills; for soil treatment against certain insects and nematodes (M. W. Stone, *U.S. Dep. Agric. Bull.*, 1969, E-786). Planting must be delayed until 8 d after soil treatment because of its phytotoxicity.

**Toxicology.**

Acute oral $LD_{50}$ for male rats 146 mg/kg. Dermal applications, if confined, will cause severe burning of the skin. Rats tolerated 7-h exposures, 5 d/week, for 0.5 y at rates $\leqslant$ 210 $mg/m^3$.

**Formulations.**

'Dowfume 85' (1.44 kg a.i./l) in an inert solvent, for soil applications. Solutions in carbon tetrachloride for mill use.

**Analysis.**

Product analysis is by glc (*CIPAC Handbook*, in press; *AOAC Methods*, 1980, 6.143-6.148). Residues may be determined by glc (R. Bielorai & E. Alumot, *J. Agric. Food Chem.*, 1966, **14**, 622; S. G. Heuser & K. A. Scudamore, *J. Sci. Food Agric.*, 1969, **20**, 566; *Pestic. Sci.*, 1970, **1**, 240). Atmospheres can be sampled by passing through 2-aminoethanol and determining bromide by Volhard titration (C. E. Castro, *Anal. Methods Pestic., Plant Growth Regul. Food Addit.*, 1964, **3**, 155; *Anal. Methods Pestic. Plant Growth Regul.*, 1972, **6**, 711).

# Ethylene dichloride

$C_2H_4Cl_2$ (98.96)
G2G

$ClCH_2CH_2Cl$

**Nomenclature and development.**

Common name EDC (JMAF);the traditional name ethylene dichloride (BSI, E-ISO);dichlorure d'ethylene (F-ISO) is accepted in lieu of a common name. Chemical name (IUPAC) 1,2-dichloroethane (I). (*C.A.*) (I) (8 & 9CI); Reg. No. *[107-06-2]*. Trivial name EDC. ENT 1656. Introduced as a component of insecticidal fumigants by R. T. Cotton & R. C. Roark (*J. Econ. Entomol.*, 1927, **20**, 636).

**Properties.**

It is a colourless liquid, with a chloroform-like odour; b.p. 83.5 °C; m.p. −36 °C; v.p. 10.4 kPa at 20 °C; $d_4^{20}$ 1.2569; $n_D^{25}$ 1.4443. Solubility at room temperature: 4.3 g/l water; soluble in most organic solvents. It is flammable, flash point (Abel Penchy) 12-15 °C, lower and upper limits of flammability in air 275 and 700 g/m$^3$.

**Uses.**

It is an insecticidal fumigant used mainly for the fumigation of stored products (W. A. Gersdorff, *U. S. Dep. Agric. Misc. Pub.*, 1932, No. 117) using 130-225 g/m$^3$.

**Toxicology.**

Acute oral $LD_{50}$: for rats 670-890 mg/kg; for mice 870-950 mg/kg; for rabbits 860-970 mg/kg.

**Formulations.**

Mixture ethylene dichloride + carbon tetrachloride (3:1) to reduce the fire hazard of the former.

**Analysis.**

Product and residue analysis is by glc (S. G. Heuser & K. A. Scudamore, *J. Sci. Food Agric.*, 1969, **20**, 566; *Analyst (London)*, 1974, **99**, 570).

$C_{17}H_{25}NO_2$ (275.4)
T C555 A DVNV IUTJ E1Y4&2

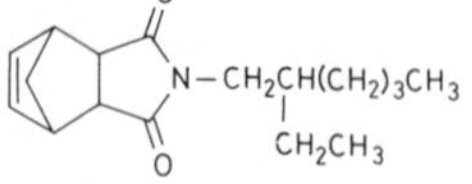

$N-CH_2CH(CH_2)_3CH_3$, $CH_2CH_3$

**Nomenclature and development.**

Chemical name (IUPAC) *N*-(2-ethylhexyl)-8,9,10-trinorborn-5-ene-2,3-dicarboximide; *N*-(2-ethylhexyl)bicyclo[2.2.1]hept-5-ene-2,3-dicarboximide. (*C.A.*) 2-(2-ethylhexyl)-3a,4,7,7a-tetrahydro-4,7-methano-1*H*-isoindole-1,3(2*H*)-dione (8 & 9CI); Reg. No. *[113-48-4]*. Trivial name *N*-octylbicycloheptenedicarboximide. ENT 8184. Its use as a synergist was described by R. H. Nelson *et al.* (*Soap Chem. Spec.*, 1949, **25**(1), 120). Introduced by Van Dyke Co. and later by McLaughlin Gormley King Co. (USP 2 476 512 to McLaughlin Gormley King Co.) as trade marks 'Van Dyke 264', 'MGK 264', formerly '*Octacide 264*' (McLaughlin Gormley King Co.).

**Properties.**

The pure compound is a liquid; f.p. $<-20$ °C; $n_D^{25}$ 1.4982-1.4988. The technical grade (98% pure, main impurity 2-ethylhexanol) has $d^{20}$ 1.040-1.060. It is practically insoluble in water; miscible with most organic solvents including petroleum oils, fluorinated hydrocarbons, also DDT and HCH. Stable in range pH6-8; to light and heat.

**Uses.**

It is used as a synergist for pyrethroids. Slight phytotoxicity is shown at concentrations $>10$ g/kg.

**Toxicology.**

Acute oral $LD_{50}$ for rats 3640 mg/kg. Acute percutaneous $LD_{50}$ for rabbits 470 mg/kg. In 2-y feeding trials no toxic effect was observed in rats receiving 1000 mg/kg diet or pigs 300 mg/kg diet. In a 3-generation teratogenic study on rats there was no effect at rates $\leqslant 36.4$ mg/kg daily, but at 364 mg/kg there was some effect on female fertility, litter size and growth ratio. $LC_{50}$ (8-d) for mallard duck and bobwhite quail $>5000$ mg/kg diet. $LC_{50}$ (96-h) is: for rainbow trout 2.6 mg/l; for bluegill 2.7 mg/l.

**Formulations.**

It is incorporated, at 5-20 g/kg, with pyrethroids (allethrin, phenothrin) in aerosol concentrates, oil concentrates and dusts.

**Analysis.**

Product and residue analysis is by glc, details of methods are available from McLaughlin Gormley King Co.

$C_9H_{21}O_3PS_3$ (304.4)
OS2&1SPS&OY1&1&OY1&1

$CH_3CH_2S(=O)CH_2SP(=S)[OCH(CH_3)_2]_2$

**Nomenclature and development.**

Common name IPSP (JMAF). Chemical name (IUPAC) *S*-ethylsulphinylmethyl *O,O*-di-isopropyl phosphorodithioate. (*C.A.*) *S*-[(ethylsulfinyl)methyl] *O,O*-bis(1-methylethyl) phosphorodithioate (9CI); *S*-ethylsulfinylmethyl *O,O*-di-isopropyl phosphorodithioate (8CI); Reg. No. *[5827-05-4]*. Its insecticidal properties were described by D. Murrayama *et al.* (*Mem. Fac. Agric. Hokkaido Univ.*, 1966, **6**, 73). Introduced by Hokko Chemical Industry Co. Ltd. (JPP 531 126; USP 3 408 426; GBP 1 068 628) as code no. 'PSP-204'; trade mark 'Aphidan'.

**Properties.**

The technical grade (90% pure) is a yellow liquid; $d_4^{20}$ 1.1696; $n_D^{20}$ 1.5260; v.p. 2 mPa at 27 °C. Solubility at 15 °C: 1.5 g/l water; 71 g/l hexane; at 20 °C freely soluble in acetone, xylene. At 25 °C 50% loss occurs in 4 d at pH 7, in 3 d at pH 9. It is stable $<$100 °C.

**Uses.**

It is a systemic insecticide effective by soil treatment against aphids on potatoes and vegetables. Granules applied at 1.5-2.5 kg a.i./ha or 50-100 mg/plant give effective control for *c*. 40 d.

**Toxicology.**

Acute oral $LD_{50}$: for male rats 25 mg a.i./kg; for male mice 320 mg/kg. Acute percutaneous $LD_{50}$ is: for male rat 28 mg/kg; for female mice 1300 mg/kg. $LC_{50}$ (48-h) for carp is 20 mg/l.

**Formulations.**

Granules (50 g a.i./kg).

**Analysis.**

Product analysis is by tlc followed by colorimetric determination. Residues may be determined by glc with FPD. Details of methods are available from Hokko Chemical Industry Co. Ltd.

$C_5H_5Cl_3N_2OS$ (247.5)
T5NS DNJ CO2 EXGGG

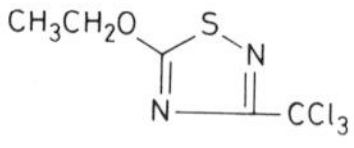

**Nomenclature and development.**

Common name etridiazole (BSI, E-ISO, F-ISO); echlomezol (JMAF). Chemical name (IUPAC) ethyl 3-trichloromethyl-1,2,4-thiadiazolyl ether; 5-ethoxy-3-trichloromethyl-1,2,4-thiadiazole. (*C.A.*) 5-ethoxy-3-(trichloromethyl)-1,2,4-thiadiazole (8 & 9CI); Reg. No. *[2593-15-9]*. Trivial names ethazol, ethazole. Introduced by Olin Chemicals as a soil fungicide (USP 3 260 588; 3 260 725) as code no. 'OM 2424'; trade mark 'Terrazole'.

**Properties.**

It is a pale yellow liquid, with a persistent odour; f.p. 20 °C; v.p. 13 mPa at room temperature. Solubility at 25 °C: 50 mg/l water; soluble in acetone, carbon tetrachloride, ethanol, saturated hydrocarbons. The technical product is a reddish-brown liquid 95-97% pure. It is stable and suffered no loss of biological activity during 3-y storage under normal conditions.

**Uses.**

It is a fungicide recommended for the control of *Phytophthora* and *Pythium* spp. and diseases of cotton, fruit, groundnuts, ornamentals, turf and vegetables at 450-900 g a.i. in 2500 l water/ha. It is also used, in combination with quintozene, as a soil fungicide and seed treatment for the control of pre-em. and post-em. cotton seedling diseases caused by *Fusarium, Phytophthora, Pythium* and *Rhizoctonia* spp.

**Toxicology.**

Acute oral $LD_{50}$ for mice 2000 mg/kg.

**Formulations.**

These include as w.p.: 'Terrazole' (Olin), 'Aaterra' (Aagrunol B.V.) (350 g a.i./kg); 'Truban' (Mallinckrodt) (300 g/kg). As e.c.: 'Terrazole 25% Emulsifiable' (Olin), 'Koban' (Mallinckrodt) (250 g/l); 'Terrazole 44% Emulsifiable' (Olin) (440 g/l); 'Pansoil EC' (Sankyo) (400 g/l). 'Pansoil 4%', dust (Sankyo) (40 g/kg). Seed treatment 'Terra-Coat L21' (Olin). Mixtures include 'Terrachlor Super X' (Olin). 'Terra-Coat L 205' (Olin), liquid seed treatment (Reg. No. *[8065-61-0*]) (58 g etridiazole + 232 g quintozene/l). 'Ban-rot' (Mallinckrodt), w.p. (150 g etridiazole + 250 g thiophanate-methyl/kg).

$C_{10}H_{17}N_2O_4PS$ (292.3)
T6N CNJ B2 DOPS&O1&O1 FO2

$CH_3CH_2O$ N $CH_2CH_3$
N
$OP(OCH_3)_2$
S

**Nomenclature and development.**

Common name etrimfos (BSI, E-ISO, F-ISO, ANSI). Chemical name (IUPAC) *O*-6-ethoxy-2-ethylpyrimidin-4-yl *O,O*-dimethyl phosphorothioate. (*C.A.*) *O*-(6-ethoxy-2-ethyl-4-pyrimidinyl) *O,O*-dimethyl phosphorothioate (9CI); Reg. No. *[38260-54-7]*. OMS 1806. Its insecticidal properties were described by H. J. Knutti & F. W. Reisser (*Proc. Br. Insectic. Fungic. Conf., 5th*, 1975, **2**, 675). Introduced by Sandoz AG (DEOS 2 209 554) as code no. 'SAN 197I'; trade marks 'Ekamet', 'Satisfar'.

**Properties.**

Pure etrimfos is a colourless oil, with a characteristic odour; m.p. −3.4 °C; v.p. 8.6. mPa at 20 °C; $d_4^{20}$ 1.195; $n_D^{20}$ 1.5068. Solubility at 23 °C: 40 mg/l water; completely miscible with acetone, chloroform, dimethyl sulphoxide, ethanol, hexane, methanol, xylene. At 25 °C, 50% loss by hydrolysis occurs in 0.4 d at pH3, 16 d at pH6, 14 d at pH9. It is unstable in its pure form, but dilute solutions in nonpolar solvents and its formulations are stable (shelf life at 20 °C: *c*. 2 y).

**Uses.**

It is a broad-range non-systemic insecticide effective at 250-750 g a.i./ha against species of Lepidoptera, Coleoptera, Diptera and, to a variable extent, Hemiptera. It is used mainly on citrus, fruit trees, grapes, lucerne, paddy maize, olives, potatoes and vegetables. Granular applications against Pyralidae in paddy rice require 1.0-1.5 kg a.i./ha. It has a moderate residual activity lasting 7-14 d. It is used at 3-10 mg/kg to protect stored products against Coleoptera, Lepidoptera, mites and Psocoptera according to duration (up to 1 y) required. It is rapidly metabolised in plants, animals and soil to 6-ethoxy-2-ethylpyrimidin-4-ol, small amounts of 2-ethylpyrimidine-4,6-diol and other hydroxylated derivatives. Also in plants to traces of the corresponding phosphate ester. Carbon dioxide is formed in soil, 50% loss of etrimfos occurring in 3-8 d.

**Toxicology.**

Acute oral $LD_{50}$ for male rats 1800 mg/kg. Acute percutaneous $LD_{50}$: for male rats >2000 mg/kg; for male rabbits >500 mg/kg. In 180 d feeding trials NEL for dogs was 12 mg/kg diet; in 90-d trials for rats it was 9 mg/kg diet. $LC_{50}$ (8-d) for quail was 740 mg/kg diet. $LC_{50}$ (96-h) for carp is 13.3 mg/l. It is toxic to honeybees by contact.

**Formulations.**

These include: 'Ekamet', e.c. (500 g/kg); granules (50 g/kg); dustable powders (10 or 20 g/kg); oil spray (50 g/kg); 'Ekamet ULV' (400 g/l); 'Satisfar' (520 g/l); 'Satisfar DP-2', dust (20 g/kg); 'Satisfar LS 3', 'Satisfar LS 5' (30 or 50 g/l).

**Analysis.**

Product analysis is by glc or by paper chromatography, followed by combustion and subsequent determination of phosphate by standard methods. Residues may be determined by glc (J. C. Karpally *et al*., *Anal. Methods Pestic. Plant Growth Regul.*, 1980, **11**, 125).

$C_{10}H_{16}NO_5PS_2$ (325.3)
1OPS&O1&OR DSWN1&1

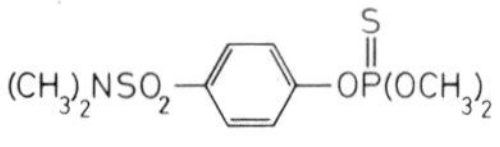

S
$(CH_3)_2NSO_2$–⟨benzene ring⟩–$\overset{\|}{\mathrm{O}}$P$(OCH_3)_2$

**Nomenclature and development.**

Common name famphur (ESA). Chemical name (IUPAC) *O*-4-dimethylsulphamoylphenyl *O,O*-dimethyl phosphorothioate. (*C.A.*) *O*-[4-[(dimethylamino)sulfonyl]phenyl] *O,O*-dimethyl phosphorothioate (9CI); *O,O*-dimethyl phosphorothioate *O*-ester with *p*-hydroxy-*N,N*-dimethylbenzenesulfonamide (8CI); Reg. No. *[52-85-7]*. Trivial name famophos. Introduced by American Cyanamid Co. (USP 3 005 004) as code no. 'CL 38 023'; trade marks 'Warbex', 'Bo-Ana'.

**Properties.**

It is a colourless crystalline powder; m.p. 52.5-53.5 °C; Solubility: sparingly soluble in water; soluble in acetone, carbon tetrachloride, chloroform, cyclohexanone, dichloromethane, toluene, xylene; sparingly soluble in aliphatic hydrocarbons.

**Uses.**

Controls cattle grubs, reindeer grub fly and the reindeer nostril fly, and reduces lice infestations.

**Toxicology.**

Acute oral $LD_{50}$: for albino rats 48 mg tech./kg; for albino mice 27 mg/kg. Acute percutaneous $LD_{50}$ for albino rabbits 2730 mg/kg.

**Formulations.**

Pour-on concentrate (143 g a.i./l); injections for reindeer.

**Analysis.**

Product analysis is by glc (N. R. Pasarela *et al.*, *J. Agric. Food Chem.*, 1967, **15**, 920). Residues may be determined by glc (P. E. Gatterdam *et al.*, *ibid.*, p.845). Details may be obtained from American Cyanamid Co.

$C_8H_{10}N_3NaO_3S$ (251.2)
WSQNUNR DN1&1 &&Na SALT

$(CH_3)_2N$—C6H4—N=N—$SO_2$.ONa

**Nomenclature and development.**

Common name fenaminosulf (BSI, E-ISO);phénaminosulf (F-ISO);DAPA (JMAF). Chemical name (IUPAC) sodium 4-dimethylaminobenzenediazosulphonate. (*C.A.*) sodium [4-(dimethylamino)phenyl]diazenesulfonate (9CI); sodium *p*-( dimethylamino)benzenediazosulfonate (8CI); Reg. No. *[140-56-7]*. Its development as a fungicide is described by E. Urbschat (*Angew. Chem.*, 1960, **72**, 981). Introduced by Bayer AG (DEAS 1 028 828) as code no. 'Bayer 22 555','Bayer 5072'; trade mark 'Lesan' (formerly '*Dexon*').

**Properties.**

Fenaminosulf is a yellowish-brown powder; stable ≤200 °C. Solubility at 20 °C 40 g/kg water; soluble in dimethylformamide; insoluble in benzene, diethyl ether, petroleum oils. Its aqueous solution is intense orange and is sensitive to light but is stabilised by sodium sulphite; it is stable in alkaline media.

**Uses.**

Fenaminosulf is a seed and soil fungicide for the control of *Pythium, Aphanomyces* and *Phytophthora* spp. in soil at 45-110 g a.i./100 kg seed (depending on the crop).

**Toxicology.**

Acute oral $LD_{50}$ for rats 60 mg a.i./kg; for guinea-pigs 150 mg/kg. Acute percutaneous $LD_{50}$ for rats >100 mg/kg. $LC_{50}$ (24-h) for goldfish >10 mg/l.

**Formulations.**

A w.p. (700 g a.i./kg).

**Analysis.**

Product analysis is by polarography; particulars from Bayer AG. Residues may be determined by colorimetry (D. MacDougall, *Anal. Methods. Pestic., Plant Growth Regul. Food Addit.*, 1964, **3**, 49). Details are available from Bayer AG.

# Fenamiphos



$C_{13}H_{22}NO_3PS$ (303.4)
1Y1&MPO&O2&OR C1 DS1

**Nomenclature and development.**

Common name fenamiphos (BSI, E-ISO); phénamiphos (F-ISO). Chemical name (IUPAC) ethyl 4-methylthio-*m*-tolyl isopropylphosphoramidate; ethyl 3-methyl-4-(methylthio)phenyl isopropylphosphoramidate. (*C.A.*) ethyl 3-methyl-4-(methylthio)phenyl 1-methylethylphosphoramidate (9CI); ethyl 4-(methylthio)-*m*-tolyl isopropylphosphoramidate (8CI); Reg. No. *[22224-92-6]*. Its nematicidal properties were described by J. H. O'Bannon & A. L. Taylor (*Plant Dis. Rep.*, 1967, **51**, 995) and B. Homeyer (*Pflanzenschutz-Nachr. (Engl. Ed.)*, 1971, **24**, 48). Introduced by Baychem Corp., Chemagro Division, (DEP 1 121 882; USP 2 978 479) as code no. 'Bay 68 138'; trade mark 'Nemacur' (formerly '*Nemacur P*').

**Properties.**

It is a colourless solid; m.p. 49 °C; v.p. 133 μPa at 30 °C. Solubility at 20 °C: *c.* 700 mg/l water; >1 kg/l dichloromethane, propan-2-ol.

**Uses.**

Fenamiphos is a systemic nematicide, active against ecto- and endo-parasitic, free-living, cyst-forming and root-knot nematodes and is recommended for overall application (with or without soil incorporation) at 5-20 kg a.i./ha; for band application at 7-40 g/100 m of row in 30-45 cm wide bands; for bare root dip at 100-400 mg/l for 5-30 min (W. M. Zeck, *ibid.*, p.114; A. Vilarbedo, *ibid.*, p.153; W. M. Burnett & I. M. Inglis, *ibid.*, p.170). Its metabolism has been studied (T. B. Waggoner & A. M. Khasawinah, *Residue Rev.*, 1974, **53**, 79).

**Toxicology.**

Acute oral $LD_{50}$: for rats 15.3-19.4 mg/kg; for dogs 10 mg/kg; for male guinea-pigs 75-100 mg/kg; for hens 12 mg/kg. Acute percutaneous $LD_{50}$ for male rats *c.* 500 mg/kg. In 2-y feeding trials rats receiving 3 mg/kg diet showed no symptom of poisoning (E. Löser & G. Kimmerle, *Pflanzenschutz-Nachr. (Engl. Ed.)*, 1971, **24**, 69). $LC_{50}$ (96-h) was: for goldfish 3.2 mg/l; for rainbow trout 0.11 mg/l.

**Formulations.**

An e.c. (400 g a.i./l); granules (50 or 100 g/kg).

**Analysis.**

Product analysis is by glc; particulars of methods are available from Bayer AG. Residue analysis is by glc (*Pestic. Anal. Man.*, 1979, **I**, 201-I; M. A. Luke *et al.*, *J. Assoc. Off. Anal. Chem.*, 1981, **64**, 1187). Details are available from Bayer AG.

$C_{17}H_{12}Cl_2N_2O$ (331.2)
T6N CNJ EXQR BG&R DG

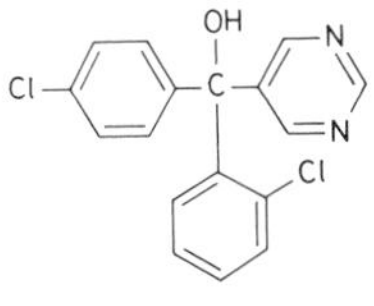

**Nomenclature and development.**

Common name fenarimol (BSI, E-ISO, F-ISO, ANSI). Chemical name (IUPAC) (±)-2,4'-dichloro-α-(pyrimidin-5-yl)benzhydryl alcohol. (*C.A.*) α-(2-chlorophenyl)-α-(4-chlorophenyl)-5-pyrimidinemethanol (9CI); Reg. No. *[60168-88-9]*. Its fungicidal properties were reported by I. F. Brown *et al.* (*Proc. Am. Phytopathol. Soc.*, 1975, **2**, 31). Introduced by Eli Lilly & Co. (GBP 1 218 623) as code no. 'EL-222'; trade marks 'Bloc', 'Rimidin', 'Rubigan'.

**Properties.**

It is a colourless crystalline solid; m.p. 117-119 °C; v.p. 13 μPa at 25 °C. Solubility at 25 °C: 13.7 mg/l water at pH7; >250 g/l acetone; 125 g/l methanol; 50 g/l xylene. It is broken down rapidly in sunlight to form many minor degradation products.

**Uses.**

It is a fungicide with a protective, curative and eradicative activity against powdery mildews, and recommended for use on apples (against *Podosphaera leucotricha* and *Venturia inaequalis*), on cucurbits (against *Erysiphe cichoracearum* and *Sphaerotheca fuliginea*), on grapes (against *Uncinula necator*) and on roses (against *S. pannosa*) and on other crops at 10-90 mg a.i./l, in sprays applied every 7-21 d.

**Toxicology.**

Acute oral $LD_{50}$ is: for mice 4500 mg/kg; for rats 2500 mg/kg; for beagle-dogs, mallard duck and bobwhite quail >200 mg/kg. Acute percutaneous $LD_{50}$ for rabbits >2000 mg/kg; no skin irritation was seen at this dose, slight irritation was observed when 0.1 ml (68 mg) was applied to their eyes. In 2-y feeding trials there was no toxic effect in: rats receiving 25 mg/kg diet, mice 600 mg/kg diet. $LC_{50}$ (96-h) for bluegill is 0.91 mg/l.

**Formulations.**

These include: e.c., s.c. and w.p. 'Rimidine Plus' (Reg. No. *[74522-90-0]*), w.p. (16 g fenarimol + 80 g carbendazim + 640 g maneb).

**Analysis.**

Product analysis is by glc with FID. Residues in soil and plant tissue may be determined by glc with ECD. Detailed analytical procedures and other information are available from Eli Lilly & Co.

# Fenbutatin oxide

$C_{60}H_{78}OSn_2$ (1053)
1X1&R&1-SN-1X1&1&R&1X1&1&R&O-SN- =
1X1&1&R&1X1&1&R&1X1&1&R

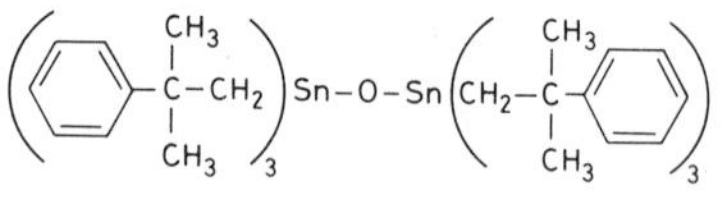

**Nomenclature and development.**

Common name fenbutatin oxide (BSI, E-ISO, JMAF);fenbutatin oxyde (F-ISO). Chemical name (IUPAC) bis[tris(2-methyl-2-phenylpropyl)tin] oxide. (*C.A.*) hexakis(2-methyl-2-phenylpropyl)distannoxane (8 & 9CI); Reg. No. *[13356-08-6]*. ENT 27 738. Introduced in USA by Shell Chemical Co., elsewhere by Shell International Chemical Company Ltd (USP 3 657 451) as code no. 'SD 14 114'; trade mark 'Vendex' (USA only), 'Torque'.

**Properties.**

Technical fenbutatin oxide is a colourless crystalline powder; m.p. 138-139 °C. Solubility at 23 °C: 5 μg/l water; 6 g/l acetone; 140 g/l benzene; 380 g/l dichloromethane. It is thermally and photochemically stable. Water causes conversion of fenbutatin oxide to tris(2-methyl-2-phenylpropyl)tin hydroxide which is reconverted to the parent compound slowly at room temperature and rapidly at 98 °C.

**Uses.**

It gives effective and long-lasting control, at 20-50 g a.i./100 l of spray, of the mobile stages of a wide range of phytophagous mites on citrus, glasshouse crops, ornamentals, top fruit, vegetables and vines. No phytotoxicity has been seen on these crops at concentrations twice those recommended. It is relatively non-toxic to predacious arthropods including *Stethorus punctum* and *Amblyseius finlandicus*.

**Toxicology.**

Acute oral $LD_{50}$: for rats 2630 mg tech./kg; for mice 1450 mg/kg; for dogs >1500 mg/kg. Acute percutaneous $LD_{50}$ for rabbits >2000 mg/kg. In feeding trials <2-y duration there was no effect of toxicological significance in rats receiving 100 mg/kg diet or dogs 30 mg/kg daily. $LC_{50}$ (8-d) for bobwhite quail 5065 mg/kg diet. The $LC_{50}$ (48-h) for rainbow trout was 0.27 mg a.i. (as w.p./l). Oral $LD_{50}$ >100 μg/bee. ADI for man 0.03 mg/kg (*JMPR* 1977).

**Formulations.**

These include: w.p. (500 g a.i./kg); s.c. (550 g/l).

**Analysis.**

Product analysis is by non-aqueous titration or i.r. spectrometry. Residues may be determined by glc with ECD of a suitable derivative (*Anal. Methods Pestic. Plant Growth Regul.*, 1978, **10**, 139). Details of the methods are available from Shell International Chemical Co. Ltd.

$C_{12}H_{11}NO_2$ (201.2)
T5OJ B1 CVMR

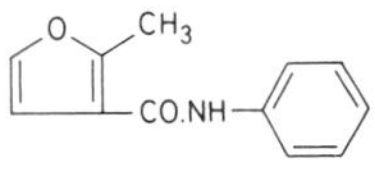

**Nomenclature and development.**

Common name fenfuram (BSI, F-ISO);fenfurame (F-ISO). Chemical name (IUPAC) 2-methyl-3-furanilide (I). (*C.A.*) 2-methyl-*N*-phenyl-3-furancarboxamide (9CI); (I) (8CI); Reg. No. *[24691-80-3]*. Its fungicidal properties were discovered by Shell Research Ltd and it was developed commercially by KenoGard AB (GBP 1 215 066 to Shell) as code no. 'WL 22 361'.

**Properties.**

Technical grade fenfuram is a cream-coloured solid; m.p. 109-110 °C; v.p. 20 μPa at 20 °C (by extrapolation). Solubility at 20 °C: 0.1 g/l water; 300 g/l acetone; 340 g/l cyclohexanone; 145 g/l methanol. It is thermally and photochemically stable, is fairly stable at neutral pH, but is hydrolysed under strongly alkaline and strongly acidic conditions.

**Uses.**

It is a fungicide highly active as a seed dressing for use against the smuts and bunts of temperate cereals. At 0.3-1.5 g a.i./kg seed, it controls *Tilletia* and *Ustilago* spp. including those within the seed, such as *U. nuda*.

**Toxicology.**

Acute oral $LD_{50}$ for rats 12 900 mg a.i./kg. In 90-d feeding trials NEL was: for dogs 300 mg/kg diet; for rats 17 mg/kg daily. The Ames test for mutagenicity was negative.

**Formulations.**

Seed treatments are available as liquids or powders. Seed treatments include: fenfuram + anthraquinone + gamma-HCH + imazalil; fenfuram + guazatine triacetate; fenfuram + gamma-HCH + guazatine triacetate + imazalil.

**Analysis.**

Product analysis is by glc with FID. Residues in barley and wheat may be determined by glc. Further information may be obtained from KenoGard AB.

$C_{13}H_{15}NO_6$ (281.3)
1VOYR&YNW1OV1 &&(1RS,2RS) FORM

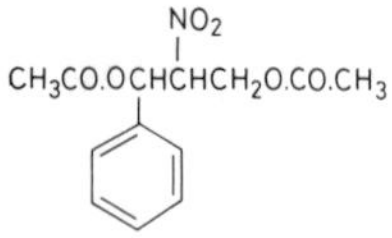

$CH_3CO.OCH(C_6H_5)CH(NO_2)CH_2O.CO.CH_3$

**Nomenclature and development.**

Common name fenitropan (BSI, draft E-ISO). Chemical name (IUPAC) (1*RS*,2*RS*)-2-nitro-1-phenyltrimethylene di(acetate). (*C.A.*) (1*RS*,2*RS*)-2-nitro-1-phenyl-1,3-propanediyl diacetate (9CI); Reg. No. *[65934-95-4]*. Its fungicidal properties were described by A. Kis-Tamás *et al.* (*Proc. Br. Crop Prot. Conf. - Pest Dis.*, 1981, **1**, 29). Introduced by EGYT Pharmacochemical Works (GBP 1 561 422; USP 4 160 035) as code no. 'EGYT 2248'; trade mark 'Volparox'.

**Properties.**

It forms colourless crystals; m.p. 70-72 °C. Solubility at 20 °C: 30 mg/l water; 1.25 kg/l chloroform; 10 g/l propan-2-ol; 350 g/l xylene.

**Uses.**

It is a contact fungicide applied as seed treatment for cereals, maize, rice and sugar beet.

**Toxicology.**

Acute oral $LD_{50}$: for male rats 3237 mg/kg, for females 3852 mg/kg. In 90-d feeding trials no effect was observed in rats receiving 2000 mg/kg diet.

**Formulations.**

Liquid (200 g a.i./l) for seed treatment.

**Analysis.**

Details of methods for product and residue analysis are available from EGYT Pharmacochemical Works.

$C_9H_{12}NO_5PS$ (277.2)
WNR B1 DOPS&O1&O1

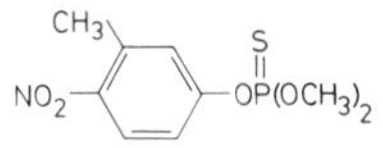

**Nomenclature and development.**

Common name fenitrothion (BSI, E-ISO, F-ISO, BPC);MEP (JMAF). Chemical name (IUPAC) *O,O*-dimethyl *O*-4-nitro-*m*-tolyl phosphorothioate. (*C.A.*) *O,O*-dimethyl *O*-(3-methyl-4-nitrophenyl) phosphorothioate (9CI); *O,O*-dimethyl *O*-(4-nitro-*m*-tolyl) phosphorothioate (8CI); Reg. No. *[122-14-5]*. OMS 43, ENT 25 715. Its insecticidal properties were described by Y. Nishizawa *et al.*, (*Bull. Agric. Chem. Soc. Jpn.*, 1960, **24**, 744; *Agric. Biol. Chem.*, 1961, **25**, 605). Introduced by the Sumitomo Chemical Co. and independently by Bayer AG and later by American Cyanamid Co. (BGP 594 669 to Sumitomo Chemical Co.;BGP 596 091 to Bayer AG) as code no. 'Bayer 41831','S-5660','S-1102A' (all to Bayer AG),'AC 47300' (American Cyanamid Co.); trade marks 'Accothion', 'Cytel', 'Cyfen' (all to American Cyanamid Co.), 'Folithion' (Bayer AG), 'Sumithion' (Sumitomo Chemical Co.)

**Properties.**

Technical grade fenitrothion is a yellow-brown liquid; b.p. 140-145 °C (decomp.)/0.1 mmHg; v.p. 18 mPa at 20 °C; $d_{25}^{25}$ 1.32-1.34, $n_D^{25}$ 1.5528. Solubility at 30 °C: 14 mg/l water; at 20-25 °C: >1 kg/kg dichloromethane, methanol, xylene; 42 g/kg hexane, 0.1-1.0 kg/kg propan-2-ol. It is hydrolysed by alkali, at 30 °C 50% loss occurs in 4.5 h in 10M sodium hydroxide.

**Uses.**

It is a potent contact insecticide, effective against a wide range of pests i.e. penetrating, chewing and sucking insect pests (coffee leafminers, locusts, rice stem borers, wheat bugs). It is also effective against household insects and all of the nuisance insects listed by WHO. Its effectiveness as a vector control agent for malaria is confirmed by WHO and has been approved as an insecticide for locust control by FAO, USAID and OCLALAV. Its properties have been summarised (*Residue Rev.*, 1976, **60**).

**Toxicology.**

Acute oral $LD_{50}$ for female rats 800 mg/kg. Acute percutaneous $LD_{50}$: for male rats 890 mg/kg, for females 1200 mg/kg. In 1.77-y feeding trials NEL for rats was 5 mg/kg diet. $LC_{50}$ (48-h) for carp is 4.1 mg/l. Its relatively low mammalian toxicity had been described earlier (J. Drabek & J. Pelikan, *Chem. Prum.*, 1956, **6**, 293).

**Formulations.**

These include: e.c. (500 g a.i./l); w.p. (400 g/kg); dusts (20, 30 or 50 g/kg); ULV (1.25 kg/l); aerosol concentrates. Also combinations (e.c. and dust): fenitrothion + 2-*sec*-butylphenyl methylcarbamate; fenitrothion + fenvalerate; fenitrothion + tetramethrin.

**Analysis.**

Product analysis is by the Averell & Norris method (*CIPAC Handbook*, 1980, **1A**, 1255; *FAO Specification* (CP/62)). Residues may be determined by glc (E. Möllhoff, *Pflanzenschutz-Nachr. (Engl. Ed.)*, 1968, 21, 331; *Pestic. Anal. Man.*, 1979, **I**, 201-A, 201-G, 201-H, 201-I; D. C. Abbott *et al.*, *Pestic. Sci.*, 1970, **1**, 10; J. Desmarchelier *et al.*, *ibid.*, 1977, **8**, 473; M. A. Luke *et al.*, *J. Assoc. Off. Anal. Chem.*, 1981, **64**, 1187; *Analyst (London)*, 1980, **105**, 515).

$C_9H_7Cl_3O_3$ (269.5)
QVY1&OR BG DG EG

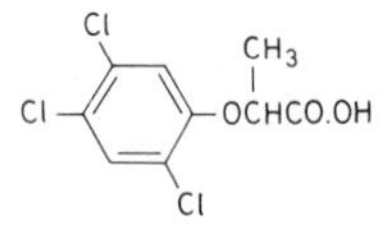

**Nomenclature and development.**

Common name fenoprop (BSI, E-ISO, F-ISO); silvex (ANSI, WSSA); 2,4,5-TP (France - as alternative to ISO name, JMAF, USSR). Chemical name (IUPAC) (±)-2-(2,4,5-trichlorophenoxy)propionic acid (I). (*C.A.*) (±)-2-(2,4,5-trichlorophenoxy)propanoic acid (9CI); (I) (8CI); Reg. No. *[93-72-1]*. Trivial name 2,4,5-TP. The plant-growth regulating properties of its salts were described by M. E. Synerholm & P. W. Zimmerman (*Contrib. Boyce Thompson Inst.*, 1945, **14**, 91). Its esters were introduced as herbicides by Dow Chemical Co. (who no longer manufacture or market it) and its salts as plant growth regulators by Amchem Products Inc. (now Union Carbide Agricultural Products Div.) as trade marks *'Kuron'* (Dow), 'Fruitone T' (Amchem).

**Properties.**

Fenoprop is a colourless powder; m.p. 179-181 °C. Solubility at 25 °C: 140 mg/l water; 180 g/kg acetone; 134 g/kg methanol; 98 g/kg diethyl ether; 860 mg/kg heptane. It is acidic p$K_a$ 2.84, forming water-soluble amine and alkali metal salts. They are soluble in acetone, lower alcohols; insoluble in aromatic and chlorinated hydrocarbons and in most non-polar organic solvents. Its lower alkyl esters are slightly volatile and its 2-butoxy-1-methylethyl esters (propylene glycol butyl ether esters) are practically non-volatile. They are sparingly soluble in water; very soluble in most organic solvents.

**Uses.**

It is a hormone-type weedkiller, absorbed by leaves and stems. It is recommended for brush control at 2-4 kg a.e./100 l; to control aquatic weeds; to control broad-leaved weeds in maize and sugarcane at lower rates. Sprays of amine salts (1.4 a.e./100 l) are used 7-14 d before harvest to reduce pre-harvest drop of apples.

**Toxicology.**

Acute oral $LD_{50}$ for rats: 650 mg a.i./kg, 500-1000 mg mixed butyl or propylene glycol butyl ether esters/kg, *c.* 3940 mg a.e. (as tris(2-hydroxyethyl)ammonium salt)/kg. Acute percutaneous $LD_{50}$ for rabbits $>$3200 mg a.e. (as tris(2-hydroxyethyl)ammonium salt)/kg. Fenoprop, this amine salt and undiluted esters are eye irritants. $LC_{50}$ (8-d) for mallard duck and quail is $>$12 800 mg a.e. (as amine salt)/kg diet.

**Formulations.**

These include: e.c. of esters; solutions of amine salts.

**Analysis.**

Product analysis is by total chlorine content. Residues may be determined by glc with ECD (R. P. Marquardt, *Anal. Methods Pestic., Plant Growth Regul. Food Addit.*, 1964, **4**, 211; *Anal. Methods Pestic. Plant Growth Regul.*, 1972, **6**, 688).

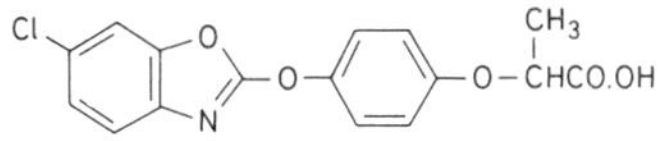

$C_{16}H_{12}ClNO_5$ (333.7)
T56 BN DOJ COR DOY1&VQ& GG
*fenoxaprop-ethyl*
$C_{18}H_{16}ClNO_5$ (361.8)
T56 BN DOJ COR DOY1&VO2& GG

**Nomenclature and development.**

Common name fenoxaprop (BSI, draft E-ISO, draft F-ISO). Chemical name (IUPAC) (±)-2-[4-(6-chlorobenzoxazol-2-yloxy)phenoxy]propionic acid. (*C.A.*) (±)-2-[4-[(6-chloro-2-benzoxazolyl)oxy]phenoxy]propanoic acid (9CI); Reg. No. *[73519-55-8]*. Its herbicidal properties were described by H. Bieringer *et al.* (*Proc. Br. Crop Prot. Conf. - Weeds*, 1982, **1,** 11). The ethyl ester, fenoxaprop-ethyl (Reg. No. *[66441-23-4]* for racemate), was introduced by Hoechst AG as code no. 'Hoe 33 171'.

**Properties.**

Fenoxaprop-ethyl is a colourless solid; m.p. 84-85 °C; v.p. 19 nPa at 20 °C. Solubility at 25 °C: 0.9 mg/l water; >500 g/kg acetone; >10 g/kg cyclohexane, ethanol, octan-1-ol; >200 g/kg ethyl acetate; >300 g/kg toluene.

**Uses.**

Fenoxaprop-ethyl is a selective post-em. herbicide controlling emerged annual and perennial grass weeds in such crops as beans, beets, cotton, groundnuts, potatoes, soyabeans and vegetables, using 180-480 g a.i./ha.

**Toxicology.**

Acute oral $LD_{50}$ for male rats 2357 mg tech./kg, for females 2500 mg/kg. For male mice 4670 mg/kg, for females 5490 mg/kg. Acute percutaneous $LD_{50}$ for rabbits >1000 mg/kg, slightly irritant to their skin and eyes; for female rats >2000 mg/kg. In 90-d feeding trials NEL was: for dogs 16 mg/kg diet; for rats 80 mg/kg diet.

**Formulations.**

These include e.c. (90 or 120 g a.i./l).

$C_{22}H_{23}NO_3$ (349.4)
L3TJ A1 A1 B1 B1 CVOYCN&R COR
&&(RS) FORM

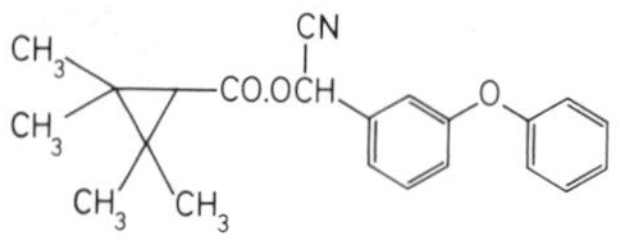

**Nomenclature and development.**

Common name fenpropathrin (BSI, E-ISO);fenpropathrine (F-ISO). Chemical name (IUPAC) (*RS*)-α-cyano-3-phenoxybenzyl 2,2,3,3-tetramethylcyclopropanecarboxylate. (*C.A.*) cyano(3-phenoxyphenyl)methyl 2,2,3,3-tetramethylcyclopropanecarboxylate (9CI); Reg. No. *[64257-84-7]* (racemate), *[39515-41-8]* (unstated stereochemistry). Its insecticidal properties were described by Fujita, *Jpn. Pestic. Inf.*, 1981, No. 38, p. 21. Introduced by Sumitomo Chemical Co. (GBP 1 356 087; USP 3 835 176) as code no. 'S-3206'; trade marks 'Rody', 'Danitrol', 'Meothrin'.

**Properties.**

Technical grade fenpropathrin is a yellow-brown solid; m.p. 45-50 °C; v.p. 730 μPa at 20 °C; $d_4^{25}$ 1.150. Solubility at 25 °C: 0.33 mg/l water; 1 kg/kg cyclohexanone, xylene; 337 g/kg methanol.

**Uses.**

It is a pyrethroid with acaricidal and insecticidal activity effective at 50-200 g a.i./ha against various species of mites and insects on cotton, ornamentals, top fruits, vine, vegetables and other field crops.

**Toxicology.**

Acute oral $LD_{50}$: for male rats 164 mg/kg, for females 107 mg/kg (vehicle: 10% gum arabic); for mallard duck 1089 mg/kg. Acute percutaneous $LD_{50}$: for male rats 1000 mg/kg, for females 870 mg/kg. $LC_{50}$ (48-h) for bluegill is 1.95 μg/l.

**Formulations.**

These include: e.c. (50, 100, 200 or 300 g a.i./l); s.c. (100 g/l).

**Analysis.**

Product analysis is by glc or hplc (S. Sakaue *et al., Agric. Biol. Chem.*, 1982, **46**, 2165). Details of the methods for residue analysis are available from Sumitomo Chemical Co.

# Fenpropimorph

$C_{20}H_{33}NO$ (303.5)
T6N DOTJ A1Y&1R DX1&1&1& C1 E1
&&CIS FORM

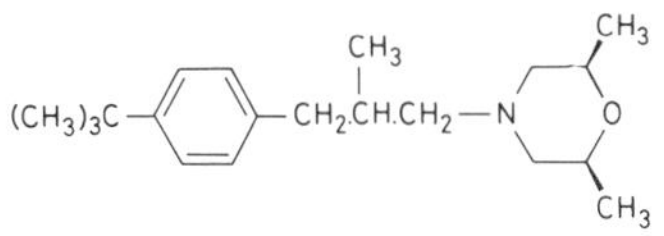

**Nomenclature and development.**

Common name fenpropimorph (BSI, draft E-ISO);fenpropimorphe (draft F-ISO). Chemical name (IUPAC) (±)-*cis*-4-[3-(4-*tert*-butylphenyl)-2-methylpropyl]-2,6-dimethylmorpholine. (*C.A.*) 4-[3-[4-(1,1-dimethylethyl)phenyl]-2-methylpropyl]-2,6-dimethylmorpholine (9CI); Reg. No. *[67564-91-4]*. Its fungicidal properties were described by K. Bohnen & A. Pfiffner (*Meded. Fac. Landbouwwet. Rijksuniv. Gent*, 1979, **44**, 487) and by E. H. Pommer & W. Himmele (*ibid.*, p. 499). Introduced as a fungicide by BASF AG, Maag AG and May & Baker Ltd (DEOS 2 752 096; GBP 1 584 290; USP 4 202 894) as code no. 'BAS 421F'; trade mark 'Corbel' (to BASF), 'Mistral' (to May & Baker).

**Properties.**

It is a colourless oil; b.p. 120 °C/0.05 mmHg; $n_D^{20}$ 1.4940. Solubility at 20 °C: 70 mg/kg water, >1 kg/kg acetone, chloroform, ethanol. It is stable under moderately acidic and alkaline conditions.

**Uses.**

It is a systemic fungicide controlling certain important fungal diseases in cereals caused by *Cercospora beticola, Erysiphe betea, E. graminis, Rhynchosporium secalis, Uromyces betea* and *Puccinia* spp.

**Toxicology.**

Acute oral $LD_{50}$: for rats 3515 mg tech./kg; for mice 5960 mg/kg. Acute percutaneous $LD_{50}$ for rats >4000 mg/kg; potential exists for skin irritation.

**Formulations.**

An e.c. (750 g a.i./l).

**Analysis.**

Product anlaysis is preferably by glc with FID. Residues in grain and cereal plants may be determined by glc with FID.

$C_{12}H_9ClO_3S$ (268.7)
WSR&OR DG

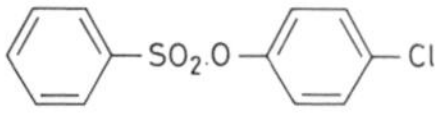

**Nomenclature and development.**

Fenson (BSI, E-ISO, F-ISO);fénizon (France). Chemical name (IUPAC) 4-chlorophenyl benzenesulphonate. (*C.A.*) 4-chlorophenyl benzensulfonate (9CI); *p*-chlorophenyl benzenesulfonate (8CI); Reg. No. *[80-38-6]*. Trivial names CPBS,PCPBS. Introduced by Murphy Chemical Ltd (who no longer manufacture or market it) as trade mark 'Murvesco' (Murphy).

**Properties.**

Pure fenson forms colourless crystals; m.p. 61-62 °C. The technical grade is a pale pink powder, m.p. 56-59 °C. Solubility: sparingly soluble in water; soluble in organic polar and aromatic solvents. It is hydrolysed by alkali.

**Uses.**

It is a non-systemic acaricide with ovicidal properties and is normally used in combination with acaricides effective against the active stages of mites. It is used to control mites on apples and glasshouse crops, including vines.

**Toxicology.**

Acute oral $LD_{50}$ for rats 1560-1740 mg/kg. Acute percutaneous $LD_{50}$ for rats and rabbits >2000 mg/kg.

**Formulations.**

A w.p. (500 a.i./kg). Mixtures include: fenson + dicofol; fenson + methomyl; fenson + parathion-methyl.

**Analysis.**

Product analysis is by hydrolysis to 4-chlorophenol which is determined by acid-base titration (*CIPAC Handbook*, 1970, **1**, 392). Residues may be determined by colorimetry of a derivative (A. H. Kutschinski & E. N. Luce, *Anal. Chem.*, 1952, **24**, 1188).

$C_{11}H_{17}O_4PS_2$ (308.3)
OS1&R DOPS&O2&O2

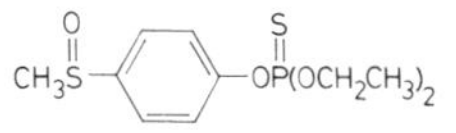

**Nomenclature and development.**

Common name fensulfothion (BSI, E-ISO, F-ISO, ESA, Society of Nematologists USA). Chemical name (IUPAC) *O,O*-diethyl *O*-4-methylsulphinylphenyl phosphorothioate. (*C.A.*) *O,O*-diethyl *O*-[4-(methylsulfinyl)phenyl] phosphorothioate (9CI); *O,O*-diethyl *O*-[*p*-(methylsulfinyl)phenyl] phosphorothioate (8CI); Reg. No. *[115-90-2]*. OMS 37, ENT 24 945. Its nematicidal properties were reported by B. Homeyer (*Mitt. Biol. Bundesanst. Land-Forstwirtsch. Berlin-Dahlem*, 1967, **121**, 50). Introduced by Bayer AG (DEP 1 101 406; USP 3 042 703) as code no. 'Bayer 25 141','S 767'; trade marks 'Dasanit', 'Terracur P'.

**Properties.**

It is a yellow oil; b.p. 138-141 °C/0.01 mmHg; $d_4^{20}$ 1.202; $n_D^{25}$ 1.540. Solubility at 25 °C 1.54 g/l water; miscible with dichloromethane, propan-2-ol and most organic solvents. It is readily oxidised to the sulphone and apparently isomerises readily to the *O,S*-diethyl isomer (E. Benjamini *et al.*, *J. Econ. Entomol.*, 1959, **52**, 94, 99).

**Uses.**

Fensulfothion is an insecticide and a nematicide active against free-living, cyst-forming and root-knot nematodes, it is recommended for soil treatment, has long persistence and some systemic activity (B. Homeyer, *Pflanzenschutz-Nachr. (Engl. Ed.)*, 1971, **24**, 367; W. Kolbe, *ibid.*, p. 431; R. Dern, *ibid.*, p. 476; H. F. Jung & K. Iwaya, *ibid.*, p. 486; C. K. Proude, *ibid.*, p. 503).

**Toxicology.**

Acute oral $LD_{50}$ for male rats 4.7-10.5 mg/kg. Acute percutaneous $LD_{50}$ (in xylene) for female rats 3.5 mg/kg, for males 30 mg/kg. In 1.34-y feeding trials rats receiving 1 mg/kg diet showed no effect (G. Löser & G. Kimmerle, *ibid.*, p. 407). $LC_{50}$ (96-h): for rainbow trout 8.8 mg/l; for golden orfe 6.8 mg/l.

**Formulations.**

These include: soluble concentrate (600 g a.i./l); w.p. (250 g/kg); dust (100 g/kg); granules (25, 50 or 100 g/kg) for foliar and soil application.

**Analysis.**

Product analysis is by indirect titration of the sulphoxide group, details from Bayer AG. Residues may be determined by glc (details from Bayer AG; *Pestic. Anal. Man.*, 1979, **I**, 201-H, 201-I; M. A. Luke *et al.*, *J. Assoc. Off. Anal. Chem.*, 1981, *64*, 1187) or by conversion to phosphate which is measured by standard colorimetric methods (G. E. Keppel, *J. Ass. Off. Anal. Chem.*, 1978, **54**, 528; C. A. Anderson, *Anal. Methods Pestic. Plant Growth Regul.*, 1973, **7**, 253).

$C_{16}H_{12}ClNO_4S$ (349.8)
T56 BN DSJ COR DOY1&VQ& GG
*fenthiaprop-ethyl*
$C_{18}H_{16}ClNO_4S$ (377.9)
T56 BN DSJ COR DOY1&VO2& GG

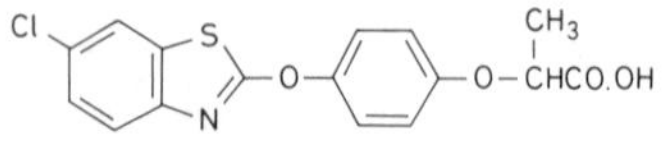

**Nomenclature and development.**

Common name fenthiaprop (BSI, draft E-ISO, draft F-ISO). Chemical name (IUPAC) (±)-2-[4-(6-chlorobenzothiazol-2-yloxy)phenoxy]propionic acid. (*C.A.*) (±)-2-[4-[(6-chloro-2-benzothiazolyl)oxy]phenoxy]propanoic acid; Reg. No. *[73519-50-3]* (no stereochemistry specified). Its herbicidal properties were described by R. Handte *et al.* (*Proc. Br. Crop Prot. Conf. - Weeds*, 1982, **1**, 19). The ethyl ester, fenthiaprop-ethyl (Reg. No. [*66441-11-0]* no stereochemistry specified), was introduced by Hoechst AG as code no. 'Hoe 35 609'.

**Properties.**

Fenthiaprop-ethyl is a crystalline solid; m.p. 56-57.5 °C; v.p. 510 nPa at 20 °C. Solubility at 25 °C: 0.8 mg/l water; >500 g/kg acetone, ethyl acetate, toluene; >40 g/kg cyclohexane; >50 g/l ethanol, octan-1-ol.

**Uses.**

Fenthiaprop-ethyl is a post-em. herbicide effective by local and systemic action. It is translocated acropetally and basipetally to roots and rhizomes, where the growing points are destroyed. It is most effective against actively growing weeds, and is active against many annual and perennial grasses. It controls *Alopecurus myosuroides, Avena* spp. and volunteer cereals at 180-240 g a.i./ha and *Agropyron repens* at 480-720 g/l in oilseed rape, potatoes and sugar beet.

**Toxicology.**

Acute oral $LD_{50}$: for male rats 977 mg/kg, for females 917 mg/kg; for male mice 1070 mg/kg, for females 1170 mg/kg. Acute percutaneous $LD_{50}$ for female rats 2000 mg/kg; slight irritant to skin and eyes of rabbits. In 90-d feeding trials NEL for rats was 50 mg/kg diet.

**Formulations.**

An e.c. (240 g fenthiaprop-ethyl/l).

**Analysis.**

Details of analytical methods are available from Hoechst AG.

$C_{10}H_{15}O_3PS_2$ (278.3)
1SR B1 DOPS&O1&O1

Nomenclature and development.

Common name fenthion (BSI, E-ISO, F-ISO, BPC);MPP (JMAF). Chemical name (IUPAC) *O,O*-dimethyl *O*-4-methylthio-*m*-tolyl phosphorothioate. (*C.A.*) *O,O*-dimethyl *O*-[3-methyl-4-(methylthio)phenyl] phosphorothioate (9CI); *O,O*-dimethyl *O*-[4-(methylthio)-*m*-tolyl] phosphorothioate (8CI); Reg. No. *[55-38-9]*. OMS 2, ENT 25 540. Developed by G. Schrader & E. Schegk (G. Schrader, *Hoefchen-Briefe (Engl. Ed.)*, 1960, **13**, 1) and introduced by Bayer AG (DEP 1 116 656; USP 3 042 703) as code no. 'Bayer 29 493','S 1752'; trade marks 'Baycid', 'Baytex', 'Entex', 'Lebaycid', 'Mercaptophos', 'Queletox', 'Tiguvon'.

**Properties.**

Pure fenthion is a colourless liquid; b.p. 87 °C/0.01 mmHg; v.p. 4 mPa at 20 °C; $n_D^{20}$ 1.5698; $d_4^{20}$ 1.250. Solubility at 20 °C: 2 mg/kg water; >1kg/kg dichloromethane, propan-2-ol. The technical grade (95-98% pure) is a brown oil, with a weak garlic-like odour. It is stable ≤210 °C, to light and to alkali.

**Uses.**

Fenthion is a contact and stomach insecticide with a useful penetrant and persistent action. It is effective, for example, against fruit flies, leaf hoppers and cereal bugs. Oxidation to the sulphoxide (mesulfenfos (7714)) and sulphone, both highly insecticidal, proceeds in plants (H. Niessen *et al.*, *Pflanzenschutz-Nachr. (Engl. Ed.)*, 1962, **15**, 125). It is also used against weaver birds (*Quelea quelea*) in the tropics.

**Toxicology.**

Acute oral $LD_{50}$: for male rats 190-315 mg/kg, for females 245-615 mg/kg. Acute percutaneous $LD_{50}$: for rats 330-500 mg/kg; more toxic to dogs and poultry. In 2-y feeding trials NEL for rats 3 mg/kg diet; in 1-y trials dogs receiving 50 mg/kg diet showed no loss of weight or in food consumption. $LC_{50}$ (96-h) for carp 2.5-3.3 mg/l; $LC_{50}$ (48-h) for goldfish 1.9 mg/l. Acute oral $LD_{50}$ for rats of the sulphoxide and sulphone was 125 mg/kg.

**Formulations.**

These include: 'Baytex', w.p. (400 g/kg), fogging concentrate (600 g/l), e.c. (500 or 1000 g/l) and granular (20 g/kg) for household use; 'Lebaycid', w.p. (250 or 400 g/kg) e.c. (500 or 1000 g/l), ULV (1000 g/l) and dust (30 g/kg) for use against crop pests; 'Queletox' for use against weaver birds; 'Tiguvon' various formulations against animal parasites.

**Analysis.**

Product analysis is by u.v. spectrometry of a derivative (F. B. Ibrahim & J. C. Cavagnol, *J. Agric. Food Chem.*, 1966, **44**, 369; *CIPAC Handbook*, 1983, **1B**, in press) - details from Bayer AG. Residues may be determined by glc (*Pestic. Anal. Man.*, 1979, **I**, 201-A, 201-G, 201-H, 201-I; D. C. Abbott *et al.*, *Pestic. Sci.*, 1970, **1**, 10; M. A. Luke *et al.*, *J. Assoc. Off. Anal. Chem.*, 1981, **64**, 1187; *Anal. Methods Pestic. Plant Growth Regul.*, 1972, **6**, 301).

$C_{20}H_{18}O_2Sn$ (409.0)
1VO-SN-R&R&R

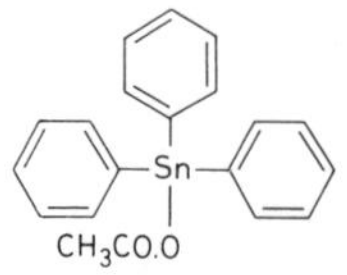

**Nomenclature and development.**

Common name fentin acetate (BSI, JMAF);the name fentin (the anion being stated e.g. acetate) (E-ISO);fentine (F-ISO);asetat fenolovo (USSR);chemical name is used in lieu of a common name (Republic of South Africa, USA). Chemical name (IUPAC) triphenyltin(IV) acetate; triphenyltin acetate. (*C.A.*) (acetyloxy)triphenylstannane (9CI); acetoxytriphenylstannane (8CI); Reg. No. *[900-95-8]*. OMS 1020, ENT 25 208. The fungicidal properties of organotin compounds were investigated by G. J. M van der Kerk & J. G. A. Luijten (*J. Appl. Chem.*, 1954, **4**, 314; 1956, **6**, 56). Fentin acetate was introduced by Hoechst AG (USP 3 499 086) as code no. 'VP 1940','Hoe 02824'; trade mark 'Brestan 60'.

**Properties.**

Pure fentin acetate forms colourless crystals; m.p. 121-123 °C; v.p. 1.9 mPa at 60 °C. Solubility at 20 °C: *c.* 9 mg/l water (pH 5); 460 g/l dichloromethane; 22 g/l ethanol; 82 g/l ethyl acetate; 5 g/l hexane; 89 g/l toluene. The technical grade (⩾94% pure) has m.p. 118-125 °C. It is stable when dry but is relatively easily decomposed on exposure to air or sunlight, finally forming insoluble tin compounds. Must *not* be mixed with oil emulsions.

**Uses.**

It is a non-systemic fungicide recommended to control *Ramularia* spp. on celery and sugar beet and *Phytophthora infestans* on potatoes at 160-260 g a.i./ha. It is non-phytotoxic to beets, celery, cocoa, potatoes and rice and may be used to control algae in paddy rice.

**Toxicology.**

Acute oral $LD_{50}$ for female rats 140-298 mg/kg. Acute percutaneous $LD_{50}$ for rats 500 mg/kg; it is irritant to mucous membranes. In 2-y feeding trials NEL was: for dogs 5 mg/kg diet; for guinea-pigs between 5 and 10 mg/kg diet. It is harmless to honeybees.

**Formulations.**

A w.p. (190 or 540 g a.i./kg). 'Brestan' (fentin acetate + maneb).

**Analysis.**

Product analysis is by hydrolysis to fentin hydroxide which is measured by non-aqueous titration (*CIPAC Handbook*, 1980, **1A**, 1263) or by glc of a derivative (*ibid.*, 1983, **1B**, in press). Residues may be determined by colorimetry of a derivative (G. A. Lloyd *et al.*, *J. Sci. Food Agric.*, 1962, **13**, 353) or by polarography of tin tetrabromide (S. Gorbach & R. Bock, *Z. Anal. Chem.*, 1958, **163**, 419).

$C_{18}H_{16}OSn$ (367.0)
Q-SN-R&R&R

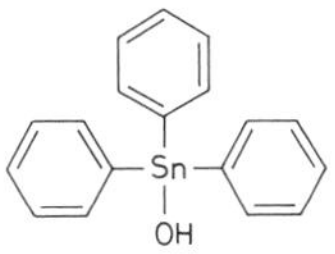

**Nomenclature and development.**

Common name fentin hydroxide (BSI, JMAF);fentin (with statement of anion present) (E-ISO);fentine (F-ISO);gidrookie fenolovo (USSR) for fentin hydroxide;*exception* (Republic of South Africa, USA (full chemical name)). Chemical name (IUPAC) triphenyltin hydroxide. (*C.A.*) hydroxytriphenylstannane (8 & 9CI); Reg. No. *[76-87-9]* OMS 1017, ENT 28 009. The fungicidal properties of organotin compounds were reported by G. J. M. van der Kerk & J. G. A. Luijten (*J. Appl. Chem.*, 1954, **4**, 314; 1956, **6**, 56) and reviewed by H. Bock (*Residue Rev.*, 1981, **79**, 1). Fentin hydroxide was introduced by N.V. Philips-Duphar (now Duphar B.V.) as trade mark 'Du-Ter'.

**Properties.**

Technical grade fentin hydroxide (≥95% pure) is a colourless crystalline solid; m.p. 116-120 °C; v.p. 47 μPa. Solubility at 20 °C: *c.* 1 mg/l water at pH7.0 (greater at lower pH-values); *c.* 50 g/l acetone; 74 g/l 1,2-dichloroethane; 171 g/l dichloromethane; 28 g/l diethyl ether; *c.* 10 g/l ethanol; 41 g [as bis(triphenyltin) oxide]/l benzene. Thermally stable ≤45 °C, above which dehydration occurs to give bis(triphenyltin) oxide the latter compound being stable ≤*c.* 250 °C. Slowly decomposed by sunlight; much quicker by u.v. light to give inorganic tin via mono- and di-phenyltin compounds. Chemically compatible with other pesticides except strongly acidic compounds.

**Uses.**

It is a non-systemic fungicide recommended for the control of early and late blights of potato at 250-350 g a.i./ha, of leaf spot on sugar beet at 350-400 g/ha, of blast diseases of rice at 220-450 g/ha, of coffee berry disease at 1 kg/ha, and of brown spot disease of tobacco at 200-350 g/ha. Other uses are control of certain leaf-eating larvae including *Spodoptera* spp. by acting as an antifeeding compound; control of algae in rice fields; control of snails in ponds. It is non-hazardous to honeybees. If used at recommended dosage rates, 'Du-Ter' products are non-phytotoxic due to special safeners in the formulations, and are in general compatible with w.p. formulations of most other pesticides. Mixing with liquid formulations is *not* recommended because of possible phytotoxicity.

**Toxicology.**

Lowest reported acute oral $LD_{50}$ values are: for male rats 171 mg/kg, for females 110 mg/kg; for male mice 245 mg/kg, for females 209 mg/kg; for male guinea-pigs 27.1 mg/kg, for females 31.1 mg/kg. In 2-y feeding trials NEL for rats was 2 mg/kg diet; for 99-d trials 20 mg/kg diet. $LC_{50}$ (8-d) for bobwhite quail was 38.5 mg/kg diet. $LC_{50}$ (48-h) is: for carp 0.05 mg/l; for Japanese killifish 0.072 mg/l; for harlequin fish 0.042 mg/l; for guppies 0.054 mg/l; for golden orfe 0.11 mg/l.

**Formulations.**

'Du-Ter' w.p. (190 g a.i./kg); 'Du-Ter Extra', w.p. (475 g/kg); 'Du-Ter Forte', w.p. (600 g/kg). 'Du-Ter M', w.p. (115 g fentin hydroxide + 340 g maneb/kg).

**Analysis.**

Product analysis is by potentiometric titration (*CIPAC Handbook*, 980, **1A**, 1266; Van Rossum *et al.*, *Anal. Methods Pest. Plant Growth Regul.*, 1980, **11**, 227) or by glc of a derivative (*idem, ibid.*). Residues may be determined by colorimetry or by glc after conversion to suitable derivatives (*idem, ibid.*; *Pestic. Anal. Man.*, 1979, **II**; P. G. Baker *et al.*, 1980, **105**, 282).

$C_9H_{12}N_2O$ (164.2)
1N1&VMR
*fenuron-TCA*
$C_{11}H_{13}Cl_3N_2O_3$ (327.6)
1N1&VMR &&$CCl_3$COOH SALT

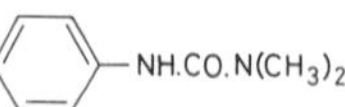


**Nomenclature and development.**

Common name fenuron (i) (BSI, E-ISO, F-ISO, ANSI, WSSA);fenidim (USSR);*exception* (Portugal, Sweden);fenuron-TCA (ii) (WSSA) for trichloroacetate salt. Chemical name (IUPAC) 1,1-dimethyl-3-phenylurea (I) (i); 1,1-dimethyl-3-phenyluronium trichloroacetate (ii). (*C.A.*) *N,N*-dimethyl-*N'*-phenylurea (i); trichloracetic acid compound with *N,N*-dimethyl-*N*-phenylurea (1:1) (ii) (9CI); (I) (i); trichloroacetic acid compound with (I) (8CI); Reg. No. *[101-42-8]* (i), *[4482-55-7]* (ii). Trivial name PDU. The herbicidal properties of fenuron were described by H. C. Bucha & C. W. Todd (*Science*, 1951, *114*, 493). It was introduced by E. I. du Pont de Nemours & Co. (Inc.) (who no longer manufacture or market it) and fenuron-TCA by Allied Chemical Corp. Agricultural Division (now Hopkins Agricultural Chemical Co.) (USP 2 655 447; GBP 691 403; 692 589 (all to du Pont for i); USP 2 782 112; 2 801 911 (both to Allied for ii)) as code no. 'GC-2603' (Allied for ii); trade marks '*Dybar*' (du Pont for i); 'Urab' (Hopkins for ii), see below for other mixtures.

**Properties.**

Fenuron is a colourless crystalline solid; m.p.133-134 °C; v.p. 21 mPa at 60 °C; $d_{20}^{20}$ 1.08. Solubility at 25 °C: 3.85 g/l water; sparingly soluble in hydrocarbons. It is stable to oxidation but is degraded microbially in soil.

Fenuron-TCA is a colourless crystalline solid; m.p. 65-68 °C. Solubility at room temperature: 4.8 g/l water; 666 g/kg 1,2-dichloroethane; 567 g/kg trichloroethylene; poorly soluble in petroleum oils.

**Uses.**

Fenuron inhibits photosynthesis and is absorbed through the roots. It is recommended to control woody plants and, in mixtures with chlorpropham, against many germinating weeds and pre-em. in vegetables against established chickweed, or post-planting in leeks and onions; also in soft fruit, top fruit and nursery trees. Fenuron-TCA combines the herbicidal actions of fenuron and trichloroacetic acid and is used for total weed control on non-crop areas and against woody plants.

**Toxicology.**

Acute oral $LD_{50}$: for rats 6400 mg fenuron/kg; 4000-5700 mg fenuron-TCA/kg. A 33% aqueous paste was practically non-irritating to the intact skin of guinea-pigs; fenuron-TCA is irritating to skin. In 90-d feeding trials no apparent effect was observed with rats receiving 500 mg fenuron/kg diet.

**Formulations.**

'Herbon Red', 'Herbon Yellow' (Cropsafe), e.c. (fenuron + chlorpropham). 'Herbon Gold' (Cropsafe), e.c. (fenuron + chlorpropham + propham). 'Urab', oil-miscible concentrate (360 g fenuron-TCA/l); granules (20 or 220 g/kg).

**Analysis.**

Product analysis of fenuron is based on hydrolysis (W. K. Lowen & H. M. Baker, *Anal. Chem.*, 1952, **24**, 1476). Residues may be determined by colorimetry (R. L. Dalton & H. L. Pease, *J. Assoc. Off. Agric. Chem.*, 1962, **45**, 377). Details of methods for fenuron-TCA are available from Hopkins Agricultural Chemical Co.

$C_{25}H_{22}ClNO_3$ (419.9)
1Y1&YR DG&VOYCN&R COR

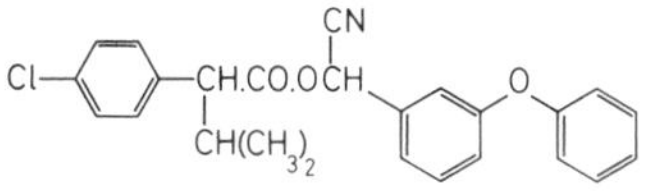

**Nomenclature and development.**

Common name fenvalerate (BSI, E-ISO, F-ISO). Chemical name (IUPAC) (*RS*)-α-cyano-3-phenoxybenzyl (*RS*)-2-(4-chlorophenyl)-3-methylbutyrate. (*C.A.*) cyano(3-phenoxyphenyl)methyl 4-chloro-α-(1-methylethyl)benzeneacetate (9CI); Reg. No. *[51630-58-1]*. OMS 2000. Its insecticidal properties were described by N. Ohno *et al.* (*Agric. Biol. Chem.*, 1974, **38**, 881) and the results of field trials by M. D. Mowlam *et al.* (*Proc. Br. Crop Prot. Conf. - Pests Dis.*, 1977, **2**, 649). Introduced by Sumitomo Chemical Co. It is also being developed in some countries by Shell International Chemical Co. (GBP 1 439 615; USP 4 062 968 both to Sumitomo) as code no. 'S-5602' (Sumitomo),'WL 43775' (Shell); trade marks 'Sumicidin' (to Sumitomo), 'Belmark', in USA 'Pydrin' (both to Shell).

**Properties.**

Technical grade fenvalerate is a viscous yellow or brown liquid, sometimes partly crystallized at room temperature; v.p. 37 μPa at 25 °C; $d_{25}^{25}$ 1.175. Solubility at 20 °C: <1 mg/l water; at 23 °C: >1 kg/kg acetone, chloroform, cyclohexanone, ethanol, xylene; 155 g/kg hexane. It is stable to heat and sunlight. It is more stable in acid than in alkaline media, with optimum stability at pH4.

**Uses.**

It is a highly active contact insecticide effective against a wide range of pests, including strains resistant to organochlorine, organophosphorus and carbamate insecticides. It controls insects that attack leaves or fruits on various crops, including cotton, fruit, vegetables and vines at 25-250 g a.i./ha, and is persistent under various field conditions. It is also used in public health and animal husbandry, controlling flies in cattle sheds for 60 d at 100 mg/m$^2$ wall, and is effective against cattle ticks at 200-300 mg/l.

**Toxicology.**

Acute oral $LD_{50}$: for rats 451 mg/kg; for domestic fowl >1600 mg/kg. Acute percutaneous $LD_{50}$ for rats >5000 mg/kg. In 2-y feeding trials NEL for rats was 250 mg/kg diet. $LC_{50}$ (96-h) for rainbow trout is 3.6 μg/l.

**Formulations.**

These include: 'Sumicidin', 'Belmark', 'Pydrin', e.c. (25-300 g a.i./l); ULV concentrates (25-75 g/l), s.c. (100 g/l). Mixtures include: 'Sumicombi'; 'Mikantop'; fenvalerate + fenitrothion.

**Analysis.**

Product anlysis is by glc or hplc. Residues may be determined by glc with ECD (Y. W. Lee, *J. Assoc. Off. Anal. Chem.*, 1978, **61**, 869; A. Ambrus *et al., ibid.*, 1981, **64**, 733; W. L. Reichel *et al., ibid.*, p. 1196; R. S. Greenberg, *J. Agric. Food Chem.*, 1981, **29**, 856). Details of methods are available from Sumitomo Chemical Co. or from Shell International Chemical Co.

$C_9H_{18}FeN_3S_6$ (416.5)
SUYSHN1&1 &&Fe SALT 3:1

$\left[(CH_3)_2NCS.S^-\right]_3 Fe^{3+}$

**Nomenclature and development.**

Common name ferbam (BSI, E-ISO, JMAF); ferbame (F-ISO); *exception* (Federal Republic of Germany). Chemical name (IUPAC) iron tris(dimethyldithiocarbamate); iron(III) dimethyldithiocarbamate; ferric dimethyldithiocarbamate. (*C.A.*) (OC-6-11)-tris(dimethylcarbamodithioato-*S,S'*)iron (9CI); tris(dimethyldithiocarbamato)iron (8CI); Reg. No. *[14484-64-1]*. ENT 14 689. Introduced by E. I. du Pont de Nemours & Co. (Inc.) (who no longer manufacture or market it) (USP 1 972 961) as trade mark '*Fermate*'.

**Properties.**

Ferbam is a black powder, decomp. >180 °C; v.p. negligible at room temperature. Solubility at room temperature: 130 mg/l water; soluble in organic solvents of high dielectric constant such as acetonitrile, chloroform, pyridine. Its chemical properties are suggestive of co-ordination, but it tends to decompose on prolonged storage or exposure to moisture and heat. It is incompatible with copper or mercury compounds or alkaline pesticides.

**Uses.**

It is mainly used for the protection of foliage against fungal pathogens including *Taphrina deformans* of peaches at 175 g/100 l.

**Toxicology.**

Acute oral $LD_{50}$ for rats >4000 mg/kg. In 2-y feeding trials NEL was: for rats 250 mg/kg diet; for dogs 5 mg/kg daily. It is not stored in the body tissues (H. C. Hodge *et al., J. Pharmacol. Exp. Ther.*, 1956, **118**, 174).

**Formulations.**

A w.p. (760 g a.i./kg).

**Analysis.**

Product analysis is either by colorimetry (*CIPAC Handbook*, 1970, **1**, 397) or by acid hydrolysis, the liberated carbon disulphide being converted to dithiocarbonate which estimated by titration with iodine (D. G. Clarke *et al., Anal. Chem.*, 1951, **23**, 1842). Residues may be determined by acid hydrolysis and suitable colorimetry of a derivative of the liberated carbon disulphide (H. L. Pease, *J. Assoc. Off. Anal. Chem.*, 1957, **40**, 1113; G. E. Keppel, *ibid.*, 1971, **54**, 528; *Analyst (London)*, 1981, **106**, 782).

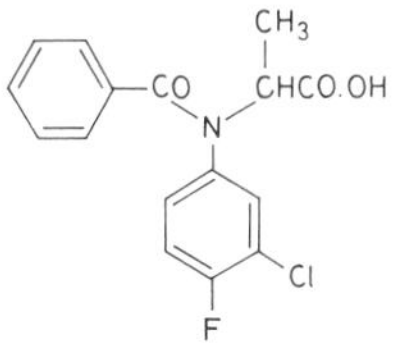

$C_{17}H_{15}ClFNO_3$ (321.7)
QVY1&NVR&R CG DF &&(DL) FORM
*flamprop-methyl* (RACEMATE)
$C_{17}H_{15}ClFNO_3$ (335.8)
GR BF ENVR&Y1&VO1 &&(DL) FORM

**Nomenclature and development.**

Common name flamprop (BSI, E-ISO, F-ISO, WSSA); the ester present, *e.g.* flamprop-methyl, should be stated. Chemical name (IUPAC) *N*-benzoyl-*N*-(3-chloro-4-fluorophenyl)-DL-alanine (I); and for the methyl ester, methyl *N*-benzoyl-*N*-(3-chloro-4-fluorophenyl)-DL-alaninate (II); methyl (±)-*N*-benzoyl-*N*-(3-chloro-4-fluorophenyl)-2-aminopropionate. (*C.A.*) (I) (8 & 9CI) for acid; (II) (8 & 9CI) for methyl ester. Reg. No. *[58667-63-3]* (acid), *[52756-25-9]* (methyl ester). The herbicidal properties of flamprop-methyl were described by E. Haddock *et al.*, (*Proc. Br. Weed Control Conf., 12th*, 1974, **1**, 9) and the reasons for the choice of this member of a series of related compounds given by B. Jeffcoat *et al.* (*Pestic. Sci.*, 1977, **8**, 1). Introduced by Shell Research Ltd. (GBP 1 437 711) as code no. 'WL 29 671'; trade marks 'Mataven' (to Shell), 'Lancer' (to ICI).

**Properties.**

Pure flamprop-methyl is a colourless crystalline solid, m.p. 84-86 °C; v.p. 47 μPa at 20 °C. The technical grade (≥93% pure) is an off-white crystalline powder, m.p. 81-82 °C. Solubility at 22 °C: *c.* 35 mg/l water; at 20 °C: >500 g/l acetone; 414 g/l cyclohexanone; 135 g/l ethanol; 7 g/l hexane. It is stable to heat, to light and to hydrolysis at $2 < pH < 7$. Hydrolysis to methanol and the parent acid occurs at pH >7.

**Uses.**

Flamprop-methyl is a selective post-em herbicide. Under European conditions it has given consistently >90% control of *Avena fatua, A. ludoviciana* and other species of wild oats at 525 g a.i./ha in the majority of both winter- and spring-sown wheats from mid-tillering to before the formation of the second node. In Australia and Mexico 450 and 600 g/ha respectively are used, applications being between mid-tillering and shooting of the crop, and in Canada 430 g/ha applied when the majority of wild oats are at the 3-leaf stage and before the crop has more than 6 leaves. Best results are achieved when crops are growing vigorously and are not subject to check or stress factors. Tests on more than 60 varieties of spring and winter wheat have indicated that, when applied as recommended, it is selective at 3-4 times the recommended dose. A useful degree of suppression of *Alopecurus myosuroides* can often be obtained in addition to control of wild oats. The selectivity of flamprop-methyl is related to the rates of hydrolysis to the parent acid and detoxication of the latter by formation of conjugates within the plant; the acid is translocated to growing points where it inhibits plant growth.

**Toxicology.**

Acute oral $LD_{50}$: for rats 1210 mg flamprop-methyl/kg; for mice 720 mg/kg; for domestic fowl >1000 mg/kg. Acute percutaneous $LD_{50}$ for rats >294 mg a.i. (as e.c.)/kg. In 2-y feeding trials no toxicological effect was noted: for rats receiving 2.5 mg a.i./kg diet; for dogs 10 mg/kg diet. $LC_{50}$ (96-h) for rainbow trout is 4.7 mg/l.

**Formulations.**

An e.c. (105 g flamprop-methyl/l).

**Analysis.**

Product analysis is by i.r. spectrometry or by glc. Residues may be determined by glc with ECD. Details of methods are available from Shell International Chemical Co. Ltd.

$C_{16}H_{13}ClFNO_3$ (321.7)
QVY1&NVR&R CG DF &&(D) FORM
*flamprop-isopropyl* (*R*)-ISOMER
$C_{19}H_{19}ClFNO_3$ (363.8)
GR BF ENVR&Y1&VOY1&1 &&(D) FORM

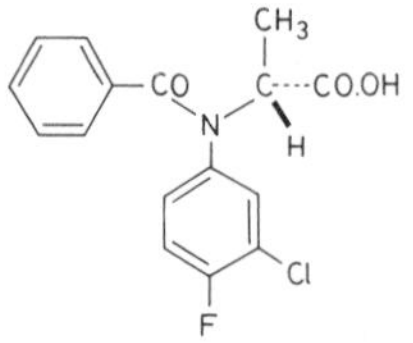

**Nomenclature and development.**

Common name flamprop (BSI, E-ISO, F-ISO) applies to the racemic acid, suitable qualification of this common name to represent the (*R*)-enantiomorph is still under discussion within ISO. The ester present, *e.g.* flamprop-isopropyl, and the stereochemistry should be stated. Chemical name (IUPAC) of the (*R*)-enantiomorph is *N*-benzoyl-*N*-(3-chloro-4-fluorophenyl)-D-alanine (I); and its isopropyl ester isopropyl *N*-benzoyl-*N*-(3-chloro-4-fluorophenyl)-D-alaninate (previously referred to in the literature as the L-alanine and L-alaninate, respectively). (*C.A.*) the acid (I); the ester 1-methylethyl *N*-benzoyl-*N*-(3-chloro-4-fluorophenyl)-D-alaninate (9CI); formerly stated to be the L-alanine and L-alaninate, respectively; Reg. No. *[57353-42-1]* (L-alanine), *[57973-67-8]* (L-alaninate), *[52756-22-6]* (DL-alaninate). Trivial name L-flamprop-isopropyl. The herbicidal properties of the (*R*)-isopropyl ester were described by R. M. Scott *et al.* (*Proc. Br. Crop Prot. Conf. - Weeds*, 1976, **2**, 723), its design has been discussed by M. A. Venis (*Pestic. Sci.*, 1982, **13**, 309) and its development by D. Jordan (*Span*, 1977, **20**, 21). Introduced by Shell Research Ltd (GBP 1 437 711) as code no. 'WL 43 425'; trade marks 'Commando' (in the U.K.), 'Suffix BW' (in most other countries). The (*R*)-isomer replaced the racemate which was less potent and less selective.

**Properties.**

The pure (*R*)-isomer is a colourless crystalline solid; m.p. 72.5-74.5 °C; v.p. 31.6 μPa at 20 °C. The technical grade (⩾93% pure) is an off-white crystalline powder, m.p. 70-71 °C. Solubility at 20 °C: *c.* 10 mg/l water; >400 g/l acetone; 677 g/l cyclohexanone; 147 g/l ethanol; 6 g/l hexane. It is stable to heat, to light and to hydrolysis at 2<pH<8. Hydrolysis to propan-2-ol and the parent acid occurs at pH>8.

**Uses.**

The (*R*)-isomer has given consistently >90% control of *Avena* spp. at 600 g a.i./ha to barley between end of tillering and the formation of the 3rd node and to wheat between end of tillering and the formation of the 4th node. Best results are achieved when the growing conditions are good and the crop is growing vigorously. When used within the above timings, the compound is selective at twice the recommended dose. A useful degree of suppression of *Alopecurus myosuroides* and *Arrhenatherum elatius* can often be obtained in addition to control of wild oats.

**Toxicology.**

Acute oral $LD_{50}$: for rats and mice >4000 mg (*R*)-ester/kg; for domestic fowl >2000 mg/kg. Acute percutaneous $LD_{50}$ for rats >1600 mg/kg. In 90-d feeding trials no toxicological effect was observed in rats receiving 50 mg/kg diet or dogs 30 mg/kg diet. $LC_{50}$ (96-h) for rainbow trout is 3.3 mg/l.

**Formulations.**

An e.c. (200 g a.i./l).

**Analysis.**

Product analysis is by optical rotation and glc. Residues can be determined by glc with ECD. Details of the methods are available from Shell International Chemical Co. Ltd.

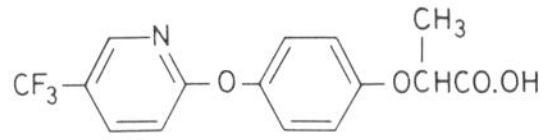

$C_{15}H_{12}F_3NO_4$ (327.3)
T6NJ BOR DOY1&VQ& EXFFF &&(RS) FORM

*fluazifop-butyl*
$C_{19}H_{20}F_3NO_4$ (383.3)
T6NJ BOR DOY1&VO4& EXFFF &&(RS) FORM

**Nomenclature and development.**

Common name fluazifop (BSI, E-ISO, ANSI). Chemical name (IUPAC) (*RS*)-2-[4-(5-trifluoromethyl-2-pyridyloxy)phenoxy]propionic acid. (*C.A.*) (±)-2-[4-[[5-(trifluoromethyl)-2-pyridinyl]oxy]phenoxy]propanoic acid (9CI); Reg. No. *[69335-91-7]* stereochemistry unstated. Discovered by Ishihara Sangyo Kaishi Ltd, the herbicidal properties of its butyl ester (fluazifop-butyl, (Reg. No. *[69806-50-4]*) stereochemistry unstated) were described by R. E. Plowman *et al.* (*Proc. British Crop Prot. Conf. - Weeds*, 1980, **1**, 29). It is being developed jointly by ICI Plant Protection Division and Ishihara Sangyo Kaishi Ltd (GBP 1 599 121 to Ishihara ) as code no.'IH 773B', 'TF1169' (both to Ishihara), PP 009 (to ICI); trade mark 'Fusilade'.

**Properties.**

Fluazifop-butyl is a pale straw-coloured liquid; m.p. *c.* 5 °C; ρ 1.21 g/cm$^3$ at 20 °C; v.p. 55 μPa at 20 °C. Solubility at room temperature: 2 mg/l water; completely miscible with acetone, cyclohexanone, dichloromethane, hexane, methanol, xylene; 24 g/l propylene glycol.

**Uses.**

Fluazifop-butyl is a potent selective herbicide used to control grass weeds in broad-leaved crops, and is most effective applied post-em. at 0.25-2.0 kg a.i./ha.

**Toxicology.**

Acute oral $LD_{50}$: for rats 3330 mg/kg; for female mice 1770 mg/kg; for guinea-pigs 2660 mg/kg; for mallard ducks >17 000 mg/kg. Acute percutaneous $LD_{50}$ for rats >6050 mg/kg; for female rabbits >2420 mg/kg; a mild irritant to their skin, practically non-irritating to their eyes. $LC_{50}$ (5-d) was: for mallard duck >25 000 mg/kg; for ring-necked pheasants >18 500 mg/kg. $LC_{50}$ (96-h) is: for rainbow trout 1.37 mg/l; for mirror carp 1.31 mg/l; for bluegill 0.53 mg/l. It has low toxicity to honeybees orally and by contact.

**Formulations.**

'Fusilade'. e.c. (250 g a.i./l); 'Fusilade W', e.c. (250 g + surfactant/l).

**Analysis.**

Residues may be determined by hplc.

$C_{12}H_{13}ClF_3N_3O_4$ (355.7)
G2N3&R BNW FNW DXFFF

**Nomenclature and development.**

Common name fluchloralin (BSI, E-ISO, ANSI, WSSA);fluchloraline (F-ISO). Chemical name (IUPAC) *N*-(2-chloroethyl)-α,α,α-trifluoro-2,6-dinitro-*N*-propyl-*p*-toluidine (I);*N*-( 2-chloroethyl)-2,6-dinitro-*N*-propyl-4-(trifluoromethyl)aniline. (*C.A.*) *N*-(2-chloroethyl)-2,6-dinitro-*N*-propyl-4-(trifluoromethyl)benzenamine (9CI); (I) (8CI); Reg. No. *[33245-39-5]*. Its herbicidal properties were reported by A. Fischer (*Pestic. Chem. Proc. Int. Congr. Pestic. Chem., 2nd*, 1972, **5**, 189). Introduced by BASF AG (DEOS 1 643 719; BEP 725 877) as code no. 'BAS 392H','BAS 3920','BAS 3921','BAS 3922'; trade mark 'Basalin'.

**Properties.**

It is an orange-yellow solid; m.p. 42-43 °C; v.p. 2.8 mPa at 20 °C. Solubility at 20 °C: <1 mg/kg water; >1 kg/kg acetone, benzene, chloroform, diethyl ether; 251 g/kg cyclohexane; 177 g/kg ethanol. Unstable in u.v. light. Technical grade has purity ⩾97%.

**Uses.**

Pre-plant or pre-em. herbicide effective against grasses and broad-leaved weeds at 475-1000 a.i./ha depending on the crop and soil type. It penetrates germinating weed seedlings through the roots and hypocotyl, preventing processes vital to development especially in the radicle. For optimum results it should be incorporated in the top 5 cm of soil by cultivation or irrigation within 5 h of application. It is used for tropical crops: e.g. cotton, groundnuts, jute, potatoes, rice, soyabeans and sunflowers. Herbicidal efficacy lasts about 10-12 weeks.

**Toxicology.**

Acute oral $LD_{50}$: for rats >6400 mg a.i./kg, 6400 mg tech./kg; for mice 730 mg tech./kg; for ducks 1300 mg/kg; for white quail 7000 mg/kg. Acute percutaneous $LD_{50}$ for rabbits >10 000 mg/kg. In 90-d feeding trials, no toxic effect was observed in rats receiving 5000 mg/kg diet or dogs 150 mg/kg diet. $LC_{50}$ (24-h): bluegill 0.031 mg/l, rainbow trout 0.027 mg/l; $LC_{50}$ (96-h): 0.016 mg/l and 0.012 mg/l, respectively.

**Formulations.**

'Basalin', e.c. (480 g a.i./l).

**Analysis.**

Product analysis is by glc.

# Flucythrinate

$C_{26}H_{23}F_2NO_4$ (451.4)
FYFOR DYY1&1&VOYCN&R COR

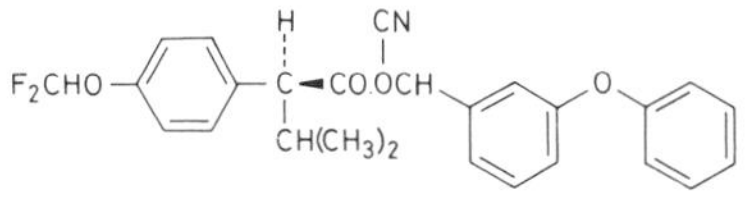

**Nomenclature and development.**

Common name flucythrinate (BSI, ANSI, draft E-ISO). Chemical name (IUPAC) (*RS*)-α-cyano-3-phenoxybenzyl (*S*)-2-(4-difluoromethoxyphenyl)-3-methylbutyrate. (*C.A.*) (*RS*)-cyano(3-phenoxyphenyl)methyl (*S*)-4-(difluoromethoxy)-α-(1-methylethyl)benzeneacetate (9CI); Reg. No. *[70124-77-5]* formerly *[71611-31-9]* (stereochemistry not stated). OMS 2007, A13-29 391. Its insecticidal properties were described by W. K. Whitney & K. Wettstein (*Proc. 1979 Br. Crop Prot. Conf. - Pests Dis.*, 1979, **2**, 387). Introduced by American Cyanamid Co. (USP 4 178 308; GBP 1 582 775) as code no. 'AC 222 705'; trade marks 'Cybolt', 'Pay-off'.

**Properties.**

Pure flucythrinate is a liquid; b.p. 108 °C/0.35 mmHg; v.p. 1.16 μPa at 25 °C; $n_D^{25}$ 1.541; $d_4^{22}$ 1.19. Solubility at 21 °C: 0.5 mg/l water; >820 g/l acetone; >560 g/l corn oil; >300 g/l cottonseed oil, soyabean oil; 90 g/l hexane; >780 g/l propan-1-ol; 1.81 kg/l xylene. On hydrolysis at 27 °C, 50% loss occurs *c.* 40 d at pH3, 52 d at pH5, 6.3 d at pH9. The technical product is a viscous, dark amber liquid with a faintly ester-like odour. It is stable for >1 y at 37 °C, >2 y at 25 °C.

**Uses.**

It is a potent insecticide effective against a wide range of pests including those which have developed resistance to other classes of insecticides. It is useful for control of insects on crops such as cereals, cotton, fruits, lucerne, maize, *Phaseolus* beans, soyabeans, sugar beet, sunflowers, tobacco. Effective dosages are 25-150 g a.i./ha depending upon pest, population pressure, crop, and frequency of treatment. Insecticidal applications often suppress phytophagous mites sufficiently to eliminate or delay the need for a separate acaricidal treatment; but flucythrinate is not recommended for use as an acaricide. It is repellent to many pests and its stability to light and resistance to wash-off by rain provide excellent residual efficacy (A. F. Saad *et al.*, *Proc. 1981 Br. Crop Prot. Conf. - Pests Dis.*, 1981, **2**, 381; K. Wettstein, *ibid.*, p. 563).

**Toxicology.**

Acute oral $LD_{50}$ for male rats 81 mg tech./kg, for females 67 mg/kg; for female mice 76 mg/kg; for bobwhite quail 2708 mg/kg; for mallard duck >2510 mg/kg. Acute percutaneous $LD_{50}$ (24-h) for rabbits >1000 mg/kg; non-irritating to skin and eyes of rabbits; undiluted formulations may be irritating. In 2-y feeding trials rats receiving 60 mg/kg diet showed no ill-effect. $LC_{50}$ (96-h): for rainbow trout 0.32 μg/l; for sheepshead minnow 1.6 μg/l. Low insecticidal dosages and immobility in soils should minimize hazards to fish. The $LD_{50}$ by topical application of a dust to honeybees is 0.078 μg/bee; treated plants are repellent to honeybees.

**Formulations.**

These include: e.c. (100 to 300 g a.i./l); ULV concentrates (10 to 25 g a.i.)/l. Combinations with other pesticides are under development.

**Analysis.**

Product analysis is by hplc and glc. Residues may be determined by glc. Details are available from American Cyanamid Co.

$C_{16}H_{12}ClF_4N_3O_4$ (421.7)
GR CF BIN2&R BNW FNW DXFFF

**Nomenclature and development.**

Common name flumetralin (BSI, draft E-ISO);flumétraline (draft F-ISO). Chemical name (IUPAC) *N*-(2-chloro-6-fluorobenzyl)-*N*-ethyl-α,α,α-trifluoro-2,6-dinitro-*p*-toluidine. (*C.A.*) 2-chloro-*N*-[2,6-dinitro-4-(trifluoromethyl)phenyl]-*N*-ethyl-6-fluorobenzenemethanamine (9CI); Reg. No. *[62924-70-3]*. Its plant growth regulating activity was reported by M. Wilcox *et al.* (*Proc. Plant Growth Regul. Working Group*, 1977, **4**, 194). Introduced by Ciba-Geigy AG. (BEP 891 327; GBP 1 531 260) as code no. 'CGA 41 065'; trade mark 'Prime'.

**Properties.**

Pure flumetralin forms yellow to orange crystals; m.p. 101-103 °C; ρ1.55 g/cm$^3$; v.p. <133 μPa. Solubility at 20 °C: <0.1 mg/l water; very soluble in benzene, dichloromethane. Decomposes exothermically >250 °C.

**Uses.**

It is applied as a topical treatment for the control of sucker growth on flue-cured, burley, dark fire, Maryland and cigar tobacco. The product performs best and can provide full season control if applied within 24 h after topping. It has local systemic effect, rain occurring later than 2 h after spraying does not reduce effectiveness.

**Toxicology.**

Acute oral $LD_{50}$ for rats >5000 mg tech./kg. Acute percutaneous $LD_{50}$ for rats >2000 mg/kg; e.c. (150 g/l) is moderate irritant to skin and extreme irritant to eyes.

**Formulations.**

'Prime+' 250 EC (in Europe), e.c. (250 g/l); 'Prime+' 150 EC (in USA), e.c. (150 g/l).

**Analysis.**

Residues may be determined by glc using ECD or FID. Particulars are available from Ciba-Geigy AG.

$C_{10}H_{11}F_3N_2O$ (232.2)
FXFFR CMVN1&1



**Nomenclature and development.**

Common name fluometuron (BSI, E-ISO, F-ISO, ANSI, WSSA). Chemical name (IUPAC) 1,1-dimethyl-3-(α,α,α-trifluoro-*m*-tolyl)urea (I). (*C.A.*) *N,N*-dimethyl-*N*'-[3-(trifluoromethyl)phenyl]urea (9CI); (I) (8CI); Reg. No. *[2164-17-2]*. Its herbicidal properties were described by C. J. Counselman *et al.*, (*Proc. South Weed Conf., 17th*, 1964, p. 189). Introduced by Ciba AG (now Ciba-Geigy AG) (BEP 594 227; GBP 914 779) as code no. 'C 2059'; trade mark 'Cotoran'.

**Properties.**

Pure fluometuron forms colourless crystals; m.p. 163-164.5 °C; v.p. 66 μPa at 20 °C; ρ 1.39 g/cm³ at 20 °C. Solubility at 20 °C: 105 mg/l water; 105 g/l acetone; 23 g/l dichloromethane; 170 mg/l hexane; 110 g/l methanol; 22 g/l octan-1-ol. On hydrolysis at 20 °C, 50% loss (calculated) occurs in 1.6 y at pH1, 2.4 y at pH5, 2.8 y at pH9.

**Uses.**

It is a herbicide mainly absorbed through the roots and, with weak activity through foliage, especially suitable for the control of broad-leaved and grass weeds in cotton at 1.0-1.5 kg a.i./ha. It is of intermediate persistence, 50% loss occurring in 60-75 d according to soil conditions.

**Toxicology.**

Acute oral $LD_{50}$ for rats 6416- >8000 mg tech./kg. Acute percutaneous $LD_{50}$ for rats >2000 mg/kg; slight irritant to skin and eyes of rabbits. In 90-d feeding trials NEL was: for rats 100 mg/kg diet (*c.* 7.5 mg/kg daily); for dogs 400 mg/kg diet (*c.* 15 mg/kg daily). $LC_{50}$ (96-h) for rainbow trout 47 mg/l; for crucian carp 170 mg/l; for bluegill 96 mg/l.

**Formulations.**

'Cotoran' 50 and 80 WP, w.p. (500 or 800 g a.i./kg); 'Cotoran' 500 FW, s.c. (500 g/l). 'Cotogard', fluometuron + prometryne (1:1). 'Cotoran multi', fluometuron + metolachlor (1:1).

**Analysis.**

Product analysis is by glc (*AOAC Methods*, 1980, 6.289-6.293; *CIPAC Handbook*, 1983, **1B**, in press). Residues may be determined by hydrolysis to α,α,α-trifluoro-*m*-toluidine, a derivative of which is measured by glc with ECD (G. Voss *et al.*, *Anal. Methods Pestic. Plant Growth Regul.*, 1973, **7**, 569).

$C_2H_4FNO$ (77.06)
ZV1F

$FCH_2CO.NH_2$

**Nomenclature and development.**

Fluoroacetamide (BSI, E-ISO, F-ISO, JMAF) is accepted in lieu of a common name. Chemical name (IUPAC) 2-fluoroacetamide (I). (*C.A.*) (I) (8 & 9CI); Reg. No. *[640-19-7]*. Its rodenticidal properties were described by C. Chapman & M. A. Phillips (*J. Sci. Food Agric.*, 1955, **6**, 231). Introduced as a rodenticide as code no. 'Compound 1081'.

**Properties.**

It is a colourless crystalline powder; m.p. 108 °C. Solubility: very soluble in water; soluble in acetone; moderately soluble in ethanol; sparingly soluble in aliphatic and aromatic hydrocarbons.

**Uses.**

It is a rodenticide and an internal mammalian poison subject to many restrictions in use. Normally it is used as a bait (20 g a.i./kg) in areas to which the public have no access such as sewers and locked warehouses. Though insecticidal, it is considered too toxic to mammals for such use.

**Toxicology.**

Acute oral $LD_{50}$: for *Rattus norvegicus c.* 13 mg/kg (C. Chapman & M. A. Phillips, *loc. cit.*; E. W. Bentley & J. H. Greaves, *J. Hyg.*, 1960, **58**, 125; E. W. Bentley *et al., ibid.*, 1961, **59**, 413).

**Formulations.**

Bait, dyed cereal base (30 g a.i./kg) which is mixed with water for use.

**Analysis.**

Product analysis is by reaction with sodium and precipitation as lead chloride fluoride (*cf. WHO Manual*, (2nd Ed.), SRT/5; *AOAC Methods*, 1980, 6.019-6.020).

$C_{13}H_7F_3N_2O_5$ (328.2)
WNR DOR BNW DXFFF

$CF_3$ — O — $NO_2$ ; $NO_2$

**Nomenclature and development.**

Common name fluorodifen (BSI, E-ISO, ANSI, WSSA, JMAF);fluorodifène (F-ISO). Chemical name (IUPAC) 4-nitrophenyl α,α,α-trifluoro-2-nitro-*p*-tolyl ether. (*C.A.*) 2-nitro-1-(4-nitrophenoxy)-4-(trifluoromethyl)benzene (9CI); *p*-nitrophenyl α,α,α-trifluoro-2-nitro-*p*-tolyl ether (8CI); Reg. No. *[15457-05-3]*. Its herbicidal properties were reported by L. Ebner *et al.*, (*Proc. Br. Weed Control Conf., 9th*, p. 1026). Introduced by Ciba AG (now Ciba-Geigy AG) (BEP 628 133; GBP 1 033 163) as code no. 'C 6989'; trade mark 'Preforan'.

**Properties.**

Pure fluorodifen forms yellowish crystals; m.p. 94.0-94.5 °C; v.p. 9.3 μPa at 20 °C; ρ 1.59 g/cm$^3$ at 20 °C. Solubility at 20 °C: 2 mg/l water; 750 g/l acetone; 680 g/l dichloromethane; 1.4 g/l hexane; 4.5 g/l methanol; 10 g/l octan-1-ol; 400 g/l toluene. No significant hydrolysis occured during 28 d, at pH values of 1, 5, 7 or 9 at 70 °C.

**Uses.**

It is a pre- and post-em. contact herbicide used at 3-4 kg a.i./ha. It is recommended pre-em. for use on drilled rice, but not for surface seeded rice.

**Toxicology.**

Acute oral $LD_{50}$ for rats 9000 mg tech./kg. Acute percutaneous $LD_{50}$ for rats >3000 mg/kg; non-irritant to skin, moderate irritant to eyes of rabbits. Inhalation $LC_{50}$ (6-h) for rats >990 mg/m$^3$ air. In 90-d feeding trials NEL was: for rats 1000 mg/kg diet (67 mg/kg daily); for dogs 316 mg/kg diet (10 mg/kg daily). $LC_{50}$ (96-h): for rainbow trout 0.18 mg/l; for bluegill 0.6 mg/l. It is practically non-toxic to birds and honeybees.

**Formulations.**

'Preforan' 300 EC, e.c. (300 g a.i./l).

**Analysis.**

Product analysis is by glc. Residues are determined by glc with ECD. Particulars are available from Ciba-Geigy AG.

$C_{22}H_{16}F_3N_3$ (379.4)
T5NN DNJ AXR&R&R CXFFF

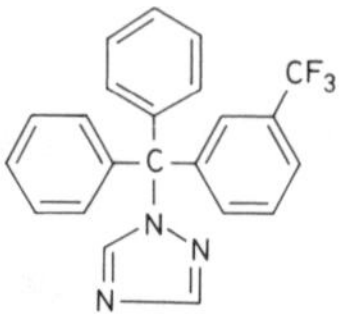

**Nomenclature and development.**

Common name fluotrimazole (BSI, E-ISO, F-ISO). Chemical name (IUPAC) 1-(3-trifluoromethyltrityl)-1*H*-1,2,4-triazole. (*C.A.*) 1-[diphenyl[3-(trifluoromethyl)phenyl]methyl]-1*H*-1,2,4-triazole (9CI); Reg. No. *[57381-79-0]*. Its fungicidal properties were described by F. Grewe & K. H. Büchel (*Mitt. Biol. Bundesanst. Land-Forstwirtsch. Berlin-Dahlem*, 1973, **151**, 208) Introduced by Bayer AG (DEP 1 795 249; USP 3 682 950) as code no. 'BAY BUE 0620'; trade mark 'Persulon'.

**Properties.**

It is a colourless crystalline solid; m.p. 132 °C. Solubility at 20 °C: 1.5 µg/l water; 200 g/kg cyclohexanone; 400 g/kg dichloromethane; 50 g/kg propan-2-ol; 100 g/kg toluene. It is stable in 0.1M sodium hydroxide, but undergoes *c.* 40% degradation in 0.2M sulphuric acid in 24 h.

**Uses.**

Fluotrimazole is a fungicide with specific action against powdery mildews on barley, cucumbers, grapes, melons and peaches.

**Toxicology.**

Acute oral $LD_{50}$: for rats >5000 mg/kg; for canaries >1000 mg/kg. Acute percutaneous $LD_{50}$ for rats >1000 mg/kg. In 2-y feeding trials NEL for rats was 50 mg/kg diet; In 90-d feeding trials NEL was: for rats 800 mg/kg diet; for dogs >5000 mg/kg diet. $LC_{50}$ (96-h) for golden orfe >100 mg/l. It is not harmful to honeybees.

**Formulations.**

These include: w.p. (500 g a.i./kg); e.c. (125 g/l).

**Analysis.**

Product analysis is by i.r. spectroscopy. Residues may be determined by glc with FID (W. Specht, *Pflanzenschutz-Nachr. (Engl. Ed.)*, 1977, **30**, 55; W. Specht & M. Tillkes *ibid.*, 1980, **33**, 61).

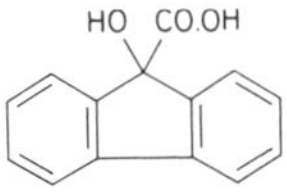

$C_{14}H_{10}O_3$ (226.2)
L B656 HHJ HVQ HQ
*flurecol-butyl*
$C_{18}H_{18}O_3$ (282.3)
L B656 HHJ HVO4 HQ

**Nomenclature and development.**

Common name flurecol (BSI, Canada, Denmark, USA);flurenol (E-ISO, F-ISO) Chemical name (IUPAC) 9-hydroxyfluorene-9-carboxylic acid (I). (*C.A.*) 9-hydroxy-9*H*-fluorene-9-carboxylic acid (9CI); (I) (8CI); Reg. No. *[467-69-6]*. The effects on plant growth of derivatives of fluorene-9-carboxylic acid were first described by G. Schneider (*Naturwissenschaften*, 1964, **51**, 416) who proposed they be called morphactins (G. Schneider *et al.*, *Nature (London)*, 1965, **208**, 1013). The butyl ester of flurecol, flurecol-butyl (flurenol-butyl), (Reg. No. *[2314-09-2]*) was introduced by E. Merck (now Celamerck GmbH Co. KG) (GBP 1 051 652; 1 051 653) as code no. 'IT 3233'.

**Properties.**

Flurecol-butyl forms colourless crystals; m.p. 71 °C; v.p. 130 μPa at 25 °C. Solubility at 20 °C: 36.5 mg/l water; 1.45 kg/l acetone; 950 g/l benzene; 550 g/l carbon tetrachloride; 35 g/l cyclohexane; 700 g/l ethanol; *c.* 7 g/l light petroleum (b.p. 50-70 °C); 1.5 kg/l methanol; 250 g/l propan-2-ol. It is compatible with herbicides.

**Uses.**

Flurecol-butyl is absorbed via leaves and roots causing a general inhibition of plant growth, but is mainly used in conjunction with phenoxyalkanoic acid herbicides, whose action it potentiates. The combination is used for weed control in cereals at rates of 2-4 l 'Aniten'/ha, which persists in the soil for about 42 d.

**Toxicology.**

Acute oral $LD_{50}$: for rats >10 000 mg/kg; for mice >5000 mg/kg. Acute percutaneous $LD_{50}$ for rats >10 000 mg/kg. In feeding trials no adverse effect was noted at 1000 mg/kg in rats (78-d) and dogs (81-d). $LC_{50}$ (96-h) is: for rainbow trout *c.* 12.5 mg/l; for carp *c.* 18.2 mg/l. Contact $LD_{50}$ for honeybees is *c.* 100 μg/bee.

**Formulations.**

An e.c. (125 g flurecol-butyl/kg). 'Aniten', (Reg. No. *[53568-85-7]*), e.c. (100 g flurecol-butyl + 400 g MCPA-isooctyl/kg).

**Analysis.**

Product analysis is by glc of an ester (E. Amadori & W. Hempt, *Anal. Methods Pestic. Plant Growth Regul.*, 1980, **11**, 319). Residues may be determined by colorimetry of a derivative or by glc of an ester (*idem, ibid.*). Details of methods for product and residue analysis are available from Celamerck GmbH & Co.

# Fluridone



$C_{19}H_{14}F_3NO$ (329.3)
T6N DVJ A1 CR CXFFF& ER

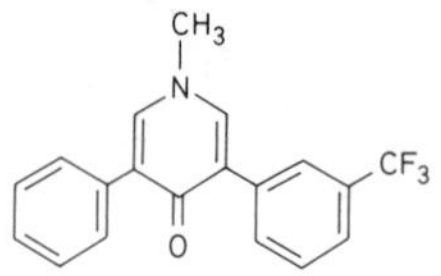

**Nomenclature and development.**

Common name fluridone (BSI, E-ISO, F-ISO, ANSI, WSSA). Chemical name (IUPAC) 1-methyl-3-phenyl-5-(α,α,α-trifluoro-*m*-tolyl)-4-pyridone. (*C.A.*) 1-methyl-3-phenyl-5-[3-(trifluoromethyl)phenyl]-4(1*H*)-pyridinone (9CI); Reg. No. *[59756-60-4]*. Its herbicidal properties were described by T. W. Waldrep & H. M. Taylor (*J. Agric. Food Chem.*, 1976, **24**, 1250). Introduced by Eli Lilly & Co. (GBP 1 521 092) as code no. 'EL-171'; trade marks 'Sonar', 'Pride'.

**Properties.**

Pure fluridone is an off-white crystalline solid; m.p. 151-154 °C; v.p. 13 $\mu$Pa at 25 °C. Solubility: *c.* 12 mg/l water at pH7; >10 g/l chloroform, methanol; >500 mg/l hexane. It is stable to hydrolysis at $3<pH<9$.

**Uses.**

It is a selective pre-em. herbicide for use particularly on cotton, controlling annual grass and broad-leaved weeds and certain perennial species. It is also used as an aquatic herbicide. Its mode of action is by inhibition of carotenoid biosynthesis; selectivity between cotton (tolerant) and maize, rice and soyabean (susceptible) being due to retention of fluridone by cotton roots whilst it is translocated to the foliage of the other species (D. F. Berard *et al., Weed Sci.*, 1978, **26**, 296).

**Toxicology.**

Acute oral $LD_{50}$: for rats and mice >10 000 mg/kg; for dogs >500 mg/kg; for cats >250 mg/kg; for bobwhite quail >2000 mg/kg. Acute percutaneous $LD_{50}$ for rabbits >500 mg/kg, no skin irritation was seen at this dose, slight irritation was observed when 26 mg was applied to their eyes. In 2-y feeding trials there was no ill-effect in rats receiving 200 mg/kg diet. $LC_{50}$ is: for rainbow trout 1157 mg/l; for bluegill 6.3 mg/l.

**Formulations.**

These include: w.p., s.c. and pellets.

**Analysis.**

Product analysis is by hplc. Residues in soil and plant tissue may be determined by hplc or glc with ECD of a suitable derivative (S. D. West, *J. Agric. Food Chem.*, 1978, **26**, 644). Details of analytical procedures and other information may be obtained from Eli Lilly & Co.

$C_{17}H_{16}F_3NO_2$ (323.3)
FXFFR BVMR COY1&1

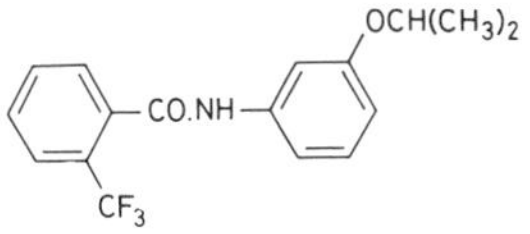

**Nomenclature and development.**

Common name flutolanil (BSI, draft E-ISO, draft F-ISO);*exception* (Hungary). Chemical name (IUPAC) α,α,α-trifluoro-*o*-toluanilide. (*C.A.*) *N*-[3-(1-methylethoxy)phenyl]-2-(trifluoromethyl)benzamide (9CI); Reg. No. *[66332-96-5]*. Its fungicidal properties were described by F. Araki & K. Yabutani (*Proc. Br. Crop Prot. Conf. Pests Dis.*, 1981, **1**, 3). Introduced by Nihon Nohyaku Co. Ltd. (JPP 1 104 514) as code no. 'NNF-136'; trade mark 'Moncut'.

**Properties.**

It forms colourless crystals; m.p. 102-103 °C; v.p. 1.77 mPa at 20 °C. Solubility at 20 °C: 9.6 mg/l water; 238 g/l chloroform; 3 g/l hexane; 606 g/l methanol; 65 g/l toluene.

**Uses.**

It is a systemic fungicide effective against some Basidiomycetes.

**Toxicology.**

Acute oral $LD_{50}$: for rats and mice >10 000 mg/kg. $LC_{50}$ (48-h) for carp 2.4 mg/l.

# Fluvalinate (D ISOMER) 6690

$C_{26}H_{22}ClF_3N_2O_3$ (502.9)
FXFFR CG DMYY1&1&VOYCN&R
COR &&(*D*) FORM

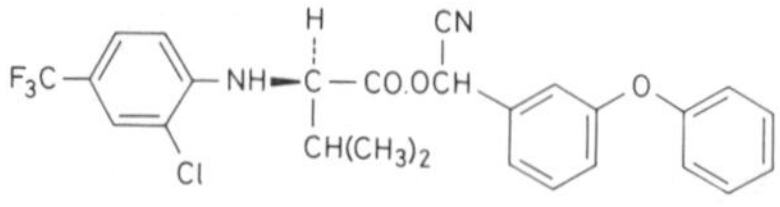

**Nomenclature and development.**

Common name fluvalinate (ANSI) for the racemate. Chemical name (*RS*)-α-cyano-3-phenoxybenzyl *N*-(2-chloro-α,α,α-trifluoro-*p*-tolyl)-D-valinate (commercial isomer); ANSI definition of fluvalinate corresponds to the DL-valinate. (*C.A.*) cyano(3-phenoxyphenyl)methyl *N*-[2-chloro-4-(trifluoromethyl)phenyl]-D-valinate (9CI) (commercial isomer); Reg. No. *[69409-94-5]* DL-isomers. The insecticidal activity of the various stereoisomers was reported by C. A. Henrick *et al*, (*Pestic. Sci.*, 1980, **11**, 224). The D-valinate was introduced by Zoecon Corp. (USA 4 243 819; 4 260 633) as code no. 'ZR-3210' (for DL-valinate); trade marks 'Mavrik-HR' (D-valinate), 'Mavrik' (DL-valinate).

**Properties.**

Fluvalinate (DL-valinate) is a viscous yellow oil; v.p. <13 μPa at 25 °C; ρ1.29 g/cm³; $n_D^{20}$ 1.5410 at 20 °C. Solubility: <5 μg/l water; readily soluble in organic solvents; partition coeff. octan-1-ol:water 7000. On hydrolysis at 25 °C, 50% loss occurred in >30 d at pH3 and pH6, 1-2 d at pH9. No significant degradation occured during 3 h at 100 °C, or 1.6 y at 42 °C in glass. Thin films exposed to sunlight suffered 50% loss in *c*. 2 d. In sandy loam 50% loss occurs in *c*. 6 d.

**Uses.**

Fluvalinate (D-valinate) is a potent, broad-range, foliar-applied insecticide which controls the major pests of cotton, field crops, fruits, vegetables and vines. It is effective against lepidopterous larvae, coleopterous, dipterous and homopterous adults. It shows promise for household, public health and veterinary uses. It has significant acaricidal activity. The D-valinate is twice as potent an insecticide as the DL-valinate, which was the product used during the early stages of development.

**Toxicology.**

Acute oral $LD_{50}$: for rats 6300 mg (tech. DL-valinate)/kg; for mice 656 mg/kg; for quail >2510 mg/kg. Acute percutaneous $LD_{50}$: for rats and rabbits >20 000 mg/kg; mild eye and skin irritant. $LC_{50}$ (8-d) was: for quail 5627 mg/kg diet; for mallard duck >5620 mg/kg. No adverse effects on reproduction were found ≤900 mg/kg diet (the highest dose tested). $LC_{50}$ (96-h) is: for sunfish 2.7 μg/l; for trout 14 μg/l. Dosages used against insects are very low, so hazards to fish are not significant under practical field conditions. It is less toxic to honeybees than to other insects.

**Formulations.**

'Mavrik HR', e.c. (240 g D-valinate/l); 'Mavrik', e.c. (240 g DL-valinate/l) was experimental formulation.

$C_{10}H_{15}OPS_2$ (246.3)
2OPS&2&SR

S
||
SPOCH$_2$CH$_3$
|
CH$_2$CH$_3$

**Nomenclature and development.**

Common name fonofos (BSI, E-ISO, F-ISO). Chemical name (IUPAC) *O*-ethyl *S*-phenyl (*RS*)-ethylphosphonodithioate (I). (*C.A.*) (I) (8 & 9CI); Reg. No. *[944-22-9]* (unstated stereochemistry), *[66767-39-3]* (racemate). OMS 410. Its insecticidal properties were described by J. J. Menn and K. Szabo (*J. Econ. Entomol.*, 1965, **58**, 734). Introduced by Stauffer Chemical Co. (USP 2 998 474) as code no. 'N-2790'; trade mark 'Dyfonate'.

**Properties.**

Fonofos (99.5% pure) is a clear colourless liquid, with an aromatic odour; b.p. *c.* 130 °C/0.1 mmHg; v.p. 28 mPa at 25 °C, $d_{20}^{20}$ 1.154; $n_D^{30}$ 1.585. Solubility: 13 mg/l water; miscible with acetone, ethanol, kerosene, 4-methylpentan-2-one, xylene. Stable $<100$ °C; corrosive to steel.

The chiral forms have been isolated (R. Allahyari *et al.*, *J. Agric. Food Chem.*, 1977, **25**, 471; P. W. Lee *et al.*, *Pestic. Biochem. Physiol.*, 1978, **8**, 146, 158). Optical rotations of the chiral forms are reversed on solution in carbon tetrachloride, cyclohexane, methanol. The (*R*)-isomer Reg. No. (*[62705-71-9]*) is more toxic to insects and mice and a more potent inhibitor of cholinesterase than the (*S*)-isomer (Reg. No. *[62680-03-9]*).

**Uses.**

Fonofos is an insecticide suitable for the control of crickets, soil pests such as rootworms, symphylids, and wireworms at 1.0-1.5 kg a.i./ha. It has caused some injury to seeds when placed in their proximity and persists in soil for *c.* 56 d. A combined granule with disulfoton is used in brassicas to extend the activity to control aphids (C. Sinclair & T. J. Purnell, *Proc. Br. Crop Prot. Conf. - Pests Dis.*, 1977, **2**, 589).

**Toxicology.**

Acute oral $LD_{50}$ of the racemate for male albino rats 24.5 mg/kg, for females 10.8 mg/kg. Acute percutaneous $LD_{50}$ for guinea-pigs 278 mg/kg; for rabbits 159 mg/kg; non-irritating to their skin and eyes.

**Formulations.**

'Dyfonate 4EC' (e.c. 480 g a.i./l); 'Dyfonate 2G', 'Dyfonate 10G', 'Dyfonate 15G', 'Dyfonate 20G', granules (20, 200, 150 or 200 g/kg). 'Doubledown', granules (fonofos + disulfoton). 'Dyfonate Tillam' 1-4E, e.c. (120 g fonofos + 480 g pebulate/l). 'Dyfonate' thiram 5-10G, granules (50 g fonofos + 100 g thiram/kg).

**Analysis.**

Product analysis is by glc (J. E. Barney *et al.*, *Anal. Methods Pestic. Plant Growth Regul.*, 1973, **7**, 269). Residues in crops and soil are determined by glc (*idem, ibid.*). Analytical methods are available from the Stauffer Chemical Co.

$CH_2O$ (30.03)
VHH

HCHO

**Nomenclature and development.**

The chemical name (IUPAC) formaldehyde (I) (BSI, E-ISO, JMAF); aldéhyde formique (F-ISO) is accepted in lieu of a common name; the name formalin applies to the aqueous solution containing 400 g a.i./l. (*C.A.*) (I) (8 & 9CI); Reg. No. *[50-00-0]*. It was introduced as a disinfectant by Loew in 1888 and was first used as a seed disinfectant by Geuther in 1896.

**Properties.**

Formaldehyde is a flammable colourless gas, with a pungent irritating odour, b.p.: −19.5 °C; m.p. −92 °C; $d_4^{-20}$ 0.815. Solubility: very soluble in water, diethyl ether, ethanol. It is very reactive chemically and a powerful reducing agent.

Formalin is a colourless liquid, with a pungent odour; $d_{25}^{25}$ 1.081-1.085; $n_D^{20}$ 1.3746. it is miscible with water, acetone, ethanol. It is chemically reactive, the formaldehyde polymerising on standing and is a powerful reducing agent.

**Uses.**

Formaldehyde is a powerful bactericide and fungicide, used as a soil fumigant and, at one time, for seed treatment though this use is limited by phytotoxicity. it is also used to fumigate glasshouse structures after cropping.

**Toxicology.**

Acute oral $LD_{50}$ for rats is 800 mg formalin/kg. Formaldehyde vapour is very irritating to the mucous membranes and toxic to animals, including man.

**Formulations.**

An aqueous solution to which methanol is added to delay polymerisation; most pharmacopoeias require a content of 370-410 g formaldehyde/kg. 'Dynoform' liquid for fogging in glasshouses and similar structures.

**Analysis.**

Product analysis is by oxidation by hydrogen peroxide to form formic acid which is determined by titration (*AOAC Methods*, 1980, 6.348-6.349; *MAFF Ref. Book*, 1958, No. 1, p. 36; *CIPAC Handbook*, in press) or by titration with thiocyanate (*AOAC Methods*, 1980, 6.351). Atmospheres may be measured by reaction with chromotropic acid and colorimetry (P. R. Ludlam & J. G. King, *Analyst (London)*, 1981, **106**, 488).

$C_{11}H_{15}N_3O_2$ (221.3)
1N1&1UNR COVM1
*formetanate hydrochloride*
$C_{11}H_{16}ClN_3O_2$ (257.8)
1N1&1UNR COVM1 &&HCl SALT

$CH_3NHCO.O$

$N{=}CHN(CH_3)_2$

**Nomenclature and development.**

Common name formetanate (BSI, E-ISO, F-ISO, ANSI). Chemical name (IUPAC) 3-dimethylaminomethyleneaminophenyl methylcarbamate. (*C.A.*) *N*,*N*-dimethyl-*N*'-[3-[(methylamino)carbonyl]oxy]phenylmethanimidamide (9CI); *N*'-(*m*-hydroxyphenyl)-*N*,*N*-dimethylformamidine methylcarbamate ester (8CI); Reg. No. *[22259-30-9]* (formetanate), *[23422-53-9]* (hydrochloride). ENT 27 566 (formetanate hydrochloride). The acaricidal properties of the hydrochloride were described by W. R. Steinhausen (*Z. Angew. Zool.*, 1968, **55**, 107). The hydrochloride was introduced by Schering AG (DBP 1 169 194) as code no. 'Schering 36 056' (formetanate); trade mark 'Dicarzol'.

**Properties.**

Formetanate is a yellowish crystalline solid; m.p. 102-103 °C; v.p. 27 μPa at 25 °C. Solubility at room temperature: 6.3 g/l water; *c.* 100 g/l acetone, chloroform; >200 g/l methanol. In aqueous methanol 50% decomposition occurs in 130 h at pH 4, in 17 h at pH 7, in 100 min at pH 9. The hydrochloride is a colourless powder decomposing at 200-202 °C. Solubility: >500 g/l water; <1 g/l acetone, chloroform, hexane; *c.* 250 g/l methanol. The hydrochloride is very slowly hydrolysed at pH <4 and unchanged when stored for 7 d at 60 °C.

**Uses.**

Formetanate is an acaricide effective against the motile stages of fruit spider mites at 420 g a.i./ha or 19 g/100 l. One spray at 25 g/100 l is recommended against glasshouse spider mite on roses; 2 sprays at 47.5 g/100 l on chrysanthemums. It is effective as an insecticide against beet fly, plant bugs and thrips.

**Toxicology.**

Acute oral $LD_{50}$ of the hydrochloride: for rats 21 mg/kg; for mice 18 mg/kg; for beagle dogs 19.1 mg/kg; for chickens 21.5 mg/kg. Acute percutaneous $LD_{50}$: for rats >5600 mg/kg; for rabbits >10 200 mg/kg. Sub-acute $LC_{50}$ (10-d) was: for pheasants and bobwhite quail >4640 mg/kg diet; for ducks 6810 mg/kg diet. In 2-y feeding trials dogs and rats receiving 200 mg/kg diet showed no significant abnormality. $LC_{50}$ (96-h) is: for rainbow trout 2.8 mg/l, for bluegill 20 mg/l, for black bullhead 75 mg/l.

**Formulations.**

These include: in France, 'Dicarzol 200', w.s.p. (250 g a.i./kg) as hydrochloride; in North America, 'Carzol SP', w.s.p. (920 g/kg) as hydrochloride. '*Fundal Forte*' (Reg. No. *[39374-41-9]*), a mixture with chlordimeform, is no longer available.

**Analysis.**

Product analysis is by titration of the hydrochloride with alkali. Residues may be determined by hydrolysis to 3-aminophenol measured as derivatives by glc or colorimetry (N. A. Jenny & K. Kossmann, *Anal. Methods Pestic. Plant Growth Regul.*, 1973, **7**, 279).

$C_6H_{12}NO_4PS_2$ (257.3)
VHN1&V1SPS&O1&O1

$(CH_3O)_2P(=S)SCH_2CO.N(CH_3)CHO$

**Nomenclature and development.**

Common name formothion (BSI, E-ISO, F-ISO, JMAF). Chemical name (IUPAC) *S*-(*N*-formyl-*N*-methylcarbamoylmethyl) *O,O*-dimethyl phosphorodithioate. (*C.A.*) *S*-[2-(formylmethylamino)-2-oxoethyl] *O,O*-dimethyl phosphorodithioate (9CI); *O,O*-dimethyl phosphorodithioate *S*-ester with *N*-formyl-2-mercapto-*N*-methylacetamide (8CI); Reg. No. *[2540-82-1]*. OMS 698, ENT 27 257. Its insecticidal properties were described by C. Klotzsche (*Mitt. Geb. Lebensmittelunters. Hyg.*, 1961, **52**, 341). Introduced by Sandoz AG (USP 3 176 035; 3 178 337) as code no. 'J-38','SAN 6913I'; trade marks 'Anthio', 'Aflix'.

**Properties.**

Pure formothion is an odourless, yellowish viscous oil or crystalline mass, which cannot be distilled without decomposition; m.p. 25-26 °C; v.p. 113 μPa at 20 °C; $d_4^{20}$ 1.361; $n_D^{20}$ 1.5541. Solubility at 24 °C: 2.6 g/l water; completely miscible with acetone, chloroform, ethanol, methanol, xylene; slightly soluble in hexane. It is rapidly hydrolysed in presence of water at 23 °C 50% loss in ⩾1 d at pH3-9 forming dimethoate (4880) and (dimethoxyphosphinthioylthio)acetic acid. It is unstable in its pure form, but dilute solutions in nonpolar solvents with stabiliser are stable as are its formulations (shelf life at 20 °C ⩾2 y).

**Uses.**

It is a contact and systemic insecticide and acaricide, effective at 250-500 g a.i./ha or at 1.0-1.5 g a.i./l, against a wide range of sucking and mining insects such as aphids, bugs, fruit flies, jassids, psyllids, scales, thrips as well as against some chewing insects (*Cydia pomonella* and *Epilachna* spp. beetles) and spider mites on a variety of field crops, fruit trees, citrus and other tropical fruits, cotton, grapes, hops, ornamentals, tobacco and vegetables. It is metabolised in plants to dimethoate, (dimethoxyphosphinthioylthio)acetic acid, bis(dimethylthiophosphoryl) disulphide, omethoate and *O,O*-dimethyl hydrogen phosphorodithioate; in animals to (dimethoxyphosphinoylthio)acetic acid and to polar metabolites. In loamy soil 50% loss occurs in about 14 d.

**Toxicology.**

Acute oral $LD_{50}$: for albino rats 365-500 mg/kg; for pigeons 630 mg/kg. Acute percutaneous $LD_{50}$ for male albino rats >1000 mg/kg. In 2-y feeding trials rats and dogs receiving 80 mg/kg diet showed no ill-effect. $LC_{50}$ (96-h) for carp is >50 mg/l.

**Formulations.**

'Anthio', e.c. (245 g or 337 g a.i./l); ULV (360 g/l). 'Anthiomix', 'Sandothion', e.c. (210 g formothion + 210 g fenitrothion/l).

**Analysis.**

Product analysis is by glc (*CIPAC Handbook*, 1980, **1A**, 1270) or by paper chromatography, followed by combustion and subsequent determination of phosphate by standard methods (details available from Sandoz AG). Residues are determined by glc (*Pestic. Anal. Man.*, 1979, **I**, 201-H, 201-I; D. C. Abbott *et al.*, *Pestic. Sci.*, 1970, **1**, 10; M. Wisson *et al.*, *Anal. Methods Pestic. Plant Growth Regul.*, 1976, **8**, 323).

$C_3H_8NO_4P$ (153.1)
ZVPQO&O2
*fosamine-ammonium*
$C_3H_{11}N_2O_4P$ (170.1)
ZVPQO&O2 &&$NH_4$ SALT

$NH_2CO-P(=O)(OH)-OCH_2CH_3$

**Nomenclature and development.**

Common name fosamine (BSI, E-ISO, F-ISO, ANSI, WSSA). Chemical name (IUPAC) ethyl hydrogen carbamoylphosphonate. (*C.A.*) ethyl hydrogen (aminocarbonyl)phosphonate (9CI); Reg. No. *[59682-52-9]* for acid, *[25954-13-6]* for ammonium salt. Its herbicidal properties were described by O.C. Zoebisch *et al.*, (*Proc. Northeast. Weed Sci. Soc.*, 1974, **28**, 347). The ammonium salt, fosamine-ammonium [ammonium ethyl carbamoylphosphonate, (*C.A.*) ammonium ethyl (aminocarbonyl)phosphonate], was introduced by E. I. du Pont de Nemours & Co. (Inc.) (USP 3 627 507; 3 846 512) as code no. 'DPX 1108'; trade mark 'Krenite'.

**Properties.**

Pure fosamine-ammonium is a colourless crystalline solid; m.p. 175 °C; *d* 1.33; v.p. 530 µPa at 25 °C. Solubility at 25 °C: 1.79 kg/kg water; 300 mg/kg acetone; 400 mg/kg benzene; 1.4 g/kg dimethylformamide; 12 g/kg ethanol; 200 mg/kg hexane; 158 g/kg methanol. Aqueous formulations and diluted sprays are stable, but it decomposes in acid media in dilute solution (5 mg/l), In soil 50% is decomposed in *c.* 7 d.

**Uses.**

It is a contact herbicide effective against woody plants during the 60-d period before autumn coloration. The parts of plants that have been sprayed tend not to grow the following season. It may be used on non-crop areas including land adjacent to water supplies. Also used for control of *Convolvulus arvensis* and *Pteridium aquilinum* and for selective removal of unwanted trees in forestry. Recommended with the addition of a non-toxic surfactant.

**Toxicology.**

Acute oral $LD_{50}$: for rats 10 125 mg a.i. (as e.c.)/kg, >2075 mg (as e.c. containing surfactant)/kg; for guinea-pigs 7380 mg/kg; for mallard duck and bobwhite quail >4150 mg (as e.c.)/kg. Acute percutaneous $LD_{50}$ for rabbits >1660 mg/kg; the e.c. with or without surfactant is not an irritant of eyes and skin of rabbits. In 90-d feeding trials in rats no adverse effects were detected at 1000 mg a.i./kg diet. $LC_{50}$ (96-h) is: for bluegill 278 mg a.i. (as e.c.)/l; for rainbow trout >415 mg a.i. (as e.c.)/l.

**Formulations.**

'Krenite' brush control agent (480 g fosamine-ammonium/l), 'Krenite S' brush control agent (480 g fosamine-ammonium + surfactant/l).

**Analysis.**

Residues in plant tissues and soil may be determined by glc of a derivative; details available from E.I. du Pont de Nemours & Co. (Inc.).

$C_2H_7O_3P$ (110.0)
QPHO&O2
*fosetyl-aluminium*
$C_6H_{18}AlO_9P_3$ (354.1)
QPHO&O2 &&Al SALT 3:1

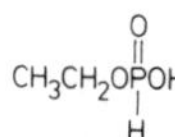

**Nomenclature and development.**

Common name fosetyl (BSI, draft E-ISO);phoséthyl (France). Chemical name (IUPAC) ethyl hydrogen phosphonate (I). (*C.A.*) (I) (9CI); Reg. No. *[39148-24-8]* (aluminium - *C.A.* aluminum - salt). Trivial name *efosite* (rejected common name proposal). The fungicidal properties of the aluminium salt were described by D. Horrière *et al.* (*Phytiatr.-Phytopharm.*, 1977, **26**, 3). The salt was introduced by Rhône-Poulenc Phytosanitaire. (FP 2 254 276) as code no. 'LS 74 783'; trade marks 'Aliette', 'Mikal'.

**Properties.**

Fosetyl-aluminium is a colourless powder; decomposing >200 °C; v.p. negligible at room temperature. Solubility at room temperature: 122 g/l water; practically insoluble in most organic solvents. It is stable under normal storage conditions, but unstable in acidic or alkaline solutions and oxidised by strong oxidising agents.

**Uses.**

It is a fungicide with systemic activity, being translocated downwards as well as upwards in plants. It is mainly effective against diseases caused by Phycomycetes.

**Toxicology.**

Acute oral $LD_{50}$: for rats 5800 mg fosetyl-aluminium/kg; for mice 3700 mg/kg. Acute percutaneous $LD_{50}$ for rats >3200 mg/kg; non-irritant to skin. In 90-d feeding trials NEL was: for rats 5000 mg/kg diet; for dogs 50 000 mg/kg diet. $LC_{50}$ (96-h) for rainbow trout is 428 mg/l. Toxicity to birds and honeybees is low.

**Formulations.**

'Aliette', w.p. (800 g fosetyl-aluminium/kg). 'Mikal', w.p. (500 g aluminium salt + 250 g folpet/kg). Fosetyl-aluminium + mancozeb.

**Analysis.**

Product analysis is by iodometric titration. Residues may be determined by glc of a derivative. Details of methods are available from Rhône-Poulenc Phytosanitaire.

$C_6H_{12}NO_3PS_2$ (241.3)
T4SYS DHJ BUNPO&O2&O2

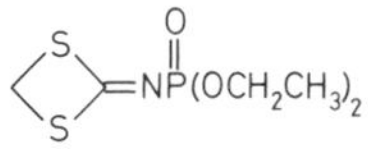

$$\text{1,3-dithietan-2-ylidene}{=}NP(O)(OCH_2CH_3)_2$$

**Nomenclature and development.**

Common name fosthietan (BSI, E-ISO, F-ISO). Chemical name (IUPAC) diethyl 1,3-dithietan-2-ylidenephosphoramidate (I). (*C.A.*) (I) (9CI); cyclic methylene diethoxyphosphinodithioimidocarbonate (8CI); 2-(diethoxyphosphinylimino)-1,3-dithietane; Reg. No. *[21548-32-3]*. Its insecticidal and nematicidal properties were described by W. K. Whitney & J. L. Aston (*Proc. Br. Insect. Fungic. Conf. 8th*, 1975, **2**, 625). Introduced by American Cyanamid Co. (USP 3 476 837) as code no. 'AC 64 475'; trade marks 'Nem-a-tak' (in USA), 'Acconem', 'Geofos'.

**Properties.**

Technical fosthietan is a yellow liquid, with a mercaptan-like odour; $d^{25}$ 1.3; v.p. 864 μPa at 25 °C. Solubility at 25 °C: 50 g/l water; soluble in acetone, chloroform, methanol, toluene. In soil 50% chemical degradation occurs in *C.* 10-42 d, depending upon environmental factors.

**Uses.**

It is a broad-range insecticide and nematicide, effective with soil applications of 1-5 kg a.i./ha.

**Toxicology.**

Acute oral $LD_{50}$ for rats 5.7 mg tech./kg. Acute percutaneous $LD_{50}$ (24-h) for rabbits 54 mg a.i./kg, 3124 mg (5% granule)/kg. Non-mutagenic to bacteria.

**Formulations.**

These include: granules (30-150 g a.i./kg); w.s.c. (250 g/l).

**Analysis.**

Details of analytical methods are available from American Cyanamid Co.

$C_{11}H_8N_2O$ (184.2)
T56 BM DNJ C- BT5OJ

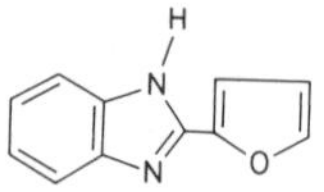

**Nomenclature and development.**

Common name fuberidazole (BSI, E-ISO, F-ISO);*exception* Canada. Chemical name (IUPAC) 2-(2-furyl)benzimidazole (I). (*C.A.*) 2-(2-furanyl)-1*H*-benzimidazole (9CI); (I) (8CI); Reg. No. *[3878-19-1]*. Its preparation had been described by R. Weidenhagen (*Ber. Dtsch. Chem. Ges.*, 1936, **69**, 2271) and its fungicidal properties by G. Schuhmann (*Nachrichtenbl. Dtsch. Pflanzenschutzdienstes (Braunschweig)*, 1968, **20**, 1). Introduced by Bayer AG (DEAS 1 209 799) as code no. 'Bayer 33 172','W VII/117'; trade mark 'Voronit'.

**Properties.**

It forms a fine crystalline powder; m.p. 286 °C (decomp.). Solubility at room temperature: 78 mg/kg water; *c.* 10 g/kg dichloromethane, toluene, light petroleum; *c.* 50 g/kg propan-2-ol; 1.6 g/kg 0.1M hydrochloric acid; 2.3g/kg 0.1M sodium hydroxide. It is unstable to light.

**Uses.**

Fuberidazole is a fungicide used for the treatment of seed against diseases caused by *Fusarium* spp., particularly *F. nivale* on rye and *F. culmorum* on peas.

**Toxicology.**

Acute oral $LD_{50}$ for rats *c.* 1100 mg/kg. Acute percutaneous $LD_{50}$ (7-d) for rats *c.* 1000 mg/kg. Acute intraperitoneal $LD_{50}$ for rats *c.* 100 mg/kg. In 90-d feeding trials rats receiving 1500 mg/kg diet showed no ill-effect.

**Formulations.**

Seed treatment. 'Neo-Voronit', (Reg. No. *[8064-30-0]*), seed treatment, fuberidazole + sodium dimethyldithiocarbamate. 'Voronit special', seed treatment (fuberidazole + quintozene). Mixture, seed treatment, fuberidazole + anthraquinone. Mixtures (Reg. No. *[39312-79-3]*), e.g. '*Voronit-C*' with hexachlorobenzene are no longer available.

**Analysis.**

Product analysis is by u.v. spectroscopy; details are available from Bayer AG. Residues may be determined by glc (W. Specht, *Pflanzenschutz-Nachr. (Engl. Ed.)*, 1977, **30**, 55; W. Specht & M. Tillkes, *ibid.*, 1980, **33**, 61).

$C_{17}H_{19}NO_4$ (301.3)
T5OJ BVNR B1 F1&Y1&VO1

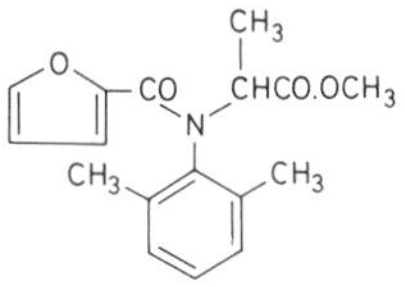

**Nomenclature and development.**

Common name furalaxyl (BSI, E-ISO, F-ISO). Chemical name (IUPAC) methyl *N*-(2-furoyl)-*N*-(2,6-xylyl)-DL-alaninate. (*C.A.*) methyl *N*-(2,6-dimethylphenyl)-*N*-(2-furanylcarbonyl)-DL-alaninate (9CI); Reg. No. *[57646-30-7]*. Its fungicidal properties were described by F. J. Schwinn *et al.*, (*Meded. Fac. Landbouwwet. Rijksuniv. Gent*, 1977, **42**, 1181). Introduced by Ciba-Geigy AG (BEP 827 419; GBP 1 448 810) as code no. 'CGA 38 140'; trade mark 'Fongarid'.

**Properties.**

Pure furalaxyl forms colourless crystals; m.p. 70 °C and 84 °C (dimorphic); v.p. 70 μPa at 20 °C; ρ 1.223 g/cm³ at 20 °C. Solubility at 20 °C: 230 mg/l water; 520 g/kg acetone; 480 g/kg benzene; 600 g/kg dichloromethane; 4 g/kg hexane; 500 g/kg methanol. On hydrolysis at 20 °C, 50% loss (calculated) occurs in >200 d at pH1 and pH9, 22 d at pH10. Stable to >300 °C.

**Uses.**

Furalaxyl is a systemic fungicide suitable for preventive and also curative control of diseases caused by soil-borne Oomycetes. It is mainly used against *Pythium* and *Phytophthora* spp. attacking ornamentals. Compost incorporation and soil drenches are suitable application methods.

**Toxicology.**

Acute $LD_{50}$ for rats 940 mg/kg. Acute percutaneous $LD_{50}$ for rats >3100 mg.kg; mild irritant to skin and minimal irritant to eyes of rabbits. In 90-d feeding trials NEL was: for rats 1250 mg/kg diet (85 mg/kg daily); for dogs 50 mg/kg diet (1.8 mg/kg daily). $LC_{50}$ (96-h) for rainbow trout 32.5 mg/l; for crucian carp 38.4 mg/l. Furalaxyl is practically non-toxic to birds.

**Formulations.**

These include: w.p. (250 and 500 g a.i./kg), granules (50 g/kg).

**Analysis.**

Product analysis is made by glc. Residues in plants and soils may be determined by glc with TID (D. J. Caverley & J. Unwin, *Analyst (London)*, 1980, **106**, 389). Particulars are available from Ciba-Geigy AG.

# Furathiocarb



$C_{18}H_{26}N_2O_5S$ (382.5)
T56 BOT&J C1 C1 IOVN1&SN1&VO4

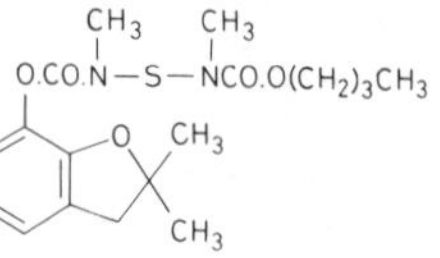

**Nomenclature and development.**

Common name furathiocarb (BSI, draft E-ISO, draft F-ISO). Chemical name (IUPAC) butyl 2,3-dihydro-2,2-dimethylbenzofuran-7-yl *N,N*-dimethyl-*N,N*-thiodicarbamate. (*C.A.*) 2,3-dihydro-2,2-dimethyl-7-benzofuranyl 2,4-dimethyl-5-oxo-6-oxa-3-thia-2,4-diazadecanoate (9CI); Reg. No. *[65907-30-4]*. Its insecticidal properties were described by F. Bachmann & J. Drabek (*Proc. Br. Crop Prot. Conf. - Pests Dis.*, 1981, **1**, 51). Introduced by Ciba-Geigy AG (BEP 865 290; GBP 1 583 713) as code no. 'CGA 73 102'; trade mark 'Deltanet'.

**Properties.**

Pure furathiocarb is a yellow liquid; b.p. 160 °C/0.01 mmHg; v.p. 84 μPa at 20 °C; ρ 1.16 g/cm³ at 20 °C. Solubility at 20 °C: 10 mg/l water; soluble in acetone, hexane, methanol, octan-1-ol, propan-2-ol, toluene. Stable up to 400 °C.

**Uses.**

It is a systemic insecticide for use against soil-dwelling insects and early-season pests. At 500-2000 g a.i./ha applied at sowing it protects seeds, seedlings and young plants of maize, oilseed rape, sugar beet and vegetables for 42 d. It is also effective as a seed treatment or as a foliar spray.

**Toxicology.**

Acute oral $LD_{50}$ for rats 137 mg a.i. (as 50% premix)/kg. Acute percutaneous $LD_{50}$ for rats >2000 mg a.i. (as 50% premix)/kg; slight irritant to skin, minimal irritant to eyes. Acute inhalation $LC_{50}$ (4-h) for rats 214 mg a.i. (as 50% premix)/m³ air.

**Formulations.**

These include: granules (50 or 100 g a.i./kg), e.c. (400 g/l) for foliar applications. Formulations for seed treatment are also available.

**Analysis.**

Product analysis is by glc. Residues may be determined by glc. Particulars are available from Ciba-Geigy AG.

$C_{14}H_{21}NO_3$ (251.3)
T5OJ B1 E1 CVNO1&- AL6TJ

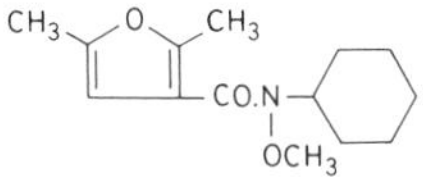

**Nomenclature and development.**

Common name furmecyclox (BSI, draft E-ISO). Chemical name (IUPAC) methyl *N*-cyclohexyl-2,5-dimethylfuran-3-carbohydroxamate. (*C.A.*) *N*-cyclohexyl-*N*-methoxy-2,5-dimethyl-3-furancarboxamide (9CI); Reg. No. *[60568-05-0]*. Its fungicidal properties were described by E. -H. Pommer & B. Zeeh, (*Pestic. Sci.*, 1977, **8**, 320). Introduced by BASF AG (DEP 2 455 082; GBP 1 517 960; USP 3 993 772) as code no. 'BAS 389F'; trade marks 'Campogran', 'Xyligen B'.

**Properties.**

Furmecyclox is a crystalline solid; m.p. 33 °C; v.p. 8.4 mPa at 20 °C. Solubility at 20°C: 1.3 mg/kg water; >1 kg/kg acetone, chloroform, ethanol. It is decomposed by sunlight and hydrolysed under strongly acidic and strongly alkaline conditions.

**Uses.**

It is used as a seed treatment for cereals, cotton, potatoes and other crops.

**Toxicology.**

Acute oral $LD_{50}$ for rats 3780 mg/kg. Acute percutaneous $LD_{50}$ for rats >5000 mg/kg. Potential exists for skin irritation.

**Formulations.**

Wet treatment (500 g a.i./l); dry treatment (400 g/kg).

**Analysis.**

Product analysis is by glc with FID or by hplc. Residues may be determined by glc with FID.

$C_6H_6Cl_6$ (290.8)
L6TJ AG BG CG DG EG FG

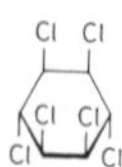


**Nomenclature and development.**

Common name either HCH or BHC (E-ISO, F-ISO);HCH (BSI);BHC (JMAF);benzene hexachloride (ESA, EPA, BPC, USA);hexachloran (USSR);hexaklor (Sweden). Chemical name (IUPAC) 1,2,3,4,5,6-hexachlorocyclohexane (mixed isomers) (I), see below. (*C.A.*) (I) (8 & 9CI); Reg. No. *[608-73-1]* (mixed isomers), *[58-89-9]* (gamma-isomer). OMS 17 (lindane), ENT 8601 (HCH), ENT 9232 (alpha-HCH), ENT 7796 (gamma-HCH), ENT 9234 (delta-HCH). Its insecticidal properties were described by A. Dupire & M. Racourt (*C. R. Hebd. Seances Acad. Agric. Fr.*, 1942, **20**, 470) and by R. E. Slade (*Chem. Ind. (London)*, 1945, p. 134). The relative configuration of the chlorine atoms allows 8 stereoisomers (in IUPAC and *C.A.* nomenclature): alpha (1,2,4/3,5,6-; 1α,2α,3β,4α,5β,6β- Reg. No. *[319-84-6]*); beta (1,3,5/2,4,6-; 1α,2β,3α,4β,5α,6β- Reg. No. *[319-85-7]*); gamma (1,2,4,5/3,6-; 1α,2α,3β,4α,5α,6β- Reg. No. *[58-89-9]*); delta (1,2,3,5/4,6-; 1α,2α,3α,4β,5α,6β Reg. No. *[319-86-8]*); epsilon (1,2,3/4,5,6; 1α,2α,3α,4β,5β,6β- Reg. No. *[6108-10-7]*); eta (1,2,3,4/5,6-; 1α,2α,3α,4α,5β,6β-); theta (1,2,3,4,5/6-; 1α,2α,3α,4α,5α,6β-); zeta (1,2,3,4,5,6/0-; 1α,2α,3α,4α,5α,6α-). Its insecticidal activity is mainly due to the gamma-isomer, common name gamma-HCH or gamma-BHC (BSI, E-ISO, F-ISO), gamma benzene hexachloride (ESA, EPA, BPC, USA), lindane (USSR - see below). The common name lindane (BSI, E-ISO, F-ISO) applies to grades of HCH/BHC in which the gamma-isomer content is ⩾99% *m/m* - *exception* (USSR). Gamma-HCH was introduced by ICI Plant Protection Ltd (now ICI Plant Protection Division) (USP 2 502 258) as trade mark 'Gammexane', numerous trade marks are held by other companies.

**Properties.**

HCH, produced by the chlorination of benzene under u.v. light, has no precise physical properties. Gamma-HCH is isolated by selective crystallisation of HCH, forming colourless crystals; m.p. 112 °C; v.p. 5.6 mPa at 20 °C. Solubility at 20 °C: 7 mg/l water; >50 g/l acetone, benzene, diethyl ether, ethanol, ethyl acetate, toluene. It is stable <180 °C, and to light. At 22 °C, 50% loss occurs in 191 d at pH7, in 11 h at pH9. It is dehydrochlorinated by alkali.

**Uses.**

Gamma-HCH acts as a stomach poison, by contact and has some fumigant action. It is effective against a wide range of soil-dwelling and phytophagous insects, those hazardous to public health, other pests and some animal ectoparasites. Except for foliar applications to cucurbits or hydrangeas, it is non-phytotoxic at insecticidal concentrations. Traces of some other isomers may produce tainting of certain crops; this drawback is less serious when the gamma-HCH content is high (as in lindane). It should not be used on crops intended for canning. It is used as a foliar spray, as soil applications, as a seed treatment and in baits for rodent control.

**Toxicology.**

Acute oral $LD_{50}$ values vary with the conditons of the test, especially the carrier: for rats 88-270 mg/kg; for mice 59-246 mg/kg; for bobwhite quail 120-130 mg/kg. Acute percutaneous $LD_{50}$ for rats 900-1000 mg/kg. In 2-y feeding trials NEL was: for rats 25 mg/kg diet; for dogs 50 mg/kg diet. $LC_{50}$ (48-h) is: for male guppies 0.16 mg/l, for females 0.3 mg/l. Regulations and side-effects of lindane have been summarised (E. Ullmann, *Lindane*, 1972, Supplements 1974, 1976).

**Formulations.**

These include: w.p., e.c., dusts, smokes. Frequently combined with fungicides or other insecticides, especially when intended for seed treatment.

**Analysis.**

Product analysis is by a cryoscopic method (*CIPAC Handbook*, 1970, **1**, 71; FAO Specification (CP/34); *Organochlorine Insecticides 1973*; *AOAC Methods*, 1980, 6.192-6.201). Residues may be determined by glc. (*AOAC Methods*, 1980, 29.001-29.018; *Pestic. Anal. Man.*, 1979, **I**, 201-A, 201-G, 201-I; P. A. Greve & W. B. F. Grevenstuk, *Meded. Fac. Landbouwwet. Rijksuniv. Gent*, 1975, **40**, 111; G.M. Telling *et al.*, *J. Chromatogr.*, 1977, **137**, 405; *Analyst (London),* 1979, **104**, 425).

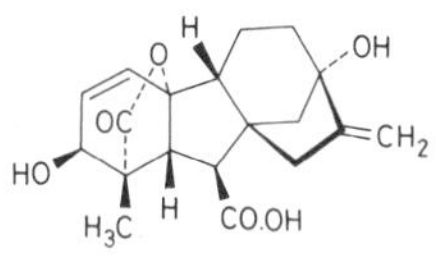

$C_{19}H_{22}O_6$ (346.4)
T C5 C6556/C-F/JP C- 3ACJ P CX EY
JXOV OUTJ BVQ EUI FQ M1 NQ

**Nomenclature and development.**

The name gibberellic acid (BSI, draft ISO); acide gibberellique (draft F-ISO) is accepted in lieu of a common name. Chemical name (IUPAC) (3*S*,3a*S*,4*S*,4a*S*,7*S*,9a*R*,9b*R*,12*S*)-7,12-dihydroxy-3-methyl-6-methylene-2-oxoperhydro-4a,7-methano-9b,3-propeno[1,2-*b*]furan-4-carboxylic acid; (3*S*,3a*S*,4*S*,4a*S*,6*S*,8a*S*,8b*S*,11*S*)-6,11-dihydroxy-3-methyl-12-methylene-2-oxo-4a,6-ethano-3,8b-prop-1-enoperhydroindeno[1,2-*b*]furan-4-carboxylic acid. (*C.A.*) (1α,2β,4aα,4bβ,10β)-2,4a,7-trihydroxy-1-methyl-8-methylenegibb-3-ene-1,10-dicarboxylic acid 1,4a-lactone (9CI); 2β,4α,7-trihydroxy-1-methyl-8-methylene-4aα,4bβ-gibb-3-ene-1α,10β-dicarboxylic acid 1,4a-lactone (8CI); Reg. No. *[77-06-5]*. Gibberellin $A_3$, $GA_3$ (ambiguous). It was discovered by E. Kurosawa (*Trans. Nat. Hist. Soc. (Formosa)*, 1926, **16**, 213) who called it gibberellin A. Later ICI Plant Protection Ltd (now ICI Plant Protection Division) began further investigations isolating a compound with biological properties similar to the Japanese one but, although chemically related, it was distinct from gibberellin A and was called gibberellic acid. It and other members of the gibberellin group (52 are known) occur naturally in a wide variety of plant species. The establishment of its chemical structure and stereochemistry have been reviewed (J. F. Grove, *Q. Rev. Chem. Soc.*, 1961, **15**, 56; B. E. Cross *et al.*, *Adv. Chem. Series*, 1961, No. 28, 13; J. McMillan & N. Takahashi, *Nature (London)*, 1968, **217**, 170, 590). Introduced by ICI as trade marks 'Berelex', 'Activol', 'Activol' GA.

**Properties.**

Produced by fermentation using the fungus *Gibberella fujikuroi*, it is a crystalline solid, m.p. 223-225 °C (decomp.). Solubility: 5 g/l water; very soluble in alcohols; soluble in ethyl acetate; slightly soluble in diethyl ether; insoluble in chloroform. It is an acid, $pK_a$ 4.0, forming water-soluble potassium (50 g potassium gibberellate/l water), sodium and ammonium salts; most other metal salts show a moderate solubility in water. Dry gibberellic acid is stable at room temperature, but is slowly hydrolysed in aqueous or aqueous-alcoholic solutions and decomposed by heat. It undergoes a rearrangement in alkali and at pH 2 and 100 °C it forms gibberellenic acid and later allogibberic acid - all these derivatives show little, if any, biological activity.

**Uses.**

Giberellic acid is a plant growth regulator its uses including: malting of barley; improved seedless and seed grape production; delayed harvesting of citrus; improved fruit set of clementines and pears: advanced cropping of artichokes and rhubarb. See also: J. N. Turner (*Outlook Agric.*, 1972, **7**, 14); (*Annu. Rev. Plant Physiol.*, 1970, **21**, 537) and J. McMillan (*Recent Adv. Phytochem.*, 1974, **7**, 1).

**Toxicology.**

Acute oral $LD_{50}$ for rats and mice $>$15 000 mg/kg. It is non-irritant to skin and eyes. In 90-d feeding trials NEL for rats and dogs was $>$1000 mg/kg diet.

**Formulations.**

These include: an effervescent water-soluble tablet (1 g a.i.); s.p. (100 g/kg); unformulated crystalline a.i.

**Analysis.**

Product and residue analysis is by tlc (V. W. Winkler, *Anal. Methods Pestic. Plant Growth Regul.*, 1978, **10**, 545).

$C_5H_{12}NO_4P$ (181.1)
QVYZ2PQO&1 &&(DL) FORM
*glufosinate-ammonium*
$C_5H_{15}N_2O_4P$ (198.2)
QVYZ2PQO&1 &&(DL) FORM $NH_4$ SALT

$$CH_3\overset{\overset{\displaystyle O}{\|}}{\underset{\underset{\displaystyle OH}{|}}{P}}CH_2CH_2\underset{\underset{\displaystyle NH_2}{|}}{C}HCO.OH$$

**Nomenclature and development.**

Common name glufosinate (BSI, draft E-ISO);*exception* Republic of South Africa. Chemical name (IUPAC) DL-homoalanin-4-yl(methyl)phosphinic acid (*C.A.*) (±)-2-amino-4-(hydroxymethylphosphinyl)butanoic acid (9CI); Reg. No. *[53369-07-6]* ((±) acid), *[51276-47-2]* (acid unstated stereochemistry). The herbicidal properties of the ammonium salt, ammonium DL-homoalanin-4-yl(methyl)phosphinate, Reg. No. *[77182-82-2]* (monoammonium salt, unstated stereochemistry), were described by F. Schwerdtle *et al.* (*Z. Pflanzenkr. Pflanzenschutz.*, 1981, Sonderheft IX, p. 431). This salt was introduced by Hoechst AG as code no. 'Hoe 39 866'; trade mark 'Basta'.

**Properties.**

Glufosinate-ammonium is very soluble in water; v.p. low.

**Uses.**

Glufosinate-ammonium is a non-selective contact herbicide effective against mono- and di-cotyledons at 1-2 kg ammonium salt/ha. It is translocated within leaves but not from them. It inhibits photosynthesis and ammonium ions are accumulated.

**Toxicology.**

Acute oral $LD_{50}$: for male rats 2000 mg glufosinate-ammonium/kg, for females 1625 mg/kg; for dogs 200-400 mg/kg; for mice 416-431 mg/kg. Acute percutaneous $LD_{50}$ for rats >2000 mg/kg. No teratogenic or neurotoxic effects have been observed. $LC_{50}$ (96-d) for rainbow trout 580 mg/l.

**Formulations.**

'Basta', aqueous solution (200 g glufosinate-ammonium/l).

**Analysis.**

Details of analytical methods are available from Hoechst AG.

$C_3H_8NO_5P$ (169.1)
QV1M1PQQO
*glyphosate-mono(isopropylammonium)*
$C_6H_{17}N_2O_5P$ (228.2)
QV1M1PQQO &&$(CH_3)_2CHNH_2$ SALT 1:1

$HO.COCH_2NHCH_2P(=O)(OH)_2$

**Nomenclature and development.**

Common name glyphosate (BSI, E-ISO, F-ISO, ANSI, WSSA, JMAF). Chemical name (IUPAC) *N*-(phosphonomethyl)glycine (I). (*C.A.*) (I) (8 & 9CI); Reg. No. *[1071-83-6]* (acid), *[38641-94-0]* (mono(isopropylammonium) salt). Its herbicidal properties were described by D. D. Baird *et al.* (*Proc. North Cent. Weed Control Conf.*, 1971, **26**, 64). Introduced by Monsanto Co. (USP 3 799 758) as code no. 'MON-0573'; trade mark 'Roundup' for the mono(isopropylammonium) salt.

**Properties.**

Pure glyphosate, which has a zwitterion structure (P. Knuuttila & H. Knuuttila, *Acta Chem. Scand.*, 1979, **33**, 623) forms colourless crystals; m.p. 200 °C; bulk density 0.5 g/cm³. Solubility at 25 °C: 12 g/l; insoluble in common organic solvents. Glyphosate-mono(isopropylammonium) is very soluble in water. Strongly absorbed by soil, in which decomposition is mainly by microbial activity, 50% loss occuring in <60 d. Solutions are corrosive to iron and galvanised steel.

**Uses.**

Glyphosate is a non-selective, non-residual post-em. herbicide which is absorbed by foliage and translocated throughout plants. It is very effective on deep-rooted perennial species, and annual and biennial species of grasses, sedges and broad-leaved weeds. Excellent control of most species has been obtained at rates of 0.34-1.12 kg a.e./ha with annual species, 1.68-2.24 kg/ha for some perennial species. In addition, better control of most weeds is obtained if applications are made at the later stage of plant maturity. Wiping devices are used where height differences allow selective removal of weeds from crops.

**Toxicology.**

Acute oral $LD_{50}$: for rats 5600 mg a.i./kg, 4050 mg a.e. (as salt)/kg; for bobwhite quail >3850 mg a.i./kg. Acute percutaneous $LD_{50}$ for rabbits >5000 mg a.i./kg, non-irritating to their skin and slightly irritating to their eyes; 3750 mg a.e. (as salt)/kg, In 2-y feeding trials on rats and dogs no ill-effect was observed at 300 mg/kg diet (highest dose treated). $LC_{50}$ (8-d) for quail and ducks was >4640 mg a.i./kg diet. $LC_{50}$ (96-h) is: for trout 86 mg a.i./l, 8.2 mg a.e. (as salt)/kg; for bluegill 120 mg/l, 10.5 mg a.e. (as salt)/kg.

**Formulations.**

'Roundup', water-based solution (480 g glyphosate-mono(isopropylammonium) ≡ 360 g a.e./l). 'Tumbleweed'.

**Analysis.**

Residues may be determined by hplc (*Pestic. Anal. Man.*, 1979, **II** ).

# Glyphosine



$C_4H_{11}NO_8P_2$ (263.1)
QPQO&1N1VQ1PQQO

$$HO.COCH_2N\left[CH_2\overset{O}{\overset{\|}{P}}(OH)_2\right]_2$$

**Nomenclature and development.**

Common name glyphosine (BSI, E-ISO, F-ISO, ANSI). Chemical name (IUPAC) *N,N*-bis(phosphonomethyl)glycine (I); *N,N*-di(phosphonomethyl)glycine. (*C.A.*) (I) (8 & 9CI); Reg. No. *[2439-99-8]*. Its effects on carbohydrate deposition in sugarcane were described by C. A. Porter & L. E. Ahlrichs (*Hawaii Sugar Tech. Rep. 1971*, 1972, **30**, 71). Introduced by Monsanto Co. (USP 3 556 762) as code no. 'CP 41 845'; trade mark 'Polaris'.

**Properties.**

Glyphosine is a colourless solid; m.p. 200 °C (decomp.); v.p. <133 nPa at 100 °C; ρ 1.8 g/cm$^3$. Solubility at 20 °C: 350 g/l water; insoluble in benzene; very slightly soluble in ethanol. Stable to light. Nonflammable. Mildly corrosive when wet.

**Uses.**

It is a plant growth regulator used to hasten ripening of sugarcane and to increase the sucrose content. Effective rates are 2.4-4.5 kg a.i./ha. It is absorbed by foliage and is translocated throughout the plant especially to growing points. It is removed from the juice during processing. It is inactivated by microbial action in soils, 50% loss occurs in *c.* 60 d.

**Toxicology.**

Acute oral $LD_{50}$ for rats 7200 mg tech./kg. Non-irritant to skin and corrosive to eyes of rabbits. In 2-y feeding trials no adverse effect was observed at rates up to 300 mg/kg diet in dogs, or 10 000 mg/kg diet in rats, and reproduction was not affected. In 1.5 y trials mice receiving 10 000 mg/kg diet showed no effect and no significant tumour frequency.

**Formulations.**

'Polaris', s.p. (850 g a.i./kg).

**Analysis.**

Residues may be determined by glc of a suitable derivative with FPD.

$C_{18}H_{41}N_7$ (355.6)
MUYZM8M8MYZUM
*guazatine triacetate*
$C_{24}H_{53}N_7O_6$ (535.7)
MUYZM8M8MYZUM &&3CH$_3$COOH
SALT

$NH_2C(=NH)NH(CH_2)_8NH(CH_2)_8NHC(=NH)NH_2$

**Nomenclature and development.**

Common name guazatine (BSI, E-ISO, F-ISO); BSI used the name *guanoctine* from 1970-1972. Chemical name (IUPAC) 1,1'-iminodi(octamethylene)diguanidine; bis(8-guanidino-octyl)amine; di-(8-guanidino-octyl)amine. (*C.A.*) *N,N'''*-(iminodi-8,1-octanediyl)bisguanidine (9CI); 1,1'-[iminobis(octamethylene)diguanidine] (8CI); Reg. No. *[13516-27-3]* (base), *[39202-40-9]* (triacetate). The fungicidal properties of its salts were described by W. S. Catling *et al.* (*Congr. Plant Pathol., 1st*, 1968, p. 27 (Abstr.)). The triacetate was introduced by Evans Medical Ltd, developed by Murphy Chemical Ltd (now part of Dow Chemical Co.) and is currently manufactured by KenoGard AB (GBP 1 114 155 to Murphy Chemical Ltd) as code no. 'EM 379' (Evans Medical Ltd), 'MC 25' (Murphy Chemical Ltd); trade mark 'Panoctine' (KenoGard AB).

**Properties.**

Produced by the amidination of technical bis(8-amino-octyl)amine. The base, which is not isolated, comprises the named isomer (*c.* 50% *m/m*), 1,1'-octamethylenediguanidine (*c.* 25% *m/m*), 1,1'-octamethylenebis(imino-octamethylene)diguanidine (*c.* 25% *m/m*), and traces of higher homologues. All of these compounds are fungicidal but, at present, the common name applies to the major component. The triacetate forms colourless crystals; m.p. 140 °C. This salt is readily soluble in water at room temperature, but is insoluble in organic solvents.

**Uses.**

Guazatine salts are fungicides, especially effective as: cereal seed dressings at 0.6-0.8 g a.i./kg seed, being repellent to hens, pigeons, pheasants and ravens; potato and sugarcane dips; and for the control of *Pyricularia oryzae* in rice by foliar application.

**Toxicology.**

Acute oral $LD_{50}$: for rats 230-260 mg guazatine triacetate/kg; for mice *c.* 300 mg/kg; for Japanese quail 263 mg/kg; for hens 125 mg/kg. Acute percutaneous $LD_{50}$ for albino rabbits *c.* 1100 mg/kg. In 90-d feeding trials for rats NEL was 800 mg/kg diet.

**Formulations.**

These include: liquids, 'Mist-O-matic Murbenine' and powders. Combinations for seed treatments include: guazatine triacetate + fenfuram + gamma-HCH + imazalil; guazatine triacetate + imazalil.

**Analysis.**

Product analysis is by potentiometric titration with acid. Residues may be determined by hydrolysis to the triamine, followed by gc-ms determination (H. Kobayashi *et al.*, *Nihon Noyaku Gakkaishi*, 1977, **2**, 427). Particulars from KenoGard AB.

$C_{10}H_5Cl_7$ (373.3)
L C555 A DU IUTJ AG AG BG FG
HG IG JG

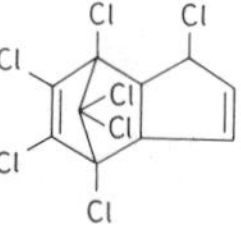


**Nomenclature and development.**

Common name heptachlor (BSI, E-ISO, JMAF);heptachlore (F-ISO). Chemical name (IUPAC) 1,4,5,6,7,8,8-heptachloro-3a,4,7,7a-tetrahydro-4,7-methanoindene (I). (*C.A.*) 1,4,5,6,7,8,8-heptachloro-3a,4,7,7a-tetrahydro-4,7-methano-1*H*-indene (9CI); (I) (8CI); Reg. No. *[76-44-8]*. OMS 193, ENT 15 152. It was isolated from technical chlordane (2250) and its insecticidal properties described by W. M. Rogoff & R. L. Metcalf (*J. Econ. Entomol.*, 1951, **44**, 910). Introduced by Velsicol Chemical Corp. (USP 3 437 664) as code no. 'E 3314','Velsicol 104'.

**Properties.**

Pure heptachlor is a crystalline solid; m.p. 95-96 °C; v.p. 53 mPa at 25 °C; solubility at 25-29 °C 56 µg/l water. The technical grade contains *c.* 72% heptachlor and 28% related compounds and is a waxy solid; m.p. 46-74 °C; $d^{25}$ 1.65-1.67; solubility at 20-30 °C: 1.65 kg/l cyclohexanone; 62.5 g/l ethanol. 263 g/l deodorised kerosene; 1.41 kg/l xylene. It is stable to light, moisture, air, and ≤261 °C. It is not readily dehydrochlorinated and is susceptible to epoxidation.

**Uses.**

It is a persistent, non-systemic contact and stomach insecticide with some fumigant action. It is non-phytotoxic at insecticidal concentrations. Heptachlor is applied as a soil treatment, a seed treatment (maize, small grains, and sorghum) or directly to foliage. It is used to control ants, cutworms, maggots, termites, thrips, weevils, wireworms and many other insect pests in both cultivated and uncultivated soils. Heptachlor also controls household insects, and pests of man and domestic animals.

**Toxicology.**

Acute oral $LD_{50}$ for rats 147-220 mg/kg. Acute percutaneous $LD_{50}$ for rabbits >2000 mg/kg; it did not irritate their skin or their eyes. Inhalation $LC_{50}$ (4-h) in rats exposed to heptachlor in an aerosol was >2 but <200 mg/l air. Numerous chronic studies have been conducted in several species. The target organ is the liver with NEL in rats at ≤5 mg/kg diet and in dogs at 0.5-1.0 mg/kg diet. A committee of the United States National Academy of Sciences has ruled that heptachlor is a carcinogen to certain mouse strains.. Heptachlor epoxide is not a teratogen in rabbits at 5 mg/kg daily. A 3-generation feeding study showed NEL in rats was 7 mg/kg diet and in dogs (2-generation reproduction study) 1 mg/kg diet. In-vivo and in-vitro mutagenicity studies have provided evidence that heptachlor is not a mutagen. There is evidence that heptachlor may act as a promoter, rather than as an initiator, of carcinogenesis.

**Formulations.**

Include: e.c. (240 g/l); w.p. (400 g/kg); granules (250 g/kg); dusts.

**Analysis.**

Product analysis is by glc (*AOAC Methods*, 1980, 6.278-6.284) or by determination of a labile chlorine substituent (*ibid.*, 6.275-6.277; *CIPAC Handbook*, 1970, **1**, 420; FAO Specification (CP/47); *Organochlorine Insecticides 1973*). Residues may be determined by glc (*AOAC Methods*, 1980, 29.001-29.018; *Pestic. Anal. Man.*, 1979, 201-A, 201-G, 201-I; M. A. Luke *et al.*, *J. Assoc. Off. Anal. Chem.*, 1981, **64**, 1187; P. A. Greve & W. B. F. Grevenstuk, *Meded. Fac. Landbouwwet. Rijksuniv. Gent*, 1975, **40**, 1115; G. M. Telling *et al.*, *J. Chromatog.*, 1977, **137**, 405; *Anal. Methods Pestic. Plant Growth Regul.*, 1972, **6**, 404; H. K. Suzuki *et al.*, *ibid.*, 1978, **10**, 73).

$C_9H_{12}ClO_4P$ (250.6)
L45 BU EUTJ BOPO&O1&O1 CG

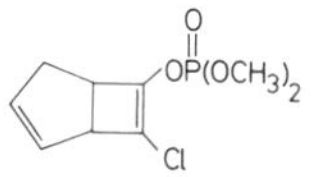

**Nomenclature and development.**

Common name heptenophos (BSI, E-ISO, F-ISO). Chemical name (IUPAC) 7-chlorobicyclo[3.2.0]hepta-2,6-dien-6-yl dimethyl phosphate (I). (*C.A.*) (I) (8 & 9CI); Reg. No. *[23560-59-0]*. OMS 1845. First described under an erroneous structure (Reg. No. *[34783-40-9], Biul. Inst. Ochr. Rosl.*, 1971, No.50, p.109), its insecticidal and acaricidal properties were reported against crop pests by R. T. Hewson (*Proc. Br. Insectic. Fungic. Conf., 8th*, 1975, **2**, 697) and against veterinary pests by W. Bonin (*ibid.*, p.705). Introduced by Hoechst AG (DEP 1 643 608; GBP 1 154 603; USP 3 600 474; 3 705 240; 3810 919) (Ciba-Geigy AG) as code no. 'Hoe 02982'; trade marks 'Hostaquick' (for crop protection use), 'Ragadan' (for veterinary use - both to Hoechst).

**Properties.**

Pure heptenophos is a pale amber liquid; b.p. 64 °C/0.075 mmHg; v.p. 170 mPa at 25 °C; $d_4^{20}$ 1.294. The technical grade ≥93% pure. Solubility at 20 °C: 2.2 g tech./l water; miscible with most organic solvents.

**Uses.**

It is an insecticide with quick initial action and short residual effect. It penetrates plant tissue and is rapidly translocated, controlling sucking insects and certain Diptera. It is also effective against ectoparasites (fleas, lice, mites and ticks) of cattle, dogs, pigs, sheep and pets.

**Toxicology.**

Acute oral $LD_{50}$: for rats 96-121 mg/kg; for dogs >500 mg/kg; for Japanese quail 17-50 mg/kg (depending on carrier and sex). Acute percutaneous $LD_{50}$ for rats *c.* 2900 mg/kg. In 2-y feeding trials NEL was: for dogs 12 mg/kg diet; for rats 15 mg/kg diet. $LC_{50}$ (96-h) for guppies 9.3 mg/l.

**Formulations.**

'Hostaquick', e.c. (550 g a.i./l) for crop use; 'Ragadan', e.c. (250 g/l) and w.p. (400 g/kg) for veterinary use.

**Analysis.**

Details of glc methods are available from Hoechst AG.

$C_{13}H_{19}NO$ (205.3)
L55 A CYTJ A1 A1 B1 CUNO2UU1
&&(E)-(1RS,4RS) FORM

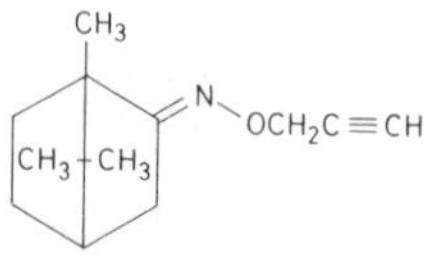

**Nomenclature and development.**

Common name heptopargil (BSI, draft E-ISO). Chemical name (IUPAC) (*E*)-(1*RS*,4*RS*)-bornan-2-one *O*-prop-2-ynyloxime. (*C.A.*) (±)-1,7,7-trimethylbicyclo[2.2.1]heptan-2-one *O*-2-propynyloxime (9CI); Reg. No. *[73886-28-9]*. Its plant growth regulating properties were described by A. Kis-Tamás *et al.* (*Proc. Br. Crop Prot. Conf. - Weeds*, 1980, **I**, 173). Introduced by EGYT Pharmacochemical Works (USP 4 244 888) as code no. 'EGYT 2250'; trade mark 'Limbolid'.

**Properties.**

It is a pale yellow oil; b.p. 95 °C/1 mmHg; $d_4^{20}$ 0.9867. Solubility at 20 °C: 1 g/l water; miscible with protic and aprotic organic solvents.

**Uses.**

It is a promising crop yield enhancer applied as seed treatment for maize, rice and sugar beet and by pre-em. or post-em. applications in beans, lucerne, maize, peas, sunflowers and various vegetables.

**Toxicology.**

Acute oral $LD_{50}$: for male rats 2100 mg/kg, for females 2141 mg/kg. Acute inhalation $LC_{50}$ for rats >1400 mg/$m^3$ air.

**Formulations.**

These include: e.c. (500 g a.i./l); film coating for seed treatment.

**Analysis.**

Details of methods for product and residue analysis are available from EGYT Pharmacochemical Works.

$C_6Cl_6$ (284.8)
GR BG CG DG EG FG

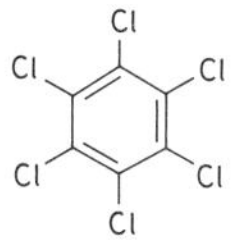

**Nomenclature and development.**

Chemical name (IUPAC) hexachlorobenzene (I) (BSI, E-ISO, F-ISO) is accepted in lieu of a common name. (*C.A.*) (I) (8 & 9CI); Reg. No. *[118-74-1]*. Trivial name HCB. Its introduction as a fungicidal seed treatment was reported by H. Yersin *et al.* (*C. R. Seances Acad. Agric. Fr.*, 1945, **31**, 24).

**Properties.**

Pure hexachlorobenzene forms colourless crystals; m.p. 226 °C; v.p. 1.45 mPa at 20 °C; $d^{23}$ 2.044. Solubility: practically insoluble in water, ethanol. The technical grade has m.p. ⩾220 °C.

**Uses.**

It is a selective fungicide used to control *Tilletia caries* on wheat (M. Lansade, *Parasitica*, 1949, **5**, 1). It was found effective against dwarf bunt, attributed to a fumigant action on the bunt spore (W. N. Siang & C. S. Holton, *Plant Dis. Rep.*, 1953,**37**, 63).

**Toxicology.**

Acute oral $LD_{50}$ for rats 10 000 mg/kg; guinea-pigs tolerated doses >3000 mg/kg. It may cause a slight irritation to the skin. Its use and presence as residues in food are banned in many countries.

**Formulations.**

Dust, powder. Previously sold in combination with other seed protectants.

**Analysis.**

Residues may be determined by glc (T. Stijve; *Mitt. Geb. Lebensmittelunters. Hyg.*, 1971, **62**, 406).

$C_{12}H_{20}N_4O_2$ (252.3)
T6NVNVNJ A1 FN1&1 C- AL6TJ

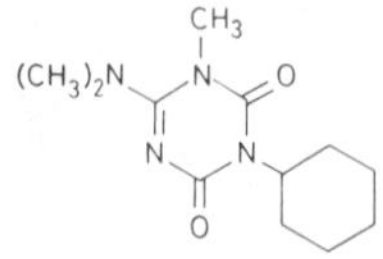

**Nomenclature and development.**

Common name hexazinone (BSI, E-ISO, F-ISO, ANSI, WSSA). Chemical name (IUPAC) 3-cyclohexyl-6-dimethylamino-1-methyl-1,3,5-triazine-2,4(1*H*,3*H*)-dione. (*C.A.*) 3-cyclohexyl-6-(dimethylamino)-1-methyl-1,3,5-triazine-2,4(1*H*,3*H*)-dione (9CI); Reg. No. *[51235-04-2]*. Its herbicidal activity was described by T. J. Hernandez *et al.* (*Proc. North Cent. Weed Control Conf.*, 1974, p. 138). Introduced by E. I. du Pont de Nemours & Co. (Inc.) (USP 3 902 997) as code no. 'DPX 3674'; trade mark 'Velpar'.

**Properties.**

Hexazinone (>98% pure) is a colourless crystalline solid; m.p. 115-117 °C; v.p. 8.5 mPa at 86 °C (27 μPa at 25 °C, extrapolated), *d* 1.25. Solubility at 25 °C: 33 g/kg water; 790 g/kg acetone; 940 g/kg benzene; 3.88 kg/kg chloroform; 836 g/kg dimethylformamide; 3 g/kg hexane; 2.65 kg/kg methanol; 386 g/kg toluene. It is stable in aqueous solution at pH 5-9 at temperatures up to 37 °C. Subject to microbial decomposition in soil, 50% loss in 30-180 d.

**Uses.**

It is a post-em. contact herbicide effective against many annual and biennial weeds and, except for *Sorghum halepense*, most perennial weeds at 6-12 kg a.i./ha. It is used for selective control in lucerne, pineapple, sugarcane and in plantations of certain coniferous species; also on non-crop areas, but *not* on sites adjacent to deciduous trees or other desirable plants. The addition of a non-ionic surfactant is recommended.

**Toxicology.**

(All results for technical grade.) Acute oral $LD_{50}$ for rats 1690 mg/kg; for guinea-pigs 860 mg/kg; for bobwhite quail 2258 mg/kg. Acute percutaneous $LD_{50}$ for rabbits >5278 mg/kg; reversible irritant to their eyes; non-irritant to skin of guinea-pigs. In 2-y feeding trials in rats NEL was 250 mg/kg diet. $LC_{50}$ (8-d) for bobwhite quail and mallard duckling >10 000 mg/kg diet. $LC_{50}$ (96-h) is: for bluegill 370-420 mg/l; for fathead minnow 274 mg/l; $LC_{50}$ (48-h) for *Daphnia magna* 151 mg/l.

**Formulations.**

'Velpar', s.p. (900 g a.i./kg); 'Velpar Gridmall', pellet (100 or 200 g/kg); 'Velpar' liquid (240 g/l). 'Velpar K4', (Reg. No. *[73347-92-9]*) (132 g hexazinone + 468 g diuron/kg).

**Analysis.**

Product analysis is by glc. Residues may be determined by glc (R. F. Holt, *J. Agric. Food Chem.*, 1981, **29**, 165) or by hplc (T. H. Byast *et al.*, *Tech. Rep. ARC Weed Res. Organ.*, No. 15, (2nd Ed.), p. 40).

# 2-Hydrazinoethanol

$C_2H_8N_2O$ (76.09)
ZM2Q

$HOCH_2CH_2NHNH_2$

**Nomenclature and development.**

Chemical name (IUPAC) 2-hydrazinoethanol (I); 2-hydroxyethylhydrazine. (*C.A.*) (I) (8 & 9CI); Reg. No. *[109-84-2]*. trivial name BOH. Introduced as a plant growth regulator by Olin Corp. trade mark 'Omaflora' (Olin Mathieson now Olin Corp.), 'Brombloom' (ACF Chemie Farma N.V.).

**Properties.**

It is a colourless liquid; m.p. −70 °C; b.p. 110-130 °C/17.5 mmHg. Solubility at room temperature: miscible with water and lower alcohols. Dilute solutions are easily oxidised, stable in the cool and in the dark.

**Uses.**

Used to promote flowering of Bromeliaceae.

**Formulations.**

Solution (10 mg a.i./l).

$CHN$ (27.03)
NCH
*calcium cyanide*
$C_2CaN_2$ (92.11)
.CA..CN2
*sodium cyanide*
$CNNa$ (49.01)
.NA..CN

HCN

**Nomenclature and development.**

The traditional names hydrogen cyanide and for the respective salts;calcium cyanide;sodium cyanide (E-ISO) are accepted in lieu of common names;acide cyanhydrique;cyanure de calcium;cyanure de sodium (F-ISO);the name hydrogen cyanide (JMAF) is used for sodium cyanide in Japan. Chemical names (IUPAC) respectively: hydrogen cyanide; calcium dicyanide; sodium cyanide. (*C.A.*) hydrocyanic acid (8 & 9CI); calcium cyanide (8 & 9CI); sodium cyanide (8 & 9CI); Reg. No. *[74-90-8]* (acid), *[529-01-8]* (calcium salt), *[143-33-9]* (sodium salt). Trivial name prussic acid (for hydrogen cyanide). Hydrogen cyanide was used as an insecticidal fumigant in 1886 by D. W. Coquillett (cited by L. O. Howard, *U.S. Dep. Agric. Yearbook*, 1899, p. 150). Introduced as the liquid acid and as salts from which the acid is liberated *in situ* by the action of moisture. Calcium cyanide was introduced by American Cyanamid Co. (who no longer manufacture or market it) and by Degesch AG, sodium cyanide by ICI Plant Protection Division, as trade marks '*Cyangas*' (to American Cyanamid formerly for calcium cyanide), 'Cymag' (to ICI for sodium cyanide).

**Properties.**

Hydrogen cyanide is a colourless liquid, with an odour of almonds; b.p. 26 °C; m.p. −14 °C; $d_4^{20}$ 0.699; soluble in water, diethyl ether, ethanol. It is a weak acid forming salts which have high m.p. (sodium cyanide 564 °C) which are very soluble in water and are readily hydrolysed, liberating hydrogen cyanide.

**Uses.**

Hydrogen cyanide is used as an insecticide and rodenticide for fumigating enclosed spaces (stored grain in warehouses etc.) and, in some countries, glasshouses. Calcium cyanide and sodium cyanide are used to fumigate rabbit burrows, rat runs etc., and owe their activity to the liberation of hydrogen cyanide by the action of atmospheric moisture on the salts. The salts are treated with acid when rapid release of hydrogen cyanide is required. Their use is normally restricted to trained personnel.

**Toxicology.**

Hydrogen cyanide is very toxic to mammals; exposure for 30 min at 360 mg/m$^3$ air is fatal to man. Acute $LD_{50}$ for rats 6.44 mg sodium cyanide/kg. The salts are very toxic to humans.

**Formulations.**

Hydrogen cyanide is packed in metal containers, a polymerisation inhibitor being included. Also absorbed on porous material such as kieselguhr. 'Cymag', powder (400 g sodium cyanide/kg).

**Analysis.**

Product analysis is by titration with silver nitrate (*AOAC Methods*, 1980, 26.135-26.136). Residues may be determined by recommended methods (*Pestic. Anal. Man,*, 1979, **II**; P. Jaulmes & R. Mestres, *Ann. Technol. Agric. (Paris)*, 1962, **11**, 249).

**7170**

$C_{17}H_{30}O_2$ (266.4)
1Y1&3Y1&2U1Y1&U1VO2 &&(E,E) FORM

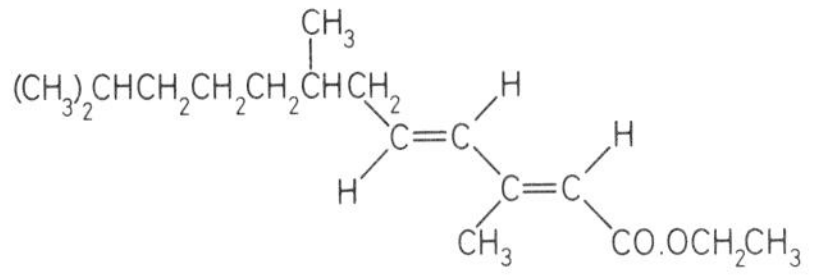

```
                          CH3
                          |
(CH3)2CHCH2CH2CH2CHCH2          H
                      \        /
                       C=C
                      /        \       H
                    H           \     /
                                 C=C
                                /      \
                             CH3        CO.OCH2CH3
```

**Nomenclature and development.**

Common name hydroprene (ANSI). Chemical name (IUPAC) ethyl (*E*,*E*)-3,7,11-trimethyldodeca-2,4-dienoate. (*C.A.*) ethyl (2*E*,4*E*)-3,7,11-trimethyl-2,4-dodecadienoate (9CI); Reg. No. *[41096-46-2]*. OMS 1696. Its properties as an insect growth regulator were described by C. A. Henrick *et al.* (*J. Agric. Food Chem.*, 1973, **21**, 354). Introduced by Zoecon Corp. (now part of Sandoz AG) as code no. 'ZR 512'; trade mark 'Altozar'.

**Properties.**

Hydroprene is an amber liquid $\rho$ 0.8955 g/cm$^3$ at 20 °C; v.p. 25 mPa at 25 °C. Solubility: 0.54 mg/l water; soluble in common organic solvents.

**Uses.**

An insect growth regulator, effective against many species of Coleoptera, Homoptera and Lepidoptera.

**Toxicology.**

Acute oral $LD_{50}$: for rats >34 600 mg/kg; for dogs >10 000 mg/kg. Acute percutaneous $LD_{50}$ for rabbits 4556 mg/kg; non-irritating to their skin, minimal irritant to their eyes. No mutagenic activity was observed in mice at 1300 mg/kg (highest dose tested).

$C_4H_5NO_2$ (99.15)
T5NOJ C1 EQ

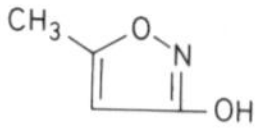

**Nomenclature and development.**

Common name hymexazol (BSI, draft E-ISO);hydroxyisozazole (JMAF). Chemical name (IUPAC) 5-methylisoxazol-3-ol. (*C.A.*) 5-methyl-3(2*H*)-isoxazolone [*C.A.* preferred tautomer (9CI)]; 5-methyl-3-isoxazolol (8CI); Reg. No. *[10004-44-1]*. Its fungicidal properties were described by I. Iwai & N. Nakamura (*Chem. Pharm. Bull.*, 1966, **14**, 1277) and its chemistry, biological properties and toxicology reviewed by K. Tomita *et al.* (*Sankyo Kenkyusho Nempo*, 1973, **25**, 1). Introduced by Sankyo Co. Ltd. (JPP 518 249; 532 202) as code no. 'F-319','SF-6505'; trade mark 'Tachigaren'.

**Properties.**

The technical grade (98% pure) forms colourless crystals; m.p. 86 °C; v.p. <133 mPa at 25 °C. Solubility at 25 °C: 85 g/l water; readily soluble in most organic solvents. It is stable and non-corrosive under alkaline conditions and comparatively stable under acidic conditions. It is stable to light and heat.

**Uses.**

It is a fungicide effective against soil-borne diseases caused by *Aphanomyces, Corticium, Fusarium*, and *Pythium* spp. and other fungi when applied as a soil drench at 30-60 g a.i./100 l to carnations, forest tree seedlings, paddy seedlings, sugar beet and other crops. It is also used as a seed treatment for sugar beet at 5-10 g/kg seed. It is effective as a plant growth promoter.

**Toxicology.**

Acute oral $LD_{50}$: for rats 3909-4678 mg/kg; for mice 1968-2148 mg/kg; for chickens >1000 mg/kg. Acute percutaneous $LD_{50}$ for rats >10 000 mg/kg; for mice >2000 mg/kg. In 90-d feeding trials NEL for rats and mice was 2500-5000 mg/kg diet. $LC_{50}$ (48-h) for carp and Japanese killifish is >40 mg/l.

**Formulations.**

These include: liquid (30 g a.i./l); dust (40 g/kg); seed treatment (700 g/kg).

**Analysis.**

Product analysis is by glc. Residues may be determined by glc of a derivative (T. Nakamura *et al.*, *Anal. Methods Pestic. Plant Growth Regul.*, 1978, **10**, 215).

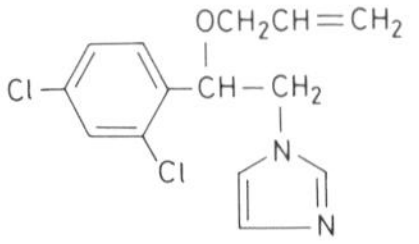

# Imazalil

$C_{14}H_{14}Cl_2N_2O$ (297.2)
T5N CNJ A1YO2U1&R BG DG

*imazalil hydrogen sulphate*
$C_{14}H_{16}Cl_2N_2O_5S$ (395.2)
T5N CNJ A1YO2U1&R BG DG
&&$H_2SO_4$ SALT 1:1

*imazalil nitrate*
$C_{14}H_{15}Cl_2N_3O_4$ (360.2)
T5N CNJ A1YO2U1&R BG DG
&&$HNO_3$ SALT 1:1

**Nomenclature and development.**

Common name imazalil (BSI, E-ISO, F-ISO, ANSI);chloramizol (Republic of South Africa);enilconazole (BPC). Chemical name (IUPAC) allyl 1-(2,4-dichlorophenyl)-2-imidazol-1-ylethyl ether; 1-(β-allyloxy-2,4-dichlorophenethyl)imidazole. (*C.A.*) 1-[2-(2,4-dichlorophenyl)-2-(2-propenyloxy)ethyl]-1*H*-imidazole (9CI); Reg. No. *[35554-44-0]* base, *[33586-66-2]* nitrate (1:1), *[60534-80-7]* sulfate (1:1). Its fungicidal properties were described by E. Laville (*Fruits*, 1973, **28**, 545). Introduced by Janssen Pharmaceutica as code no. 'R 23 979' for base, 'R 27 180' for hydrogen sulphate, 'R 18 531' for nitrate.

**Properties.**

Imazalil is a slightly yellowish to brownish oil; $d^{23}$ 1.2429; $n_D^{20}$ 1.5643; v.p. 9.3 μPa at 20 °C. Solubility: slightly soluble in water; freely in organic solvents. It is chemically stable at room temperature in the absence of light, and stable ≤*c.* 285 °C. Imazalil hydrogen sulphate is an almost colourless to beige-coloured powder, freely soluble in water, alcohols and slightly soluble in apolar organic solvents.

**Uses.**

Imazalil is a systemic fungicide effective against a wide range of fungi affecting fruit, ornamentals and vegetables. Having high activity against *Fusarium, Helminthosporium* and *Septoria* spp. it is recommended as a seed treatment for the control of cereal diseases. Storage decay on bananas, citrus and other fruit can be controlled by post-harvest spray or dip in water or wax emulsions. Imazalil is highly effective against benzimidazole-resistant strains of *Penicillium* spp. Typical use rates are: for seed treatment 4-5 g a.i./100 kg seed; for ornamentals and vegetables 5-30 g/100 l and for post-harvest treatment 2-4 g/t fruit.

**Toxicology.**

Acute oral $LD_{50}$ for rats 320 mg/kg. Acute percutaneous $LD_{50}$ for rats 4200-4880 mg/kg.

**Formulations.**

These include: 'Fungaflor' e.c. (200, 500 or 700 g a.i./l); imazalil hydrogen sulphate, s.p. (750 g base/kg). Liquid and powder seed treatments as mixtures: imazalil + anthraquinone + fenfuram + gamma-HCH; imazalil + fenfuram + gamma-HCH + guazatine triacetate.

**Analysis.**

Product analysis is by glc. Residues may be determined by glc with ECD or FID (R. Greenberg & C. Reswick, *Pestic. Sci.*, 1977, **8**, 59; G. R. Cayley *et al.*, *ibid.*, 1981, **12**, 103) hplc may also be used (*idem, ibid.*; R. Norman, *J. Assoc. Off. Anal. Chem.*, 1978, **61**, 1469; J. Wijnants, *Proc. Int. Soc. Citriculture, Orlando*, 1977, **3**, H-38).

It is the aim of the editors of *The Pesticide Manual* to present accurate and up-to-date information and its contents have been stored in a computer. This will simplify modification of the text for future editions. The contents and periodic updates of the *Manual* will be available on-line as the Pesticide Databank from 1984.

The easiest way to find a given compound is by its ***Entry Number***. Because the majority of pesticides are known by various synonyms it is recommended that the reader ascertains the required ***Entry Number*** from one or more of the Indexes.

**Index 1** Wiswesser Line-Formula Notation (for description of this see pp. 613).

**Index 2** Molecular formulae.

**Index 3** Code numbers used by official bodies, the manufacturers and suppliers. This index is arranged numerically and whether the qualifying letters are, for example, BAY or 'Bayer' is immaterial.

**Index 4** Chemical, common, and trivial names and trade marks.
Each of these give you the ***Entry Number***. A simple number (e.g. 1230) indicates that the entry is in the ***Main Section***, while a preceding asterisk (*2400) shows it is in the ***Superseded Compounds Section*** printed on coloured paper so that it is easy to find.

**Read the Section on How to Use Part I of the Manual,** page xvi.

Remember: an equals sign ( = ) at the end of a line means run on to the next line (omitting = ); chemical names ending in -, ( or [ run straight on. Otherwise the word finishes on a given line.

$C_{10}H_9NO_2$ (175.2)
T56 BMJ D1VQ

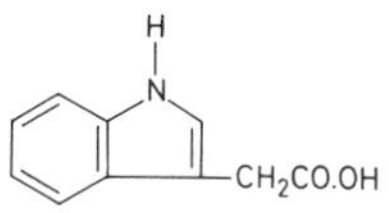

**Nomenclature and development.**

Chemical name (IUPAC) indol-3-ylacetic acid; β-indoleacetic acid. (*C.A.*) 1*H*-indole-3-acetic acid (9CI); indole-3-acetic acid (8CI); Reg. No. *[87-51-4]*. Trivial name IAA. It is the main naturally-occurring hormone in higher plants. trade mark 'Rhizipon A' (ACF Chemie Farma N.V.).

**Properties.**

It forms crystals; m.p. 168-170 °C. Solubility at room temperature: sparingly soluble in water; soluble in acetone, diethyl ether, ethanol; sparingly soluble in chloroform. Acidic p$K_a$ 4.75. Unstable in light.

**Uses.**

It is used to induce rooting in cuttings.

**Formulations.**

'Rhizopon A', tablets (each contains 50 mg a.i.); 'Rhizopon A', dust (5-10 g/kg).

$C_{12}H_{13}NO_2$ (203.2)
T56 BMJ D3VQ

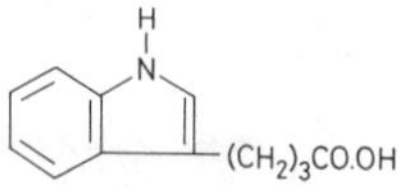

**Nomenclature and development.**

Chemical name (IUPAC) 4-indol-3-ylbutyric acid (draft BSI, E-ISO) is accepted in lieu of a common name. (*C.A.*) 1*H*-indole-3-butanoic acid (9CI); indole-3-butyric acid (8CI); Reg. No. *[133-32-4]*. Trivial name IBA. The ability of the acid and its simple esters to stimulate root formation in cuttings was reported by P. W. Zimmerman & F. Wilcoxon (*Contrib. Boyce Thompson Inst.*, 1935, **7**, 209) and by P. W. Zimmerman & A. E. Hitchcock (*ibid*, p. 439). Introduced by Union Carbide Corp. and by May & Baker Ltd (USP 3 051 723) as trade mark 'Seradix' (May & Baker); 'Rootone F' (Union Carbide) for mixture with 1-naphthylacetic acid.

**Properties.**

The technical acid forms colourless or slightly yellow crystals; m.p. 123-125 °C. Solubility: practically insoluble in water; soluble in acetone, diethyl ether, ethanol. It is non-flammable and non-corrosive.

**Uses.**

4-Indol-3-yl-butyric acid is used as an aid to rooting cuttings (J. van Overbeek, *Encyclopedia of Plant Physiology*, 1961, **14**, 1145).

**Toxicology.**

The intraperitoneal $LD_{50}$ for mice is 100 mg/kg (H. H. Anderson *et al., Proc. Soc. Exp. Biol. Med.*, 1936, **34**, 138).

**Formulations.**

These include the following powders: 'Seradix No. 1' (1 g a.i./kg) for soft wood, 'Seradix No. 2' (3 g/kg) for semi-hardwoods, 'Seradix No. 3' (8 g/kg) for hardwood. 'Rootone F', in combination with 1-naphthylacetic acid.

**Analysis.**

Methods include glc of a suitable volatile derivative (S. Spice, *Analyst (London)*, 1951, **76**, 664).

$C_8H_8Cl_2IO_3PS$ (413.0)
IR BG EG DOPS&O1&O1

**Nomenclature and development.**

Common name iodofenphos (BSI, F-ISO, BPC, ESA, Canada, New Zealand);jodfenphos (E-ISO). Chemical name (IUPAC) *O*-2,5-dichloro-4-iodophenyl *O,O*-dimethyl phosphorothioate. (*C.A.*) *O*-(2,5-dichloro-4-iodophenyl) *O,O*-dimethyl phosphorothioate (8 & 9CI); Reg. No. *[18181-70-9]*. OMS 1211. Its insecticidal and acaricidal properties were reported by B. C. Haddow & T. G. Marks (*Proc. Br. Insectic. Fungic. Conf., 5th*, 1969, **2**, 531). Introduced by Ciba AG (now Ciba-Geigy AG) (BEP 672 431; GBP 1 057 609) as code no. 'C 9491'; trade mark 'Nuvanol N'.

**Properties.**

Pure iodofenphos forms colourless crystals; m.p. 76 °C; v.p. 106 μPa at 20 °C; ρ 2.0 g/cm$^3$ at 20 °C. Solubility at 20 °C: <2 mg/l water; 450 g/l acetone, toluene; 810 g/l dichloromethane; 30 g/l methanol, octan-1-ol. Technical grade stable to >160 °C.

**Uses.**

It is a contact and stomach insecticide and acaricide for the control of flies and mosquitoes (adults and larvae) in public hygiene, indoor pests such as ants, bedbugs, cockroaches, fleas and warehouse pests. Also recommended for farm use, especially in poultry houses against mealworm beetles, mites and flies.

**Toxicology.**

Acute oral $LD_{50}$ for rats 2100 mg tech./kg. Acute percutaneous $LD_{50}$ for rats >2000 mg/kg; slight irritant to skin, non-irritant to eyes of rabbits. Inhalation $LC_{50}$ (6-h) for rats >246 mg/m$^3$ air. In 90-d feeding trials NEL was: for rats 5 mg/kg diet (0.38 mg/kg daily); for dogs 15 mg/kg diet (0.45 mg/kg daily). $LC_{50}$ (96-h) for rainbow trout 0.06-0.10 mg/l; for bluegill 0.42-0.75 mg/l. In laboratory trials it is slightly toxic to honeybees and non-toxic to birds.

**Formulations.**

'Nuvanol N' WP, w.p. (500 g a.i./kg); 'Nuvanol N' 500 FW, s.c. (500 g/l); 'Nuvanol N' 20 U, e.c. (200 g/l); 'Nuvanol N' 5 P ready-to-use powder (50 g/kg). Products sold under the trade mark '*Alfracon*' are no longer available.

**Analysis.**

Product analysis is by glc. Residues may be determined by glc with ECD or FPD (N. Burkhard & G. Voss, *Pestic. Sci.*, 1972, **3**, 183). Particulars are available from Ciba-Geigy AG.

$C_7H_3I_2NO$ (370.9)
QR BI FI DCN
*ioxynil-sodium*
$C_7H_2I_2NNaO$ (392.9)
QR BI FI DCN &&Na SALT
*ioxynil octanoate*
$C_{15}H_{17}I_2NO_2$ (497.1)
NCR CI EI DOV7

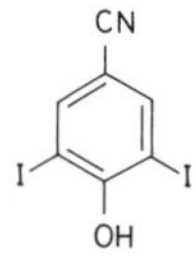

**Nomenclature and development.**

Common name ioxynil for parent compound (BSI, E-ISO, F-ISO, WSSA);ioxynil for octanoate ester (JMAF). Chemical name (IUPAC) 4-hydroxy-3,5-di-iodobenzonitrile (I); 4-hydroxy-3,5-di-iodophenyl cyanide. (*C.A.*) (I) (8 & 9CI); Reg. No. *[1689-83-4]* (ioxynil), *[2961-62-8]* (ioxynil-sodium), *[3861-47-0]* (ioxynil octanoate). The herbicidal properties of ioxynil were reported independently by R. L. Wain (*Nature (London)*, 1963, **200**, 28), by K. Carpenter & B. J. Heywood (*ibid.*, p. 28) and by R. D. Hart *et al.* (*Proc. Br. Weed Control Conf., 7th*, 1964. p. 3). Its development was reviewed by B. J. Heywood (*Chem. Ind. (London)*, 1966, p. 1946). Introduced by May & Baker Ltd. and by Amchem Products Inc. (now Union Carbide Agricultural Products Co. Inc.) (GBP 1 067 033 to May & Baker Ltd; USP 3 397 054 to Union Carbide) as code no. 'ACP 63-303' (Amchem for ioxynil),'M & B 8873' (May & Baker for ioxynil),'M & B 11 641' (May & Baker for octanoate),'15 830 RP' (Rhône-Poulenc for octanoate); trade marks 'Certrol' (Amchem for ioxynil and its esters), 'Mylone' (Union Carbide for the esters), 'Actril' (May & Baker for ioxynil and its octanoate), 'Totril' (May & Baker for octanoate).

**Properties.**

Pure ioxynil is a colourless solid; m.p. 209 °C. Solubility at 25 °C: 50 mg/l water; 70 g/l acetone; 20 g/l methanol; 340 g/l tetrahydrofuran. The technical grade (*c.* 95% pure) is a cream powder with a faint phenolic odour; m.p. 200 °C. It is stable and non-corrosive.

It is acidic p$K_a$ 3.96, forming salts (m.p. *c.* 360 °C): ioxynil-sodium (solubility at 20-25 °C: 140 g a.e./l water; 120 g/l acetone; 670 g/l 20% *m/V* aqueous acetone; 640 g/l 2-methoxyethanol; 650 g/l tetrahydrofurfuryl alcohol); ioxynil-potassium (solubility: 107 g a.e./l water, 60 g/l acetone; 560 g/l 20% *m/V* aqueous acetone; 770 g/l 2-methoxyethanol; 750 g/l tetrahydrofurfuryl alcohol).

Ioxynil forms stable esters of which the octanoate [4-cyano-2,6-di-iodophenyl octanoate (IUPAC and *C.A.* 9CI)], the butyrate and, occasionally, the heptanoate are used. Ioxynil octanoate is a cream, waxy solid; m.p. 59-60 °C. Solubility at 20-25 °C: practically insoluble in water; 100 g/l acetone; 650 g/l benzene, chloroform; 500 g/l cyclohexanone, xylene; 700 g/l dichloromethane; 150 g/l ethanol. It is stable to storage but is readily hydrolysed by alkali.

**Uses.**

Ioxynil and its salts and esters are contact herbicides with some systemic activity; ioxynil itself is the toxicant, inhibiting photosynthesis and uncoupling oxidative phosphorylation. Mainly they are used post-em. on annual broad-leaved weeds in cereals, leeks, onions, sugarcane and newly-sown turf. Frequently mixtures with other herbicides are used to extend the range of weeds controlled. Ioxynil is decomposed in soils (50% loss in *c.* 10 d). In plants hydrolysis of both the ester and nitrile groups occurs with some de-iodination.

$C_7H_3I_2NO$ (370.9)
QR BI FI DCN
*ioxynil-sodium*
$C_7H_2I_2NNaO$ (392.9)
QR BI FI DCN &&Na SALT
*ioxynil octanoate*
$C_{15}H_{17}I_2NO_2$ (497.1)
NCR CI EI DOV7

**Toxicology.**

Acute oral $LD_{50}$ for rats 110 mg a.i./kg, 120 mg formulated salt/kg, 190 mg formulated octanoate/kg; for mice 230 mg a.i./kg, 190 mg formulated salt/kg, 240 mg formulated octanoate/kg; for pheasants 75 mg a.i./kg, 35 mg formulated sodium salt/kg, 1000 mg octanoate/kg; for hens 200 mg a.i./kg, 120 mg formulated sodium salt/kg. Acute percutaneous $LD_{50}$ for rats >2000 mg a.i./kg, 210 mg formulated ioxynil salt/kg, >912 mg formulated octanoate/kg; transient irritation of the eyes of rabbits was observed. In 90-d feeding trials NEL for rats was 5.5 mg ioxynil-sodium/kg daily, 4.0 mg ioxynil octanoate/kg daily. $LC_{50}$ (48-h) for harlequin fish 3.3 mg ioxynil-sodium/l, 4.0 mg ioxynil octanoate/l. A spray of ioxynil sodium showed negligible contact toxicity to honeybees.

**Formulations.**

These include aqueous concentrates of salts 'Actrilawn'; e.c. of esters: 'Mate', 'Totril'. Mixtures include as salts: 'Actril 3' (ioxynil + dichlorprop + MCPA); 'Actril 4', 'Actril S', 'Dantril', 'Tetroxone', (ioxynil + bromoxynil + dichlorprop + MCPA); 'Certrol-Lin Onions', w.p. (ioxynil + linuron), 'Certrol PA' (ioxynil + dichlorprop + MCPA); 'Actril C', 'Iotox', (ioxynil + mecoprop). As octanoates and esters: 'Actril DS' (ioxynil + 2,4-D-isooctyl); 'Brittox' (ioxynil + bromoxynil + mecoprop-isooctyl); 'Certrol H' (ioxynil + mecoprop); 'Oxytril CM' (ioxynil + bromoxynil); 'Oxytril P' (ioxynil + bromoxynil + dichlorprop-isooctyl).

**Analysis.**

Product analysis is by titration or by estimation of iodine (H. S. Segal & M. L. Sutherland, *Anal. Methods Pestic., Plant Growth Regul. Food Addit*., 1967, **5**, 423; *Anal. Methods Pestic. Plant Growth Regul*., 1972, **6**, 654). Residues may be determined by glc of derivatives (*loc. cit*., T. H. Byast *et al*., *Tech. Rep. ARC Weed Res. Organ*., No. 15 (2nd Ed.), p. 13) or by i.r. spectrometry as ioxynil.

$C_{13}H_{13}Cl_2N_3O_3$ (330.2)
T5NVNV EHJ AVMY1&1 CR CG EG

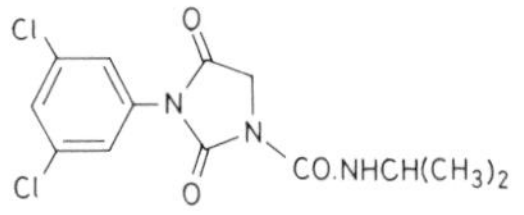

Cl O N N CO.NHCH(CH3)2 Cl O

**Nomenclature and development.**

Common name iprodione (BSI, E-ISO, F-ISO, ANSI). Chemical name (IUPAC) 3-(3,5-dichlorophenyl)-*N*-isopropyl-2,4-dioxoimidazolidine-1-carboxamide. (*C.A.*) 3-(3,5-dichlorophenyl)-*N*-(1-methylethyl)-2,4-dioxo-1-imidazolidinecarboxamide (9CI); Reg. No. *[36734-19-7]*. *Glycophene* rejected common name proposal. Its fungicidal properties were described by L. Lacroix *et al.* (*Phytiatr. Phytopharm.*, 1974, **23**, 165). Introduced by Rhône-Poulenc Phytosanitaire (GBP 1 312 536; USP 3 755 350; FP 2 120 222) as code no. '26 019 RP', 'ROP 500F', 'NRC 910', 'LFA 2043', 'FA 2071'; trade mark 'Rovral'.

**Properties.**

Iprodione forms colourless crystals; m.p. *c.* 136 °C; v.p. <133 μPa at 20 °C. Solubility at 20 °C: 13 mg/l water; 300 g/l acetone, acetophenone, anisole; 500 g/l dichloromethane, dimethylformamide, 1-methyl-2-pyrrolidone; 25 g/l ethanol, methanol.

**Uses.**

It is a fungicide which is particularly effective against *Botrytis, Monilia* and *Sclerotium* spp. It is also active against other fungi including *Alternaria, Corticium, Fusarium, Helminthosporium, Rhizoctonia, Typhula* spp. It is used mainly on cereals, soft fruit, deciduous tree fruits, ornamentals, vegetables, and vines at 0.5-1.0 kg a.i./ha, and on turf at 0.3-1.2 g/$m^2$. The dosage for seed treatment varies with the crop.

**Toxicology.**

Acute oral $LD_{50}$: for rats 3500 mg/kg; for mice 4000 mg/kg; for bobwhite quail 930 mg/kg; for mallard duck 10 400 mg/kg. No toxic effect was observed in dermal tests: for rats at 2500 mg/kg; for rabbits 1000 mg/kg. In 1.5-y feeding trials no ill-effect was observed: for rats at 1000 mg/kg diet; for dogs at 2400 mg/kg daily. It is practically non-toxic to honeybees.

**Formulations.**

'Rovral', w.p. (500 g a.i./kg).

**Analysis.**

Product analysis is by hplc or glc (L. Lacroix *et al.*, *Anal. Methods Pestic. Plant Growth Regul.*, 1980, **11**, 247). Residues may be determined by glc with ECD (*idem, ibid.*; R. Mestres *et al., Trav. Soc. Pharm. Montpellier*, 1979, **39**, 323).

$C_9H_{17}ClN_3O_3PS$ (313.7)
T5NN DNJ AY1&1 COPS&O2&O2 EG

**Nomenclature and development.**

Common name isazofos (BSI, E-ISO F-ISO). Chemical name (IUPAC) *O*-5-chloro-1-isopropyl-1*H*-1,2,4-triazol-3-yl *O,O*-diethyl phosphorothioate. (*C.A.*) *O*-[5-chloro-1-(1-methylethyl)-1*H*-1,2,4-triazol-3-yl] *O,O*-diethyl phosphorothioate (9CI); Reg. No. *[42509-80-8]*. Its insecticidal and nematicidal properties were described by D. Dawes *et al.*, (*Meded. Fac. Landbouwwet. Rijksuniv. Gent*, 1974, **39**, 727) and F. Bachmann & D. Dawes (*ibid.*, p.801). Developed by Ciba-Geigy AG (BEP 792 452; GBP 1 419 131; 1 419 132) as code no. 'CGA 12 223'; trade mark 'Miral'.

**Properties.**

Pure isazofos is a yellow liquid; b.p. 100 °C/0.001 mmHg; v.p. 4.3 mPa at 20 °C; ρ 1.22 g/cm$^3$ at 20°C. Solubility at 20 °C: 250 mg/l water; soluble in benzene, chloroform, hexane and methanol. At 20 °C, 50% hydrolysis (calculated) occurs in 85 d at pH5, 48 d at pH7 and 19 d at pH9. Decomposes at 210 °C.

**Uses.**

Isazofos is a soil-applied nematicide with contact and systemic properties for use on bananas and in turf against soil-dwelling insects.

**Toxicology.**

Acute oral $LD_{50}$ for rats 40-60 mg tech./kg. Acute percutaneous $LD_{50}$ for male rats >3100 mg/kg, for females 118 mg/kg; mild irritant to skin and minimal irritant to eyes of rabbits. Inhalation $LC_{50}$ (4-h) for rats 245 mg/m$^3$ air. In 90-d feeding trials NEL was: for rats 2 mg/kg diet (0.2 mg/kg daily); for dogs 2 mg/kg diet (0.05 mg/kg daily). $LC_{50}$ (96-h): for trout 0.008 mg/l; for carp 0.22 mg/l; for bluegill 0.01 mg/l. It is toxic to birds and honeybees.

**Formulations.**

These include 'Miral' 5G, 'Miral' 10G, granules (50 or 100 g a.i./kg).

**Analysis.**

Product analysis is by glc. Residues may be determined by glc with TID. Particulars are available from Ciba-Geigy AG.

# Isocarbamid 7380

$C_8H_{15}N_3O_2$ (185.2)
T5MVNTJ CVM1Y1&1

HN N–CO.NHCH$_2$CH(CH$_3$)$_2$ O

**Nomenclature and development.**

Common name isocarbamid (BSI, E-ISO);isocarbamide (F-ISO, WSSA). Chemical name (IUPAC) *N*-isobutyl-2-oxoimidazolidine-1-carboxamide. (*C.A.*) *N*-(2-methylpropyl)-2-oxo-1-imidazolidinecarboxamide (9CI); Reg. No. *[30979-48-7]*. Its herbicidal properties were reported by L. Eue *et al.*, (*Proc. Br. Weed Control Conf., 10th*, 1970, **2**, 610). Introduced by Bayer AG (DEAS 1 795 117; USP 3 875 180) as code no. 'BAY MNF 0166'; trade mark 'Merpelan AZ' (as mixture with lenacil).

**Properties.**

It is a colourless crystalline solid; m.p. 95-96 °C; v.p. <13 Pa at 50 °C. Solubility at 20 °C: 1.3 g/l water; 130 g/kg cyclohexanone; 281 g/kg dichloromethane. It is stable in alkaline and acid media.

**Uses.**

Isocarbamid is a selective herbicide for use in sugar beet and fodder beet at rates of 3-4 kg 'Merpelan AZ'/ha for pre-em. application.

**Toxicology.**

Acute oral $LD_{50}$: for male rats >2500 mg/kg; for female dogs >500 mg/kg. Acute percutaneous $LD_{50}$ for male rats >500 mg/kg. In 90-d feeding trials NEL was: for rats 800 mg/kg diet; for dogs >5000 mg/kg diet. Its toxicity to birds and fish is low; it is not harmful to honeybees.

**Formulations.**

'Merpelan AZ' ('Terratop' in UK), (Reg. No. *[57762-44-4]*), w.p. (630 g isocarbamid + 130 g lenacil/kg).

**Analysis.**

Details of product analysis are available from Bayer AG. Residues may be determined by glc (H. J. Jarczyk, *Pflanzenschutz Nachr. (Engl. Ed.)*, 1974, **27**, 130).

$C_{15}H_{24}NO_4PS$ (345.4)
1Y1&OVR BOPS&O2&MY1&1

$(CH_3)_2CHO.OC$ S
OPOCH$_2$CH$_3$
NHCH(CH$_3$)$_2$

**Nomenclature and development.**

Common name isofenphos (BSI, E-ISO);isophenphos (F-ISO). Chemical name (IUPAC) isopropyl *O*-[ethoxy-*N*-isopropylamino(thiophosphoryl)]salicylate; *O*-ethyl *O*-2-isopropoxycarbonylphenyl isopropylphosphoramidothioate. (*C.A.*) 1-methylethyl-2-[[ethoxy[(1-methylethyl)amino]phosphinothioyl]oxy]benzoate (9CI); isopropyl salicylate *O*-ester with *O*-ethyl isopropylphosphoramidothioate (8CI); Reg. No. *[25311-71-1]*. Its insecticidal properties were reported by B. Homeyer (*Meded. Fac. Landbouwwet., Rijksuniv. Gent*, 1974, **39**, 789). Introduced by Bayer AG (DEP 1 668 047) as code no. 'BAY SRA 12 869'; trade marks 'Oftanol', 'Amaze' (Mobay Chemical Corp.).

**Properties.**

It is a colourless oil; v.p. 530 μPa at 20 °C; $n_D^{20}$ 1.5165; $d_4^{20}$ 1.13. Solubility at 20 °C: 23.8 mg/kg water; >600 g/kg cyclohexanone, dichloromethane.

**Uses.**

Isofenphos is a contact and stomach insecticide and is translocated from roots to a limited extent in plants. It is effective against soil-dwelling insects when applied overall at 5 kg a.i./ha, against leaf-eating pests at 50-100 g/100 l, and is used on maize, rape and vegetables.

**Toxicology.**

Acute oral $LD_{50}$: for rats 28-38.7 mg/kg; for mice 91.3-127 mg/kg. Acute percutaneous $LD_{50}$ for rats >1.1 mg/kg. In 2-y feeding trials NEL for rats was 1 mg/kg diet. $LC_{50}$ for Japanese quail is 5-12.5 mg/kg diet. $LC_{50}$ (96-h) for goldfish is 2 mg/l.

**Formulations.**

These include: e.c. (500 g a.i./l); w.p. (400 g/kg); seed treatment; granules.

**Analysis.**

Product analysis is by glc. Residues may be determined by glc with FID (K. Wagner, *Pflanzenschutz-Nachr. (Engl. Ed.)*, 1976, **29**, 67).

# Isoprocarb



$C_{11}H_{15}NO_2$ (193.2)
1Y1&R BOVM1

$OCO.NHCH_3$
$CH(CH_3)_2$

**Nomenclature and development.**

Common name isoprocarb (BSI, E-ISO);isoprocarbe (F-ISO);MIPC (JMAF). Chemical name (IUPAC) *o*-cumenyl methylcarbamate (I); 2-isopropylphenyl methylcarbamate (*C.A.*) 2-(1-methylethyl)phenyl methylcarbamate (9CI); (I) (8CI); Reg. No. *[2631-40-5]*. OMS 32,ENT 25 670. Its insecticidal properties have been described (*Jpn. Pestic. Info.*, 1969, No. 1, p. 22). Introduced by Bayer AG and Mitsubishi Chemical Industries Ltd as code no. 'Bayer 105 807'; trade mark 'Etrofolan' (Bayer AG), 'Mipcin' (Mitsubishi).

**Properties.**

Technical grade isoprocarb is a colourless crystalline solid; m.p. 88-93 °C. Solubility: insoluble in water; readily soluble in acetone, methanol. Unstable in alkaline media.

**Uses.**

It is a contact insecticide effective against pests of rice at 0.5-1.0 kg a.i./ha, or 1.0-1.5 kg/ha for granular application to the water surface. Also for the control of aphids, capsid bugs, leaf hoppers and other pests of deciduous fruit and other crops.

**Toxicology.**

Acute oral $LD_{50}$: for rats 403-485 mg/kg; for mice 487-512 mg/kg; for guinea-pigs and rabbits *c.* 500 mg/kg. Acute percutaneous $LD_{50}$ for male rats $>$500 mg/kg; it is non-irritant to the eyes and ears of rabbits. In 90-d feeding trials NEL for rats was 300 mg/kg diet. $LC_{50}$ (48-h) for carp is 4.2 mg/l. It is harmful to honeybees.

**Formulations.**

These include: e.c. (200 g a.i./l); w.p. (500 g/kg); concentrate (150 g/l) for fogging; granules (50 g/kg); dusts.

**Analysis.**

Product analysis is by u.v. spectroscopy; details of methods are available from Bayer AG. Residues may be determined by glc (*Nihon Noyaku Gakkaishi*, 1978, **3**, 119).

$C_{15}H_{23}N_3O_4$ (309.4)
3N3&R BNW FNW DY1&1

**Nomenclature and development.**

Common name isopropalin (BSI, E-ISO, ANSI, WSSA);isopropaline (F-ISO). Chemical name (IUPAC) 4-isopropyl-2,6-dinitro-*N*,*N*-dipropylaniline. (*C.A.*) 4-(1-methylethyl)-2,6-dinitro-*N*,*N*-dipropylbenzenamine (9CI); 2,6-dinitro-*N*,*N*-dipropylcumidine (8CI); Reg. No. *[33820-53-0]*. Its herbicidal properties were described by L. R. Guse (*Proc. North Cent. Weed Control Conf.*, 1969, p. 44; G. J. Shoop, *ibid.*, p. 19). Introduced by Eli Lilly & Co. (USP 3 257 190) as code no. 'EL-179'; trade mark 'Paarlan'.

**Properties.**

Technical grade isopropalin is a red-orange liquid. Solubility at 25 °C: 0.1 mg/l water; >1 kg/l acetone, hexane and methanol. It is stable in use though susceptible to decomposition by u.v. irradiation in laboratory studies.

**Uses.**

It is a pre-plant soil-incorporated herbicide used at 1-2 kg a.i./ha to control broad-leaved and grass weeds in direct-seeded peppers and tomatoes and in transplanted tobacco.

**Toxicology.**

Acute oral $LD_{50}$ for mice and rats >5000 mg tech./kg. Chickens, dogs, mallard duck and rabbits receiving oral doses of 2000 mg/kg (highest tested) or bobwhite quail and Japanese quail receiving 1000 mg/kg suffered no fatality. Acute percutaneous $LD_{50}$ for rabbits >2000 mg/kg; it caused slight skin and eye irritation. In 90-d feeding trials NEL for rats and dogs was >250 mg/kg diet. $LC_{50}$ (96-h) is: for fathead minnow >0.1 mg/l; for goldfish >0.15 mg/l.

**Formulations.**

An e.c. (720 g a.i./l).

**Analysis.**

Analysis is by glc (W. S. Johnson & O. D. Decherer, *Anal. Methods Pestic. Plant Growth Regul.*, 1976, **8**, 369); particulars from Eli Lilly & Co.

$C_{12}H_{18}O_4S_2$ (290.4)
T5SYSTJ BUYVOY1&1&VOY1&1

$$\text{1,3-dithiolane ring}=C(CO.OCH(CH_3)_2)_2$$

**Nomenclature and development.**

Common name isoprothiolane (BSI, JMAF, draft E-ISO). Chemical name (IUPAC) di-isopropyl 1,3-dithiolan-2-ylidenemalonate. (*C.A.*) bis(1-methylethyl) 1,3-dithiolan-2-ylidenepropanedioate (9CI); Reg. No. *[50512-35-1]*. Trivial name IPT. Its fungicidal properties were described by F. Araki *et al.* (*Proc. Insectic. Fungic. Conf., 8th*, 1975, **2**, 715). Introduced by Nihon Nohyaku Co. Ltd as code no. 'SS 11 946'; trade mark 'Fuji-one'.

**Properties.**

It forms colourless crystals; m.p. 50-54.5 °C; b.p. 167-169 °C/0.5 mmHg. Solubility at 20 °C 48 mg/l water. The mobility of various analogues in rice plants has been reported (M. Uchida *Pestic. Biochem. Physiol.* 1980, **14**, 249).

**Uses.**

It is a systemic fungicide effective against *Pyricularia oryzae* of rice at 3.2-4.8 kg a.i./ha. The population of planthoppers on the treated rice was markedly decreased by an 'insectistatic' action (*idem, ibid.*).

**Toxicology.**

Acute oral $LD_{50}$: for male rats 1190 mg/kg; for male mice 1340 mg/kg. $LC_{50}$ (48-h) for carp is 6.7 mg/l.

**Formulations.**

Granules (120 g a.i./kg), for application in paddy water.

**Analysis.**

Product analysis is by glc with FID (T. Hattori & M. Kanauchi, *Anal. Methods Pestic. Plant Growth Regul.*, 1978, **10**, 229). Residues may be determined by glc with ECD (*idem, ibid.*).

$C_{12}H_{18}N_2O$ (206.3)
1Y1&R DMVN1&1

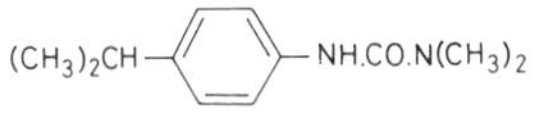

$(CH_3)_2CH-C_6H_4-NH.CO.N(CH_3)_2$

**Nomenclature and development.**

Common name isoproturon (BSI, E-ISO, F-ISO). Chemical name (IUPAC) 3-*p*-cumenyl-1,1-dimethylurea (I); 3-(4-isopropylphenyl)-1,1-dimethylurea. (*C.A.*) *N*,*N*-dimethyl-*N*'-[4-(1-methylethyl)phenyl]urea (9CI); (I) (8CI); Reg. No. *[34123-59-6]*. Introduced as a herbicide by Hoechst AG, Ciba-Geigy AG and Rhône-Poulenc Phytosanitaire (GBP 1 407 587 to Ciba-Geigy AG) as code no. 'Hoe 16 410','CGA 18 731'; trade marks 'Alon', 'Arelon' (both to Hoechst AG), 'Graminon' (to Ciba-Geigy AG).

**Properties.**

Isoproturon is a colourless powder; m.p. 155-156 °C; v.p. 3.3 μPa at 20 °C. Solubility at 20 °C: 55 mg/l water; soluble in most organic solvents. It is stable to light, acid and alkali. The technical grade is 97% pure.

**Uses.**

It controls annual grasses, (*Alopecurus myosuroides, Apera spica-venti, Avena fatua* and *Poa annua*) and broad-leaved weeds in barley, rye, wheat at 1.0-1.5 kg a.i./ha.

**Toxicology.**

Acute oral $LD_{50}$ for rats 1826-2417 mg/kg; for Japanese quail 3042-7926 mg/kg. Acute percutaneous $LD_{50}$ for rats >3170 mg/kg. In 90-d feeding trials NEL was: for rats 400 mg/kg diet; for dogs 50 mg/kg diet. $LC_{50}$ (96-h) is: for carp 193 mg/l; for rainbow trout 240 mg/l.

**Formulations.**

These include: w.p. (500-800 g a.i./kg); 500 FW, s.c. (500 g/l). Combinations include: 'Arelon P' (Reg. No. *[55602-90-9]*) (isoproturon + mecoprop); 'Twin-Tak' (isoproturon + bromoxynil + ioxynil).

**Analysis.**

Product analysis is by titration. Residues may be determined by glc. Details of these methods are available from Hoechst AG or Ciba-Geigy AG.

$C_7H_{17}O_2PS_3$ (260.4)
1Y1&S2SPS&O1&O1

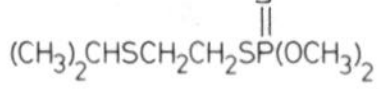

$(CH_3)_2CHSCH_2CH_2SP(=S)(OCH_3)_2$

**Nomenclature and development.**

Common name isothioate (BSI, JMAF, draft E-ISO). Chemical name (IUPAC) *S*-2-isopropylthioethyl *O,O*-dimethyl phosphorodithioate. (*C.A.*) *O,O*-dimethyl *S*-[2-[(1-methylethyl)thio]ethyl] phosphorodithioate (9CI); *O,O*-dimethyl *S*-[(2-isopropylthio)ethyl] phosphorodithioate (8CI); Reg. No. *[36614-38-7]*. Introduced by Nihon Nohyaku Co. Ltd (JPP 624 714) as trade mark 'Hosdon'.

**Properties.**

Isothioate is a light yellowish-brown liquid, with an aromatic odour; v.p. 293 mPa at 20 °C. Solubility at 25 °C 97 mg/l water.

**Uses.**

It is a systemic insecticide also active in the vapour phase. It is effective against aphids when used as a seed dressing or when applied to foliage at 1.0-1.5 kg/ha.

**Toxicology.**

Acute oral $LD_{50}$: for rats 150-170 mg/kg; for mice 50-80 mg/kg. Acute percutaneous $LD_{50}$ for male mice 240 mg/kg.

**Formulations.**

Seed treatment (330 g a.i./l); granules (50 g/kg).

**Analysis.**

Product analysis is by tlc with estimation of the phosphorus in the spots by standard colorimetric methods (*CIPAC Handbook*, 1983, **1B**, in press; T. Nakagawa & M. Kanauchi, *Anal. Methods Pestic. Plant Growth Regul.*, 1978, **10**, 75) or by glc (*idem, ibid.*). Residue analysis is by glc with FID (*idem, ibid.*).

$C_{13}H_{16}NO_4PS$ (313.3)
T5NOJ CR& EOPS&O2&O2

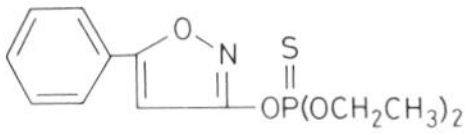

**Nomenclature and development.**

Common name isoxathion (BSI, JMAF, draft E-ISO). Chemical name (IUPAC) *O,O*-diethyl *O*-5-phenylisoxazol-3-yl phosphorothioate. (*C.A.*) *O,O*-diethyl *O*-(5-phenyl-3-isoxazolyl) phosphorothioate (8 & 9CI); Reg. No. *[18854-01-8]*. Its insecticidal properties were described by N. Sampei *et al.* (*Sankyo Kenkyusho Nempo*, 1970, **22**, 221). Introduced by the Sankyo Co. Ltd (JPP 525 850) as code no. 'E-48','SI-6711'; trade mark 'Karphos'.

**Properties.**

It is a yellowish liquid; b.p. 160 °C/0.15 mmHg; v.p. <133 μPa at 25 °C. Solubility at 25 °C: 1.9 mg/l water; readily soluble in organic solvents. It is unstable to alkali and decomposes at high temperatures.

**Uses.**

It is a contact insecticide with a wide range of action, controlling aphids and scale insects at 35-50 g a.i./100 l; also effective against borers, hoppers and gall midges of paddy rice and against beetles, caterpillars and mites on many crops.

**Toxicology.**

Acute oral $LD_{50}$: for rats 112 mg/kg; for mice 98.4 mg/kg. Acute percutaneous $LD_{50}$: for rats >450 mg/kg; for mice 193 mg/kg. In 90-d feeding trials rats and mice receiving 3.2 mg/kg diet showed no ill-effect. $LC_{50}$ (48-h) for carp is 2.13 mg/l.

**Formulations.**

An e.c. (500 g a.i./l); w.p. (400 g/kg); microgranule (30 g/kg); dusts (20 or 30 g/kg).

**Analysis.**

Product analysis is by glc with FID (T. Nakamura & K. Yamaoka, *Anal. Methods Pestic. Plant Growth Regul.*, 1978, **10**, 83). Residues may be determined by glc with FPD or FTD (*idem, ibid.*).

# Karbutilate 7600

$C_{14}H_{21}N_3O_3$ (279.3)
1X1&1&MVOR CMVN1&1

$OCO.NHC(CH_3)_3$

$NHCO.N(CH_3)_2$

**Nomenclature and development.**

Common name karbutilate (BSI, E-ISO, F-ISO, ANSI, WSSA). Chemical name (IUPAC) 3-(3,3-dimethylureido)phenyl *tert*-butylcarbamate. (*C.A.*) 3-[[(dimethylamino)carbonyl]aminophenyl] (1,1-dimethylethyl)carbamate (9CI); *tert*-butylcarbamic acid ester with 3-(*m*-hydroxyphenyl)-1,1-dimethylurea (8CI); Reg. No. *[4849-32-5]*. Its herbicidal properties were described by J. H. Dawson (*Bull. Wash. Agric. Exp. Stn.*, 1967, No. 691). Introduced by Agricultural Division of FMC Corp., who no longer manufacture or market it, and later by Ciba-Geigy AG (USP 3 434 822; 3 532 738; 3 801 625 to FMC Corp.) as code no. 'FMC 11 092','CGA 61 837'; trade mark 'Tandex' (Ciba-Geigy AG and formerly FMC Corp.).

**Properties.**

Pure karbutilate is a colourless solid; m.p. 176-176.5/187.3 °C; ρ 1.2 g/$cm^3$ at 20 °C; v.p. 6.0 nPa at 20 °C. Solubility at 20 °C: 325 mg/l water; <30 g/kg acetone, propan-2-ol, 3,5,5-trimethylcyclohex-2-enol, xylene; >250 g/kg dimethylformamide, dimethyl sulphoxide. On hydrolysis at 22 °C 50% loss occurs in 4.6 d at pH 8, stable in acid media.

**Uses.**

It is a non-selective herbicide for residual control of most annual and perennial broad-leaved weeds and grasses. It is primarily absorbed through the roots. For eradication weed control rates of 4-12 kg a.i./ha are recommended; for subsequent annual maintenance reduced rates and mixtures with simazine can be applied. It may also be used for the control of undesirable woody species in pastures and rangeland at 4-16 g/bush.

**Toxicology.**

Acute oral $LD_{50}$ for rats 3000 mg/kg. Acute percutaneous $LD_{50}$ for rats >15 400 mg/kg; slight irritant to eyes and non-irritant to skin of rabbits. In 90-d feeding trials NEL was: for rats 1000 mg/kg diet (70 mg/kg daily); for dogs 15 mg/kg daily. $LC_{50}$ is: for rainbow trout >135 mg/l; for bluegill >75 mg/l.

**Formulations.**

'Tandex' 80 WP (800 g a.i./kg). 'Erbotan plus' 20 G, granules (10 g karbutilate + 10 g thiazafluron/kg).

**Analysis.**

Product analysis is made by hplc. Residues are determined by hplc of karbutilate or the corresponding phenol (R. F. Cook, *Anal. Methods Pestic. Plant Growth Regul.*, 1976, **8**, 381; J. F. Lawrence, *ibid.*, 1982, **12**, 196).

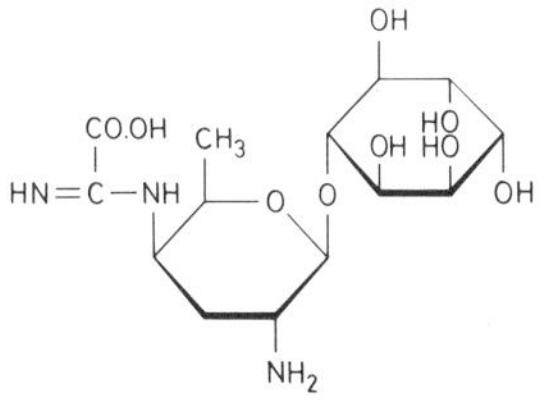

$C_{14}H_{25}N_3O_9$ (379.4)
T6OTJ B1 CMYVQUM EZ FO- AL6TJ
BQ CQ DQ EQ FQ
*kasugamycin hydrochloride hydrate*
$C_{14}H_{28}ClN_3O_{10}$ (433.8)
T6OTJ B1 CMYVQUM EZ FO- AL6TJ
BQ CQ DQ EQ FQ &&HCl SALT $H_2O$

**Nomenclature and development.**

Common name kasugamycin (JMAF). Chemical name (IUPAC) 1L-1,3,4/2,5,6-1-deoxy-2,3,4,5,6-pentahydroxycyclohexyl-2-amino-2,3,4,6-tetradeoxy-4-(α-iminoglycino)-α-D-*arabino*-hexopyranoside; [5-amino-2-methyl-6-(2,3,4,5,6-pentahydroxycyclohexyloxy)tetrahydropyran-3-yl]amino-α-iminoacetic acid. (*C.A.*) D-3-*O*-[2-amino-4-[(1-carboxyiminomethyl)amino]-2,3,4,6-tetradeoxy-α-D-*arabino*-hexopyranosy]-D-*chiro*-inositol (9CI); 3-*O*-[2-amino-4-[(1-carboxyformidoyl)amino]-2,3,4,6-tetradeoxy-α-D-arabino-hexopyranosyl]inositol (8CI); Reg. No. *[6980-18-3]* (kasugamycin), *[19408-46-9]* (kasugamycin monohydrochloride). Kasugamycin was dicovered by H. Umezawa *et al.* (*J. Antibiot. (Tokyo)*, 1965, **18**, 101) and its fungicidal properties described by T. Ishiyama *et al.* (*ibid.*, p. 115). Introduced by the Institute of Microbial Chemistry and by Hokko Chemical Industry Co. Ltd (JPP 6818(67); BEP 657 659; GBP 1 094 566) as trade mark 'Kasumin'.

**Properties.**

Produced by fermentation of *Streptomyces kasugaensis*. Kasugamycin hydrochloride hydrate forms colourless crystals, m.p. 202-204 °C (decomp.). Solubility at 25 °C: 125 g/l water; <1 mg/kg acetone, xylene; 2.76 g/kg methanol. At 50 °C 50% loss occurs in 47 d at pH5, in 14 d at pH9.

**Uses.**

It is a fungicide with specific action against *Pyricularia oryzae* on rice at 20 g a.i./ha; also effective against many plant diseases such as *Cercospora beticola* of sugar beet, *Erwinia carotovora* of potato, *Fulvia fulva* of tomato, *Pseudomonas phaseolicola* of kidney bean, *Pseudomonas lachrymans* of cucumber, *Pseudomonas glumae* of rice and *Xanthomonas citri* of citrus (in combination with copper oxychloride).

**Toxicology.**

Acute oral $LD_{50}$: for male rats 22 g a.i./kg; for male mice 21 g/kg; for male Japanese quail >4 g/kg. Acute percutaneous $LD_{50}$: for rats >4 g/kg; for mice >10 g/kg. Applications (24-h) of tech. to eyes and skin of rabbits caused no irritation. In 90-d feeding trials NEL was 100 mg/kg daily. In rats there was no teratogenicity nor effect on reproduction and no mutagenic activity. $LC_{50}$ (48-h) for carp and goldfish >40 mg/l; $LC_{50}$ (6-h) for waterflea >40 mg/l.

**Formulations.**

These include: dust (2 g a.i./kg); w.p. (20 g/kg); liquid (20 g/l); granules (20 g/kg). Mixtures include: 'Kasuran', 'Kasumin' - Bordeaux, w.p. (50 g kasugamycin + 756 g copper oxychloride). 'Kasumiron', 'Hokumiron', 'Kasubaron', dust 1 g kasugamycin + 20 g bis(2,4-dichlorophenyl) ethyl phosphate/kg; w.p. (10 g kasugamycin + 250 phosphate ester/kg).

**Analysis.**

Product analysis is by cup assay with *Pseudomonas fluorescens* (NIHJ B-254), and residues are determined by cup assay with *Piricularia oryzae* (P2) (*J. Antibiot. (Tokyo)*, 1968, **21**, 49).

$C_{18}H_{28}O_2$ (276.4)
1Y1&3Y1&2U1Y1&U1VO2UU1 &&(E,E)
FORM

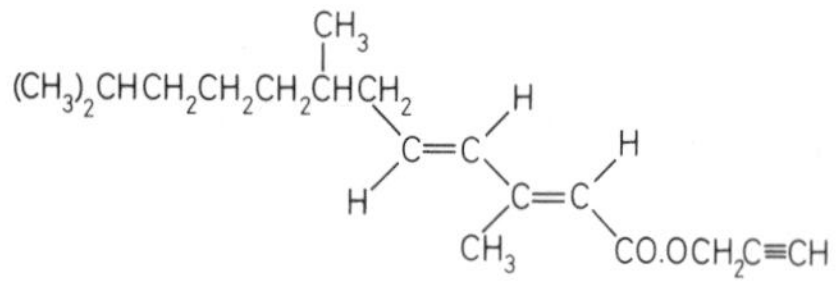

$(CH_3)_2CHCH_2CH_2CH_2CH(CH_3)CH_2$–CH=CH–C($CH_3$)=CH–CO.OCH$_2$C≡CH

**Nomenclature and development.**

Common name kinoprene (BSI, E-ISO, F-ISO);*exception* (France). Chemical name (IUPAC) prop-2-ynyl (±)-(*E,E*)-3,7,11-trimethyldodeca-2,4-dienoate. (*C.A.*) 2-propynyl (*E,E*)-3,7,11-trimethyl-2,4-dodecadienoate (8 & 9CI); Reg. No. *[42588-37-4]*. Introduced by Zoecon Corp. (now owned by Sandoz AG) (USP U 021 461) as code no. 'ZR 777'; trade mark 'Enstar'.

**Properties.**

Kinoprene is an amber liquid; b.p. 115-116 °C/0.04 mmHg; v.p. 956 μPa at 20 °C; ρ 0.921 g/cm$^3$; flash point (Tag closed cup) 40.5 °C. Solubility: 5.2 mg/l water; soluble in most organic solvents. It is stable on storage but not persistent in the environment and is decomposed by sunlight.

**Uses.**

It is an insect growth regulator effective against Homoptera, including aphids, mealybugs, scale insects and whiteflies. It is used in glasshouses against these pests, which are often difficult to control by conventional means, on ornamental or vegetable seed crops. It is phytotoxic to some open flowers but safer to foliage.

**Toxicology.**

Acute oral $LD_{50}$ for rats 4900 mg/kg. Acute percutaneous $LD_{50}$ for rabbits 9000 mg/kg.

**Formulations.**

'Enstar', e.c. (653 g/l).

$C_{13}H_{18}N_2O_2$ (234.3)
T56 FMVNVT&J H- AL6TJ

**Nomenclature and development.**

Common names lenacil (BSI, E-ISO, ANSI, WSSA, JMAF);lenacile (F-ISO). Chemical name (IUPAC) 3-cyclohexyl-1,5,6,7-tetrahydrocyclopentapyrimidine-2,4(3*H*)-dione. (*C.A.*) 3-cyclohexyl-6,7-dihydro-1*H*-cyclopentapyrimidine-2,4(3*H*,5*H*)-dione (8 & 9CI); Reg. No. *[2164-08-1]*. 3-Cyclohexyl-5,6-trimethyleneuracil. Its herbicidal properties were described by G. W. Cussans (*Proc. Br. Weed Control Conf., 7th*, 1964, **2**, 671). Introduced by E. I. du Pont de Nemours & Co. (Inc.) (USP 3 235 360) as code no. 'Du Pont 634'; trade mark 'Venzar'.

**Properties.**

Pure lenacil forms colourless crystals; m.p. 315.6-316.8 °C; *d* 1.32. Solubility at 25 °C: 6 mg/l water; *c*. 4 g/kg cyclohexanone; *c*. 8 g/kg dimethylformamide; *c*. 6 g/kg dimethyl sulphoxide; *c*. 2 g/kg xylene. Stable ≤ 316 °C, in water and aqueous acids. With sodium hydroxide forms salt of limited solubility; decomposed by hot alkali. Undergoes microbial decomposition in moist soil.

**Uses.**

It is a herbicide absorbed by the roots and inhibits photosynthesis. It is used at pre-planting with incorporation or as a pre-em. treatment of fodder, red and sugar beets at 0.6-1.2 kg a.i./ha, treatment is broadcast or band (the latter based on actual soil area treated), shallow incorporation into the soil improves efficacy under dry conditions and allows reduction of the rates; for pre-em. treatment of flax at 0.4-1.0 kg/ha, and of spinach, strawberries and various ornamentals at 1.0-2.0 kg/ha. Used in tank mixes with other herbicides for sugar beet.

**Toxicology.**

Acute oral $LD_{50}$ for rats >11 000 mg/kg (max. feasible dose, at which no deaths occurred). Acute percutaneous $LD_{50}$ for rabbits >5000 mg/kg (max. feasible dose, no clinical sign of toxicity or gross pathological change); w.p. produced a slight transient irritation to eyes of rabbits. In 2-y feeding trials with rats no adverse effect was observed. $LC_{50}$ (8-d) for bobwhite quail 2300 mg/kg diet. $Lc_{50}$ (96-h) for carp is 10 mg/l.

**Formulations.**

'Venzar', w.p. (800 g a.i./kg). 'Merpelan', (Reg. No. *[12738-05-5]*) w.p. (125 g lenacil + 600 g benzthiazuron/kg).

**Analysis.**

Residues may be determined by glc (H. L. Pease, *J. Sci. Food Agric.* 1966, **17**, 121) or by hplc (T. H. Byast *et al., Tech. Rep. ARC Weed Res. Organ.*, No. 15 (2nd Ed.), p. 49).

$C_9H_{10}Cl_2N_2O_2$ (249.1)
GR BG DMVN1&O1

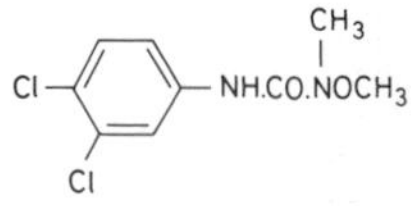

**Nomenclature and development.**

Common name linuron (BSI, E-ISO, F-ISO, ANSI, WSSA, JMAF). Chemical name (IUPAC) 3-(3,4-dichlorophenyl)-1-methoxy-1-methylurea (I). (*C.A.*) *N'*-(3,4-dichlorophenyl)-*N*-methoxy-*N*-methylurea (9CI); (I) (8CI); Reg. No. *[330-55-2]*. Its herbicidal properties were described by K. Härtel (*Meded. Landbouwhogesch. Opzoekingsstn. Staat Gent*, 1962, **27**, 1275). Introduced by E. I. du Pont de Nemours & Co. (Inc.) and by Hoechst AG (DEP 1 028 986; GBP 852 422 both to Hoechst AG) as code no. 'Du Pont Herbicide 326','Hoe 02810' (Hoechst AG); trade marks 'Lorox' (du Pont), 'Afalon' (Hoechst AG).

**Properties.**

Pure linuron forms colourless crystals; m.p. 93-94 °C; v.p. 2.0 mPa at 24 °C. Solubility at 24 °C: 81 mg/l water; soluble in acetone; moderately soluble in ethanol, aromatic hydrocarbons; sparingly soluble in aliphatic hydrocarbons. It is stable at its m.p. and in neutral aqueous media; hydrolysed in acid or alkaline media or at elevated temperatures. Microbial decomposition occurs in moist soils. The technical grade is $\geqslant$94% pure.

**Uses.**

It is a pre-em. and post-em. selective herbicide which inhibits photosynthesis. It is recommended for pre-em. use in asparagus, beans, cotton, maize, peas, potatoes, soyabeans, and for pre-em. or post-em. use in carrots and winter wheat.

**Toxicology.**

Acute oral $LD_{50}$: for rats 4000 mg/kg; for dogs *c.* 500 mg/kg. Acute percutaneous $LD_{50}$ for rabbits >5000 mg a.i. (as 50% w.p.)/kg. In 2-y feeding trials NEL for rats and dogs was 125 mg/kg diet (H. C. Hodge *et al.*, *Food Cosmet. Toxicol.*, 1968, **6**, 171), but in a recent study increases in the incidence of testicular tumours were noted in rats receiving 125 and 625 mg/kg diet but *not* at 50 mg/kg diet. $LC_{50}$ (8-d) was: for mallard duckling 3083 mg/kg diet; for Japanese quail >5000 mg/kg diet. $LC_{50}$ (96-h) for bluegill and rainbow trout is 16 mg/l.

**Formulations.**

'Lorox', w.p. (500 g a.i./kg); liquid formulations are also available. Mixtures include: 'Bronox', w.p. (linuron + trietazine); 'Certrol-Lin' (linuron + ioxynil); 'Chandor', 'Mudekan', 'Trinulan' (Reg. No. *[8070-92-6]*) (linuron + trifluralin); 'Profalon' (linuron + chlorpropham); linuron + cyanazine; linuron + diuron + terbacil; linuron + lenacil; linuron + monolinuron; linuron + terbacil.

**Analysis.**

Product analysis is by hydrolysis to 2,4-dichloroaniline, a derivative of which is measured colorimetrically, or by non-aqueous titration with perchloric acid (*CIPAC Handbook*, 1980, **1A**, 1281). Residues may be determined by hplc or glc (T. H. Byast *et al.*, *Tech. Rep. ARC Weed Res. Organ.*, No. 15 (2nd Ed.), p. 49; H. L. Pease *et al.*, *Anal. Methods Pestic., Plant Growth Regul. Food Addit.*, 1967, **5**, 433; *Anal. Methods Pestic. Plant Growth Regul.*, 1972, **6**, 659).

$C_{10}H_{19}O_6PS_2$ (330.3)
2OV1YVO2&SPS&O1&O1

$$(CH_3O)_2\overset{S}{\overset{\|}{P}}S\underset{COOCH_2CH_3}{\underset{|}{C}}HCH_2CO.OCH_2CH_3$$

**Nomenclature and development.**

Common name malathion (BSI, E-ISO, F-ISO, ESA, BPC);maldison (Australia, New Zealand);malathon (JMAF);mercaptothion (Republic of South Africa);carbofos (USSR);*exception* (Federal Republic of Germany). Chemical name (IUPAC) diethyl (dimethoxythiophosphorylthio)succinate; *S*-1,2-bis(ethoxycarbonyl)ethyl *O,O*-dimethyl phosphorodithioate. (*C.A.*) diethyl (dimethoxyphosphinothioyl)thiobutanedioate (9CI); diethyl mercaptosuccinate *S*-ester with *O,O*-dimethyl phosphorodithioate (8CI); Reg. No. *[121-75-5]*. OMS 1, ENT 17 034. Introduced by American Cyanamid Co. (USP 2 578 652) as code no. 'EI 4049'; trade mark 'Cythion' for deodorised grade, other marks include 'Malathion' in countries where this is not the common name.

**Properties.**

Technical grade malathion (*c.* 95% pure) is a clear amber liquid; m.p. 2.85 °C; b.p. 156-157 °C/0.7 mmHg; v.p. 5.3 mPa at 30 °C; $d_4^{25}$ 1.23; $n_D^{25}$ 1.4985. Solubility at room temperature: 145 mg/l water; miscible with most organic solvents; of limited solubility in petroleum oils (350 g light petroleum/l malathion).

**Uses.**

It is a non-systemic insecticide and acaricide of low mammalian toxicity. It is generally non-phytotoxic but may damage cucumber, string bean and squash under glasshouse conditions. In addition to a wide range of agricultural and horticultural uses, it is used to control animal ectoparasites, flies, household insects, human head and body lice, and mosquitoes. Metabolism is by hydrolysis of the carboxylate and phosphorodithioate esters and/or oxidation to the phosphorothioate (sometimes known as malaxon); methods for their separation have been reported (N. W. H. Houx *et al., Pestic. Sci.*, 1979, **10**, 185).

**Toxicology.**

Acute oral $LD_{50}$ for rats 2800 mg/kg. Acute percutaneous $LD_{50}$ (24-h) for rabbits 4100 mg/kg. In 1.75-y trials rats receiving 100 mg tech./kg diet showed normal weight gain. $LC_{50}$ (5-d) is: for bobwhite quail 3497 mg/kg diet; for ring-necked pheasant 4320 mg/kg diet (*USDI Wildlife Rep.*, 1972, No. 152, p. 36). $LC_{50}$ (96-h) is: for bluegill 0.103 mg/l; for largemouth bass 0.285 mg/l (*USDI Resource Pub.*, 1980, No. 157, p. 47). Topical toxicity for honeybees, $LD_{50}$ 0.71 µg/bee.

**Formulations.**

These include: e.c. (25-1000 g a.i./l); w.p. (250 or 500 g/kg); dusts (40 g/kg); ULV concentrates (920 g/l).

**Analysis.**

Product analysis is by glc (*CIPAC Handbook*, 1983, **1B**, in press). Residues on a wide range of crops may be determined by glc (*AOAC Methods*, 1980, 29.001-29.018; *Pestic. Anal. Man.*, 1979, **I**, 201-A, 201-G, 201-H, 201-I; D. C. Abbott *et al., Pestic. Sci.*, 1970, **1**, 10; J. Desmarchelier *et al., ibid.*, 1977, **8**, 473; *Analyst (London)*, 1973, **98**, 19; 1977, **102**, 858; 1980, **105**, 515; J. E. Boyd, *Anal. Methods Pestic. Plant Growth Regul.*, 1972, **6**, 418). Details are available from American Cyanamid Co.

$C_4H_4N_2O_2$ (112.1)
T6NMVJ FQ

**Nomenclature and development.**

Common name MH (WSSA);the trivial name maleic hydrazide (BSI, E-ISO, JMAF);hydrazide maléique (F-ISO) is accepted in lieu of a common name. Chemical name (IUPAC) 6-hydroxy-2*H*-pyridazin-3-one (the actual structure - P. D. Cradwick, *J. Chem. Soc. Perkin Trans. 2*, 1976, p. 1836); 1,2-dihydropyridazine-3,6-dione. (*C.A.*) 1,2-dihydro-3,6-pyridazinedione (I) (8 & 9CI); 6-hydroxy-3(2*H*)-pyridazinone (II) (8 & 9CI); Reg. No. *[123-33-1]* (I), *[10071-13-3]* (II). Its plant growth regulating properties were described by D. L. Schoene & O. L. Hoffmann (*Science*, 1949, **109**, 588). Introduced by U.S. Rubber Co. (now Uniroyal Chemical Inc.) (USP 2 575 954; 2 614 916; 2 614 917; 2 805 926) as trade mark 'MH-30'.

**Properties.**

The dry technical grade (>99% pure) is a colourless crystalline solid; m.p. 292-298 °C; $d^{25}$ 1.60. Solubility at 25 °C: 6 g/kg water; 24 g/kg dimethylformamide; 1 g/kg ethanol. It is a monobasic acid forming water-soluble alkali metal and amine salts but the calcium salt is precipitated by hard water. Maleic hydrazide is stable to hydrolysis but is decomposed by concentrated oxidising acids with formation of nitrogen.

**Uses.**

It is translocated in plants and inhibits cell division but not cell extension. It is used to retard the growth of grass, hedges and trees; to inhibit sprouting of beets, carrots, onions, potatoes and rutabagas; to prevent sucker development in tobacco. A mixture with 2,4-D is used to control broad-leaved weeds.

**Toxicology.**

Acute oral $LD_{50}$: for rats >5000 mg/kg; for mallard duck and bobwhite quail >10 000 mg/kg. Acute percutaneous $LD_{50}$ for rabbits >2000 mg/kg. Acute inhalation $LC_{50}$ for rats >20 mg/l air. In 2-y feeding trials, including a 3-generation reproduction study, rats receiving 50 000 mg sodium salt/kg diet showed no ill-effect. There was no oncogenic effect by the potassium salt in mice; in a 1-y trial on dogs with 20 000 mg sodium salt/kg diet mild reversible effects were noted. Maleic hydrazide gave negative results in mutagenicity tests. $LC_{50}$ (96-h) is: for bluegill 1608 mg/l; for rainbow trout 1435 mg/l; for *Daphnia* spp. 107 mg/l.

**Formulations.**

Royal 'MH-30', Royal 'Slo-Gro', aqueous solutions (160 g sodium salt/l), other liquids include: 'Burtolin', 'Regulox' (Diamond Shamrock); 'Mazide' (Synchemicals). 'Royal MH-30 SG', soluble granule (600 g maleic hydrazide/kg as potassium salt). 'BH 43' (Diamond Shamrock) solution (maleic hydrazide + 2,4-D).

**Analysis.**

Product analysis is by differential titration, details from Uniroyal Chemical. Residues may be determined by hydrolysis to hydrazine which is determined by colorimetry (*AOAC Methods*, 1980, 29.129-29.135; J. R. Lane, *J. Assoc. Off. Anal. Chem.*, 1963, **46**, 261; 1965, **48**, 744; *Anal. Methods Pestic., Plant Growth Regul. Food Addit.*, 1964, **4**, 147).

$$\left[-SCS.NHCH_2CH_2NHCS.S.Mn-\right]_x (Zn)_y$$

**Nomenclature and development.**

Common name mancozeb (BSI, E-ISO); mancozèbe (F-ISO); manzeb (JMAF). ISO definition is 'a complex of zinc and maneb containing 20% of manganese and 2.55% of zinc, the salt present being stated (for instance mancozeb chloride)'. Chemical name (IUPAC) manganese ethylenebis(dithiocarbamate) (polymeric) complex with zinc salt. (*C.A.*) [[1,2-ethanediylbis[carbamodithioato]](2-)]manganese mixture with [[1,2-ethanediylbis[carbamodithioato]](2-)]zinc (9CI); [ethylenebis[dithiocarbamato]]mang = anese mixture with [ethylenebis[dithiocarbamato]]zinc (8CI); Reg. No. *[8018-01-7]* (formerly *[8065-67-5]*). Its fungicidal properties were described in *Fungic. Nematic. Tests*, 1961, **17**. Introduced by Rohm & Haas Co. (GBP 996 264; USP 3 379 610; 2 974 156) as trade mark 'Dithane M-45'. Various other combination products are available based on zinc and manganese ethylenebis(dithiocarbamate)s, some with proportions close to the ISO definition of mancozeb; indexing policy of *Chemical Abstracts* is to list all mixtures of the same components under the same Registry Number.

**Properties.**

Mancozeb is a greyish-yellow powder, which decomposes without melting; flash point (Tag open cup) 137.8 °C. solubility: practically insoluble in water and most organic solvents. It is stable under normal storage conditions but decomposed at high temperatures by moisture, and by acid.

**Uses.**

It is a protectant fungicide generally used at 1.4-1.9 kg a.i./ha and effective against a wide range of foliage fungal diseases including those caused by *Phytophthora infestans* on potatoes and *Fulvia fulva* on tomatoes. It is used, in combined formulation, with zineb, against a wide range of foliage fungal diseases such as those caused by *Venturia* spp. on pome fruit and against various rust diseases. It is also used in combination with some systemic fungicides to increase the duration of protection given to the crop.

**Toxicology.**

Acute oral $LD_{50}$ for rats >8000 mg/kg. It may cause skin irritation on repeated exposure.

**Formulations.**

These include: 'Dithane M-45', 'Dithane 945' w.p. (≥800 g mancozeb/kg). 'Karamate' (mancozeb + zineb).

**Analysis.**

Product analysis is by decomposition with acid and measuring the carbon disulphide liberated, either by glc or by a titrimetric method (*CIPAC Handbook*, 1980, **1A**, 1288). Residues may be determined by reaction with acid of the liberated carbon disulphide being measured by glc or by colorimetry (*Analyst (London)*, 1981, **106**, 782; G. E. Keppel, *J. Assoc. Off. Anal. Chem.*, 1971, **54**, 528).

$(C_4H_6MnN_2S_4)_x$ $(265.3)x$
SUYSHM2MYUS&SH &&Mn COMPLEX

$\left[ -SCS.NHCH_2CH_2NHCS.S-Mn- \right]_x$

**Nomenclature and development.**

Common name maneb (BSI, E-ISO, JMAF);manèbe (F-ISO). Chemical name (IUPAC) manganese ethylenebis(dithiocarbamate) (polymeric). (*C.A.*) [1,2-ethanediylbis[carbamodithioato](2-)]manganese (9CI); [ethylenebis(dithiocarbamato)manganese] (8CI); Reg. No. *[12427-38-2]*. Trivial name MEB. ENT 14 875. Introduced by E. I. du Pont de Nemours & Co. (Inc.) and by Rohm & Haas Co. (USP 2 504 404; 2 710 822 both to du Pont) as trade marks 'Manzate' (to du Pont), 'Dithane M-22' (to Rohm & Haas).

**Properties.**

Maneb is a yellow crystalline solid which decomposes without melting on heating; *d* 1.92; v.p. negligible. Solubility: virtually insoluble in water and common organic solvents; dissolves in chelating agents (*e.g.* sodium salts of ethylenediaminetetra-acetic acid) but it cannot be recovered from such solutions. Decomposes on prolonged exposure to air or moisture and rapidly on contact with acids. Etem (*5660) is one of the products formed on contact with moisture (R. A. Ludwig *et al., Can. J. Bot.*, 1955, **33**, 42; C. W. Pluijgers *et al., Tetrahedron Lett.*, 1971, p. 1371).

**Uses.**

It is a protectant fungicide used at 1.5-2.0 g/l to control many fungal diseases of field crops, fruits, nuts, ornamentals, turf and vegetables, especially blights of potatoes and tomatoes and downy mildews of lettuce and vines. Frequently used in combination with other fungicides to prolong the duration of protection they would otherwise provide.

**Toxicology.**

Acute oral $LD_{50}$ for male rats 6750 mg tech./kg; ALD for rats and guinea-pigs 6400 mg/kg. In 2-y feeding trials no ill-effect was observed in rats receiving 250 mg/kg diet, 2500 mg/kg diet showed signs of toxicity; in 1-y trials with dogs, no effect was observed at 20 mg/kg daily but toxicity was observed at 75 mg/kg daily. $LC_{50}$ (8-d) for mallard duckling and bobwhite quail was $>$10 000 mg/kg diet. $LC_{50}$ (48-h) for carp is 1.8 mg/l.

**Formulations.**

A w.p. (800 g a.i./kg). Other metal salts, especially zinc, are included in some products *e.g.* 'Manzate D', 'Teresan'. Various mixtures with systemic fungicides are available.

**Analysis.**

Product analysis is by decomposition with acid and measurement of the liberated carbon disulphide, either by glc or by colorimetry of a derivative (*CIPAC Handbook*, 1970, **1**, 463; 1980, **1A**, 1293). Residues may be determined by reaction to form carbon disulphide which is measured by standard methods (G. E. Keppel *J. Assoc. Off. Anal. Chem.*, 1969, **52**, 162; 1971, **54**, 528; *Analyst (London)*, 1981, **106**, 782; *Pestic. Anal. Man.*, 1979, **II**).

$C_9H_9ClO_3$ (200.6)
QV1OR DG B1

Cl—C6H3(CH3)—OCH2CO.OH

**Nomenclature and development.**

Common name MCPA (BSI, E-ISO, F-ISO, WSSA);2,4-MCPA (France);metaxon (USSR);MCP (JMAF). Chemical name (IUPAC) 4-chloro-*o*-tolyloxyacetic acid; (4-chloro-2-methylphenoxy)acetic acid (I). (*C.A.*) (I) (9CI); [(4-chloro-*o*-tolyl)oxy]acetic acid (8CI); Reg. No. *[94-74-6]*. The discovery of its plant-growth regulating activity was described by R. E. Slade (*Nature* (*London*), 1945, **155**, 498). Introduced as a herbicide by ICI Plant Protection Division and, later, numerous other firms; trade marks include 'Agroxone' (ICI Plant Protection Division), 'Agritox' (May & Baker Ltd), 'Phenoxylene Plus' (FBC Limited), formerly '*Cornox M*' (The Boots Co. Ltd).

**Properties.**

Produced by condensation of sodium chloroacetate with 4-chloro-*o*-cresol (which may contain traces of 6-chloro-*o*-cresol). Pure MCPA is a colourless crystalline solid; m.p. 118-119 °C; v.p. 200 μPa at 21 °C. Solubility at room temperature: 825 mg/l; 1.5 kg/ethanol; 5 g/l heptane; 62 g/l toluene; 49 g/l xylene. The technical grade (85-99% pure) has m.p. 100-115 °C. The acid, $pK_a$ 3.07 forms water-soluble alkali metal and amine salts [MCPA-sodium, MCPA-potassium, MCPA-(dimethyl)ammonium]; though precipitation of calcium or magnesium salts may occur with hard water. Solutions of the alkali metal salts are alkaline in reaction and will corrode aluminium and zinc. Esters of MCPA [MCPA-(2-butoxyethyl)], are readily soluble in organic solvents.

**Uses.**

It is a systemic hormone-type selective herbicide, readily absorbed by leaves and roots. Its uses include the control of annual and perennial weeds in cereals, grassland and turf at 0.28-2.25 kg a.i./ha.

**Toxicology.**

Acute oral $LD_{50}$: for rats 700 mg/kg; for mice 550 mg/kg. Acute percutaneous $LD_{50}$ for rats >1000 mg/kg. In 210-d feeding trials rats receiving 100 mg/kg diet suffered slight kidney enlargement but no other ill-effect. Safe for fish at 10 mg/l.

**Formulations.**

These include aqueous concentrates of the salts (200-500 g/l); s.p. of salts (750-800 g a.e./l); e.c. of esters (200-500 g/l). 'Agritox' MCPA-potassium (250 g/l). There are numerous mixtures usually as salts or esters. 'Actril 4', 'Actril S', 'Dantril', 'Oxytril 4' (MCPA + bromoxynil + dichlorprop + ioxynil) as salts. 'Banlene Plus' (MCPA + dicamba + mecoprop). 'Buctril M' (MCPA + bromoxynil) as esters. 'Cambilene' (MCPA + dicamba + 2,3,6-TBA). 'Embutox Plus' (MCPA + 2,4-DB). 'Graslam' (MCPA + asulam + mecoprop) as salts. 'Harness' (MCPA + bromoxynil + mecoprop) as esters. 'Legumex Extra' (MCPA + benazolin + 2,4-DB). 'Mayclene' (MCPA + 3,6-dichloropicolinic acid + dichlorprop). 'New Legumex', 'Tropotox Plus (MCPA + MCPB).

**Analysis.**

Product analysis is by i.r. spectrometry (*CIPAC Handbook*, 1970, **1**, 483; 1980, **1A**, 1295), by liquid chromatography (*ibid.*, 1970, **1**, 477) or by glc of a derivative (*ibid.*, in press); FAO Specification (CP/48); *Herbicides 1977*, pp. 22-29. Residues may be determined by glc of a derivative (T. H. Byast *et al.*, *Tech. Rep. ARC Weed Res. Organ.*, No. 15 (2nd Ed.), p. 71; J. Lest, *Anal. Methods Pestic., Plant Growth Regul. Food Addit.*, 1967, **5**, 439; *Anal. Methods Pestic. Plant Growth Regul.*, 1972, **6**, 663).

$C_{11}H_{13}ClO_2S$ (244.7)
GR C1 DO1VS3

Cl–C6H3(CH3)–$OCH_2CO.SCH_2CH_3$

**Nomenclature and development.**

Common name MCPA-thioethyl (BSI, E-ISO, F-ISO);phenothiol (JMAF);*exception* (USA). Chemical name (IUPAC) *S*-ethyl 4-chloro-*o*-tolyloxythioacetate. (*C.A.*) *S*-ethyl (4-chloro-2-methylphenoxy)ethanethioate (9CI); *S*-ethyl [(4-chloro-*o*-tolyl)oxy]thioacetate (8CI); Reg. No. *[25319-90-8]*. Its herbicidal properties were described by T. Ohi *et al.* (*Zassu Kenkyo* 1969, **9**, 46). Introduced by Hokko Chemical Industry Co. Ltd (USP 3 708 278; GBP 1 263 169) as code no. 'HOK-7501'; trade marks 'Zero One', 'Herbit'.

**Properties.**

The technical grade (92% pure) forms brown crystals; m.p. 41-42 °C; b.p. 165 °C/7 mmHg; v.p. 21 mPa at 20 °C. Solubility at 25 °C: 2.3 mg/l water; >1 kg/l acetone, xylene; 290 g/l hexane. At 25 °C 50% loss occurs in 22 d at pH 7, in 2 d at pH 9. It is stable <200 °C.

**Uses.**

It is a hormone-type selective herbicide for the post-em. control of annual and perennial weeds, including *Chenopodium album, Convolvulus arvensis, Cyperus* spp., *Monochoria vaginalis, Polygonum avicular* and *Sagittaria pygmaea*, in orchards, rice paddies and wheat fields at 400-800 g a.i./ha.

**Toxicology.**

Acute oral $LD_{50}$: for male rats 790 mg a.i./kg, for females 880 mg/kg; for male mice 810 mg/kg, for females 750 mg/kg; for Japanese quail >1000 mg/kg. Acute intraperitoneal $LD_{50}$: for male rats 530 mg/kg, for females 570 mg/kg. Applications (24-h) of tech. to skin and eyes of rabbits produced no irritation. In 90-d feeding trials NEL for rats and mice was 300 mg/kg diet. No effect was observed on reproduction of or teratogenicity in rats. Not mutagenic. $LC_{50}$ (48-h) for carp is 2.5 mg/l; $LC_{50}$ (6-h) for waterfleas 40 mg/l.

**Formulations.**

These include: 'Zero One', granules (14 g a.i./kg); 'Herbit', e.c. (200 g/l). 'Grakill 1.5', granules (7 g MCPA-thioethyl + 15 g simetryne/kg).

**Analysis.**

Product and residue analysis is by glc. Details of methods are available from Hokko Chemical Industry Co. Ltd.

$C_{11}H_{13}ClO_3$ (228.7)
QV3OR DG B1
*MCPB-sodium*
$C_{11}H_{12}ClNaO_3$ (250.7)
QV3OR DG B1 &&Na SALT

Cl—(benzene ring)—O(CH2)3CO.OH, CH3

**Nomenclature and development.**

Common name MCPB (BSI, E-ISO, F-ISO, WSSA, JMAF);2,4-MCPB (France);2M-4Kh-M (USSR). Chemical name (IUPAC) 4-(4-chloro-*o*-tolyloxy)butyric acid (I); 4-(4-chloro-2-methylphenoxy)butyric acid. (*C.A.*) 4-(4-chloro-2-methylphenyl)butanoic acid (9CI); (I) (8CI); Reg. No. *[94-81-5]* (MCPB), *[6062-26-6]* (MCPB-sodium). Its herbicidal properties were reported by R. L. Wain & F. Wightman (*Proc. Roy. Soc.*, 1955, **142B**, 525). Introduced by May & Baker Ltd (GBP 758 980) as code no. 'MB 3046'; trade mark 'Tropotox'.

**Properties.**

Pure MCPB forms colourless crystals; m.p. 100 °C. Solubility at room temperature: 44 mg/l water; >200 g/l acetone; 150 g/l ethanol. The acid (p$K_a$ 4.84) forms water-soluble amine and alkali metal salts, though precipitation occurs with hard water. Technical grade MCPB (*c.* 92% pure) has m.p. 99-100 °C.

**Uses.**

It is a selective herbicide used post-em. to control broad-leaved annual and perennial weeds in undersown cereals, peas and established grassland at 1.7-3.4 kg a.e./ha. It owes its selectivity to the ability of susceptible plants to translocate it (R. C. Kirkwood *et al.*, *Pestic. Sci.*, 1972, **3**, 307) and to oxidise it to MCPA which is the real toxicant.

**Toxicology.**

Acute oral $LD_{50}$ for rats: 680 mg a.i./kg, 690 mg MCPB-sodium/kg.

**Formulations.**

'Tropotox', solution (400 g a.e./l). Mixtures include: 'Tropotox Plus' solution, MCPB + MCPA (300 g total a.e./l, as sodium/potassium salts); 'Tropotox Plus' 400 solution, same components (400 g total a.e./l).

**Analysis.**

Product analysis is by titration of extractable acids (*CIPAC Handbook*, in press). Residues may be determined by glc of a derivative with ECD (T. H. Byast *et al.*, *Tech. Rep. ARC Weed Res. Organ.*, No. 15 (2nd Ed.), p. 71; A. Guardiglii, *Anal. Methods Pestic. Plant Growth Regul.*, 1976, **8**, 397).

$C_{10}H_{20}NO_5PS_2$ (329.4)
2OVN1&V1SPS&O2&O2

$CH_3CH_2O.CON(CH_3)CO.CH_2SP(S)(OCH_2CH_3)_2$

**Nomenclature and development.**

Common name mecarbam (BSI, E-ISO, JMAF); mécarbame (F-ISO); *exception* (France). Chemical name (IUPAC) *S*-(*N*-ethoxycarbonyl-*N*-methylcarbamoylmethyl) *O,O*-diethyl phosphorodithioate. (*C.A.*) ethyl 6-ethoxy-2-methyl-3-oxo-7-oxa-5-thia-2-aza-6-phosphanonanoate 6-sulfide (9CI); formerly ethyl [[(diethoxyphosphinothioyl)thio]acet = yl]methylcarbamate (9CI); ethyl (mercaptoacetyl)methylcarbamate *S*-ester with *O,O*-diethyl phosphorodithioate (8CI); Reg. No. *[2595-54-2]*. Its insecticidal properties were described by M. Pianka (*Chem. Ind. (London)*, 1961, p. 324). Introduced by Murphy Chemical Ltd (now part of Dow Chemical Co.) and, later, by Takeda Chemical Industries Ltd (GBP 867 780) as code no. 'P 474','MC 474' (Murphy Chemical Ltd); trade marks 'Murfotox' (Murphy Chemical Ltd), 'Pestan' (Takeda Chemical Industries Ltd), and formerly '*Afos*'.

**Properties.**

Pure mecarbam is a light brown to pale yellow oil; b.p. 144 °C/0.02 mmHg; v.p. negligible at room temperature; $d_{20}^{20}$ 1.223; $n_D^{20}$ 1.5138. Solubility at room temperature: <1 g/l water; <50 g/kg in aliphatic hydrocarbons; miscible with alcohols, aromatic hydrocarbons, esters, ketones. The technical grade is ⩾85% pure, $d_{20}^{20}$ 1.223. Though subject to hydrolysis, it is compatible with all but highly alkaline pesticides. The technical product slowly attacks metals.

**Uses.**

It is an acaricide and insecticide with slight systemic properties. It is used to control scale insects and other Hemiptera, olive fly and other fruit flies at 60 g a.i./100 l; against leaf hoppers and stem flies of rice as a dust (15 g/kg); and against root fly larvae of cabbage, carrots, celery and onions. At recommended rates of use it persists in soil for 28-42 d.

**Toxicology.**

Acute oral $LD_{50}$: for rats 36-53 mg/kg; for mice 106 mg/kg. Acute percutaneous $LD_{50}$ for rats >1220 mg/kg. In 0.5-y feeding trials rats receiving 1.6 mg/kg daily suffered no ill-effect, but at 4.56 mg/kg daily there was a slight depression of growth rate.

**Formulations.**

These include: 'Murfotox', e.c. (500, 680 or 900 g a.i./l), w.p. (250 g/kg), dust (15 or 40 g/kg); 'Murfotox Oil' (50 g/l) in petroleum oil; 'Pestan', e.c. (400 g/l).

**Analysis.**

Product analysis is by glc with FID (V. P. Lynch, *Anal. Methods Pestic. Plant Growth Regul.*, 1976, **8**, 135). Residues may be determined by glc with ECD (*idem, ibid.*). Details of both methods are available from Murphy Chemical Ltd.

$C_{10}H_{11}ClO_3$ (214.6)
QVY1&OR DG B1

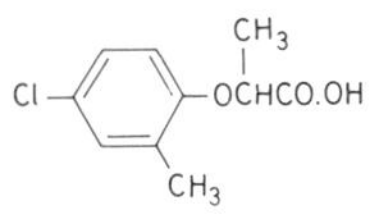

**Nomenclature and development.**

Common name mecoprop (BSI, E-ISO, F-ISO, WSSA); mechlorprop (Denmark); MCPP (JMAF). Chemical name (IUPAC) (±)-2-(4-chloro-*o*-tolyloxy)propionic acid; (±)-2-(4-chloro-2-methylphenoxy)propionic acid. (*C.A.*) 2-(4-chloro-2-methylphenoxy)propanoic acid (9CI); (±)-2-[(4-chloro-*o*-tolyl)oxy]propionic acid (8CI); Reg. No. *[7085-19-0]* (racemate), *[93-65-2]* (unstated stereochemistry). Trivial name CMPP. Its plant growth regulating activity was described by C. H. Fawcett *et al.* (*Ann. Appl. Biol.*, 1953, **40**, 232) and its uses as a herbicide by G. B. Lush & E. L. Leafe (*Proc. Br. Weed Control Conf., 3rd*, 1956, pp. 625, 633). Introduced by The Boots Co. Ltd (now FBC Limited) (GBP 820 180; 822 973; 825 875) as code no. 'RD 4593'; trade mark 'Iso-Cornox'.

**Properties.**

The commercial product is the racemate (in which doses are normally expressed) though only the (+)-form is active as a herbicide. Pure mecoprop forms colourless crystals; m.p. 94-95 °C. Solubility at 20 °C: 620 mg/l water; readily soluble in most organic solvents. It readily forms salts, many of which are very soluble in water: mecoprop-sodium (460 g/l at 15 °C); mecoprop-potassium (795 g/l at 0 °C); mecoprop-[bis(2-hydroxyethyl)]ammonium (580 g/l at 20 °C). Technical grade mecoprop has m.p. $\geq$90 °C. It is stable to heat, and resistant to hydrolysis, oxidation, reduction, and atmospheric oxidation. Mecoprop is corrosive to metals in the presence of moisture but solutions of mecoprop-potassium $<$80 °C and pH$\geq$8.6 do not corrode brass, iron or mild steel.

**Uses.**

It is a systemic hormone-type herbicide used post-em. to control chickweed, cleavers and other weeds in cereals at 1.5-2.7 kg/ha. It is mainly used in combination with other herbicides to extend the range of weeds controlled.

**Toxicology.**

Acute oral $LD_{50}$ for rats 930 mg/kg; for mice 650 mg/kg. In 21-d feeding trials rats receiving 65 mg/kg daily showed no ill-effect; those receiving 100 mg/kg diet for 210 d suffered only a slight enlargement of the kidneys.

**Formulations.**

These include aqueous solutions of mecoprop-potassium, mecoprop-sodium or amine salts; e.c. of esters; e.c. based on dried non-hygroscopic mecoprop salts. Mixtures include: 'Banlene Plus', 'Cornox Plus' (mecoprop + dicamba + MCPA). 'Cambilene' (mecoprop + dicamba + MCPA + 2,3,6-TBA). 'Springclene' (mecoprop + bromoxynil + ioxynil + linuron). Mecoprop + cyanazine. Mecoprop + 2,4-D. Mecoprop + 3,6-dichloropicolinic acid. Mecoprop + dichlorprop. Mecoprop + dichlorprop + 2,3,6-TBA. Mecoprop + fenoprop. Mecoprop + ioxynil.

**Analysis.**

Product analysis is by glc of a derivative (details from FBC Limted) or by titration of extractable acids (CIPAC Handbook, 1980, **1A**, 1297; 1983, **1B**, in press). Residues may be determined by glc of a derivative (T. H. Byast *et al.*, *Tech. Rep. ARC Weed Res. Organ.*, No. 15, (2nd Ed.), p. 71).

# Mefluidide



$C_{11}H_{13}F_3N_2O_3S$ (310.3)
FXFFSWMR B1 D1 EMV1

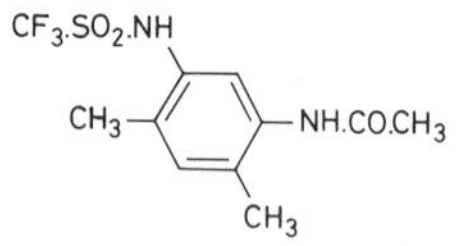

$CF_3.SO_2.NH$
$CH_3$ — NH.CO.$CH_3$
$CH_3$

**Nomenclature and development.**

Common name mefluidide (BSI, E-ISO, F-ISO, ANSI, WSSA). Chemical name (IUPAC) 5'-(trifluoromethanesulphonamido)acet-2',4'-xylidide. (*C.A.*) *N*-[2,4-dimethyl-5-[[(trifluoromethyl)sulfonyl]amino]phenyl]acetamide (9CI); Reg. No. *[53780-34-0]*. Its plant growth regulating properties were reported (*Proc. North Cent. Weed Control Conf.*, 1974). Introduced by the 3M Company (USP 3 894 078) as code no. 'MBR 12 325'; trade mark 'Embark'.

**Properties.**

Mefluidide is a colourless crystalline solid; m.p. 183-185 °C; v.p. <13 mPa at 25 °C; p$K_a$ 4.6. Solubility at 23 °C: 180 mg/l water; 350 g/l acetone; 310 mg/l benzene; 2.1 g/l dichloromethane; 310 g/l methanol; 17 g/l octan-1-ol. It is stable at elevated temperatures, but the acetamido moiety is hydrolysed on refluxing mefluidide in acidic or alkaline solutions. It is degraded in aqueous solutions exposed to u.v. radiation; aqueous solutions and suspensions are mildly corrosive to metals on prolonged exposure.

**Uses.**

It is a plant growth regulator and herbicide used: for the suppression of growth and seed production of turf grasses, trees and woody ornamentals; as an agent for enhancing sucrose content of sugarcane; for the control of growth and seed production of weeds in various crops, particularly rhizomatous *Sorghum halepense* in soyabeans. The optimum rates for these uses range from 0.3 to 1.1 kg a.i./ha.

**Toxicology.**

Acute oral $LD_{50}$: for mice >1920 mg tech./kg; for rats >4000 mg/kg; for mallard ducks >4640 mg/kg. Acute percutaneous $LD_{50}$ for rabbits >4000 mg/kg; it is mildly irritating to the eyes. No teratogenic effect was observed in rabbits and no mutagenic effect was observed in *Salmonella typhimurium*. $LC_{50}$ (5-d) for mallard duck and bobwhite quail (observed on the 8th d) was >10 000 mg/kg diet. $LC_{50}$ (96-h) for rainbow trout and bluegill is >100 mg/l.

**Formulations.**

'Embark' Plant Growth Regulator/Herbicide 2-S, solution (240 g a.i./l).

**Analysis.**

Analysis is by glc of a derivative; details may be obtained from 3M Company.

# Mephosfolan

$C_8H_{16}NO_3PS_2$ (269.3)
T5SYSTJ BUNPO&O2&O2 D1

**Nomenclature and development.**

Mephosfolan (BSI, E-ISO); méphospholan (F-ISO). Chemical name (IUPAC) diethyl 4-methyl-1,3-dithiolan-2-ylidenephosphoramidate; 2-(diethoxyphosphinylimino)-4-methyl-1,3-dithiolane. (*C.A.*) diethyl (4-methyl-1,3-dithiolan-2-ylidene)phosphoramidate (9CI); *P,P*-diethyl cyclic propylene ester of phosphonodithioimidocarbonic acid (8CI); Reg. No. *[950-10-7]*. ENT 25 991. Introduced by American Cyanamid Co. (GBP 974 138) as code no. 'EI 47 470'; trade mark 'Cytrolane'.

**Properties.**

Technical grade is a yellow to amber liquid; b.p. 120 °C/1 × $10^{-3}$ mmHg; $n_D^{26}$ 1.539. Solubility at 25 °C: 57 mg/l water; soluble in acetone, benzene, 1,2-dichloroethane, ethanol. Stable in water under neutral conditions; hydrolysed $2 < pH > 9$.

**Uses.**

It is a contact and stomach insecticide, with systemic activity following root or foliar absorption, and is used for the control of aphids, bollworms, mites, *Spodoptera* spp., stem borers and whiteflies on such major crops as cotton, maize, field crops, fruit and vegetables.

**Toxicology.**

Acute oral $LD_{50}$: for rats 3.9-8.9 mg tech./kg; for albino mice 11.3 mg/kg; for Japanese quail 12.8 mg/kg. Acute percutaneous $LD_{50}$ (24-h) for male albino rabbits: 28.7 mg tech./kg, >5000 mg (as 2% granules)/kg. In 90-d feeding trials with male albino rats receiving ≤15 mg/kg diet, no significant effect was observed on weight gain but there was a reduction in erythrocyte and brain cholinesterase activity. $LC_{50}$ (96-h) was: for trout 2.12 mg/l; for carp 54.5 mg/l. $LC_{50}$ (topical application) to honeybees is 3.51 μg/bee.

**Formulations.**

These include: e.c. (250 or 750 g a.i./l); granules (20-100 g/kg).

**Analysis.**

Product analysis is by u.v. spectroscopy. Residues may be determined by glc (N. R. Pasarela & E. J. Orloski, *Anal. Methods Pestic. Plant Growth Regul.*, 1973, **7**, 231; R. C. Blinn & J. E. Boyd, *J. Assoc. Off. Agric. Chem.*, 1964, **47**, 1106).

$C_7H_{16}N$ (114.2)
T6KTJ A1 A1
*mepiquat chloride*
$C_7H_{16}ClN$ (149.7)
T6KTJ A1 A1 &&Cl SALT

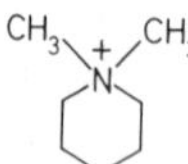


**Nomenclature and development.**

Common name mepiquat (BSI, E-ISO, F-ISO). Chemical name (IUPAC) 1,1-dimethylpiperidinium ion (I). (*C.A.*) 1,1-dimethylpiperidinium (9CI); (I) (8CI); Reg. No. *[15302-91-7]* (ion), *[24307-26-4]* (chloride). The plant growth regulating properties of mepiquat chloride (1,1-dimethylpiperidinium chloride) were described by B. Zeeh *et al.*, (*Kem-Kemi*, 1974, **1**, 621). It was introduced as a plant growth regulator by BASF AG (DEOS 2 207 575) as code no. 'BAS 08 300W'; trade marks 'Pix'; 'Terpal' (mixture with ethephon).

**Properties.**

Mepiquat chloride forms colourless hygroscopic crystals, decomposing at 285 °C. Solubility at 20 °C: >1kg/kg water; 20 g/kg acetone; 10.5 g/kg chloroform; 162 g/kg ethanol.

**Uses.**

It is used on cotton at early bloom or when plants are *c.* 60 cm high (whichever comes first) to reduce vegetative growth with beneficial, if any, effect on quantity or quality of fibre yield. Mixtures with ethephon strengthen the straw and reduce its length in winter barley.

**Toxicology.**

Acute oral $LD_{50}$ for rats 1490 mg tech. mepiquat chloride/kg. Acute percutaneous $LD_{50}$ for rats >7800 mg/kg. Non-toxic to wild fowl or fish.

**Formulations.**

'Pix' aqueous solution (50 g mepiquat chloride/l); 'Pix' ULV, concentrate (25 g/l). 'Terpal', (Reg. No. *[71587-73-0]*), mepiquat chloride + ethephon.

# Mercuric chloride

$Cl_2Hg$ (271.5)
.HG..G2

$HgCl_2$

**Nomenclature and development.**

The traditional chemical name mercuric chloride (E-ISO); chlorure mercurique (F-ISO) is accepted in lieu of a common name. Chemical name (IUPAC) mercury(II) chloride; mercury dichloride. (*C.A.*) mercury chloride ($HgCl_2$) (8 & 9CI); Reg. No. *[7487-94-7]*. Trivial name corrosive sublimate. Its use in crop protection was described by H. L. Bolley (*N. D. Agric. Exp. Stn. Bull.*, 1891, No. 4).

**Properties.**

It is a colourless crystalline powder; m.p. 277 °C; v.p. 18.6 mPa at 35 °C; *d* 5.32. Solubility at 20 °C: 69 g/l water; soluble in ethanol, diethyl ether, pyridine. It is unstable to alkalis which precipitate mercury chloride oxide; it is readily reduced chemically or by sunlight to mercury(II) chloride and to metallic mercury.

**Uses.**

In Canada it is used, in combination with mercurous chloride, as a fungicide on turf. Owing to its toxicity to animals, it is no longer used in most other countries.

**Toxicology.**

Acute oral $LD_{50}$ for rats 1-5 mg/kg.

**Formulations.**

'Merfusan', 'Mersil' (mercuric chloride + mercurous chloride).

**Analysis.**

Product analysis is by AOAC Methods; *CIPAC Handbook*, in press.

HgO (216.6)
.HG..O

HgO

**Nomenclature and development.**

The traditional chemical name mercuric oxide (E-ISO) is accepted in lieu of a common name;oxyde mercurique (F-ISO). Chemical name (IUPAC) mercury oxide. (*C.A.*) mercury oxide (HgO) (8 & 9CI); Reg. No. *[21908-53-2]*, formerly *[1344-45-2]*. Introduced by Sandoz AG as trade mark 'Santar'.

**Properties.**

Pure mercuric oxide is a red powder (yellow when finely divided) which darkens *c.* 400 °C and decomposes into its constituent elements at *c.* 500 °C. Solubility at 25 °C: 50 mg/l water; it dissolves in acids forming the corresponding mercury salts.

**Uses.**

Mercuric oxide is used as a paint against apple canker and as a protective seal on bark injuries and pruning cuts on fruit trees and ornamental shrubs and trees.

**Toxicology.**

It is extremely poisonous orally to all animals.

**Formulations.**

'Santar' paint (30 g a.i./kg).

**Analysis.**

Product analysis is by standard methods (titration with potassium iodide after dissolution in acid).

$Cl_2Hg_2$ (472.1)
.HG2.G2

$Hg_2Cl_2$

**Nomenclature and development.**

The traditional name mercurous chloride (E-ISO);chlorure mercureux (F-ISO) is accepted in lieu of a common name. Chemical name (IUPAC) mercury(I) chloride; dimercury dichloride. (*C.A.*) mercury chloride ($Hg_2Cl_2$) (8 & 9CI); Reg. No. *[7546-30-7]*. Trivial name calomel. It has long been used as an insecticide and, having a lower toxicity to mammals, has largely replaced mercuric chloride as recommended by H. Glasgow (*J. Econ. Entom.*, 1929, **22**, 335).

**Properties.**

It is a colourless powder, which sublimes at 400-500 °C; *d* 7.15. Solubility at 18 °C: 2 mg/l water; soluble in cold dilute acids, ethanol, most organic solvents. In the presence of water, it slowly dissociates to form mercury and mercuric chloride, a reaction accelerated by alkali.

**Uses.**

Mercurous chloride is limited to soil applications in crop protection use because of its phytotoxicity. It is recommended to control root maggots; club root of brassicas; white rot of onions; as a fungicide and moss-killer on turf.

**Toxicology.**

Acute oral $LD_{50}$ for rats is 210 mg/kg.

**Formulations.**

'Cyclosan' (May & Baker Ltd), Synchemicals 'M-C Turf Fungicide', dust (40 g/kg). 'Merfusan', 'Mersil' (mercurous chloride + mercuric chloride).

**Analysis.**

Product analysis is by oxidation with sulphuric acid and potassium nitrate followed by titration with potassium thiocyanate (*CIPAC Handbook*, 1970, **1**, 514; *AOAC Methods,* 1980, 6.132-6,135) or by titration with iodine and sodium thiosulphate (*ibid.*, 36.085-36.087). Residues may be determined by colorimetry of the complex with dithizone.

$C_{15}H_{21}NO_4$ (279.3)
1OVY1&NV1O1&R B1 F1 &&(DL) FORM

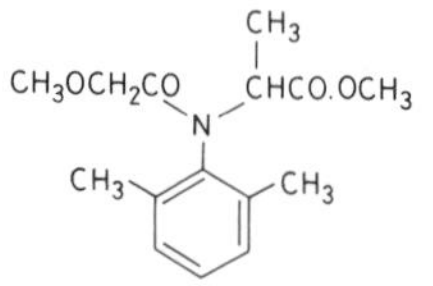

**Nomenclature and development.**

Common name metalaxyl (BSI, E-ISO, F-ISO, ANSI). Chemical name (IUPAC) methyl *N*-(2-methoxyacetyl)-*N*-(2,6-xylyl)-DL-alaninate. (*C.A.*) methyl *N*-(2,6-dimethylphenyl)-*N*-(2-methoxyacetyl)-DL-alaninate (9CI); Reg. No. *[57837-19-1]*. Its fungicidal properties were described by F. J. Schwinn *et al.* (*Mitt. Biol. Bundesanst. Land-Fortswirtsch. Berlin-Dahlem*, 1977, **178**, 145) and by P. A. Urech, (*Proc. 1977 Br. Crop. Prot. Conf. - Pests Dis.*, 1977, **2**, 623). Introduced by Ciba-Geigy AG (BEP 827 671; GBP 1 500 581) as code no. 'CGA 48 988'; trade marks 'Ridomil', 'Apron'; 'Fubol' (mixture with mancozeb).

**Properties.**

Pure metalaxyl forms colourless crystals; m.p. 71.8-72.3 °C; v.p. 293 μPa at 20 °C; ρ 1.21 g/cm³ at 20 °C. Solubility at 20 °C: 7.1 g/l water; 550 g/l benzene; 750 g/l dichloromethane; 650 g/l methanol; 130 g/l octan-1-ol; 270 g/l propan-2-ol. At 20 °C, 50% hydrolysis (calculated) occurs in >200 d at pH1, 115 d at pH9, 12 d at pH10. Stable ≤300 °C.

**Uses.**

It is a systemic fungicide suitable for the control of diseases caused by air and soil-borne Peronosporales on a wide range of temperate, subtropical and tropical crops. *Foliar sprays* with mixtures of metalaxyl and conventional protectant fungicides are recommended for the control of airborne diseases caused by *Pseudoperonospora humuli* on hops, *Phytophthora infestans* on potatoes, *Peronospora tabacina* on tobacco and *Plasmopara viticola* on vines. Metalaxyl alone is used as a *soil application* for the control of soil-borne pathogens causing root and lower stem rots on crops such as avocado and citrus. This method of use is also recommended for primary systemic infections of downy mildew on hops and in tobacco seed beds. It is used as a *seed treatment* for the control of systemic downy mildews on crops such as maize, peas, sorghum and sunflowers as well as damping-off (*Pythium spp.*) of various crops.

**Toxicology.**

Acute oral $LD_{50}$ for rats 669 mg/kg. Acute percutaneous $LD_{50}$ for rats >3100 mg/kg; slight irritant to eyes and skin of rabbits. In 0.5 y feeding trials NEL for dogs was 250 mg/kg diet (7.6 mg/kg daily); in 90-d trials NEL for rats was 250 mg/kg diet (17 mg/kg daily). $LC_{50}$ (96-h) is >100 mg/l for rainbow trout. carp, bluegill. In laboratory trials it is practically non-toxic to honeybees and birds.

**Formulations.**

'Ridomil' 5G, granules (50 g a.i./kg) for soil application. Seed treatments: 'Apron SD 35', (350 g/kg); 'Apron FW 350', s.c. (350 g/l). Combinations for foliar use include: 'Acylon Super F', 'Ridomil combi' (metalaxyl + folpet); 'Apron 70 SD' (350 g metalaxyl + 350 g captan/kg); 'Fubol', 'Ridomil MZ' (metalaxyl + mancozeb); 'Ridomil M' (metalaxyl + maneb); 'Ridomil plus', (Reg. No. *[71873-89-7]*), (metalaxyl + copper oxychloride).

**Analysis.**

Product analysis is by glc. Residues in plants and soil may be determined by glc with TID (D. J. Caverley & J. Unwin, *Analyst (London)*, 1980, **106**, 389). Particulars are available from Ciba-Geigy AG.

$C_8H_{16}O_4$ (176.2)
T8O CO EO GOTJ B1 D1 F1 H1

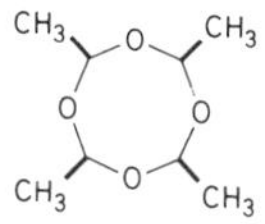

**Nomenclature and development.**

The name metaldehyde (BSI, E-ISO, F-ISO, JMAF) is accepted in lieu of a common name. Chemical name (IUPAC) *r*-2,*c*-4,*c*-6,*c*-8-tetramethyl-1,3,5,7-tetroxocane is the main component; former name 2,4,6,8-tetramethyl-1,3,5,7-tetraoxacyclo-octane (I). (*C.A.*) (I) or acetaldehyde homopolymer; Reg. No. *[108-62-3]* (for tetramer), *[9002-91-9]* (homopolymer). Trivial name metacetaldehyde. Its slug-killing properties were described by G. W. Thomas (*Gard. Chron.*, 1936, **100**, 453). Available from numerous suppliers under various trade marks: 'Slugit Pellets' (Murphy Chemical Co.), 'Slug Pellets', 'Mini Slug Pellets'.

**Properties.**

Produced by polymerisation of acetaldehyde, the technical grade contains higher oligomers but mainly the tetramer mentioned above. The pure tetramer forms colourless crystals; m.p. 246 °C (sealed tube), subliming at 110-120 °C. Solubility at 17 °C: 200 mg/l water; soluble in benzene, chloroform; sparingly soluble in diethyl ether, ethanol. It is subject to depolymerisation and should be stored with care, containers of soldered tinplate are unsuitable. It is flammable. It does not give the chemical tests for an aldehyde group.

**Uses.**

It is a molluscicide, effective against slugs which are immobilised and death follows exposure to low r.h. Neither acetaldehyde nor paraldehyde (a trimer) share the biological action.

**Toxicology.**

Acute oral $LD_{50}$ for dogs 600-1000 mg/kg. It is toxic to animals and humans so the bait, which is attractive to them, should be stored with care.

**Formulations.**

Slug baits (25-40 g a.i./kg) in a protein-rich base.

**Analysis.**

Product analysis is by conversion to acetaldehyde, which is estimated by reaction with sodium hydrogen sulphite and titration with iodine (*CIPAC Handbook*, 1970, **1**, 532) or by reaction with hydroxyammonium chloride and subsequent acid-base titration (*MAFF Ref. Bk.*, 1958, No. 1, p. 58).

# Metamitron

$C_{10}H_{10}N_4O$ (202.2)
T6NN DNVJ C1 DZ FR

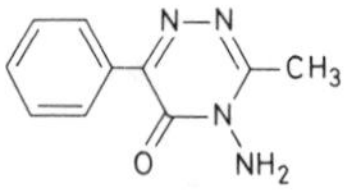

**Nomenclature and development.**

Common name metamitron (BSI, E-ISO); métamitrone (F-ISO); methiamitron (Belgium). Chemical name (IUPAC) 4-amino-3-methyl-6-phenyl-1,2,4-triazin-5(4*H*)-one (I); 4-amino-4,5-dihydro-3-methyl-6-phenyl-1,2,4-triazin-5-one. (*C.A.*) (I) (9CI); Reg. No. *[41394-05-2]*. Its herbicidal properties were described by R. R. Schmidt *et al.* (*3rd Int. Meeting Selective Weed Control in Beet*, 1975, **1**, 713). and H. Hack (*ibid.*, p. 729) and reviewed by H. Lembrich (*Pflanzenschutz-Nachr. (Engl. Ed.)*, 1978, **31**, 197). Introduced by Bayer AG (BEP 799 854; GBP 1 368 416) as code no. 'BAY DRW 1139'; trade mark 'Goltix'.

**Properties.**

Pure metamitron is a crystalline solid; m.p. 166.6 °C, v.p. 13 mPa up to 70 °C. Solubility at 20 °C: 1.8 g/l water; 10-50 g/kg cyclohexanone, dichloromethane. It is stable to acid media but unstable at pH>10.

**Uses.**

It is a herbicide with a high selectivity to sugar and fodder beets in which crops it is used to control broad-leaved and grass weeds. It is generally recommended at 3.5-5 kg a.i./ha. Selectivity may be due to the ability of some crop plants to deaminate metamitron (R. R. Schmidt & C. Fedtke, *Pestic. Sci.*, 1977, **8**, 611).

**Toxicology.**

Acute oral $LD_{50}$: for male rats 3343 mg/kg, for females 1832 mg/kg; for mice 1450-1463 mg/kg; for canaries >1000 mg/kg. Acute percutaneous $LD_{50}$ for rats >1000 mg/kg. In 2-y feeding trials NEL for rats was 250 mg/kg diet; in 90-d feeding trials NEL for dogs 500 mg/kg diet. $LC_{50}$ (96-h) for goldfish is >100 mg/l. It is not harmful to honeybees.

**Formulations.**

A w.p. (700 g a.i./kg).

**Analysis.**

Product analysis is by i.r. spectroscopy, details from Bayer AG. Residues in plants may be determined by glc, details from Bayer AG, and in soil by hplc (T. H. Byast *et al.*, *Tech. Rep. ARC Weed Res. Organ.*, No. 15 (2nd Ed.), p. 84).

$C_{14}H_{16}ClN_3O$ (277.8)
T5NNJ A1NV1GR B1 F1

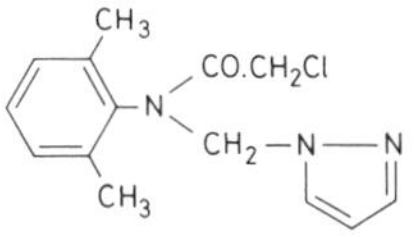

**Nomenclature and development.**

Common name metazachlor (BSI, draft E-ISO). Chemical name (IUPAC) 2-chloro-*N*-(pyrazol-1-ylmethyl)acet-2',6'-xylidide. (*C.A.*) 2-chloro-*N*-(2,6-dimethylphenyl)-*N*-(1*H*-pyrazol-1-ylmethyl)acetamide (9CI); Reg. No. *[67129-08-2]*. Introduced as a herbicide by BASF AG as code no. 'BAS 47 900H'; trade mark 'Butisan S'.

**Properties.**

Technical metazachlor is a beige solid; m.p. *c.* 85 °C. For the pure compound m.p. 85 °C; v.p. 49 μPa at 20 °C. Solubility at 20 °C: 17 mg/kg water; >1 kg/kg acetone, chloroform; 200 g/kg ethanol.

**Uses.**

It is a pre-em. herbicide for the control of grass and dicotyledonous weeds in potatoes, rape, soyabeans and tobacco.

**Toxicology.**

Acute oral $LD_{50}$ for rats 2150 mg/kg. Acute percutaneous $LD_{50}$ for rats >6810 mg/kg.

**Formulations.**

'Butisan S', s.c. (500 g a.i./l).

**Analysis.**

Details of methods for product and residue analysis are available from BASF AG.

$C_{10}H_{11}N_3OS$ (221.3)
T56 BN DSJ CN1&VM1

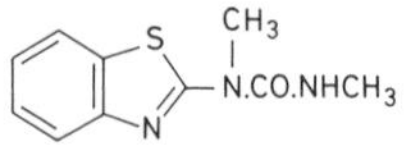

**Nomenclature and development.**

Common name methabenzthiazuron (BSI, E-ISO, F-ISO);*exception* Belgium, Canada, USA. Chemical name (IUPAC) 1-benzothiazol-2-yl-1,3-dimethylurea. (*C.A.*) *N*-2-benzothiazoly-*N*,*N*'-dimethylurea (9CI); 1-(2-benzothiazolyl)-1,3-dimethylurea (8CI); Reg. No. *[18691-97-9]*. Its herbicidal properties were described by H. Hack (*Pflanzenschutz-Nachr. (Engl. Ed.)*, 1969, **22**, 341). Introduced by Bayer AG (GBP 1 085 430) as code no. 'Bayer 74 283'; trade mark 'Tribunil'.

**Properties.**

It is a colourless crystalline solid; m.p. 119-120 °C; v.p. *c.* 133 μPa at 20 °C. Solubility at 20 °C: 59 mg/l water; 116 g/l acetone; *c.* 100 g/l dimethylformamide; 65.9 g/l methanol.

**Uses.**

It is a selective herbicide recommended for the control of blackgrass, meadow grass and certain other annual weeds pre-em. or early post-em. in winter cereals at 2.0-3.5 kg a.i./ha, in spring wheat and oats at 1.75-2.0 kg/ha, in broad beans and peas at 2.0-2.8 kg/ha. Safety to winter barley varieties and timing of applications have been reviewed (W. Kolbe, *ibid.*, 1974, **27**, 90; H. Hack *ibid.*, p.167). It is not recommended on crops under-sown with clover. The soil should be moist at the time of treatment for methabenzthiazuron is absorbed mainly through the roots; weeds die within 14-20 d after treatment but root-propagated weeds survive. It is broken down in the soil and has no after-effects on subsequent crops.

**Toxicology.**

Acute oral $LD_{50}$: for guinea-pigs, male rats >2500 mg/kg; for cats, dogs, female mice, rabbits >1000 mg/kg. Acute percutaneous $LD_{50}$ for female rats >500 mg/kg. In 2-y feeding trials rats receiving 150 mg/kg diet showed no ill-effect. $LC_{50}$ (96-h) is: for rainbow trout 15.9 mg/l; for golden orfe 29 mg/l.

**Formulations.**

'Tribunil', w.p. (700 g a.i./kg). 'Tribunil combi', (Reg. No. *[39283-72-2]*), methabenzthiazuron + dichlorprop.

**Analysis.**

Details of product analysis are available from Bayer AG. Residues in plants and soil may be determined by glc with FID (H. J. Jarczyk, *ibid.*, 1972, **25**, 21; T. H. Byast *et al.*, *Tech. Rep. ARC Weed Res. Organ.*, No. 15 (2nd Ed.), p.49) or by hplc (*idem, ibid.*).

$C_7H_{13}O_5PS$ (240.2)
1OVY1&U1PS&O1&O1

S
||
$(CH_3O)_2PO$ $CH_3$
C=C
H $CO.OCH_3$

**Nomenclature and development.**

Common name methacrifos (BSI, E-ISO F-ISO). Chemical name (IUPAC) methyl (*E*)-3-(dimethoxyphosphinothioyloxy)-2-methylacrylate; *O,O*-dimethyl (*E*)-*O*-2-methoxycarbonylprop-1-enyl phosphorothioate. (*C.A.*) methyl (*E*)-3-[(dimethoxyphosphinothioyl)oxy]-2-methyl-2-propenoate (9CI); methyl (*E*)-3-hydroxy-2-methylacrylate *O*-ester with *O,O*-dimethyl phosphorothioate (8CI); Reg. No. *[30864-28-9]* (unstated stereochemistry), *[62610-77-9]* ((*E*)-isomer). OMS 2005. Its insecticidal properties were reported by R. Wyniger *et al.*, (*Proc. 1977 Br. Crop Prot. Conf. - Pests Dis.*, 1978, **3**, 1033). Introduced by Ciba-Geigy AG (BEP 766 000; GBP 1 342 630) as code no. 'CGA 20 168' trade mark 'Damfin'.

**Properties.**

Pure methacrifos is a colourless liquid; b.p. 90 °C/0.01 mmHg; v.p. 160 mPa at 20 °C; ρ 1.225 g/cm³ at 20 °C. Solubility at 20 °C; 400 mg/l water; very soluble in benzene, dichloromethane, hexane, methanol. On hydrolysis at 20 °C, 50% loss (calculated) occurs in 66 d at pH1, 29 d at pH7, 9.5 d at pH9. Decomposes *c.* 200 °C.

**Uses.**

Methacrifos is an insecticide and acaricide with a vapour, contact and stomach action. Its main use is for the control of arthropod pests in stored products by mixture or surface treatment.

**Toxicology.**

Acute oral $LD_{50}$ for rats 678 mg/kg; practically non-toxic to Japanese quail. Acute percutaneous $LD_{50}$ for rats >3100 mg/kg; non-irritant to eyes and slight irritant to skin of rabbits. Inhalation $LC_{50}$ (6-h) for rats 2200 mg/m³ air.

**Formulations.**

These include e.c. (950 g a.i./l); ready-to-use spray (50 g/l); dust (20 g/kg).

**Analysis.**

Product analysis is by glc. Residues in grain may be determined by glc with FPD or TID. (J. Desmarchelier *et al.*, *Pestic. Sci.*, 1977, **8**, 473). Details of methods are available from Ciba-Geigy AG.

$C_2H_5NS_2$ (107.2)
SUYSHM1
*metham-sodium*
$C_2H_4NNaS_2$ (129.2)
SUYSHM1 &&Na SAI

$CH_3NHCS.SH$

**Nomenclature and devel nent.**

Common name met m (BSI);metam (Canada, New Zealand) in these cases for the free acid - the salt to be ed. The corresponding sodium salt is named as metam-sodium (E-ISO, F-ISO);karbat n (USSR);metham (WSSA);carbam (JMAF, for *both* the ammonium and sodium salts). Chemical name (IUPAC) methyldithiocarbamic acid (I); salt, sodium methyldithiocarbamate (II). (*C.A.*) methylcarbamodithioic acid (9CI), (I) (8CI); salt, sodium methylcarbamodithioate (9CI), (II) (8CI); Reg. No. *[144-54-7]* acid, *[137-42-8]* sodium salt. Trivial name SMDC (sodium salt). The fungicidal properties of metham-sodium were described by H. L. Klopping (Thesis, University of Utrecht, 1951) and by A. J. Overman & D. S. Burgis (*Proc. Fla. St. Hortic. Soc.*, 1956, **69**, 250). Metham-sodium was introduced by Stauffer Chemical Co. and later by E. I. du Pont de Nemours & Co. (Inc.) and ICI Plant Protection Division (neither of whom now market it) (USP 2 766 554; 2 791 605; GBP 789 690) all Stauffer Chemical Co. as code no. 'N-869' (Stauffer); trade marks 'Vapam' (Stauffer), '*VPM*' (formerly du Pont), 'Sistan' (ICI).

**Properties.**

Metham-sodium forms a colourless crystalline dihydrate (Reg. No. *[6734-80-1]*). Solubility at 20 °C: 722 g dihydrate/l; moderately soluble in ethanol; sparingly soluble in most other organic solvents. It is stable in concentrated solution but unstable when diluted, decomposition being promoted by soil, acids and heavy metal salts. The aqueous solution is corrosive to brass, copper and zinc.

**Uses.**

Metham-sodium is a soil fungicide, nematicide and herbicide with a fumigant action, applied at *c.* 1100 l 32.7% solution in 2000-12 000 l water/ha. Its activity is due to decomposition to methyl isothiocyanate (see 8390). It is phytotoxic and planting in treated soil must be delayed until decomposition and aeration are complete as shown by the normal germination of cress seed sown on a sample of treated soil (W. M. Morgan & M. S. Ledieu, *Pest Disease Control Handbook*, 1979, p. 7.4). Under moist conditions this occurs within 14 d.

**Toxicology.**

Acute oral $LD_{50}$: for male rats 1800 mg metham-sodium/kg, for females 1700 mg/kg. Acute percutaneous $LD_{50}$ for rabbits 1300 mg/kg; mildly irritating to skin, corrosive to eyes. Any contact with skin or organs should be treated as a burn.

**Formulations.**

Solutions: 'Polefume', 'Vapam' (382 g metham-sodium/l); 'Vapam B' (334 g/l); 'Sistan' (Universal Crop Protection).

**Analysis.**

Product analysis is by hydrolysis to liberate carbon disulphide, which is reacted and measured by titration with iodine (*CIPAC Handbook*, 1970, **1**, 537). Residues in crops and soils may be determined by analysing them for methyl isothiocyanate by glc. Analytical methods are available from the Stauffer Chemical Co.; see also R. A. Gray (*Anal. Methods Pestic., Plant Growth Regul. Food Addit.*, 1964, **3**, 177; *Anal. Methods Pestic. Plant Growth Regul.*, 1972, **6**, 717).

$C_2H_8NO_2PS$ (141.1)
1SPZO&O1

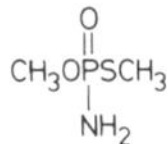


**Nomenclature and development.**

Common name methamidophos (BSI, E-ISO, F-ISO, ANSI). Chemical name (IUPAC) *O,S*-dimethyl phosphoramidothioate (I). (*C.A.*) (I) (8 & 9CI); Reg. No. *[10265-92-6]*. ENT 27 396. Its insecticidal properties were described by I. Hammann (*Pflanzenschutz-Nachr. (Engl. Ed.)*, 1970, **23**, 133). Introduced by Chevron Chemical Co. and by Bayer AG (USP 3 309 266; DEP 1 210 835 to Bayer AG) as code no. 'Ortho 9006' (Chevron), 'Bayer 71 628', 'SRA 5172' (Bayer); trade marks 'Monitor' (to Chevron), 'Tamaron' (to Bayer).

**Properties.**

Pure methamidophos is a solid; m.p. 44.5 °C; v.p. 40 mPa at 30 °C; $n_D^{40}$ 1.5092; $d^{44.5}$ 1.31. Solubility at 20 °C: >2 kg/l water; <100 g/l benzene, xylene; 20-25 g/l chloroform, dichloromethane, diethyl ether; <10 g/l kerosene. It is stable at ambient temperature but 50% is decomposed in 140 h at 40 °C and pH 2, in 120 h at 37 °C and pH 9. It decomposes before its b.p. is reached. The technical grade and concentrates are corrosive to mild steel and copper-containing alloys. Incompatible with alkaline pesticides.

**Uses.**

It is an acaricide and insecticide effective against a broad range of insect pests on many crops. It is systemic in action when applied to the base or trunk of deciduous trees; defoliation has occurred when applied as a foliage spray to deciduous fruit. It is especially useful on brassica crops, cotton, head lettuce and potatoes. At 0.5-1.0 kg/ha its contact effectiveness persists for 7-21 d.

**Toxicology.**

Acute oral $LD_{50}$: for rats and mice 30 mg/kg; for guinea-pigs 30-50 mg/kg; for rabbits 10-30 mg/kg; for hens 25 mg/kg; for bobwhite quail 57.5 mg/kg. Acute percutaneous $LD_{50}$ for male rats is 50-110 mg/kg. In 2-y feeding trials: dogs receiving 0.75 mg/kg daily showed no significant abnormalities; rats receiving 10 mg/kg diet showed no effect. $LC_{50}$ (96-h) is: for trout 51 mg/l; for guppies 46 mg/l.

**Formulations.**

These include: 'Monitor Spray Concentrate', water miscible conc. (717 g/l); 'Tamaron', w.s.c. (600 g/l).

**Analysis.**

Product analysis is by i.r. spectrometry or by glc (J. B. Leary, *Anal. Methods Pestic. Plant Growth Regul.*, 1973, **7**, 339). Residues may be determined by glc with FID (*idem, ibid.*; E. Möllhoff, *Pflanzenschutz-Nachr. (Engl. Ed.)*, 1971, **24**, 252; *Pestic. Anal. Man.*, 1979, **I**, 201-H, 201-I). Details of these methods are available from Chevron Chemical Co. and from Bayer AG.

$C_9H_6Cl_2N_2O_3$ (261.1)
T5NOVNVJ AR CG DG& D1

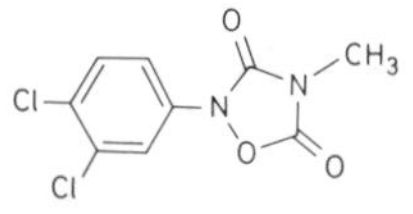

**Nomenclature and development.**

Common name methazole (BSI, ANSI, WSSA). Chemical name (IUPAC) 2-(3,4-dichlorophenyl)-4-methyl-1,2,4-oxadiazolidine-3,5-dione (I). (*C.A*) (I) (8 & 9CI); Reg. No. *[20354-26-1]*. Its herbicidal properties were described by W. Furness (*Proc. Int. Congr. Plant Prot., 7th*, Paris, 1970, p. 314). Introduced by Velsicol Chemical Corp. (USP 3 437 664) as code no. 'VCS-438'; trade mark 'Probe' (Velsicol Chemical Corp).

**Properties.**

Technical grade methazole (95% pure) is a light tan solid; m.p. 123-124 °C; $d^{25}$ 1.24; v.p. 133 μPa at 25 °C; it decomposes before the b.p. is reached. Solubility at 25 °C: 1.5 mg/l water; 40 g/l acetone; 6.5 g/l methanol; 55 g/l xylene. Solutions of methazole in methanol are decomposed by u.v. light but it is more stable in water when exposed to sunlight (G. W. Ivie *et al., J. Agric. Food Chem.*, 1973, **21**, 386).

**Uses.**

It is a selective herbicide effective against certain grasses and many broad-leaved weeds: pre-em. in garlic and potatoes (1-2 kg a.i./ha); as a directed spray on to soil or emerged weeds in citrus, stone-fruits, nuts, tea and established vines (6 kg/ha) and on cotton 15 cm tall (1-2 kg/ha), onions (>2 leaves) (2.5 kg/ha), newly seeded or winter-dormant lucerne (>1 y old) (2.25 kg/ha). It is preferably applied to moist soil; lower rates are applicable to light sandy soils.

**Toxicology.**

Acute oral $LD_{50}$ for rats 2500 mg/kg. Acute percutaneous $LD_{50}$ for rabbits >12 500 mg/kg; mildly irritating to their skin and eyes and chloracne was observed in rabbit-ear studies and among certain workers at manufacturing plants. It does not produce skin sensitisation in guinea-pigs. Acute inhalation $LD_{50}$ (4-h) in rats exposed to dust >200 mg/l. In 2-y feeding trials in rats and mice a yellow-brown pigment (of unknown significance) occurred in the spleen or liver at levels >100 mg/kg diet. $LC_{50}$ (8-d) was: for mallard duck 11 500 mg/kg diet; for bobwhite quail 1825 mg/kg diet. It is not a mutagen and not a teratogen in rabbits ⩽ 60 mg/kg daily, but foetal toxicity was observed ⩾ 30 mg/kg daily. In a 3-generation study in rats NEL was 50 mg/kg diet, cataracts were observed ⩾ 100 mg/kg diet but no other ill-effect; in a 1-generation study NEL for bobwhite quail and mallard duck was 3 mg/kg diet. $LC_{50}$ (96-h) for bluegill and rainbow trout is 4 mg/l.

**Formulations.**

A w.p. (750 g a.i./kg); granules (50 g/kg).

**Analysis.**

Product analysis is by i.r. spectrometry (D. M. Whiteacre *et al., Anal. Methods Pestic. Plant Growth Regul.*, 1978, **10**, 367). Residues may be determined by glc with ECD (*idem, ibid.*) or by hplc (*idem, ibid.*; T. H. Byast *et al., Tech. Rep. ARC Weed Res. Organ.*, No. 15 (2nd Ed.), p. 85).

$C_6H_{11}N_2O_4PS_3$ (302.3)
T5NNVSJ B1SPS&O1&O1 EO1

$CH_3O$ S O S
N—N—$CH_2SP(OCH_3)_2$

**Nomenclature and development.**

Common name methidathion (BSI, E-ISO, F-ISO, ANSI); DMTP (JMAF). Chemical name (IUPAC) *S*-2,3-dihydro-5-methoxy-2-oxo-1,3,4-thiadiazol-3-ylmethyl *O,O*-dimethyl phosphorodithioate. (*C.A.*) *S*-[(5-methoxy-2-oxo-1,3,4-thiadiazol-3(2*H*)-yl)methyl] *O,O*-dimethyl phosphorodithioate (9CI); *O,O*-dimethyl phosphorodithioate *S*-ester with 4-(mercaptomethyl)-2-methoxy-$\Delta^2$-1,3,4-thiadiazolin-5-one (8CI); Reg. No. *[950-37-8]*. OMS 844, ENT 27 193. Its insecticidal properties were described by H. Grob *et al.* (*Proc. Br. Insectic. Fungic. Conf., 3rd*, 1965, p. 451). Introduced by J. R. Geigy S.A. (now Ciba-Geigy AG) (BEP 623 246; GBP 1 008 451) as code no. 'GS 13 005'; trade marks 'Supracide', 'Ultracide Ciba-Geigy'.

**Properties.**

Pure methidathion forms colourless crystals; m.p. 39-40 °C; v.p. 186 μPa at 20 °C; ρ 1.495 g/cm$^3$ at 20 °C. Solubility at 20 °C: 250 mg/l water; 690 g/kg acetone; 850 g/kg cyclohexanone; 260 g/kg ethanol; 53 g/kg octan-1-ol; 600 g/kg xylene. Relatively stable to hydrolysis in neutral, or slightly acidic media, less stable in more acidic (pH1), or in alkaline media (pH13, 50% loss in 30 min at 25 °C).

**Uses.**

It is a non-systemic insecticide controlling a wide range of sucking and leaf-eating insects with a specific use against scale insects. The application rate for sucking insects is 30-60 g a.i./100 l on fruit, 250-800 g/ha on field crops. Foliar penetration enables it to be used against leafrollers. It is rapidly metabolised by plants and animals and excreted by the latter (H. O. Esser & P. W. Müller, *Experientia*, 1966, **22**, 36; H. O. Esser *et al., Helv. Chim. Acta*, 1968, **51**, 513).

**Toxicology.**

Acute oral $LD_{50}$ for rats 25-54 mg tech./kg. Acute percutaneous $LD_{50}$ for rats 1546 mg/kg; non-irritant to eyes, slight irritant to skin of rabbits. In 2-y feeding trials NEL was: for rats 4 mg/kg diet (0.15 mg/kg daily); for dogs 0.25 mg/kg daily. Human volunteers tolerated daily oral doses of up to 0.11 mg/kg for at least 42 d without reaction. $LC_{50}$ (96-h) is: for rainbow trout 0.01 mg/l; for bluegill 0.0022 mg/l. In laboratory trials it is toxic to birds in chronic oral application; slightly toxic to honeybees.

**Formulations.**

These include: 'Supracide 40'/'Ultracide 40 Ciba-Geigy', e.c. (400 g a.i./l), w.p. (400 g/kg); 'Supracide 20'/'Ultracide 20 Ciba-Geigy', e.c. (200 g/l), w.p. (200 g/kg); 'Supracide/'Ultracide Ulvair' 250. 'Supracide combi'/'Ultracide combi' 40 EC, e.c. (150 g methidathion + 250 g DDT/l).

**Analysis.**

Product analysis is by glc. Residues may be determined by tlc or glc (*Pestic. Anal. Manual*, 1979, **I**, 201-A, 201-G, 201-H, 201-I; M. A. Luke *et al., J. Assoc. Off. Anal. Chem.*, 1981, **64**, 1187; D. O. Eberle & R. Suter, *Anal. Methods Pestic. Plant Growth Regul.*, 1976, **8**, 141).

$C_{11}H_{15}NO_2S$ (225.3)
1SR B1 F1 DOVM1

OCO.NHCH3
CH3 CH3
SCH3

**Nomenclature and development.**

Common name methiocarb (BSI, Canada, New Zealand, Republic of South Africa, Turkey, ESA, E-ISO);methiocarbe (F-ISO);or mercaptodimethur (E-ISO, F-ISO, France, Federal Republic of Germany);*exception* (USA). Chemical name (IUPAC) 4-methylthio-3,5-xylyl methylcarbamate. (*C.A.*) 3,5-dimethyl-4-(methylthio)phenyl methylcarbamate (9CI); 4-(methylthio)-3,5-xylyl methylcarbamate (8CI); Reg. No. *[2032-65-7]*. OMS 93, ENT 25 726. Its insecticidal properties were reported by G. Unterstenhöfer (*Pflanzenschutz-Nachr. (Engl. Ed.)*, 1962, **15**, 181). Introduced by Bayer AG (FP 1 275 658; DEP 1 162 352) as code no. 'Bayer 37 344','H 321'; trade marks 'Mesurol', 'Draza'.

**Properties.**

Methiocarb is a colourless crystalline powder; m.p. 117-118 °C; v.p. 15 mPa at 60 °C. Solubility at 20 °C: 30 mg/l water; 500 g/l dichloromethane; 80 g/l propan-2-ol. It is hydrolysed in alkaline media.

**Uses.**

It is a non-systemic acaricide and insecticide with a broad range of action and good residual activity, used at 50-100 g a.i./100 l. It is also a powerful molluscicide used as pellets at 200 g a.i./ha; and bird repellent when used as a seed treatment (G. Hermann & W. Kolbe, *ibid.*, 1971, **24**, 279).

**Toxicology.**

Acute oral $LD_{50}$: for male rats 100 mg/kg; for guinea-pigs 40 mg/kg. Acute percutaneous $LD_{50}$ for male rats 350-400 mg/kg. In 1.67-y feeding trials no symptom was noted in rats receiving 100 mg/kg diet. $LC_{50}$ (96-h) is: for rainbow trout 0.64 mg/l; for carp 1.0-10 mg/l.

**Formulations.**

These include: 'Mesurol', w.p. (500 or 750 g a.i./kg); dust (30 g/kg); pellets for slug control (40 g/kg).

**Analysis.**

Product analysis is by u.v. spectrometry. Residues may be determined by u.v. spectrometry after hydrolysis to give 4-methylthio-3,5-xylenol (H. Niessen & H. Frehse, *ibid.*, 1963, **16**, 205). Residues of the parent compound and some of its metabolites including the corresponding sulphoxide and sulphone, may be determined by glc (M. C. Bowman & M. Beroza, *J. Assoc. Off. Anal. Chem.*, 1969, **52**, 1054).

$C_5H_{10}N_2O_2S$ (162.2)
1SY1&UNOVM1

$$\begin{matrix} CH_3S \\ \quad\diagdown \\ \quad\quad C{=}N{-}OCO.NHCH_3 \\ \quad\diagup \\ CH_3 \end{matrix}$$

**Nomenclature and development.**

Common name methomyl (BSI, E-ISO, F-ISO, ANSI, JMAF). Chemical name (IUPAC) *S*-methyl *N*-(methylcarbamoyloxy)thioacetimidate. (*C.A.*) methyl *N*-[[(methylamino)carbonyl]oxy]ethanimidothioate (9CI); methyl *N*-[(methylcarbomoyl)oxy]thioacetimidate. (8CI); Reg. No. *[16752-77-5]*. OMS 1196. Its insecticidal properties were described by G. A. Roodhans & N. B. Joy (*Meded. Rijksfac. Landbouwwet. Gent*, 1968, **33**, 833). Introduced by E. I. du Pont de Nemours & Co. (Inc.) (USP 3 576 834; 3 639 633) as code no. 'Du Pont 1179'; trade mark 'Lannate'.

**Properties.**

Methomyl, a mixture of (*Z*)- and (*E*)-isomers (the former predominating), forms colourless crystals, with a slight sulphurous odour; m.p. 78-79 °C; $d_4^{25}$ 1.2946; v.p. 6.65 mPa at 25 °C. Solubility at 25 °C: 58 g/kg water; 720 g/kg acetone; 420 g/kg ethanol; 1 kg/kg methanol; 30 g/kg toluene. Aqueous solutions decompose slowly at room temperature, more rapidly on aeration, in sunlight, in alkaline media or at higher temperatures. Decomposes rapidly in soil.

**Uses.**

It is used as a foliar spray and controls many insects including: aphids, *Bacoulutrix thurberiella, Heliothis* spp., *Spodoptera* spp. and *Trichoplusia ni* on field crops, certain fruit crops, ornamentals and vegetables. Following soil treatment, it is taken up by roots and translocated but this method is generally less effective than foliar applications.

**Toxicology.**

Acute oral $LD_{50}$: for male rats 17 mg a.i./kg, for females 24 mg/kg, for males 130 mg a.i. (as liquid formulation)/kg; for mallard duck 15.9 mg a.i./kg; for pheasant 15.4 mg/kg. Acute percutaneous $LD_{50}$ for male rabbits >5000 mg a.i./kg, 5880 mg a.i. (as liquid formulation)/kg; irritant to eyes of rabbits but not to skin of guinea-pigs. In 2-y feeding trials NEL for rats and dogs was 100 mg/kg diet. $LC_{50}$ (8-d): for bobwhite quail 3680 mg/kg diet; for Pekin duck 1890 mg/kg diet. $LC_{50}$ (96-h) is: for rainbow trout 3.4 mg/l; for bluegill 0.87 mg/l. Relatively non-toxic to honeybees once the spray had dried.

**Formulations.**

'Lannate', s.p. (900 g a.i./kg); 'Lannate L', water soluble liquid (220 g/l).

**Analysis.**

Product analysis is by hplc (J. E. Thean *et al.*, *J. Assoc. Off. Anal. Chem.*, 1978, **61**, 15; R. E. Leitch & H. L. Pease, *Anal. Methods Pestic. Plant Growth Regul.*, 1973, **7**, 331). Residues may be determined by glc with FPD, details available from E. I. du Pont de Nemours & Co. (Inc.).

# Methoprene



$C_{19}H_{34}O_3$ (310.5)
1Y1&OV1UY1&1U2Y1&3X1&1&O1
&&(E)-(E)-(RS) FORM

**Nomenclature and development.**

Common name methoprene (BSI, E-ISO, F-ISO, ANSI). Chemical name (IUPAC) isopropyl (*E-E*)-(*RS*)-11-methoxy-3,7,11-trimethyldodeca-2,4-dienoate. (*C.A.*) 1-methylethyl (*E,E*)-11-methoxy-3,7,11-trimethyl-2,4-dodecadienoate (9CI); Reg. No. *[40596-69-8]*. OMS 1697. Its insect growth-regulating properties were described by C. A. Hendrick *et al.* (*J. Agric. Food Chem.*, 1973, **21**, 354). Introduced by Zoecon Corp. (now part of Sandoz AG) (USP 3 904 662; 3 912 815) as code no. 'ZR 515'; trade marks 'Altosid', 'Apex', 'Kabat', 'Pharorid', 'Precor'.

**Properties.**

Methoprene is an amber liquid; b.p. 100 °C/0.05 mmHg; ρ 0.9261 g/cm$^3$ at 20 °C; v.p. 3.15 mPa at 25 °C. Solubility: 1.4 mg/l water; miscible with all common organic solvents. Flash point 187 °C (open cup). The technical grade is stable >4 y in glass in the dark at 43 °C.

**Uses.**

It is used as an insect growth regulator, exhibiting morphological rather than direct toxicity against target species. It is effective against numerous insect pests of plants, stored products, farm animals and in public health, especially Diptera, and is used against cattle-horn flies, indoor fleas, *Megaselia* spp., *Lasioderma serricorne* and *Ephestia elutella*.

**Toxicology.**

Acute oral $LD_{50}$: for rats >34 600 mg/kg; for dogs 5000-10 000 mg/kg. Acute percutaneous $LD_{50}$ for rabbits 3500 mg/kg; non-irritant to their eyes or skin. In 2-y feeding trials no methoprene-related effects were observed: with rats at 5000 mg/kg diet; with mice at 250 mg/kg diet. No effect was observed at the highest rates tested: in 3-generation reproduction studies in rats (2500 mg/kg diet); teratogenicity in rabbits (500 mg/kg) or rats (1000 mg/kg); mutagenicity in rats (2000 mg/kg).

**Formulations.**

These include: 'Altosid' SR-10, s.c. (103 g/l); 'Altosid Briquet', charcoal briquet (80 g/kg); 'Apex' 5E, 'Precor' 5E, e.c. (600 g/l); 'Kabat', alcohol-based concentrate (41 g/l); 'Precor' aerosol concentrate (1.5 g/l).

$C_{11}H_{21}N_5OS$ (271.4)
T6N CN ENJ BS1 DMY1&1 FM3O1

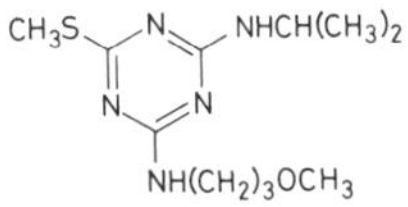

$CH_3S$, $NHCH(CH_3)_2$, $NH(CH_2)_3OCH_3$

**Nomenclature and development.**

Common name methoprotryne (BSI, E-ISO);métoprotryne (F-ISO). Chemical name (IUPAC) *N*-isopropyl-*N*'-(3-methoxypropyl)-6-methylthio-1,3,5-triazine-2,4-diyldiamine; 2-isopropylamino-4-(3-methoxypropylamino)-6-methylthio-1,3,5-triazine. (*C.A.*) *N*-(3-methoxypropyl)-*N*'-(1-methylethyl)-6-(methylthio)-1,3,5-triazine-2,4-diamine (9CI); 2-(isopropylamino)-4-[(3-methoxypropyl)amino]-6-(methylthio)-*s*-triazine (8CI); Reg. No. *[841-06-5]*. Its herbicidal properties were reported by A. Gast *et al.* (*Z. Pflanzenkr. Pflanzenpathol. Pflanzenschutz*, 1965, **72**, 325). Introduced by J. R. Geigy S.A. (now Ciba-Geigy AG) (BEP 584 306; GBP 927 348) as code number 'G 36 393'; trade mark 'Gesaran'.

**Properties.**

Pure methoprotryne is a colourless powder; m.p. 68-70 °C; v.p. 38 µPa at 20 °C; ρ 1.186 g/cm$^3$ at 20 °C. Solubility at 20 °C: 320 mg/l water; 450 g/l acetone; 650 g/l dichloromethane; 5 g/l hexane; 400 g/l methanol; 150 g/l octan-1-ol; 380 g/l toluene. At 21 °C p*K* is 4.0.

**Uses.**

Methoprotryne is a herbicide used post-em. for the control of weed grasses in winter-sown cereals at 1.5-2.0 kg a.i./ha.

**Toxicology.**

Acute oral $LD_{50}$ for rats >5000 mg tech./kg. Acute percutaneous $LD_{50}$ for rats >150 mg/kg. In 90-d feeding trials NEL was: for rats 3750 mg a.i./kg diet (*c.* 250 mg a.i./kg daily); for dogs 11 250 mg/kg diet (*c.* 375 mg/kg daily). $LC_{50}$ (96-h) is: for rainbow trout 8 mg/l; for crucian carp 31 mg/l; for bluegill 9.8 mg/l.

**Formulations.**

'Gesaran 25', w.p. (250 g a.i./kg).

**Analysis.**

Product analysis is by glc (*CIPAC Handbook*, 1980, **1A**, 1302). Residues may be determined by glc (K. Ramsteiner *et al., J. Assoc. Off. Anal. Chem.*, 1974, **57**, 192). Particulars are available from Ciba-Geigy AG.

$C_8H_9O_3PS$ (216.2)
T66 BOPO EHJ CS CO1

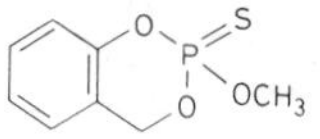

**Nomenclature and development.**

Common name salithion (JMAF). Chemical name (IUPAC) 2-methoxy-4*H*-1,3,2$\lambda^5$-benzodioxaphosphorine 2-sulphide; 2-methoxy-4*H*-benzo-1,3,2-dioxaphosphorin 2-sulphide. (*C.A.*) 2-methoxy-4*H*-1,3,2-benzodioxaphosphorin 2-sulfide (9CI); cyclic *O*,*O*-(methylene-*o*-phenylene) phosphorothioate *O*-methyl ester (8CI); Reg. No. *[3811-49-2]*. Its insecticidal properties were described by M. Eto & Y. Oshima (*Agric. Biol. Chem.*, 1962, **26**, 452). Introduced by Sumitomo Chemical Co. (JPP 467 645; 446 747; GBP 987 378; FP 1 360 130).

**Properties.**

The technical grade is a light yellow powder; m.p. 52.5-54 °C; v.p. 420 mPa at 20 °C. Solubility at 20 °C: 43 mg/l water; at 21-23 °C: >1 kg/kg acetonitrile, cyclohexanone; 1 kg/kg xylene. It is stable in acidic to weakly alkaline media (pH2-8).

**Uses.**

It is an insecticide having a vapour action; it is effective at 25 g a.i./100 l against a wide range of boring, chewing and sucking insects (aphids, borers, diamond moth, scales, weevils and worms).

**Toxicology.**

Acute oral $LD_{50}$: for male rats 125 mg/kg, for females 180 mg/kg. Acute percutaneous $LD_{50}$ for mice >1250 mg/kg. $LC_{50}$ (48-h) for carp is 7.8 mg/l.

**Formulations.**

These include: e.c. (250 g a.i./l); w.p. (250 g/kg); granules (50 or 100 g/kg).

**Analysis.**

Product analysis is by glc. Residues may be determined by glc with FPD (M. Eto & J. Miyamoto, *Anal. Methods Pestic. Plant Growth Regul.*, 1973, **7**, 431).

$C_{16}H_{15}Cl_3O_2$ (345.7)
GXGGYR DO1&R DO1

$CH_3O$ — CH — $OCH_3$
$CCl_3$

**Nomenclature and development.**

Common name methoxychlor (BSI, E-ISO, JMAF); méthoxychlore (F-ISO). Chemical name (IUPAC) 1,1,1-trichloro-2,2-bis(4-methoxyphenyl)ethane; 1,1,1-trichloro-2,2-di(4-methoxyphenyl)ethane. (*C.A.*) 1,1'-(2,2,2-trichloroethylidene)bis[4-methoxybenzene] (9CI); 1,1,1-trichloro-2,2-bis(*p*-methoxyphenyl)ethane (8CI); Reg. No. *[72-43-5]*. Trivial name DMTD. OMS 466; ENT 1716. Its insecticidal properties were described by P. Läuger *et al.* (*Helv. Chim. Acta*, 1944, **27**, 892). Introduced by J. R. Geigy AG (now Ciba-Geigy AG who no longer manufacture it) and by E. I. du Pont de Nemours & Co. (Inc.) (CHP 226 180; GBP 547 871 both to Ciba-Geigy AG; USP 2 420 928; 2 477 655; GBP 624 561 to du Pont) as trade mark 'Marlate' (to du Pont).

**Properties.**

Condensation of trichloroacetaldehyde with anisole (a reaction reported by K. Elbs, *J. Prakt. Chem.*, 1893, **47**, 44) produces the technical product, a grey powder, setting point 77°C, $d^{25}$ 1.41. It contains $\geqslant$ 88% methoxychlor and $\leqslant$ 12% related isomers. Solubility at 25 °C 0.1 mg/l water; at 22 °C: 440 g/kg chloroform, xylene; 50 g/kg methanol. Pure methoxychlor forms colourless crystals m.p. 89 °C. Stable to u.v. light.

**Uses.**

It is a contact and stomach insecticide effective against a wide range of pests, but not aphids, in field and forage, fruit and vegetable crops. It is also used against certain household and industrial insects. It shows little tendency to be stored in the body fat or to be excreted in milk and so is recommended to control flies in dairy barns and on lactating cows (F. Kunze *et al.*, *Fed. Proc., Fed. Am. Soc. Exp. Biol.*,1950, **9**, 293).

**Toxicology.**

Acute oral $LD_{50}$: for rats 6000 mg tech./kg; for mallard duck >2000 mg/kg. Application to skin of rabbits of 2820 mg (in dimethyl sulphoxide) produced no symptom. In 1-y feeding trials no toxic effect was observed in dogs receiving 300 mg/kg daily; in 2-y trials in rats there was no effect at 200 mg/kg diet, but some reduction in growth occurred at 1600 mg/kg diet. $LC_{50}$ (8-d) for bobwhite quail and ring-necked pheasant >5000 mg/kg diet. $LC_{50}$ (24-h) is: for rainbow trout 52 μg/l; for bluegill 67 μg/l; $LC_{50}$ (48-h) for water flea 0.78 μg/l.

**Formulations.**

These include: w.p. (500 g tech./kg); e.c.; dusts; granules; aerosol concentrates.

**Analysis.**

Product analysis is by total chlorine content (H. L. Pease, *J. Assoc. Off. Anal. Chem.*, 1976, **58**, 40). Residues may be determined by glc (*AOAC Methods* 1980, 29.001-29.028, 29.137; J. Solomon & W. L. Lockhart, *ibid.*, 1978, **60**, 690; W. K. Lowen, *et al.*, *Anal. Methods Pestic., Plant Growth Regul. Food Addit.*, 1964, **2**, 303; *Anal. Methods Pestic. Plant Growth Regul.*, 1972, **6**, 441).

$C_{16}H_{16}O_2$ (240.3)
1OR B1 DVR C1

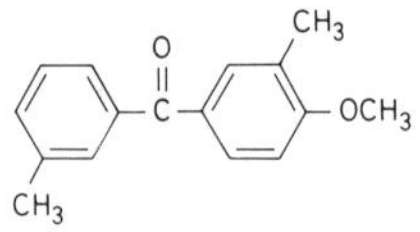

**Nomenclature and development.**

Common name methoxyphenone (JMAF). Chemical name (IUPAC) 4-methoxy-3,3'-dimethylbenzophenone (I). (*C.A.*) (4-methoxy-3-methylphenyl)(3-methylphenyl)methanone (9CI); (I) (8CI); Reg. No. *[41295-28-7]*. Its herbicidal properties were described (*Proc. Asian-Pacific Weed Sci. Soc. Conf., 4th*, 1973, p. 215). Introduced by Nippon Kayaku Co. Ltd. (JPP 51-5446; BP 1 355 926; USP 3 873 304) as code no. 'NK-049'; trade mark 'Kayametone'.

**Properties.**

It forms colourless crystals; m.p. 62.0-62.5 °C. Solubility at 20 °C: 2 mg/l water; soluble in most organic solvents. It is stable under acid and alkaline conditions, but is slowly decomposed by sunlight.

**Uses.**

It is a selective pre-em. herbicide effective against annual grasses and broad-leaved weeds in paddy rice and vegetable crops at 3-5 kg a.i./ha; also used in rice pre-em. in combination with bensulide at 30-50 kg granules/ha. It produces chlorosis and inhibits photosynthesis. It is biodegradable, leaving no residue problems.

**Toxicology.**

Acute oral $LD_{50}$ for rats and mice >4000 mg/kg. In 90-d feeding trials NEL was: for rats 1500 mg/kg diet; for mice 1000 mg/kg diet. $LC_{50}$ (48-h) is: for carp 3.2 mg/l; for goldfish 10 mg/l.

**Formulations.**

'Kayametone', w.p. (500 g a.i./kg) for use on vegetables and paddy rice. 'Kayaphenone', granules (80g 4-methoxy-3,3'-dimethylbenzophenone + 30 g bensulide/kg).

**Analysis.**

Product and residue analysis is by glc.

# 2-Methoxyethylmercury chloride

$C_3H_7ClHgO$ (295.1)
G-HG-2O1

$CH_3OCH_2CH_2HgCl$

**Nomenclature and development.**

Chemical name (IUPAC) 2-methoxyethylmercury chloride (BSI, New Zealand, Republic of South Africa, JMAF) is accepted in lieu of a common name (*C.A.*) chloro(2-methoxyethyl)mercury (8 & 9CI); Reg. No. *[123-88-6]*. Introduced by I. G. Farbenindustrie AG (W. Bonrath, *Nachr. Schaedlingsbekaempfung I. G. Farbenindustrie (Leverkusen)*, 1935, **10**, 23) (now Bayer AG) as trade mark 'Agallol'; 'Aretan'; 'Ceresan Universal Liquid Sterilant'.

**Properties.**

It is a colourless crystalline powder; m.p. 65 °C; v.p. 133 mPa (G. F. Phillips *et al.*, *J. Sci. Food Agric.*, 1959, **10**, 604). Solubility at room temperature: 50 g/l water; readily soluble in acetone, ethanol. 2-Methoxyethylmercury salts, though stable to alkali, are decomposed by hydrochloric acid (J. Chatt, *Chem. Rev.*, 1951, **48**, 7).

**Uses.**

It is effective as a seed treatment against the major seedborne diseases of cereals and as a dip for seed potato against *Phoma solanicola* f. *foveata*.

**Toxicology.**

Acute oral $LD_{50}$: for rats 30 mg/kg; for mice 47 mg/kg.

**Formulations.**

These include: 'Ceresan Universal Liquid Seed Treatment', (20, 25 or 36 g mercury/kg) for use as a slurry seed treatment; 'Agallol' (30 g mercury/kg) for potato dips; 'Aretan 6' (60 g mercury/kg) for dips.

**Analysis.**

Product analysis is by digestion with acids, dilution and titration with potassium iodide. Particulars may be obtained from Bayer AG. See also: *CIPAC Handbook*, 1970, **1**, 503.

$CH_3OCH_2CH_2Hg$.silicate

**Nomenclature and development.**

The trivial name 2-methoxyethylmercury silicate (New Zealand, Republic of South Africa) is accepted in lieu of a common name. The compound is an organomercury silicate but there is some doubt about whether it is an orthosilicate or a metasilicate and therefore about its true molecular composition. Introduced by I. G. Farbenindustrie AG (now Bayer AG) as trade mark 'Ceresan Universal Dry Seed Treatment'.

**Properties.**

It is a colourless crystalline powder; v.p. 440 mPa at 35 °C. It is practically insoluble in water. 2-Methoxyethylmercury salts, though stable to alkali, are decomposed by halogen acids (J. Chatt, *Chem. Rev.* 1951 **48**, 7).

**Uses.**

2-Methoxyethylmercury silicate is used in seed treatments against various seed-borne diseases of cereals.

**Toxicology.**

Acute oral $LD_{50}$ for rats 1140 mg (formulated product)/kg.

**Formulations.**

'Ceresan Special'. Also combinations: 'Ceresan-Gamma M Special' (with gamma-HCH + anthraquinone); 'Ceresan Morkit Special' (with anthraquinone).

**Analysis.**

Product analysis is by digestion with acids and measurement of mercury by titration with potassium iodide (*CIPAC Handbook*, 1970, **1**, 503). Details available from Bayer AG.

$CH_3As(O)(OH)_2$

$CH_5AsO_3$ (140.0)
Q-AS-QO&1
*MSMA*
$CH_4AsNaO_3$ (162.0)
Q-AS-QO&1 &&Na SALT
*DSMA*
$CH_3AsNa_2O_3$ (183.9)
Q-AS-QO&1 &&2Na SALT

**Nomenclature and development.**

Common names MAA (WSSA for acid); methylarsonic acid (E-ISO) is draft proposal in lieu of common name;MSMA (WSSA for monosodium salt);DSMA (WSSA, JMAF for disodium salt);CMA (WSSA for calcium acid salt);MAMA (WSSA for monoammonium salt). Chemical name (IUPAC) methylarsonic acid (I); salts are sodium hydrogen methylarsonate; disodium methylarsonate; calcium bis(hydrogen methylarsonate); ammonium hydrogen methylarsonate. (*C.A.*) methylarsonic acid (9CI); methanearsonic acid (8CI); monosodium methylarsonate (9CI); disodium methylarsonate (9CI); monosodium methanearsonate (8CI); disodium methanearsonate (8CI); Reg. No. *[124-58-3]* (acid), *[2163-80-6]* (MSMA), *[144-21-8]* (DSMA), *[5902-95-4]* (calcium hydrogen salt). MSMA and DSMA were introduced by Ansul Chemical Co. (which no longer exists), by Diamond Shamrock Co. and by Vineland Chemical Co. (USP 2 678 265; 2 889 347 to Ansul) as trade marks 'Ansar', 'Daconate', 'Bueno' (to Diamond Shamrock Corp.) for MSMA or DSMA, 'Calcar' for CMA.

**Properties.**

Methylarsonic acid is a strong dibasic acid; m.p. 161 °C; $pK_1$ −0.44, $pK_2$ −0.09. Freely soluble in water; soluble in ethanol.

MSMA forms a colourless crystalline hydrate MSMA $1.5H_2O$; m.p. 113-116 °C. Solubility at 20 °C: 1.4 kg (anhydrous salt)/kg water; soluble in methanol; insoluble in most organic solvents.

DSMA forms colourless crystals as DSMA $6H_2O$; m.p. 132-139 °C, losing water at elevated temperatures. Solubility at 20 °C: 279 g (anhydrous salt)/kg water; soluble in methanol; practically insoluble in most organic solvents.

These salts are stable to hydrolysis, MSMA being formed from DSMA at pH6-7. They are decomposed by strong oxidising and reducing agents; non-flammable.

**Uses.**

MSMA and DSMA are selective pre-em. contact herbicides with some systemic properties. MSMA is recommended to control grass weeds: in cotton at 2.24 kg in 150 l water/ha; in sugarcane at 3.3 kg in 200 l/ha; as directed sprays under tree crops almost to run-off at 460 g/100 l; on non-crop areas at 0.55-1.1 kg/100 l. The addition of a surfactant is advisable for unformulated products (octylammonium and decylammonium hydrogen methylarsonate are also used); air temperatures should be >21 °C. DSMA is recommended for control of grass weeds in cotton at 2.5 kg in 375 l/ha; for turf treatment and for weed control on uncropped land.

**Toxicology.**

Acute oral $LD_{50}$ for young albino rats: 900 mg MSMA/kg; 1800 mg DSMA/kg. Applications of methylarsonic acid are only mildly irritating to the skin of rabbits. In feeding trials dogs receiving 100 mg acid/kg showed no ill-effect. $LC_{50}$ (48-h) for bluegill is >1000 mg MSMA or DSMA/l, but the addition of surfactants increases the toxicity.

**Formulations.**

These include for MSMA: 'Ansar 529', 'Daconate', 'Bueno', w.s.c. (480 g/l); 'Ansar 529 HC', 'Daconate 6', 'Bueno 6' (720 g/l); 'Arsonate Liquid' (791 g/l); 'Super Arsonate' (959 g/l). For DSMA: 'Ansar 8100', s.p. (810 g/kg); 'Diamond Shamrock DSMA Liquid' (270 g/l).

**Analysis.**

Product analysis is by acid-base titration; total arsenic is determined by wet oxidation, (E. A. Dietz & L. O. Moore, *Anal. Methods Pestic. Plant Growth Regul.*, 1978, **10**, 385). Residues may be determined by estimation of total arsenic reduction to arsine and colorimetry (*idem, ibid.*).

$CH_3Br$ (94.94)
E1

$CH_3Br$

**Nomenclature and development.**

The traditional chemical name methyl bromide (I) (BSI, E-ISO, JMAF);bromure de méthyle (F-ISO) is accepted in lieu of a common name. Chemical name (IUPAC) bromomethane (II); (I). (*C.A.*) (II) (8 & 9CI); Reg. No. *[74-83-9]*. Its insecticidal properties were described by Goupil (*Rev. Pathol. Veg. Entomol. Agric. Fr.*, 1932, **19**, 169). Introduced by Dow Chemical Co. trade mark 'Dowfume MC2'.

**Properties.**

Methyl bromide is a colourless liquid; b.p. 4.5 °C; $d^{0}$ 1.73; specific heat at 0 °C 0.5 J/g. It forms a colourless odourless gas, which has a chloroform-like odour at high concentrations, f.p. −93 °C. Solubility at 25 °C 13.4 g/kg water, it forms a crystalline hydrate with ice water; soluble in most organic solvents. It is stable, non-flammable, corrosive to aluminium, magnesium and their alloys.

**Uses.**

It is a potent insectide with some acaricidal properties and is used for space fumigation and for the fumigation of plants and plant products in stores, mills and ships. It is a soil fumigant used for the control of fungi, nematodes and weeds. Residues of bromide ion and methyl bromide in soils fumigated with the latter and their effects were reviewed by G. A. Maw & R. J. Kempton (*Soils Fert.*, 1973, **36**, 41).

**Toxicology.**

It is highly toxic to man, the threshold limit value being 65 mg/m$^3$, above which concentration respirators must be worn. In many countries its use is restricted to trained personnel.

**Formulations.**

It is packed in glass ampoules (up to 50 ml), in metal cans and cylinders for direct use. Chloropicrin is sometimes added, up to 2% as a warning agent.

**Analysis.**

Methyl bromide may be detected by the halide lamp, a non-specific test given by all volatile halides; by commercially available detector tubes (H. K. Heseltine, *Pest. Tech.*, 1959, **1**, 253). Mixtures with chloropicrin, *e.g.* 'Dowfume MC-2' are analysed by i.r. spectroscopy, details are available from the Dow Chemical Co. For its estimation in air a commercial thermal conductivity meter may be used (H. K. Heseltine *et al.*, *Chem. Ind. (London)*, 1958, p. 1287); or glc (S. G. Heuser & K. A. Scudamore, *J. Sci. Food Agric.*, 1969, **20**, 566; *idem, Analyst (London)*, 1968, **93**, 252; P. A. Greve & E. A. Hogendoorn, *Meded. Fac. Landbouwwet. Rijksuniv. Gent*, 1979, **44**, 877; R. Mestres *et al.*, *Ann. Falsif. Expert. Chim.*, 1980, **73**, 407; H. J. Kolbezen & F. J. Abu-El Haj, *Pestic. Sci.*, 1972, **3**, 73) also glc of a derivative (R. J. Fairall & K. A. Scudamore, *Analyst (London)*, 1980, **105**, 251). It reacts with soil or organic material to form bromide ion. Residues of methyl bromide and bromide ion may be distinguished (S. G. Heuser & K. A. Scudamore, *Pestic. Sci.*, 1970, **1**, 244).

$C_{16}H_{23}NO_3$ (277.4)
3VN1&VOR C1 EY1&1

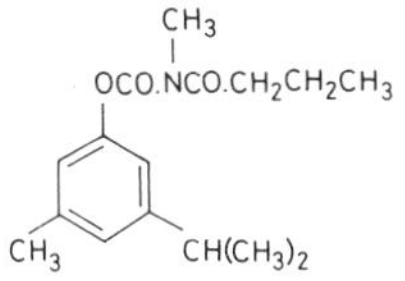

$CH_3$
$OCO.NCO.CH_2CH_2CH_3$
$CH_3$ $CH(CH_3)_2$

**Nomenclature and development.**

Common name promacyl (Australia). Chemical name (IUPAC) 5-methyl-*m*-cumenyl butyryl(methyl)carbamate. (*C.A.*) 3-methyl-5-(1-methylethyl)phenyl methyl(1-oxobutyl)carbamate (9CI); *m*-cym-5-yl butyrylmethyl carbamate (8CI); Reg. No. *[34264-24-9]*. Trivial name promecarb A. Introduced by ICI Australia (AUP 441 004; 454 280) as code no. 'CRC 7320'; trade mark 'Promicide'.

**Properties.**

The technical grade is a clear amber to dark brown liquid; b.p. 158 °C/5 mmHg; $d_4^{20}$ 0.996; $n_D^{24}$ 1.5052; v.p. 400 Pa at 149 °C. Solubility: sparingly soluble in water; completely miscible with aliphatic and aromatic hydrocarbons, alcohols, esters, ethers, ketones. Stable $\leqslant$200 °C; also unchanged after 300 d at 50 °C.

**Uses.**

It is effective for controlling *Boophilus microplus* on cattle by sprays or plunge dips at 1.5 g a.i./l applied at intervals of *c.* 21 d. Also controls *Haemaphysalis longicornis, Haematobia irritans exigua* and *Ixodes holocyclus* on cattle and horses.

**Toxicology.**

Acute oral $LD_{50}$: for female rats 1220 mg tech./kg; female mice 2000-4000 mg/kg; female guinea-pigs 250 mg/kg; female rabbits 8000 mg/kg; hens 4000-8000 mg/kg. Acute percutaneous $LD_{50}$ for female rats >4000 mg/kg; slightly irritating to their skin. In 2-y feeding trials NEL for rats was 500 mg/kg diet.

**Formulations.**

'Promicide', e.c. (600 g/l).

**Analysis.**

Product analysis is by D. Gunew, *Anal. Methods Pestic. Plant Growth Regul.*, 1980, **11**, 139.

$C_3H_{10}N_6S_2$ (194.3)
SUYZMM1MMYZUS

$NH_2CS.NHNHCH_2NHNHCS.NH_2$

**Nomenclature and development.**

Common name bisthiosemi (JMAF). Chemical name (IUPAC) 1,1'-methylenedi(thiosemicarbazide). (*C.A.*) 2,2'-methylenebis(hydrazinecarbothioamide) (9CI); Reg. No. *[39603-48-0]*. Introduced as a rodenticide by Nippon Kayaku Co. Ltd (JPP 50 20 126; GBP 1 351 710; USP 3 826 841) as code no. 'NK-15 561'; trade mark 'Kayanex'.

**Properties.**

It is a colourless crystalline powder, which decomposes at 171-174 °C. Solubility: practically insoluble in water and in common organic solvents; soluble in dimethyl sulphoxide. It is slowly decomposed in water, more rapidly under acid or alkaline conditions.

**Uses.**

It is a quick-acting rodenticide effective against *Rattus norvegicus, R. rattus* and the Japanese field vole, *Microtus montebelli.*

**Toxicology.**

Acute oral $LD_{50}$: for mice 30.4-50 mg/kg; for guinea-pigs 32-36 mg/kg; for chickens 120-150 mg; for cats *c.* 150 mg/kg; for dogs >1500 mg/kg. It induces vomiting in cats and dogs. In 90-d feeding trials NEL was: for mice 100 mg/kg diet; for rats 50 mg/kg diet.

**Formulations.**

Baits: (5, 10 or 20 g a.i./kg).

**Analysis.**

Product analysis is by hydrolysis to thiosemicarbazide which is measured colorimetrically as a derivative.

$C_2H_3NS$ (73.11)
SCN1

$CH_3NCS$

**Nomenclature and development.**

Chemical name (IUPAC) methyl isothiocyanate (I) (BSI, E-ISO, JMAF); isothiocyanate de méthyle (F-ISO) accepted in lieu of common name. (*C.A.*) isothiocyanatomethane (9CI); (I) (8CI); Reg. No. *[556-61-6]*. Trivial name MIT. Its nematicidal activity was described by E. A. Pieroh *et al.* (*Anz. Schaedlingsskd.*, 1959, **32**, 183). Introduced by Schering AG (USP 3 113 908) as trade mark 'Trapex'; 'Di-Trapex' for mixture with 1,2-dichloropropane + 1,3-dichloropropene.

**Properties.**

It has a pungent horse-radish-like odour, and forms colourless crystals; m.p. 35 °C; v.p. 2.7 kPa at 20 °C; $d_4^{37}$ 1.069. Solubility at 20 °C: 7.6 g/l water; readily soluble in acetone, benzene, cyclohexanone, dichloromethane, ethanol, light petroleum, methanol. The technical grade (>96% pure) has b.p. 117-119 °C.

**Uses.**

It is a soil fumigant used at a rate of 17.5-25 g/m$^2$ for the control of soil fungi, insects and nematodes, and against weed seeds. It is phytotoxic and planting must be delayed until decomposition is complete, *c.* 21 d at soil temperatures of 15-18 °C, *c.* 45 d at 5 °C; the cress germination test (W. M. Morgan & M. S. Ledieu, in *Pest Disease Control Handbook*, 1979, p. 7.4) should be used to check the absence of residue before planting. Methyl isothiocyanate is also generated in moist soil from metham-sodium (8080) and from dazomet (3810).

**Toxicology.**

Acute oral $LD_{50}$: for male rats 175 mg/kg; for male mice 90 mg/kg; for ducks 136 mg/kg. Acute percutaneous $LD5_0$: for male mice 1870 mg/kg; prolonged skin contact causes irritation. In 2-y feeding trials NEL for rats was 10 mg/l drinking water, $LC_{50}$ (96-h) is: for bluegill 0.13 mg/l; for trout and carp 0.37 mg/l.

**Formulations.**

'Trapex', e.c. (175 g a.i./l). 'Di-Trapex' ('Vorlex' in North America), (Reg. No. *[8066-01-1]*), (235 g/l 1,3-dichloropropane + 1,2-dichloropropene mixture) do *not* store in aluminium containers. 'Di-Trapex CP' (Reg. No. *[8069-50-9]*), as 'Di-Trapex', with added chloropicrin. These preparations are moderately corrosive and appliances should be cleaned with mineral oil and not water.

**Analysis.**

Product analysis is by glc or by reaction with 1-butylamine measured by potentiometric titration (M. Ottnad *et al.*, *Anal. Methods Pestic. Plant Growth Regul.*, 1978, **10**, 563). Residues may be determined by glc with FPD (*idem, ibid.*). Details are availabe from Schering AG.

$C_9H_{11}BrN_2O_2$ (259.1)
ER DMVN1&O1

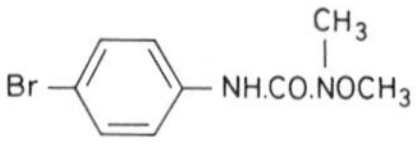

**Nomenclature and development.**

Common name metobromuron (BSI, E-ISO, F-ISO, ANSI, WSSA). Chemical name (IUPAC) 3-(4-bromophenyl)-1-methoxy-1-methylurea. (*C.A.*) *N'*-(4-bromophenyl)-*N*-methyoxy-*N*-methylurea (9CI); 3-(*p*-bromophenyl)-1-methoxy-1-methylurea (8CI); *[3060-89-7]*. Its herbicidal properties were described by J. Schuler & L. Ebner (*Proc. Br. Weed Control Conf., 7th*, 1964, p. 450) Introduced by Ciba AG (now Ciba-Geigy AG) (BEP 662 268; GBP 965 313) as code no. 'C 3126'; trade mark 'Patoran'.

**Properties.**

Pure metobromuron is a crystalline powder; m.p. 95.5-96 °C; v.p. 400 µPa at 20 °C; ρ 1.60 g/cm³ at 20 °C. Solubility at 20 °C: 330 mg/l water; 500 g/l acetone; 550 g/l dichloromethane; 2.6 g/l hexane; 240 g/l methanol; 70 g/l octan-1-ol; 100 g/l toluene. On hydrolysis at 20 °C, 50% loss occurs in: 150 d at pH1, >200 d at pH9, 83 d at pH13.

**Uses.**

It is a herbicide absorbed by roots and leaves and recommended for pre-em. use on beans (*Phaseolus* spp.), potatoes, soyabeans at 1.5-2.0 kg/ha, on tobacco and tomatoes, prior to transplanting, at 1.5-2.5 kg/ha. At these dosages it is degraded to non-phytotoxic levels within 90 d. Used in combination with metolachlor (see below) on hemp, paprika and sunflowers.

**Toxicology.**

Acute oral $LD_{50}$ for rats 2603 mg tech./kg. Acute percutaneous $LD_{50}$ for rats >3000 mg/kg; slight irritant to skin and eyes of rabbits. Inhalation $LC_{50}$ (4-h) for rats >1100 mg/m³ air. In 2-y feeding trials NEL was: for rats 250 mg/kg diet (17 mg/kg daily); for dogs 100 mg/kg diet (3 mg/kg daily). $LC_{50}$ (96-h) is: for rainbow trout 36 mg/l; for crucian carp, bluegill 40 mg/l. It is slightly toxic to birds and practically non-toxic to honeybees.

**Formulations.**

'Patoran' 50 WP, w.p. (500 g a.i./kg). 'Galex', metobromuron + metolachlor (1:1); 'Tobacron' (1:2). 'Igrater' (Reg. No. *[52080-96-3]*), metobromuron + terbutryne (1:1).

**Analysis.**

Product analysis is by glc. Residues may be determined by alkaline hydrolysis to 4-bromoaniline, derivatives of which are measured by colorimetry or by glc with ECD (G. Voss *et al., Anal. Methods Pestic. Plant Growth Regul.*, 1973, **7**, 569).

$C_{15}H_{22}ClNO_2$ (283.8)
2R C1 BNV1GY1&1O1

$CH_2CH_3$, $CO.CH_2Cl$, N, $CHCH_2OCH_3$, $CH_3$, $CH_3$

**Nomenclature and development.**

Common name metachlor (BSI, E-ISO, ANSI, WSSA);métolachlore (F-ISO). Chemical name (IUPAC) 2-chloro-6'-ethyl-*N*-(2-methoxy-1-methylethyl)acet-*o*-toluidide. (*C.A.*) 2-chloro-*N*-(2-ethyl-6-methylphenyl)-*N*-(2-methoxy-1-methylethyl)acetamide (9CI); Reg. No. *[51218-45-2]*. Its herbicidal properties were described by H. R. Gerber *et al.* (*Proc. Br. Weed Control Conf., 12th*, 1974, **2**, 787). Introduced by Ciba-Geigy AG (BEP 800 471; GBP 1 438 311; 1 438 312) as code no. 'CGA 24 705'; trade mark 'Dual'.

**Properties.**

Pure metolachlor is a colourless liquid; b.p. 100 °C/0.001 mmHg; v.p. 1.7 mPa at 20 °C; $\rho$ 1.12 g/cm$^3$ at 20 °C. Solubility at 20 °C: 530 mg/l water; very soluble in benzene, dichloromethane, hexane, methanol, octan-1-ol. On hydrolysis at 20 °C, 50% loss (calculated) in $>$200 d at $1 \leqslant pH \leqslant 9$. Stable $\leqslant$300 °C.

**Uses.**

It is a germination inhibitor active mainly on grasses at 1.0-2.5 kg a.i./ha, rates depending on soil type and climatic conditions. It is selective in cotton, groundnuts, maize, potatoes, sorghum, sugar beet, sugarcane, sunflowers and most other broad-leaved crops. Mixtures with other herbicides (see below) are also used in broad beans, carrots, hemp, lentils, paprika.

**Toxicology.**

Acute oral $LD_{50}$ for rats 2780 mg tech./kg. Acute percutaneous $LD_{50}$ for rats $>$3170 mg/kg; non-irritant to eyes, slight irritant to skin of rabbits. Inhalation $LC_{50}$ (6-h) for rats $>$1750 mg/m$^3$ air. In 90-d feeding trials NEL was: for rats 1000 mg/kg diet (*c.* 90 mg/kg daily); for dogs 500 mg/kg diet (*c.* 17 mg/kg daily). $LC_{50}$ (96-h) is: for rainbow trout 2 mg/l; for carp 4.9 mg/l; for bluegill 15 mg/l. It is practically non-toxic to birds and honeybees.

**Formulations.**

'Dual', e.c. (500 or 720 g a.i./l). 'Bicep' (Reg. No. *[59316-87-9]*), metolachlor + atrazine (1.25:1); 'Primextra' (2:1 or 1.5:1); 'Primagram' (1:1). 'Codal', metolachlor + prometryne (1:1 or 1.5:1). 'Cotoran multi', (Reg. No. *[72878-73-0]*), metolachlor + fluometuron (1:1). 'Galex', metolachlor + metobromuron (1:1); 'Tobacron' (2:1). 'Gardomil', metolachlor + terbuthylazine (1:3). 'Milocep', metolachlor + propazine (2:1). 'Cotodon', metolachlor + dipropetryn (1:1.5).

**Analysis.**

Product analysis is by glc. Residues may be determined by glc with TID or MCD. Particulars are available from Ciba-Geigy AG.

$C_9H_{11}NO_2$ (165.2)
1MVOR C1

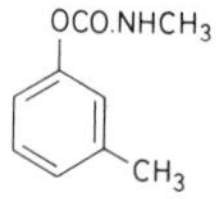

OCO.NHCH$_3$
CH$_3$

**Nomenclature and development.**

Common name metolcarb (BSI, draft E-ISO);métholcarb (draft F-ISO);MTMC (JMAF). Chemical name (IUPAC) *m*-tolyl methylcarbamate (I). (*C.A.*) 3-methylphenyl methylcarbamate (9CI); (I) (8CI); Reg. No. *[1129-41-5]*. Introduced as an insecticide by Nihon Nohyaku Co. and by Sumitomo Chemical Co. code no. 'C-3' (Nihon Nohyaku); trade marks 'Tsumacide' (to Nihon Nohyaku), 'Metacrate' (to Sumitomo Chemical Co.).

**Properties.**

Technical grade metolcarb is a colourless solid; m.p. 74-75 °C; v.p. 145 mPa at 20 °C. Solublity at 30 °C: 2.6 g/l water; 790 g/kg cyclohexanone; 100 g/kg xylene; at room temperature 880 g/kg methanol.

**Uses.**

It is an insecticide mainly used to control sucking insects on rice *i.e.* leafhoppers and planthoppers.

**Toxicology.**

Acute oral $LD_{50}$: for male rats 580 mg/kg, for females 498 mg/kg. Acute percutaneous $LD_{50}$ for rats >2000 mg/kg.

**Formulations.**

These include: e.c. (300 g a.i./l); w.p. (500 g/kg); dust (20 g or 30 g/kg). Combinations with fungicides and other insecticides as dusts are available.

**Analysis.**

Product analysis is by glc or by hplc (S. Sakaue *et al., Nippon Nogei Kagaku Kaishi.*, 1981, **55**, 1237). Residues may be determined by glc of a derivative with ECD (J. Miyamoto *et al., Nihon Noyaku Gakkaishi*, 1978, **3**, 119).

$C_{10}H_{13}ClN_2O_2$ (228.7)
1OR BG DMVN1&1

$CH_3O$—[3-chloro-4-methoxyphenyl]—$NH.CO.N(CH_3)_2$, Cl

**Nomenclature and development.**

Common name metoxuron (BSI, E-ISO, F-ISO). Chemical name (IUPAC) 3-(3-chloro-4-methoxyphenyl)-1,1-dimethylurea (I). (*C.A.*) *N*'-(3-chloro-4-methoxyphenyl)-*N,N*-dimethylurea (9CI); (I) (8CI); Reg. No. *[19937-59-8]*. Its herbicidal properties were described by W. Berg (*Z. Pflanzenkr. Pflanzenpathol. Pflanzenschutz*., 1968, **75**, 233). Introduced by Sandoz AG (GBP 1 165 160; FP 1 497 868) as code no. 'SAN 6915H','SAN 7102H'; trade mark 'Dosanex'.

**Properties.**

Pure metoxuron is a colourless crystalline solid; m.p. 126-127 °C; v.p. 4.3 mPa at 20 °C. Solubility at 24 °C: 678 mg/l water; soluble in acetone, cyclohexanone, hot ethanol; moderately soluble in diethyl ether, toluene; practically insoluble in petroleum spirit. At 50 °C, 50% hydrolysis occurs in: 18 d at pH3, 21 d at pH5, 24 d at pH7, >30 d at pH9, 26 d at pH11. It is stable during storage (shelf life at 20 °C: ≥4 y). It is sensitive to light.

**Uses.**

It is a selective herbicide for use in cereals and carrots, particularly against blackgrass, canary grass, ryegrass, silky bent grass, wild oats, and most annual broad-leaved weeds. It is applied to winter and some spring wheats, winter barley and winter rye at early or late post-em. On irrigated wheat in light to medium soils good results against canary grass and wild oats are obtained. Most winter-wheat, winter-barley and winter-rye, and some spring-wheat varieties show a high tolerance but some damage has occurred on a few varieties. On carrots it is used pre- and post-em. the rate depending on the soil. Other uses are as a defoliant for hemp, flax, tomatoes and as a potato-haulm killer. It inhibits the Hill reaction. A 50% loss of metoxuron in soil occurs in 10-30 d. In plants, the main metabolic reactions are *N*-demethylation and hydrolysis of the urea moiety.

**Toxicology.**

Acute oral $LD_{50}$ for rats 3200 mg/kg. Acute percutaneous $LD_{50}$ for albino rats >2000 mg/kg. In 90-d feeding trials dogs receiving 2500 mg/kg diet showed no toxic symptom. In 42-d trials chicks receiving 1250 mg/kg diet showed no significant abnormality. $LC_{50}$ (96-h) for rainbow trout is 19.8 mg/l. Safe to honeybees.

**Formulations.**

These include: 'Dosanex', w.p. (800 g a.i./kg); 'Dosanex FL' and 'Dosaflo', s.c. (500 g/l); 'Dosanex G' and 'Dosamet', granules (400 g/kg); 'Dosanex Instant', 'Sulerex', water dispersible granules (800 g/kg). 'Purivel' and 'Piruvel', w.p. (800 g/kg) are special formulations for pre-harvest defoliation of hemp and potatoes. 'Certosan', s.c. (285 g metoxuron + 240 g DNOC/l). 'Dosamix', (Reg. No. *[53126-75-3]*), w.p. (720 g metoxuron + 80 simazine/kg). 'Dosater', s.c. (335 g metoxuron + 195 g mecoprop/l). 'Riflex', granules (140 g metoxuron + 60 g tri-allate/kg).

**Analysis.**

Product analysis is by hydrolysis and titration of the liberated amine (*CIPAC Handbook*, 1980, **1A**, 1304) or by tlc and subsequent spectrometric determination of the eluted compound (M. Wisson, *Anal. Methods Pestic. Plant Growth Regul*., 1960, **8**, 417). Residues may be determined by a colorimetric method (*idem, ibid*., p. 425).

$C_8H_{14}N_4OS$ (214.3)
T6NN DNVJ CS1 DZ FX1&1&1

$(CH_3)_3C$ — 1,2,4-triazin-5-one ring with $SCH_3$, O, $NH_2$

**Nomenclature and development.**

Common name metribuzin (BSI, E-ISO, WSSA);métribuzine (F-ISO). Chemical name (IUPAC) 4-amino-6-*tert*-butyl-3-methylthio-1,2,4-triazin-5(4*H*)-one; 4-amino-6-*tert*-butyl-4,5-dihydro-3-methylthio-1,2,4-triazin-5-one. (*C.A.*) 4-amino-6-(1,1-dimethylethyl)-3-(methylthio)-1,2,4-triazin-5(4*H*)-one (9CI); 4-amino-6-*tert*-butyl-3-(methylthio)-*as*-triazin-5(4*H*)-one (8CI); Reg. No. *[21087-64-9]*. Its herbicidal properties were described by W. Draber *et al.* (*Naturwissenschaften*, 1968, **55**, 446) and reviewed by L. Eue (*Pflanzenschutz-Nachr. (Engl. Ed.*), 1972, **25**, 175). Introduced by Bayer AG and E. I. du Pont de Nemours & Co. (Inc.) (BEP 697 083; DEPS 1 795 784 both to Bayer AG; USP 3 905 801 to du Pont) as code no. 'Bayer 94 337','Bayer 6159H','Bayer 6443H','DIC 1468' (all to Bayer); trade mark 'Sencor', 'Sencorex' in Great Britain, 'Sencoral' in France (all to Bayer), 'Lexone' (to du Pont).

**Properties.**

Pure metribuzin forms colourless crystals; m.p. 125.5-126.5 °C; v.p. $<$1.3 mPa at 20 °C. Solubility at 20 °C: 1.2 g/l water; 820 g/kg acetone; 220 g/kg benzene; 850 g/kg chloroform; 1 kg/kg cyclohexanone; 190 g/kg ethanol; 2 g/kg hexane; 120 g/kg toluene. The technical grade has a slight sulphurous odour. Undergoes microbial decomposition in moist soil.

**Uses.**

It is used pre-em. or post-em. to control weeds in asparagus, lucerne, potatoes, soyabeans, sugar beet, tomatoes and other crops.

**Toxicology.**

Acute oral $LD_{50}$: for rats 2200-2345 mg/kg; for mice 698-711 mg/kg; for bobwhite quail 164-168 mg/kg. Acute percutaneous $LD_{50}$ for rats $>$20 000 mg/kg. In 2-y feeding trials NEL for rats and dogs was 100 mg/kg diet. $LC_{50}$ (96-h) is: for rainbow trout 76 mg/l; for bluegill 80 mg/l.

**Formulations.**

These include w.p. (350, 500, 700, 750 g a.i./kg); aqueous suspension (420 g/kg).

**Analysis.**

Product analysis is by i.r. spectrometry (J. W. Betker *et al.*, *J. Assoc. Off. Anal. Chem.*, 1976, **59**, 278). Residues may be determined by glc (T. H. Byast *et al.*, *Tech. Rep. ARC Weed Res. Organ.*, No. 15 (2nd Ed.), pp. 40, 67; C. A. Anderson, *Anal. Methods Pestic. Plant Growth Regul.*, 1976, **8**, 453). Details of both methods are available from Bayer AG.

$C_7H_{13}O_6P$ (224.1)
1OV1UY1&OPO&O1&O1

$(CH_3O)_2P(=O)-O(CH_3)C=CH\,CO.OCH_3$ (*E*)

$(CH_3O)_2P(=O)-O(CH_3)C=CH\,CO.OCH_3$ (*Z*)

**Nomenclature and development.**

Common name mevinphos (BSI, E-ISO, F-ISO, ESA) the isomer present (*E*)- and (*Z*)- [or *cis* and *trans* (with respect to carbon chain)] should be stated;*exception* (USSR). Chemical name (IUPAC) methyl 3-(dimethoxyphosphinyloxy)but-2-enoate; 2-methoxycarbonyl-1-methylvinyl dimethyl phosphate. (*C.A.*) methyl 3-[(dimethoxyphosphinyl)oxyl]-2-butenoate (9CI); methyl 3-hydroxycrotonate dimethyl phosphate ester (8CI); Reg. No. *[26718-65-0]* (formerly *[298-01-1]*) (*E*)-isomer, *[338-45-4]* (*Z*)-isomer, *[7786-34-7]* mixed isomers. ENT 22 374. Its insecticidial properties were described by R. A. Corey *et al.* (*J. Econ. Entomol.*, 1953, **45**, 386). Introduced by Shell Chemical Co., USA. (USP 2 685 552) as code no. 'OS-2046'; trade mark 'Phosdrin'.

**Properties.**

Produced by the reaction of trimethyl phosphite with methyl 2-chloro-3-oxybutyrate, the technical product contains >60% *m/m* of the (*E*)-isomer and *c.* 20% *m/m* of the (*Z*)-isomer. It is a pale yellow to orange liquid; b.p. 99-103 °C/0.3 mmHg; v.p. 17 mPa at 20 °C; $d_{20}^{20}$ 1.24. The (*E*)-isomer has m.p. 21 °C, $n_D^{20}$ 1.4452, $d^{20}$ 1.2345; the (*Z*)-isomer, m.p. 6.9 °C, $n_D^{20}$ 1.4524, $d^{20}$ 1.245. Technical mevinphos is completely miscible with water, alcohols, ketones, chlorinated hydrocarbons, aromatic hydrocarbons; only slightly soluble in aliphatic hydrocarbons. It is stable when stored at ambient temperatures but is hydrolysed in aqueous solution, 50% loss occurring after 120 d at pH6, 35 d at pH7, 3 d at pH9, 1.4 h at pH11. It should not be used with Bordeaux mixture, lime sulphur or other alkaline products. It is corrosive to cast iron, mild and some stainless steels and brass; it is non-corrosive to glass and many plastics but passes slowly through thin films of polythene.

**Uses.**

It is a contact and systemic acaricide and insecticide with short residual activity. It is effective against beetles and mites at 200-300 g/ha, caterpillars at 250-500 g/ha, sap-feeding insects at 125-250 g/ha. Its metabolism and degradation have been reviewed (K. I. Beynon *et al., Residue Rev.* 1973, **47**, 55).

**Toxicology.**

Acute oral $LD_{50}$: for rats 3-12 mg tech./kg; for mice 7-18 mg/kg; for birds 0.75-7.0 mg/kg. Acute percutaneous $LD_{50}$: for rats 4-90 mg/kg; for rabbits 16-33 mg/kg. In 2-y feeding trials no effect was observed in rats receiving 4 mg/kg diet or dogs 5 mg/kg diet. $LC_{50}$ (48-h) for fish in range 0.02-31.6 mg/l. It has little effect on wildlife in practice because it is rapidly broken down to less toxic decomposition products. ADI for man 0.0015 mg/kg (*JMPR* 1972).

**Formulations.**

These include; e.c. (100 or 240 g tech./l); water-soluble concentrates (100 or 240 g/l).

**Analysis.**

Product analysis is by glc with FID (*CIPAC Handbook*, in press). Residues may be determined using glc with FPD (*Pestic. Anal. Man.*, 1979, **I**, 201-H, 201-I; R. Mestres *et al., Trav. Soc. Pharm. Montpellier*, 1979, **39**, 323; D. C. Abbott *et al., Pestic. Sci.*, 1970, **1**, 10; P. E. Porter *et al., Anal. Methods Pestic., Plant Growth Regul. Food Addit.*, 1964, **2**, 351; *Anal. Methods Pestic. Plant Growth Regul.*, 1972, **6**, 450). Details can be obtained from Shell International Chemical Co. Ltd.

$C_9H_{17}NOS$ (187.3)
T7NTJ AVS2

**Nomenclature and development.**

Common name molinate (BSI, E-ISO, F-ISO, JMAF) *exception* Federal Republic of Germany. Chemical name (IUPAC) *S*-ethyl azepane-1-carbothioate; perhydroazepin-1-carbothioate; *S*-ethyl *N,N*-hexamethylenethiocarbamate; *S*-ethyl perhydroazepine-1-thiocarboxylate. (*C.A.*) *S*-ethyl hexahydro-1*H*-azepine-1-carbothioate (8 & 9CI); Reg. No. *[2212-67-1]*. Introduced by Stauffer Chemical Co. (USP 3 198 786; 3 573 031) as code no. 'R-4572'; trade mark 'Ordram'.

**Properties.**

Molinate (99.5% pure) is a clear liquid, with an aromatic odour; b.p. 202 °C/10 mmHg; v.p. 746 mPa at 25 °C; $d_{20}^{20}$ 1.063; $n_D^{30}$ 1.5124. Solubility at 20 °C: 880 mg/l water; miscible with acetone, ethanol, kerosene, 4-methylpentan-2-one, xylene. In soil 50% is degraded in 14-35 d, mostly through microbial activity. It is stable $<$200 °C.

**Uses.**

Molinate is toxic to germinating broad-leaved and grassy weeds and is particularly useful for the control of *Echinochloa* spp. in rice at 2-4 kg a.i./ha. It is applied either before planting to water-seeded or shallow soil-seeded rice or post-flood, post-em. on other types of rice culture. It is rapidly taken up by plant roots.

**Toxicology.**

Acute oral $LD_{50}$: for male rats 369 mg/kg for females 450 mg/kg. Acute percutaneous $LD_{50}$ for rabbits $>$4640 mg/kg; non-irritating to skin and moderately irritating to their eyes. It is rapidly metabolized by rats, about 50% to carbon dioxide, 25% excreted in urine and 7-20% in the faeces in 3 d. $LC_{50}$ (96-h)was: for rainbow trout 1.3 mg/l; for goldfish 30 mg/l. At recommended rates, it has had no detectable effects on fish in ditches draining water from treated rice fields in California.

**Formulations.**

'Ordram E', e.c. (720 or 960 g a.i./l); 'Ordram G', granules (50 or 100 g/kg). 'Arrosolo 3-3E', 'Arrozan 36-36E', (Reg. No. *[50934-16-2]*), e.c. (360 g molinate + 360 g propanil/l).

**Analysis.**

Product analysis is by glc (*AOAC Methods*, 1980, 6.426-6.430; *CIPAC Handbook*, in press). Residues in crops and soil are determined by glc or colorimetry after conversion to a suitable derivative. See also: G. R. Patchett & G. H. Batchelder, *Anal. Methods Pestic., Plant Growth Regul. Food Addit.*, 1967, **5**, 469; G. R. Patchett *et al.*, *Anal. Methods Pestic. Plant Growth Regul.*, 1972, **6**, 668. Analytical methods are available from the Stauffer Chemical Co.

$C_{13}H_{18}ClNO$ (239.7)
GR DMVX3&1&1

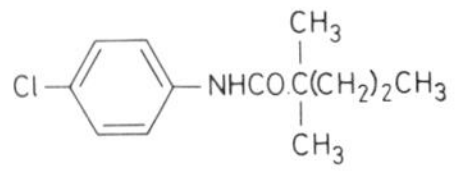

$Cl-C_6H_4-NHCO.C(CH_3)_2(CH_2)_2CH_3$

**Nomenclature and development.**

Common name monalide (BSI, E-ISO, F-ISO). Chemical name (IUPAC) 4'-chloro-2,2-dimethylvaleranilide (I); 4'-chloro-α,α-dimethylvaleranilide; *N*-(4-chlorophenyl)-2,2-dimethylvaleramide. (*C.A.*) *N*-(4-chlorophenyl)-2,2-dimethylpentanamide (9CI); (I) (8CI); Reg. No. *[7287-36-7]*. Its herbicidal properties were described by F. Arndt (*Z. Pflanzenkr. Pflanzenpathol. Pflanzenschutz*, 1965, Sonderheft III, p. 277). Introduced by Schering AG (GBP 971 819) as code no. 'Schering 35 830'; trade mark 'Potablan'.

**Properties.**

Monalide is a colourless crystalline solid; m.p. 87-88 °C; v.p. 240 μPa at 25 °C. Solubility at 23 °C: 22.8 mg/l water; *c.* 500 g/l cyclohexanone; <10 g/l light petroleum; *c.* 100 g/l xylene. It is stable to high temperatures and to hydrolysis.

**Uses.**

It is a post-em. herbicide, absorbed by leaves and roots and effective at 4 kg a.i. in 400-800 l water/ha against broad-leaved weeds in umbelliferous crops.

**Toxicology.**

Acute oral $LD_{50}$ for rats >4000 mg/kg. Acute percutaneous $LD_{50}$ for rats and rabbits >800 mg a.i. (as e.c.)/kg. In 28-d feeding trials NEL for rats was 150 mg/kg daily.

**Formulations.**

'Potablan', e.c. (200 g a.i./l). '*Potablan S*', (monalide + linuron) discontinued.

**Analysis.**

Residues may be determined by tlc or glc. Details may be obtained from Schering AG.

$C_7H_{14}NO_5P$ (223.2)
1OPO&O1&OY1&U1VM1

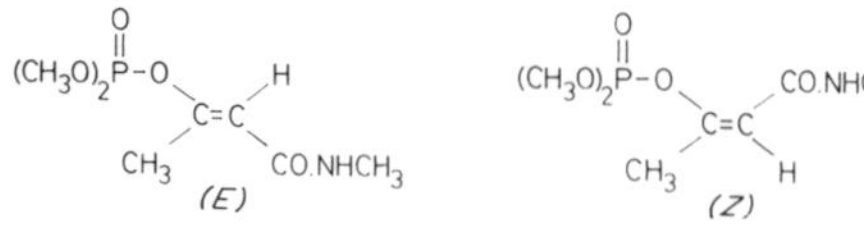

**Nomenclature and development.**

Common name monocrotophos (BSI, E-ISO, F-ISO, JMAF). Chemical name (IUPAC) dimethyl (*E*)-1-methyl-2-(methylcarbamoyl)vinyl phosphate; 3-(dimethoxyphosphinyloxy)-*N*-methylisocrotonamide. (*C.A.*) (*E*)-dimethyl 1-methyl-3-(methylamino)-3-oxo-1-propenyl phosphate (9CI); dimethyl phosphate ester with (*E*)-3-hydroxy-*N*-methylcrotonamide (8CI); Reg. No. *[6923-22-4]* (formerly *[919-44-8]*) (*E*)-isomer, *[919-44-8]* for *Z*-isomer, *[2157-98-4]* for (*E*)- + (*Z*)-isomers. OMS 834; ENT 27 129. Introduced by Ciba AG (now Ciba-Geigy AG) and by Shell Chemical Co. USA (BEP 552 284; GBP 829 276 both to Ciba-Geigy AG) as code no. 'C 1414' (Ciba-Geigy AG),'SD 9129' (Shell); trade marks 'Nuvacron' (to Ciba-Geigy AG), 'Azodrin' (to Shell).

**Properties.**

Pure monocrotophos forms colourless crystals; m.p. 54-55 °C; b.p. 125 °C/0.0005 mmHg; v.p. 293 μPa at 20 °C; ρ 1.33 g/cm³ at 20°C. Solubility at 20 °C: 1 kg/kg water; 700 g/kg acetone; 800 g/kg dichloromethane; 1 kg/kg methanol; 250 g/kg octan-1-ol; 60 g/kg toluene. On hydrolysis at 20 °C, 50% loss (calculated) occurs in 96 d at pH5, 66 d at pH7, 17 d at pH9. The technical grade (*c.* 80% pure) is a reddish-brown semi-solid, m.p. 25-30 °C. It is unstable in short-chain alcohols and glycols. It is corrosive to black iron, drum steel, stainless steel 304 and brass.

**Uses.**

It is a fast-acting insecticide with both systemic and contact action, used against a wide range of pests including mites, sucking insects, leaf-eating beetles, bollworms and other lepidopterous larvae on a variety of crops. Dosage rates against mites and sucking insects are 250-500 g/ha and against lepidopterous larvae 500-1000 g/ha. It persists for 7-14 d. It has caused phytotoxicity when applied under cool conditions to apples (Red Delicious and Ananas Reinette) to cherries, peaches and to sorghum varieties related to Red Swazi.

**Toxicology.**

Acute oral $LD_{50}$ for rats 14 mg tech./kg; for birds 1.0-6.5 mg/kg. Acute percutaneous $LD_{50}$ for rats 336 mg/kg; non-irritant to skin and eyes of rabbits. Inhalation $LC_{50}$ (4-h) for rats *c.* 80 mg/m³ air. In 2-y feeding trials NEL was estimated by WHO as: for rats 0.5 mg/kg diet (0.025 mg/kg daily); for dogs 0.5 mg/kg diet (0.0125 mg/kg daily). In a 30-d trial with human volunteers a daily oral dose of 0.006 mg/kg was without effect. This provides the basis for the WHO estimates of ADI as 0.0006 mg/kg. $LC_{50}$ (24-h) is: for rainbow trout 12 mg/l; for bluegill 23 mg/l. It is highly toxic to honeybees ($LD_{50}$ 33-84 μg/bee).

Its metabolism and breakdown have been reviewed (K. I. Beynon *et al.*, *Residue Rev.*, 1973, **47**, 55).

**Formulations.**

These include: concentrates (200, 400 or 600 g a.i./l); w.s.c. (400, 552 or 600 g/l); ULV concentrate (250 g/l). Mixtures include: monocrotophos + mevinphos; monocrotophos + parathion-methyl; monocrotophos + parathion-methyl + sulphur.

**Analysis.**

Product analysis is by i.r. spectrometry or by glc: details can be obtained from Shell International Chemical Co. Ltd. Residues are determined by glc using phosphorus-sensitive detectors (*Pestic. Anal. Man.*, 1979, **I**, 201-H, 201-I; *ibid*, 1979, **II**; B. Y. Giang and H. F. Beckman, *ibid.*, 1968, **16**, 899; *Anal. Methods Pestic. Plant Growth Regul.*, 1972, **6**, 287).

$C_9H_{11}ClN_2O_2$ (214.6)
GR DMVN1&O1

$$\mathrm{Cl{-}C_6H_4{-}NH.CO.N(CH_3)OCH_3}$$

**Nomenclature and development.**

Common name monolinuron (BSI, E-ISO, F-ISO, WSSA). Chemical name (IUPAC) 3-(4-chlorophenyl)-1-methoxy-1-methylurea. (*C.A.*) *N'*-(4-chlorophenyl)-*N*-methoxy-*N*-methylurea (9CI); 3-(*p*-chlorophenyl)-1-methoxy-1-methylurea (8CI); Reg. No. *[1746-81-2]*. Its herbicidal properties were described by K. Härtel (*Meded. Landbouwhogesch. Opzoekingsstn. Staat Gent*, 1962, **27**, 1275); Its history was reported by H. Maier-Bode & K. Härtel (*Residue Rev.*, 1981, **77**, 1). Introduced by Hoechst AG (DEP 1 028 986; GBP 852 422) as code no. 'Hoe 02 747'; trade mark 'Aresin'.

**Properties.**

Pure monolinuron forms colourless crystals; m.p. 80-83 °C; v.p. 6.4 Pa at 65 °C. Solubility at 25 °C: 735 mg tech./l water; soluble in acetone, dioxane, ethanol, xylene. It is stable at its m.p. and in solution, but slowly decomposes in acids and bases, and in moist soil. The technical grade is ⩾95% pure. Decomposes at 220 °C.

**Uses.**

It is absorbed by leaves and roots and is used as a pre-em. herbicide effective against annual grasses and broad-leaved weeds. It is recommended for pre-em. use on beans and maize at 0.5-1.0 kg a.i./ha, on potatoes at 1.0-1.5 kg/ha, on asparagus at 1.0-2.0 kg/ha, in vineyards at 2.0-3.0 kg/ha; also on black currants and ornamental shrubs before bud burst at 1.0-2.0 kg/ha. It is degraded in soil (H. Börner, *Z. Pflanzenkr. Pflanzenpathol. Pflanzenschutz*, 1965, **72**, 516).

**Toxicology.**

Acute oral $LD_{50}$ for rats 2100 mg tech./kg. In 2-y feeding trials NEL for rats was 250 mg/kg diet.

**Formulations.**

'Aresin', w.p. (475 g a.i./kg); 'Arresin Emulsion' (liquid). 'Broadside' (Hoechst), 'Super PWK' (FBC Limited), monolinuron + linuron. 'Gramonol' (Hoechst, ICI Plant Protection Division), (Reg. No. *[69312-59-6]*), monolinuron + paraquat. 'Ivorin' (Hoechst), monolinuron + dinoseb acetate.

**Analysis.**

Residues may be determined by colorimetry (R. L. Dalton & H.L. Pease, *J. Assoc. Agric. Chem.*, 1962, **45**, 377).

$C_9H_{11}ClN_2O$ (198.7)
GR DMVN1&1
*monuron-TCA*
$C_{11}H_{12}Cl_4N_2O_3$ (326.0)
GR DMVN1&1 &&$CCl_3COOH$ SALT

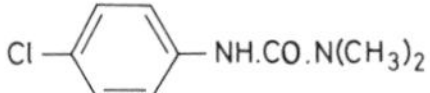

**Nomenclature and development.**

Common name monuron (i) (BSI, E-ISO, F-ISO, ANSI, WSSA);chlorfenidim (USSR);CMU (JMAF);*exception* (Portugal);monuron-TCA (WSSA) for the trichloroacetate salt (ii). Chemical name (IUPAC) 3-(4-chlorophenyl)-1,1-dimethylurea; 3-(4-chlorophenyl)-1,1-dimethyluronium trichloroacetate (ii). (*C.A.*) *N'*-(4-chlorophenyl)-*N,N*-dimethylurea (i); trichloroacetic acid compound with *N'*-(4-chlorophenyl)-*N,N*-dimethylurea (1:1) (9CI); 3-(*p*-chlorophenyl)-1,1-dimethylurea (ii); trichloroacetic acid compound with 3-(*p*-chlorophenyl)-1,1-dimethylurea (1:1) (8CI); Reg. No. *[150-68-5]* (i), *[140-41-0]* (ii). The herbicidal properties of monuron were described by H. C. Bucha & C. W. Todd (*Science*, 1951, **114**, 493). Introduced by E. I. du Pont de Nemours & Co. (Inc.) (who no longer manufacture or market it) and as the trichloroacetate by Allied Chemical Corp. Agricultural Division (now Hopkins Agricultural Chemical Co.) (USP 2 655 445; GBP 691 403; 692 589 (all to du Pont); USP 2 782 112; 2 801 911 (both to Allied)) as code no. 'GC-2996' (Allied) for (ii); trade marks '*Telvar*' (du Pont) for (i); 'Urox' (Allied) for (ii).

**Properties.**

Monuron is a colourless crystalline solid; m.p. 174-175 °C; v.p. 67 µPa at 25 °C; $d^{20}_{20}$ 1.27. Solubility at 25 °C: 230 mg/l water; at 27 °C: 52 g/kg acetone; sparingly soluble in petroleum oils and in polar organic solvents. The rate of hydrolysis at room temperature and pH7 is negligible, but increases at elevated temperatures and under acidic or alkaline conditions. It decomposes at 185-200 °C. It is slowly decomposed in moist soils.

Monuron-TCA is a crystalline solid; m.p. 78-81 °C. Solubility at room temperature: 918 mg/l water; 400 g/kg 1,2-dichloroethane; 177 g/kg methanol; 91 g/kg xylene. It is acidic in reaction and incompatible with alkaline materials.

**Uses.**

Monuron is an inhibitor of photosynthesis and is absorbed via the roots. It is recommended at 10-30 kg a.i./ha for total weed control in non-crop areas; 5-10 kg/ha suffices for annual maintenance. Monuron-TCA is used at 10-15 kg/ha for total weed control of uncropped areas.

**Toxicology.**

Acute oral $LD_{50}$: for rats 3600 mg monuron/kg, 2300-3700 mg monuron-TCA (in corn oil)/kg. Application of monuron to intact or abraded skin of guinea-pigs produced no irritation or sensitisation; monuron-TCA is irritating to skin and mucous membranes. In feeding trials NEL for rats and dogs was 250-500 mg/kg diet (M. C. Hodge *et al., AMA Arch. Ind. Health*, 1958, **17**, 45).

**Formulations.**

A w.p. (800 g monuron/kg). 'Urox' weedkiller: granules (55, 110 or 220 g monuron-TCA/kg); oil-miscible concentrate; oil/water miscible concentrate. Oil/water miscible concentrate (monuron-TCA + 2,4-D).

**Analysis.**

Product analysis of monuron is by hydrolysis and titration of the liberated amine (*CIPAC Handbook*, 1980, **1A**, 1310) and of monuron-TCA by i.r. spectrometry (details from Hopkins Agricultural Chemical Co.). Residues of monuron may be determined by glc (*Anal. Methods Pestic. Plant Growth Regul.*, 1972, **6**, 664; T. H. Byast *et al., Tech. Rep. ARC Weed Res. Organ.*, No. 15 (2nd Ed.), p. 49) or by hplc (*idem, ibid.*).

$C_{12}H_{11}Cl_2NO_4$ (304.1)
T5OVNV EHJ CR CG EG& E1O1 E1

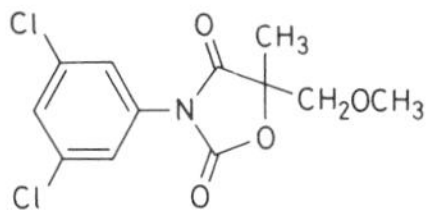

**Nomenclature and development.**

Common name myclozolin (BSI, draft E-ISO); myclozoline (draft F-ISO). Chemical name (IUPAC) (±)3-(3,5-dichlorophenyl)-5-methoxymethyl-5-methyl-1,3-oxazolidine-2,4-dione. (*C.A.*) 3-(3,5-dichlorophenyl)-5-(methoxymethyl)-5-methyl-2,4-oxazolidinedione (9CI); Reg. No. *[54864-61-8]*. Its fungicidal properties were described by E. H. Pommer & B. Zeeh (*Meded. Fac. Landbouwwet. Rijksuniv. Gent*, 1982, **47**, 935). Introduced by BASF AG. (DEOS 2 324 591) as code no. 'BAS 436F'.

**Properties.**

Myclozolin is a colourless crystalline solid; m.p. 111 °C; v.p. 59 µPa at 20 °C. Solubility at 20 °C: 6.7 mg/kg water; 400 g/kg chloroform; 20 g.kg ethanol. It is hydrolysed under alkaline conditions.

**Uses.**

It is a contact fungicide. Diseases caused by *Botrytis, Monilia, Sclerotinia* spp. are controlled well on beans, cucumbers, grapes, groundnuts, lettuce, oilseed rape, ornamentals, stonefruit, strawberries, sunflowers, tomatoes.

**Toxicology.**

Acute oral $LD_{50}$ for rats >5000 mg/kg. Acute percutaneous $LD_{50}$ for rats >2000 mg/kg.

**Formulations.**

A w.p. 330 g a.i./kg; s.c. 330 g/l.

**Analysis.**

Product analysis is by glc with FID. Residues may be determined by glc with ECD.

$C_4H_6N_2Na_2S_4$ (256.3)
SUYSHM2MYUS&SH &&2Na SALT

$Na^+ \ ^-S.CS.NHCH_2CH_2NHCS.S^- \ Na^+$

**Nomenclature and development.**

Common name nabam (BSI, E-ISO);nabame (F-ISO). Chemical name (IUPAC) disodium ethylenebis(dithiocarbamate) (I). (*C.A.*) disodium 1,2-ethanediylbis(carbamodithioate) (9CI); (I) (8CI); Reg. No. *[142-59-6]*. Its fungicidal activity was described by A. E. Dimond *et al.* (*Phytopathology*, 1943, **33**, 1095). Introduced by E. I. du Pont de Nemours & Co. (Inc.) and by Rohm & Haas Co. (though neither company now manufactures it) (USP 2 317 765) as trade marks '*Parzate*' (du Pont), '*Dithane D-14*' (Rohm & Haas).

**Properties.**

Nabam forms colourless crystals of the hexahydrate. Solubility at room temperature: *c.* 200 g (as anhydrous salt)/l water, forming a yellow solution. On aeration, aqueous solutions deposit yellow mixtures of which the main fungicidal components are sulphur and etem (C. W. Pluijgers *et al., Tetrahedron Lett.*, 1971, p. 1371).

**Uses.**

It has been superseded as a protectant fungicide by zineb though the latter may be prepared as a tank mix from solutions of nabam and zinc sulphate, similarly maneb with manganese(II) sulphate. Nabam is too phytotoxic for general use on foliage; soil applications were reported to have a systemic action on *Phytophthora fragariae* (E. M. Stoddard, *Phytopathology*, 1951, **41**, 858). It is also used to control algae in paddy fields.

**Toxicology.**

Acute oral $LD_{50}$ for rats 395 mg/kg. A goitrogenic effect was noted in rats receiving 1000-2500 mg/kg diet for 10 d (R. B. Smith *et al., J. Pharmacol. Exp. Ther.*, 1953, **109**, 159).

**Formulations.**

'Campbell's X-Spor' (aqueous solution).

**Analysis.**

Product analysis is by hydrolysis to carbon disulphide which is converted to dithiocarbonate and this measured by titration with iodine (*CIPAC Handbook*, 1970, **1**, 539; 1983, **1B**, in press).

$C_4H_7Br_2Cl_2O_4P$ (380.8)
GXGEYEOPO&O1&O1

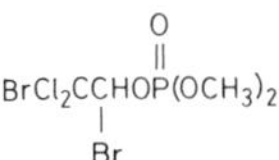

**Nomenclature and development.**

Common name naled (BSI, E-ISO, F-ISO, ANSI); bromchlophos (Republic of South Africa); dibrom (Denmark); BRP (JMAF); *exception* (Sweden). Chemical name (IUPAC) 1,2-dibromo-2,2-dichloroethyl dimethyl phosphate (I). (*C.A.*) (I) (8 & 9CI); Reg. No. *[300-76-5]*. OMS 75, ENT 24 988. Its insecticidal properties were reported by J. M. Grayson & B. D. Perkins (*Pest Control*, 1960, **28**(6), 9). Introduced by Chevron Chemical Co. (GBP 855 157; USP 2 971 882) as code no. 'RE-4355'; trade mark 'Dibrom'.

**Properties.**

Technical grade naled (purity *c.* 93%) is a yellow liquid, with a slightly pungent odour; b.p. 110 °C/0.5 mmHg; v.p. 266 mPa at 20 °C; $d_{20}^{20}$ 1.97; $n_D^{28}$ 1.5108. Solubility: practically insoluble in water; slightly soluble in aliphatic solvents; readily soluble in aromatic solvents. Pure naled has m.p. 26 °C. It is stable under anhydrous conditions, but is rapidly hydrolysed in water ($\geqslant$90% in 48 h at room temperature) and by alkali. It is degraded by sunlight. It is stable in brown glass containers but, in the presence of metals and reducing agents, rapidly loses bromine forming dichlorvos (4490).

**Uses.**

It is a fast-acting, non-systemic, contact and stomach acaricide and insecticide with some fumigant action. It is recommended for use against flies on many crop plants, also in glasshouses and mushroom houses at *c.* 8.1 g a.i./100 m$^3$; and against adult mosquitoes.

**Toxicology.**

Acute oral $LD_{50}$ for rats 430 mg/kg. Acute percutaneous $LD_{50}$ for rabbits 1100 mg/kg. In 2-y feeding trials no ill-effect was observed in albino rats receiving 100 mg tech. (91% pure)/kg diet. $LC_{50}$ (24-h) is: for goldfish 2-4 mg/l; for crabs 0.33 mg/l; there was no mortality to mosquito fish or tadpoles when it was applied at 560 g/ha.

**Formulations.**

An e.c. (960 g a.i./l); dust (40 g/kg).

**Analysis.**

Product analysis is by glc (D. E. Pack *et al.*, *Anal. Methods Pestic., Plant Growth Regul. Food Addit.*, 1964, **2**, 125; J. B. Leary, *Anal. Methods Pestic. Plant Growth Regul.*, 1972, **6**, 350). Residues may be determined by glc (D. E. Pack *et al.*, *loc. cit.*; J. B. Leary, *loc. cit.*). Details of methods are available from Chevron Chemical Co.

$C_{10}H_8$ (129.2)
L66J

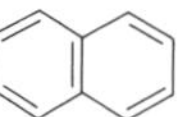

**Nomenclature and development.**

Chemical name (IUPAC) naphthalene (I) (BSI, E-ISO); naphtalène (F-ISO) is accepted in lieu of a common name. (*C.A.*) (I) (8 & 9CI); Reg. No. *[91-20-3]*. It has long been used as a household fumigant against clothes moths.

**Properties.**

Pure naphthalene forms colourless flaky crystals; m.p. 80 °C; b.p. 218 °C; v.p. 6.5 Pa at 20 °C; $d_4^{15}$ 1.517. Solubility at room temperature: 30 mg/l water; 285 g/l benzene, toluene; 500 g/l carbon tetrachloride, chloroform; 77 g/l ethanol, methanol; very soluble in 1,2-dichloroethane, diethyl ether. It is flammable, flash point 79 °C (open cup), 88 °C (closed cup) but otherwise stable.

**Uses.**

Its effectiveness as a household fumigant against clothes moths has been questioned (W. S. Abbott & S. C. Billings, *J. Econ. Entomol.*, 1935, **28**, 493). It is also used as a fumigant against soil fungi, but is rapidly decomposed by soil organisms.

**Toxicology.**

Acute oral $LD_{50}$ for rats 2200 mg/kg.

**Analysis.**

Product analysis is by distillation from mixtures in the presence of ethanol and precipitation as the picrate (D. S. Binnington & W. F. Giddes, *Ind. Eng. Chem. Anal. Ed.*, 1934, **6**, 461; W. L. Millar, *J. Assoc. Off. Agric. Chem.*, 1934, **17**, 308).

$C_{12}H_6O_3$ (198.2)
T666 1A M CVOVJ

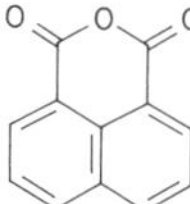

**Nomenclature and development.**

The trivial name naphthalic anhydride (I) (draft E-ISO) is used in lieu of a common name. Chemical name (IUPAC) naphthalene-1,8-dicarboxylic anhydride. (*C.A.*) 1*H*,3*H*-naphtho[1,8-*cd*]pyran-1,3-dione (9CI); (I) (8CI); Reg. No. *[81-84-5]*. Its ability to increase the selectivity of thiocarbamate herbicides to maize was described by O. L. Hoffman (*Weed Sci. Soc. Am., Abstr.*, 1969, No. 12) and reviewed by G. R. Stephenson & F. Y. Chang (*Chemistry and Mode of Action of Herbicide Antidotes*, p. 35). Introduced by Gulf Oil Corp. (USP 3 131 509; 3 564 768) as trade mark 'Protect'.

**Properties.**

Pure naphthalic anhydride is a light tan crystalline solid; m.p. 270-274 °C. Solubility: relatively insoluble in water and most non-polar solvents; 13.9 g/l dimethylformamide. It is stable under normal storage conditions, and is non-hygroscopic. The technical grade is ⩾98% pure.

**Uses.**

It is used as a seed coating, at 5 g/kg seed, prior to planting and protects maize from damage by EPTC, vernolate and related herbicides. Germination of maize seed planted 2 y after treatment was not affected.

**Toxicology.**

Acute oral $LD_{50}$: for rats 12 300 mg/kg; for mallard duck >6810 mg/kg; for bobwhite quail 4100 mg/kg. Acute percutaneous $LD_{50}$ for rabbits >2025 mg/kg; it is moderately irritating to eyes of albino rabbits. Inhalation of the dust by rats for 4 h at 820 mg/$m^3$ air revealed no toxic symptom. In 90-d feeding trials no ill-effect was noted for rats and dogs receiving 500 mg/kg diet. $LC_{50}$ (96-h) is: for rainbow trout 6.0 mg/l; for bluegill 4.9 mg/l.

**Formulations.**

'Protect' technical grade (980 g/kg).

**Analysis.**

Residues in crops may be determined by glc with ECD (J. R. Riden, *Anal. Methods Pestic. Plant Growth Regul.*, 1976, **8**, 483).

**Nomenclature and development.**

The traditional name naphthenic acid (BSI, E-ISO);acide naphténique (F-ISO) is accepted in lieu of a common name. Common salts include copper naphthenate (I); zinc naphthenate (II) (E-ISO); naphténate de cuivre; naphtenate de zinc (F-ISO). Chemical name (IUPAC) naphthenic acid is a group of carboxylic acids derived from a fraction of crude petroleum oils and includes alkylcyclopentanecarboxylic acids and alkylcyclohexanecarboxylic acids containing 9 or 10 carbon atoms. Their salts include (I) and (II). (*C.A.*) (I) and (II) (8 & 9CI); Reg. No. *[1338-02-9]* for (I). Copper naphthenate and zinc naphthenate have long been used to protect wood from fungal attack. Trade marks include 'Cuprinol'.

**Properties.**

Copper naphthenate is a deep greenish viscous oil; v.p. <133 mPa at 100 °C. The zinc salt is similar but almost colourless. Solubility: they are practically insoluble in water; moderately soluble in petroleum oils; soluble in most organic solvents. Naphthenic acid has been reviewed (E. R. Littleman & J. R. M. Klotz, *Chem. Rev.*, 1942, **30**, 97).

**Uses.**

Copper and zinc naphthenates are generally toxic to fungi, on account of both their metal and their acid content. Naphthenic acid and its salts are toxic to growing plants. The salts may be used to disinfect and to preserve wooden seed boxes, benches, stakes etc.

**Formulations.**

'Cuprinol green', e.c. (copper naphthenate in oil, 6-8% copper); 'Cuprinol clear', e.c. (zinc naphthenate in oil).

**Analysis.**

Product analysis is by estimating the copper electrolytically (*MAFF Ref. Book*, 1958, No. 1, p. 16; *AOAC Methods*, 1980, 6.066) or by iodometric titration (*ibid.*, 6.065).

$C_{12}H_{11}NO$ (185.2)
L66J B1VZ

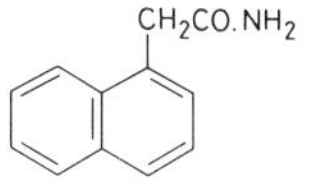

$CH_2CO.NH_2$

**Nomenclature and development.**

Chemical name (IUPAC) 2-(1-naphthyl)acetamide (draft BSI, E-ISO) in lieu of a common name. (*C.A.*) 1-naphthaleneacetamide (8 & 9CI); Reg. No. *[86-86-2]*. Also known as α-naphthaleneacetamide,NAD. Introduced as a thinning agent for apples and pears by Amchem Products, Inc. (now Union Carbide Agricultural Products Co., Inc.) as trade mark 'Amid-Thin'.

**Properties.**

It forms colourless crystals; m.p. 184 °C. Solubility: sparingly soluble in water; soluble in acetone, ethanol, propan-2-ol; insoluble in kerosene. It is stable under normal storage conditions and non-flammable.

**Uses.**

It is used for thinning many apple and pear varieties at 2.5-5.0 g a.i./100 l, a few days after petal fall. It induces formation of an abscission zone in the peduncle.

**Toxicology.**

Acute oral $LD_{50}$ for rats *c.* 6400 mg/kg. Acute percutaneous $LD_{50}$ for rabbits >5000 mg/kg; it is not a skin or eye irritant.

**Formulations.**

'Amid-Thin' W, w.p. (84 g a.i./kg).

**Analysis.**

Product and residue analysis is by glc (details of methods available from Union Carbide Agricultural Products Co., Inc.).

$C_{12}H_{10}O_2$ (186.2)
L66J B1VQ
*ethyl 2-(1-naphthyl)acetate*
$C_{14}H_{14}O_2$ (214.3)
L66J B1VO2

$CH_2CO.OH$

**Nomenclature and development.**

The names 1-naphthylacetic acid or 1-naphthaleneacetic acid (I) (BSI, E-ISO), acide naphtylacétique (F-ISO) are accepted in lieu of a common name. Chemical name (IUPAC) 2-(1-naphthyl)acetic acid. (*C.A.*) (I) (8 & 9CI); Reg. No. *[86-87-3]*. Trivial names NAA, α-naphthaleneacetic acid. Its plant growth-regulating activity was described by F. E. Gardiner *et al.* (*Science*, 1939, **90**, 208). Introduced by Amchem Products Inc. (now Union Carbide Agricultural Products Co., Inc.) and ICI Plant Protection Division as trade marks 'NAA-800', 'Fruitone-N', 'Rootone' (all to Amchem), 'Phyomone' (to ICI).

**Properties.**

It is a colourless crystalline powder; m.p. 134-135 °C. Solubility at 20 °C: 420 mg/kg water; at 26 °C: 10.6 g/l carbon tetrachloride; 55 g/l xylene; very soluble in acetone, ethanol, propan-2-ol. It forms water-soluble salts with bases. The technical grade has m.p. 125-128 °C.

The ethyl ester (Reg. No. *[2122-70-5]*) is a colourless liquid; b.p. 158-160 °C/3mmHg; $d^{25}$ 1.106. Solubility: sparingly soluble in water; very soluble in acetone, ethanol, propan-2-ol; slightly soluble in kerosene, diesel oil.

**Uses.**

2-(1-Naphthyl)acetic acid is a plant growth regulator used at 0.5-1.0 g a.i./100 l to prevent pre-harvest drop of apples, mangoes and pears after the calyx period or after petal fall. It is also used to stimulate rooting of cuttings, sometimes in combination with 4-indol-3-ylbutyric acid.

Ethyl 2-(1-naphthyl)acetate is used as a bark paint to reduce water-shoot production in pruned apple and pear trees and regrowth of unwanted shoots of some ornamental trees.

**Toxicology.**

Acute oral $LD_{50}$: for rats *c.* 1000-5900 mg acid/kg, *c.* 3580 mg ethyl ester/kg; for mice 670 mg a.e. (as sodium salt)/kg. Acute percutaneous $LD_{50}$ for rabbits: >5000 mg acid/kg, >5000 mg ethyl ester/kg; slight to moderate irritation to their skins has been reported on prolonged contact. $LC_{50}$ (8-d) for mallard duck and bobwhite quail was >10 000 mg/kg diet.

**Formulations.**

These include 'Phyomone' (ICI), w.p. (182 g acid/kg); 'NAA-800' (Union Carbide), s.p. (201 g/kg); 'Fruitone-N', w.s.c. (35 g sodium salt/l). 'Tre-Hold Sprout Inhibitor A112', (ethyl ester). Mixtures include: 'Keriroot' (ICI), (acid + captan); 'Rootone-F' (Union Carbide), (methyl ester + 4-indol-3-ylbutyric acid + 2-(1-naphthyl)acetamide + thiram); 'Tre-Hold Tree Wound Dressing', paint (ethyl ester + phenylmercury acetate).

**Analysis.**

Product analysis is by glc of a derivative or by acid-base titration. Residues may be determined by u.v. spectrometry or by colorimetry of a derivative (C. A. Bache *et al.*, *J. Agric. Food Chem.*, 1962, **10**, 365). Details are available from ICI Plant Protection Division and Union Carbide Agricultural Products Co., Inc.

# (2-Naphthyloxy)acetic acid

$C_{12}H_{10}O_3$ (202.2)
L66J CO1VQ

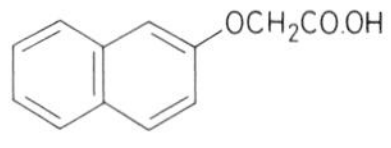

Nomenclature and development.

Chemical name (IUPAC) (2-naphthyloxy)acetic acid (I) (BSI, E-ISO); acide naphtyloxyacétique (F-ISO) is accepted in lieu of a common name. (*C.A.*) 2-naphthalenyloxyacetic acid (9CI); (I) (8CI); Reg. No. *[120-23-0]*. Its effect of increasing fruit setting was reported by S. C. Bausor (*Am. J. Bot.*, 1939, **26**, 415). Introduced by Synchemicals Ltd as trade mark 'Betapal', the methyl ester '*Kamillemittel*' Reg. No. *[1929-87-9]* is no longer marketed.

**Properties.**

The technical product forms green crystals. The pure acid is a colourless crystalline solid, m.p. 156 °C. Solubility at room temperature: sparingly soluble in water; soluble in ethanol, diethyl ether. It forms water-soluble alkali metal and amine salts.

**Uses.**

It is used as a fruit setting spray on grapes, holly, pineapples, strawberries and tomatoes. The normal rate is 40-60 mg a.e./l.

**Toxicology.**

Acute oral $LD_{50}$ for rats 600 mg/kg.

**Formulations.**

'Betapal', solution of the tris(2-hydroxyethyl)ammonium salt (16 g a.e./l) with added surfactant.

**Analysis.**

Residues may be deterimined by hplc (T. E. Archer & J. D. Stokes, *J. Agric. Food Chem.*, 1978. **26**, 452).

# Napropamide 

$C_{17}H_{21}NO_2$ (271.4)
L66J BOY1&VN2&2

$CH_3CHCON(CH_2CH_3)_2$
O

**Nomenclature and development.**

Common name napropamide (BSI, E-ISO, F-ISO, WSSA, JMAF). Chemical name (IUPAC) (*RS*)-*N*,*N*-diethyl-2-(1-naphthyloxy)propionamide (I). (*C.A.*) *N*,*N*-diethyl-2-(1-naphthalenyloxy)propanamide (9CI); I (8CI); Reg. No. *[15299-99-7]*. Its herbicidal properties were described by B. J. van den Brink *et al.* (*Symp. New Herbic., 3rd*, 1969, p. 35). Introduced by Stauffer Chemical Co. (USP 3 480 671; 3 718 455) as code no. 'R-7465'; trade mark 'Devrinol'.

**Properties.**

Napropamide (99.7% pure) is a colourless crystalline solid; m.p. 75 °C; v.p. 530 μPa at 25 °C. Solubility at 20 °C: 73 mg/l water; *c.* 60 g/l kerosene; *c.* 500 g/l xylene; miscible with acetone, ethanol, 4-methylpentan-2-one. No decomposition was observed at 40 °C in 63-d at $4 < pH < 10$. Technical grade (92-96% pure) is a brown solid, m.p. 68-70 °C.

**Uses.**

It is a herbicide effective against annual grasses and certain annual broad-leaved weeds and used in asparagus, brassicas, citrus, grapevines, lima beans, oilseed rape, peppers, sunflowers, tobacco, tomatoes, tree fruits. Also effective against perennial grass weeds. Soil treatments (2-4 kg a.i./ha) should be incorporated into the soil within 2-d of application; for pre-em. surface treatments (3-6 kg/ha) applied to weed-free soil irrigate if no rain falls within 2-d. The (*R*)-(-)-isomer, Reg. No. *[41643-35-0]*, is 8 times as toxic to 3 weed species as the (*S*)-(+)-isomer, Reg. No. *[41643-36-1]* (J. H. Chan *et al.*, *J. Agric. Food Chem.*, 1975, **23**, 1008).

**Toxicology.**

Acute oral $LD_{50}$: for male rats >5000 mg tech./kg, for females 4680 mg/kg. Acute percutaneous $LD_{50}$ for rabbits >4640 mg/kg; not a primary eye nor skin irritant.

**Formulations.**

'Devrinol 2E', e.c. (240 g a.i./l); 'Devrinol 50WP' w.p. (500 g/kg); 'Devrinol 10G' granules (100 g/kg). 'Devrinol T', e.c. (120 g napropamide + 120 g trifluralin/l); 'Tillam Devrinol 4-1E', w.p.. (120 g napropamide/l + 480 g pebulate). Mixture (280 g napropamide + 320 g nitralin/kg).

**Analysis.**

Product analysis is by glc. Crop and soil residues are determined by glc (G. G. Patchett *et al.*, *Anal. Methods Pestic. Plant Growth Regul.*, 1976, **8**, 347). Details of analytical methods are available from the Stauffer Chemical Co.

$C_{18}H_{13}NO_3$ (291.3)
L66J BMVR BVQ
*naptalam-sodium*
$C_{18}H_{12}NNaO_3$ (313.3)
L66J BMVR BVQ &&Na SALT

**Nomenclature and development.**

Common name naptalam (BSO, E-ISO, WSSA);naptalame (F-ISO);alanap (Turkey);NPA for sodium salt (JMAF). Chemical name (IUPAC) *N*-1-naphthylphthalamic acid (I). (*C.A.*) 2-[(1-naphthalenylamino)carbonyl]benzoic acid (9CI); (I) (8CI); Reg. No. *[132-66-1]*. The plant growth-regulating activity of *N*-arylphthalamic acids was reported by O. L. Hoffman & A. E. Smith (*Science*, 1949, **109**, 588). The sodium salt, naptalam-sodium (Reg. No. *[132-67-2]*) was introduced by Uniroyal Inc. (USP 2 556 664; 2 556 665) as trade mark 'Alanap'.

**Properties.**

Naptalam is a crystalline solid, m.p. 185 °C; v.p. $<$133 Pa at 20 °C. Solubility at room temperature 200 mg/l water. Solubility of naptalam-sodium 300 g/kg water. Naptalam is hydrolysed in solutions of pH $>$9.5, and is unstable at elevated temperatures, forming *N*-(1-naphthyl)phthalimide.

**Uses.**

Naptalam inhibits seed germination and is recommended as a pre-em. herbicide for use in cucurbits, groundnuts, soyabeans at 2.0-5.5 kg/ha. Also used as post-em. herbicide in combination with dinoseb or 2,4-DB to control certain dicotyledonous weeds in soyabeans.

**Toxicology.**

Acute oral $LD_{50}$ for rats 8200 mg a.i./kg; 1770 mg/naptalam-sodium/kg. Acute percutaneous $LD_{50}$ for rabbits is $>$2000 mg/kg. Acute inhalation $LC_{50}$ for rats $>$2.07 mg/l air. In 90-d feeding trials no ill-effect was observed in rats and dogs receiving 1000 mg naptalam-sodium/kg diet. $LC_{50}$ (8-d) for mallard duck and bobwhite quail is $>$10 000 mg/kg diet. Long term studies on rats have shown no carcinogenic effect $\leqslant$3000 mg/kg diet. Naptalam was negative in a bacterial mutagenic test. $LC_{50}$ (96-h) is: for bluegill 354 mg/l; for rainbow trout 76 mg/l.

**Formulations.**

These include: 'Alanap-L', naptalam-sodium (240 g/l). 'Rescue' (naptalam + 2,4-DB). 'Dyanap' (Reg. No. *[8075-57-8]*), (naptalam + dinoseb).

**Analysis.**

Product analysis is by u.v. spectrometry, details from Uniroyal Inc. Residues may be determined by hydrolysis to 1-naphthylamine, a derivative of which is measured by colorimetry (A. E. Smith & G. M. Stone, *Anal. Methods Pestic., Plant Growth Regul. Food Addit.*, 1964, **4**, 1).

$C_{12}H_{16}Cl_2N_2O$ (275.2)
GR BG DMVN4&1

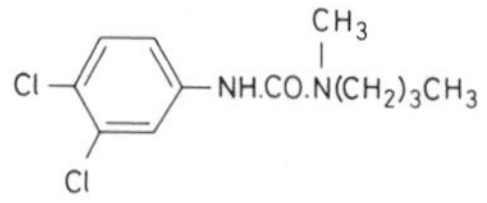

CH3
Cl—C6H3(Cl)—NH.CO.N(CH2)3CH3

**Nomenclature and development.**

Common name neburon (BSI, E-ISO, F-ISO, ANSI, WSSA);neburea (Republic of South Africa). Chemical name (IUPAC) 1-butyl-3-(3,4-dichlorophenyl)-1-methylurea (I). (*C.A.*) *N*-butyl-*N'*-(3,4-dichlorophenyl)-*N*-methylurea (9CI); (I) (8CI); Reg. No. *[555-37-3]*. Its herbicidal properties were described by H. C. Bucha & C. W. Todd (*Science*, 1951, **144**, 493). Introduced by E. I. du Pont de Nemours & Co. (Inc.) (USP 2 655 444; 2 655 445) as trade mark 'Kloben'.

**Properties.**

Pure neburon forms colourless crystals; m.p. 102-103 °C. Solubility at 24 °C: 4.8 mg/l; sparingly soluble in common hydrocarbon solvents. Hydrolysis is negligible at ambient temperatures and in neutral pH range. Hydrolysis rate increases at higher temperatures or in acid or alkaline media.

**Uses.**

It inhibits photosynthesis and is absorbed through plant roots. It is recommended at 2-3 kg a.i./ha for the pre-em. control of annual weeds and grasses in lucerne, strawberries, wheat and nursery plantings of certain woody ornamentals.

**Toxicology.**

Acute oral $LD_{50}$ for rats >11 000 mg/kg. A suspension (150 g a.i./l dimethyl phthalate) applied to the shaved backs of guinea-pigs produced only mild irritation and no sensitisation. Chronic toxicity is low.

**Formulations.**

'Kloben' weedkiller, w.p. (600 g a.i./kg). Mixtures include: neburon + isoproturon; neburon + linuron + trifluralin; neburon + pendimethalin; neburon + terbutryne;

**Analysis.**

Residues may be determined by glc with ECD (T. H. Byast *et al.*, *Tech. Rep. ARC Weed Res. Organ.*, No. 15 (2nd Ed.), p. 49; *Anal. Methods Pestic., Plant Growth Regul. Food Addit.*, 1964, **4**, 157; *Anal. Methods Pestic. Plant Growth Regul.*, 1972, **6**, 664).

$C_{13}H_8Cl_2N_2O_4$ (327.1)
WNR CG DMVR BQ EG
*niclosamide-(2-hydroxyethyl)ammonium*
$C_{15}H_{15}Cl_2N_3O_5$ (388.2)
WNR CG DMVR BQ EG
&&$HOCH_2CH_2NH_2$ SALT

**Nomenclature and development.**

Common name niclosamide (BSI, E-ISO, F-ISO, BPC);*exception* niclosamide for veterinary use, clonitralid (for 2-hydroxyethylammonium salt) for public health use (Federal Republic of Germany). Chemical name (IUPAC) 2',5-dichloro-4'-nitrosalicylanilide (I); 5-chloro-*N*-(2-chloro-4-nitrophenyl)salicylamide. (*C.A.*) 5-chloro-*N*-(2-chloro-4-nitrophenyl)-2-hydroxybenzamide (9CI); (I) (8CI); Reg. No. *[50-65-7]* (free base), *[1420-04-8]* (2-hydroxyethylammonium salt). Its molluscicidal properties were described by R. Gönnert & E. Schraufstätter (*Proc. Int. Conf. Trop. Med. Malar.*, 1958, **2**, 5) and its development discussed by R. Gönnert *et al.* (*Z. Naturforsch. Teil B*, 1961, **16**, 95). The 2-hydroxyethylammonium salt was introduced as a molluscicide by Bayer AG (DEPS 1 126 374; USP 3 079 297; 3 113 067) as code no. 'Bayer 25 648','Bayer 73','SR 73'; ready-to-use formulations have the trade marks 'Bayluscid', 'Bayluscide'.

**Properties.**

Niclosamide is an almost colourless solid; m.p. 230 °C; v.p. <1 mPa at 20 °C. Solubility at room temperature 5-8 mg/l water.

The 2-hydroxyethylammonium salt is a yellow solid; m.p. 216 °C. Solubility at room temperature 180-280 mg/l water. It is stable to heat and is hydrolysed by concentrated acid or alkali. Its biological activity is not affected by hard water.

**Uses.**

Niclosamide is a powerful molluscicide giving a total kill of *Australorbis glabratus* at 0.3 mg/l. It is non-phytotoxic at field concentrations. It is used in the treatment of tapeworm infestation as 'Yomesan'.

**Toxicology.**

Rats and rabbits survived single oral doses of 5000 mg niclosamide/kg; rats 5000 mg 2-hydroxyethylammonium salt/kg; cats vomited a dose of 500 mg salt/kg but tolerated 250 mg salt/kg without ill-effect. Acute intraperitoneal $LD_{50}$ for rats 250 mg salt/kg. $LC_{50}$ (48-h) is: for rainbow trout 50 μg salt/l; for carp 235 μg/l.

**Formulations.**

These include: e.c. (250 g a.i./l); w.p. (700 g/kg).

**Analysis.**

Product analysis is by redox titration. Residues in water may be measured by colorimetry (R. Strufe, *Pflanzenschutz-Nachr. (Engl. Ed.)*, 1962, **15**, 42) or by titration (details from Bayer AG).

$C_{10}H_{14}N_2$ (162.2)
T6NJ C- BT5NTJ A1

Nomenclature and development.

The trivial names nicotine (BSI, E-ISO, F-ISO);nicotine sulphate (BSI, E-ISO);sulfate de nicotine (F-ISO);nicotine sulfate (JMAF) for its sulphate are accepted in lieu of a common name. Chemical name (IUPAC) (*S*)-3-(1-methylpyrrolidin-2-yl)pyridine. (*C.A.*) (*S*)-3-(1-methyl-2-pyrrolidinyl)pyridine (9CI); nicotine (8CI); Reg. No. *[54-11-5]*. Extracts of tobacco have long been used against sucking insects but have been replaced by technical nicotine and nicotine sulphate.

**Properties.**

Pure nicotine is a colourless liquid; b.p. 247 °C; m.p. −80 °C; v.p. 5.65 Pa at 25 °C; $d_4^{20}$ 1.009; $n_D^{22.4}$ 1.5239; $[\alpha]_D^{20}$ −161.55°. Solubility: miscible with water below 60 °C, forming a hydrate, and above 210 °C; miscible with ethanol, diethyl ether; readily soluble in most organic solvents. It darkens slowly and becomes viscous on exposure to air. It is a base, $pK_1$ 6.16, $pK_2$ 10.96 forming salts with acids. The predominant component of the crude alkaloid extract is (*S*)-(−)-nicotine, small amounts of related alkaloids may be present.

**Uses.**

Nicotine is a non-persistent, non-systemic contact insecticide. It is used as a fumigant in closed spaces, *e.g.* in glasshouses.

**Toxicology.**

Acute oral $LD_{50}$ for rats 50-60 mg/kg. Acute percutaneous $LD_{50}$ (single application) for rabbits 50 mg/kg. It is toxic to man by inhalation and by dermal contact.

**Formulations.**

Nicotine is marketed as the technical alkaloid (950 g a.i./l) or as nicotine sulphate (400 g/kg); the addition of soap or alkali is required to dilutions of the latter to liberate the nicotine; also as dusts (30-50 g/kg). For fumigation, nicotine is applied to a heated metal surface or nicotine 'shreds' are burnt.

**Analysis.**

Product analysis is by steam distillation and precipitation as silicotungstate (*AOAC Methods*, 1980, 6.175-6.176; *MAFF Ref. Book*, No.1, 1958, p. 59) (*CIPAC Handbook*, 1970, **1**, 543; 1983, **1A**, 1316). Residues may be determined by glc (R. J. Martin, *J. Assoc. Off. Anal. Chem.*, 1967, **50**, 939).

$C_{13}H_{19}N_3O_6S$ (345.4)
WS1&R CNW ENW DN3&3

$NO_2$
$CH_3SO_2$ — — $N[(CH_2)_2CH_3]_2$
$NO_2$

**Nomenclature and development.**

Common name nitralin (BSI, E-ISO, F-ISO, WSSA, JMAF). Chemical name (IUPAC) 4-methylsulphonyl-2,6-dinitro-*N*,*N*-dipropylaniline. (*C.A.*) 4-(methylsulfonyl)-2,6-dinitro-*N*,*N*-dipropylbenzenamine (9CI); 4-(methylsulfonyl)-2,6-dinitro-*N*,*N*-dipropylaniline (8CI); Reg. No. *[4726-14-1]*. Its herbicidal properties were described by J. B. Regan *et al.* (*Proc. Northeast. Weed Control Conf.*, 1966, p. 36). Introduced by Shell Research Ltd (BEP 672 199) as code no. 'SD 11 831'; trade mark 'Planavin'.

**Properties.**

Technical nitralin (*c.* 94% pure) is a yellow powder; m.p. 151-152 °C; v.p. 2 μPa at 25 °C; ρ 1.39 g/cm$^3$ at 20 °C. Solubility at 22 °C: 600 μg/l water; 360 g/l acetone; 330 g/l dimethyl sulphoxide; 250 g/l 2-nitropropane; sparingly soluble in common hydrocarbon and aromatic solvents and alcohols. Contact of nitralin with concentrated bases or temperatures >200 °C should be avoided.

**Uses.**

It is a selective herbicide acting by disruption of primary cell wall formation during cell division. Pre-em. control of annual grasses and many broad-leaved weeds is successful in cotton, groundnuts, rape, soyabeans and transplanted crops such as brassicas, tobacco, tomatoes using 0.5-0.75 kg a.i./ha on light soils, 1.0-1.5 kg/ha on heavier soils (<5% o.m.), not used in highly organic or peat soils. Shallow mechanical incorporation into the upper 2-4 cm of soil is recommended. It is relatively immobile in soil and 50% loss occurs in *c.* 30-50 d.

**Toxicology.**

Acute oral $LD_{50}$ for rats and mice >2000 mg/kg; for mallard duck >2000 mg/kg. Acute percutaneous $LD_{50}$ for rabbits >2000 mg/kg. In 2-y feeding trials NEL for rats and dogs was 2000 mg/kg diet. $LC_{50}$ (48-h) for rainbow trout 31 mg/l.

**Formulations.**

A w.p. (750 g/kg).

**Analysis.**

Product analysis is by i.r. spectrometry or by glc (*Anal. Methods Pestic. Plant Growth Regul.*, 1973, **7**, 625). Residues may be determined by glc with ECD (*ibid.*). Details of methods are available from Shell International Chemical Co. Ltd.

$C_6H_3Cl_4N$ (230.9)
T6NJ BXGGG FG

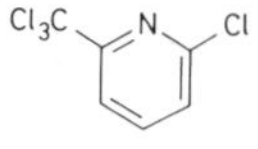

**Nomenclature and development.**

Common name nitrapyrin (BSI, E-ISO, ANSI);nitropyrine (F-ISO). Chemical name (IUPAC) 2-chloro-6-trichloromethylpyridine. (*C.A.*) 2-chloro-6-( trichloromethyl)pyridine (8 & 9CI); Reg. No. *[1929-82-4]*. Introduced as a soil bactericide by Dow Chemical Co. (USP 3 135 594; GBP 960 109) as code no. 'Dowco 163'; trade mark 'N-Serve'.

**Properties.**

Pure nitrapyrin is a colourless crystalline solid; m.p. 62-63 °C; v.p. 370 mPa at 23 °C. Solubility at 22 °C: 40 mg/kg water; 540 g/kg anhydrous ammonia; 300 g/kg ethanol; at 20 °C: 1.98 kg/kg acetone; 1.85 kg/kg dichloromethane; at 26 °C, 1.04 kg/kg xylene.

**Uses.**

It is a nitrogen stabiliser at rates from 480 g a.i./ha because of its highly selective action as a soil bactericide against *Nitrosomonas* spp., the micro-organisms that oxidise ammonium ions in soil. It is hydrolysed in soil to 6-chloropicolinic acid (Reg. No. *[4684-94-0]*) which is absorbed by plants and is the principal metabolite (C. T. Redeman *et al., J. Agric. Food Chem.*, 1964, **12**, 207; 1965, **13**, 518).

**Toxicology.**

Acute oral $LD_{50}$: for rats 1072-1232 mg/kg; for chickens 235 mg/kg. Acute percutaneous $LD_{50}$ for rabbits 2830 mg/kg. In 94-d feeding trials NEL was: for rats 300 mg/kg diet; for dogs 600 mg/kg diet. In 2-y feeding studies on the metabolite (6-chloropicolinic acid) NEL was: for rats 15 mg/kg daily; for dogs 50 mg/kg daily. The $LC_{50}$ (8-d) is: for mallard duck 1466 mg/kg diet; for Japanese quail 820 mg/kg diet. $LC_{50}$ for channel catfish is 5.8 mg/l; it caused no mortality of Ramshorn snails or *Daphnia* at 10 mg/l.

**Formulations.**

These include: tech. grade (>90% pure); non-emulsifiable concentrates and e.c. (240 g a.i./l).

**Analysis.**

Product analysis is by glc, details may be obtained from Dow Chemical Co. Residues of 6-chloropicolinic acid, the main degradation product, may be determined by glc of the methyl ester (D. J. Jensen, *ibid.*, 1971, **19**, 897; C. T. Redeman, *Bull. Environ. Contam. Toxicol.*, 1967, **2**, 289).

$C_{12}H_7Cl_2NO_3$ (284.1)
WNR DOR BG DG

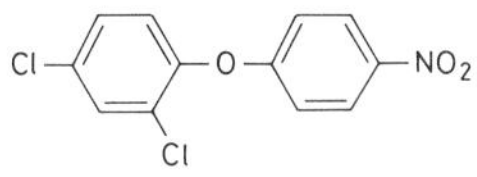

**Nomenclature and development.**

Common name nitrofen (BSI, E-ISO, WSSA);nitrofène (F-ISO);niclofen (Canada);NIP (JMAF);*exception* Federal Republic of Germany. Chemical name (IUPAC) 2,4-dichlorophenyl 4-nitrophenyl ether. (*C.A.*) 2,4-dichloro-1-(4-nitrophenoxy)benzene (9CI); 2,4-dichlorophenyl *p*-nitrophenyl ether (8CI); Reg. No. *[1836-75-5]*. Introduced by Rohm & Haas Co. (GBP 974 475; USP 3 080 225) as code no. 'FW-925'; trade marks 'Tok E-25', 'Tokkorn'.

**Properties.**

It is a crystalline solid; m.p. 70-71 °C; v.p. 1.06 mPa at 40 °C. Solubility at 22 °C 0.7-1.2 mg/l water.

**Uses.**

It is a selective herbicide, toxic to a number of broad-leaved and grass weeds, and effective when left as a thin layer on the soil surface; activity is rapidly lost on incorporation in soil. It is used for weed control in cereals pre-em. at 2 kg a.i./ha. Many vegetable crops tolerate applications pre-em. $\leqslant$3 kg/ha and post-em. $\leqslant$1 kg/ha.

**Toxicology.**

Acute oral $LD_{50}$: for rats 638-888 mg a.i. (as e.c.)/kg; for rabbits 300-510 mg a.i. (as e.c.)/kg. Neither the a.i. nor the formulations caused irritation to skin of rabbits or visible toxic effects. Women of child-bearing age should not be exposed to the product.

**Formulations.**

These include: 'Tok E-25', 'Tokkorn', e.c. (250 g a.i./kg; 240 g/l); 'Tok WP-50', w.p. (500 g/kg).

**Analysis.**

Product analysis is by glc with FID (I. L. Adler & B. M. Jones, *Anal. Methods Pestic. Plant Growth Regul.*, 1978, **10**, 403). Residues may be determined by glc with ECD of a derivative (*idem., ibid.*).

$C_{14}H_{17}NO_6$ (295.3)
1Y1&OVR CNW EVOY1&1

CO.OCH(CH$_3$)$_2$

$O_2N$ CO.OCH(CH$_3$)$_2$

**Nomenclature and development.**

Common name nitrothal-isopropyl (BSI, E-ISO, F-ISO);nitrothale-isopropyl (France). Chemical name (IUPAC) di-isopropyl 5-nitroisophthalate (I). (*C.A.*) bis(1-methylethyl) 5-nitro-1,3-benzenedicarboxylate (9CI); (I) (8CI); Reg. No. *[10552-74-6]*. Its herbicidal properties were reported by W. H. Phillips *et al.* (*Proc. Br. Insectic. Fungic. Conf., 7th,* 1973, **2**, 673). Introduced by BASF AG as code no. 'BAS 30 000F','BAS 38 501F' (mixture with sulphur),'BAS 37 900F' (mixture with zinc ammoniate ethylenebisdithiocarbamate-polyethylenethiuram disulphide); trade marks 'Kumulan' (mixture with sulphur), 'Pallinal' (mixture with zinc ammoniate ethylenebisdithiocarbamate-polyethylenethiuram disulphide).

**Properties.**

It forms yellow crystals; m.p. 65 °C; v.p. $<$10 μPa at 20 °C. Solubility at 20 °C: 390 μg/l water; $>$1 kg/kg acetone, benzene, chloroform, ethyl acetate.

**Uses.**

Nitrothal-isopropyl is a non-systemic fungicide effective against powdery mildews. Sprays of 50 g a.i./100 l every $\leqslant$14 days have controlled *Podosphaera leucotricha* without effect on *Panonychus ulmi*. It is used in combinations with other fungicides.

**Toxicology.**

Acute oral $LD_{50}$: for rats 12 800 mg a.i. (as 50% w.p.)/kg; for rabbits $>$10 000 mg/kg. In 90-d feeding trials NEL was: for rats 500 mg/kg diet, for dogs 20 000 mg/kg diet.

**Formulations.**

Mixtures include: 'Kumulan', (Reg. No. *[61583-33-3]*), (167 g nitrothal-isopropyl + 533 g sulphur/kg). 'Pallinal', (Reg. No. *[55257-78-8]*), (125 g nitrothal-isopropyl + 600 g zinc ammoniate ethylenedithiocarbamate-poly(ethylenethiuram disulphide)/kg).

**Analysis.**

Product analysis is by glc. Residues may be determined by glc with ECD. Details of methods are available from BASF AG.

$C_{33}H_{25}N_{3}O_{3}$ (511.6)
T C555 A AY DVMV IUTJ AUYR&-
BT6NJ& 1XQR&- BT6NJ

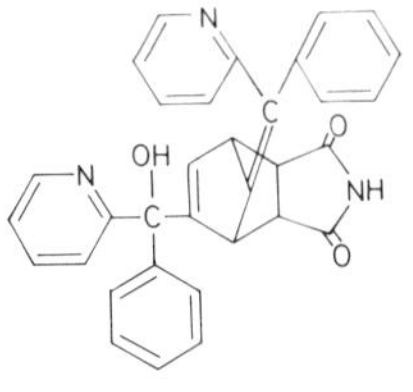

**Nomenclature and development.**

Common name norbormide (E-ISO, ANSI);nobormide (F-ISO). Chemical name (IUPAC) 5-(α-hydroxy-α-2-pyridylbenzyl)-7-(α-2-pyridylbenzylidene)-8,9,10-trinorborn-5-ene-2,3-dicarboximide; 5-(α-hydroxy-α-2-pyridylbenzyl)-7-(α-2-pyridylbenzylidene)bicyclo[2.2.1]hept-5-ene-2,3-dicarboximide. (*C.A.*) 3a,4,7,7a-tetrahydro-5-(hydroxyphenyl-2-pyridyl-methyl)-7-(phenyl-2-pyridylmethylene)-4,7-methano-1*H*-isoindole-1,3(2*H*)-dione (9CI); 5-(α-hydroxy-α-2-pyridylbenzyl)-7-(α-2-pyridylbenzylidene)-5-norbornene-2,3-dicarboximide (8CI); Reg. No. *[991-42-4]*. ENT 51 762. Its rodenticidal properties were described by A. P. Roszkowski *et al., Science*, 1964, **144**, 412. Introduced by the McNeil Laboratories Inc. (GBP 1 059 405) as code no. 'McN-1025'; trade marks 'Shoxin', 'Raticate'.

**Properties.**

It is a colourless to off-white crystalline powder; m.p. >160 °C. Solubility at room temperature: 60 mg/l water; at 30 °C: 14 mg/l ethanol; >150 mg/l chloroform; 1 mg/l diethyl ether; 29 mg/l 0.1M hydrochloric acid. It is a mixture of stereoisomers, stable at room temperature when dry, and to boiling water; hydrolysed by alkali.

**Uses.**

It is a selective rodenticide lethal to all members of the genus *Rattus* that have been tested; the acute oral $LD_{50}$ for *R. rattus* is 52 mg/l, for *R. norvegicus* 11.5 mg/kg, for *R. hawaiiensis c.* 10 mg/kg. it is relatively non-lethal to other rodents.

**Toxicology.**

Acute oral $LD_{50}$: for mice 2250 mg/kg; for hamsters 140 mg/kg; for prairie dogs >1000 mg/kg; a single oral dose of 1000 mg/kg had no effect on cats, dogs, monkeys, chickens, turkeys; nor was it lethal to any of 40 other animal species tested (A. P. Roszkowski, *J. Pharmacol. Exp. Ther.*, 1965, *149*, 288). Three adult male volunteers received 15 mg/kg daily for 3 d without ill-effect.

**Formulations.**

'Raticate', concentrate (5-10 g a.i./kg) for bait preparation in cereal offals, fish meal, etc.

**Analysis.**

Product analysis is by u.v. spectrometry (C. A. Janick *et al., J. Pharm. Sci.*, 1966, **55**, 1077).

$C_{12}H_9ClF_3N_3O$ (303.7)
T6NNVJ BR CXFFF& DG EM1

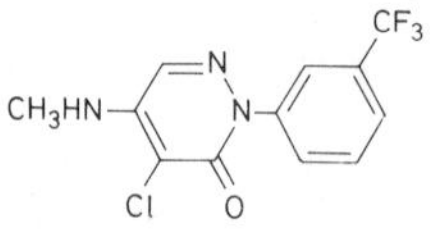

**Nomenclature and development.**

Common name norflurazon (BSI, E-ISO, ANSI, WSSA);norflurazone (F-ISO). Chemical name (IUPAC) 4-chloro-5-methylamino-2-(α,α,α-trifluoro-*m*-tolyl)pyridazin-3(2*H*)-one; 4-chloro-5-methylamino-2-(3-trifluoromethylphenyl)pyridazin-3-one. (*C.A.*) 4-chloro-5-(methylamino)-2[3-(trifluoromethyl)phenyl]-3(2*H*)-pyridazinone (9CI); 4-chloro-5-methylamino-2-(α,α,α-trifluoro-*m*-tolyl)-3(2*H*)-pyridazinone (8CI); Reg. No. *[27314-13-2]*. Introduced by Sandoz AG (USP 3 644 355) as code no. 'H 52 143','H 9789'; trade marks 'Zorial', 'Evital', 'Solicam'.

**Properties.**

Pure norflurazon is a colourless crystalline solid; m.p. 174-180 °C; v.p. 2.8 μPa at 20 °C. Solubility at 25 °C: 28 mg/l water; 50 g/l acetone; 142 g/l ethanol; 2.5 g/l xylene. It is stable in aqueous solution at pH 3-9 (<8% loss within 24 d). It is stable upon storage (shelf life at 20 °C ≥4 y). It is rapidly degraded by sunlight (F. A. Eder *et al.*, *Proc. Eur. Weed Res. Com. Symp. Herbic. - Soil*, 1973, p. 202).

**Uses.**

It inhibits photosynthesis by reduction of carotenoid biosynthesis. It is a selective herbicide used in citrus, cotton, cranberries, nuts, pome fruits, soyabeans, stone fruits. It is effective against many annual broad-leaved weeds at 1-4 kg a.i./ha; and also suppresses perennial grass and sedge species (*Agropyron, Cynodon, Cyperus* and *Sorghum* spp.). In soil 50% loss occurs in 21-28 d according to soil type and method of application. In plants a major metabolic pathway is *N*-demethylation.

**Toxicology.**

Acute oral $LD_{50}$: for rats >8000 mg/kg; for bobwhite quail and mallard duck >1250 mg/kg. Acute percutaneous $LD_{50}$ for rabbits >20 000 mg/kg. In 2-y feeding trials NEL for rats was 375 mg/kg diet; in 90-d feeding trials for dogs 12.5 mg/kg daily. $LC_{50}$ for catfish and goldfish is >200 mg/l.

**Formulations.**

These include: 'Solicam', 'Zorial', w.p. (800 g a.i./kg); 'Evital', granules (50 g/kg). 'Telok', granules (40 g norflurazon + 20 g simazine/kg).

**Analysis.**

Product analysis is by glc with FID (S. S. Brady *et al.*, *Anal. Methods Pestic. Plant Growth Regul.*, 1978, **10**, 415), or by tlc, followed by u.v. spectromety of the eluted compound (details from Sandoz AG). Residues may be determined by glc with ECD (*idem, ibid.*).

$C_{17}H_{12}ClFN_2O$ (314.7)
T6N CNJ EXQR BG&R DF

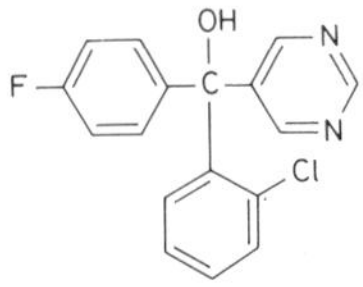

**Nomenclature and development.**
Common name nuarimol (BSI, E-ISO, F-ISO, ANSI). Chemical name (IUPAC) (±)-2-chloro-4'-fluoro-α-(pyrimidin-5-yl)benzhydryl alcohol. (*C.A.*) α-(2-chlorophenyl)-α-(4-fluorophenyl)-5-pyrimidinemethanol (9CI); Reg. No. *[63284-71-9]*. Introduced by Eli Lilly & Co. (GBP 1 218 623) as 'EL-228'; trade marks 'Trimidal', 'Triminol', 'Gauntlet', 'Murox'.

**Properties.**
It is a colourless crystalline solid; m.p. 126-127 °C; v.p. < 2.7 μPa at 25 °C. Solubility at 25 °C: 26 mg/l water at pH 7; 170 g/l acetone; 55 g/l methanol; 20 g/l xylene. It breaks down rapidly in sunlight to form a large number of minor degradation products.

**Uses.**
It is a systemic fungicide with activity against a wide range of plant-pathogenic fungi. It is used as a foliar spray at 40 g a.i./ha to control *Erysiphe graminis* on barley and as a seed treatment at 100-200 mg/kg for barley and wheat to control *E. graminis, Leptosphaeria nodorum, Pyrenophora graminea* and *Ustilago* spp.

**Toxicology.**
Acute oral $LD_{50}$: for male rats 1250 mg/kg, for females 2500 mg/kg; for male mice 2500 mg/kg, for females 3000 mg/kg; for beagle dogs 500 mg/kg; for bobwhite quail 200 mg/kg. Acute percutaneous $LD_{50}$ for rabbits >2000 mg/kg; no skin irritation was seen at this dose, slight irritation was observed when 0.1 ml (71 mg) was applied to their eyes. In 2-y feeding trials NEL for rats and mice was 50 mg/kg diet. In 7-d trials no observable effect was produced in bluegill exposed to 1.1 mg/l in a continuous flow-through system.

**Formulations.**
These include: e.c., s.c. and solution. 'Trimisem Total', w.p. (65 g nuarimol + 165 g anthraquinone + 165 g gamma-HCH + 265 g maneb). 'Trimidal GT'.

**Analysis.**
Product analysis is by glc and FID. Residues in plant tissue and soils may be measured by glc with ECD. Detailed analytical procedures and information on nuarimol may be obtained from Eli Lilly & Co.

$C_{13}H_{16}O_2$ (204.3)
T B656 HO DU KUTJ GVH

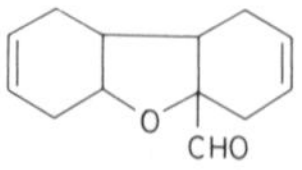


**Nomenclature and development.**

Chemical name (IUPAC) 1,4,4a,5a,6,9,9a,9b-octahydrodibenzofuran-4a-carbaldehyde. (*C.A.*) 1,5a,6,9,9a,9b-hexahydro-4a(4*H*)-dibenzofurancarboxaldehyde (8 & 9CI); Reg. No. *[126-15-8]*. Butadiene-furfural copolymer; 2,3:4,5-bis(2-butylene)tetrahydrofurfural. ENT 17 596. Its repellent action on cockroaches was described by L. D. Goodhue & C. Linnaid (*J. Econ. Entomol.*, 1952, **45**, 133). Introduced by Phillips Petroleum Co. and by McLaughlin Gormley King Co. (USP 2 683 511; 2 934 471; 2 683 151) as 'Phillips Repellent 11'; 'MGK Repellent 11' (McLaughlin Gormley King Co.).

**Properties.**

It is a pale yellow liquid, with a fruity odour; b.p. 307 °C; $n_D^{20}$ 1.5420, the technical grade has $d_{20}^{20}$ 1.120. It is practically insoluble in water and dilute alkali; miscible with ethanol, petroleum oils, toluene, xylene. Stable for long periods in drums.

**Uses.**

It is an insect repellent effective against cockroaches, gnats, mosquitoes, stable and horse flies. Its main uses are in combination with other materials in pet sprays and repellants for personal use.

**Toxicology.**

Acute oral $LD_{50}$ for rats 2500 mg/kg. Acute percutaneous $LD_{50}$ for rabbits >2000 mg/kg. In 90-d feeding trials rats receiving 20 000 mg/kg diet suffered no gross ill-effect. $LC_{50}$ (8-d) for mallard duck and bobwhite quail >5000 mg/kg diet. $LC_{50}$ (96-h) is: for rainbow trout 22.8 mg/l; for bluegill 18.1 mg/l.

**Formulations.**

Oil solution (10-50 g a.i./l); aerosol concentrates.

**Analysis.**

Product analysis is by glc, details available from McLaughlin Gormley King Co.

$C_{10}H_{22}OS$ (190.3)
Q2S8

$CH_3(CH_2)_7SCH_2CH_2OH$

**Nomenclature and development.**

Chemical name (IUPAC) 2-(octylthio)ethanol (I). (*C.A.*) (I) (8 & 9CI); Reg. No. *[3547-33-9]*. 2-Hydroxyethyl octyl sulfide. Its insect-repelling properties were described by L. D. Goodhue (*J. Econ. Entomol.*, 1960, **53**, 805). Introduced by Phillips Petroleum Co. and by the McLaughlin Gormley King Co. (USP 2 863 799 to Phillips Petroleum Co.) as code no. 'Phillips R-874'; trade mark 'MGK Repellent R-874'.

**Properties.**

It is a pale amber liquid, with a mild mercaptan-like odour; b.p. 98 °C/0.1 mmHg; m.p. 0 °C; $d_4^{20}$ 0.925-0.935; $n_D^{20}$ 1.470-1.478. It is slightly soluble in water; miscible with most organic solvents including refined kerosene, though a co-solvent such as propan-2-ol is required with the latter to maintain solution at low temperatures. It is stable at pH 6-8; and to heat and light.

**Uses.**

Its chief use is as a cockroach repellent at 1 g a.i./m$^2$. Also as a repellent using pressurised 'patio' foggers against ants, flies and crawling insects. It is not normally applied to plants.

**Toxicology.**

Acute oral $LD_{50}$ for rats 8530 mg/kg. Acute percutaneous $LD_{50}$ for albino rabbits 13 590 mg/kg. The application of 0.05 mg to the cornea of albino rats produced corneal necrosis in 2 of the 5 eyes tested but they healed without opacities. In 90-d feeding trials rats receiving 20 000 mg/kg diet showed no adverse effect. $LC_{50}$ (8-d) for mallard duck and bobwhite quail was >5000 mg/kg diet. $LC_{50}$ (96-h) is: for rainbow trout 3.1 mg/l; for bluegill 5.0 mg/l.

**Formulations.**

An e.c. or oil solution (10-50 g a.i./l).

**Analysis.**

Details of a glc method are available from McLaughlin Gormley King Co.

$C_5H_{12}NO_4PS$ (213.2)
1OPO&O1&S1VM1

$$CH_3NHCO.CH_2SP(O)(OCH_3)_2$$

**Nomenclature and development.**

Common name omethoate (BSI, E-ISO, F-ISO);*exception* (Italy). Chemical name (IUPAC) *O,O*-dimethyl *S*-methylcarbamoylmethyl phosphorothioate. (*C.A.*) *O,O*-dimethyl *S*-[2-(methylamino)-2-oxoethyl] phosphorothioate (9CI); *O,O*-dimethyl phosphorothioate *S*-ester with 2-mercapto-*N*-methylacetamide (8CI); *[1113-02-6]*. Its insecticidal properties described by R. Santi & P. de Pietri-Tonelli (*Nature (London)*, 1959, **183,** 398). Introduced by Bayer AG (DEAS 1 251 304) as code no. 'Bayer 45 432', 'S 6876'; trade mark 'Folimat'.

**Properties.**

It is a colourless to yellow oil, decomposing *c.* 135 °C; v.p. 3.3 mPa at 20 °C; $d_4^{20}$ 1.32; $n_D^{20}$ 1.4987. Miscible with water, acetone, ethanol and many hydrocarbons; slightly soluble in diethyl ether; almost insoluble in light petroleum. On hydrolysis at 24 °C, 50% decomposition occurs in 2.5 d at pH7.

**Uses.**

Omethoate is a systemic acaricide and insecticide effective against aphids, beetles, caterpillars, *Eriosoma lanigerum*, scale insects, thrips, especially on hops and against *Delia coarotata* on wheat. The usual rate is 50 g a.i./100 l.

**Toxicology.**

Acute oral $LD_{50}$ for rats *c.* 50 mg/kg. Acute dermal $LD_{50}$ (7-d) for rats 700 mg/kg. In 90-d feeding trials NEL for rats was 1 mg/kg diet. $LC_{50}$ (96-h) for goldfish is 10-100 mg/l.

**Formulations.**

Soluble concentrates of various a.i. content; concentrates for ULV application.

**Analysis.**

Details of product analysis are available from Bayer AG. Residues may be determined by glc with FID (K. Wagner & H. Frehse, *Pflanzenschutz-Nachr. (Engl. Ed.)*, 1976, **29**, 54; *Pestic. Anal. Man.*, 1979, **I**, 201-H, 201-I; D. C. Abbott *et al., Pestic. Sci.*, 1970, **1**, 10; *Analyst (London)*, 1977, **102**, 858; M. A. Luke *et al., J. Assoc. Off. Anal. Chem.*, 1981, **64**, 1187) or by oxidation to phosphoric acid which is measured colorimetrically by standard methods. Details are available from Bayer AG.

$C_{12}H_{18}N_4O_6S$ (346.4)
ZSWR CNW ENW DN3&3

$NO_2$
$H_2NSO_2$ — $N[(CH_2)_2CH_3]_2$
$NO_2$

**Nomenclature and development.**

Common name oryzalin (BSI, E-ISO, F-ISO, ANSI, WSSA). Chemical name (IUPAC) 3,5-dinitro-*N*$^4$,*N*$^4$-dipropylsulphanilamide. (*C.A.*) 4-(dipropylamino)-3,5-dinitrobenzenesulfonamide (9CI); 3,5-dinitro-*N*$^4$,*N*$^4$-dipropylsulfanilamide (8CI); Reg. No. *[19044-88-3]*. Its herbicidal properties were described by J. V. Gramlich *et al.* (1969 abstr. WSSA). Introduced by Eli Lilly & Co. (USP 3 367 949) as code no. 'EL-119'; trade marks 'Dirimal', 'Ryzelan', 'Surflan'.

**Properties.**

Technical grade oryzalin is a yellow-orange crystalline solid; m.p. 141-142 °C. Solubility at 25 °C: 2.4 mg/l water; >500 g/l acetone; >50 g/l methanol; 1.6 g/l xylene.

**Uses.**

It is recommended for use, alone or with other herbicides, as a pre-em. herbicide on cotton, groundnuts, winter oilseed rape, soyabeans and sunflower at 1.0-2.0 kg a.i./ha.

**Toxicology.**

Acute oral $LD_{50}$: for rats and gerbils >10 000 mg tech./kg; for cats and chickens 1000 mg/kg; for dogs >1000 mg/kg; for bobwhite quail and mallard duck >500 mg/kg. Acute percutaneous $LD_{50}$ for rabbits >2000 mg/kg; it caused a slight irritation to their skin but not eyes. In 2-y feeding trials NEL for rats was 300 mg/kg diet. $LC_{50}$ (96-h) for goldfish fingerlings is >1.4 mg/l.

**Formulations.**

A w.p. (750 g a.i./kg). 'Dirimal Extra' (Reg. No. *[52452-96-7]*) (oryzalin + diuron).

**Analysis.**

Product analysis is by glc. Residues may be determined by glc of a suitable derivative (O. D. Decker & W. S. Johnson, *Anal. Methods Pestic. Plant Growth Regul.*, 1976, **8**, 433). Particulars from Eli Lilly & Co.

$C_{15}H_{18}Cl_2N_2O_3$ (345.2)
T5NNVOJ BR BG DG EOY1&1&
EX1&1&1

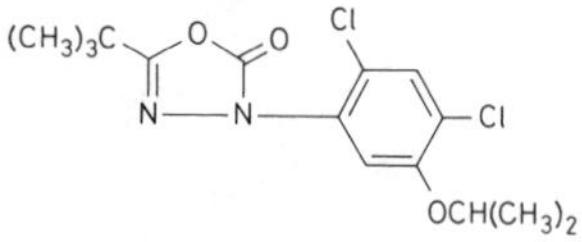

**Nomenclature and development.**

Common name oxadiazon (BSI, E-ISO, F-ISO, ANSI, WSSA, JMAF). Chemical name (IUPAC) 5-*tert*-butyl-3-(2,4-dichloro-5-isopropoxyphenyl)-1,3,4-oxadiazol-2(3*H*)-one. (*C.A.*) 3-[2,4-dichloro-5-(1-methylethoxy)phenyl]-5-(1,1-dimethylethyl)-1,3,4-oxadiazol-2(3*H*)-one (9CI); 2-*tert*-butyl-4-(2,4-dichloro-5-isopropoxyphenyl)-$\Delta^2$-1,3,4-oxadiozolin-5-one (8CI); Reg. No. *[19666-30-9]*. Its herbicidal properties were described by L. Burgaud *et al.* (*Symp. New Herbic., 3rd*, 1969, p. 201). Introduced by Rhône-Poulenc Phytosanitaire (GBP 1 110 500; USP 3 385 862) as code no. '17 623 RP'; trade mark 'Ronstar'.

**Properties.**

Oxadiazon forms colourless crystals; m.p. *c.* 90 °C; v.p. <133 μPa at 20 °C. Solubility at 20 °C: 0.7 mg/l water; 600 g/l acetone, acetophenone, anisole; 1 kg/l benzene, chloroform, toluene; 100 g/l ethanol, methanol. It is stable under normal storage conditions.

**Uses.**

It is a selective herbicide, effective against mono- and di-cotyledonous weeds in rice at *c.* 1 kg/ha; in orchards and vineyards at 2 kg/ha post-em. or 4 kg/ha pre-em. Under temperate conditions at these rates 50% loss from soil occurs in 90-180 d.

**Toxicology.**

Acute oral $LD_{50}$: for rats >8000 mg/kg; for bobwhite quail 6000 mg/kg; for mallard duck 1000 mg/kg. Acute percutaneous $LD_{50}$ for rats >8000 mg/kg. In feeding trials rats and dogs receiving 25 mg/kg daily were not affected.

**Formulations.**

These include e.c. (250 g a.i./l); granules (20 g/kg).

**Analysis.**

Product analysis is by glc (J. Desmoras *et al., Anal. Methods Pestic. Plant Growth Regul.*, 1973, **7**, 595). Residues may be determined by glc (*idem, ibid.*).

$C_7H_{13}N_3O_3S$ (219.3)
1N1&VYS1&UNOVM1

$(CH_3)_2NCO.C(SCH_3)=NO.CO.NHCH_3$

**Nomenclature and development.**

Common name oxamyl (BSI, ANSI, draft E-ISO);oxamil (JMAF). Chemical name (IUPAC) *N*,*N*-dimethyl-2-methylcarbamoyloxyimino-2-(methylthio)acetamide; *N*,*N*-dimethyl-α-methylcarbamoyloxyimino-α-(methylthio)acetamide; *S*-methyl *N*'*N*'-dimethyl-*N*-(methylcarbamoyloxy)-1-thio-oxamimidate (I). (*C.A.*) methyl 2-( dimethylamino)-*N*-[[(methylamino)carbonyl]oxy]-2-oxoethanimidothioate (9CI); (I) (8CI); Reg. No. *[23135-22-0]*. Introduced by E.I. du Pont de Nemours & Co. (Inc.) (USP 3 530 220; 3 658 870) as code no. 'DPX 1410'; trade mark 'Vydate'.

**Properties.**

Pure oxamyl forms colourless crystals, with a slight sulphurous odour; m.p. 100-102 °C; changing to a dimorphic form m.p. 108-110 °C; v.p. 31 mPa at 25 °C. Solubility at 25 °C: 280 g/kg water; 670 g/kg acetone; 330 g/kg ethanol; 1.44 kg/kg methanol; 10 g/kg toluene. The solid and formulations are stable; aqueous solutions decompose slowly, a change accelerated by aeration, sunlight, alkali, and elevated temperatures. Decomposes in soil to non-toxic substances, 50% loss in *c*. 7 d.

**Uses.**

It controls insects, mites and nematodes on many field crops, fruits, ornamentals and vegetables. A foliar spray (0.28-1.12 kg a.i./ha) is effective against a broad range of these pests; soil applications can utilise its systemic activity. It is translocated from the foliage of certain plants to their roots (C. A. Peterson *et al.*, *Pestic. Biochem. Physiol.*, 1978, **8**, 1).

**Toxicology.**

Acute oral $LD_{50}$: for rats 5.4 mg a.i./kg, 8.9 mg a.i. (as liquid formulation)/kg; for quail 4.18 mg a.i./kg. Acute percutaneous $LD_{50}$ for male rabbits 710 mg a.i. (as liquid formulation)/kg; non-irritating to skin of guinea-pigs. Inhalation $LC_{50}$ (1-h) for male rats 0.17 mg a.i./l air using a dust cloud, 0.035 mg a.i. (as liquid formulation)/kg for spray mist (head exposures only). In 2-y feeding trials NEL was: for rats 50 mg a.i./kg diet; for dogs 100 mg/kg diet. $LC_{50}$ (8-d) was: for bobwhite quail 54 mg a.i. (as liquid formulation)/kg diet; for mallard duck 369 mg a.i. (as liquid formulation)/kg diet. $LC_{50}$ (96-h) is: for bluegill 5.6 mg a.i./l; for rainbow trout 4.2 mg/l; for goldfish 27.5 mg/l.

**Formulations.**

These include: 'Vydate' L, water soluble liquid (240 g/l); 'Vydate' G, granules (100 g/kg).

**Analysis.**

Product analysis is by hplc (R. F. Holt & R. E. Leitch, *Anal. Methods Pestic. Plant Growth Regul.*, 1978. **10**, 111). Residues may be determined by glc with FPD (*idem, ibid.*; R. F. Holt & H. L. Pease, *J. Agric. Food Chem.*, 1976, **24**, 263) or by hplc (J. E. Thean *et al.*, *J. Assoc. Off. Anal. Chem.*, 1978, **61**, 15).

$C_{18}H_{12}CuN_2O_2$ (351.9)
D566 1A L BND-CU-OJ C-& CD566 1A
L BND-CU-OJ

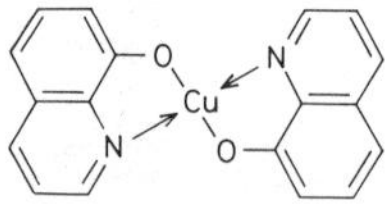

**Nomenclature and development.**

Common name oxine-copper (JMAF);oxine-Cu (either name BSI, E-ISO);oxine-cuivre or oxine-Cu (F-ISO);oxyquinoléate de cuivre (France). Chemical name (IUPAC) bis(quinolin-8-olato)copper (I); cupric 8-quinolinoxide. (*C.A.*) bis(8-quinolinato-*N*',*O*[8])copper (9CI); (I) (8CI); Reg. No. *[10380-28-6]*. The fungicidal properties of this complex of copper with quinolin-8-ol, the latter known by the trivial name oxine, were described by D. Powell (*Phytopathology*, 1946, **36**, 572).

**Properties.**

Oxine-copper is a greenish-yellow powder, stable ≤200 °C and non-volatile. Solubility: insoluble in water and common organic solvents; slightly soluble in pyridine. It is stable, chemically inert, and not decomposed by u.v. light.

**Uses.**

It is a protectant fungicide used against foliar diseases of soft fruit, top fruit and nuts. Mixtures with insecticides and other fungicides are used as seed treatments. A mixture with bitumen is effective in sealing wounds and pruning cuts on trees.

**Toxicology.**

Acute oral $LD_{50}$ for rats *c.* 10 000 mg/kg. It is non-irritating to skin. $LC_{50}$ (48-h) for brown and rainbow trout is 0.2-0.3 mg/l (J. S. Alabaster, *Proc. Br. Weed Control Conf., 4th*, 1958, p. 84). It is non-toxic to honeybees.

**Formulations.**

'Quinolate 400', e.c. (400 g a.i./l); 'Arbrex 805', mixture with bitumen. Seed treatments include: oxine-copper + anthraquinone + carboxin + gamma-HCH; oxine copper + anthraquinone + gamma-HCH; oxine-copper + anthraquinone + endosulfan + gamma-HCH; oxine-copper + endosulfan + gamma-HCH; oxine-copper + gamma-HCH;

**Analysis.**

Product analysis is by reaction with sulphuric acid and standard methods for estimation of copper (*AOAC Methods*, 1980, 6.015-6.016; *CIPAC Handbook*, 1970, **1**, 226).

$C_{12}H_{13}NO_4S$ (267.3)
T6O DSW BUTJ B1 CVMR

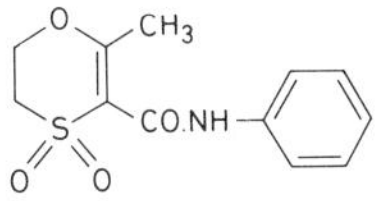

**Nomenclature and development.**

Common name oxycarboxin (BSI, E-ISO, ANSI, JMAF);oxycarboxine (F-ISO). Chemical name (IUPAC) 5,6-dihydro-2-methyl-1,4-oxathi-ine-3-carboxanilide 4,4-dioxide; 5,6-dihydro-2-methyl-1,4-oxathi-in-3-carboxanilide 4,4-dioxide (I); 2,3-dihydro-6-methyl-5-phenylcarbamoyl-1,4-oxathi-in 4,4-dioxide. (*C.A.*) 5,6-dihydro-2-methyl-*N*-phenyl-1,4-oxathiin-3-carboxamide 4,4-dioxide (9CI); (I) (8CI); Reg. No. *[5259-88-1]*. Its fungicidal properties were described by B. von Schmeling & M. Kulka (*Science*, 1966, **152**, 659). Introduced by Uniroyal Inc. (USP 3 399 214; 3 402 241; 3 454 391) as code no. 'F 461'; trade mark 'Plantvax'.

**Properties.**

It forms colourless crystals; m.p. 127.5-130 °C; v.p. <133 Pa at 20°C. Solubility at 25 °C: 1 g/l water; 180 g/kg acetone; 1530 g/kg dimethyl sulphoxide; 17 g/kg ethanol; 33 g/kg methanol. Compatible with all except highly acidic or highly alkaline pesticides.

**Uses.**

Oxycarboxin is a systemic fungicide used for the treatment of rust diseases of cereals, ornamentals and vegetables at 200-400 g a.i./ha.

**Toxicology.**

Acute oral $LD_{50}$ for rats 2000 mg/kg. Acute dermal $LD_{50}$ for rabbits >16 000 mg/kg. In 2-y feeding trials with rats and dogs, levels at up to 3000 mg/kg diet had no adverse effect. $LC_{50}$ (8-d) for bobwhite quail was >10 000 mg/kg diet; for mallard duck >4640 mg/kg diet. $LC_{50}$ (96-h) is: for bluegill 28.1 mg/l; for rainbow trout 19.9 mg/l and water flea 69.1 mg/l.

**Formulations.**

These include: e.c. (200 g a.i./l); w.p. (750 g/kg).

**Analysis.**

Product analysis is by hplc or by i.r. spectrometry. Details of methods are available from Uniroyal Chemical Co. Residue analysis is by hydrolysis and determination of the aniline so formed, either by glc (H. R. Siskin & J. E. Newell, *J. Agric. Food Chem.*, 1971, **19**, 738) or by colorimetry of a derivative (J. R. Lane, *ibid.*, 1970, **18**, 409).

$C_6H_{15}O_4PS_2$ (246.3)
OS2&2SPO&O1&O1

$CH_3CH_2\overset{O}{\overset{\|}{S}}CH_2CH_2S\overset{O}{\overset{\|}{P}}(OCH_3)_2$

**Nomenclature and development.**

Common name oxydemeton-methyl (BSI, E-ISO, F-ISO);metilmerkaptofosoksid (USSR). Chemical name (IUPAC) *S*-2-ethylsulphinylethyl *O,O*-dimethyl phosphorothioate. (*C.A.*) *S*-[2-(ethylsulfinyl)ethyl] *O,O*-dimethyl phosphorothioate (8 & 9CI); Reg. No. *[301-12-2]*. ENT 24 964. Its insecticidal properties were described by G. Schrader (*Die Entwicklung neuer insektizider Phosphorsäure-Ester)*. Introduced by Bayer AG (DEP 947 368; USP 2 963 505) as code no. 'Bayer 21 097','R 2170'; trade mark 'Metasystox-R'.

**Properties.**

It is a clear amber-coloured liquid; m.p. $<-10$ °C; b.p. 106 °C/0.01 mmHg; v.p. 3.8 mPa at 20 °C; $d_4^{20}$ 1.289; $n_D^{20}$ 1.5216. Miscible with water; solubility at 20 °C: 100-1000 g/l dichloromethane, propan-2-ol; sparingly soluble in light petroleum. It is hydrolysed in alkaline media.

**Uses.**

It is a systemic and contact insecticide suitable for the control of sap-feeding insects and mites, with a range of action similar to that of demeton-S-methyl (3920) of which it is a metabolic product.

**Toxicology.**

Acute oral $LD_{50}$ for rats 65-80 mg/kg. Acute intraperitoneal $LD_{50}$ for rats 20 mg/kg. Acute percutaneous $LD_{50}$ for male rats 250 mg/kg. In 84-d feeding trials rats receiving 20 mg/kg diet suffered a slight depression of cholinesterase. $LC_{50}$ (24-h) for rainbow trout and bluegill is 10 mg/l.

**Formulations.**

Include water-soluble concentrate (500 g a.i./l); e.c. (250 g/l with emulsifier); aerosol concentrate.

**Analysis.**

Product analysis is by oxidation to phosphoric acid which is measured by standard colorimetric methods (*CIPAC Handbook*, 1983, **1B**, in press), or by reduction of the sulphoxide group by titanium(III) sulphate and titration of the excess (details from Bayer AG). Residues in plants, soil or water may be determined, after oxidation to the corresponding sulphone (demeton-S-methyl sulphone, 3930), by glc with FID (K. Wagner & J. S. Thornton, *Pflanzenschutz-Nachr. (Engl. Ed.)*, 1977, **30**, 1; J. H. van der Merwe & W. B. Taylor, *ibid.*, 1971, **24**, 259; *Pestic. Anal. Man.*, 1979, **I**, 201-H, 201-I; D. C. Abbott *et al.*, *Pestic. Sci.*, 1970, **1**, 10; *Anal. Methods Pestic. Plant Growth Regul.*, 1972, **6**, 432).

# Oxydeprofos

$C_7H_{17}O_4PS_2$ (260.3)
OS2&1Y1&SPO&O1&O1

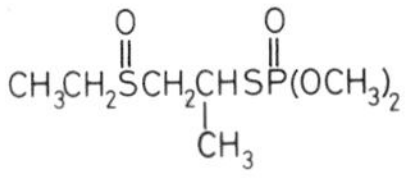

$$\underset{\displaystyle CH_3}{CH_3CH_2\overset{O}{\overset{\|}{S}}CH_2\underset{|}{C}HS\overset{O}{\overset{\|}{P}}(OCH_3)_2}$$

**Nomenclature and development.**

Common name oxydeprofos (BSI, draft E-ISO);ESP (JMAF). Chemical name (IUPAC) *S*-2-ethylsulphinyl-1-methylethyl *O,O*-dimethyl phosphorothioate. (*C.A.*) *S*-[2-(ethylsulfinyl)-1-methylethyl] *O,O*-dimethyl phosphorothioate (8 & 9CI); Reg. No. *[2674-91-1]*. ENT 25 647. Its insecticidal properties were described by G. Schrader (*Die Entwicklung neuer insektizider Phosphorsäure-Ester*, 3rd Ed.). Introduced by Bayer AG. (DEP 1 035 958; USP 2 952 700) as code no. 'Bayer 23 655','S 410'; trade marks 'Metasystox S', 'Estox'.

**Properties.**

It is a yellow oil; b.p. 115 °C/0.02 mmHg; v.p. 625 μPa at 20 °C; $d_4^{20}$ 1.257; $n_D^{25}$ 1.5149. It is soluble in water, alcohols, chlorinated hydrocarbons, ketones; sparingly soluble in light petroleum.

**Uses.**

It is a systemic and contact insecticide and acaricide, effective against sap-feeding insects and mites at 25 g a.i./100l.

**Toxicology.**

Acute oral $LD_{50}$ for rats 105 mg a.i./kg. Acute percutaneous $LD_{50}$ for male rats 800 mg/kg. Intraperitoneal $LD_{50}$ for guinea-pigs 100 mg/kg. In 50-d feeding trials the growth of rats receiving 10 mg/kg daily was not affected. $LC_{50}$ (48-h) for carp and Japanese killifish is >40 mg/l.

**Formulations.**

An e.c. (500 g a.i./l).

**Analysis.**

Product analysis is by titration after reduction by titanium disulphate. Residues may be determined by glc. Details of both methods are available from Bayer AG.

$C_{12}H_{14}N_2$ (186.3)
T6KJ A1 D- DT6KJ A1
*paraquat dichloride*
$C_{12}H_{14}Cl_2N_2$ (257.2)
T6KJ A1 D- DT6KJ A1 &&2Cl SALT

**Nomenclature and development.**

Common name paraquat (BSI, E-ISO, F-ISO, ANSI, WSSA, JMAF);*exception* (Federal Republic of Germany). Chemical name (IUPAC) 1,1'-dimethyl-4,4'-bipyridinium ion (I); 1,1'-dimethyl-4,4-bipyridyldiylium ion. (*C.A.*) 1,1'-dimethyl-4,4'-bipyridinium (8 & 9CI); Reg. No. *[4685-14-7]* (ion), *[1910-42-5]* (dichloride), *[2074-50-2]* (di(methyl sulphate)). Methyl viologen,the dichloride was introduced as an oxidation-reduction indicator by L. Michaelis - *Biochem. Z.*, **1932, 250, 564.** The herbicidal properties of the dichloride and *di(methyl sulphate)* were described by R. C. Brian (*Nature (London)*, 1958, **181**, 446) and their properties reviewed by A. Calderbank (*Adv. Pest Control Res.*, 1968, **8**, 127). Both salts (though only the former is still sold) were introduced by ICI Plant Protection Division (GBP 813 531) as code no. 'PP 148' for dichloride, 'PP 910' for di(methyl sulphate); trade marks 'Gramoxone', 'Dextrone X', 'Esgram'; and for various mixtures 'Cleansweep' (Shell), 'Weedol', 'Dexuron', 'Tota-Col', 'Gramuron', 'Para-Col', 'Pathclear', 'Gramonol' (all ICI).

**Properties.**

The dichloride forms colourless crystals decomposing *c.* 300 °C; v.p. negligible at room temperature. Very soluble in water; slightly in short-chain alcohols; insoluble in hydrocarbons. Neither salt is normally isolated from the technical products which are >95% pure. The salts are stable in neutral and acid media but are oxidised under alkaline conditions. They are inactivated by inert clays and by anionic surfactants. The unformulated salts are corrosive to common metals; when diluted they present no significant hazard to spray equipment.

**Uses.**

Paraquat destroys green plant tissue by contact action with some translocation. It is rapidly inactivated on contact with soil. Uses include stubble cleaning (140-840 g a.i./ha); pasture renovation (140-2210 g/ha), *i.e.* the killing of unproductive grass in such a way that the ground can be resown without ploughing; inter-row weed control in vegetable crops (560-1120 g/ha); desiccation of various crops (140-1680 g/ha); weed control in plantation crops (280-560 g/sprayed ha).

**Toxicology.**

Acute oral $LD_{50}$: for rats 150 mg paraquat ion/kg; for mice 104 mg/kg; for dogs 25-50 mg/kg; for Rhode Island hens 262 mg/kg; for sheep 70 mg/kg. Acute percutaneous $LD_{50}$ for rabbits *c.* 236 mg/kg; paraquat is absorbed through human skin only after prolonged exposure; shorter exposures can cause irritation and a delay in the healing of cuts and wounds; it is an irritant to eyes, can cause temporary damage to nails and, if inhaled, may cause nose bleeding. In 2-y feeding trials NEL was: for dogs 34 mg/kg diet; for rats 170 mg/kg diet. Toxicity to fish depends on the formulation and wetter used, $LC_{50}$ (96-h) is: for rainbow trout 32 mg/l; for brown trout 2.5-13 mg/l.

**Formulations.**

These include paraquat dichloride as: aqueous concentrates (100-240 g paraquat/l); water-soluble granules (25 g/kg). 'Weedol', granules (25 g paraquat + 25 g diquat/kg, as dichloride and dibromide respectively). 'Dexuron' (Chipman), paraquat + diuron. 'Gramonol', (Reg. No. *[69312-67-0]*), paraquat + monolinuron.

**Analysis.**

Product analysis is by colorimetry after reduction (*AOAC Methods*, 1980, 6.332-6.337; *CIPAC Handbook*, 1970, **1**, 547); impurities are determined by glc (*ibid.*, 1980, **1A**, 1317; FAO Specification (CP/50), *Herbicides 1977*, pp. 52, 54). Residues may be determined by colorimetry after reduction (*Pestic. Anal. Man.*, 1979, **II**; A. Calderbank & S. H. Yuen, *Analyst (London)*, 1965, **90**, 99; T. H. Byast *et al., Tech. Rep. ARC Weed Res. Organ.*, No. 15 (2nd Ed.), p. 35; P. F. Lott *et al., J. Chromatogr. Sci.*, 1978, **16**, 390; J. B. Leary, *Anal. Methods Pestic. Plant Growth Regul.*, 1978, **10**, 321; M. G. Ashley, *Pestic. Sci.*, 1970, **1**, 101). Details of methods are available from ICI Plant Protection Division.

$C_{10}H_{14}NO_5PS$ (291.3)
WNR DOPS&O2&O2

**Nomenclature and development.**

Common name parathion (BSI, E-ISO, F-ISO, JMAF); thiophos (USSR); *exception* parathion-ethyl (in some countries). Chemical name (IUPAC) *O*,*O*-diethyl *O*-4-nitrophenyl phosphorothioate. (*C.A.*) *O*,*O*-diethyl *O*-(4-nitrophenyl) phosphorothioate (9CI); *O*,*O*-diethyl *O*-*p*-nitrophenyl phosphorothioate (8CI); Reg. No. *[56-38-2]*. OMS 19, ENT 15 108. Discovered by G. Schrader (cited by H. Martin & H. Shaw, *BIOS, Final Report*, 1946, No. 1095). Introduced by, in turn, American Cyanamid Co. (who no longer manufacture or market it), Bayer AG, ICI Plant Protection Ltd (now ICI Plant Protection Division), Monsanto Chemical Co. (DEP 814 152; USP 1 893 018; 2 842 063) as code no. 'ACC 3422' to Cyanamid, 'E-605' to Bayer; trade marks '*Thiophos*' (to Cyanamid), 'Bladan', 'Folidol' (both to Bayer), 'Fosferno' (to ICI), 'Niran' (to Monsanto).

**Properties.**

Pure parathion is a pale yellow liquid; b.p. 157-162 °C/0.6 mmHg; v.p. 5.0 mPa at 20°C; $d_4^{25}$ 1.265; $n_D^{25}$ 1.5370. Solubility at 25°C: 24 mg/l water; slightly soluble in petroleum oils; completely miscible with most organic solvents. The technical product is a brown liquid with a garlic-like odour. It is rapidly hydrolysed in alkaline media; at 25 °C pH5-6, 1% in 62 d. On heating it isomerises to the *O*,*S*-diethyl isomer.

**Uses.**

It is a non-systemic contact and stomach-acting insecticide and acaricide with some fumigant action. It is non-phytotoxic except to some ornamentals and, under certain weather conditions, to pears and some apple varieties. It is very effective for controlling soil-dwelling insect pests.

**Toxicology.**

Acute oral $LD_{50}$: for male rats 13 mg/kg, for females 3.6 mg/kg; for mallard duck 1.9-2.1 mg/kg; for pheasant 12.4 mg/kg; for pigeon 2.5 mg/kg. Acute percutaneous $LD_{50}$: for male rats 21 mg/kg, for females 6.8 mg/kg. $LC_{50}$ (96-h) is: for rainbow trout 1.5 mg/l; for golden orfe 0.57 mg/l; for fathead minnow 1.4-2.7 mg/l. Its toxicity is enhanced by metabolic oxidation to diethyl 4-nitrophenyl phosphate.

**Formulations.**

These include: w.p. and e.c. of various a.i. content; also dusts, smokes, and aerosol concentrates. Mixtures with various other insecticides are available.

**Analysis.**

Product analysis is by glc or hplc (*CIPAC Handbook*, 1983, **1B**, in press; *AOAC Methods*, 1980, 6.379-6.387) or by redox titration (*ibid*., 1980, 6.388-6.394; WHO Specifications for Pesticides 1961, 2nd Ed., p. 70; FAO Specification (CP/32)) or by alkaline hydrolysis and spectrometry of the liberated 4-nitrophenol (*CIPAC Handbook*, 1970, **1**, 550; *AOAC Methods*, 1980, 6.395-6.399). Residues may be determined by glc (*ibid*., 1980, 29.001-29.018, 29.039-29.043; *Pestic. Anal. Man.*, **I**, 201-A, 201-G, 201-H, 201-I; D.C. Abbott *et al., Pestic. Sci.*, 1970, **1**, 10; *Analyst (London)*, 1977, **102**, 858).

$C_8H_{10}NO_5PS$ (263.2)
WNR DOPS&O1&O1

$NO_2-C_6H_4-OP(=S)(OCH_3)_2$

**Nomenclature and development.**

Common name parathion-methyl (BSI, E-ISO, F-ISO);methyl parathion (ESA, JMAF);metaphos (USSR). Chemical name (IUPAC) *O,O*-dimethyl *O*-4-nitrophenyl phosphorothioate. (*C.A.*) *O,O*-dimethyl *O*-(4-nitrophenyl) phosphorothioate (9CI); *O,O*-dimethyl *O*-(*p*-nitrophenyl) phosphorothioate (8CI); Reg. No. *[298-00-0]*. OMS 213, ENT 17 292. Its insecticidal activity was described by G. Schrader (*Angew. Chem. Monograph* No. 52 (2nd Ed.), 1952). Introduced by Bayer AG (DEP 814 142) as trade marks 'Folidol-M', 'Metacide', 'Bladan M', 'Nitrox 80', formerly '*Dalf*'.

**Properties.**

Pure parathion-methyl forms colourless crystals; m.p. 35-36 °C; v.p. 1.3 mPa at 20 °C; $d_4^{20}$ 1.358; $n_D^{35}$ 1.5515 Solubility: at 25 °C 55-60 mg/l water; at 20 °C >1 kg/l dichloromethane; sparingly soluble in light petroleum; 100-1000 g/l propan-2-ol. The technical grade (*c.* 80% pure) is a light to dark tan-coloured liquid; f.p. *c.* 29 °C; $d^{20}$ 1.20-1.22. It is hydrolysed by alkali and readily isomerises to the *O,S*-dimethyl analogue on heating.

**Uses.**

It is a non-systemic contact and stomach insecticide with some fumigant action. It is generally recommended at 15-25 g a.i./100 l. It is non-persistent.

**Toxicology.**

Acute oral $LD_{50}$: for male rats 14 mg/kg, for females 24 mg/kg. Acute percutaneous $LD_{50}$ for rats 67 mg/kg. In 2-y feeding trials NEL for rats was 2 mg/kg diet. $LC_{50}$ (96-h) is: for rainbow trout 2.7 mg/l; for golden orfe 6.9 mg/l.

**Formulations.**

These include e.c., w.p. and dusts of various a.i. content.

**Analysis.**

Product analysis is by glc or hplc (*CIPAC Handbook*, 1983, **1B**, in press) or by hydrolysis to 4-nitrophenol which is determined colorimetrically (*ibid.*, 1970, **1**, 568; 1980, **1A**, 568). Residues may be determined by glc (E. Möllhoff, *Pflanzenschutz-Nachr. (Engl. Ed.)*, 1968, **21**, 331; *AOAC Methods*, 1980, 29.001-29.018; *Pestic. Anal. Man.*, 1979, **I**, 201-A, 201-G, 201-H, 201-I; D. C. Abbott *et al., Pestic. Sci.*, 1970, **1**, 10).

$C_{10}H_{21}NOS$ (203.3)
4N2&VS3

$CH_3(CH_2)_3$—NCO.S$CH_2CH_2CH_3$
|
$CH_2CH_3$

**Nomenclature and development.**

Common name pebulate (BSI, E-ISO, F-ISO, WSSA, JMAF). Chemical name (IUPAC) *S*-propyl butyl(ethyl)thiocarbamate (I). (*C.A.*) *S*-propyl butylethylcarbamothioate (9CI); (I) (8CI); Reg. No. *[1114-71-2]*. Its herbicidal properties were described by E. O. Burt (*Proc. South. Weed Conf., 12th*, 1959, p.19). Introduced by Stauffer Chemical Co. (USP 3 175 897) as code number 'R-2061'; trade mark 'Tillam'.

**Properties.**

A clear colourless liquid, with an aromatic odour; b.p. 142 °C/21 mmHg; v.p. 4.7 Pa at 25 °C; $d_4^{20}$ 0.956; $n_D^{30}$ 1.474. Solubility at 20 °C: 60 mg/l water; miscible with acetone, ethanol, kerosene, 4-methylpentan-2-one, xylene. At 40 °C in water, 50% loss occurs in 11 d. at $4<pH<10$. Stable $<200$ °C.

**Uses.**

A pre-em. herbicide controlling annual grasses, nut sedges and broad-leaved weeds in sugar beet, tomatoes and transplanted tobacco by soil incorporation at 4-6 kg a.i./ha. Rapidly taken up by roots, translocated throughout the plant and metabolised to carbon dioxide.

**Toxicology.**

Acute oral $LD_{50}$: for rats 1120 mg tech./kg. Acute percutaneous $LD_{50}$ for rabbits 4640 mg/kg. Rapidly metabolised in animals: about 50% of the dose administered to rats was expired as carbon dioxide in 3 d, about 25% excreted in the urine and 5% in the faeces. $LC_{50}$ (7-d) for bobwhite quail was 8400 mg/kg diet. $LC_{50}$ (96-h) for rainbow trout and bluegill is *c.* 7.4 mg/l.

**Formulations.**

'Tillam E', e.c. (720 g a.i./l); 'Tillam G', granules (100 or 250 g a.i./kg). 'Tillam Devrinol 4-1 E', (Reg. No. *[51052-53-0]*), e.c. (480 g pebulate + 120 g napropamide/l).

**Analysis.**

Product analysis is by glc (*AOAC Methods*, 1980, 6.426-6.430; *CIPAC Handbook*, in press). Residues in crops and soils are determined by glc (G. G. Patchett *et al.*, *Anal. Methods Pestic., Plant Growth Regul. Food Addit.*, 1964, **4**, 343; W. Y. Ja *et al.*, *Anal. Methods Pestic. Plant Growth Regul.*, 1972, **6**, 698). Analytical methods are available from the Stauffer Chemical Co.

# Pencycuron

$C_{19}H_{21}ClN_2O$ (328.8)
L5TJ ANVMR&1R DG

**Nomenclature and development.**

Common name pencycuron (BSI, draft E-ISO). Chemical name (IUPAC) 1-(4-chlorobenzyl)-1-cyclopentyl-3-phenylurea. (*C.A.*) *N*-[(4-chlorophenyl)methyl]-*N*-cyclopentyl-*N*'-phenylurea (9CI); Reg. No. *[66063-05-6]*. Its herbicidal properties were described by P. -E. Frohberger & F. K. Grossman, (*Mitt. Biol. Bundesanst. Land-Forstwirtsch., Berlin-Dahlem*, 1981, **203**, 230). Introduced by Bayer AG (BEP 856 922; DEOS 2 732 257) as code no. 'BAY NTN 19 701'; trade mark 'Monceren'.

**Properties.**

Pure pencycuron forms colourless crystals; v.p. $<1$ mPa at 20 °C. Solubility at 20 °C: 0.4 mg/l water; 0.1-1.0 kg./l dichloromethane; 1-10 g/l propan-2-ol.

**Uses.**

It is a non-systemic fungicide with specific action against diseases caused by *Rhizoctonia solani*, in particular sheath blight of rice. Diseases effectively controlled include black scurf of potatoes and damping-off of ornamentals.

**Toxicology.**

Acute oral $LD_{50}$ for rats $>5000$ mg/kg. Acute percutaneous $LD_{50}$ (24-h) for rats $>2000$ mg/kg. No symptom of poisoning was observed at these doses. $LC_{50}$ (96-h) is: 8.8 mg/l for carp; 5-10 mg/l for guppies.

**Formulations.**

These include: water-dispersible powder; dust. Dry seed treatment, (pencycuron + captan).

**Analysis.**

Details of product analysis may be obtained from Bayer AG. Residues may be determined by glc.

$C_{13}H_{19}N_3O_4$ (281.3)
WNR B1 C1 ENW FMY2&2

$NO_2$
$CH_3$ — NHCH($CH_2CH_3$)$_2$
$CH_3$ $NO_2$

**Nomenclature and development.**

Common name pendimethalin (BSI, E-ISO, ANSI, WSSA);pendiméthaline (F-ISO);*penoxalin* (WSSA former name). Chemical name (IUPAC) *N*-(1-ethylpropyl)-2,6-dinitro-3,4-xylidine. (*C.A.*) *N*-(1-ethylpropyl)-3,4-dimethyl-2,6-dinitrobenzenamine (9CI); Reg. No. *[40487-42-1]*. Its herbicidal properties were described by P. L. Sprankle (*Proc. Br. Weed Control Conf., 12th*, 1974, **2**, 825). Introduced by American Cyanamid Co. (BEP 816 837; USP 4 199 669) as code no. 'AC 92 553'; trade marks 'Prowl', 'Stomp', 'Herbadox'.

**Properties.**

It forms orange-yellow crystals; m.p. 54-58 °C; v.p. 4.0 mPa at 25 °C. Solubility at 20 °C: 0.3 mg/l water; 700 g/l acetone; 148 g/l corn oil; 77 g/l propan-2-ol; 628 g/l xylene. It is stable to alkaline and acidic conditions. Technical grade is 90% pure *m/m*.

**Uses.**

It is a selective herbicide effective against most annual grasses and several small-seeded annual broad-leaved weeds. It can be applied pre-em. after seeding in cereals, maize and rice; or with shallow soil incorporation before seeding beans, cotton, groundnuts and soyabeans. In vegetable crops it can be applied pre-em. or pre-transplanting. It is also used to control suckers in tobacco.

**Toxicology.**

Acute oral $LD_{50}$: for albino rats 1050-1250 mg tech./kg; for albino mice 1340-1620 mg/kg; for beagle dogs >5000 mg/kg, Acute percutaneous $LD_{50}$ for albino rabbits >5000 mg/kg. In 2-y feeding trials rats receiving 100 mg/kg diet showed no ill-effect. $LC_{50}$ (8-d) are: for bobwhite quail 4187 mg tech./kg diet: for mallard duck 10 388 mg/kg diet. $LC_{50}$ (96-h) for channel catfish is 0.42 mg a.i./l. Non-toxic to honeybees; the $LD_{50}$ for topical application >49.8 μg tech./bee.

**Formulations.**

These include: e.c. (330 or 500 g a.i./l); granules (30, 50 or 100 g/kg).

**Analysis.**

Product analysis is by glc with FID (J. C. Wyckoff, *Anal. Methods Pestic. Plant Growth Regul.*, 1978, **10**, 461). Residues of the 4-hydroxymethyl analogue (after formation of a derivative) and of pendimethalin may be determined by glc with ECD (T. H. Byast *et al., Tech. Rep. ARC Weed Res. Organ.*, No.15, (2nd Ed.), p. 77; J. C. Wyckoff, *loc. cit.*). Details of analytical methods are available from American Cyanamid Co.

# Pentachlorophenol 9420

$C_6HCl_5O$ (266.3)
QR BG CG DG EG FG
*sodium pentachlorophenoxide*
$C_6Cl_5NaO$ (288.3)
QR BG CG DG EG FG &&Na SALT

**Nomenclature and development.**

Common name PCP (WSSA, JMAF);the chemical names pentachlorophenol (I) (BSI, E-ISO, F-ISO);sodium pentachlorophenoxide (II) (BSI, E-ISO);pentachlorophénate de sodium (F-ISO) are accepted in lieu of common names. Chemical name (IUPAC) (I); (II); sodium pentachlorophenate. (*C.A.*) (I) (8 & 9CI); Reg. No. *[608-93-5]* (pentachlorophenol), *[131-52-2]* (sodium salt). Trivial name PCP. Introduced *c.* 1936 as a timber preservative and later used as a general disinfectant. Trade marks include '*Dowicide EC7*' (Dow), '*Penta*' for pentachlorophenol, '*Dowicide G*' (Dow), '*Santobrite*' (Monsanto) for the sodium salt - neither company now manufactures or markets these products.

**Properties.**

Pure pentachlorophenol forms colourless crystals, with a phenolic odour; m.p. 191 °C; v.p. 16 Pa at 100 °C. Solubility at 30 °C: 20 mg/l water; soluble in most organic solvents; slightly soluble in carbon tetrachloride, paraffins. The technical grade is dark grey, m.p. 187-189 °C. It is non-flammable and non-corrosive in the absence of moisture. Its solutions in oil cause natural rubber to deteriorate, but synthetic rubbers may be used in equipment and for protective clothing.

It is a weak acid, $pK_a$ 4.71. The sodium salt crystallises from water as a monohydrate. Solubility at 25 °C: 330 g/l water; insoluble in petroleum oils. The calcium and magnesium salts are soluble in water.

**Uses.**

Pentachlorophenol is used to control termites, and to protect wood from fungal rots and wood-boring insects, as a pre-harvest defoliant and as a general herbicide. The sodium salt is used as a general disinfectant, e.g. for trays in mushroom houses. The uses have been reviewed (*Pentachlorophenol*).

**Toxicology.**

Acute oral $LD_{50}$ for rats 210 mg/kg. It irritates mucous membranes; the solid and aqueous solutions ($>$10 g/l) cause skin irritation. No deaths occurred among dogs and rats receiving 3.9-10 mg/daily for 70-190 d. $LC_{50}$ for rainbow and brown trout is 0.17 mg sodium pentachlorophenoxide/l (J. S. Alabaster, *Proc. Br. Weed Control Conf., 4th*, 1958, p. 84).

**Formulations.**

It is used as such: as solutions in oil; as technical sodium pentachlorophenoxide.

**Analysis.**

Product analysis is by titration with alkali (*MAFF Ref. Book*, 1958, No. 1, p. 64). Residues may be determined by colorimetry of derivatives (W. W. Kilgore & K. W. Cheng, *Anal. Methods Pestic., Plant Growth Regul. Food Addit.*, 1967, **5**, 313; *Anal. Methods Pestic. Plant Growth Regul.*, 1972, **6**, 581).

# Pentanochlor

$C_{13}H_{18}ClNO$ (293.7)
GR B1 EMVY3&1

$CH_3$ — $CH_3$ — NH.CO.CH$(CH_2)_2CH_3$ — Cl

**Nomenclature and development.**

Common name pentanochlor (BSI, E-ISO, F-ISO);solan (WSSA, Canada, ex ANSI);CMMP (JMAF);*exception* Turkey. Chemical name (IUPAC) 3'-chloro-2-methylvaler-*p*-toluidide; *N*-(3-chloro-*p*-tolyl)-2-methylvaleramide. (*C.A.*) *N*-(3-chloro-4-methylphenyl)-2-methylpentanamide (9CI); 3'-chloro-2-methyl-*p*-valerotoluidide; Reg. No. *[2307-68-8]*. Its herbicidal properties were described by D. H. Moore (*Proc. Northeast. Weed Control Conf.*, 1960, p. 86). Introduced by the Agricultural Division of FMC Corp. (who no longer manufacture or market it) (GBP 869 169; USP 3 020 142) as code no. 'FMC 4512'.

**Properties.**

Pure pentanochlor is a colourless solid, m.p. 85-86 °C. The technical grade is a colourless to pale cream powder; m.p. 82-86 °C; $d_{20}^{20}$ 1.106. Solubility at room temperature: 8-9 mg/l water; 460 g/kg di-isobutyl ketone; 550 g/kg dimethylcyclohexenone; 520 g/kg 4-methylpentan-2-one; 410 g/kg pine oil; 200-300 g/kg xylene. It is stable to hydrolysis at room temperature.

**Uses.**

It is a selective herbicide used post-em. to control annual weeds at <4 kg a.i. in 500-1000 l water/ha in carrots, celery, parsley, strawberries, tomatoes and as a directed contact spray on carnations, chrysanthemums, roses, tomatoes, fruit and ornamental trees. It is also used pre-em. in combination with chlorpropham in narcissus and tulips in addition to many of the crops listed above. There is only a limited and short-lived residual activity in soil.

**Toxicology.**

Acute oral $LD_{50}$ for rats >10 000 mg/kg. In 140-d feeding trials rats receiving 20 000 mg/kg diet suffered no effect on body weight or survival, but histopathological changes were found in the liver; these effects were not observed at 2000 mg/kg diet.

**Formulations.**

'Herbon Solan' (Cropsafe), e.c. (400 g a.i./l). 'Herbon Brown' (Cropsafe), e.c. (pentanochlor + chlorpropham).

**Analysis.**

Residues may be determined by hydrolysis to 3-chloro-*p*-toluidine which is estimated by glc.

$C_{14}H_{12}F_3NO_4S_2$ (379.4)
WSR&R C1 DMSWXFFF

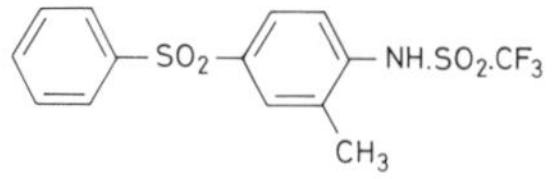

**Nomenclature and development.**

Common name perfluidone (BSI, E-ISO, F-ISO, ANSI, WSSA). Chemical name (IUPAC) 1,1,1-trifluoro-2'-methyl-4'-(phenylsulphonyl)methanesulphonanilide; 1,1,1-trifluoro-*N*-(4-phenylsulphonyl-*o*-tolyl)methanesulphonamide. (*C.A.*) 1,1,1-trifluoro-*N*-[2-methyl-4-(phenylsulfonyl)phenyl]methanesulfonamide (9CI); 1,1,1-trifluoro-4'-(phenylsulfonyl)methanesulfono-*o*-toluidide (8CI); Reg. No. *[37924-13-3]*. Its herbicidal properties were described by W. A. Gentner (*Agric. Res. (Wash. D.C.)*, 1971, **20**(2), 5). Introduced by 3M Company (GBP 1 306 564; BEP 765 558) as code no. 'MBR-8251'; trade mark 'Destun'.

**Properties.**

Perfluidone is a colourless crystalline solid; m.p. 142-144 °C; v.p. <1.3 mPa at 25 °C; $pK_a$ 2.5. Solubility at 22 °C; 60 mg/l water; 750 g/l acetone; 11 g/l benzene; 162 g/l dichloromethane; 595 g/l methanol. It is stable at 100 °C to thermal degradation and to both acid and alkaline hydrolysis. In aqueous environments it is susceptible to degradation under u.v. radiation. Aqueous solutions and suspensions are mildly corrosive to metals on prolonged exposure.

**Uses.**

It is a selective herbicide which, at 1.1-4.5 kg a.i./ha, provides pre-em. control of *Cyperus esculentus*, many grass and several broad-leaved weed species in emerging cotton, container-grown ornamentals, transplanted tobacco and established turf. Other crops at various growth stages show a high degree of tolerance and include sugarcane seed pieces and ratoons, transplanted and established rice in water, and groundnuts at ⩾4-leaf growth stage.

**Toxicology.**

Acute oral $LD_{50}$: for rats 633 mg tech./kg; for mice 920 mg/kg. Acute percutaneous $LD_{50}$ for rabbits >4000 mg/kg; it is mildly irritating to their skin and irritating to their eyes. $LC_{50}$ (96-h) is: for bluegill 318 mg/l; for rainbow trout 312 mg/l.

**Formulations.**

These include: 'Destun', w.p. (500 g a.i./kg); 'Destun 4-S', solution (480 g/l); 'Destun 3.75 G' and 'Destun 5.0 G', granules (37.5 or 50 g/kg).

**Analysis.**

Product analysis is by glc of a derivative with FID (C. D. Green, *Anal. Methods Pestic. Plant Growth Regul.*, 1978, **10**, 437). Residues may be determined by glc of a derivative with ECD (*idem, ibid.*).

$C_{21}H_{20}Cl_2O_3$ (391.3)
L3TJ A1 A1 BVO1R COR&& C1UYGG
&&(1RS)-CIS-TRANS FORM

**Nomenclature and development.**

Common name permethrin (the ratio of isomers should be stated) (BSI, E-ISO, ANSI);perméthrine (F-ISO). Chemical name (IUPAC) 3-phenoxybenzyl (1*RS*,3*RS*; 1*RS*,3*SR*)-3-(2,2-dichlorovinyl)-2,2-dimethylcyclopropanecarboxylate; 3-phenoxybenzyl (1*RS*)-*cis-trans*-3-(2,2-dichlorovinyl)-2,2-dimethylcyclopropanecarboxylate. (*C.A.*) (3-phenoxyphenyl)methyl 3-(2,2-dichloroethenyl)-2,2-dimethylcyclopropanecarboxylate (9CI); Reg. No. *[52645-53-1]* formerly *[57608-04-5]* and *[63364-00-1]*. OMS 1821. Its insecticidal properties were described by M. Elliott *et al.* (*Proc. Br. Insectic. Fungic. Conf., 7th*, 1973, **2**, 721; *Nature (London)*, 1973, **246**, 169). It has been developed by FMC Corp., ICI Plant Protection Division, Mitchell Cotts Chemicals, Penick Corp., Shell International Chemical Co. Ltd, Sumitomo Chemical Co. and the Wellcome Foundation (GBP 1 413 491 to NRDC) as code no. 'NRDC 143', 'FMC 33 297', 'PP 557' (ICI), 'WL 43 479' (Shell); trade marks 'Ambush', 'Ambushfog', 'Kafil', 'Perthrine', 'Picket', 'Picket G' (all to ICI), 'Pounce', 'Pramex' (both to Penick), 'Talcord' (for agronomic use), 'Outflank' and 'Stockade' (for veterinary use) (all to Shell), 'Eksmin' (for Sumitomo), 'Coopex', 'Peregin', 'Stomoxin', 'Stomoxin P', 'Qamlin' (all to Wellcome Foundation), 'Permasect 25'.

**Properties.**

Technical grade permethrin is a yellow-brown to brown liquid, which sometimes tends to crystallise partly at room temperature; v.p. 261 mPa at 30 °C; $d^{25}_{25}$ 1.214. Solubility at 30 °C: 0.2 mg/l water; at 25 °C: >1 kg/kg hexane; 258 g/kg methanol; ⩾1 kg/kg xylene. It is stable to heat (⩾2 y at 50 °C), is more stable in acid than alkaline media with optimum stability *c.* pH 4. Some photochemical degradation has been observed in laboratory studies (F. Barlow, *Pestic. Sci.*, 1977, **8**, 291) but field data indicate this does not adversely effect biological performance. Pure permethrin has m.p. 34-39 °C, the *cis*-isomers m.p. 63-65 °C, the *trans*-isomers m.p. 44-47 °C.

**Uses.**

It is a contact insecticide effective against a broad range of pests. It controls leaf- and fruit-eating lepidopterous and coleopterous pests in cotton at 100-150 g a.i./ha, in fruit at 25-50 g/ha, in tobacco, vines and other crops at 50-200 g/ha, in vegetables at 40-70 g/ha. It has good residual activity on treated plants. It is effective against a wide range of animal ectoparasites, provides <60 d residual control of biting flies in animal housing at 200 mg a.i. (as e.c.)/m$^2$ wall or 30 mg a.i. (as w.p.)/m$^2$ wall and is effective as a wool preservative at 200 mg/kg wool (P. A. Duffield, *ibid.*, p. 279). It provides >120 d control of cockroaches and other crawling insects at 100 mg a.i. (as w.p.)/m$^2$.

**Toxicology.**

Oral $LD_{50}$ values of permethrin depend on such factors as: carrier, *cis/trans* ratio of the sample, the test species, its sex, age and degree of fasting. Values reported sometimes differ markedly. Typical oral $LD_{50}$ values for a *cis/trans* ratio of *c.* 40:60 are: for rats 430-4000 mg/kg; for mice 540-2690 mg/kg; for chickens >3000 mg/kg; for Japanese quail >13 500 mg/kg. In 2-y feeding trials rats receiving 100 mg/kg diet showed no ill-effect. $LC_{50}$ (96-h) for rainbow trout is 9 µg/l. Metabolism in rats (M. Elliott *et al.*, *J. Agric. Food Chem.*, 1976, **24**, 270; L. C. Gaughan *et al.*, *ibid.*, 1977, **25**, 9) has been studied.

**Formulations.**

These include: e.c. (100-500 g a.i./l), solution (50 g/l) for fogging, ULV concentrates (50 or 100 g/l), all for agronomic use; dusts (2.5-10 g/kg) and e.c. (25-200 g/l) for veterinary use; w.p. (100-500 g/kg) for industrial and public hygiene use; aerosol concentrates and fumigants for household use. Many commercial formulations contain a *cis/trans* isomer ratio of 40-50:60-50 *m/m*, but formulations with a 25:75 ratio are also available.

**Analysis.**

Product analysis is by glc (M. Horiba *et al.*, *Agric. Biol. Chem.*, 1977, **41**, 581). Residues may be determined by glc with ECD (I. H. Williams, *Pestic. Sci.*, 1976, **7**, 336). Details may be obtained from Shell International Chemical Co.

**Nomenclature and development.**

Petroleum oils are also known as mineral oils; refined grades have been called white oils. They consist largely of aliphatic hydrocarbons, both saturated and unsaturated, the content of the latter being reduced by refinement. Oils of higher distillation range came into use *c.* 1922 and later highly refined oils. Recently, highly refined oils have been used as adjuvants to increase the effectiveness of some other pesticides. Trade marks include: 'Volck' (Chevron) for traditional use; 'Actipron' (British Petroleum), and 'Fyzol' (FBC Limited) for adjuvant use.

**Properties.**

Produced by the distillation and refinement of crude mineral oils, those used as pesticides generally distil >310 °C. They may be classified by the proportion distilling at 335 °C, namely: 'light' (67-79%), 'medium' (40-49%), and 'heavy' (10-25%). Viscosity and density vary according to the geographical area from which the crude oils came; density rarely exceeds 0.92 g/cm$^3$ at 15.5 °C.

Adjuvant oils are highly refined self-emulsifying oils which form a quick-breaking emulsion, spread quickly and help penetration of the active ingredient into the plant and pest. 'Actipron' is a pale yellow liquid; dynamic viscosity *c.* 30 cSt at 40 °C; flash point ⩾204 °C; $\rho$ 0.86-0.88 g/cm$^3$ at 15 °C.

**Uses.**

Petroleum oils are effective against certain insects such as scales and red spider mites especially under glass on cucumbers, tomatoes and vines; they are ovicidal. Their use is limited by their phytotoxic properties; a semi-refined oil can be used as an ovicide on dormant trees; for foliage use a refined oil of narrow viscosity range and high content of unsulphonated residue is required. Such oils should not be applied to foliage bearing sulphur residues. An oil of lower distillation range containing some aromatic hydrocarbons is suitable for herbicidal use in umbelliferous crops.

'Actipron' alone has little biological activity but it will destroy summer eggs and newly hatched nymphs of spider mites, is effective against scale insects and has an effect on powdery mildews. It acts as a humectant, so increasing the time available for uptake by plants and pests.

**Toxicology.**

Acute oral $LD_{50}$ for rats and mice >4300 mg 'Actipron'/kg. No toxilogical problem due to petroleum oils has been reported in practice. Tests have shown there is no risk of polynuclear aromatic compounds entering the food chain.

**Formulations.**

Stock emulsions; e.c. (solutions of a suitable surfactant in the oil); 'Actipron', ready for mixing with specified fungicides, herbicides and insecticides.

**Analysis.**

Formulation analysis is by standard methods (*MAFF Ref. Book*, 1958, No. 1, p. 66; *CIPAC Handbook*, 1970, **1**, 582; *ibid.*, in press).

$C_{19}H_{22}N_2O_4$ (342.4)
2NR&VOR CMVOY1&1

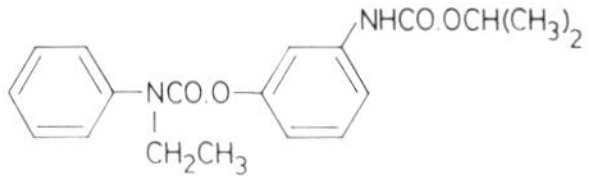

**Nomenclature and development.**

Common name phenisopham (BSI, draft E-ISO);phenisophame (draft F-ISO). Chemical name (IUPAC) isopropyl 3-(*N*-ethyl-*N*-phenylcarbamoyloxy)phenylcarbamate. (*C.A.*) 3-[[(1-methylethoxy)carbonyl]amino]phenyl ethylphenylcarbamate (9CI); Reg. No. *[57375-63-0]*. its herbicidal properties were described in *Schering AG Tech. Inf.* 1977. Introduced by Schering AG (DEP 2 413 933) as code no. 'SN 58 132'; trade marks 'Diconal', 'Verdinal'.

**Properties.**

Phenisopham is a colourless solid; m.p. 109-110 °C; v.p. 665 nPa at 25 °C. Solubility at 25 °C: 3 mg/l water; 500 g/l dichloromethane; 98 g/l ethanol; 60 g/l methanol; 26 g/l propan-2-ol; 36 g/l toluene. Unstable under alkaline conditions, 50% loss occurring in 35 d at pH9, 29 d at pH12, 2 d at pH13, 7 h at pH14.

**Uses.**

It is a selective herbicide mainly used to control broad-leaved weeds in cotton. It acts mainly by contact but there also appears to be some activity through the soil. It should be applied at 1-2 kg a.i. in 300-400 l water/ha as soon as most of the weeds have emerged and not later than their 2-4 true leaf stage.

**Toxicology.**

Acute oral $LD_{50}$: for rats and ducks >4000 mg/kg; for mice >5000 mg/kg. Acute percutaneous $LD_{50}$ for rabbits >1000 mg/kg. In 90-d feeding trials NEL for rats was 1 mg/kg daily; in 182-d trials for dogs 3 mg/kg daily. $LC_{50}$ (96-h) is: for trout 3 mg/l; for carp and bluegill 4 mg/l.

**Formulations.**

An e.c. (138 g a.i./l).

**Analysis.**

Product analysis is by tlc. Residues may be determined by hydrolysis to *N*-ethylaniline a derivative being measured by glc with ECD. Particulars available from Schering AG.

# Phenmedipham 

$C_{16}H_{16}N_2O_4$ (300.3)
1OVMR COVMR C1

**Nomenclature and development.**

Common name phenmedipham (BSI, E-ISO, ANSI, WSSA, JMAF);phenmédiphame (F-ISO). Chemical name (IUPAC) methyl 3-(3-methylcarbaniloyloxy)carbanilate; 3-methoxycarbonylaminophenyl 3-methylcarbanilate. (*C.A.*) 3-[(methoxycarbonyl)amino]phenyl (3-methylphenyl)carbamate (9CI); methyl *m*-hydroxycarbanilate *m*-methylcarbanilate (ester) (8CI); Reg. No. *[13684-63-4]*. Its herbicidal properties were described by F. Arndt & C. Kötter (*Abstr. Int. Congr. Plant Prot., 6th, Vienna*, 1967, p. 433). Introduced by Schering AG (GBP 1 127 050) as code no. 'Schering 38 584'; trade mark 'Betanal'.

**Properties.**

Pure phenmedipham forms colourless crystals; m.p. 143-144 °C; ρ 0.25-0.30 g/cm$^3$ at 20 °C. Solubility at room temperature 4.7 mg/l water; *c.* 200 g/kg acetone, cyclohexanone; 2.5 g/kg benzene; 20 g/kg chloroform; *c.* 500 mg/kg hexane; *c.* 50 g/kg methanol. The technical grade (>95% pure) has m.p. 140-144 °C; v.p. 1.3 nPa at 25 °C. No change was observed on storage at 50 °C for 6 d. On hydrolysis at 22 °C in buffer solutions, 50% loss occured in 70 d at pH5, 24 h at pH7, 10 min at pH9.

**Uses.**

It is a post-em. herbicide used in beet crops, especially sugar beet, at 1 kg a.i. in 200-300 l water/ha after the emergence of most broad-leaved weeds and before they develop >2-4 true leaves. It is absorbed through the leaves but has little action when applied to the soil. In soil, 71-86% of the amount determined 1 d after treatment was degraded in 90 d, mainly to methyl 3-hydroxycarbanilate.

**Toxicology.**

Acute oral $LD_{50}$: for rats and mice >8000 mg/kg; for dogs and guinea-pigs >4000 mg/kg; for chickens >3000 mg/kg. Acute percutaneous $LD_{50}$ for rats >4000 mg/kg. In 120-d feeding trials rats receiving up to 500 mg a.i./kg daily survived, but there was a reduced food intake dependent on the dose. No effect on trout at 1.6 mg a.i. (as 'Betanal')/l and carp 2.4 mg a.i. (as 'Betanal')/l.

**Formulations.**

'Betanal' (157 g a.i./l), in the UK 'Betanal E' (120 g/l). 'Betanal AM 11', (80 g phenmedipham + 80 g desmedipham/l).

**Analysis.**

Product analysis is by bromometric titration (*CIPAC Handbook*, in press) or by quantitative tlc (K. Kossmann & N. A. Jenny, *Anal. Methods Pestic. Plant Growth Regul.*, 1973, **7**, 611). Residues in soil may be determined by hplc (T. H. Byast *et al.*, *Tech. Rep. ARC Weed Res. Organ.*, No. 15 (2nd Ed.), p. 87) or by hydrolysis to *m*-toluidine, derivatives of which are determined by glc with ECD or by colorimetry (K. Kossmann, *Weed Res.*, 1970, **10**, 340).

$C_{23}H_{26}O_3$ (350.5)
L3TJ A1 A1 BVO1R COR&& C1UY1&1
&&(1R)-CIS-TRANS FORM

$(CH_3)_2C{=}CH$ — cyclopropane — $CO.OCH_2$ — $C_6H_4$ — O — $C_6H_5$; $CH_3$ $CH_3$

**Nomenclature and development.**

Common name phenothrin (BSI, E-ISO); phénothrine (F-ISO). Chemical name (IUPAC) 3-phenoxybenzyl (1*RS*,3*RS*; 1*RS*,3*SR*)-2,2-dimethyl-3-(2-methylprop-1-enyl)cyclopropanecarboxylate; 3-phenoxybenzyl (1*RS*)-*cis-trans*-2,2-dimethyl-3-(2-methylprop-1-enyl)cyclopropanecarboxylate; 3-phenoxybenzyl (1*RS*)-*cis-trans*-chrysanthemate. (*C.A.*) (3-phenoxyphenyl)methyl 2,2-dimethyl-3-(2-methyl-1-propenyl)cyclopropanecarboxylate (9CI); *m*-phenoxybenzyl 2,2-dimethyl-3-(2-methylpropenyl)cyclopropanecarboxylate (8CI); Reg. No. *[26002-80-2]* as defined above, *[51186-88-0]* (1*R*)-*cis*-isomer, *[26046-85-5]* (1*R*)-*trans*-isomer. OMS 1809 for phenothrin, OMS 1810 for (1*R*)-*cis-trans*-isomers, ENT 27 972 for (1*R*)-*cis-trans*-isomers. The insecticidal activity of the (1*R*)-*cis-trans*-isomers was described by K. Fujimoto *et al.* (*Agric. Biol. Chem.*, 1973, **37**, 2681). They were introduced by Sumitomo Chemical Co. (JPP 1 027 088; USP 3 934 028) as code no. 'S-2539 Forte'; trade marks 'Sumithrin' (*cis*/*trans* ratio 20:80 *m*/*m*), 'Pesguard'.

**Properties.**

The technical (1*R*)-*cis-trans*-isomeric mixture is a pale yellow to yellow-brown liquid; $n_D^{20}$ 1.547-1.552; $d_{25}^{25}$ 1.061; v.p. 160 μPa at 20 °C. Solubility at 30 °C: 2 mg/l; at 25 °C: >1 kg/kg hexane, methanol, xylene. It is stable under irradiation, on inorganic mineral diluents, and in most organic solvents, but is hydrolysed by alkali.

**Uses.**

The (1*R*)-*cis-trans*-mixture is a non-systemic insecticide, effective by contact and as a stomach poison, used to control injurious and nuisance insects of public health and for controlling human lice. It is also used to protect stored grain.

**Toxicology.**

Acute oral $LD_{50}$: for rats >10 000 (1*R*)-*cis-trans*-mixture /kg; for bobwhite quail >2510 mg/kg. Acute percutaneous $LD_{50}$ for rats >10 000 mg/kg. In 0.5-y feeding trials NEL for rats was 2500 mg/kg diet. $LC_{50}$ (96-h) is: for rainbow trout 16.7 μg/l; for bluegill 18 μg/l.

**Formulations.**

'Pesguard' applies to the (1*R*)-*cis-trans*-isomers when used in household and public health. 'Pesguard-A NS' and 'Pesguard-A NS W' are aerosol concentrates; 'Pesguard NS' liquid for ULV spraying or thermal fogging, e.c. for power sprayer or mist blower. Also powders, shampoos, lotions which usually include piperonyl butoxide. Mixtures, as above, containing tetramethrin or the (1*R*)-*cis-trans* isomers of allethrin.

**Analysis.**

Product analysis is by glc or hplc (S. Sakaue, *et al.*, *ibid.*, 1981, **45**, 1135). Residues may be determined by glc with ECD. Details of these methods are available from Sumitomo Chemical Co.

$C_{12}H_{17}O_4PS_2$ (320.4)
2OVYR&SPS&O1&O1

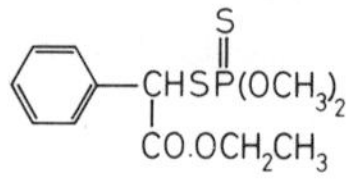


**Nomenclature and development.**

Common name phenthoate (BSI, E-ISO, F-ISO);PAP (JMAF). Chemical name (IUPAC) ethyl 2-dimethoxythiophosphorylthio-2-phenylacetate; *S*-α-ethoxycarbonylbenzyl *O,O*-dimethyl phosphorodithioate. (*C.A.*) ethyl α-[(dimethoxyphosphinothioyl)thio]ben = zeneacetate (9CI); ethyl mercaptophenylacetate *S*-ester with *O,O*-dimethyl phosphorodithioate (8CI); Reg. No. *[2597-03-7]*. OMS 1075, ENT 27 386. Introduced by Montecatini S.p.A. (now Montedison S.p.A.) (GBP 834 814; USP 2 947 662) as code no. 'L 561'; trade marks 'Cidial', 'Elsan'.

**Properties.**

Pure phenthoate is a colourless crystalline solid; m.p. 17-18 °C; v.p. 5.3 mPa at 40 °C; $d_4^{20}$ 1.226, $n_D^{20}$ 1.5550. Solubility at 24 °C: 11 mg/l water. The technical grade (90-92% pure) is a reddish-yellow oil; its solubility at 20 °C is: 200 mg/l water; >200 g/l 1,2-dimethoxyethane; 120 g/l hexane; 100-170 g/l petroleum spirit. It is stable in buffer solutions $3.9<pH<7.8$, but *c.* 25% is hydrolysed in 20 d in buffer solutions at pH9.7.

**Uses.**

It is a non-systemic insecticide, with contact and stomach action. Applied at 0.5-1.0 kg a.i./ha, it protects citrus, fruit, cotton, rice, vegetables and other crops from Lepidoptera, aphids, jassids and soft scales. It is also effective against adults and larvae of mosquitoes. It may be phytotoxic to some fig, grape and peach varieties and may discolour red-skinned apple varieties.

**Toxicology.**

Acute oral $LD_{50}$: for rats 300-400 mg tech./kg; for mice 350-400 mg/kg; for pheasants 218 mg/kg; for quail 300 mg/kg. In 4-h dermal tests application to the skins of rats of 4800 mg/kg caused no toxic symptom. In 1.67-y feeding trials the NEL for rats was 10 mg/kg diet. $LC_{50}$ (96-h) is: for goldfish 2.9 mg/l; for minnows 2.5 mg/l; for mosquito fish 0.12 mg/l; for guppies 0.69 mg/l. The $LD_{50}$ for honeybees is 0.12 μg/bee.

**Formulations.**

For agricultural use: 'Cidial 50L', liquid (500 g tech./kg); 'Cidial Oil', (50 g tech. + 800 g mineral oil); 'Cidial WP40', w.p. (400 g/kg); 'Cidial D', dust (30 g/kg); 'Cidial Granules', granules (20 g/kg); 'Cidial AS', liquid (850 g/kg); 'Cidial ULV', ULV concentrate (1 kg/kg). For mosquito control: 'Tanone 50L', liquid (500 g/kg), 'Tanone AS', liquid (850 g/kg); 'Tanone ULV', ULV concentrate (1 kg/kg).

**Analysis.**

Product analysis is by glc (B. Bazzi *et al.*, *Anal. Methods Pestic. Plant Growth Regul.*, 1976, **8**, 159) or by tlc with colorimetric estimation of phosphorus in the appropriate spots (*CIPAC Handbook* 1983, **1B**, in press). Residues on apples or citrus fruits may be determined by glc (B. Bazzi *et al.*, *loc. cit.*). Details of these methods are available from Montedison S.p.A.

$C_8H_8HgO_2$ (336.7)
1VO-HG-R

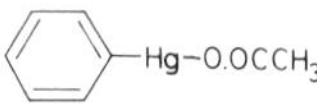


**Nomenclature and development.**

Common name PMA (JMAF). Chemical name (IUPAC) phenylmercury acetate (BSI, E-ISO); acétate de phénylmercure (F-ISO) is accepted in lieu of a common name. (*C.A.*) (acetato-*O*)phenylmercury (9CI); (acetato)phenylmercury (8CI); Reg. No. *[62-38-4]*. Trivial name PMA. The fungicidal properties of the corresponding chloride were described by E. Riehm (*Zentralbl. Bakteriol. Parasitenkd. Infektionskr. Hyg. Abt. 2*, 1914, **40**, 424); the acetate was marketed as a seed treatment ('Ceresan') by I. G. Farbenindustrie (now Bayer AG) in 1932 and its toxicity to crabgrass reported by J. A. De France (*Greenkeepers Rep.*, 1947, **15**(1), 30). Trade marks include: 'Agrosan D', 'Ceresol' (ICI Plant Protection Division), 'Mist-O-Matic Mercury Liquid Seed Treatment' (Murphy Chemical Ltd).

**Properties.**

It forms colourless crystals m.p. 149-153 °C; v.p. 1.2 mPa at 35 °C. Solubility at room temperature: 4.37 g/l water; soluble in acetone, benzene, ethanol; very soluble in 2-(2-ethoxyethoxy)ethanol.

**Uses.**

It is a powerful eradicant fungicide mainly used as a treatment for cereal seed, often in combination with insecticides or fungicides. It is used as a selective herbicide to control crabgrass in lawns.

**Toxicology.**

Acute oral $LD_{50}$: for rats 24 mg/kg; for mice 70 mg/kg; for chicks 60 mg/kg. Minimum lethal dose by intravenous injection: for rats 20 mg/kg; for mice 20 mg/kg; for dogs 5 mg/kg. There is some vapour hazard, so conditions under which seed is treated are controlled by law.

**Formulations.**

Powder seed treatments (10-300 g mercury/kg). Mixtures include: phenylmercury acetate + carboxin.

**Analysis.**

Product analysis is by digestion with concentrated sulphuric and nitric acids and titration of the mercury with a thiocyanate. Residues may be determined by a similar oxidative digestion with colorimetry of a complex with dithizone (*AOAC Methods*, 1980, 25.117-25.123; *EPPO Recommended Methods, Series A*, No. 37) or by atomic absorption spectroscopy (*AOAC Methods*, 1980, 25.110-25.112).

$C_{12}H_{10}O$ (170.2)
QR BR
*sodium 2-phenylphenoxide*
$C_{12}H_9NaO$ (192.2)
QR BR &&Na SALT

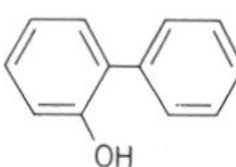


**Nomenclature and development.**

The traditional chemical name 2-phenylphenol (BSI, E-ISO);phényl-2 phenol (F-ISO) is accepted in lieu of a common name. Chemical name (IUPAC) biphenyl-2-ol. (*C.A.*) [1,1'-biphenyl]-2-ol (9CI); 2-biphenylol (8CI); Reg. No. *[90-43-7]* 2-phenylphenol, *[132-27-4]* sodium 2-phenylphenoxide. Trivial names 2-hydroxybiphenyl,*o*-phenylphenol. Its fungicidal properties were described by R. G. Tomkins (*Rep. Food Investigation Board 1936*, p. 149).

**Properties.**

2-Phenylphenol forms colourless to pinkish crystals, of a mild odour; m.p. 57 °C; b.p. 286 °C; $d_{25}^{25}$ 1.217. Solubility at 25 °C: 700 mg/kg water; soluble in most organic solvents. it forms salts of which those of the alkali metals are water-soluble; the sodium salt crystallises as a tetrahydrate, solubility *c.* 1.1 kg/kg water, giving a solution of pH 12.0-13.5 at 35 °C.

**Uses.**

2-Phenylphenol is a powerful disinfectant and fungicide for the impregnation of fruit wrappers so inhibiting post-harvest microbial decay and for the disinfection of seed boxes. The tendency to scald citrus fruits is reduced by the addition of hexamine. It is also applied during the dormant period to control apple canker.

**Toxicology.**

Acute oral $LD_{50}$ for white rats 2480 mg/kg. in 2-y feeding trials rats receiving 2g/kg diet showed no ill-effect (H. C. Hodge *et al., J. Pharmacol. Exp. Ther.*, 1952, **104**, 202). It can cause skin irritation.

**Formulations.**

These include: 'Dowicide 1', 2-phenylphenol; 'Dowicide A', sodium 2-phenylphenoxide; 'Nectryl', canker paint (Stanhope).

**Analysis.**

Residues in fruit wrappers or citrus peel may be determined by colorimetry of derivatives (P. H. Caulfield & R. J. Robinson, *Anal. Chem.*, 1953, **25**, 982; R. G. Tompkins & F. A. Isherwood, *Analyst (London)*, 1945, **70**, 330; C. D. Schiffman, *J. Assoc. Off. Agric. Chem.*, 1957, **40**, 238).

$C_7H_{17}O_2PS_3$ (260.4)
2S1SPS&O2&O2

$$CH_3CH_2SCH_2S\overset{S}{\overset{\|}{P}}(OCH_2CH_3)_2$$

**Nomenclature and development.**

Common name phorate (BSI, E-ISO, F-ISO, ANSI);timet (USSR). Chemical name (IUPAC) *O,O*-diethyl *S*-ethylthiomethyl phosphorodithioate. (*C.A.*) *O,O*-diethyl *S*-[(ethylthio)methyl]phosphorodithioate (8 & 9CI); Reg. No. *[298-02-2]*. ENT 24 042. Introduced by American Cyanamid Co. (USP 2 586 655; 2 596 076; 2 970 080 (Cyanamid); 2 759 010 (Bayer AG)) as code no. 'EI 3911'; trade mark 'Thimet'.

**Properties.**

Technical phorate (>90% pure) is a clear mobile liquid; f.p. <−15 °C; b.p. 118-120 °C/0.8 mmHg; v.p. 85 mPa at 25 °C: $n_D^{25}$ 1.5349; $d^{25}$ 1.167. Solubility at room temperature: 50 mg/l water; miscible with alcohols, carbon tetrachloride, dioxane, esters, ethers, vegetable oils and xylene. Stable at 25 °C for 2 y. Stability is optimum in the range pH5-7. Hydrolysis occurs at rates dependent upon the temperature and pH, at 2>pH>9.

**Uses.**

It is a systemic and contact insecticide and acaricide used to protect primarily brassicas, coffee, cotton, field and root crops, from sucking and biting insects, mites and certain nematodes. Also used in maize and sugar beet against soil-dwelling pests. In plants and animals it is metabolically-oxidised giving the sulphoxide and sulphone and their phosphorothioate analogues; as both the parent compound and the oxidation products are readily hydrolysed only a small proportion of the sulphone results (J. S. Bowman & J. E. Casida, *J. Agric. Food Chem.*, 1957, **5**, 192: R. L. Metcalf *et al., J. Econ. Entomol.*, 1957, **50**, 338). Behaviour in soils is similar but the sulphones can persist under some conditions (D. L. Suett, *Pestic. Sci.*, 1975, **6**, 385)

**Toxicology.**

Acute oral $LD_{50}$: for rats 1.6-3.7 mg tech./kg; for mallard duck 0.62 mg/kg; for ringnecked pheasant 7.1 mg/kg, Acute percutaneous $LD_{50}$: for rats 2.5-6.2 mg/kg; for guinea-pigs 20-30 mg/kg; values for granular formulations depend on a.i. content, carrier, test method and animal species - typical values include: for male rats 137 mg a.i. (as 5% granules)/kg, 98 mg a.i. (as 10% granules)/kg; for male rabbits 93-245 mg a.i. (as 5% granules)/kg, 116 mg a.i. (as 10% granule)/kg. In 90-d feeding trials rats receiving 6 mg tech./kg diet showed no ill-effect other than depression of cholinesterase levels. $LC_{50}$ (96-h) is: for rainbow trout 0.013 mg/l; for channel catfish 0.28 mg/l. The $LD_{50}$ for topical application to honeybees is 10 μg/bee.

**Formulations.**

These include: e.c. of various a.i. content including 'Thimet LC-8' (960 g tech./l) also (200 or 250 g a.i./l); granules (50, 100 or 150 g/kg).

**Analysis.**

Product analysis is by i.r. spectrometry (*CIPAC Handbook*, in press). Residues of phorate and its oxidation products may be determined by glc (*Pestic. Anal. Man.*, 1979, **I**, 201-A, 201G, 201-H, 201-I; D. C. Abbott *et al., Pestic. Sci.*, 1970, **1**, 10; *Analyst (London)*, 1980, **105**, 515; J. E. Boyd, *Anal. Methods Pestic. Plant Growth Regul.*, 1972, **6**, 493). Details of methods are available from American Cyanamid Co.

$C_{12}H_{15}ClNO_4PS_2$ (367.8)
T56 BNVOJ B1SPS&O2&O2 GG

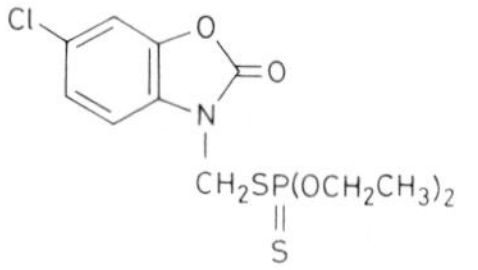

**Nomenclature and development.**

Common name phosalone (BSI, E-ISO, F-ISO, ANSI, JMAF);benzofos (USSR). Chemical name (IUPAC) *S*-6-chloro-2,3-dihydro-2-oxobenzoxazol-3-ylmethyl *O,O*-diethyl phosphorodithioate. (*C.A.*) *S*-[(6-chloro-2-oxo-3(2*H*)-benzoxazolyl)methyl] *O,O*-diethyl phosphorodithioate (9CI); *O,O*-diethyl phosphorodithioate *S*-ester with 6-chloro-3-(mercaptomethyl)-2-benzoxazolinone (8CI); Reg. No. *[2310-17-0]*. ENT 27 163. Its insecticidal properties were described by J. Desmoras *et al.* (*Phytiatr. Phytopharm.*, 1963, **12**, 199). introduced by Rhône-Poulenc Phytosanitaire (GBP 1 005 372; BEP 609 209; FP 1 482 025) as code no. '11 974 RP'; trade mark 'Zolone'.

**Properties.**

Phosalone forms colourless crystals, with a slight garlic odour; m.p. 48 °C; v.p. negligible at room temperature. Solubility at room temperature 10 mg/l water; sparingly soluble in cyclohexane, light petroleum; soluble in acetone, acetonitrile, benzene, chloroform, dioxane, ethanol, methanol, toluene, xylene.

**Uses.**

It is a non-systemic acaricide and insecticide used at 300-600 g a.i./100 l against aphids, apple maggot, codling moth, oriental fruit moth, pear psylla, plum curculio, red spider mites, tortrix moths on deciduous tree fruits; at 400-800 g/ha on field and market garden crops; against aphids, bollworms, jassids, red spider mites and thrips on cotton; aphids and tuber moths on potatoes; pollen beetles and weevils on rape. It persists on plants *c.* 14 h, being converted to the corresponding phosphorothioate which is rapidly hydrolysed.

**Toxicology.**

Acute oral $LD_{50}$: for male rats 120-170 mg/kg; for mice 180 mg/kg; for guinea-pigs 380 mg/kg; for pheasants 290 mg/kg. Acute percutaneous $LD_{50}$: for rats 1500 mg/kg; for rabbits $>$1000 mg/kg. In 2-y feeding trials rats receiving 250 mg/kg diet and dogs 290 mg/kg diet suffered no ill-effect. Rates of 700 g/ha were without hazard to honeybees, provided the worker bees were not actively foraging at the time of spraying.

**Formulations.**

These include e.c. 'Zolone Liquid', in Japan 'Rubitox' (300-350 g a.i./l); w.p. (300 g/kg); dusts (25 or 40 g/kg). Mixtures include phosalone + parathion-methyl.

**Analysis.**

Product analysis is by glc (J. Desmoras *et al., Anal. Methods Pestic. Plant Growth Regul.*, 1973, **7**, 385). Residues may be determined by glc (*idem, ibid.*; *Pestic. Anal. Man.*, 1979, **I**, 201-A, 201-G, 201-H, 201-I; D. C. Abbott *et al., Pestic. Sci.*, 1970, **1**, 10; M. A. Luke *et al., J. Assoc. Off. Anal. Chem.*, 1981, **64**, 1187).

$C_7H_{14}NO_3PS_2$ (255.3)
T5SYSTJ BUNPO&O2&O2

O
‖
S
$=NP(OCH_2CH_3)_2$
S

**Nomenclature and development.**

Common name phosfolan (BSI, E-ISO);phospholan (F-ISO). Chemical name (IUPAC) diethyl 1,3-dithiolan-2-ylidenephosphoramidate (I); 2-(diethoxyphosphinylimino)-1,3-dithiolan. (*C.A.*) (I) (9CI); *P,P*-diethyl cyclic ethylene ester of phosphonodithiomidocarbonic acid (8CI); Reg. No. *[947-02-4]*. OMS 646, ENT 25 830. Introduced by American Cyanamid Co. (GBP 974 138; FP 1 327 386) as code no. 'EI 47 031'; trade marks 'Cyolane', 'Cyolan', 'Cyalane', 'Cylan'.

**Properties.**

Technical phosfolan is a colourless to yellow solid; m.p. 37-45 °C; b.p. 115-118 °C/0.001 mmHg. Solubility: 650 g/l water; soluble in acetone, benzene, cyclohexane, ethanol, toluene; slightly soluble in diethyl ether; sparingly soluble in hexane. The aqueous solution is stable under neutral and slightly acidic conditions but is hydrolysed at 2<pH>9.

**Uses.**

It is a systemic insecticide for use against sucking insects, mites and lepidopterous larvae. It controls *Prodenia* spp. and *Spodoptera littoralis* on cotton at 0.75-1.0 kg/ha; jassids and whiteflies at 0.5 kg/ha; *Trichoplusia ni* at 1 kg/ha; also *Thrips tabaci* at 30 g/100 l. It is non-persistent in soil; in plants and animals it is metabolised at the N-P bond to less toxic, water-soluble compounds.

**Toxicology.**

Acute oral $LD_{50}$ for male rats 8.9 mg tech./kg. Acute percutaneous $LD_{50}$ for guinea-pigs 54 mg tech./kg; for granular formulations from 24 mg a.i. (as 10% granule)/kg to >100 g a.i. (as 2% granule)/kg. In 90-d feeding trials dogs receiving 1 mg/kg daily showed no clinical symptom.

**Formulations.**

An e.c. (250 g a.i./l); granules (20-50 g/kg).

**Analysis.**

Residues may be determined by glc (details are available from American Cyanamid Co.; N. R. Pasarela & E. J. Orloski, *Anal. Methods Pestic. Plant Growth Regul.*, 1973, **7**, 231); or by colorimetry (R. C. Blinn & J. E. Boyd, *J. Assoc. Off. Agric. Chem.*, 1964, **47**, 1106).

$C_{11}H_{12}NO_4PS_2$ (317.3)
T56 BVNVJ C1SPS&O1&O1

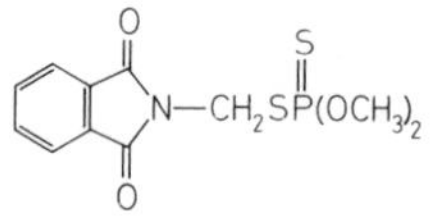

**Nomenclature and development.**

Common name phosmet (BSI, E-ISO, F-ISO, ESA);phtalofos (USSR);PMP (JMAF). Chemical name (IUPAC) *O,O*-dimethyl *S*-phthalimidomethyl phosphorodithioate. (*C.A.*) *S*-[(1,3-dihydro-1,3-dioxo-2*H*-isoindol-2-yl)methyl] *O,O*-dimethyl phosphorodithioate (9CI); *O,O*-dimethyl phosphorodithioate *S*-ester with *N*-(mercaptomethyl)phthalimide (8CI); Reg. No. *[732-11-6]*. ENT 25 705. Its insecticidal properties were reported by B. A. Butt & J. C. Keller (*J. Econ. Entomol.*, 1961, **54**, 813). It was introduced by Stauffer Chemical Co. (USP 2 767 194) as code no. 'R-1504'; trademark 'Imidan'.

**Properties.**

Phosmet (99.5% pure) is a colourless crystalline solid; m.p. 72.5 °C; v.p. 133 mPa at 50 °C. Solubility at 25 °C: 22 mg/l water; 650 g/l acetone; 600 g/l benzene; 5 g/l kerosene; 50 g/l methanol; 300 g/l toluene; 250 g/l xylene. In buffered solutions (20 mg/l) at 20 °C, 50% is hydrolysed in 13 d at pH 4.5, in <12 h at pH7, in <4 h at pH 8.3. It decomposes rapidly >100 °C. Incompatible with pesticides under alkaline conditions. Technical grade (94-96% pure) is an off-white or pink solid, with an offensive odour; m.p. 67-70 °C.

**Uses.**

Phosmet is a non-systemic acaricide and insecticide, used on top fruit (*e.g.* apples, pears, peaches, apricots, cherries), citrus, grapes, potatoes and in forestry at rates (0.5-1.0 kg a.i./ha) such that it is safe for a range of predators of mites and therefore useful in integrated control programmes.

**Toxicology.**

Acute oral $LD_{50}$ for male and female rats 113 mg/kg. Acute percutaneous $LD_{50}$ for albino rabbits >5000 mg/kg; non-irritating to their skin and moderately irritating to their eyes. In 2-y feeding trials NEL for rats and dogs was 40 mg/kg diet. Readily degraded both in laboratory animals and the environment.

**Formulations.**

'Prolate E', e.c. (120 g a.i./l); 'Prolate 8-OS', solution (80 g/l). 'Imidan WP', w.p. (125 or 500 g/kg). 'Imidan 5 Dust', 'Prolate 5 Dust', dust (50 g/kg).

**Analysis.**

Product analysis is by glc or hplc. Residues in crops are determined by glc (G. H. Batchelder *et al.*, *Anal. Methods Pestic., Plant Growth Regul. Food. Addit.*, 1967, **5**, 257; J. E. Barney *et al.*, *Anal. Methods Pestic. Plant Growth Regul.*, 1972, **6**, 408). Analytical methods are available from the Stauffer Chemical Co.

$C_{10}H_{19}ClNO_5P$ (299.7)
2N2&VYGUY1&OPO&O1&O1

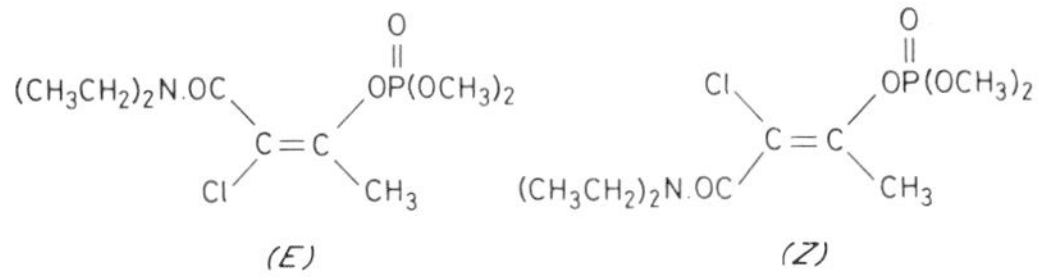

$(CH_3CH_2)_2N.OC(Cl)C{=}C(CH_3)OP(O)(OCH_3)_2$ *(E)*

$Cl(( CH_3CH_2)_2N.OC)C{=}C(CH_3)OP(O)(OCH_3)_2$ *(Z)*

**Nomenclature and development.**

Common name phosphamidon (BSI, E-ISO, F-ISO, ANSI);*exception* (Italy, Turkey). Chemical name (IUPAC) 2-chloro-2-diethylcarbamoyl-1-methylvinyl dimethyl phosphate. (*C.A.*) 2-chloro-3-(diethylamino)-1-methyl-3-oxo-1-propenyl dimethyl phosphate (9CI); dimethyl phosphate ester with 2-chloro-*N*,*N*-diethyl-3-hydroxycrotonamide (8CI); Reg. No. *[13171-21-6]* (*E*)- + (*Z*)-isomers, *[23783-98-4]* (*Z*)-isomer, *[297-99-4]* (*E*)-isomer. OMS 1325, ENT 25 515. Its insecticidal properties were described by F. Bachmann & J. Meierhans (*Bull. Cent. Int. Antiparasit.*, 1956, Nov. p. 18). Introduced by Ciba AG (now Ciba-Geigy AG) (BEP 552 284; GBP 829 576) as code no. 'C 570'; trade mark 'Dimecron'.

**Properties.**

Pure phosphamidon is a yellow liquid; b.p. 94 °C/0.04 mmHg; v.p. 3.3 mPa at 20 °C; ρ 1.21 g/cm³ at 20 °C; $n_D^{25}$ 1.4721. Solubility at 20 °C: completely miscible in water, acetone, dichloromethane, octan-1-ol, toluene. At 20 °C (calculated) 50% hydrolysis occurs in 60 d at pH5, 54 d at pH7 and 12 d at pH9.

The commercial compound contains 70% *m/m* (*Z*)-isomer (β-isomer) and 30% (*E*)-isomer (α-isomer) which are indistinguishable chemically, but may be separated by glc or countercurrent distribution (Ciba AG, 1964; *Residue Rev.*, 1971, **37**, 1; D. L. Bull *et al., J. Econ. Entomol.*, 1967, **60**, 332). The (*Z*)-isomer has the greater insecticidal activity.

**Uses.**

Phosphamidon is a systemic insecticide which is rapidly absorbed by plants and has little contact action. It is effective against sap-feeding insects at 300-600 g/ha, and other pests including rice and sugarcane stem borers and rice leaf beetles at 500-1000 g/ha. it is non-phytotoxic except to some cherry varieties and sorghum varieties related to Red Swazi. In plants 50% loss occurs in *c.* 2 d (R. Schuppen, *Chim. Ind. (Paris)*, 1961, **85**, 421).

**Toxicology.**

Acute oral $LD_{50}$ for rats 17.4 mg/kg. Acute percutaneous $LD_{50}$ for rats 374 mg/kg; slight irritant to skin and eyes of rabbits. Acute inhalation $LC_{50}$ (4-h) for rats *c.* 180 mg/m³ air. In 2-y feeding trials NEL was: for rats 1.25 mg/kg daily; for dogs 0.1 mg/kg daily. It is highly toxic to birds and honeybees.

**Formulations.**

'Dimecron 20', 'Dimecron 50', solution (200 or 500 g/l).

**Analysis.**

Product analysis is by glc (A. A. Carlstrom, *J. Assoc. Off. Anal. Chem.*, 1972, **55**, 1331) or by reaction with iodine under conditions to differentiate from impurities (R. Anlicker *et al., Helv. Chim. Acta*, 1961, **44**, 1622). Residues may be determined by glc with FPD (*Pestic. Anal. Man.*, 1979, **I**, 201-H, 201-I; D. C. Abbott, *Pestic. Sci.*, 1970, **1**, 10; M. A. Luke *et al., J. Assoc. Off. Anal. Chem.*, 1981, **64**, 1187; G. Voss *et al., Residue Rev.*, 1971, **37**, 101).

$C_{12}H_{15}N_2O_3PS$ (298.3)
NCYR&UNOPS&O2&O2

$C_6H_5-C(CN)=NOP(=S)O(CH_2CH_3)_2$

**Nomenclature and development.**

Common name phoxim (BSI, E-ISO, ESA);phoxime (F-ISO). Chemical name (IUPAC) *O,O*-diethyl α-cyanobenzylideneamino-oxyphosphonothioate; 2-( diethoxyphosphinothioyloxyimino)-2-phenylacetonitrile. (*C.A.*) α-[[(diethoxyphosphinothioyl)oxy]imino]benzeneacetonitrile (9CI); phenylglyoxylonitrile oxime *O,O*-diethyl phosphorothioate (8CI); Reg. No. *[14816-18-3]*. OMS 1170. Its insecticidal properties were described by A. Wybou & I. Hammann (*Meded. Rijksfac. Landbouwwet. Gent*, 1968, **33**, 817). Introduced by Bayer AG (BEP 678 139; DEP 1 238 902) as code no. 'Bay 5621','Bayer 77 488'; trade marks 'Baythion', 'Volaton'.

**Properties.**

Pure phoxim is a yellowish liquid; m.p. 5-6 °C; $d_4^{20}$ 1.176. Solubility at 20 °C: 7 mg/l water; >500 g/kg dichloromethane; >600 g/kg propan-2-ol; less soluble in light petroleum. The technical product is a reddish oil at room temperature. It decomposes on attempted distillation; is stable to water and to acids, 50% hydrolysis occurs at room temperature in 700 h at pH7, 170 min at pH11.6.

**Uses.**

It is an insecticide effective against a broad range of insects, used particularly to control pests of man or of stored products. Soil applications have controlled dipterous larvae at 50 mg a.i./plant; rootworms and wireworms at 5 kg/ha (P. Villeroy & P. Poucharesse, *Pflanzenschutz-Nachr. (Engl. Ed.)*, 1974, **27**, 284; G. Zoebelein, *ibid.*, 1975, **28**, 162, 178). ULV foliar applications have controlled grasshoppers. It is of brief persistence and has no systemic action. Photochemical breakdown and metabolism in cotton plants involve isomerisation to *O,O*-diethyl *S*-α-cyanobenzylideneaminothiophosphonoate, TEPP being among the other products (G. Dräger, *ibid.*, 1971, **24**, 239).

**Toxicology.**

Acute oral $LD_{50}$: for rats 1976-2170 mg/kg; for mice 1935-2340 mg/kg; for female guinea-pigs *c.* 600 mg/kg; for female cats and dogs 250-500 mg/kg; for female rabbits 250-375 mg/kg. Acute percutaneous $LD_{50}$ for male rats 1000 mg/kg. In 2-y feeding trials NEL for rats was 15 mg/kg diet; in 90-d trials rats receiving 50 mg/kg diet showed no symptom apart from mild depression of cholinesterase activity. $LC_{50}$ is: for trout and carp 0.1-1.0 mg/l; for goldfish 1-10 mg/l. Toxic to honeybees by contact and vapour action.

**Formulations.**

Include 'Volaton' for agricultural uses: e.c. (500 g a.i./l), granules (50 or 100 g/kg), seed treatment (200 g/kg), concentrate for ULV application. 'Baythion' for use against pests of man and of stored products, e.c. (100, 200 or 500 g/l), w.p. (500 g/kg), dust (30 g/kg), cold fog solution (50 g/l).

**Analysis.**

Details of product analysis are available from Bayer AG. Residues may be determined by glc (*idem, ibid.*, 1969, **22**, 301).

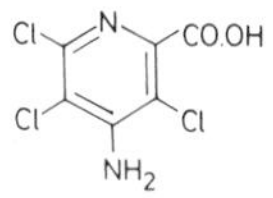

$C_6H_3Cl_3N_2O_2$ (241.5)
T6NJ BVQ CG DZ EG FG
*picloram-potassium*
$C_6H_2Cl_3KN_2O_2$ (280.6)
T6NJ BVQ CG DZ EG FG &&K SALT

**Nomenclature and development.**

Common name picloram (BSI, E-ISO, ANSI, WSSA, JMAF);piclorame (F-ISO). Chemical name (IUPAC) 4-amino-3,5,6-trichloropyridine-2-carboxylic acid; 4-amino-3,5,6-trichloropicolinic acid (I). (*C.A.*) 4-amino-3,5,6-trichloro-2-pyridinecarboxylic acid (9CI); (I) (8CI); Reg. No. *[1918-02-1]*. Its herbicidal properties were described by J. W. Hamaker *et al.* (*Science*, 1963, **141**, 363). Introduced by Dow Chemical Co. (USP 3 285 925) as trade mark 'Tordon'.

**Properties.**

Picloram is a colourless powder, with a chlorine-like odour, decomposing at *c.* 215 °C without melting; v.p. 82 μPa at 35 °C. Solubility at 25 °C: 430 mg/l water; 19.8 g/l acetone; 600 mg/l dichloromethane; 5.5 g/l propan-2-ol. It is acidic, $pK_a$ 3.6, and forms water-soluble alkali metal and amine salts, *e.g.* picloram-potassium, Reg. No. *[2545-60-0]* (1:1 salt) *[11562-68-2]* (unstated potassium content), solubility at 25 °C: 400 g/l water.

**Uses.**

Picloram and its salts are herbicides that produce epinasty and leaf curling, are rapidly absorbed by leaves and roots and translocated, accumulating in new growth. Most broad-leaved crops, except crucifers, are sensitive; most grasses are resistant. Rates as low as 17 g a.e./ha are effective in controlling annual weeds; it is used at 2.2-3.3 kg a.e./ha alone or at 0.3-1.8 kg/ha in combination with 2,4-D against deep-rooted perennials on non-crop land; as pellets 2.2-9.5 kg picloram/ha or 0.3-2.4 kg/ha in combination with 2,4-D or 2,4,5-T for brush control. At the higher doses 50% loss from soil occurs in 30-330 d, depending upon soil conditions.

**Toxicology.**

Acute oral $LD_{50}$: for rats 8200 mg/kg; for mice 2000-4000 mg/kg; for rabbits *c.* 2000 mg/kg; for guinea-pigs *c.* 3000 mg/kg; for sheep >100 mg/kg; for cattle >750 mg/kg. Acute percutaneous $LD_{50}$ for rabbits >4000 mg/kg; no serious hazard from contact with their skin or eyes was observed. In 2-y feeding trials NEL for rats was 150 mg/kg daily. $LC_{50}$ (8-d) for mallard duck, pheasant, Japanese quail and bobwhite quail was >5000 mg/kg. $LC_{50}$ (96-h) is: for rainbow trout 19.3 mg/l; for fathead minnow 55.3 mg/l. $LC_{50}$ (48-h) for *Daphnia* is 50.7 mg/l. It is relatively non-toxic to honeybees ($LC_{50}$ >1000 mg/kg).

**Formulations.**

These include: pellets (20 or 100 g a.i./kg); aqueous concentrates (240 g a.e. picloram-potassium/l). Mixtures include aqueous concentrates of salts: picloram + aminotriazole + atrazine + simazine; picloram + aminotriazole + bromacil + dalapon + diuron; picloram + bromacil + diuron; picloram + 2,4-D; picloram + 2,4-D + dichlorprop; picloram + 2,4-D + tebuthiuron; picloram + 2,4-D + MCPA + mecoprop; also e.c. picloram as amine salt + triclopyr-(2-butoxyethyl); granules picloram + bromacil.

**Analysis.**

Product analysis is by glc of suitable esters (P. A. Hargreaves & S. H. Rapkins, *Pestic. Sci.*, 1976, **7**, 515) or by hplc (*CIPAC Handbook*, 1983, **1B**, in press). Residues may be determined by glc of derivatives (T. H. Byast *et al.*, *Tech. Rep. ARC Weed Res. Organ.*, No. 15 (2nd Ed.), p. 33; J. R. Ramsey, *Anal. Methods Pestic., Plant Growth Regul. Food Addit.*, 1967, **5**, 507; *Anal. Methods Pestic. Plant Growth Regul.*, 1972, **6**, 700). Details of methods are available from Dow Chemical Co.

$C_{33}H_{47}NO_{13}$ (665.7)
T F3-24-6 A AO GO KVO IU OU QU
SU UUTJ BQ DQ M1 C&VQ D&Q WO-
BT6OTJ CQ DZ EQ F1

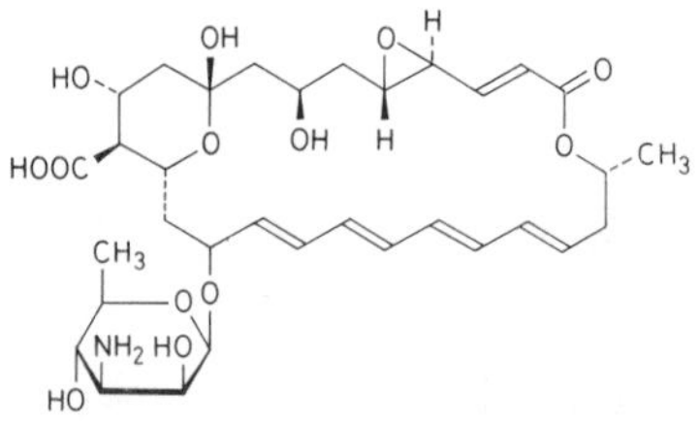

**Nomenclature and development.**

The traditional names pimaricin;tennecetin;natamycin (BPC) are applied to an antifungal antibiotic isolated from *Streptomyces natalensis*. Chemical name (IUPAC) (8*E*,14*E*,16*E*,18*E*,20*E*)-(1*S*,3*R*,5*S*,7*S*,12*R*,24*R*,25*S*,26*R*)-22-(3-amino-3,6-dideoxy-β-D-mannopyranosyloxy)-1,3,26-trihydroxy-12-methyl-10-oxo-6,11,28-trioxatricyclo[22.3.1.$0^{5,7}$]octacosa-8,14,16,18,20-pentaene-25-carboxylic acid. (*C.A.*) 22-[(3-amino-3,6-dideoxy-β-D-mannopyranosyl)oxy]-1,3,26-trihydroxy-12-methyl-10-oxo-6,11,28-trioxatricyclo-[22.3.1.$0^{5,7}$]octacosa-8,14,16,18,20-pentaene-25-carboxylic acid (9CI); 16-[(3-amino-3,6-dideoxy-β-D-mannopyranosyl)oxy]-18,20,24-trihydroxy-6-methyl-4,22-dioxo-5,27-dioxabicyclo[24.1.0]heptacosa-2,8,10,12,14-pentaene-19-carboxylic acid (8CI); Reg. No. *[7681-93-8]*. Its structure was established by B. T. Golding *et al.* (*Tetrahedron Lett.*, 1966, p. 3551), W. E. Meyer (*Chem. Commun.*, 1968, p. 470) and G. Gaudiano *et al.* (*Chem. Ind. (Milan)*, 1966, **48**, 1327). Introduced as a fungicide by Gist-Brocades N.V. (GBP 712 547; 844 289; USP 3 892 850) as trade marks 'Delvolan' (for crop protection use), 'Pimafucin' (for medical use), 'Delvocid', 'Delvopos' (for food additive use).

**Properties.**

Produced by fermentation of *S. natalensis* and *S. chattanoogensis*, it is isolated as pale yellow crystals which decompose *c.* 200 °C. Solubility at 20-22 °C: 4.1 g/l water; 850 mg/l acetone; 50 mg/l benzene; >200 g/l dimethyl sulphoxide; 5.5 g/l ethanol; 97 g/l methanol; 100 mg/l petroleum spirit (J. R. Marsh & P. J. Weiss, *J. Assoc. Off. Anal. Chem.*, 1967, **50**, 457). It forms water-soluble salts with acid or alkali. It is light-sensitive but otherwise stable when dry.

**Uses.**

Pimaricin is used as a fungicide to control diseases of bulbs, especially basal rot of daffodils at 200 mg a.i./l - preferably combined with hot-water treatment of the bulbs.

**Toxicology.**

Acute oral $LD_{50}$ for rats 2730-4670 mg/kg (G. J. Levinskas, *Toxicol. Appl. Pharmacol.*, 1966, **8**, 97).

**Formulations.**

'Delvolan', w.p. (68 g a.i./kg).

**Analysis.**

Product analysis is either by bioassay with a suitable microorganism, and confirmed by the effect of the enzyme pimaricinase, or by u.v. spectrometry. Residues may be determined by bioassay.

$C_{14}H_{14}O_3$ (230.3)
L56 BV DV CHJ CVX1&1&1

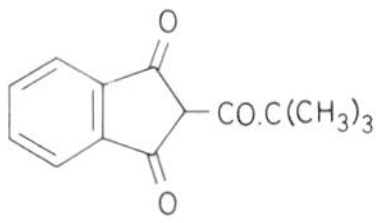

O
CO.C(CH3)3
O

**Nomenclature and development.**

Common name pindone (BSI, E-ISO, F-ISO);pivaldione (France);pival (Turkey);*exception* (Portugal). Chemical name (IUPAC) 2-pivaloylindan-1,3-dione. (*C.A.*) 2-(2,2-dimethyl-1-oxopropyl)-1*H*-indene-1,3(2*H*)-dione (9CI); 2-pivaloyl-1,3-indandione (8CI); Reg. No. *[83-26-1]*. Its insecticidal properties were described by L. B. Kilgore *et al.* (*Ind. Eng. Chem.*, 1942, **34**, 494). Introduced by Kilgore Chemical Co. (USP 2 310 949) as trade marks 'Pivalyl Valone', 'Pival', 'Pivalyn' (all to Kilgore Chemical Co.).

**Properties.**

Pure pindone is a yellow crystalline solid; m.p. 108.5-110.5 °C. Solubility at 25 °C: 18 mg/l water; soluble in aqueous alkali or ammonia to give bright yellow salts; soluble in most organic solvents.

**Uses.**

It was originally suggested as a substitute for the greater part of the pyrethrins in pyrethrum sprays. It is used as an anticoagulant rodenticide in baits containing 250 mg/kg.

**Toxicology.**

Acute $LD_{50}$ by injection for rats is *c.* 50 mg/kg, but it is more toxic when given in small daily doses of 15-35 mg/kg. Dogs are killed by daily doses of 2.5 mg/kg (J. R. Beauregard *et al.*, *J. Agric. Food Chem.*, 1955, **3**, 124; J. P. Saunders, *et al.*, *ibid.*, p. 762).

**Formulations.**

These include: 'Pival' powder (5 g a.i./kg) for use in rat baits; 'Pivalyn', pindone-sodium with chelating agent.

**Analysis.**

Analysis is by colorimetry (J. B. La Clair, *J. Assoc. Off. Agric. Chem.*, 1955, **38**, 299).

$C_{19}H_{30}O_5$ (338.4)
T56 BO DO CHJ G3 H1O2O2O4

$CH_3(CH_2)_3OCH_2CH_2OCH_2CH_2OCH_2$ / $CH_3(CH_2)_2$

**Nomenclature and development.**

The trivial name piperonyl butoxide (BSI, E-ISO, BPC);piperonyl butoxyde (F-ISO) is accepted in lieu of a common name. Chemical name (IUPAC) 2-(2-butoxyethoxy)ethyl 6-propylpiperonyl ether; 5-[2-(2-butoxyethoxy)ethoxymethyl]-6-propyl-1,3-benzodioxole (I). (*C.A.*) (I) (9CI); α-[2-(2-butoxyethoxy)ethoxy]-4,5-(methylenedioxy)-2-propyltoluene (8CI); Reg. No. *[51-03-6]*. ENT 14 250. Its activity as a synergist for pyrethrum was described by H. Wachs (*Science*, 1947, **105**, 530) (USP 2 485 681; 2 550 737).

**Properties.**

The technical grade comprises ⩾85% *m/m* piperonyl butoxide and ⩾15% *m/m* related compounds, and is a pale yellow oil, b.p. 180 °C/1 mmHg; $d^{25}$ 1.05-1.07; $n_D^{20}$ 1.497-1.512. It is stable to light, resistant to hydrolysis.

**Uses.**

Piperonyl butoxide is a synergist for the pyrethrins and related insecticides.

**Toxicology.**

Acute oral $LD_{50}$ for rats and rabbits *c.* 7500 mg/kg. In 2-y feeding trials rats receiving 100 mg/kg diet suffered no ill-effect. It is non-carcinogenic and the human tolerance for chronic ingestion is estimated at 42 mg/kg diet (M. P. Sarles & W. B. Vandergrift, *Am. J. Trop. Hyg. Med.*, 1952, **1**, 862; J. R. M. Innes, *J. Nat. Cancer Res.*, 1969, **42**, 1101).

**Formulations.**

It is used in conjunction with pyrethrins in ratios of 5:1 to 20:1 *m/m*, in solutions, as aerosol concentrates, emulsions or dusts.

**Analysis.**

Product analysis is by glc (*Br. Pharmacopoeia Vet.*, 1977, p. 64) or by colorimetry of a derivative (R. A. Jones *et al., J. Assoc. Off. Agric. Chem.*, 1952, **35**, 771; *AOAC Methods*, 1980, 6.166-6.168; *CIPAC Handbook, 1980,* **1A**, 1325; *Anal. Methods Pestic. Plant Growth Regul.*, 1972, **6**, 458). Residues may be determined by colorimetry of a derivative (*idem, ibid.*; *AOAC Methods*, 1980, 29.161-29.164; *Pestic. Anal. Man.*, 1979, **II**; W. Specht & M. Tillkes, *Fresenius' Z. Anal. Chem.*, 1980, **301**, 300).

$C_{14}H_{28}NO_3PS_2$ (353.5)
T6NTJ AV1SPS&O3&O3 B1

N—CO.$CH_2SP(O)(OCH_2CH_2CH_3)_2$ (2-methylpiperidine ring, $CH_3$)

**Nomenclature and development.**

Common name piperophos (BSI, E-ISO, F-ISO). Chemical name (IUPAC) *S*-2-methylpiperidinocarbonylmethyl *O,O*-dipropyl phosphorodithioate. (*C.A.*) *S*-[2-(2-methyl-1-piperidinyl)-2-oxoethyl] *O,O*-dipropyl phosphorodithioate (9CI); *O,O*-dipropyl phosphorodithioate *S*-ester with 1-mercaptoacetyl-2-pipecoline (8CI); Reg. No. *[24151-93-7]*. Its herbicidal properties were described by D. H. Green & L. Ebner (*Proc. Br. Weed Control Conf., 11th*, 1972, **2**, 822). Introduced by Ciba-Geigy AG (BEP 725 992; GBP 1 255 946) as code no. 'C 19 490'; trade marks 'Rilof'; 'Avirosan' (mixture with dimethametryn).

**Properties.**

Pure piperophos is a liquid which decomposes on attempted distillation; v.p. 32 μPa at 20 °C; ρ 1.130 g/cm³ at 20 °C. Solubility at 20 °C: 25 mg/l water; completely miscible with acetone, benzene, dichloromethane, hexane, methanol, octan-1-ol. On hydrolysis at 20 °C, 50% loss (calculated) occurs in >200 d at $5 \leqslant pH \leqslant 7$, 178 d at pH9.

**Uses.**

It is a selective herbicide active against annual grasses and sedges in seeded or flooded rice. Used in combination with dimethametryn, for the control of both mono- and di-cotyledonous weeds, both herbicides being taken up by young plants through roots, coleoptiles and leaves. In tropical regions piperophos is applied in combination with 2,4-D to control moncotyledonous weeds.

**Toxicology.**

Acute oral $LD_{50}$ for rats 324 mg tech./kg. Acute percutaneous $LD_{50}$ for rats >2150 mg/kg; non-irritant to skin and slight irritant to eyes of rabbits. Inhalation $LC_{50}$ (1-h) for rats >1960 mg/m³ air. In 90-d feeding trials NEL was: for rats 10 mg/kg diet (0.8 mg/kg daily); for dogs 5 mg/kg diet (0.15 mg/kg daily). $LC_{50}$ (96-h) is: for rainbow trout 6 mg/l; for crucian carp 5 mg/l. It is practically non-toxic to birds and honeybees.

**Formulations.**

'Rilof' 500 EC, e.c. (500 g a.i./l). 'Avirosan' (Reg. No. *[12701-69-8]*), piperophos + dimethametryn (4:1). 'Rilof H', piperophos + 2,4-D (*c*.2:1 or 3:2).

**Analysis.**

Product analysis is by glc. Residues may be determined by glc with TID. Details of both methods are available from Ciba-Geigy AG.

$C_{18}H_{36}N$ (266.5)
T6KTJ A2Y1&3Y1&1 A2U1
*piproctanyl bromide*
$C_{18}H_{36}BrN$ (346.4)
T6KTJ A2Y1&3Y1&1 A2U1 &&Br SALT

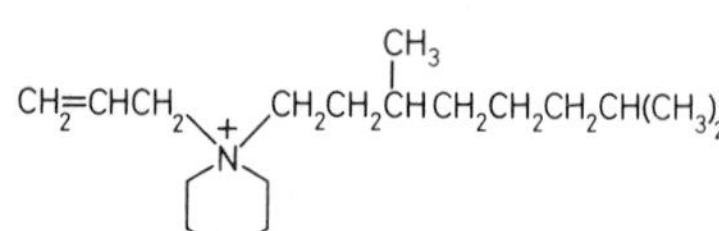

**Nomenclature and development.**

Common name piproctanyl (BSI, E-ISO, F-ISO). Chemical name (IUPAC) 1-allyl-1-(3,7-dimethyloctyl)piperidinium ion. (*C.A.*) 1-(3,7-dimethyloctyl)-1-(2-propenyl)piperidinium (9CI); Reg. No. *[69309-47-3]* (ion), *[56717-11-4]* (bromide salt). Its plant growth-regulating properties were described by G. A. Hüppi *et al.* (*Experientia*, 1976, **32**, 37). The bromide was introduced by R. Maag Ltd (DEOP 2 459 129 to Hoffman-La Roche AG) as code no. 'Ro 06-0761'; trade marks 'Alden', 'Stemtrol' (Maag).

**Properties.**

Piproctanyl bromide is a pale yellow wax; m.p. 75 °C. Solubility: highly soluble in water and alcohols. It is non-corrosive in aqueous solution, insensitive to light and stable at temperatures ≤100 °C, and ≤3 y under normal conditions.

**Uses.**

It is a plant growth regulator which reduces internodal elongation, decreasing plant height, producing stronger stems and peduncles and a deeper green foliage. it is taken up through leaves and roots but is not readily translocated between shoots. A surfactant is included in sprays. It is used on chrysanthemums at 75-150 mg a.i./l, the spray concentration depends on cultivar. It is also used on *Begonia elatior, Calceolaria rugosa, Fuchsia hybrida* and *Petunia* spp.

**Toxicology.**

Acute oral $LD_{50}$: for rats 820-990 mg a.i./kg; for mice 182 mg/kg. Acute percutaneous $LD_{50}$ for rats 115-240 mg/kg; a solution (30 g/l acetone) caused no irritation to the skin of guinea-pigs; rabbits' eyes were not affected by 3 g/l water. Acute inhalation $LC_{50}$ for rats 1.5 g/$m^3$ air. In 90-d feeding trials there was no significant effect for rats receiving 150 mg/kg daily or dogs 25 mg/kg daily. $LC_{50}$ (8-d) for bobwhite quail and mallard duck was >10 000 mg/kg diet. $LC_{50}$ (96-h) is: for rainbow trout 12.7 mg/l; for bluegill 62 mg/l.

**Formulations.**

'Alden' and 'Stemtrol', water-miscible concentrates (50 g a.i./l, plus surfactant).

**Analysis.**

Product analysis is by reaction with sodium phenylsulphide to form allyl phenyl sulphide which is determined by glc. Residues may be determined in the same way.

$C_{11}H_{18}N_4O_2$ (238.3)
T6N CNJ BN1&1 DOVN1&1 E1 F1



**Nomenclature and development.**

Common name pirimicarb (BSI, E-ISO, ANSI, JMAF);pyrimicarbe (F-ISO). Chemical name (IUPAC) 2-dimethylamino-5,6-dimethylpyrimidin-4-yl dimethylcarbamate. (*C.A.*) 2-(dimethylamino)-5,6-dimethyl-4-pyrimidinyl dimethylcarbamate (8 & 9 CI); Reg. No. *[23103-98-2]*. ENT 27 766. Its aphicidal properties were described by F. L. C. Baranyovits & R. Ghosh *(Chem. Ind. (London)*, 1969, p. 1018). Introduced by ICI Plant Protection Division (GBP 1 181 657) as code no. 'PP 062'; trade marks 'Pirimor', 'Aphox'.

**Properties.**

A colourless solid; m.p. 90.5 °C; v.p. 4.0 mPa at 30 °C. Solubility at 25 °C; 2.7 g/l water; 4.0 g/l acetone; 3.2 g/l chloroform; 2.5 g/l ethanol; 2.9 g/l xylene. Decomposed by prolonged boiling with acid or alkali. Aqueous solutions unstable to light. Forms well defined crystalline salts with acids, which are readily soluble in water; the hydrochloride is deliquescent.

**Uses.**

A selective aphicide, effective against organophosphorus-resistant aphid strains. It is fast acting and has fumigant and translaminar properties; taken up by roots and translocated in the xylem system. It has neither acaricidal nor fungicidal properties, and can be used to control aphids in integrated control programmes.

**Toxicology.**

Acute oral $LD_{50}$: for rats 147 mg/kg; for mice 107 mg/kg; for poultry 25-50 mg/kg; for dogs 100-200 mg/kg; for mallard duck 17.2 mg/kg; for bobwhite quail 8.2 mg/kg. Daily applications of 500 mg/kg (for 24 h) to rabbit skin over a 14-d period produced no toxic symptom; a solution (5 g tech./l) was not an irritant to rabbit eyes. Rats exposed for 6 h/d (5 d/week) for 21 d, to air which had been passed over technical pirimicarb at room temperature developed no toxic sign, nor was there inhibition of cholinesterase. In 2-y feeding trials NEL was: for dogs 1.8 mg/kg daily; for rats 250 mg/kg diet (12.5 mg/kg daily). $LC_{50}$ (96-h) is: for bluegill 55 mg/l; for rainbow trout 29 mg/l.

**Formulations.**

These include: e.c. (80 g a.i./l); w.p. (500 g/kg); dispersible grains (500 g/kg) and smoke generators all for use in agriculture and horticulture; an aerosol generator for home/garden use.

**Analysis.**

Product analysis is by glc (J. E. Bagness & W. G. Sharples, *Analyst (London)*, 1974, **99**, 225). Residues may be determined by glc with FID or by colorimetry (D. J. W. Bullock, *Anal. Methods Pestic. Plant Growth Regul.*, 1973, **7**, 399). Details available from ICI Plant Protection Division.

$C_{13}H_{24}N_3O_3PS$ (333.4)
T6N CNJ BN2&2 DOPS&O2&O2 F1

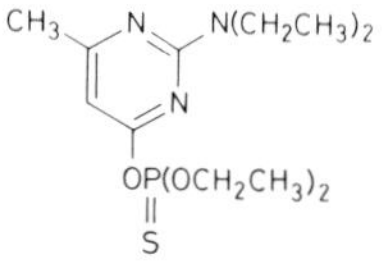

**Nomenclature and development.**

Common name pirimiphos-ethyl (BSI, E-ISO, ANSI);pyrimiphos-éthyl (F-ISO). Chemical name (IUPAC) *O*-2-diethylamino-6-methylpyrimidin-4-yl *O,O*-diethyl phosphorothioate. (*C.A.*) *O*-[2-(diethylamino)-6-methyl-4-pyrimidinyl] *O,O*-diethyl phosphorothioate (8 & 9CI); Reg. No. *[23505-41-1]*. Introduced as an insecticide by ICI Plant Protection Division (GBP 1 019 227; 1 205 000) as code no. 'PP 211'; trade marks 'Pirimicid', 'Fernex'.

**Properties.**

It is a straw-coloured liquid; v.p. 39 mPa at 25 °C; $d^{20}$ 1.14; $n_D^{25}$ 1.520. Decomposition begins >130 °C so no b.p. is available. Solubility at 30 °C: <1 mg/l water; miscible with, or very soluble in most organic solvents. Stable ⩾5 d at 80 °C and ⩾1 y at room temperature. Corrosive to iron and unprotected tinplate.

**Uses.**

A broad-range insecticide effective against dipterous and coleopterous pests living in the soil or on the soil surface. It is relatively non-phytotoxic and may be used as a seed treatment, sometimes in combination with fungicides. It is also used in compost to prevent infestation of mushrooms by sciarids and phorids.

**Toxicology.**

Acute oral $LD_{50}$: for rats 140-200 mg/kg; for guinea-pigs 50-100 mg/kg; mallard duck 2.5 mg/kg; bobwhite quail 10-20 mg/kg. Acute percutaneous $LD_{50}$ for rats 1000-2000 mg/kg; no irritation followed the application of 100 mg/kg to shorn rabbit skin for 24-h periods on alternate days over a 10-d period; it is not a sensitiser. In 90-d feeding trials NEL was: for rats 1.6 mg/kg diet (0.08 mg/kg daily); for dogs 0.2 mg/kg daily. For rats receiving 27 mg/kg diet (1.6 mg/kg daily) and dogs 2 mg/kg daily the only effect was on the cholinesterase levels. From inhalation tests on rats the maximum allowable concentration is calculated to be 14 mg/m³. $LC_{50}$ (96-h) is: for brown trout 0.02 mg/l; for common carp 0.22 mg/l.

**Formulations.**

These include: encapsulated (200 g a.i./l); e.c. (250 or 500 g/l); granules (50 or 100 g/kg). 'Pirimicid', seed treatment (pirimiphos-ethyl + drazoxolon).

**Analysis.**

Product analysis is by glc (J. E. Bagness & W. G. Sharples, *Analyst (London)*, 1974, **99**, 225), or by u.v. spectrometry (S. H. Yuen, *ibid*,, 1976, **101**, 533). Residues may be determined by glc with FPD or FTD; a colorimetric method is also available. Details of these methods are available from ICI Plant Protection Division. See also: D. J. W. Bullock, *Anal. Methods Pestic. Plant Growth Regul.*, 1976, **8**, 171.

# Pirimiphos-methyl

$C_{11}H_{20}N_3O_3PS$ (305.3)
T6N CNJ BN2&2 DOPS&O1&O1 F1

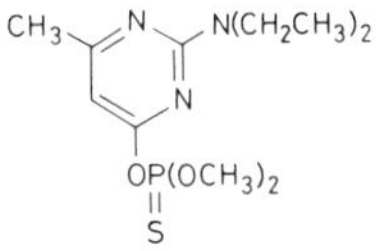

**Nomenclature and development.**

Common name pirimiphos-methyl (BSI, E-ISO, ANSI);pyrimiphos-méthyl (F-ISO). Chemical name (IUPAC) *O*-2-diethylamino-6-methylpyrimidin-4-yl O,O-dimethyl phosphorothioate. (*C.A.*) *O*-[2-(diethylamino)-6-methyl-4-pyrimidinyl] *O,O*-dimethyl phosphorothioate (8 & 9CI); Reg. No. *[29232-93-7]*. OMS 1424. Introduced as an insecticide by ICI Plant Protection Division (GBP 1 019 227; 1 204 552) as code no. 'PP 511'; trade marks 'Actellic', 'Actellifog', 'Silo San', 'Blex', 'Sybol 2'.

**Properties.**

It is a straw-coloured liquid; v.p. *c.* 13 mPa at 30 °C; $d^{30}$ 1.157; $n_D^{25}$ 1.527. Solubility at 30 °C: *c.* 5 mg/l water; miscible with, or very soluble in most organic solvents. It is hydrolysed by concentrated acid and alkali. Slightly corrosive to unprotected tinplate and mild steel.

**Uses.**

A fast-acting insecticide and acaricide with both contact and fumigant action, able to penetrate leaf tissue to give a translaminar action. Its range of activity includes many crop pests and it is also used against pests of stored products, in public health, domestic and amenity use.

**Toxicology.**

Acute oral $LD_{50}$: for female rats 2050 mg/kg; for male mice 1180 mg/kg; for female guinea-pigs 1000-2000 mg/kg; for male rabbits 1150-2300 mg/kg; for hens 30-60 mg/kg; for bobwhite quail *c.* 140 mg/kg. Acute percutaneous $LD_{50}$ for rabbits is $>$2000 mg/kg. In 2-y feeding trials NEL for rats 10 mg/kg diet; in 90-d trials NEL: for dogs 20 mg/kg diet; for rats 8 mg/kg diet. At 360 mg/kg diet the only effect was on cholinesterase levels. Rats were not affected when exposed to saturated vapour (40 mg/$m^3$) for 6 h/d for 5 d/week over a period of 21 d. $LC_{50}$ (24-h) for mirror carp is 1.6 mg/l; $LC_{50}$ (48-h) 1.4 mg/l.

**Formulations.**

These include: e.c. (80, 250 or 500 g a.i./l); ULV concentrate (500 g/l); encapsulated (200 g/l); dust (20 g/kg); solvent-free formulations (900 g/kg); smoke generator. Domestic sprays, (pirimiphos-methyl + synergised pyrethrins).

**Analysis.**

Product analysis is by glc (J. E. Bagness & W. G. Sharples, *Analyst (London)*, 1974, **99**, 225) or by u.v. spectrometry (S. H. Yuen, *ibid.*, 1976, **101**, 533). Residues may be determined by glc with FPD or FTD (*Analyst (London)*, 1980, **105**, 515; J. Desmarchelier *et al.*, *Pestic. Sci.*, 1977, **8**, 473). A colorimetric method is also available (D. J. W. Bullock, *Anal. Methods Pestic. Plant Growth Regul.*, 1976, **8**, 181). Details are also available from ICI Plant Protection Division.

$C_{17}H_{25}N_5O_{13}$ (507.4)
T6NVMVJ E1Q A- BT5OTJ CQ DQ
EYVQMVYZYQYQ1OVZ

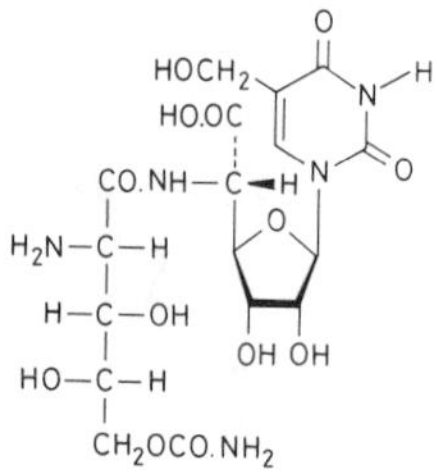

**Nomenclature and development.**

Common name polyoxins (JMAF). Chemical name (IUPAC) for polyoxin B is 5-(2-amino-5-*O*-carbamoyl-2-deoxy-L-xylonamido)-5-deoxy-1-(1,2,3,4-tetrahydro-5-hydroxymethyl-2,4-dioxopyrimidinyl)-β-D-allofuranuronic acid. (*C.A.*) 5-[[2-amino-5-*O*-(aminocarbonyl)-2-deoxy-L-xylonyl]amino]-1,5-dideoxy-1-[3,4-dihydro-5-(hydroxymethyl)-2,4-dioxo-1(2*H*)-pyrimidinyl]-β-D-allofuranuronic acid (9CI); 5-(2-amino-2-deoxy-L-xylonamido)-1-[3,4-dihydro-5-hydroxymethyl-2,4-dioxo-1(2*H*)-pyrimidinyl]allofuranuronic acid β-D-monocarbamate (8CI); Reg. No. *[19396-06-6]*. It was isolated by S. Suzuki *et al.* (*J. Antibiot. Ser. A.*, 1965, **18**, 131). Introduced by Nihon Nohyaku Ltd and by Kumiai Chemical Industry Co. Ltd.

**Properties.**

Produced by fermentation of *S. cacaoi* it is an amorphous powder, m.p. 160 °C. Solubility: soluble in water; insoluble in acetone, benzene, chloroform, ethanol, hexane, methanol. It is stable over the range pH1-8.

**Uses.**

Polyoxin B is a fungicide effective at 50-100 mg a.i./l against *Alternaria* spp., causing an abnormal germination, and loss of phytopathogenicity and impeding hyphal development. It is also effective against *Botrytis* spp., *Pellicularia sasaki* of rice and several diseases of fruit, *e.g. Podosphaera leucotricha* on apple, *Fulvia fulva* on tomato, and cucumber scab. It is compatible with highly alkaline pesticides.

Polyoxin A is effective against brown spot of rice, polyoxin D against *Pellicularia sasaki* of rice.

**Toxicology.**

Oral doses of 1500 mg polyoxin B/kg caused no deaths in mice, nor did intravenous injections of 800 mg/kg. It is non-irritant to mucous membranes or skin. Japanese killifish are unaffected by 100 mg/l for 72 h.

**Formulations.**

A w.p. (100 g polyoxin B/l).

$C_{17}H_{26}ClNO_2$ (311.9)
G1VN2O3&R B2 F2

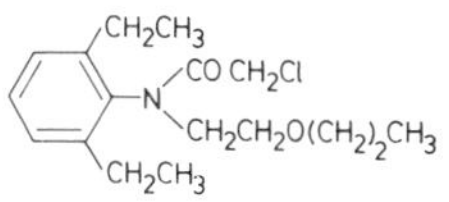

$CH_2CH_3$, $N$, $CO\,CH_2Cl$, $CH_2CH_2O(CH_2)_2CH_3$, $CH_2CH_3$

**Nomenclature and development.**

Common name pretilachlor (BSI, E-ISO); prétilachlore (F-ISO). Chemical name (IUPAC) 2-chloro-2',6'-diethyl-*N*-(2-propoxyethyl)acetanilide. (*C.A.*) 2-chloro-*N*-(2,6-diethylphenyl)-*N*-(2-propoxyethyl)acetamide (9CI); Reg. No. *[51218-49-6]*. Introduced by Ciba-Geigy AG (BEP 800 471; GBP 1 438 311; 1 438 312) as code no. 'CGA 26 423'; trade mark 'Rifit'.

**Properties.**

Pure pretilachlor is a colourless liquid; b.p. 135 °C/0.001 mmHg; v.p. 133 μPa at 20 °C; ρ 1.076 g/cm$^3$ at 20 °C. Solubility at 20 °C: 50 mg/l water; very soluble in benzene, dichloromethane, hexane, methanol. At 20 °C, 50% hydrolysis (calculated) occurs in >200 d at pH1-9, 14 d at pH13.

**Uses**

An experimental herbicide effective against broad-leaved weeds and sedges in transplanted rice.

**Toxicology.**

Acute oral $LD_{50}$ for rats 6099 mg tech./kg. Acute percutaneous $LD_{50}$ for rats >3100 mg/kg; moderate irritant to skin, minimal irritant to eyes of rabbits. Inhalation $LC_{50}$ (4-h) for rats >2800 mg/m$^3$ air. In 0.5-y feeding trials NEL for dogs was 300 mg/kg diet (*c.* 7.5 mg/kg daily). $LC_{50}$ (96-h) is: for rainbow trout 0.9 mg/l; for crucian carp 2.3 mg/l; for catfish 2.7 mg/l. It is slightly toxic to birds and toxic to honeybees.

**Analysis.**

Product analysis is by glc. Residues are determined by glc with MCD or TID. Particulars are available from Ciba-Geigy AG.

$C_{15}H_{16}Cl_3N_3O_2$ (376.7)
T5N CNJ AVN3&2OR BG DG FG

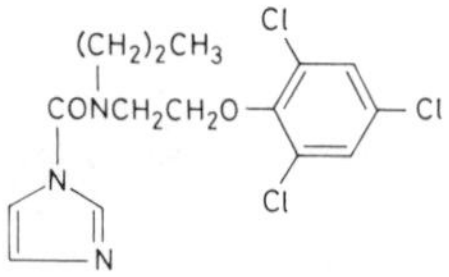

**Nomenclature and development.**

Common name prochloraz (BSI, E-ISO, F-ISO, ANSI). Chemical name (IUPAC) *N*-propyl-*N*-[2-(2,4,6-trichlorophenoxy)ethyl]imidazole-1-carboxamide; 1-*N*-propyl-*N*-[2-(2,4,6-trichlorophenoxy)ethyl]carbamoylimidazole. (*C.A.*) *N*-propyl-*N*-[2-(2,4,6-trichlorophenoxy)ethyl]-1*H*-imidazole-1-carboxamide (9CI); Reg. No. *[67747-09-5]*. Its fungicidal properties were described by R. J. Birchmore *et al.* (*Proc. Br. Crop. Prot. Conf.-Pests Dis.*, 1977, **2**, 593). Introduced by The Boots Co. Ltd (now FBC Limited) (GBP 1 469 772; USP 3 991 071; 4 080 462) as code no. 'BTS 40 542'; trade mark 'Sportak'.

**Properties.**

Pure prochloraz is a colourless crystalline solid; m.p. 38.5-41.0 °C; v.p. 80 nPa at 20 °C. Solubility at 23 °C 47.5 mg/l water; approximate values at 25 °C: 3.5 kg/l acetone, 2.5 kg/l chloroform, diethyl ether, toluene, xylene. It is stable in water (50% loss in 5 y (extrapolated) at 20 °C and pH 7); unstable to concentrated acid or alkali and to sunlight. The technical grade (*c.* 97% pure) is a golden brown liquid that tends to solidify on cooling.

**Uses.**

It is a protectant and eradicant fungicide effective against a wide range of diseases affecting field crops, fruit, turf, and vegetables. An e.c. is recommended for use on cereals at 400-600 g a.i./ha against *Erysiphe, Pseudocercosporella, Pyrenophora, Rhynochosporium*, and *Septoria* spp. and also on oilseed rape against several foliar and stem diseases. Good activity is also shown against storage or transit diseases of citrus and tropical fruits e.g. banana, mango when applied as a dip treatment at 0.5-0.7 g a.i./l; shower treatments are also recommended. It will control a range of important foliar diseases on turf at 0.75-3.0 kg/ha. It is active against *Alternaria* on potato, and *Cercospora* on sugar beet, at 0.5-1.5 kg/ha.

A w.p. is recommended for topical use on certain broad-leaved crops. On mushrooms, three sprays after casing give effective control of *Verticillium fungicola* with good crop safety.

A 0.2-0.5 g/kg seed treatment will control several cereal diseases caused by *Cochliobolus, Fusarium, Pyrenophora* and *Septoria* spp.

**Toxicology.**

Acute oral $LD_{50}$: for rats *c.* 1600 mg/kg; for mice 2400 mg/kg; for mallard duck 3132 mg/kg. Acute percutaneous $LD_{50}$ for rats >5000 mg/kg. In 2-y feeding trials NEL for dogs was 30 mg/kg diet. The $LC_{50}$ (96-h) for rainbow trout is 1 mg/l.

**Formulations.**

These include: 'Sportak' (400 or 450 g a.i./l); e.c. (400 or 450 g/kg); w.p. (500 g/kg); a seed treatment and liquid (200 g/l).

**Analysis.**

Product analysis is by hplc and glc. Residue analysis is by glc. Details available from FBC Limited.

# Procymidone

$C_{13}H_{11}Cl_2NO_2$ (284.1)
T35 DVNVTJ A1 C1 ER CG EG

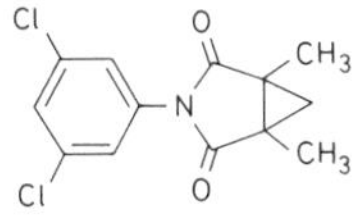

**Nomenclature and development.**

Common name procymidone (BSI, E-ISO, F-ISO, JMAF). Chemical name (IUPAC) *N*-( 3,5-dichlorophenyl)-1,2-dimethylcyclopropane-1,2-dicarboximide. (*C.A.*) 3-(3,5-dichlorophenyl)-1,5-dimethyl-3-azabicyclo[3.1.0]hexane-2,4-dione (9CI); Reg. No. *[32809-16-8]*. Its fungicidal activities were described by Y. Hisada *et al.* (*J. Pestic. Sci.*, 1976, **1**, 145). Introduced by Sumitomo Chemical Co. (GBP 1 298 261; USP 3 903 090) as code no. 'S-7131'; trade marks 'Sumisclex', 'Sumilex'.

**Properties.**

The technical grade is a light brown solid; m.p. 164-166 °C; v.p. 10.5 mPa at 20 °C. Solubility at 25 °C: 4.5 mg/l water; 16 g/kg methanol; 45 g/kg xylene.

**Uses.**

Procymidone is a preventive, curative and persistent fungicide with moderate systemic activity. It is effective against *Botrytis, Cochliobolus, Helminthosporium* and *Sclerotinia* spp. on field crops, fruits, ornamentals and vegetables particularly in vineyards and greenhouses. It is usually applied at 0.5-1.0 kg a.i./ha. It can also be used for the control of storage rots of fruits and vegetables.

**Toxicology.**

Acute oral $LD_{50}$: for male rats 6,800 mg/kg, for females 7,700 mg/kg. Acute percutaneous $LD_{50}$ for rats $>$ 2,500 mg/kg; no skin or eye irritation was observed in rabbits. $LC_{50}$ (48-h) for carp is $>$ 10 mg/l.

**Formulations.**

These include: w.p. (500 g a.i./kg); fumigant (300 g/kg); flowable dust (250 g/kg).

**Analysis.**

Product analysis is by glc (M. Horiba, *Agric. Biol. Chem.*, 1982, **46**, 1095). The details of methods for residue analysis are available from Sumitomo Chemical Co.

$C_{11}H_{15}BrClO_3PS$ (373.6)
GR CE FOPO&S3&O2

Br— (2-Cl-phenyl) —OP(=O)—$OCH_2CH_3$, $SCH_2CH_2CH_3$

**Nomenclature and development.**

Common name profenofos (BSI, E-ISO, F-ISO, ANSI). Chemical name (IUPAC) *O*-4-bromo-2-chlorophenyl *O*-ethyl *S*-propyl phosphorothioate. (*C.A.*) *O*-(4-bromo-2-chlorophenyl) *O*-ethyl *S*-propyl phosphorothioate (9CI); Reg. No. *[41198-08-7]*. OMS 2004. Its insecticidal properties were reported by F. Buholzer (*Proc. Br. Insectic. Fungic. Conf., 8th*, 1975, **2**, 659). Introduced by Ciba-Geigy AG (BEP 789 937; GBP 1 417 116) as code no. 'CGA 15 324s'; trade mark 'Curacron', 'Selecron'.

**Properties.**

Pure profenofos is a pale yellow liquid; b.p. 110 °C/0.001 mmHg; v.p. 1.3 mPa at 20 °C; ρ 1.455 g/cm³ at 20 °C; $n_D^{25}$ 1.5493-1.5495. Solubility at 20 °C: 20 mg/l water; miscible with most organic solvents. On hydrolysis at 20 °C, 50% loss (calculated) occurs in 93 d at pH5, 14.6 d at pH7, 5.7 h at pH9.

**Uses.**

It is a non-systemic broad spectrum insecticide for use against insect pests and mites on cotton. It has contact and stomach action and due to its translaminar effect it kills lepidopterous larvae on the untreated side of the leaves. Rates of application for sucking insects and mites are 250-500 g a.i./ha; for chewing insects 400-1200 g/ha. The separate optical isomers, due to the phosphorus atom, have been prepared and show different types of insecticidal activity and ability to inhibit acetylcholinsterase (H. Leader & J. E. Casida, *J. Agric. Food Chem.*, 1982, **30**, 546).

**Toxicology.**

Acute oral $LD_{50}$ for rats 358 mg tech./kg. Acute percutaneous $LD_{50}$ for rats *c.* 3300 mg/kg; slight irritant to skin and eyes of rabbits. Inhalation $LC_{50}$ (4-h) for rats *c.* 3000 mg/m³ air. In feeding trials NEL for e.c. formulation (380 g a.i.)/l were: for rats (2-y) 0.38 mg a.i./kg diet (0.02 mg/kg daily); for mice (1.5-y) 0.08 mg/kg diet (0.01 mg/kg daily). $LC_{50}$ (96-h) is: for rainbow trout 0.08 mg/l; for crucian carp 0.09 mg/l; for bluegill 0.3 mg/l. It is toxic to honeybees and birds.

**Formulations.**

These include: e.c. 'Curacron' 500 EC (500 g a.i./l); 400 EC (400 g/l).

**Analysis.**

Product analysis is by glc. Residues may be determined by glc with TID. Particulars are available from Ciba-Geigy AG.

$C_{14}H_{16}F_3N_3O_4$ (347.3)
L3TJ A1N3&R BNW FNW DXFFF

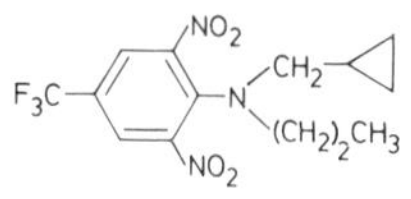

$F_3C$–[2,6-dinitrophenyl]–N($CH_2$-cyclopropyl)($(CH_2)_2CH_3$)

**Nomenclature and development.**

Common name profluralin (BSI, E-ISO, ANSI, WSSA);profluraline (F-ISO). Chemical name (IUPAC) *N*-(cyclopropylmethyl)-α,α,α-trifluoro-2,6-dinitro-*N*-propyl-*p*-toluidine (I); *N*-cyclopropylmethyl-2,6-dinitro-*N*-propyl-4-trifluoromethylaniline. (*C.A.*) *N*-( cyclopropylmethyl)-2,6-dinitro-*N*-propyl-4-(trifluoromethyl)benzenamine (9CI); (I) (8CI); Reg. No. *[26399-36-0]*. Its herbicidal properties were described by T. D. Taylor *et al.* (*Annu. Meet. Weed Sci. Soc. Am.*, 1973, Abstr. 169). Introduced by Ciba-Geigy AG (USP 3 546 295) as code no. 'CGA 10 832'; trade marks 'Tolban' (in USA), 'Pregard'.

**Properties.**

Pure profluralin forms yellow to orange crystals; m.p. 32-33 °C; v.p. 8.4 mPa at 20 °C; ρ 1.38 g/cm$^3$ at 20 °C. Solubility at 20 °C: 0.1 mg/l water; 220 g/l octan-1-ol. On hydrolysis at 100 °C of a solution (2.5 g/l), 50% loss occurs in 6h at pH3, pH7 and pH10.

**Uses.**

Profluralin is a soil-incorporated herbicide used pre-planting at 0.75-1.5 kg a.i./ha for annual and perennial weed and grass control in cotton, soyabeans and many other crops.

**Toxicology.**

Acute oral $LD_{50}$ for rats *c.* 10 000 mg/kg. Acute percutaneous $LD_{50}$ for rats >3170 mg/kg. In 90-d feeding trials NEL was: for rats 200 mg/kg diet (*c.* 13 mg/kg daily); for dogs 600 mg/kg diet (*c.* 20 mg/kg daily). $LC_{50}$ (96-h) is: for trout 0.015 mg/l; for bluegill 0.023 mg/l. It is practically non-toxic to birds but toxic to honeybees.

**Formulations.**

An e.c. (480 g a.i./l).

**Analysis.**

Product analysis is by glc with FID (R. A. Kahrs, *Anal. Methods Pestic. Plant Growth Regul.*, 1978, **10**, 451). Residues may be determined by glc with ECD or MCD (*idem, ibid.*). Particulars are available from Ciba-Geigy AG.

$C_8H_{12}ClN_5O_2$ (245.6)
T6N CN ENJ BMY1&1 DM1VQ FG
*proglinazine-ethyl*
$C_{10}H_{16}ClN_5O_2$ (273.3)
T6N CN ENJ BMY1&1 DM1VO2 FG

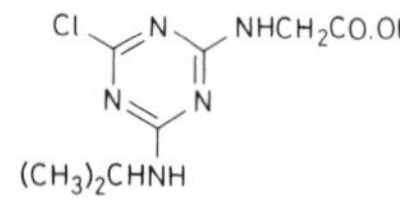

Cl N $NHCH_2CO.OH$ N N $(CH_3)_2CHNH$

**Nomenclature and development.**

Common name proglinazine (BSI, E-ISO, F-ISO, WSSA). Chemical name (IUPAC) *N*-(4-chloro-6-isopropylamino-1,3,5-triazin-2-yl)glycine. (*C.A.*) *N*-[4-chloro-6-(1-methylethylamino)-1,3,5-triazine-2-yl]glycine (9CI); *N*-[4-chloro-6-(isopropylamino)-*s*-triazin-2-yl]glycine (8CI); Reg. No. *[68228-20-6]* (proglinazine), *[68228-18-2]* (proglinazine-ethyl). The ethyl ester, proglinazine-ethyl, was introduced as a herbicide by Nitrokémia Ipartelepek as code no. 'MG-07'.

**Properties.**

Pure proglinazine-ethyl forms colourless crystals; m.p. 110-112 °C; v.p. 267 μPa at 20 °C. Solubility at 25 °C: 750 mg/l water; 500 g/l acetone; 35 g/l hexane; 100 g/l xylene. It is stable at room temperature at pH5-8, but on heating it is hydrolysed by acid or alkali to the corresponding herbicidally-inactive hydroxytriazine. It is stable to light. It decomposes at 160 °C. It is decomposed in soil, 50% loss occurring in 56-70 d. The technical grade is *c.* 94% pure.

**Uses.**

Proglinazine-ethyl is used as a pre-em. herbicide in maize at 4 kg a.i./ha, and is especially effective for the control of seedlings of dicotyledonous weeds.

**Toxicology.**

Acute oral $LD_{50}$: for rats and mice >8000 mg/kg; for guinea-pigs 857-923 mg/kg, for rabbits >3000 mg/kg. Acute percutaneous $LD_{50}$: for rats > 1500 mg/kg; for rabbits > 4000 mg/kg; it is not an irritant to their skin or eyes. The sensitising action for guinea-pigs is moderate. Acute intraperitoneal $LD_{50}$: for rats 829-891 mg/kg; for mice 720-1080 mg/kg.

**Formulations.**

A w.p. (500 g a.i./kg).

**Analysis.**

Product analysis is by glc. Residues may be determined by glc with TID.

$C_{12}H_{17}NO_2$ (207.3)
1Y1&R C1 EOVM1

OCO.NHCH3
CH3 CH(CH3)2

**Nomenclature and development.**

Common name promecarb (BSI, E-ISO, JMAF);promécarbe (F-ISO). Chemical name (IUPAC) 5-methyl-*m*-cumenyl methylcarbamate; 3-isopropyl-5-methylphenyl methylcarbamate. (*C.A.*) 3-methyl-5-(1-methylethyl)phenyl methylcarbamate (9CI); *m*-cym-5-yl methylcarbamate (8CI); Reg. No. *[2631-37-0]*. OMS 716, ENT 27 300a. Its insecticidal properties were described by A. Formigoni & G. P. Bellini (*Congr. Int. Degli Antiparassitari, Naples*, 1965) and by A. Jäger (*Z. Angew. Entomol.*, 1966, **58**, 188). Introduced by Schering AG (DEP 1 156 272; GBP 916 707) as code no. 'Schering 34 615'; trade mark 'Carbamult'.

**Properties.**

Pure promecarb is a colourless crystalline solid; m.p. 87-88 °C; b.p. 117 °C/0.01 mmHg; v.p. 4 mPa at 25 °C. Solubility at 25 °C: 91 mg/l water; 400-600 g/l acetone, 1,2-dichloroethane, dimethylformamide; 100-200 g/l carbon tetrachloride, xylene; 200-400 g/l cyclohexanol, cyclohexanone, methanol, propan-2-ol. The technical grade is >98% pure. No change was observed on storage at 50 °C for 5.8 d. At 22 °C it is stable at pH5, 50% loss occurs: in 5.2 d at pH7, in 36 h at pH9.

**Uses.**

Promecarb is a non-systemic contact insecticide effective against coleopterous pests such as post-embryonic stages of *Leptinotarsa decemlineata* at 450 g a.i./ha, against lepidopterous pests and leaf miners of fruit at 50-100 g/100 l.

**Toxicology.**

Acute oral $LD_{50}$ for rats 74-118 mg/kg; for ducks 3.5 mg/kg; for quail 78 mg/kg. Acute percutaneous $LD_{50}$ for rats and rabbits >1000 mg a.i. (as 50% w.p.)/kg. In 1.5-y feeding trials rats receiving 5 mg a.i./kg daily suffered no ill-effect. $LC_{50}$ (96-h) is: for trout 0.3 mg/l; for carp 4.3 mg/l.

**Formulations.**

These include: e.c. (200 g a.i./l); w.p. (375 or 500 g/kg).

**Analysis.**

Product analysis is by hydrolysis, isolation of methylamine which is estimated by titration. Residues may be determined by glc or by measurement of 5-methyl-*m*-cumenol produced on hydrolysis.

$C_{10}H_{19}N_5O$ (225.3)
T6N CN ENJ BO1 DMY1&1 FMY1&1

**Nomenclature and development.**

Common name premeton (BSI, E-ISO, ANSI, WSSA);prométone (F-ISO);*exception* (Federal Republic of Germany). Chemical name (IUPAC) *N,N'*-diisopropyl-6-methoxy-1,3,5-triazine-2,4-diyldiamine; 2,4-bis(isopropylamino)-6-methoxy-1,3,5-triazine. (*C.A.*) 6-methoxy-*N,N'*-bis(1-methylethyl)-1,3,5-triazine-2,4-diamine (9CI); 2,4-bis(isopropylamino)-6-methoxy-*s*-triazine (8CI); Reg. No. *[1610-18-0]*. Its herbicidal properties were described by H. Gysin & E. Knüsli (*Adv. Pest Control Res.*, 1960, **3**, 289). Introduced by J. R. Geigy S.A. (now Ciba-Geigy AG) (CHP 337 019; GBP 814 948) as code no. 'G 31 435'; trade marks 'Gesagram'; 'Primatol', in USA only 'Pramitol'.

**Properties.**

Pure prometon is a colourless powder; m.p. 91-92 °C; v.p. 306 μPa at 20 °C; ρ 1.088 g/cm$^3$ at 20 °C. Solubility at 20 °C: 620 mg/l water; 300 g/l acetone; 350 g/l dichloromethane; 600 g/l methanol; 150 g/l octan-1-ol; 250 g/l toluene. Stable to hydrolysis at 20 °C in neutral, alkaline or slightly acidic media; p*K* 4.3 at 21 °C.

**Uses.**

It is a non-selective herbicide used for the control of most annual and perennial broad-leaved, grass and brush weeds on non-crop areas at 10-20 kg a.i./ha and may be applied to the ground before laying asphalt.

**Toxicology.**

Acute oral $LD_{50}$ for rats 3000 mg tech./kg. Acute percutaneous $LD_{50}$ for rabbits >2000 mg/kg; non-irritant to eyes and minimal irritant to skin of rabbits. Inhalation $LC_{50}$ (4-h) for rats >3260 mg/m$^3$ air. In 90-d feeding trials NEL for rats was 5.4 mg/kg daily. $LC_{50}$ (96-h) is: for rainbow trout 12 mg/l; for crucian carp 70 mg/l; for bluegill 40 mg/l. It is slightly toxic to birds and practically non-toxic to honeybees.

**Formulations.**

These include 80 WP, w.p. (800 g a.i./kg); 25E, e.c. (250 g/l). 'Atrol', prometon + atrazine (1:15). 'Pramitol', prometon + sodium chlorate + sodium metaborate (1:8:10) + simazine (0.75%).

**Analysis.**

Product analysis is by glc (*AOAC Methods*, 1980, 6.431-6.435; K. Hofberg *et al., J. Assoc. Off. Anal. Chem.*, 1973, **56**, 586; B. G. Tweedy & R. A. Kahrs, *Anal. Methods Pestic. Plant Growth Regul.*, 1978, **10**, 493) or by acidimetric titration. Residues may be determined by glc with ECD (*idem, ibid.*; E. Knüsli, *ibid.*, 1972, **6**, 679) or FID (K. Ramsteiner *et al., J. Assoc. Off. Anal. Chem.*, 1974, **57**, 192).

$C_{10}H_{19}N_5S$ (241.4)
T6N CN ENJ BS1 DMY1&1 FMY1&1

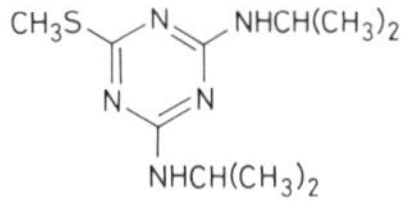

**Nomenclature and development.**

Common name prometryne (BSI, F-ISO, JMAF);prometryn (E-ISO, ANSI, WSSA). Chemical name (IUPAC) *N,N'*-di-isopropyl-6-methylthio-1,3,5-triazine-2,4-diyldiamine; 2,4-bis(isopropylamino)-6-methylthio-1,3,5-triazine. (*C.A.*) *N,N'*-bis(1-methylethyl)-6-( methylthio)-1,3,5-triazine-2,4-diamine (9CI); 2,4-bis(isopropylamino)-6-(methylthio)-*s*-triazine (8CI); Reg. No. *[7287-19-6]*. Its herbicidal properties were described by H. Gysin (*Chem. Ind. (London)*, 1962, p. 1393). Introduced by J. R. Geigy S.A. (now Ciba-Geigy AG) (CHP 337 019; GBP 814 948) as code no. 'G 34 161'; trade marks 'Caparol' (in USA), 'Gesagard'.

**Properties.**

Pure prometryne is a colourless powder; m.p. 118-120 °C; v.p. 133 μPa at 20 °C; ρ 1.157 g/cm$^3$ at 20 °C. Solubility at 20 °C: 33 g/l water; 240 g/l acetone; 300 g/l dichloromethane; 5.5 g/l hexane; 160 g/l methanol; 100 g/l octan-1-ol; 170 g/l toluene. Stable to hydrolysis at 20 °C in neutral, slightly acid or slightly alkaline media at 20 °C; p*K* 4.1 at 21 °C.

**Uses.**

It is a herbicide used either pre-em or post-em. for selective weed control in broad beans, carrots, celery, cotton, leeks, lentils, parsley, peas, potatoes, sunflowers at 1.0-1.5 kg a.i./ha pre-em. and 0.5-1.0 kg/ha post-em. At the higher rates it persists in soils for 30-90 d.

**Toxicology.**

Acute oral $LD_{50}$ for rats 5233 mg tech./kg. Acute percutaneous $LD_{50}$ for rats >3100 mg/kg; non-irritant to skin and slight irritant to eyes of rabbits. In 2-y feeding trials NEL was: for rats 1250 mg/kg diet (83 mg/kg daily); for dogs 150 mg/kg diet (4 mg/kg daily). $LC_{50}$ (96-h) is: for rainbow trout 2.5 mg/l; for bluegill 10 mg/l. It is practically non-toxic to birds and honeybees.

**Formulations.**

'Gesagard' 50 and 80 WP, w.p. (500 or 800 g a.i./kg); 500 FW, s.c. (500 g/l). 'Codal', prometryne + metolachlor (1:1 or 1:1.5). 'Cotogard', prometryne + fluometuron (1:1). 'Gesatene', prometryne + ametryne (1:1).

**Analysis.**

Product analysis is by glc (*CIPAC Handbook*, 1980, **1A**, 1328; *AOAC Methods*, 1980, 6.431-6.435; R. T. Murphy *et al., J. Assoc. Off. Anal. Chem.*, 1971, **54**, 703). Residues may be determined by glc with ECD or FID (K. Ramsteiner *et al., ibid.*, 1974, **57**, 192; E. Knüsli, *Anal. Methods Pestic. Plant Growth Regul.*, 1972, **6**, 680; B. G. Tweedy & R. A. Hahrs, *ibid.*, 1978, **10**, 493; T. H. Byast *et al., Tech. Rep. ARC Weed Res. Organ.*, No. 15 (2nd Ed.) p. 40).

# Propachlor 10100

$C_{11}H_{14}ClNO$ (211.7)
G1VNR&Y1&1

$CO.CH_2Cl$ / N / $CH(CH_3)_2$

**Nomenclature and development.**

Common name propachlor (BSI, E-ISO, WSSA);propachlore (F-ISO). Chemical name (IUPAC) 2-chloro-*N*-isopropylacetanilide (I); α-choro-*N*-isopropylacetanilide. (*C.A.*) 2-chloro-*N*-(1-methylethyl)-*N*-phenylacetamide (9CI); (I) (8CI); Reg. No. *[1918-16-7]*. Its herbicidal activity was described by D. D. Baird *et al.* (*Proc. North Cent. Weed Control Conf.*, 1964). Introduced by Monsanto Chemical Co. (USP 2 863 752) as code no. 'CP 31 393'; trade mark 'Ramrod'.

**Properties.**

Pure propachlor is a light tan solid; m.p. 77 °C; b.p. 110 °C/0.03 mmHg; v.p. 30.6 mPa at 25 °C; $d_{25}^{25}$ 1.242. Solubility at 25 °C: 613 mg/kg water; 448 g/kg acetone; 737 g/kg benzene; 602 g/kg chloroform; 408 g/kg ethanol; 239 g/kg xylene. Stable to u.v. light; decomposes at 170 °C; nonflammable.

**Uses.**

It is a pre-em. or pre-planting incorporated or early post-em. herbicide effective against annual grasses and some broad-leaved weeds in beans, brassicas, cotton, groundnuts, leeks, maize, onions, peas, roses, ornamental trees and shrubs, soyabeans and sugarcane at 3.36-6.72 kg a.i. broadcast/ha. Absorbed mainly by germinating seedling shoots, less so by roots; translocated throughout the plant being concentrated in vegetative rather than reproductive parts; rapidly metabolised in plants. Persists in soil 28-42 d, main loss being by microbial degradation.

**Toxicology.**

Acute oral $LD_{50}$: for rats 1800 mg tech./kg; for bobwhite quail 91 mg/kg. Acute percutaneous $LD_{50}$: for rabbits >20 000 mg/kg; slightly irritating to eyes, moderately to skin of rabbits. Formulations may produce skin sensitisation in susceptible humans. In 2-y feeding trials no adverse effect in dogs at 1000 mg/kg diet (highest dose); in rats at 100 mg/kg diet. There was no evidence of carcinogenic, teratogenic or reproductive effects in rats or rabbits at 300 mg/kg diet. In 8-d feeding trials $LC_{50}$ for bobwhite quail and mallard ducks was >5000 mg/kg diet. $LC_{50}$ (96-h) is: for bluegill >1.4 mg/l; for rainbow trout 0.17 mg/l.

**Formulations.**

'Ramrod' (Monsanto) and 'Bexton' (Dow Chemical Co.), s.c. (480 g a.i./l); 'Ramrod', granules (200 g/kg); 'Ramrod', w.p. (650 g/kg). Mixture, s.c. (propachlor + atrazine).

**Analysis.**

Product analysis is by glc. Residues may be determined by glc with ECD. Details are available from Monsanto Co.

$C_9H_{20}N_2O_2$ (188.3)
3OVM3N1&1
*propamocarb hydrochloride*
$C_9H_{21}ClN_2O_2$ (224.7)
3OVM3N1&1 &&HCl SALT

$(CH_3)_2NCH_2CH_2CH_2NHCO.OCH_2CH_2CH_3$

**Nomenclature and development.**

Common name propamocarb (BSI, E-ISO);propamocarbe (F-ISO). Chemical name (IUPAC) propyl 3-(dimethylamino)propylcarbamate. (*C.A.*) propyl [3-( dimethylamino)propyl]carbamate (9CI); Reg. No. *[24579-73-5]* (propamocarb), *[25606-41-1]* (hydrochloride). Its fungicidal properties were described by E. A. Pieroh *et al.* (*Meded. Fac. Landbouwwet. Rijksuniv. Gent*, 1978, **43**, 933). The hydrochloride was introduced by Schering AG (DEP 1 567 169; 1 643 040) as code no. 'SN 66 752'; trade mark 'Previcur N'.

**Properties.**

Propamocarb hydrochloride is a colourless crystalline solid; m.p. 45-55 °C; v.p. 800 mPa at 25 °C. It is hygroscopic, solubility at 25 °C: >700 g/l water, methanol; 430 g/l dichloromethane; 23 g/l ethyl acetate; <100 mg/l hexane, toluene.

**Uses.**

Propamocarb hydrochloride is a soil-applied systemic fungicide, but is also suitable for use as a dip treatment (bulbs and tubers) and as a seed protectant. Its action is specific against Phycomycetes, including the following genera of fungi; *Aphanomyces, Bremia, Peronospora, Phytophthora, Pseudoperonospora* and *Pythium*. It is recommended as a preventive treatment.

**Toxicology.**

Acute oral $LD_{50}$ of hydrochloride: for rats 8600 mg/kg; for mice 1600-2000 mg/kg; for pheasant 3050 mg/kg; for ducks >6000 mg/kg. Acute percutaneous $LD_{50}$: for rats and rabbits >3.5 mg (70% solution)/kg. In 2-y feeding trials NEL was: for rats 40 mg/kg daily; for mice 50 mg/kg daily. $LC_{50}$ (96-h) is: for carp 235 mg/l; for trout 410-600 mg/l; for bluegill 415 mg/l.

**Formulations.**

'Previcur N', aqueous solution (722 g propamocarb hydrochloride/l).

**Analysis.**

Product analysis is by potentiometric titration with perchloric acid in a non-aqueous medium. Residues may be determined by glc with FID. Particulars available from Schering AG.

$C_9H_9Cl_2NO$ (218.1)
GR BG DMV2

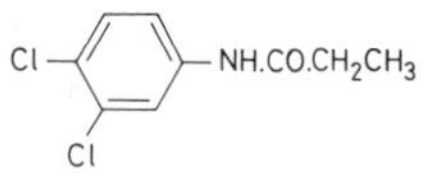

**Nomenclature and development.**

Common name propanil (BSI, E-ISO, F-ISO, WSSA);DCPA (JMAF);*exception* (Austria, Federal Republic of Germany). Chemical name (IUPAC) 3',4'-dichloropropionanilide (I); *N*-(3,4-dichlorophenyl)propionamide. (*C.A.*) *N*-(3,4-dichlorophenyl)propanamide (9CI); (I) (8CI); Reg. No. *[709-98-8]*. Its herbicidal properties were described (*Proc. South. Weed Control Conf.*, 1960, p.20). Introduced by Rohm & Haas Co., later by Bayer AG and formerly by Monsanto Chemical Co. (DEAS 1 039 779; GBP 903 766 both to Bayer AG) as code no. FW-734' (Rohm & Haas Co.),'Bayer 30 130'; trade marks 'Stam F-34' (Rohm & Haas Co.), 'Surcopur' (Bayer AG), 'Riselect' (Farmoplant S.p.A.), and formerly *'Rogue'* (Monsanto Chemical Co.).

**Properties.**

Pure propanil is a colourless solid; m.p. 92-93 °C; v.p. 12 mPa at 60 °C. Solubility at room temperature 225 mg/l water; at 25 °C: 540 g/kg ethanol; 600 g/kg 3,5,5-trimethylcyclohex-2-enone. The technical grade is a brown crystalline solid, m.p. 88-91 °C. It is stable in emulsion concentrates, but is hydrolysed in acid and alkaline media to 3,4-dichloroaniline and propionic acid. The formulations should be packed in steel drums coated with 'Unichrome B-124-17' or glass; polyethylene coatings are not suitable.

**Uses.**

It is a contact herbicide recommended for post-em. use in rice and potatoes at 1-4 kg/ha, at which dose its persistence in soils is brief. If applied to some crops that have been treated with organophosphorus insecticides, severe phytotoxicity may result.

**Toxicology.**

Acute oral $LD_{50}$ for rats 1285-1483 mg tech./kg. Acute percutaneous $LD_{50}$ for rabbits 7080 mg (in corn oil)/kg; non-irritant to skin of rabbits. In 2-y feeding trials NEL for rats 400 mg/kg diet. $LC_{50}$(48-h) for carp 13 mg/l; for goldfish and Japanese killifish 14 mg/l; for *Daphnia* 4.8 mg/l indicative that contamination of streams and lakes should be avoided.

**Formulations.**

These include; 'Stam F-34, e.c. (360 g a.i./l); e.c. 'Riselect' (370 g/l), 'Surcopur' (350 g/l); ULV (600 g/l).

**Analysis.**

Product analysis is by hydrolysis and titration with nitrite of the resulting 3,4-dichloroaniline. Residues may be determined by glc (*Anal. Methods Pestic. Plant Growth Regul.*, 1972, **6**, 692) or colorimetry of the 3,4-dichloroaniline formed on hydrolysis (C. F. Gordon *et al., Anal. Methods Pestic., Plant Growth Regul. Food Addit.*, 1964, **4**, 235).

$C_{13}H_{21}O_4PS$ (304.3)
3OPO&O3&OR DS1

$CH_3S-C_6H_4-OP(=O)(OCH_2CH_2CH_3)_2$

**Nomenclature and development.**

Common name propaphos (BSI, JMAF, draft E-ISO);propafos (draft F-ISO). Chemical name (IUPAC) 4-(methylthio)phenyl dipropyl phosphate (I). (*C.A.*) (I) (8 & 9CI); Reg. No. *[7292-16-2]*. Its insecticidal properties were described (*Jpn. Pestic. Inf.*, 1970, No. 4, p. 7). Introduced by Nippon Kayaku Co. Ltd. (JPP 482 500; 462 729) as code no. 'NK-1158'; trade mark 'Kayaphos'.

**Properties.**

Propaphos is a colourless liquid; b.p. 175-177 °C/0.85 mmHg. Solubility at 25 °C: 125 mg/l water; soluble in most organic solvents. It is stable in neutral and acid media but is slowly decomposed in alkaline media.

**Uses.**

It is a systemic insecticide effective at 600-800 g a.i./ha in controlling the rice pests *Chilo suppressalis, Laodelphax striatella, Nephotettix cincticeps, Oulema oryzae* in paddy rice. It is effective against strains resistant to carbamate and other organophosphorus insecticides.

**Toxicology.**

Acute oral $LD_{50}$: for rats 70 mg/kg; for mice 90 mg/kg; for rabbits 82.5 mg/kg. Acute percutaneous $LD_{50}$ for mice 156 mg/kg. In 90-d feeding trials NEL was: for rats 100 mg/kg diet; for mice 5 mg/kg diet. $LC_{50}$ (48-h) for carp is 4.8 mg/l.

**Formulations.**

These include: e.c. (500 g a.i./l); dust (20 g/kg); granules (50 g/kg).

**Analysis.**

Product and residue analysis is by glc.

$C_{19}H_{26}O_4S$ (350.5)
L6TJ AOSO&O2UU1 BOR DX1&1&1

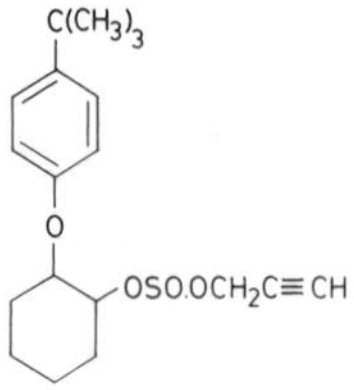

**Nomenclature and development.**

Common name propargite (BSI, E-ISO, F-ISO, ANSI); BPPS (JMAF). Chemical name (IUPAC) 2-(4-*tert*-butylphenoxy)cyclohexyl prop-2-ynyl sulphite. (*C.A.*) 2-[4-(1,1-dimethylethyl)phenoxy]cyclohexyl 2-propynyl sulfite (9CI); 2-(*p-tert*-butylphenoxy)cyclohexyl 2-propynyl sulfite (8CI); Reg. No. *[2312-35-8]*. ENT 27 226. Introduced by Uniroyal Inc. (USP 3 272 854; 3 463 859) as code no. 'DO 14'; trade mark 'Omite'.

**Properties.**

The technical grade (>85% *m/m* pure) is a dark amber viscous liquid; $d_4^{25}$ 1.085-1.115; v.p. 400 Pa at 20 °C. Practically insoluble in water; miscible with most organic solvents.

**Uses.**

Propargite is an acaricide effective for the control of phytophagous mites on cotton, fruit, hops, maize, nuts, vegetables and other crops at 0.75 to 5.5 kg a.i. in 750-1900 l water/ha.

**Toxicology.**

Acute oral $LD_{50}$ for rats 220 mg tech./kg. Acute percutaneous $LD_{50}$ for rabbits >300 mg/kg. Acute inhalation $LD_{50}$ for rats >2.5 mg/l. In 2-y feeding trials NEL was: for rats and dogs 900 mg/kg diet; for mice 1000 mg/kg diet. It is not mutagenic or carcinogenic. $LC_{50}$ (8-d) is: for mallard duck >4640 mg/kg diet; for bobwhite quail 3401 mg/kg diet. $LC_{50}$ (96-h) is: for bluegill 0.1 mg/l; for rainbow trout 0.12 mg/l; $LC_{50}$ (48-h) for water fleas 0.12 mg/l.

**Formulations.**

These include: e.c. (720 g/l or 790 g/l); w.p. (300 g/kg).

**Analysis.**

Product analysis is by i.r. spectroscopy; details are available from Uniroyal Chemical Co. Residues may be determined by glc (*Pestic. Anal. Man.*, 1979, **I**, 201-I; M. A. Luke *et al.*, *J. Assoc. Off. Anal. Chem.*, 1981, **64**, 1187; G. M. Stone, *Anal. Methods Pestic. Plant Growth Regul.*, 1973, **7**, 355).

$C_9H_{16}ClN_5$ (229.7)
T6N CN ENJ BMY1&1 DMY1&1 FG

**Nomenclature and development.**

Common name propazine (BSI, E-ISO, F-ISO, ANSI, WSSA, JMAF);*exception* (Germany, Sweden). Chemical name (IUPAC) 6-chloro-*N*,*N*'-di-isopropyl-1,3,5-triazine-2,4-diyldiamine; 2-chloro-4,6-bis(isopropylamino)-1,3,5-triazine. (*C.A.*) 6-chloro-*N*,*N*'-bis(1-methylethyl)-1,3,5-triazine-2,4-diamine (9CI); 2-chloro-4,6-bis(isopropylamino)-*s*-triazine (8CI); Reg. No. *[139-40-2]*. Its herbicidal properties were reported by H. Gysin & E. Knüsli (*Proc. Int. Congr. Crop Protect., 4th*, 1957, **1**, 549). Introduced by J. R. Geigy S.A. (now Ciba-Geigy AG) (BEP 540 947; GBP 814 947) as code no. 'G 30 028'; trade marks 'Gesamil', 'Milogard'.

**Properties.**

Pure propazine is a colourless powder; m.p. 212-214 °C; v.p. 3.9 μPa at 20 °C; ρ 1.162 g/cm$^3$ at 20 °C. Solubility at 20 °C 5.0 mg/l water; at 22 °C: 6.2 g/kg benzene, toluene; 2.5 g/kg carbon tetrachloride. Stable to hydrolysis in neutral, slightly acidic or slightly alkaline media.

**Uses.**

It is a pre-em herbicide recommended for the control of broad-leaved and grass weeds in sorghum and umbelliferous crops at 0.5-3.0 kg a.i./ha. It is metabolised in tolerant plants to the corresponding 2-hydroxy compound.

**Toxicology.**

Acute oral $LD_{50}$ for rats >7700 mg tech./kg. Acute percutaneous $LD_{50}$ for rats >3100 mg/kg; non-irritant to eyes and mild irritant to skin of rabbits. In 90-d feeding trials with a w.p. (800 g a.i./kg) NEL was: for rats 200 mg a.i./kg diet (13 mg/kg daily); for dogs 200 mg a.i./kg diet (7 mg/kg daily). $LC_{50}$ (96-h) is: for rainbow trout 17.5 mg/l; for bluegill >100 mg/l. It is slightly toxic to birds and practically non-toxic to honeybees.

**Formulations.**

'Gesamil' 50 and 80 WP, w.p. (500 or 800 g a.i./kg); 90 WDG, water-dispersible granule (900 g/kg); 4 L, liquid (480 g/l). 'Milocep', propazine + metolachlor (1:2).

**Analysis.**

Product analysis is by glc (*CIPAC Handbook*, 1980, **1A**, 1333; *AOAC Methods*, 1980, 6.431-6.435; B. G. Tweedy & R. A. Kahrs, *Anal. Methods Pestic., Plant Growth Regul.*, 1978, **10**, 493). Residues may be determined by glc with ECD or FID (E. Knüsli, *ibid.*, 1972, **6**, 234, 679; K. Ramsteiner *et al., J. Assoc. Off. Anal. Chem.*, 1974, **57**, 192; T. H. Byast *et al., Tech. Rep. ARC Weed Res. Organ.*, No. 15 (2nd Ed.), p. 40).

$C_{10}H_{20}NO_4PS$ (281.3)
2MPS&O1&OY1&U1VOY1&1
&&(E) FORM

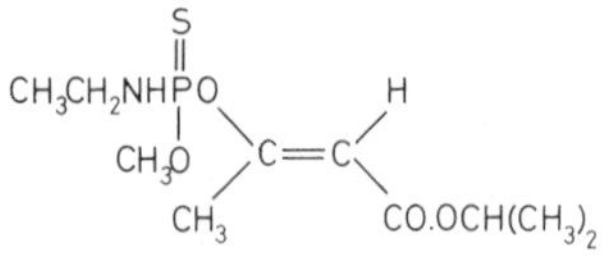

S
CH$_3$CH$_2$NHPO H
CH$_3$O C=C
CH$_3$ CO.OCH(CH$_3$)$_2$

**Nomenclature and development.**

Common name propetamphos (BSI, E-ISO, F-ISO, ANSI);propetamfos (Austria). Chemical name (IUPAC) isopropyl (3-[ethylamino(methoxy)phosphinothioy = loxy]isocrotonate; (*E*)-*O*-2-isopropoxycarbonyl-1-methylvinyl *O*-methyl ethylphosphoramidothioate. (*C.A.*) (*E*)-1-methylethyl 3-[[(ethylamino)methoxyphosphinothioyl]oxy]-2-butenoate (9CI); Reg. No. *[31218-83-4]*. OMS 1502. Its insecticidal properties were described by J. P. Leber, (*Proc. Int. IUPAC Congr., 2nd*, 1974, **1**, 381). Introduced by Sandoz AG (DEP 2 035 103; CHP 526 585; BEP 753 579) as code no. 'SAN 52 139I'; trade mark 'Safrotin'.

**Properties.**

Pure propetamphos is a colourless oil, b.p. 87-89 °C/0.005 mmHg at 20 °C; v.p. 1.9 mPa at 20 °C; $d_4^{20}$ 1.1294; $n_D^{20}$ 1.495. Solubility at 23 °C: 110 mg/l water; completely miscible with acetone, chloroform, diethyl ether, dimethyl sulphoxide, ethanol, hexane, methanol, xylene. At 25 °C, 50% loss by hydrolysis occurs in 11 d at pH3, 1 y at pH6, 41 d at pH9. It is stable during storage (shelf life at 20 °C: $\geqslant$2 y) and stable to light.

**Uses.**

Propetamphos is a contact insecticide with stomach activity, effective against household and public health pests, notably cockroaches, flies and mosquitoes; also against animal ectoparasites.

**Toxicology.**

Acute oral $LD_{50}$ for male rats 119 mg/kg. Acute percutaneous $LD_{50}$ for rats 2825 mg/kg. In 2-y feeding trials NEL for rats was 6 mg/kg diet.

**Formulations.**

These include: for ectoparasite control 'Blotic' (300 g a.i./kg); public health use 'Safrotin', e.c. (500 g/kg); 'Safrotin Dust' (20 g/kg); 'Safrotin', w.p. (200 g/kg); 'Safrotin', lacquer (40 g/kg). Also 'Safrotin Aerosol', (Reg. No. *[77491-30-6]*), (20 g propetamphos + 5 g dichlorvos/kg); 'Safrotin Liquid' (10 g propetamphos + 5 g dichlorvos/kg).

**Analysis.**

Product analysis is by glc or by paper chromatography with subsequent phosphorus determination by standard colorimetric methods. Residues may be determined by glc. Details are available from Sandoz AG.

$C_{10}H_{13}NO_2$ (179.2)
1Y1&OVMR

**Nomenclature and development.**

Common name propham (BSI, E-ISO, WSSA);prophame (F-ISO);IFC (USSR). Chemical name (IUPAC) isopropyl carbanilate (I); isopropyl phenylcarbamate. (*C.A.*) 1-methylethyl phenylcarbamate (9CI); (I) (8CI); Reg. No. *[122-42-9]*. Trivial name IPC. Its plant growth regulating properties were described by W. G. Templeman & W. A. Sexton (*Nature (London)*, 1945, **156**, 630). Introduced as a herbicide by ICI Plant Protection Division (GBP 574 995).

**Properties.**

Pure propham forms colourless crystals; m.p. 87.0-87.6 °C. Solubility at 20-25 °C: 32-250 mg/l water; soluble in most organic solvents. The technical grade (99% pure) has m.p. 86.5-87.5 °C. It is stable <100 °C.

**Uses.**

It is mainly used at 2.3-5.0 kg a.i./ha to control annual grass weeds in peas and sugar beet, or in combination with other herbicides for weed control in beetroot, fodder beet, lettuces and mangels. It is absorbed by roots but not by foliage. It is also used, in combination with chlorpropham, to inhibit sprouting of potatoes.

**Toxicology.**

Acute oral $LD_{50}$: for rats 5000 mg/kg; for mice 3000 mg/kg. Intraperitoneal $LD_{50}$: for mice 1000 mg/kg; for rats 600 mg/kg. In 30-d feeding trials rats receiving 10 000 mg/kg diet suffered no ill-effect and there was no evidence of carcinogenic activity (W. C. Heuper, *Ind. Med.*, 1952, **21**, 71). Fish exposed to 5 mg/l suffered no ill-effect.

**Formulations.**

A w.p. These include e.c. or s.c.: propham + chloridazon + fenuron; propham + chloridazon + chlorpropham + fenuron; propham + chlorpropham; propham + chlorpropham + fenuron.

**Analysis.**

Product analysis is by hydrolysis, the liberated carbon dioxide being determined titrimetrically (*CIPAC Handbook*, 1970, **1**, 593). Residues may be determined by hydrolysis to aniline, a derivative of which is estimated by glc (L. N. Gard & C. E. Ferguson, *Anal. Methods Pestic., Plant Growth Regul. Food Addit.*, 1964, **4**, 139; *Anal. Methods Pestic. Plant Growth Regul.*, 1972, **6**, 657).

$C_{15}H_{17}Cl_2N_3O_2$ (342.2)
T5O COTJ BR BG DG& D3 B1- AT5NN
DNJ &&(±) FORM

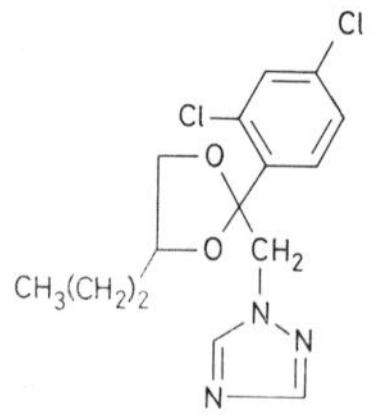

**Nomenclature and development.**

Common name propiconazole (BSI, draft E-ISO). Chemical name (IUPAC) (±)-1-[2-(2,4-dichlorophenyl)-4-propyl-1,3-dioxolan-2-ylmethyl]-1*H*-1,2,4-triazole. (*C.A.*) 1-[[2-(2,4-dichlorophenyl)-4-propyl-1,3-dioxolan-2-yl]methyl]-1*H*-1,2,4-triazole (9CI); Reg. No. *[60207-90-1]*. Its fungicidal properties were described by Janssen Pharmaceutica. Developed by Ciba-Geigy AG and described by P. A. Urech *et al.* (*Proc. Br. Crop Prot. Conf.*, 1979, **2**, 508) (GBP 1 522 657; BEP 895 579 to Janssen Pharmaceutica) as code no. 'CGA 64 250'; trade marks 'Tilt', in Federal Republic of Germany 'Desmel' (both to Ciba-Geigy AG), 'Radar' (to ICI Plant Protection Division).

**Properties.**

Pure propiconazole is a yellowish viscous liquid; b.p. 180 °C/0.1 mmHg; v.p. 133 μPa at 20 °C; $n_D^{20}$ 1.5468; ρ 1.27 g/cm$^3$ at 20 °C. Solubility at 20 °C: 110 mg/l water; completely miscible with acetone, methanol, propan-2-ol; 60 g/kg hexane. Hydrolysis is not significant; stable below 320 °C.

**Uses.**

It is a systemic foliar fungicide with a broad range of activity. On cereals it controls, at 125 g a.i./ha with 1-2 applications, diseases caused by *Erysiphe graminis, Leptosphaeria nodorum, Pseudocerosporella herpotrichoides, Puccinia* spp., *Pyrenophora teres, Rhynchosporium secalis, Septoria* spp. In some countries it is also recommended against *Uncinula necator* on grapes.

**Toxicology.**

Acute oral $LD_{50}$, for rats 1517 mg/kg. Acute percutaneous $LD_{50}$ for rats >4000 mg/kg; slight irritant to skin and eyes of rabbits. $LC_{50}$ (96-h) in laboratory trials is: for brown trout 20 mg/l water; for carp >100 mg/l.

**Formulations.**

These include: 'Tilt' 100, 'Tilt' 250, e.c. (100 or 250 g a.i./l); 'Tilt' 125, w.s.c. (125 g/l). Mixtures include: 'Tilt C' 275, s.c. (125 g propiconazole + 150 g carbendazim/l). 'Tilt CB' 45, w.p. (250 g propiconazole + 200 g carbendazim/kg). 'Tilt CF' 72.5, w.p. (125 g propiconazole + 600 g captafol/kg).

**Analysis.**

Product analysis is by glc with TID. Residue analysis is by glc with FID. Particulars are available from Ciba-Geigy AG.

$(C_5H_8N_2S_4Zn)_\chi$ (289.8)*x*
SUYSHMY1&1MYUS&SH &&Zn
COMPLEX

$$\left[-S.CS.NHCH_2\overset{CH_3}{\overset{|}{C}}HNHCS.S.Zn-\right]_x$$

**Nomenclature and development.**

Common name propineb (BSI, E-ISO, JMAF);propinèbe (F-ISO). Chemical name (IUPAC) polymeric zinc propylenebis(dithiocarbamate). (*C.A.*) [[(1-methyl-1,2-ethanediyl)bis[carbamodithioato]](2-)]zinc homopolymer (9CI); [propylenebis[dithiocarbamato]]zinc polymer (8CI); Reg. No. *[9016-72-2]*. Its fungicidal properties were described by H. Goeldner (*Pflanzenschutz-Nachr. (Engl. Ed.)*, 1963, **16**, 49). Introduced by Bayer AG (BEP 611 960) as code no. 'Bayer 46 131','LH 30/Z'; trade mark 'Antracol'.

**Properties.**

It is a white to yellowish powder which decomposes >160 °C; at *c*. 300 °C only a slight residue remains. Practically insoluble in all common solvents. It is decomposed in strongly alkaline or acid media. It is stable when dry. Its degradation on apples and grapes has been studied (K. Vogeler, *et al.*, *ibid.*, 1977, **30**, 72).

**Uses.**

It is a protective fungicide with a long residual activity; suitable for the control of *Pseudoperonospora humuli* on hops, *Phytophthora infestans* on potatoes and tomatoes and *Venturia inaequalis* on apples. It has some inhibitory action on powdery mildews and on red spider mites.

**Toxicology.**

Acute oral $LD_{50}$ for male rats 8500 mg/kg. Acute percutaneous $LD_{50}$ (7-h) for rats >1000 mg/kg. In 2-y feeding trials no ill-effect was caused in rats receiving 50 mg/kg diet. $LC_{50}$ (96-h) is: for rainbow trout 1.9 mg/l; for golden orfe 133 mg/l. It is harmless to honeybees.

**Formulations.**

'Antracol', w.p. (700 g/kg); w.p. (650-750 g a.i./kg); dusts of various a.i. content.

**Analysis.**

Product analysis is by titration with iodine, after hydrolysis and conversion of the liberated carbon disulphide to dithiocarbonate (D. G. Clarke *et al.*, *Anal. Chem.*, 1951, **23**, 1842). Residues may be determined by colorimetry of a complex formed with the carbon disulphide produced on hydrolysis (G. E. Keppel, *J. Assoc. Off. Anal. Chem.*, 1971, **54**, 528; K. Vogeler, *Pflanzenschutz-Nachr. (Engl. Ed.)*, 1967, **20**, 525) or by polarography (*idem., ibid.*).

$C_{11}H_{15}NO_3$ (209.2)
1Y1&OR BOVM1

OCO.NHCH$_3$
OCH(CH$_3$)$_2$

**Nomenclature and development.**

Common name propoxur (BSI, E-ISO, F-ISO, ESA);PHC (JMAF);original BSI name was *arprocarb*. Chemical name (IUPAC) 2-isopropoxyphenyl methylcarbamate. (*C.A.*) 2-(1-methylethoxy)phenyl methylcarbamate (9CI); *o*-isopropoxyphenyl methylcarbamate (8CI); Reg. No. *[114-26-1]*. OMS 33, ENT 25 671; Its insecticidal properties were described by G. Unterstenhöfer (*Meded. Landbouwhogesch. Opzoekingsstn. Staat Gent,* 1963, **28**, 758). Introduced by Bayer AG (USP 3 111 539; DEP 1 108 202) as code no. 'Bayer 39 007', '58 12 315'; trade marks 'Baygon', 'Blattanex' (both for household and public health use), 'Unden', 'Undene' (both for agricultural uses).

**Properties.**

It is a colourless crystalline powder; m.p. 84-87 °C; v.p. 1.3 Pa at 120 °C. Solubility at 20 °C: 2 g/l water; soluble in most organic solvents. Unstable in highly alkaline media, 50% loss at 20 °C in 40 min at pH10.

**Uses.**

It is a non-systemic insecticide with rapid knock-down, effective against ants, aphids, bugs, cockroaches, flies, jassids, millipedes, mosquitoes and other household pests.

**Toxicology.**

Acute oral $LD_{50}$ for rats 90-128 mg/kg; for male mice 100-109 mg/kg; for male guinea-pigs 40 mg/kg; for red-winged blackbirds 2-6 mg/kg; for starlings 15-20 mg/kg. Acute percutaneous $LD_{50}$ for male rats 800-1000 mg/kg. In 2-y feeding trials male and female rats receiving 250 mg a.i./kg diet showed no ill-effect; at 750 mg/kg diet the liver weight of female rats increased, otherwise there was no ill-effect. $LC_{50}$ (96-h) is: for bluegill 6.6 mg/l; for rainbow trout 4-14 mg/l. It is highly toxic to honeybees.

**Formulations.**

These include: e.c., w.p., dusts, granules, pressurised sprays, smokes and baits of different a.i. concentrations.

**Analysis.**

Product analysis is by hydrolysis and titration of the liberated methylamine (*CIPAC Handbook*, 1980, **1A**, 1338), by u.v. spectrometry after hydrolysis to 2-isopropoxyphenol or by i.r. spectrometry (C. A. Anderson, *Anal. Methods Pestic. Plant Growth Regul.*, 1973, **7**, 163). Residues may be determined by glc (*idem., ibid.*; *AOAC Methods*, 1980, 29.058-29.063; M. A. Luke *et al.*, *J. Assoc. Off. Anal. Chem.*, 1981, **64**, 1187) or, after hydrolysis, by colorimetric estimation of 2-isopropoxyphenol (H. Niessen & H. Frehse, *Pflanzenschutz-Nachr. (Engl. Ed.)*, 1964, **17**, 25).

$C_{15}H_{22}O_3$ (250.3)
3OV1OR CX1&1&1

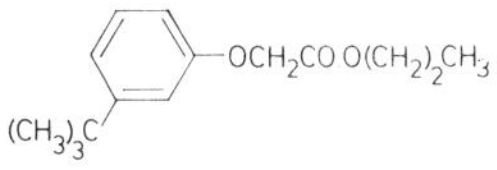

$(CH_3)_3C$ — $OCH_2CO\,O(CH_2)_2CH_3$

**Nomenclature and development.**

Chemical name (IUPAC) propyl 3-*tert*-butylphenoxyacetate. (*C.A.*) propyl 3-(1,1-dimethylethyl)phenoxyacetate (9CI); propyl *m-tert*-butylphenoxyacetate (8CI); Reg. No. *[66227-09-6]*. Its plant growth-regulating properties were described by C. J. Hibbitt & J. A. Hardisty (*Meded. Fac. Landbouwwet. Rijksuniv. Gent*, 1979, **44**, 835). Introduced by May & Baker Ltd (GBP 1 524 320) as code no. 'M&B 25 105'; trade mark 'M&B 25-105'.

**Properties.**

It is a colourless liquid with a characteristic odour; b.p. 162 °C/20mmHg. It is sparingly soluble in water.

**Uses.**

It is a plant growth regulator used to promote lateral branching in maiden and young non-cropping apple and pear trees by temporarily inhibiting the growth of the apical meristem.

**Toxicology.**

Acute oral $LD_{50}$: for rats 1800 mg/kg; for Japanese quail 2160 mg/kg. Acute percutaneous $LD_{50}$ for rats $>$ 2000 mg/kg; moderate irritation to the eyes and skin of rabbits has been reported. It is practically non-toxic to honeybees and earthworms.

**Formulations.**

An e.c. (750 g a.i./l).

**Analysis.**

Product and residue analysis is by glc.

$C_{12}H_{11}Cl_2NO$ (256.1)
GR CG EVMX1&1&1UU1

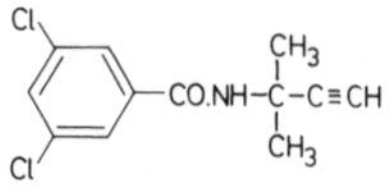

**Nomenclature and development.**

Common name propyzamide (BSI, E-ISO, F-ISO, JMAF);pronamide (WSSA). Chemical name (IUPAC) 3,5-dichloro-*N*-(1,1-dimethylpropynyl)benzamide (I). (*C.A.*) (I) (8 & 9CI); Reg. No. *[23950-58-5]*. Introduced by Rohm & Haas Co. (GBP 1 209 068; USP 3 534 098; 3 640 699) as code no. 'RH 315'; trade mark 'Kerb'.

**Properties.**

Pure propyzamide is a colourless solid; m.p. 155-156 °C; v.p. 11.3 mPa at 25 °C. Solubility at 25 °C: 15 mg/l water; soluble in many aliphatic and aromatic solvents. The technical grade is 94-95% pure.

**Uses.**

It is a selective herbicide for post-em. use in new and established crops of lucerne and other small-seeded legumes grown for fodder, and for pre-em. treatment of lettuces, related leafy crops and many species of trees, ornamental shrubs, top fruit and some species of soft fruit (depending on season and soil type). It is also effective in controlling *Poa annua* in Bermuda grass, Zoysia and certain other specific turf species. Rates vary from 0.56-2.2 kg a.i./ha, depending on weed species, environmental conditions and the residual activity needed.

**Toxicology.**

Acute oral $LD_{50}$: for male rats 8350 mg tech./kg, for female rats 5620 mg/kg. Acute percutaneous $LD_{50}$ for rabbits >3160 mg/kg. The w.p. formulation is only mildly irritating to the eyes and skin.

**Formulations.**

'Kerb 50-W', w.p. (500 g a.i./kg); granules (40 g/kg). Mixtures include: propyzamide + diuron.

**Analysis.**

Product analysis is by glc (I. L. Adler *et al., J. Assoc. Off. Anal. Chem.*, 1972, **55**, 802; *idem, Anal. Methods Pestic. Plant Growth Regul.*, 1976, **8**, 443). Residues may be determined by glc of a derivative (*idem, loc. cit.*) or by glc with ECD (T. H. Byast *et al., Tech. Rep. ARC Weed Res. Organ.*, No. 15 (2nd Ed.), pp. 9, 73).

$C_8H_{18}N_2OS$ (190.3)
2SVM3N1&1
*prothiocarb hydrochloride*
$C_8H_{19}ClN_2OS$ (226.8)
2SVM3N1&1 &&HCl SALT

$(CH_3)_2NCH_2CH_2CH_2NHCO.SCH_2CH_3$

**Nomenclature and development.**

Common name prothiocarb (BSI, E-ISO);prothiocarbe (F-ISO). Chemical name (IUPAC) *S*-ethyl (3-dimethylaminopropyl)thiocarbamate (I). (*C.A.*) *S*-ethyl [3-(dimethylamino)propyl]carbamothioate (9CI); (I) (8CI); Reg. No. *[19622-08-3]*. Its fungicidal properties were described by M. G. Bastiaansen *et al.* (*Meded. Fac. Landbouwwet. Rijksuniv. Gent*, 1974, **39**, 1019). The hydrochloride, Reg. No. *[19622-19-6]*, was introduced by Schering AG (DEP 1 567 169) as code no. 'SN 41 703'; trade mark 'Previcur', formerly '*Dynone*' in UK and Republic of South Africa.

**Properties.**

Pure prothiocarb hydrochloride is a colourless, odourless crystalline solid; m.p. 120-121 °C; v.p. 1.9 μPa at 25 °C. It is hygroscopic; solubility at 23 °C: 890 g/l water; <150 mg/l benzene, hexane; 100 g/l chloroform; 680 g/l methanol. The technical grade (95% pure) has a strong odour.

**Uses.**

Prothiocarb hydrochloride is a soil-applied systemic fungicide with a specific action against Phycomycetes, *e.g. Phytophthora, Pythium* spp. It is taken up by the roots and translocated to the aerial parts. It is only recommended for ornamental crops, mainly as a protective fungicide.

**Toxicology.**

Acute oral $LD_{50}$ of the hydrochloride: for rats 1300 mg/kg; for mice 660-1200 mg/kg; for ducks 1754 mg salt (as formulation)/kg. Acute percutaneous $LD_{50}$: for rats >1470 mg a.i. (as formulation)/kg; for rabbits >980 mg a.i. (as formulation)/kg. In 90-d feeding trials NEL was: for rats 0.5 mg/kg diet; for dogs 1.8 mg/kg diet. $LC_{50}$ (96-h) is: for trout 328 mg a.i./l; for bluegill 258 mg/l.

**Formulations.**

'Previcur S70', aqueous solution (745 g prothiocarb hydrochloride/l).

**Analysis.**

Product analysis is by argentometric titration of the ethanethiol liberated by hydrolysis. Residues may be determined by the fluorimetric measurement of the *N*,*N*-dimethylpropane-1,3-diamine liberated by alkaline hydrolysis. Particulars available from Schering AG.

$C_{11}H_{15}Cl_2O_2PS_2$ (345.2)
GR CG DOPS&S3&O2

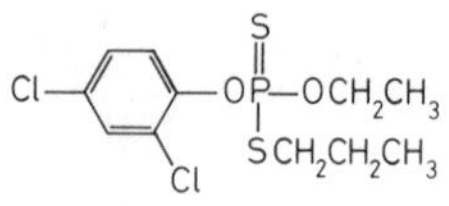

**Nomenclature and development.**

Common name prothiofos (BSI, E-ISO, F-ISO, JMAF);*exception* (Eire). Chemical name (IUPAC) *O*-2,4-dichlorophenyl *O*-ethyl *S*-propyl phosphorodithioate. (*C.A.*) *O*-(2,4-dichlorophenyl) *O*-ethyl *S*-propyl phosphorodithioate (9CI); Reg. No. *[34643-46-4]*. Its insecticidal properties were described by A. Kadamatsu (*Jpn. Pestic. Inf.*, 1976, No. 26, p. 14). Introduced by Bayer AG (DEOS 2 111 414) as 'BAY NTN 8629'; trade marks 'Tokuthion' (agricultural use), 'Bideron' (public health use).

**Properties.**

It is a colourless liquid; b.p. 125-128 °C/0.1 mmHg; v.p. 1.0 mPa at 20 °C; $d_4^{20}$ 1.3. Solubility at 20 °C *c.* 1.7 mg/kg water; completely miscible with cyclohexanone, toluene.

**Uses.**

It is an insecticide for use against leaf-eating caterpillars, generally recommended for vegetables at 50-75 g a.i./100 l. Also public health use against flies.

**Toxicology.**

Acute oral $LD_{50}$ for male rats 925-966 mg/kg. Acute percutaneous $LD_{50}$ for male rats >1.0 ml (1300 mg)/kg. In 90-d feeding trials NEL for rats was 8 mg/kg diet. $LC_{50}$ (96-h) for carp 1.8 mg/l.

**Formulations.**

An e.c. (500 g/l); w.p. (400 g/kg).

**Analysis.**

Product analysis is by glc. Residues may be determined by glc (E. Möllhoff, *Pflanzenschutz-Nachr. (Engl. Ed.)*, 1975, **28**, 382).

$C_9H_{20}NO_3PS_2$ (285.4)
2OPS&O2&S1VMY1&1

$(CH_3)_2CHNHCO.CH_2SP(S)(OCH_2CH_3)_2$

**Nomenclature and development.**

Common name prothoate (BSI, E-ISO, F-ISO). Chemical name (IUPAC) *O,O*-diethyl *S*-isopropylcarbamoylmethyl phosphorodithioate. (*C.A.*) *O,O*-diethyl *S*-[2-(1-methylethyl)amino-2-oxoethyl] phosphorodithioate (9CI); *O,O*-diethyl phosphorodithioate *S*-ester with *N*-isopropyl-2-mercaptoacetamide (8CI); Reg. No. *[2275-18-5]*. ENT 24 652. Its insecticidal properties were reported (*Ital. Agric.*, 1955, **99**, 747). Discovered by American Cyanamid Co. and later introduced by Montecatini S.p.A. (now Montedison S.p.A.) (USP 2 494 283 to American Cyanamid Co; GBP 791 824 to Montedison) as code no. 'E.I. 18 682' to Cyanamid,'L 343' to Montedison; trade mark 'Fac' (to Montedison).

**Properties.**

Pure prothoate is a colourless crystalline solid, with a camphor-like odour; m.p. 28.5 °C; v.p. 13 mPa at 40 °C; $d^{32}$ 1.151, $n_D^{32}$ 1.5128; Solubility at 20 °C 2.5 g/l water. The technical grade is an amber to yellow semi-solid; f.p. 21-24 °C. Solubility at 20 °C: $<$30 g/kg cyclohexane, hexane and higher petroleum fractions; $<$10 g/kg glycerol; $<$20 g/kg light petroleum; miscible with most organic solvents. It is stable at $4.0<pH<8.2$, but it is decomposed at 50 °C in *c.* 48 h at pH 9.2. The technical grade is stable to light and to temperatures $\leqslant$60 °C.

**Uses.**

It is a systemic acaricide and insecticide used at 20-30 g a.i./100 l (300-500 g/ha) to protect citrus, fruit, and vegetable crops from Tetranychid and some Eryophyid mites and some insects, notably aphids, Thysanoptera, Tingidae, and Psyllidae.

**Toxicology.**

Acute oral $LD_{50}$: for rats 8.0-8.9 mg tech./kg; for mice 19.8-20.3 mg/kg (R. Schuppon, *Chim. Ind. (France)*, 1961, **85**, 421). In 90-d feeding trials NEL was: for rats 0.5 mg/kg daily; for mice 1 mg/kg daily. NEL (10-d) was: for goldfish 6-8 mg/l; for mosquito fish 2-3 mg/l.

**Formulations.**

These include: 'Fac 20', 'Fac 40', liquid (200 or 400 g tech./kg); 'Fac 40WP', w.p. (400 g/kg); 'Fac P', solid (30 g/kg); 'Fac Granules', granules (50 g/kg).

**Analysis.**

Product analysis is by glc (B. Bazzi, *Anal. Methods Pestic. Plant Growth Regul.*, 1976, **8**, 213) or by tlc and colorimetric estimation of phosphorus in the appropriate spots (*CIPAC Handbook*, 1983, **1B**, in press). Residues on crops are determined by glc (B. Bazzi, *loc. cit.*; B. Bazzi *et al.*, *Pestic. Sci.*, 1974, **5**, 511). Details of the methods are available from Montedison S.p.A.

$C_{13}H_{15}NO_2$ (217.3)
T6O BUTJ B1 CVMR

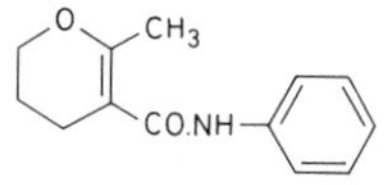

**Nomenclature and development.**

Common name pyracarbolid (BSI, E-ISO);pyracarbolide (F-ISO). Chemical name (IUPAC) 3,4-dihydro-6-methyl-2*H*-pyran-5-carboxanilide (I). (*C.A.*) 3,4-dihydro-6-methyl-*N*-phenyl-2*H*-pyran-5-carboxamide (9CI); (I) (8CI); Reg. No. *[24691-76-7]*. Its fungicidal properties were described by H. Stingle *et al.* (*Int. Congr. Plant Prot., 7th, Paris.*, 1970, p. 205 (Abstr.)) and B. Jank & F. Grossman (*Pestic. Sci.*, 1971, **2**, 43). Introduced by Hoechst AG (DEP 1 668 899; GBP 1 194 526) as code no. 'Hoe 13 764', formerly '*Hoe 02 989*', '*Hoe 6052*', '*Hoe 6053*'; trade mark 'Sicarol'.

**Properties.**

It is a colourless solid; m.p. 110-111 °C; v.p. 16 μPa at 25 °C. Solubility at 40 °C 0.6 g/l water; at 25 °C: 366 g/l chloroform; 89 g/l ethanol; 86 g/l ethyl acetate; 100 mg/l hexane; 13 g/l xylene. It is stable to light and heat and in alkaline media, but decomposed by acids. Technical grade is ⩾98% pure.

**Uses.**

It is a systemic fungicide, effective against Basidiomycetes. It controls rusts (Uredinales), smuts (Ustilaginales), and damping off disease caused by *Rhizoctonia solani*. It is absorbed by plants via roots and leaves. It is used in bean, cereal, coffee, and tea crops. Rates and concentrations vary considerably depending on disease/crop combination and formulation.

**Toxicology.**

Acute oral $LD_{50}$ for female rats >15 000 mg (in starch mucilage)/kg. Acute percutaneous $LD_{50}$ for female rats >1000 mg/kg. Acute intraperitoneal $LD_{50}$ for female rats 1600 mg/kg. In 2-y feeding trials NEL was: for rats 400 mg/kg diet; for dogs 1600 mg/kg diet. $LC_{50}$ (96-h) is: for carp 42.3 mg/l; for rainbow trout 45.5 mg/l.

**Formulations.**

These include: w.p. (500 g a.i./kg); dispersion (135 g/l); dry seed treatment (750 g/kg).

**Analysis.**

Product analysis is by titration. Residues may be determined by colorimetric methods. Details are available from Hoechst AG.

# Pyrazophos

$C_{14}H_{20}N_3O_5PS$ (373.4)
T56 ANN FNJ COPS&O2&O2 G1 HVO2

$CH_3CH_2O.CO$ N—N S
$CH_3$ N $OP(OCH_2CH_3)_2$

**Nomenclature and development.**

Common name pyrazophos (BSI, E-ISO, F-ISO). Chemical name (IUPAC) ethyl 2-diethoxythiophosphoryloxy-5-methylpyrazolo[1,5-*a*]pyrimidine-6-carboxylate; *O*-6-ethoxycarbonyl-5-methylpyrazolo[1,5-*a*]pyrimidin-2-yl *O,O*-diethyl phosphorothioate. (*C.A.*) ethyl 2-[(diethoxyphosphinothioyl)oxy]-5-methylpyrazolo[1,5-*a*]pyrimidine-6-carboxylate (9CI); ethyl 2-hydroxy-5-methylpyrazolo[1,5-*a*]pyrimidine-6-carboxylate *O*-ester with *O,O*-diethyl phosphorothioate (8CI); Reg. No. *[13457-18-6]*. Its fungicidal properties were described by F. M. Smit (*Meded. Rijksfac. Landbouwwet. Gent*, 1969, **34**, 763). Introduced by Hoechst AG (DEP 1 545 790; GBP 1 145 306) as code no. 'Hoe 02 873'; trade marks 'Afugan', 'Curamil'.

**Properties.**

It forms colourless crystals; m.p. 50-51 °C; v.p. (Antoine) 220 μPa at 50 °C. Solubility at 20 °C 4.2 mg/l water; at 25 °C: 1.2 kg/l acetone; 95 g/l ethanol; 898 g/l ethyl acetate; 11 g/l hexane; >980 g/l toluene. Technical grade ⩾92.6% pure. Decomposed by acids and alkali; unstable unless diluted, then stable ⩾2 y under normal storage conditions.

**Uses.**

It is a systemic fungicide controlling powdery mildews on a wide range of crops at 10-30 g a.i./100 l, and on cereals at 500-700 g/ha. It has only a limited insecticidal and acaricidal activity. It is absorbed by foliage and green stems and translocated within the plant. When applied to the soil or as a seed treatment uptake by roots is insufficient for effective fungicidal action within the plant.

**Toxicology.**

Acute oral $LD_{50}$ is (depending on carrier and sex): for rats 151-778 mg tech./kg; for quail 118-480 mg/kg. Acute percutaneous $LD_{50}$ for rats >2000 mg/kg. In 2-y feeding trials NEL for rats was 5 mg/kg diet; a 4-generation test on rats showed no effect at 50 mg/kg diet. $LC_{50}$ (96-h) is: for carp 6.1 mg/l; for rainbow trout 0.48 mg/l. It is non-toxic to honeybees up to 1 g/l when applied by high volume spraying and up to 2 l/ha when applied by mist-blowers (low volume spray).

**Formulations.**

These include: e.c. (295 g a.i./l); w.p. (300 g/kg) for HV spraying.

**Analysis.**

Product and residue analysis is by glc; details are available from Hoechst AG. See also J. Asshauer *et al.* (*Anal. Methods Pestic. Plant Growth Regul.*, 1978, **10**, 237).

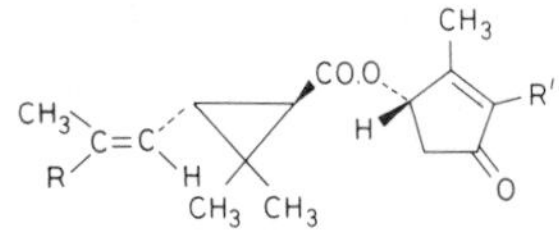

**Nomenclature and development.**

The term pyrethrins (BSI, E-ISO, ESA, JMAF); pyrèthres (F-ISO) is used collectively for the six insecticidal constituents with BSI and E-ISO (F-ISO) names: cinerin I (cinérine I); cinerin II; (cinérine II); jasmolin I; (jasmoline I); jasmolin II; (jasmoline II); pyrethrin I; (pyréthrine I); pyrethrin II; (pyréthrine II) present in extracts of the flowers *Pyrethrum cinerariaefolium* and other species. Reg. No. *[8003-34-7]*. Further details of the individual components are listed opposite. The dried flower heads of *P. cinerariae= folium* (once known as the Dalmatian Insect Flower) were introduced into Europe from western Asia *c* 1820, but their use is now replaced by that of extracts of crops grown in Ecuador, Kenya, and Tanzania.

**Properties.**

The ratio of pyerthrin:cinerin:jasmolin is generally 71:21:7; most commercial extracts contain 20-25% pyrethrins and are pale yellow, the plant waxes and pigments having been removed. They are unstable in light and are rapidly hydrolysed by alkali with loss of insecticidal properties.

**Uses.**

The pyrethrins are potent, non-systemic, contact insecticides causing a rapid paralysis or 'knockdown', death occurring at a later stage. Insecticidal activity is markedly increased by the addition of synergists, *e.g.* piperonyl butoxide.

**Toxicology.**

Acute oral $LD_{50}$ for rats 584-900 mg/kg. Acute percutaneous $LD_{50}$ for rats >1500 mg/kg (J. C. Malone & N. C. Brown, *Pyrethrum Post*, 1968, **9**(3), 3). Constituents of the flowers may cause dermatitis to sensitised individuals (J. T. Martin & K. H. C. Hester, *Br. J. Dermatol.*, 1941, **53**, 127; F. E. Rickett *et al., Pestic. Sci.*, 1972, **3**, 57; 1973, **4**, 801) but are removed during the preparation of refined extracts. There is no evidence that synergists increase toxicity of the pyrethrins to mammals. Pyrethrins are highly toxic to fish.

**Formulations.**

Dusts of various a.i. content on non-alkaline carriers. For aerosol concentrates (up to 20 g/l) the extract is dissolved in refined odourless kerosene, occasionally with dichloromethane, plus suitable propellants. Synergists (usually piperonyl butoxide) and organophosphorus insecticides (especially diazinon) may be added to increase the rapid kill of the pest.

**Analysis.**

Commercial extracts of flowers may be analysed by glc (D. B. McClellan, *Anal. Methods Pestic. Plant Growth Regul.*, 1972, **6**, 461; J. Sherma, *ibid.*, 1976, **8**, 225) or by a titrimetric method. Due to interference by certain adjuvants, only approximate results on formulations may be obtained; an empirical factor and the determined pyrethrin I content may be used (*CIPAC Handbook*, 1970, **1**, 598; D. B. McClellan, *Anal. Methods Pestic., Plant Growth Regul. Food Addit.*, 1964, **2**, 399). Residues may be determined by colorimetry or glc (*Pestic. Anal. Man*, 1979, **II**; J. Sherma, *loc. cit.*; R. Mestres *et al., Ann. Falsif. Expert. Chim.*, 1979, **72**, 577; W. Specht & M. Tillkes, *Fresenius' Z. Anal. Chem.*, 1980, **301**, 300).

*Pyrethrin I* $C_{21}H_{28}O_3$; (328.4)
L5V BUTJ B2U2U1 C1 DOV- BL3TJ A1 A1 C1UY1&1
(R -CH$_3$; R′ -CH$_2$CH = CH-CH = CH$_2$)

(IUPAC) (1*S*)-2-methyl-4-oxo-3-[(*Z*)-penta-2,4-dienyl]cyclopent-2-enyl (1*R*,3*R*)-2,2-dimethyl-3-(2-methylprop-1-enyl)cyclopropanecarboxylate; (1*S*)-2-methyl-4-oxo-3-[*Z*)-penta-2,4-dienyl]cyclopent-2-enyl (1*R*)-*trans*-2,2-dimethyl-3-(2-methylprop-1-enyl)cyclopropanecarboxylate; (1*S*)-2-methyl-4-oxo-3-[(*Z*)-penta-2,4-dienyl]cyclopent-2-enyl (1*R*)-*trans*-chrysanthemate. (*C.A.*) [1*R*-[1α[*S**(*Z*)],3β]]-2-methyl-4-oxo-3-(2,4-pentadienyl)cyclopenten-1-yl 2,2-dimethyl-3-(2-methyl-1-propenyl)cyclopropanecarboxylate (9CI); 2,2-dimethyl-3-(2-methylpropenyl)cyclopropanecarboxylic acid, ester with 4-hydroxy-3-methyl-2-(2,4-pentadienyl)-2-cyclopenten-1-one (8CI); Reg. No. [*121-21-1*].

*Pyrethrin II* $C_{22}H_{28}O_5$ (372.4)
L5V BUTJ B2U2U1 C1 DOV- BL3TJ A1 A1 C1UY1&VO1
(R -CO.OCH$_3$; R′ CH$_2$CH = CH-CH = CH$_2$)

(IUPAC) (1*S*)-2-methyl-4-oxo-3-[(*Z*)-penta-2,4-dienyl]cyclopent-2-enyl (1R,3*R*)-3-[(*E*)-2-methoxycarbonylprop-1-enyl]-2,2-dimethylcyclopropanecarboxylate; (1*S*)-2-methyl-4-oxo-3-[(*Z*)-penta-2,4-dienyl]cyclopent-2-enyl (1*R*)-*trans*-3-[(*E*)-2-methoxycarbonylprop-1-enyl]-2,2-dimethylcyclopropane=carboxylate; (1*S*)-2-methyl-4-oxo-3-[(*Z*)-penta-2,4-dienyl]cyclopent-2-enyl pyrethrate. (*C.A.*) [1*R*-[1α[*S**(*Z*)],3β(*E*)]]-2-methyl-4-oxo-3-(2,4-pentadienyl)-2-cyclopenten-1-yl 3-(3-methoxy-2-methyl-3-oxo-1-propenyl)-2,2-dimethylcyclopropanecarboxylate (9CI); 3-carboxy-α,2,2-trimethylcyclopropanecarboxylic acid 1-methyl ester, ester with 4-hydroxy-3-methyl-2-(2,4-pentadienyl)-2-cyclopenten-1-one (8CI); Reg. No. *[121-29-9]*.

*Cinerin I* $C_{20}H_{28}O_3$ (316.4)
L5V BUTJ B2U2 C1 DOV- BL3TJ A1 A1 C1UY1&1
(R -CH$_3$; R′ -CH$_2$CH = CHCH$_3$)

(IUPAC) (1*S*)-3-[(*Z*)-but-2-enyl]-2-methyl-4-oxocyclopent-2-enyl (1*R*,3*R*)-2,2-dimethyl-3-(2-methylprop-1-enyl)cyclopropanecarboxylate; (1*S*)-3-[(*Z*)-but-2-enyl]-2-methyl-4-oxocyclopent-2-enyl (1*R*)-*trans*-2,2-dimethyl-3-(2-methylprop-1-enyl)cyclopropanecarboxylate; (1*S*)-3-[(*Z*)-but-2-enyl]-2-methyl-4-oxocyclopent-2-enyl (1*R*)-*trans*-chrysanthemate. (*C.A.*) [1*R*-[1α[*S**(*Z*)],3β]]-3-(2-butenyl)-2-methyl-4-oxo-2-cyclopenten-1-yl 2,2-dimethyl-3-(2-methyl-1-propenyl)cyclopropanecarboxylate (9CI); 2,2-dimethyl-3-(2-methylpropenyl)cyclopropanecarboxylic acid, ester with 2-(2-butenyl)-4-hydroxy-3-methyl-2-cyclopenten-1-one (8CI); Reg. No. *[25402-06-6]*.

*Cinerin II* $C_{21}H_{28}O_5$ (360.4)
L5V BUTJ B2U2 C1 DOV- BL3TJ A1 A1 C1UY1&VO1
(R -CO.OCH$_3$; R′ -CHCH = CHCH$_3$)

(IUPAC) (1*S*)-3-[(*Z*)-but-2-enyl]-2-methyl-4-oxocyclopent-2-enyl (1*R*,3*R*)-3-[(*E*)-2-methoxycarbonylprop-1-enyl]-2,2-dimethylcyclopropanecarboxylate; (1*S*)-3-[(*Z*)-but-2-enyl]-2-methyl-4-oxocyclopent-2-enyl (1*R*)-*trans*-3-[(*E*)-2-methoxycarbonylprop-1-enyl]-2,2-dimethylcyclopropanecarboxylate; (1*S*)-3-[(*Z*)-but-2-enyl]-2-methyl-4-oxocyclopent-2-enyl pyrethrate. (*C.A.*) [1*R*-[1α[*S**(*Z*)],3β(*E*)]]-3-(2-butenyl)-2-methyl-4-oxo-2-cyclopenten-1-yl 3-(3-methoxy-2-methyl-3-oxo-1-propenyl)-2,2-dimethylcyclopropanecarboxylate (9CI); 3-carboxy-α,2,2-trimethylcyclopropanecarboxylic acid 1-methyl ester, ester with 2-(2-butenyl)-4-hydroxy-3-methyl-2-cyclopenten-1-one (8CI); Reg. No. *[121-20-0]*.

*Jasmolin I* $C_{21}H_{30}O_3$ (330.4)
L5V BUTJ B2U3 C1 DOV- BL3TJ A1 A1 C1UY1&1
(R -CH$_3$; R′ -CH$_2$CH = CH-CH$_2$CH$_3$)

(IUPAC) (1*S*)-2-methyl-4-oxo-3-[(*Z*)-pent-2-enyl]cyclopent-2-enyl (1*R*,3*R*)-2,2-dimethyl-3-(2-methylprop-1-enyl)cyclopropanecarboxylate; (1*S*)-2-methyl-4-oxo-3-[(*Z*)-pent-2-enyl]cyclopent-2-enyl (1*R*)-*trans*-2,2-dimethyl-3-(2-methylprop-1-enyl)cyclopropanecarboxylate; (1*S*)-2-methyl-4-oxo-3-[(*Z*)-pent-2-enyl]cyclopent-2-enyl (1*R*)-*trans*-chrysanthemate. (*C.A.*) [(1*R*-[1α[*S**(*Z*)],3β]]-2-methyl-4-oxo-3-(2-pentenyl)-2-cyclopenten-1-yl 2,2-dimethyl-3-(2-methyl-1-propenyl)cyclopropanecarboxylate (9CI); 2,2-dimethyl-3-(2-methylpropenyl)cyclopropanecarboxylic acid, ester with 4-hydroxy-3-methyl-2-(2-pentenyl)-2-cyclopenten-1-one (8CI); Reg. No. *[4466-14-2]*.

*Jasmolin II* $C_{22}H_{30}O_5$ (374.4)
L5V BUTJ B2U3 C1 DOV- BL3TJ A1 A1 C1UY1&VO1
(R -CO.OCH$_3$; R′ -CH$_2$CH = CH-CH$_2$CH$_3$)

(IUPAC) (1*S*)-2-methyl-4-oxo-3-[(*Z*)-pent-2-enyl]cyclopent-2-enyl (1*R*,3*R*)-3-[(*E*)-2-methoxy=carbonylprop-1-enyl]-2,2-dimethylcyclopropanecarboxylate; (1*S*)-2-methyl-4-oxo-3-[(*Z*)-pent-2-enyl]cyclopent-2-enyl (1*R*)-*trans*-3-[(*E*)-2-methoxycarbonylprop-1-enyl]-2,2-dimethyl=cyclopropanecarboxylate; (1*S*)-2-methyl-4-oxo-3-[(*Z*)-pent-2-enyl]cyclopent-2-enyl pyrethrate. (*C.A.*) [1*R*-[1α[*S**(*Z*)],3β(*E*)]]-2-methyl-4-oxo-3-(2-pentenyl)-2-cyclopenten-1-yl 3-(3-methoxy-2-methyl-3-oxo-1-propenyl)-2,2-dimethylcyclopropanecarboxylate (9CI); 3-carboxy-α,2,2-trimethylcyclopropanecarboxylic acid 1-methyl ester, ester with 4-hydroxy-3-methyl-2-(2-pentenyl)-2-cyclopenten-1-one (8CI); Reg. No. *[1172-63-0]*.

$C_{19}H_{23}ClN_2O_2S$ (378.9)
T6NNJ CR& DOVS8 FG

Cl N N
$CH_3(CH_2)_7S.CO.O$

**Nomenclature and development.**

Common name pyridate (BSI, E-ISO, F-ISO, WSSA). Chemical name (IUPAC) 6-chloro-3-phenylpyridazin-4-yl *S*-octyl thiocarbonate. (*C.A.*) *O*-(6-chloro-3-phenyl-4-pyridazinyl) *S*-octyl carbonothioate (9CI); Reg. No. *[55512-33-9]*. Its herbicidal properties were described by A. Diskus *et al.* (*Proc. Br. Crop Prot. Conf. - Weeds*, 1976, **2**, 717). Introduced by Chemie Linz AG (ATP 326 409) as code no. 'CL 11 344'; trade mark 'Lentagran'.

**Properties.**

Pure pyridate is a colourless crystalline solid; m.p. 27 °C. The technical grade is a brown oil, b.p. 220 °C/0.1 mmHg; v.p. 133 nPa; $d^{20}$ 1.16; $n_D^{20}$ 1.568. Solubility: 90 mg/l water; highly soluble in organic solvents.

**Uses.**

It is a foliar-acting herbicide with contact activity on annual dicotyledonous plants, especially *Galium aparine* and *Amaranthus retroflexus* (atrazine-resistant biotypes), and some grassy weeds. It controls weeds selectively in cereals, maize, rice and other crops at 1.0-1.5 kg a.i./ha. Effectiveness depends on weed species and stage of weed development. There is a strong indication that the action depends upon inhibition of the Hill reaction.

**Toxicology.**

Acute oral $LD_{50}$ : for rats *c.* 2000 mg/kg; for 10-d-old pheasants and 5-d-old Pekin ducks ≥10 000 mg/kg; for bobwhite quail 1500 mg/kg. Acute percutaneous $LD_{50}$ for rabbits ≥3400 mg/kg; it is moderately irritant to their skin but not to their eyes. It produced evidence of sensitivity in guinea-pigs, but no symptom has been observed in man (applicators, technicians, workers). Five tests for mutagenicity were negative. $LC_{50}$ (96-h) is: for rainbow trout 81 mg/l; for bluegill and carp >100 mg/l. It is non-toxic to honeybees.

**Formulations.**

These include: 'Lentagran', w.p. (450 g a.i./kg); e.c. (640 g/l).

**Analysis.**

Product analysis is by hplc. Residues may be determined by hplc. Details of methods are available from Chemie Linz AG.

$C_{12}H_{15}N_2O_3PS$ (298.3)
T66 BN ENJ COPS&O2&O2

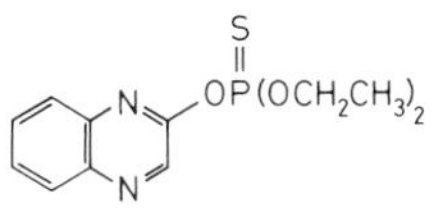

**Nomenclature and development.**

Common name quinalphos (BSI, E-ISO, F-ISO);chinalphos (France). Chemical name (IUPAC) *O,O*-diethyl *O*-quinoxalin-2-yl phosphorothioate. (*C.A.*) *O,O*-diethyl *O*-2-quinoxalinyl phosphorothioate (8 & 9CI); Reg. No. *[13593-03-8]*. ENT 27 394. Its insecticidal properties were described by K. -J. Schmidt & L. Hammann (*Pflanzenschutz-Nachr. (Engl. Ed.)*, 1969, **22**, 314). Introduced by Bayer AG and Sandoz AG (BEP 681 443; DEAS 1 545 817 to Bayer AG) as code no. 'Bayer 77 049', 'Sandoz 6538 e.c.', 'Sandoz 6626 g'; trade marks 'Bayrusil' (to Bayer AG), 'Ekalux' (to Sandoz AG).

**Properties.**

Pure quinalphos is a colourless crystalline solid; m.p. 31-32 °C; b.p. 142 °C (decomp.)/3 × $10^{-4}$ mmHg at 20 °C; v.p. 346 µPa at 20 °C; $d_4^{20}$ 1.235, $n_D^{25}$ 1.5624 (supercooled melt). Solubility at 23-24 °C: 22 mg/l water; completely miscible with acetone, chloroform, diethyl ether, dimethyl sulphoxide, ethanol, methanol, xylene; *c.* 250 g/l hexane. At 23 °C (extrapolated) 50% hydrolysis occurs in 56 d at pH5, 40 d at pH7, 30 d at pH9. The pure crystalline a.i. is stable at 20 °C for *c.* 1 y; technical liquid material is less stable, but can be stabilised by suitable nonpolar solvents. Formulations are stable (shelf life at 20 °C *c.* 2 y).

**Uses.**

It is a contact and stomach insecticide and acaricide with good penetrative properties. It is used at 190-500 g a.i. (as e.c.)/ha against caterpillars on cotton, groundnuts and vegetables, also scales and caterpillars on fruit trees; at 250-500 g a.i. (as e.c.)/ha or 0.75-1.0 kg a.i. (as granules)/ha against the pest complex on rice. It is rapidly metabolised in plants, animals and soil to quinoxalin-2-ol (free and as its conjugates), polar metabolites and carbon dioxide (soil).

**Toxicology.**

Acute oral $LD_{50}$ for rats 62-137 mg/kg. Acute percutaneous $LD_{50}$ for rats 1250-1400 mg/kg. In 2-y feeding trials NEL (based on cholinesterase inhibition) for rats was 3 mg/kg diet. $LC_{50}$ (8-d) was: for quail 150 mg/kg diet; for mallard duck 220 mg/kg diet. $LC_{50}$ (96-h) is: for carp 2.8 mg/l; for goldfish 1.0-10 mg/l.

**Formulations.**

These include: 'Bayrusil', e.c. (200 g a.i./l); 'Ekalux', e.c. (250 g/l); 'Ekalux Forte', e.c. (480 g/l); 'Ekalux', w.p. (250 g/kg); 'Ekalux', spray oil (30 g/l); 'Ekalux', ULV (300 g/kg); 'Ekalux', granules (50 g/kg); 'Ekalux', dust (15 g/kg). 'Tombel', e.c. (160 g quinalphos + 160 g thiometon/l).

**Analysis.**

Product analysis is by glc or by tlc and subsequent u.v. spectrometric determination of the eluted compound. Residues are determined by glc (M. Wisson *et al.*, *Anal. Methods Pestic. Plant Growth Regul.*, 1980, **11**, 147).

$C_{10}H_6N_2OS_2$ (234.3)
T C566 BN DSVS HNJ K1

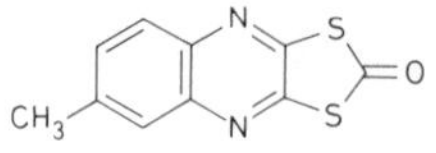

**Nomenclature and development.**

Common name chinomethionat (E-ISO);chinomethionate (F-ISO);quinomethionate (BSI);oxythioquinox (Australia, ESA);quinoxalines (JMAF). Chemical name (IUPAC) 6-methyl-1,3-dithiolo[4,5-*b*]quinoxalin-2-one (I); *S,S*-(6-methylquinoxaline-2,3-diyl) dithiocarbonate. (*C.A.*) (I) (9CI); cyclic *S,S*-(6-methyl-2,3-quinoxalinediyl) dithiocarbonate (8CI); Reg. No. *[2439-01-2]*. ENT 25 606. Its biological properties were described by K. Sasse (*Hoefchen-Briefe (Engl. Ed.)*, 1960, **13**, 197; K. Sasse *et al., Angew. Chem.*, 1960, **72**, 973). Introduced by Bayer AG (DEAS 1 100 372; BEP 580 478) as code no. 'Bayer 36 205','Bayer Ss 2074'; trade mark 'Morestan'.

**Properties.**

Quinomethionate forms yellow crystals; m.p. 169.8-170 °C; v.p. 27 μPa at 20 °C. Solubility: practically insoluble in water; 18 g/kg cyclohexanone; 10 g/kg dimethylformamide; 4 g/kg petroleum oils.

**Uses.**

It is a selective non-systemic acaricide and fungicide specific to powdery mildews on fruits, ornamentals and vegetables at 7.5-12.5 g a.i./100 l.

**Toxicology.**

Acute oral $LD_{50}$ for rats 2500-3000 mg/kg. Acute percutaneous $LD_{50}$ (7-d) for rats >500 mg/kg. In 2-y feeding trials rats receiving 60 mg/kg diet suffered no ill-effect. It is non-toxic to honeybees.

**Formulations.**

A w.p. (250 g a.i./kg); smokes; dusts.

**Analysis.**

Product analysis is by u.v. spectrometry, details are available from Bayer AG. Residues may be determined by glc (K. Vogeler & H. Niessen, *Pflanzenschutz-Nachr. (Engl. Ed.)*, 1967, **20**, 550; *Pestic. Anal. Man.*, 1979, **I**, 201-I), or by colorimetry after conversion to a derivative (H. Tietz *et al., ibid.*, 1962, **15**, 166; R. Havens *et al., J. Agric. Food Chem.*, 1964, **12**, 247; C. A. Anderson, *Anal. Methods Pestic., Plant Growth Regul. Food Addit.*, 1967, **5**, 277).

$C_{12}H_6Cl_3NO_3$ (318.5)
L66 BV EJV CMVYGG DG

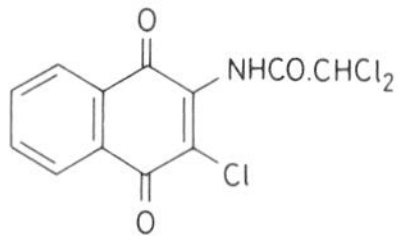

**Nomenclature and development.**

Common name quinonamid (BSI, E-ISO);quinonamide (F-ISO). Chemical name (IUPAC) 2,2-dichloro-*N*-(3-chloro-1,4-naphthoquinon-2-yl)acetamide. (*C.A.*) 2,2-dichloro-*N*-(3-chloro-1,4-dihydro-1,4-dioxo-2-naphthalenyl)acetamide (9CI); 2,2-dichloro-*N*-(3-chloro-1,4-dihydro-1,4-dioxo-2-naphthyl)acetamide (8CI); Reg. No. *[27541-88-4]*. Its algicidal activity was reported by P. Hartz *et al.* (*Meded. Fac. Landbouwwet. Rijksuniv. Gent*, 1972, **37**, 699). Introduced by Hoechst AG (DEP 1 768 447; GBP 1 263 625) as code no. 'Hoe 02 997'; trade marks 'Alginex', formerly '*Nosprasit*'.

**Properties.**

Pure quinonamid forms yellow needles; m.p. 214-216 °C; 11 µPa at 20 °C. Solubility at 23 °C: 3.0 mg/l water (pH4.6), 60 mg/l (pH7); at 20 °C: *c.* 7 g/l acetone; *c.* 3 g/l ethanol; 40 mg/l hexane; 8 g/l xylene. Decomposed by acid or alkali. Technical grade 97% pure.

**Uses.**

Quinonamid is effective against algae in the open as well as algae and mosses under glass. It can be used as a seed treatment or spray for control of algae in paddies, as a dip for clay pots, and for treating benches, etc., in greenhouses.

**Toxicology.**

Acute oral $LD_{50}$: for rats 11 700-15 000 mg/kg; for Japanese quail 11 136-15 542 mg/kg (depending on carrier and sex). In 90-d feeding trials NEL for rats was 2000 mg/kg diet. $LC_{50}$ (96-h) for rainbow trout is 0.45 mg/l.

**Formulations.**

A w.p. (500 g a.i./kg); granules (100 g/kg).

**Analysis.**

Product and residue analysis is by glc. Details are available from Hoechst AG.

$C_6Cl_5NO_2$ (295.3)
WNR BG CG DG EG FG



**Nomenclature and development.**

Common name quintozene (BSI, E-ISO, F-ISO);terrachlor (Turkey);PKhNB (USSR);PCNB (JMAF). Chemical name (IUPAC) pentachloronitrobenzene (I). (*C.A.*) (I) (8 & 9CI); Reg. No. *[82-68-8]*. Trivial name PCNB. Introduced as a fungicide by I. G. Farbenindustrie AG (now Bayer AG who no longer manufacture or market it). (DERP 682 048) as trade marks 'Botrilex' (ICI Plant Protection Division), 'Tritisan' (Hoechst AG), 'Folosan', 'Terrachlor' (Olin Mathieson Chemical Corp.), 'Brassicol'.

**Properties.**

Pure quintozene forms colourless needles; m.p. 146 °C; v.p. 1.8 Pa at 25 °C; $d^{25}$ 1.718. Solubility at 25 °C: practically insoluble in water; soluble in benzene, chloroform; *c.* 20 g/kg ethanol. It is highly stable in soil and compatible with pesticides at $\leqslant$pH7. The technical grade is $\leqslant$98.5% pure, m.p. 142-145°.

**Uses.**

It is a fungicide of specific use for seed and soil treatment, effective against *Botrytis, Rhizoctonia* and *Sclerotinia* spp. on brassicas, vegetable and ornamental crops, and *Tilletia caries* of wheat.

**Toxicology.**

Acute oral $LD_{50}$ for rats >12 000 mg/kg. In 2-y feeding trials NEL was: for rats 25 mg/kg diet; for dogs 30 mg/kg diet.

**Formulations.**

These include: 'Brassicol', 'Tritisan' (200 g a.i./kg; in Canada 600 g/kg); 'Terrachlor', w.p. (750 g/kg); e.c. (240 g/l). 'Terrachlor Super X', 'Terra-Coat L 205' (Reg No. *[8065-61-0]*) (quintozene + etridiazole). 'Acti-dione RZ' (quintozene + cycloheximide).

**Analysis.**

Product analysis is by AOAC Methods. Residues may be determined by glc (*Pestic. Anal. Man.* 1979, **I**, 201A, 201G, 201I; M. A. Luke *et al.*, *J. Assoc. Off. Anal. Chem.*, 1981, **64**, 1187; *Anal. Methods Pestic. Plant Growth Regul.*, 1972, **6**, 577).

# Resmethrin

$C_{22}H_{26}O_3$ (338.4)
T5OJ B1R& D1OV- BL3TJ A1 A1
C1UY1&1 &&(1RS)-CIS-TRANS FORM

$(CH_3)_2C{=}CH$— —CO.OCH$_2$—
CH$_3$ CH$_3$ O CH$_2$—

**Nomenclature and development.**

Common name resmethrin (the ratio of isomers should be stated) (BSI, E-ISO, ANSI, JMAF); resméthrine (F-ISO). Chemical name (IUPAC) 5-benzyl-3-furylmethyl (1*RS*,3*RS*; 1*RS*,3*SR*)-2,2-dimethyl-3-(2-methylprop-1-enyl)cyclopropanecarboxylate; 5-benzyl-3-furylmethyl (1*RS*)-*cis-trans*-2,2-dimethyl-3-(2-methylprop-1-enyl)cyclopropanecarboxylate; 5-benzyl-3-furylmethyl (*RS*)-*cis-trans*-chrysanthemate. (*C.A.*) [5-(phenylmethyl)-3-furanyl]methyl 2,2-dimethyl-3-(2-methyl-1-propenyl)cyclopropanecarboxylate (9CI); (5-benzyl-3-furyl)methyl 2,2-dimethyl-3-(2-methylpropenyl)cyclopropanecarboxylate (8CI); Reg. No. *[10453-86-8]*. OMS 1206. The insecticidal properties of this mixture of isomers was described by M. Elliott *et al.* (*Nature (London)*, 1967, **213**, 493). Selective isomers are also available, see bioresmethrin (1000). Resmethrin was introduced by FMC Corp., Mitchell Cotts Chemicals, Penick Corp., Sumitomo Chemical Co. (GBP 1 168 797; 1 168 798; 1 168 799 all to NRDC) as code no. 'NRDC 104', 'FMC 17 370', 'SBP 1382' (Penick Corp.); trade marks 'Benzyfuroline', 'Chryson' (to Sumitomo Chemical Co.).

**Properties.**

Technical grade resmethrin is a colourless waxy solid; m.p. 43-48 °C; v.p. 1.46 μPa at 30 °C. Solubility at 30 °C: <1 mg/l water; 220 g/kg hexane; 81 g/kg methanol; >1kg/kg xylene. It contains 20-30% (1*RS*)-*cis*- and 80-70% (1*RS*)-*trans*-isomers. It is decomposed rapidly on exposure to air and light (J. H. Fales *et al.*, CSMA *Proc. 55th Annu. Meeting*, December 1968; A. B. Hadaway *et al., Bull. W.H.O.*, 1970, **42**, 387). Unstable in alkaline media.

**Uses.**

It is a potent contact insecticide effective against a wide range of insects (I. C. Brooks *et al., Soap. Chem. Spec.*, 1969, **45**(3), 62). The toxicity to normal *Musca domestica* is 20× that of natural pyrethrum (M. Elliott *et al., loc. cit.*) but it is not synergised by pyrethrum synergists. Toxicity to plants is low. It is used to control agricultural, horticultural, household and public health insect pests, often in combination with other more persistent insecticides.

**Toxicology.**

Acute oral $LD_{50}$ for rats >2500 mg/kg. Acute percutaneous $LD_{50}$ for rats >3000 mg/kg. In 90-d feeding trials rats receiving 3000 mg/kg diet suffered no ill-effect. No teratogenic effect was seen in rats receiving 25 mg/kg daily, or in mice at 50 mg/kg daily.

**Formulations.**

These include: aerosol concentrates, e.c., w.p. concentrate for ULV application. Combinations with other insecticides or synergists: resmethrin + tetramethrin; resmethrin + malathion.

**Analysis.**

Product analysis is by glc with FID (B. B. Brown, *Anal. Methods Pestic. Plant Growth Regul.*, 1973, **7**, 441). The *cis*- and *trans*- isomers can be separated and estimated by hplc (J. H. Zehner & R. A. Simonaitis, *J. Assoc. Off. Anal. Chem.*, 1976, **59**, 1101) or by glc (A. Murano *et al., Agric. Biol. Chem.*, 1971, **35**, 1200). Residues may be determined by glc (R. Mestres *et al., Annal. Falsif. Expert. Chim.*, 1979, **72**, 577).

$C_{23}H_{22}O_6$ (394.4)
T G5 D6 B666 CV HO MO POT&TT&J
1Y1&U1 SO1 TO1

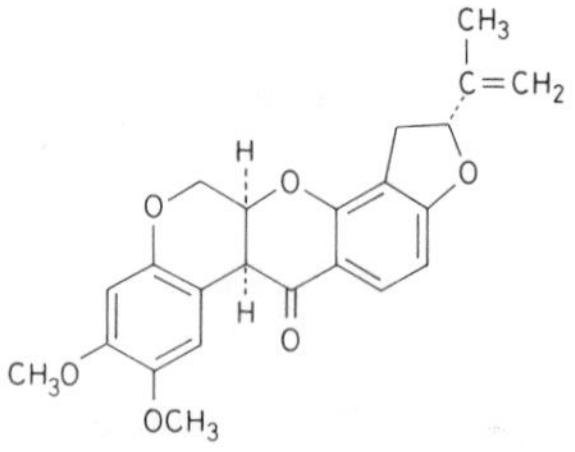

**Nomenclature and development.**

The traditional name rotenone (BSI, E-ISO, F-ISO);derris (JMAF) is accepted in lieu of a common name for the main insecticidal compound present in *Derris elliptica* and certain other *D.* spp., *Lonchocarpus utilis, L. urucu* and *L. nicou*. Chemical name (IUPAC) (2*R*,6a*S*,12a*S*)-1,2,6,6a,12,12a-hexahydro-2-isopropenyl-8,9-dimethoxychromeno[3,4-*b*]furo[2,3-*h*]chromen-6-one. (*C.A.*) (*R*)-1,2-dihydro-8,9-dimethoxy-2-(1-methylethenyl)[1]benzopyrano[3,4-*b*]furo[2,3-*h*][1]benzopyran-6,12-dione (9CI); 1,2,12,12aα-tetrahydro-2a-isopropenyl-8,9-dimethoxy[1]benzopyranol[3,4-*b*]furo[2,3-*h*][1]benzopyran-6(6a*H*)-one (8CI); Reg. No. *[83-79-7]*. Trivial names for the plant extract include derris root,tuba-root,aker-tuba; those for the plants barbasco, cube, haiari, nekoe, timbo. ENT 133. Derris root has long been used as a fish poison and its insecticidal properties were known to the Chinese well before it was isolated in 1895 by E. Geoffrey (*Ann. Inst. Colon. (Marseilles)*, 1895, **2**, 1); its structure was established by E. B. LaForge *et al.* (*Chem. Rev.*, 1933, **12**, 181).

**Properties.**

Isolated from roots of *Derris* and *Lonchocarpus* spp. by crystallisation from carbon tetrachloride, it forms colourless crystals, m.p. 163 °C, a dimorphic form m.p. 181 °C; $[\alpha]_D^{20}$ −231° (benzene). Solubility at 100 °C: 15 mg/l water; at room temperature: slightly soluble in carbon tetrachloride, petroleum oils; soluble in polar organic solvents. It crystallises with solvent of crystallisation (H. A. Jones, *J. Am. Chem. Soc., 1931,* **53**, 2738). It is readily racemised by alkali and readily oxidised, especially in the presence of light and alkali, to less insecticidal products, a change accelerated in certain solvents, *e.g.* pyridine.

**Uses.**

Rotenone is a selective non-systemic insecticide with some acaricidal properties. It is of low persistence in spray or dust residues.

**Toxicology.**

Acute oral $LD_{50}$: for white rats 132-1500 mg/kg; for white mice 350 mg/kg. It is toxic to pigs (A. A. Kinscote *et al., Annu. Rep. Entomol. Soc. Ont.*, 1951, p. 37). it is highly toxic to fish.

**Formulations.**

These include dusts from the ground root on a non-alkaline carrier, and stabilised with phosphoric acid.

**Analysis.**

Product analysis is by i.r. spectrometry (*AOAC Methods*, 1980, 6.160-6.165); the purity of a benzene solution of rotenone may be checked by polarimetry (*CIPAC Handbook*, 1970, **1**, 610). Residues may be determined by hplc (M. C. Bowman *et al., J. Assoc. Off. Anal. Chem.*, 1978, **61**, 1445) or by colorimetry (C. R. Gross & C. M. Smith, *ibid.*, 1934, **17**, 336; L. D. Goodhue, *ibid.*, 1936, **19**, 118).

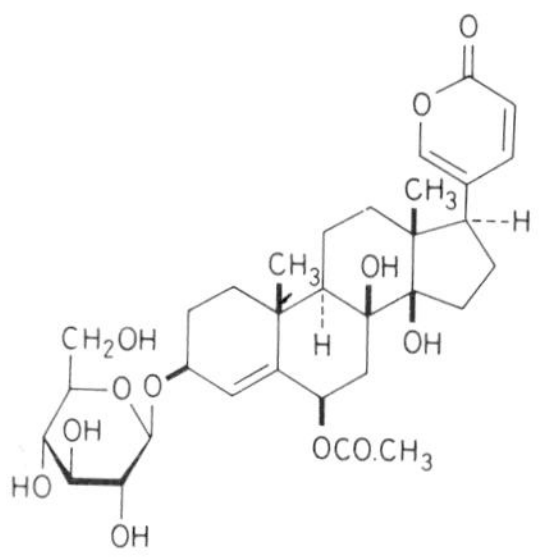

$C_{32}H_{44}O_{12}$ (620.7)
L E5 B666 MUTJ A1 E1 IQ JQ LOV1 F-
ET6OVJ& OO- BT6OTJ CQ DQ EQ F1Q

**Nomenclature and development.**

Common name scilliroside (JMAF) is used for a glycoside contained in the extract of the powdered bulbs of the red squill, *Urginea (Scilla) maritima*. Chemical name (IUPAC) 3β-(β-D-glucopyranosyl)-17β-(2-oxo-2*H*-pyran-5-yl)-14β-androst-4-ene-6β,8,14-triol 6-acetate. (*C.A.*) (3β,6β)-6-(acetyloxy)-3-(β-D-glucopyranosyloxy)-8,14-dihydroxybufa-4,20,22-trienolide (9CI); 3β-(β-D-glucopyropyranosyloxy)-6β,8,14-trihydroxybufa-4,20,22-trienolide 6-acetate (8CI); scilliroside (8CI); Reg. No. *[507-60-8]*. Scilliroside, the toxic principle, of red sqill was isolated and identified by A. Stoll & J. Renz (*Helv. Chim. Acta*, 1942, **25**, 377; 1943, **26**, 648). The chemistry and toxicology of cardiac glycosides was reviewed by A. Stoll (*Experientia*, 1954, **10**, 282). The two varieties of *U. maritima*, red squill and white squill, contain glycosides but only the former is used against rats. The toxicity of white squill to rats is lost on drying (H. Roques, *Bull. Fil. Soc. Biol. Paris*, 1942-1944, p. 21). The activity of red squill is thought to be protected by the red pigment, but is lost on heating, hence the necessity for drying bulbs at <80 °C.

**Properties.**

Scilliroside forms hydrated prisms (from aqueous methanol) which lose *c.* 8% *m/m in vacuo* giving a hemihydrate, m.p. 168-170 °C (decomposing at 200 °C); $[\alpha]^{20}_{D}$ −59° to −60° (methanol). Solubility: sparingly soluble in water, acetone, chloroform, ethyl acetate; soluble in alcohols, ethylene glycol, dioxane, glacial acetic acid; practically insoluble in diethyl ether, petroleum spirit.

**Uses.**

Extracts of red squill are used in baits to control rats. it is claimed that its specific toxicity to rats is due to the inability of the rodent to vomit, a reaction squill induces in other animals.

**Toxicology.**

Acute oral $LD_{50}$ of scilliroside is: for male rats 0.7 mg/kg, for females 0.43 mg/kg (A. Stoll & J. Renz, *loc. cit.*). Pigs and cats survived 16 mg/kg, fowls survived 400 mg/kg (S. A. Barnett *et al.*, *J. Hyg.*, 1949, **47**, 431).

**Analysis.**

Product analysis is by tlc (details from Sandoz AG) or by chromatographic separation on paper and, after elution, photometric assay (M. Wichtl & L. Fuchs, *Arch. Pharm.*, 1962, **295**, 361).

$C_{10}H_{19}N_5O$ (225.3)
T6N CN ENJ BO1 DMY2&1 FM2

**Nomenclature and development.**

Common name secbumeton (BSI, E-ISO, F-ISO, ANSI, WSSA). Chemical name (IUPAC) *N*-*sec*-butyl-*N*'-ethyl-6-methoxy-1,3,5-triazine-2,6-diyldiamine; 2-*sec*-butylamino-4-ethylamino-6-methoxy-1,3,5-triazine. (*C.A.*) *N*-ethyl-6-methoxy-*N*'-(1-methylpropyl)-1,3,5-triazine-2,4-diamine (9CI); 2-(*sec*-butylamino)-4-(ethylamino)-6-methoxy-*s*-triazine (8CI); Reg. No. *[26259-45-0]*. Its herbicidal properties were described by A. Gast & E. Fankhauser (*Proc. Br. Weed Control Conf., 8th*, 1966, p. 485). Introduced by J. R. Geigy S.A. (now Ciba-Geigy AG) (CHP 337 019; GBP 814 948) as code no. 'GS 14 254'; trade marks 'Etazine', formerly '*Sumitol*' in USA.

**Properties.**

Pure secbumeton forms colourless crystals; m.p. 86-88 °C; v.p. 971 µPa at 20 °C; ρ 1.105 g/cm³ at 20 °C. Solubility at 20 °C: 600 mg/l water; 400 g/l acetone; 600 g/l dichloromethane; 22 g/l hexane; 500 g/l methanol; 200 g/l octan-1-ol; 350 g/l toluene. On hydrolysis at 20 °C 50% loss (calculated) occurs in: 30 d at pH1, 175 d at pH13. Acidic, $pK_a$ 4.4.

**Uses.**

Secbumeton is a herbicide taken up by leaves and roots and controls mono- and dicotyledonous weeds both annual and perennial. It is used in established lucerne either at 1-3 kg a.i./ha or in combination with simazine. In combination with terbuthylazine it is suitable for non-cropped land.

**Toxicology.**

Acute oral $LD_{50}$ for rats 2680 mg tech./kg. In 90-d feeding trials NEL was: for rats 640 mg/kg diet (*c.* 43 mg/kg daily); for dogs 1600 mg/kg diet (40 mg/kg daily). $LC_{50}$ (96-h) is: for rainbow trout 18 mg/l; for crucian carp 43 mg/l; for bluegill 30 mg/l. It is practically non-toxic to birds and honeybees.

**Formulations.**

'Etazine' 50 WP, w.p. (500 g a.i./kg). 'Etazine 3585' WP (Reg. No. *[61952-98-5]*), w.p. secbumeton + simazine (7:3); 'Etazine 3947' 500 FW, s.c. (500 g/l, similar composition). 'Primatol 3588', secbumeton + terbuthylazine (1:1).

**Analysis.**

Product analysis is by glc or by acidimetric titration - details are available from Ciba-Geigy AG. Residues may be determined by glc with DMC or FPD (K. Ramsteiner *et al.*, *J. Assoc. Off. Anal. Chem.*, 1974, **57**, 192; E. Knüsli, *Anal. Methods Pestic., Plant Growth Regul. Food Addit.*, 1964, **4**, 13; T. H. Byast *et al., Tech. Rep. ARC Weed Res. Organ.*, No. 15 (2nd Ed.), p. 40).

$C_{17}H_{29}NO_3S$ (327.5)
L6V BUTJ B1UN3&O2 CQ E1Y1&S2
&&(±) FORM

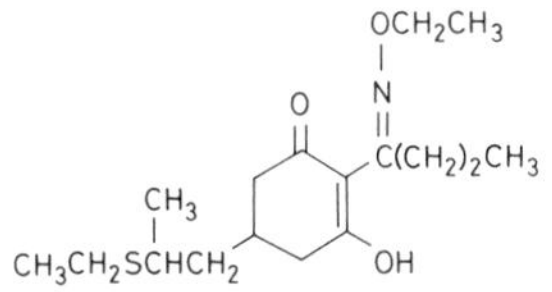

L6V CVTJ B1UN3&O2 E1Y1&S2
&&(±) FORM

**Nomenclature and development.**

Common name sethoxydim (BSI, draft E-ISO). Chemical name (IUPAC) (±)-2-(1-ethoxyiminobutyl)-5-[2-(ethylthio)propyl]-3-hydroxycyclohex-2-enone. (*C.A.*) (±)-2-[1-(ethoxyimino)butyl]-5-[2-(ethylthio)propyl]-3-hydroxy-2-cyclohexen-1-one (I) (9CI); first publication on the compound presented the structure as a tautomer, 2-[(1-ethoxyamino)butylidene]-5-[2-(ethylthio)propyl]-1,3-cyclohexanedione (II) (9CI); Reg. No. *[74051-80-2]* (I), *[71441-80-0]* (II). Introduced by Nippon Soda Co., Ltd. (JPA 112 945-77) as code no. 'NP-55', 'BAS 90 520 H' (BASF AG), 'SN 81 742'; trade marks 'Nabu', 'Poast', 'Fervinal'.

**Properties.**

Sethoxydim is a mobile liquid; b.p. >90 °C/3 × $10^{-5}$ mmHg; $d^{25}$ 1.043 at 25 °C. Solubility at 20 °C: 25-4700 mg/l water (pH 4-7); miscible with organic solvents.

**Uses.**

It is a post-em. herbicide that acts predominantly on grass weeds by absorption through their foliage and roots, and has no herbicidal effect on broad-leaved weeds. It is recommended for use on broad-leaved crops, e.g. cotton, oilseed rape, soyabeans, sugar beet, sunflower.

**Toxicology.**

Acute oral $LD_{50}$ for rats 3200 mg/kg. Acute percutaneous $LD_{50}$ for rats >5000 mg/kg.

**Formulations.**

'Poast', e.c. (184 g a.i./l); 'Nabu', 'Fervinal', e.c. (200 g/kg).

**Analysis.**

Product analysis is by hplc. Residues may be determined by hplc.

$C_{14}H_{20}N_2O$ (232.3)
L6TJ AMVMR& B1

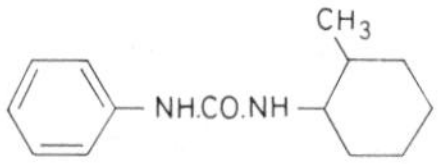

Nomenclature and development.

Common name siduron (BSI, E-ISO, F-ISO, ANSI, WSSA, JMAF);*exception* (Austria). Chemical name (IUPAC) 1-(2-methylcyclohexyl)-3-phenylurea (I). (*C.A.*) *N*-(2-methylcyclohexyl)-*N*'-phenylurea (9CI); (I) (8CI); Reg. No. *[1982-49-6]*. Its herbicidal properties were described by R. W. Varner *et al.* (*Proc. Br. Weed Control Conf., 7th*, 1964, p. 38). Introduced by E. I. du Pont de Nemours & Co. (Inc.) (USP 3 309 192) as code no. 'Du Pont 1318'; trade mark 'Tupersan'.

**Properties.**

Pure siduron forms colourless crystals; m.p. 133-138 °C; *d* 1.08. Solubility at 25 °C: 18 mg/l water; ⩾100 g/kg dichloromethane, dimethylacetamide, dimethylformamide, 3,5,5-trimethylcyclohex-2-enone. It is stable up to its m.p. and in water at neutral pH, but is slowly decomposed in acid and alkaline media.

**Uses.**

It is a selective herbicide used pre-em. against *Digitaria* spp. and annual weed grasses, but tolerated by cereals, turf species and many broad-leaved crop plants. It is recommended for the treatment of grass at 2-6 kg a.i./ha for new seedlings, 8-12 kg/ha for established turf. It resists leaching in soil, 50% loss in 120-150 d mainly by microbial degradation.

**Toxicology.**

Acute oral $LD_{50}$ for rats >7500 mg/kg. In percutaneous tests on rabbits 5500 mg/kg (max. feasible dose) applied to their intact or abraded skin caused no sign of toxicity. In 2-y feeding trials NEL was: for rats 500 mg/kg diet; for dogs 2500 mg/kg diet. $LC_{50}$ (8-d) for mallard duckling and bobwhite quail was >10 000 mg/kg diet. $LC_{50}$ (48-h) is: for carp 18 mg/l; for Japanese goldfish 10-40 mg/l.

**Formulations.**

'Tupersan', w.p. (500 g a.i./kg).

**Analysis.**

Product analysis is by hplc. Residues may be determined by colorimetry (R. L. Dalton & H. L. Pease, *J. Assoc. Off. Agric. Chem.*, 1962, **45**, 377).

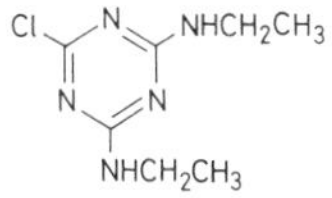

$C_7H_{12}ClN_5$ (201.7)
T6N CN ENJ BM2 DM2 FG

**Nomenclature and development.**

Common name simazine (BSI, E-ISO, F-ISO, ANSI, WSSA); CAT (JMAF); *exception* (Turkey). Chemical name (IUPAC) 6-chloro-*N*,*N*'-diethyl-1,3,5-triazine-2,4-diyldiamine; 2-chloro-4,6-bis(ethylamino)-1,3,5-triazine. (*C.A.*) 2-chloro-*N*,*N*'-diethyl-1,3,5-triazine-4,6-diamine (9CI); 2-chloro-4,6-bis(ethylamino)-*s*-triazine (8CI); Reg. No. *[122-34-9]*. Its herbicidal properties were reported by A. Gast *et al.* (*Experientia*, 1956, **12**, 146). Introduced by J. R. Geigy S.A. (now Ciba-Geigy AG) (BEP 540 590; GBP 894 947) as code no. 'G 27 692'; trade marks 'Gesatop', 'Weedex', in USA 'Aquazine'.

**Properties.**

Pure simazine is a colourless powder; m.p. 225-227 °C (decomp.); v.p. 810 nPa at 20 °C; $\rho$ 1.302 g/cm$^3$ at 20 °C. Solubility at 20 °C: 5 mg/l water; 900 mg/l chloroform; 300 mg/l diethyl ether; 2 mg/l light petroleum; 400 mg/l methanol. Slow hydrolysis occurs at 70 °C in neutral media, the rate increases with either increasing or decreasing pH. At 21 °C p*K* is 1.7.

**Uses.**

It is the pre-em. herbicide recommended for the control of broad-leaved and grass weeds in deep-rooted crops such as artichokes, asparagus, berry crops, broad beans, citrus, cocoa, coffee, forestry, hevea, hops, oil palms, olives, orchards (pome and stone fruits), ornamentals, sisal, sugarcane, tea, tree nurseries, turf, vineyards and non-crop areas. A major use is on maize which can convert simazine to the herbicidally-inactive 6-hydroxy analogue. In the USA it is also used to control vegetation and algae in farm ponds, fish hatcheries, etc.

**Toxicology.**

Acute oral $LD_{50}$ for rats >5000 mg tech./kg. Acute percutaneous $LD_{50}$ for rats >3100 mg/kg; non-irritant to skin and eyes of rabbits. In 2-y feeding trials NEL was: for rats 100 mg/kg diet (7 mg/kg daily); for dogs 150 mg/kg diet (5 mg/kg daily). $LC_{50}$ (96-h) is: for rainbow trout and crucian carp >100 mg/l; for bluegill 90 mg/l. It is practically non-toxic to birds and honeybees.

**Formulations.**

'Gesatop' as: 50 WP and 80 WP, w.p. (500 or 800 g a.i./kg); 4, 8 and 10G, granules (40, 80 and 100 g/kg); 90WDG, water-dispersible granule (900 g/kg); 4L, liquid (480 g/l); 500FW, s.c. (500 g/l). 'Princep' 4L, liquid (480 g/l). 'Weedex' S2G, granules. Other brand names include: 'Simadex' (FBC Limited), 'Simflow' (Diamond Shamrock), 'Herbazin' (Fisons Horticulture). 'Etazine 3583' WP, (Reg. No. *[61952-98-5]*), w.p. simazine + secbumeton (3:7); 'Etazine 3947' 500 FW, s.c. (500 g/l, similar composition). 'Gesaprim S', (Reg. No. *[39331-45-8]*), simazine + atrazine (1:1). 'Gesatop Z', simazine + ametryne (1:1). 'Primatol', simazine + prometon + sodium chlorate + sodium metaborate. 'Primatol SE' 500 L, simazine + aminotriazole. 'Midox Forte', simazine + aminotriazole + MCPA. 'Herbon Blue' (Cropsafe), simazine + 2,4-DES. 'Fylene Flowable' (FBC Limited), simazine + metoxuron. 'Aventox' (Murphy) & 'Rental' (FBC Limited), simazine + trietazine.

**Analysis.**

Product analysis is by glc (*CIPAC Handbook*, 1980, **1A**, 1343; B. G. Tweedy & R. A. Kahrs, *Anal. Methods Pestic. Plant Growth Regul.*, 1979, **10**, 493; *AOAC Methods*, 1980, 6.431-6.435) or by titration of liberated chloride (H. P. Bosshardt *et al.*, *J. Assoc. Off. Anal. Chem.*, 1971, **54**, 749). Residues may be determined by glc with DMC or FPD, (K. Ramsteiner *et al.*, *ibid.*, 1974, **57**, 192; B. G. Tweedy & R. A. Kahrs, *loc. cit.*; T. H. Byast *et al.*, *Tech. Rep. ARC Weed Res. Organ.*, No. 15 (2nd Ed.) p. 79) or by hplc (*idem, ibid.*, p. 40).

$C_8H_{15}N_5S$ (213.3)
T6N CN ENJ BS1 DM2 FM2

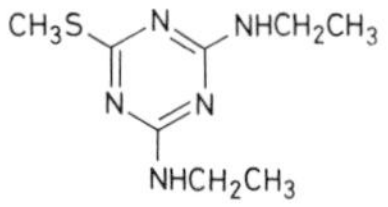

$CH_3S$–triazine–$NHCH_2CH_3$; $NHCH_2CH_3$

**Nomenclature and development.**

Common name simetryne (BSI, F-ISO, Canada, JMAF);simetryn (E-ISO, WSSA). Chemical name (IUPAC) *N*,*N*'-diethyl-6-methylthio-1,3,5-triazine-2,4-diyldiamine; 2,4-bis(ethylamino)-6-methylthio-1,3,5-triazine; 2,4-di(ethylamino)-6-methylthio-1,3,5-triazine. (*C.A.*) *N*,*N*'-diethyl-6-(methylthio)-1,3,5-triazine-2,4-diamine (9CI); 2,4-bis(ethylamino)-6-(methylthio)-*s*-triazine (8CI); Reg. No. *[1014-70-6]*. First reported as a herbicide by J. R. Geigy S.A. (now Ciba-Geigy AG). Introduced in Japan by Hokko Chemical Industry Co. Ltd, Nihon Nohyaku Co. Ltd, Nippon Kayaku Co., Sankyo Co. Ltd (CHP 337 019 to Ciba-Geigy AG) as code no. 'G 32 911' (to Ciba-Geigy AG); trade mark 'Gy-bøn' (to Japanese companies listed above).

**Properties.**

Simetryne forms crystals; m.p. 82-83 °C. Solubility at room temperature 450 mg/l water.

**Uses.**

Simetryne is used as a mixture with thiobencarb to control broad-leaved weeds in rice.

**Toxicology.**

Acute oral $LD_{50}$ for rats 1830 mg/kg.

**Formulations.**

These include: 'Saturn' (simetryne + thiobencarb).

**Analysis.**

Product analysis is by titration or by glc (*CIPAC Handbook*, 1983, **1B**, in press).

$ClNaO_3$ (106.4)
.NA..G-O3

$NaClO_3$

**Nomenclature and development.**

Chemical name (IUPAC) sodium chlorate (I) (E-ISO); chlorate de sodium (F-ISO); is used in lieu of a common name. (*C.A.*) (I) (8 & 9CI); Reg. No. *[7775-09-9]*. It has been in use as a herbicide since *c*. 1910. Trade marks include 'Altacide'.

**Properties.**

It is a colourless powder; m.p. 248 °C, decomposing with evolution of oxygen *c*. 300 °C. Solubility at 0 °C: 790 g/l water; soluble in ethanol, glycerol. It is a potent oxidising agent reacting with organic materials so creating a serious fire hazard for example with splashed clothing. Corrosive to zinc and to mild steel. Marketed formulations contain a fire depressant.

**Uses.**

It is a non-selective herbicide used against established vegetation at 200-600 kg/ha and against annual weeds at 100-200 kg/ha. At >300 kg/ha it gives control for 0.5 y but is leached by high rainfall.

**Toxicology.**

Acute oral $LD_{50}$ for rats 1200 mg/kg (E. F. Edson, *Pharm. J*., 1960, **185**, 361). It can cause local irritation to skin and mucous membranes.

**Formulations.**

Mixtures with fire depressants to reduce fire hazard. Combinations with other herbicides include: sodium chlorate + atrazine; sodium chlorate + atrazine + 2,4-D; sodium chlorate + bromacil + sodium metaborate; sodium chlorate + sodium metaborate.

**Analysis.**

Product analysis is by titration (*CIPAC Handbook*, 1970, **1**, 626; *Herbicides 1977*, p. 41). Residues, in soil may be determined by colorimetry of a derivative (T. H. Byast, *et al.*, *Tech. Rep. ARC Weed Res. Organ*., No. 15 (2nd Ed.), p. 38).

FNa (41.99)
.NA..F

NaF

**Nomenclature and development.**

Chemical name (IUPAC) sodium fluoride (E-ISO); fluorure de sodium (F-ISO); is accepted in lieu of a common name. (*C.A.*) sodium fluoride (8 & 9CI); Reg. No. *[7681-49-4]*. Introduced in baits against insects in stores.

**Properties.**

It is a colourless powder; m.p. 993 °C; *d* 2.8; v.p. negligible at room temperature. Solubility at 18 °C: 42.2 g/l water; slightly soluble in ethanol.

**Uses.**

It is a potent stomach-acting insecticide with some contact activity. It is highly phytotoxic, and is used in insect baits and as a timber preservative.

**Toxicology.**

Acute oral $LD_{50}$: for rats 180 mg/kg (*Am. Ind. Hyg. Assoc. J.*, 1969, **30**, 470); Highly toxic, the lowest oral lethal dose being 28-100 mg/kg for a range of vertebrates.

**Formulations.**

Commercial grade (purity 93-99%) is used to prepare baits.

**Analysis.**

Product analysis is by determination of the fluorine content by titrimetric methods (*AOAC Methods*, 1980, 25.049-25.060).

$C_2H_2FNaO_2$ (100.0)
QV1F &&Na SALT

$FCH_2COONa$

**Nomenclature and development.**

Chemical names (IUPAC) sodium fluoroacetate (I) (E-ISO); fluoroacétate de sodium (F-ISO); fluoroacetic acid (BSI) for parent acid are used in lieu of common name. (*C.A.*) (I) (8 & 9CI); Reg. No. *[62-74-8]*. Trivial name Compound 1080. Its rodenticidal properties were described by E. R. Kalmbeck (*Science*, 1945, **102**, 232).

**Properties.**

Sodium fluoroacetate is a colourless non-volatile powder, decomposing *c.* 200 °C. It is hygroscopic. Solubility: very soluble in water; poorly soluble in acetone, ethanol, petroleum oils.

**Uses.**

It is a potent mammalian poison used in baits as a rodenticide. Its use, in many countries is restricted to trained personnel. It is a systemic insecticide (W. A. L. David, *Nature (London)*, 1950, **165**, 493) but is considered too toxic to mammals to be used for this purpose. Its toxicity is due to its interruption of the tricarboxylic acid cycle.

**Toxicology.**

Acute oral $LD_{50}$ for *Rattus norvegicus* 0.22 mg/kg (S. H. Dieke & C. P. Richter, *U.S. Public Health Rep.*, 1946, **61**, 672).

**Formulations.**

An aqueous solution for bait preparation.

**Analysis.**

Product analysis is by reaction with sodium to form sodium fluoride and precipitation as lead chloride fluoride (*AOAC Methods*, 1980, 29.142-29.148); sodium fluoride is a usual contaminant (*WHO Manual* (2nd Ed.), SRT/5).

$C_{21}H_{22}N_2O_2$ (334.4)
T6 G656 B7 C6 E5 D 5ABCEF A& FX
MNV QO VN SU AHT&&TTTTJ

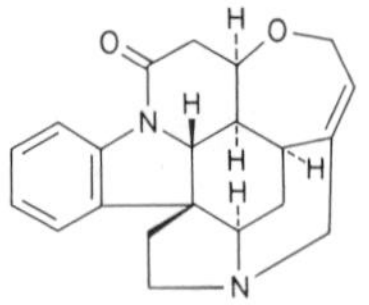

**Nomenclature and development.**

The traditional name strychnine (I) (BSI, E-ISO, F-ISO) is accepted in lieu of a common name. Chemical name (*C.A.*) strychnidin-10-one (9CI); (I) (8CI); Reg. No. *[57-24-9]*. The physiological properties of the alkaloids in extract of *Strychnos nux-vomica* seeds have long been known, strychnine being the most important of these.

**Properties.**

Produced by extracting seeds of *Strychnos* or *Loganiaceae* spp., strychnine forms a colourless crystalline powder; m.p. 270-280 °C (decomp.); $[\alpha]_D^{18}$ −139 ° (chloroform). Solubility: 143 mg/l water; 5.6 g/l benzene; 200 g/l chloroform; 6.7 g/l ethanol; sparingly soluble diethyl ether, petroleum spirit.

It is a base, forming water-soluble salts with acids: the hydrochloride, colourless prisms with 1.5-2.0 mol of water of crystallisation which are lost at 110 °C; the sulphate, colourless crystals (as pentahydrate) becoming anhydrous at 110 °C, m.p. >199 °C, solubility at 15 °C: 30 g/l water; soluble in ethanol.

**Uses.**

It is a potent mammalian poison, sometimes used to control moles.

**Toxicology.**

Lethal dose: for rats 1-30 mg/kg; for man 30-60 mg/kg.

**Formulations.**

Baits are usually prepared from strychnine sulphate (5-10 g/kg)

**Analysis.**

Product analysis is by acid-base titration (*AOAC Methods*, 1980, 38.002).

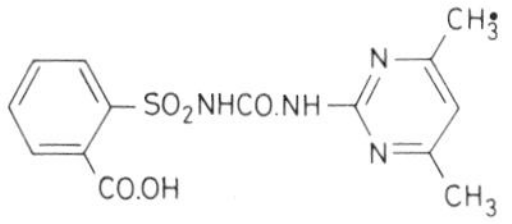

$C_{14}H_{14}N_4O_5S$ (350.3)
T6N CNJ BMVMSWR BVQ& D1 F1
*sulfometuron-methyl*
$C_{15}H_{16}N_4O_5S$ (364.4)
T6N CNJ BMVMSWR BVO1& D1 F1

SO₂NHCO.NH CO.OH N N CH₃ CH₃

**Nomenclature and development.**

Common name sulfometuron (BSI, ANSI, draft E-ISO). Chemical name (IUPAC) 2-[3-(4,6-dimethylpyrimidin-2-yl)ureidosulphonyl]benzoic acid. (*C.A.*) 2-[[[[(4,6-dimethyl-2-pyrimidinyl)amino]carbonyl]amino]sulfonyl]benzoic acid (9CI); Reg. No. *[74222-97-2]* (methyl ester). The methyl ester, sulfometuron-methyl, was introduced by E. I. du Pont de Nemours & Co. (Inc.) as code no. 'DPX 5648'; trade mark 'Oust'.

**Properties.**

Technical grade sulfometuron-methyl is a colourless solid; m.p. 203-205 °C; v.p. 8 mPa at 25 °C; *d* 1.48. Solubility at 25 °C: 10 mg/l water at pH5, 300 mg/l at pH7; 2.4 g/kg acetone; 1.5 g/kg acetonitrile; 32 mg/kg diethyl ether; 137 mg/kg ethanol; 37 mg/kg xylene. Aqueous suspensions are stable to hydrolysis at pH 7-9, 50% loss occurs in *c.* 14 d at pH5. It is a weak acid $pK_a$ 5.3. Loss from soil is by hydrolysis and by microbial action; 50% loss takes *c.* 28 d.

**Uses.**

It is a broad range pre-em. and post-em. herbicide controlling annual and perennial grasses and broad-leaved weeds, particularly effective against *Sorghum halepense*. It suppresses plant growth by arresting cell division in the growing tips of roots and plants. It is rapidly taken up by the foliage and roots of sensitive herbaceous species, being translocated throughout the plant.

**Toxicology.**

Acute oral $LD_{50}$ for rats >5000 mg/kg. Mild skin irritation, but no sensitisation occurs with guinea-pigs. Temporary mild eye irritation occurs with rabbits.

**Formulations.**

A water dispersible granule (750 g sulfometuron-methyl/kg).

$C_8H_{20}O_5P_2S_2$ (322.3)
2OPS&O2&OPS&O2&O2

$(CH_3CH_2O)_2P(=S)-O-P(=S)(OCH_2CH_3)_2$

**Nomenclature and development.**

Common name sulfotep (BSI, E-ISO, F-ISO);sulfotepp (ESA). Chemical name (IUPAC) *O,O,O',O'*-tetraethyl dithiopyrophosphate. (*C.A.*) tetraethyl thiodiphosphate (9CI); tetraethyl thiopyrophosphate (8CI); Reg. No. *[3689-24-5]*. Trivial names dithio, dithione, thiotep. ENT 16 273. Its insecticidal properties were discovered in 1944 by G. Schrader & H. Kükenthal (cited by G. Schrader, *Die Entwicklung insektizider Phosphorsaüre-Ester*, 3rd Ed.). Introduced by Bayer AG (DEP 848 812) as code no. 'Bayer E 393', 'ASP-47' (Victor Chemical Works); trade mark 'Bladafum' (Bayer AG).

**Properties.**

Pure sulfotep is a pale yellow mobile liquid; b.p. 136-139 °C/2 mmHg; v.p. 22.6 mPa at 20 °C; $d_4^{25}$ 1.196; $n_D^{25}$ 1.4753. Solubility at room temperature: 25 mg/l water; completely miscible with chloromethane and most organic solvents. The technical grade is a dark-coloured liquid, b.p. 131-135 °C/2 mmHg; $n_D^{25}$ 1.4725.

**Uses.**

It is a non-systemic insecticide effective against a wide range of pests but of brief persistence on foliage.

**Toxicology.**

Acute oral $LD_{50}$ for rats *c.* 5 mg a.i./kg. Acute percutaneous $LD_{50}$ (4-h) for rats 262 mg/kg. In 90-d feeding trials NEL was: for rats 10 mg/kg diet; for dogs 0.5 mg/kg diet. Inhalation $LC_{50}$ (1-h) for male rats 330 mg/m$^3$, for females 160 mg/m$^3$.

**Formulations.**

'Bladafum', for smoke generation.

**Analysis.**

Product analysis is by i.r. spectroscopy (details from Bayer AG). Residues may be determined by glc (G. Dräger, *Pflanzenschutz-Nachr. (Engl. Ed.)*, 1968, **21**, 359; *Anal. Methods Pestic. Plant Growth Regul.*, 1972, **6**, 483).

S (32.06)
.S

S

**Nomenclature and development.**

The chemical name (IUPAC) sulphur (E-ISO); soufre (F-ISO); sulfur (ESA, JMAF) is accepted in lieu of a common name. (*C.A.*) sulfur (8 & 9CI); Reg. No. *[7704-34-9]*. It has been used as a pesticide for many years. Typical trade marks include 'Elosal', 'Kumulus S', 'Thiovit'.

**Properties.**

It is a yellow solid and can exist as various allotropic forms. Rhombic sulphur is stable at normal temperatures but forms other allotropes on heating to between 94-119 °C, the observed melting point depending on the rate of heating and extent of conversion. Rhombic sulphur has *d* 2.06; v.p. 527 μPa at 30.4 °C. Solubility: practically insoluble in water; the crystalline forms are soluble in carbon disulphide whereas the amorphous are not. It is incompatible with petroleum oils.

**Uses.**

Sulphur is a non-systemic contact and protectant acaricide and fungicide, normally applied as sprays or a dust. It is generally non-phytotoxic, except to certain varieties known as 'sulphur-shy'. It has also been vaporised in glasshouses to control powdery mildews, care being taken to avoid ignition with the formation of the highly phytotoxic sulphur dioxide.

**Toxicology.**

It is relatively non-toxic to mammals, but can cause irritation of skin and the mucous membranes.

**Formulations.**

Dusts (inert material is usually added during manufacture to prevent electrostatic 'balling'); w.p.; finely-ground 'colloidal' suspensions. 'Kumulan', w.p. (sulphur + nitrothal-isopropyl).

**Analysis.**

Product analysis is by conversion to sodium thiosulphate which is determined by titration (*CIPAC Handbook*, 1970, **1**, 632; *FAO Plant Prot. Bull.*, 1961, **9**(5), 80). Residues may be determined by colorimetry (H. A. Ory *et al.*, *Analyst (London)*, 1957, **82**, 189).

$F_2O_2S$ (102.1)
WSFF

$SO_2F_2$

**Nomenclature and development.**

Chemical name (IUPAC) sulphuryl fluoride (E-ISO) is accepted in lieu of a common name. (*C.A.*) Sulfuryl fluoride (8 & 9CI); Reg. No. *[2699-79-8]*. Its insecticidal properties were described by E. E. Kenaga (*J. Econ. Entomol.*, 1957, **40**, 1). Introduced by Dow Chemical Co.. (USP 2 875 127) trade mark 'Vikane'.

**Properties.**

Sulphuryl fluoride is a colourless gas; b.p. −55.2 °C; v.p. 1.7 MPa at 25 °C. Solubility: 40-50 ml gaseous sulphuryl fluoride/l water; 1.36-1.38 l/l carbon tetrachloride; 240-270 ml/l ethanol; 2.1-2.2 l/l toluene. Hydrolysed by aqueous alkali, but not by water. It is stable, non-corrosive and harmless to fabrics.

**Uses.**

It is an insecticide used for fumigating structures, vehicles and wood products to control drywood termites, wood-infesting beetles and certain other insects. It has poor ovicidal activity, is phytotoxic but with little effect on the germination of weed and crop seeds. Its toxic effect may be due to formation of fluoride ions (R. W. Meikle *et al., J. Agric. Food Chem.*, 1963, **11**, 226).

**Toxicology.**

The adopted time-weighted average exposure value for the commercial gas for humans is 20 mg/m$^3$ for 8 h/d, 5 d/week repeated exposure.

**Formulations.**

It is used as the 99% *m/m* technical product.

**Analysis.**

Atmospheres may be analysed by trapping in aqueous alkali and titration of the fluorosulphate produced, or by use of a thermal conductivity meter (S. G. Heuser, *Anal. Chem.*, 1963, **35**, 1476). Details of methods for total fluoride residues are available from Dow Chemical Co.

$C_{12}H_{19}O_2PS_3$ (322.4)
3SPS&O2&OR DS1

$CH_3S-C_6H_4-OP(=S)(OCH_2CH_3)SCH_2CH_2CH_3$

**Nomenclature and development.**

Common name sulprofos (BSI, E-ISO, F-ISO). Chemical name (IUPAC) *O*-ethyl *O*-4-methylthiophenyl *S*-propyl phosphorodithioate. (*C.A.*) *O*-ethyl *O*-[4-(methylthio)phenyl] *S*-propyl phosphorodithioate (9CI); Reg. No. *[35400-43-2]*. Its insecticidal activity was described by G. Zoebelein (*Proc. Conf. Pest Control, Cairo, 4th*, 1978, p. 456). Introduced by Bayer AG as code no. 'BAY NTN 9306'; trade marks 'Bolstar', 'Helothion'.

**Properties.**

It is a colourless oil; b.p. 210 °C; v.p. <100 μPa at 20 °C; $d_4^{20}$ 1.20. Solubility at 29 °C: <5 mg/kg water; 120 g/kg cyclohexanone; 400-600 g/kg propan-2-ol; >1.2 kg/kg toluene.

**Uses.**

Sulprofos is an insecticide with specific effect against Lepidoptera and is used on cotton.

**Toxicology.**

Acute oral $LD_{50}$ for male rats 304 mg/kg. Acute percutaneous $LD_{50}$ for rats >1200 mg/kg. In 2-y feeding trials NEL for rats was 6 mg/kg diet. $LC_{50}$ (96-h) is: for carp 5.2 mg/l; for rainbow trout 23 mg/l.

**Formulations.**

An e.c. (720 g a.i./l); ULV (720 g/l).

**Analysis.**

Product analysis is by glc, details of which and of methods for determination of residues are available from Bayer AG.

$C_8H_5Cl_3O_3$ (255.5)
QV1OR BG DG EG

**Nomenclature and development.**

Common name 2,4,5-T (BSI, E-ISO, F-ISO, WSSA, JMAF);JMAF also apply it to the corresponding 2-butoxyethyl ester. Chemical name (IUPAC) (2,4,5-trichlorophenoxy)acetic acid (I). (*C.A.*) (I) (8 & 9CI); Reg. No. *[93-76-5]*. Its herbicidal properties were reported by C. L. Hamner & H. B. Tukey (*Science*, 1944. **100**, 154). Introduced by Amchem Products Inc. (now Union Carbide Agricultural Products Co. Inc.) as trade mark 'Weedone'.

**Properties.**

It is produced by condensation of sodium chloroacetate with sodium 2,4,5-trichlorophenoxide. At high temperatures the action of alkali on 2,4,5-trichlorophenol can produce some 2,3,7,8-tetrachlorodibenzo-*p*-dioxin. Technical grade 2,4,5-T (94% pure) forms colourless crystals; m.p. 153-156 °C; $d_{20}^{20}$ 1.80; v.p. 700 nPa at 25 °C. Solubility at 25 °C: 150 mg/l water; >50 g/l diethyl ether, ethanol, methanol, toluene; 400 mg/l heptane. It is stable in aqueous solution at pH5-9.

Its salts with alkali metals and amines are water-soluble, precipitation occurs with hard water in the absence of sequestering agents. Its tris(2-hydroxyethyl)ammonium salt (Reg. No. *[3813-14-7]*) has m.p. 113-115 °C; the triethylammonium (Reg. No. *[57213-69-1]*) and other salts are soluble in water but insoluble in oils. Esters of 2,4,5-T are insoluble in water but soluble in oils. Examples include: 2-butoxyethyl ester (Reg. No. *[2545-59-7]*); butyl ester (Reg. No. *[93-79-8]*) (tech. grade solidifies at 20 °C); iso-octyl ester (Reg. No. *[25168-15-4]*); (2-butoxy-1-methylethyl) ester (Reg. No. *[7173-98-0]*).

**Uses.**

2,4,5-T is used post-em. alone or with 2,4-D, for the control of shrubs and trees. It is applied as a foliage, dormant shoot or basal bark spray. It is also used for girdling, injection or cut-stump treatment. It is absorbed through roots, foliage and bark. Esters of low volatility are used for ULV application.

**Toxicology.**

Acute oral $LD_{50}$: for rats 300-1700 mg/kg (depending on vehicle and rat strain); 389-1380 mg/kg for mice. Acute percutaneous $LD_{50}$ for rats >5000 mg/kg. In 2-y feeding trials no effect was observed in rats receiving 30 mg/kg diet; nor in 90-d trials in beagle dogs at 60 mg/kg diet. $LC_{50}$ (8-d) was: for bobwhite quail 2776 mg/kg diet; for ducks >4650 mg/kg diet. $LC_{50}$ (96-h) is: for rainbow trout 350 mg/l; for carp 355 mg/l. A contaminant, 2,3,7,8-tetrachlorodibenzo-*p*-dioxin, produced foetal deaths in hamsters at 0.0091 mg/kg and it causes serious acne in man. Modern methods of manufacture of 2,4,5-T limit the amount of the contaminant to <0.01 mg/kg 2,4,5-T.

**Formulations.**

Typical formulations include: e.c. (500 g a.e./l) as ester; e.c. (480 g a.e./l) as 2-butoxyethyl ester; water-soluble concentrate (480 g a.e./l) of amine salts. Mixtures of similar derivatives of 2,4,5-T + 2,4-D.

**Analysis.**

Product analysis is by glc of a suitable ester (*CIPAC Handbook*, 1970, **1**, 642; 1980, **1A**, 1347; FAO Specification (CP/45), *Herbicides 1977*, p.32), or by hplc (*J. Assoc. Off. Anal. Chem.*, 1980, **63**, 379). Residues may be determined by glc of a suitable ester (T. H. Byast *et al., Tech. Rep. ARC Weed Res. Organ.*, No. 15 (2nd Ed.), p. 31; W. Specht & M. Tillkes, *Fresenius' Z. Anal. Chem.*, 1980, **301**, 300; 1981, **307**, 257; E. L. Bjerke *et al., J. Agric. Food Chem,*, 1972, **20**, 963; D. E. Clark *et al., ibid.*, 1975, **23**, 573; *Anal. Methods Pestic. Plant Growth Regul.*, 1972, **6**, 702; H. Løkke & P. Odgaard, *Pestic. Sci.*, 1981, **12**, 375; H. E. Munro, *ibid.*, 1977, **8**, 157). Estimation of 2,3,7,8-tetrachlorodibenzo-*p*-dioxin content has been studied (J. W. Edmunds *et al., ibid.*, 1973, **4**, 101; P. G. Baker *et al., ibid.*, 1981, **12**, 297).

**Nomenclature and development.**

Tar oils are produced by the distillation of tars resulting from the high-temperature carbonisation of coal and of coke oven and blast furnace tars. They consist mainly of aromatic hydrocarbons, but contain components soluble in aqueous alkali: the 'phenols' or 'tar acids'; and nitrogenous bases soluble in dilute mineral acids: 'tar bases'. Although they have been used for wood preservation since 1890, the introduction of the formulated products known as tar oil washes for crop protection, using the heavy creosote and anthracene oil ranges, dates from *c*. 1920.

**Properties.**

The oils are brown to black liquids distilling from 230 °C to the temperature at which the residue is pitch, $d^{15}$ 1.05-1.11. Solubility: insoluble in water; soluble in organic solvents and in dimethyl sulphate.

**Uses.**

Tar oils are toxic to the eggs of many insect species, particularly of aphids and psyllids. They are highly phytotoxic and their use on fruit trees is limited to the dormant season. They also kill moss and lichen on the tree trunks.

**Toxicology.**

Liable to cause dermatitis to operators, especially in sunlight.

**Formulations.**

An e.c. (solution of surfactants in the oil with higher phenols and petroleum oils as mutual solvents); stock emulsions, concentrated emulsions using aqueous solutions of a suitable emulsifier.

**Analysis.**

Product analysis is by standard methods (*MAFF Ref. Book*, 1958, No. 1, p. 81).

$C_7H_3Cl_3O_2$ (225.5)
QVR BG CG FG

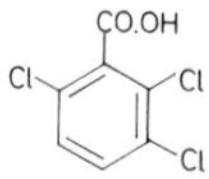

**Nomenclature and development.**

Common name 2,3,6-TBA (BSI, E-ISO, F-ISO, WSSA); acide trichlorobenzoique (optional alternative in France); TCBA (JMAF). Chemical name (IUPAC) 2,3,6-trichlorobenzoic acid (I). (*C.A.*) (I) (8 & 9CI); Reg. No. *[50-31-7]*. Its herbicidal properties were described by H. J. Miller (*Weeds*, 1952, **1**, 185). Introduced by the Heyden Chemical Corp. and later by E. I. du Pont de Nemours & Co. (Inc.) (neither now manufacture or market it) and other firms and is currently available in mixtures with other herbicides (USP 2 848 470; 3 081 162) as code no. 'HC-1281' (Heyden); trade marks '*Trysben*' (du Pont); 'Cambilene' (FBC Limited) for mixture with dicamba, MCPA and mecoprop.

**Properties.**

Produced by the chlorination of 2-chlorotoluene, giving a mixture of 2,3,6- and 2,4,5-trichlorotoluene, which is oxidised to the corresponding benzoic acids (*c.* 60% 2,3,6-TBA). The technical grade is a colourless to buff crystalline powder; m.p. 87-99 °C; v.p. 3.2 Pa at 100 °C. Solubility at 22 °C: 7.7 g/l water; readily soluble in most organic solvents. Pure 2,3,6-TBA has m.p. 125-126 °C. It forms water-soluble alkali metal and amine salts (at 25 °C 440 g 2,3,6-TBA-sodium/kg water). 2,3,6-TBA is stable to light and ⩽60 °C. It is compatible with the phenoxyalkanoic acid herbicides.

**Uses.**

2,3,6-TBA is used post-em., in combinations with other growth regulator herbicides, in cereals and grass seed crops to control broad-leaved annual and perennial weeds including black bindweed, cleavers, knotgrass, mayweeds and redshank.

**Toxicology.**

Acute oral $LD_{50}$: for rats 1500 mg/kg; for guinea-pigs and hens >1500 mg/kg; for rabbits 600 mg/kg. Acute percutaneous $LD_{50}$ for rats >1000 mg/kg. In 64-69-d feeding trials rats receiving 10 000 mg/kg diet suffered a minor disturbance of water metabolism, no trace of which was apparent at 1000 mg/kg diet. It is largely excreted unchanged in the urine.

**Formulations.**

Mixtures include: 'Cambilene', (25 g 2,3,6-TBA + 19 g dicamba + 100 g MCPA + 150 g mecoprop/l) solution of mixed potassium/sodium salts; 2,3,6-TBA + dichlorprop + mecoprop (solution of mixed potassium/sodium salts); 2,3,6-TBA + MCPA + mecoprop (solution of mixed potassium/sodium salts).

**Analysis.**

Product analysis is by glc *(CIPAC Handbook*, in press). Residues may be determined by glc of a suitable derivative (J. J. Kirkland & H. L. Pease, *J. Agric. Food Chem.*, 1964, **12**, 468).

$C_2Cl_3NaO_2$ (185.4)
QVXGGG &&Na SALT
*trichloroacetic acid*
$C_2HCl_3O_2$ (163.4)
QVXGGG

$Cl_3C.CO.O^- \ Na^+$

**Nomenclature and development.**

Common names TCA (BSI, E-ISI, F-ISO with exceptions outlined below) for sodium trichloroacetate;TCA applies to trichloroacetic acid (Australia, Canada, New Zealand, USA, JMAF) the sodium salt being indicated as usual, TCA-sodium;trichloroacétate de sodium (France) as option to common name. Chemical name (IUPAC) trichloroacetic acid (I), sodium trichloroacetate (II) for the appropriate salt. (*C.A.*) (I) (8 & 9CI) for acid; (II) (8 & 9CI) for sodium salt; Reg. No. *[76-03-9]* (acid), *[650-51-1]* (sodium salt). The herbicidal properties of the sodium salt were described by K. C. Barrons & A. J. Watson (*North Cent. Weed Control Conf., Res. Rep.*, 1947, pp. 43, 284). Introduced by E. I. du Pont de Nemours & Co. (Inc.) and by Dow Chemical Co. (neither of whom now manufacture or market it) and currently available from several suppliers.

**Properties.**

Trichloroacetic acid forms colourless hygroscopic crystals; m.p. 55-58 °C; b.p. 196-197 °C; $d_4^{25}$ 1.62. Solubility at 25 °C: 10 kg/l water; soluble in diethyl ether, ethanol. It tends to decompose to chloroform under alkaline conditions but is stable in the absence of moisture. It is corrosive to aluminium, iron and zinc.

Technical sodium trichloroacetate is a yellowish deliquescent powder. Solubility at room temperature 1.2 kg/l water; soluble in ethanol and many organic solvents. It is less corrosive than the acid.

**Uses.**

TCA is a pre-em. herbicide acting via the soil. It is used against: couch grass (at 15-30 kg a.e./ha); wild oats (at 7.8 kg/ha) prior to planting kale, peas or sugar beet. Persistence in soil is from 14 to *c.* 90 d, depending on soil, moisture and temperature.

**Toxicology.**

Acute oral $LD_{50}$: for rats 3200-5000 mg sodium salt/kg, 400 mg acid/kg; for mice 5640 mg sodium salt/kg. The acid is extremely corrosive to skin; its sodium salt is a skin and eye irritant.

**Formulations.**

Water-soluble solid (790-870 g a.e./kg) as sodium salt. Mixtures include: TCA + aminotriazole + 2,4-D + MCPA; TCA + aminotriazole + 2,4-D + sodium chlorate; TCA + atrazine.

**Analysis**

Product analysis is by decarboxylation with sulphuric acid followed by titration of the excess of acid (*CIPAC Handbook*, 1970, **1**, 697). Residues may be determined by glc of a derivative with ECD (T. H. Byast *et al.*, *Tech. Rep. ARC Weed Res. Organ.*, No. 15 (2nd Ed.), p.29).

$C_{15}H_{23}NO$ (233.4)
1Y1&N1R&VX1&1&1

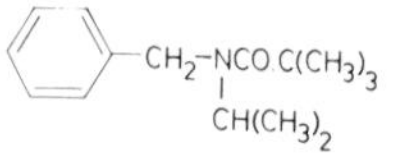

$C_6H_5-CH_2-N(CH(CH_3)_2)CO.C(CH_3)_3$

**Nomenclature and development.**

Common name tebutam (BSI, draft E-ISO);butam (ANSI, WSSA). Chemical name (IUPAC) *N*-benzyl-*N*-isopropylpivalamide. (*C.A.*) 2,2-dimethyl-*N*-(1-methylethyl)-*N*-(phenylmethyl)propanamide (9CI); Reg. No. *[35256-85-0]*. Its herbicidal properties were described by R. A. Schwartzbeck (*Proc. Br. Crop Prot. Conf. - Weeds*, 1976, **2**, 739). Introduced as an experiment by Gulf Oil Chemicals Co. and by Midox Ltd (USP 3 974 218; 3 707 366) as code no. 'GCP-5544' (Gulf); trade mark 'Comodor' (Midox).

**Properties.**

Tebutam is a colourless oil; b.p. 86-87 °C/0.07 mmHg; $n_D^{25}$ 1.5074. Solubility: almost insoluble in water; very soluble in benzene, ethanol, toluene.

**Uses.**

It is effective as a pre-em. herbicide controlling broad-leaved and annual grass weeds in winter oilseed rape. It is frequently used as a tank-mix with TCA.

**Toxicology.**

Acute oral $LD_{50}$: for albino rats 6210 mg a.i./kg; for guinea-pigs 2025 mg/kg. Acute percutaneous $LD_{50}$ for albino rabbits >2000 mg/kg.

**Formulations.**

An e.c., 'Comodor'.

# Tebuthiuron

$C_9H_{16}N_4OS$ (228.3)
T5NN DSJ CX1&1&1 EN1&VM1

$CH_3$
$(CH_3)_3C$ S N.CO.NHCH$_3$
N—N

**Nomenclature and development.**

Common name tebuthiuron (BSI, E-ISO, F-ISO, ANSI, WSSA). Chemical name (IUPAC) 1-(5-*tert*-butyl-1,3,4-thiadiazol-2-yl)-1,3-dimethylurea (I). (*C.A.*) *N*-[5-(1,1-dimethylethyl)-1,3,4-thiadiazol-2-yl]-*N*,*N*'-dimethylurea (9CI); (I) (8CI); Reg. No. *[34014-18-1]*. Its herbicidal properties were described by J. F. Schwer (*Proc. Br. Weed Control Conf., 12th*, 1974, **2**, 847). Introduced by Eli Lilly & Co. (GBP 1 266 172) as code no. 'EL-103'; trade marks 'Spike', 'Perflan'.

**Properties.**

Technical grade tebuthiuron is a colourless solid; m.p. 161.5-164 °C; v.p. 267 μPa at 25 °C. Solubility at 25 °C: 2.5 g/l water; 70 g/l acetone; 60 g/l acetonitrile; 6.1 g/l hexane; 170 g/l methanol; 60 g/l 2-methoxyethanol. It is stable to light.

**Uses.**

It is a broad spectrum herbicide for control of herbaceous and woody plants. Uses include: total vegetation control in non-crop areas; control of undesirable woody plants in pastures and rangeland; control of grass and broad-leaved weeds in sugarcane.

**Toxicology.**

Acute oral $LD_{50}$: for rats 644 mg/kg; for mice 579 mg/kg; for rabbits 286 mg/kg; for cats >200 mg/kg; for dogs, bobwhite quail, mallard duck and chickens >500 mg/kg. There was no irritation to rabbits by dermal application of 200 mg/kg or eye application of 71 mg/kg. In 2-y feeding trials rats and mice receiving 1600 mg/kg diet suffered no ill-effect. $LC_{50}$ (96-h) is: for goldfish and fathead minnow >160 mg/l; for trout 144 mg/l; for bluegill 112 mg/l.

**Formulations.**

These include: w.p., pellets and granules.

**Analysis.**

Product analysis is by hplc or by glc with FID (A. Loh *et al., Anal. Methods Pestic. Plant Growth Regul.*, 1980, **11**, 351). Residues may be determined by glc with FPD (*idem, ibid.*). Details may be obtained from Eli Lilly & Co.

$C_6HCl_4NO_2$ (260.9)
WNR BG CG EG FG

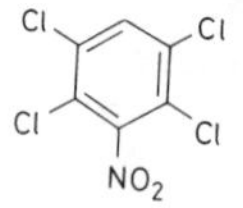

**Nomenclature and development.**

Common name tecnazene (BSI, E-ISO, F-ISO). Chemical name (IUPAC) 1,2,4,5-tetrachloro-3-nitrobenzene (I). (*C.A.*) (I) (8 & 9CI); Reg. No. *[117-18-0]*. Trivial name TCNB. Introduced as a fungicide *c.* 1946 by Bayer AG (who no longer manufacture or market it) (USP 2 615 801) as trade marks '*Folosan*' (Bayer), 'Fusarex' (ICI Plant Protection Division).

**Properties.**

Tecnazene forms colourless crystals; m.p. 99 °C; appreciably volatile at room temperature. Solubility: at 25 °C practically insoluble in water; *c.* 40 g/kg ethanol; readily soluble in benzene, carbon disulphide, chloroform.

**Uses.**

Tecnazene is a fungicide selective for the control of *Fusarium caerulum* of potato tubers. Smoke formulations are used against *Botrytis* spp. on various glasshouse crops. It inhibits the sprouting of seed potatoes (W. Brown, *Ann. Appl. Biol.*, 1947, **34**, 801).

**Toxicology.**

No ill-effect was noted in rats receiving 57 mg/kg daily or mice 215 mg/kg daily (G. A. H. Buttle & F. J. Dyer, *J. Pharm. Pharmacol.*, 1950, **2**, 371).

**Formulations.**

Dusts (30-60 g a.i./kg), granules, smokes. 'Fumite Tecnalin', smoke (tecnazene + gamma-HCH).

**Analysis.**

Product analysis is by polarography (*CIPAC Handbook*, 1970, **1**, 663). Residues may be determined by glc (*Pestic. Anal. Man.*, 1979, **I**, 201-A, 201-G).

# Temephos

$C_{16}H_{20}O_6P_2S_3$ (466.5)
1OPS&O1&OR DSR DOPS&O1&O1

**Nomenclature and development.**

Common name temephos (BSI, E-ISO, F-ISO, ANSI). Chemical name (IUPAC) *O,O,O',O'*-tetramethyl *O,O'*-thiodi-*p*-phenylene bis(phosphorothioate); *O,O,O',O'*-tetramethyl *O,O'*-thiodi-*p*-phenylene diphosphorothioate. (*C.A.*) *O,O'*-(thiodo-4,1-phenylene) bis(*O,O*-dimethyl phosphorothioate) (9CI); *O,O'*-thiodi-*p*-phenylene *O,O,O',O'*-tetramethyl di(phosphorothioate) (8CI); Reg. No. *[3383-96-8]*. OMS 786, ENT 27 165. Introduced by American Cyanamid Co. (BEP 648 531; GBP 1 039 238; USP 3 317 636) as code no. 'AC 52 160'; trade marks 'Abate', 'Abathion', 'Abat', 'Swebat', 'Nimitex', 'Biothion'.

**Properties.**

Pure temephos is a colourless crystalline solid; m.p. 30.0-30.5 °C. The technical product is a brown viscous liquid >90% pure, $n_D^{25}$ 1.586-1.588. Almost insoluble in: water, hexane, methylcyclohexane; soluble in: acetonitrile, carbon tetrachloride, diethyl ether, 1,2-dichloroethane, toluene, lower alkyl ketones. It is stable at 25 °C and in natural fresh and saline waters; optimum stability is at pH5-7, hydrolysis occurs at 2>pH>9 at rates depending on the temperature and pH of the medium.

**Uses.**

Temephos is used in public health programmes to control the larvae of biting midges (Ceratopogonidae), blackfly (Simuliidae), chironomid midges, mosquitoes and moth and sand flies (Psychodidae). It is also effective in controlling fleas on dogs and cats and lice on humans. On crops, it is effective in controlling cutworms, thrips on citrus and Lygus bugs.

**Toxicology.**

Acute oral $LD_{50}$: for male rats 8600 mg/kg, for females 13 000 mg/kg. Acute percutaneous $LD_{50}$ (24-h): for rabbits 1300-1930 mg/kg; for rats >4000 mg/kg. In 2-y feeding trials rats receiving 300 mg/kg diet showed no observable clinical effect. No toxic symptom was felt by humans receiving 256 mg/man for 5 d, or 64 mg/man for 28 d (R. L. Laws *et al., Arch. Environ. Health*, 1967, **14**, 289). In 5-d feeding tests the $LC_{50}$ is: for mallard ducks 1200 mg/kg diet; for ring-necked pheasants 170 mg/kg diet. $LC_{50}$ for rainbow trout is 31.8 mg/l. $LD_{50}$ by topical application to honeybees is 1.55 μg/bee.

**Formulations.**

These include: e.c. (200 or 500 g a.i./l); w.p. (500 g/kg); granules (10, 20 or 50 g/kg); dusting powder (20 g/kg); lotion (20 g/kg).

**Analysis.**

Residues may be determined by colorimetry (R. C. Blinn *et al., J. Agric. Food Chem.*, 1966, **14**, 152) or by glc (N. R. Pasarela & E. J. Orloski, *Anal. Methods Pestic. Plant Growth Regul.*, 1973, **7**, 119).

$C_8H_{20}O_7P_2$ (290.2)
2OPO&O2&OPO&O2&O2

$(CH_3CH_2O)_2P(=O)-O-P(=O)(OCH_2CH_3)_2$

**Nomenclature and development.**

Common name TEPP (BSI, E-ISO, F-ISO, JMAF); ethyl pyrophosphate (BPC). Chemical name (IUPAC) tetraethyl pyrophosphate (I). (*C.A.*) tetraethyl diphosphate (9CI); (I) (8CI); Reg. No. *[107-49-3]*. ENT 18 771. Its aphicidal properties were discovered in 1938 by G. Schrader & H. Kükenthal (cited by G. Schrader, *Die Entwickling neuer insektizider Phosphorsaüre-Ester*, 3rd Ed.). In 1943 I. G. Farbenindustrie AG introduced a derivative then thought to be hexaethyl tetraphosphate (known as HETP) but since shown to contain TEPP as the main insecticidal component. Trade marks include '*Nifos T*' (Monsanto Chemical Co. who no longer manufacture or market it), 'Vapotone' (Chevron Chemical Co.).

**Properties.**

Pure TEPP is a colourless hygroscopic liquid, b.p. 124 °C/1 mmHg; v.p. 21 mPa at 20 °C; $d_4^{20}$ 1.185; $n_D^{20}$ 1.4196. Solubility: miscible with water and most organic solvents; sparingly soluble in petroleum oils. The technical grade is a dark amber-coloured mobile liquid, $d_{25}^{25}$ *c.* 1.2. TEPP is rapidly hydrolysed by water, 50% decomposition occurs in 6.8 h at pH7 and 25 °C. It decomposes at 170 °C with the formation of ethylene. It is corrosive to most metals.

**Uses.**

TEPP is a non-systemic acaricide and aphicide with a very brief persistence.

**Toxicology.**

Acute oral $LD_{50}$ for rats 1.12 mg/kg. Acute percutaneous $LD_{50}$ for male rats 2.4 mg/kg. It is rapidly metabolised by animals.

**Formulations.**

Aerosol concentrate (solution in methyl chloride).

**Analysis.**

Product analysis is by selective hydrolysis (*CIPAC Handbook*, 1970, **1**, 667; *AOAC Methods*, 1980, 6.420-6.422). Residues may be determined by glc (J. Crossley, *Anal. Methods Pestic. Plant Growth Regul.*, 1973, **7**, 471).

$C_9H_{13}ClN_2O_2$ (216.7)
T6MVNVJ CX1&1&1 EG F1

H
CH3 N O
Cl N C(CH3)3
O

**Nomenclature and development.**

Common name terbacil (BSI, E-ISO, F-ISO, ANSI, WSSA, JMAF). Chemical name (IUPAC) 3-*tert*-butyl-5-chloro-6-methyluracil (I). (*C.A.*) 5-chloro-3-(1,1-dimethylethyl)-6-methyl-2,4(1*H*,3*H*)-pyrimidinedione (9CI); (I) (8CI); Reg. No.*[5902-51-2]*. The herbicidal activity of certain substituted uracils was reported by H. C. Bacha *et al.*, (*Science*, 1962, **137**, 537). Introduced by E. I. du Pont de Nemours & Co. (Inc.) (USP 3 235 357; BEP 625 897) as code no. 'Du Pont Herbicide 732'; trade mark 'Sinbar'.

**Properties.**

Pure terbacil forms colourless crystals; m.p. 175-177 °C; v.p. 62.5 μPa at 29.5 °C; *d* 1.34. Solubility at 25 °C: 710 mg/kg water; 220 g/kg cyclohexanone; 337 g/kg dimethylformamide; 65 g/kg xylene. Stable up to its m.p. and in water and aqueous alkali at room temperature. Undergoes microbial decomposition in moist soil.

**Uses.**

It inhibits photosynthesis and is absorbed by plant roots. It is used for the selective control of many annual and some perennial weeds in apples, citrus, lucerne, peaches, sugarcane at 0.5-4 kg/ha (as area actually treated), also *Cynodon dactylon* and *Sorghum halepense* in citrus at 4-8 kg/ha.

**Toxicology.**

Acute oral $LD_{50}$ for rats >5000 mg/kg. In percutaneous tests on rabbits no toxic effect was observed at 5000 mg/kg (max. feasible dose). In 2-y feeding trials NEL for rats and dogs was 250 mg/kg diet. $LC_{50}$ (8-d) was: for pheasant chicks >31 450 mg/kg diet; for Pekin ducklings >56 000 mg/kg. $LC_{50}$ (48-h) is: for pumpkinseed sunfish 86 mg/l; for fiddler crab >1000 mg/l.

**Formulations.**

'Sinbar', w.p. (800 mg a.i./kg). Mixtures include: terbacil + diuron + linuron; terbacil + linuron; terbacil + linuron + monolinuron.

**Analysis.**

Product analysis is by hplc (H. L. Pease *et al.*, *Anal. Methods Pestic. Plant Growth Regul.*, 1978, **10**, 483) or by i.r. spectrometry, details from E. I. du Pont de Nemours & Co. (Inc.). Residues may be determined by glc (H. L. Pease *et al.*, *loc cit.*; R. F. Holt & H. L. Pease, *J. Agric. Food Chem.*, 1977, **25**, 373; T. H. Byast *et al.*, *Tech. Rep. ARC Weed Res. Organ.*, No. 15 (2nd Ed.), p.47).

$C_9H_{21}O_2PS_3$ (288.4)
2OPS&O2&S1SX1&1&1

$$(CH_3)_3CSCH_2SP(S)(OCH_2CH_3)_2$$

**Nomenclature and development.**

Common name terbufos (BSI, E-ISO, F-ISO, ANSI). Chemical name (IUPAC) *S-tert*-butylthiomethyl *O,O*-diethyl phosphorodithioate. (*C.A.*) *S*-[[(1,1-dimethylethyl)thio]methyl] *O,O*-diethyl phosphorodithioate (9CI); *S*-(*tert*-butylthio)methyl *O,O*-diethyl phosphorodithioate (8CI); Reg. No. *[13071-79-9]*. ENT 27 920. Its insecticidal properties were described by E. B. Fagan (*Proc. Br. Insectic. Fungic. Conf. 7th,* 1973, **2**, 695). Introduced by American Cyanamid Co. as code no. 'AC 92 100'; trade mark 'Counter'.

**Properties.**

Technical grade terbufos is a colourless to pale yellow liquid, ≥85% pure; b.p. 69 °C/0.01 mmHg; m.p. -29.2 °C; v.p. 34.6 mPa at 25 °C; $n_D^{23}$ 1.52; $d^{24}$ 1.105; flash point 88 °C (tag open cup). Solubility at ordinary temperature *c.* 10-15 mg/l water; soluble in acetone, alcohols, aromatic hydrocarbons, chlorinated hydrocarbons. It decomposes at >120 °C or in the presence of acid (pH<2) or alkali (pH>9).

**Uses.**

It has effective initial and residual activity against soil-dwelling arthropods. Soil application of granules controls *Diabrotica* spp. larvae on maize, *Tetanops myopaeformis* on sugar beet, *Erioischia brassicae* on cabbages, millipedes, onion maggots, symphylids, wireworms and other soil-dwelling arthropods. Various above-ground pests also are controlled on plants grown in terbufos-treated soil. It has nematicidal activity.

**Toxicology.**

Acute oral $LD_{50}$: for male albino rats 1.6 mg tech./kg; for female albino mice 5.4 mg/kg. Acute percutaneous $LD_{50}$: for rabbits 1.0 mg tech./kg; for rats 7.4 mg tech./kg, for 24-h contact 27.5 mg a.i. (as 10% granule)/kg, >200 mg a.i. (as 2% granule)/kg. In 2-y feeding trials rats receiving 1 mg/kg diet showed no adverse effect other than cholinesterase depression and associated syndrome. In 8-d feeding trials $LC_{50}$ was: for mallard ducks 185 mg/kg diet; for ring-necked pheasants 145 mg/kg diet. $LC_{50}$ (96-h) is: for bluegill 0.004 mg/l; for rainbow trout 0.01 mg/l. Topical $LD_{50}$ for honeybees 4.09 μg/bee.

**Formulations.**

Granules (20-150 g a.i./kg).

**Analysis.**

Formulations and residues may be determined by glc (E. J. Orloski, *Anal. Methods Pestic. Plant Growth Regul.*, 1980, **11**, 165).

$C_{10}H_{19}N_5O$ (225.3)
T6N CN ENJ BO1 DMX1&1&1 FM2



**Nomenclature and development.**

Common name terbumeton (BSI, E-ISO, F-ISO). Chemical name (IUPAC) *N-tert*-butyl-*N'*-ethyl-6-methoxy-1,3,5-triazine-2,4-diyldiamine; 2-*tert*-butylamino-4-ethylamino-6-methoxy-1,3,5-triazine. (*C.A.*) *N*-(1,1-dimethylethyl)-*N'*-ethyl-6-methoxy-1,3,5-triazine-2,4-diamine (9CI); 2-(*tert*-butylamino)-4-(ethylamino)-6-methoxy-*s*-triazine (8CI); Reg. No. *[33693-04-8]*. Its herbicidal properties were described by A. Gast & E. Fankhauser (*Proc. Br. Weed Control Conf., 8th*, 1966, **2**, 485). Introduced by J. R. Geigy S.A. (now Ciba-Geigy AG) (CHP 337 019; GBP 814 948) as code no. 'GS 14 259'; trade mark 'Caragard'.

**Properties.**

Pure terbumeton forms colourless crystals; m.p. 123-124 °C; v.p. 267 μPa at 20°C; ρ 1.081 g/cm³ at 20 °C. Solubility at 20 °C: 130 mg/l water; 130 g/l acetone; 360 g/l dichloromethane; 220 g/l methanol; 90 g/l octan-1-ol; 110 g/l toluene. At 20 °C, 50% hydrolysis (calculated) occurs in: 29 d at pH1, 1.6 y at pH13.

**Uses.**

It is a herbicide absorbed by leaves and roots and effective against grasses and broad-leaved weeds, both annual and perennial species, used in citrus orchards after 3rd year. In combination with terbuthylazine, it is used for post-em. weed control in apple orchards, citrus, forestry and vineyards at 3-10 kg total a.i./ha.

**Toxicology.**

Acute oral $LD_{50}$ for rats 483-657 mg tech./kg. Acute percutaneous $LD_{50}$ for rats >3170 mg/kg. In 90-d feeding trials using 500 g a.i./kg w.p. NEL was: for rats 150 mg a.i./kg diet (10 mg/kg daily); for dogs 750 mg/kg diet (25 mg/kg daily). $LC_{50}$ (96-h) is: for rainbow trout 14 mg/l; for bluegill and crucian carp 30 mg/l.

**Formulations.**

'Caragard', 50 WP, w.p. (500 g a.i./kg). 'Caragard combi A', (Reg. No. *[8072-81-9]*), terbumeton + terbuthylazine (1:1); 'Caragard combi B', same components (1:2).

**Analysis.**

Product analysis is by glc or by acidimetric titration; particulars available from Ciba-Geigy AG. Residues may be determined by glc with MCD (K. Ramsteiner *et al., J. Assoc. Off. Anal. Chem.*, 1974, **57**, 192; E. Knüsli, *Anal. Methods Pestic., Plant Growth Regul. Food Addit.*, 1964, **4**, 13).

$C_9H_{16}ClN_5$ (229.7)
T6N CN ENJ BMX1&1&1 DM2 FG

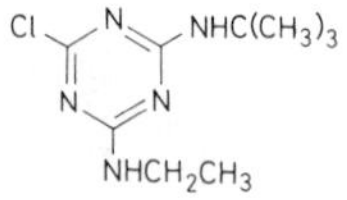

**Nomenclature and development.**

Common name terbuthylazine (BSI, E-ISO, F-ISO, ANSI, WSSA). Chemical name (IUPAC) *N-tert*-butyl-6-chloro-*N'*-ethyl-1,3,5-triazine-2,4-diyldiamine; 2-*tert*-butylamino-4-chloro-6-ethylamino-1,3,5-triazine. (*C.A.*) 6-chloro-*N*-(1,1-dimethylethyl)-*N'*-ethyl-1,3,5-triazine-2,4-diamine (9CI); 2-(*tert*-butylamino)-4-chloro-6-(ethylamino)-*s*-triazine (8CI); Reg. No. *[5915-41-3]*. Its herbicidal properties were described by A. Gast & E. Fankhauser (*Proc. Br. Weed Control Conf., 8th*, 1966, p. 485). Introduced by J. R. Geigy S.A. (now Ciba-Geigy AG) (BEP 540 590; GBP 814 947) as code no. 'GS 13 529'; trade mark 'Gardoprim'.

**Properties.**

Pure terbuthylazine is a colourless powder; m.p. 177-179 °C; v.p. 146 μPa at 20 °C; ρ 1.188 g/cm$^3$ at 20 °C. Solubility at 20 °C: 8.5 mg/l water; 100 g/l dimethylformamide; 40 g/l ethyl acetate; 14.3 g/l octan-1-ol. At 20 °C, 50% hydrolysis (calculated) occurs in: 8 d at pH1; 86 d at pH5; >200 d at pH9; 12 d at pH13.

**Uses.**

It is a herbicide which controls a wide range of weeds, and is absorbed mainly by plant roots. It remains largely in the top soil. It is used pre-em. in sorghum at 1.2-1.8 kg a.i./ha; also for selective weed control in citrus, maize, pod forests and vineyards. Mixed with terbumeton it controls perennial weeds in established stands of apples, citrus and grapes. It is used for non-selective weed control when mixed with secbumeton and it is also used in combination with bromofenoxim at total a.i. rates of 0.8-1.25 kg/ha as a broad-spectrum broad-leaved herbicide in winter and spring cereals.

**Toxicology.**

Acute oral $LD_{50}$ for rats 2000 mg tech./kg. Acute percutaneous $LD_{50}$ for rats >3000 mg/kg; non-irritant to eyes and slight irritant to skin of rabbits. Inhalation $LC_{50}$ for rats >3510 mg/m$^3$. In 90-d feeding trials NEL for beagle dogs was 3.5 mg/kg daily. $LC_{50}$ (96-h) is: for rainbow trout 4.6 mg/l; for crucian carp 66 mg/l; for bluegill 52 mg/l.

**Formulations.**

'Gardoprim' 50WP or 80WP, w.p. (500 or 800 g a.i./kg); 'Gardoprim' 5G, granules (50 g/kg); 'Gardoprim' 500FW, s.c. (500 g/l). Mixtures include the following. 'Caragard combi A' (Reg. No. *[8072-81-9]*), terbuthylazine + terbumeton (1:1); 'Caragard combi B', same ingredients (2:1). 'Faneron combi' (Reg. No. *[39380-46-6]*), terbuthylazine + bromofenoxim (*c.* 1:2); 'Faneron multi' and 'Mofix', same ingredients (*c.* 1:3) and (1:1.6). 'Faneron plus', terbuthylazine + bromofenoxim + mecoprop (1:3:6). 'Gardomil', terbuthylazine + metolachlor (3:1). 'Gardopax', terbuthylazine + ametryne (2:1 or 1:1). 'Piotal', terbuthylazine + chlortoluron + terbutryne (4:5:1).'Primatol 3588', terbuthylazine + secbumeton (1:0). 'Sorgoprim' (Reg. No. *[8066-11-3]*), terbuthylazine + terbutryne (1:1) and 'Topogard', same ingredients (3:7).

**Analysis.**

Product analysis is by glc (*CIPAC Handbook*, 1983, **1B**, in press; FAO Specification CP/61); particulars also available from Ciba-Geigy AG. Residues may be determined by glc with DMC or FPD. (K. Ramsteiner *et al., J. Assoc. Off. Anal. Chem.*, 1974, **57**, 192; E. Knüsli, *Anal. Methods Pestic., Plant Growth Regul. Food Addit.*, 1964, **4**, 13).

$C_{10}H_{19}N_5S$ (241.4)
T6N CN ENJ BS1 DMX1&1&1 FM2

$CH_3S$, N, $NHC(CH_3)_3$, N, N, $NHCH_2CH_3$

**Nomenclature and development.**

Common name terbutryne (BSI, F-ISO); terbutryn (E-ISO, ANSI, WSSA). Chemical name (IUPAC) *N-tert*-butyl-*N*'-ethyl-6-methylthio-1,3,5-triazine-2,4-diyldiamine; 2-*tert*-butylamino-4-ethylamino-6-methylthio-1,3,5-triazine. (*C.A.*) *N*-(1,1-dimethylethyl)-*N*'-ethyl-6-(methylthio)-1,3,5-triazine-2,4-diamine (9CI); 2-(*tert*-butylamino)-4-( ethylamino)-6-(methylthio)-*s*-triazine (8CI); Reg. No. *[886-50-0]*. Its herbicidal properties were described by A. Gast *et al.* (*Proc. Symp. New Herbic., 2nd*, 1965, p. 305). Introduced by J. R. Geigy (now Ciba-Geigy AG) (CHP 337 019; GBP 814 948) as code no. 'GS 14 260'; trade marks 'Igran', 'Clarosan', and in Great Britain only 'Prebane'.

**Properties.**

Pure terbutryne is a white powder; m.p. 104-105 °C; v.p. 128 μPa at 20 °C; ρ 1.115 g/$cm^3$ at 20 °C. Solubility at 20 °C: 25 mg/l water; 280 g/l acetone, methanol; 300 g/l dichloromethane; 9 g/l hexane; 130 g/l octan-1-ol; 45 g/l toluene. No significant hydrolysis occurred at pH 5, 7 and 9 at 30 °C or 70 °C.

**Uses.**

It is a selective herbicide for pre-em. use in winter cereals at 1-2 kg a.i./ha for the control of blackgrass and annual meadow grass. Among the autumn-germinating broad-leaved weeds controlled are chickweed, mayweed, poppies and speedwell, but cleavers are rather resistant. Other pre-em. uses are on sugarcane and sunflowers; and in mixture with terbuthylazine on peas and potatoes. It is used post-em. directed on maize. Also used as 'Clarosan' for control of algae and submerged vascular plants in waterways, reservoirs and fish ponds.

**Toxicology.**

Acute oral $LD_{50}$ for rats 2000 mg tech./kg. Acute percutaneous $LD_{50}$ for rats >2000 mg/kg. In 90-d feeding trials NEL for rats was 600 mg/kg diet (50 mg/kg daily); in 0.5-y trials for dogs 10 mg/kg daily. $LC_{50}$ (96-h) is: for rainbow trout 1.8-3.0 mg/l; for crucian carp 1.4-4.0 mg/l; for bluegill 4 mg/l. It is practically non-toxic to birds and honeybees.

**Formulations.**

'Igran' 50WP and 80WP, w.p. (500 or 800 g a.i./kg); 'Igram' 500FW, 'Clarosan 500FW, s.c. (500 g/l); 'Clarosan' 1G, granules (10 g/kg). 'Dicuran extra', terbutryne + chlortoluron (1:6). 'Gesaprim combi' (Reg. No. *[8063-10-2]*), terbutryne + atrazine (1:1). 'Igrater' (Reg. No. *[52080-96-3]*), terbutryne + metobromuron (1:1). 'Piotal', terbutryne + terbuthylazine + chlortoluron (1:4:5). 'Sorgoprim' (Reg. No. *[8066-11-3]*), terbutryne + terbuthylazine (1:1); 'Topogard', same ingredients (7:3).

**Analysis.**

Product analysis is by glc (*CIPAC Handbook*, 1980, **1A**, 1351; *AOAC Methods*, 1980, 6.431-6.435; FAO Specification (CP/61); A. H. Hofberg *et al., J. Assoc. Off. Anal. Chem.*, 1973, **56**, 586). Residues may be determined by glc with DMC or FPD (K. Ramsteiner *et al.*, 1974, **57**, 192; T. H. Byast *et al., Tech. Rep. ARC Weed Res. Organ.*, No. 15 (2nd Ed.), pp. 40, 80; B. G. Tweedy & R. A. Kahrs, *Anal. Methods Pestic. Plant Growth Regul.*, 1978, **10**, 493; E. Knüsli, *Anal. Methods Pestic., Plant Growth Regul. Food Addit.*, 1964, **4**, 13), or by hplc (T. H. Byast *et al., loc. cit.*, p. 40).

$C_{13}H_{12}ClN_5$ (273.7)
T C45 H55 A DNN INNN DU IUTJ KR
DG &&(1R,2S,6S,7R,8R,11S) FORM

**Nomenclature and development.**

Common name tetcyclacis (BSI, draft E-ISO); tetcyclasis (draft F-ISO). Chemical name (IUPAC) *rel*-(1*R*,2*R*,6*S*,7*R*,8*R*,11*S*)-5-(4-chlorophenyl)-3,4,5,9,10-penta-azatetracyclo[5.4.1.$0^{2,6}$.$0^{8,11}$]dodeca-3,9-diene. (*C.A.*) usage (3aα,4β,4aα,6aα,7β,7aα)-1-(4-chlorophenyl)-3a,4,4a,6a,7,7a-hexahydro-4,7-methano-1*H*-[1,2]diazeto[3,4-*f*]benzotriazole (9CI); Reg. No. *[77788-21-7]*. Its plant growth-regulating properties were described by J. Jung *et al.* (*Z. Acker-Pflanzenbau*, 1980, **149**, 128). Introduced by BASF AG (DEOS 2 615 878; GBP 1 573 161; USP 4 189 434) as code no. 'BAS 106W'; trade mark 'Ken byo'.

**Properties.**

It is a colourless crystalline solid, m.p. 190 °C (decomp.). Solubility at 20 °C: 3.7 mg/kg water; 42 g/kg chloroform; 2 g/kg ethanol. It is decomposed by sunlight and concentrated acid.

**Uses.**

It is used for improvement of rice seedlings. Treated seedlings are more compact and shorter than untreated ones and they show enhanced rooting and tillering capacity after transplanting into the field.

**Toxicology.**

Acute oral $LD_{50}$ for rats 261 mg/kg. Acute percutaneous $LD_{50}$ for rats $>$4640 mg/kg.

**Formulations.**

A w.s.p. (10 g a.i./kg).

**Analysis.**

Product analysis is by hplc with u.v. detection. Residues may be determined by glc of a derivative with ECD.

# 4,5,6,7-Tetrachlorophthalide

$C_8H_2Cl_4O_2$ (271.9)
T56 BVO DHJ FG GG HG IG

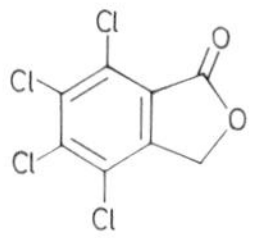

**Nomenclature and development.**

Common name phthalide (JMAF);spelling fthalide also used (JMAF). Chemical name (IUPAC) 4,5,6,7-tetrachlorophthalide (I). (*C.A.*) 4,5,6,7-tetrachloro-1(3*H*)-isobenzofuranone (9CI); (I) (8CI); Reg. No. *[27355-22-2]*. Trivial name TCP. Its fungicidal properties were first described by K. Nambu (*Jpn. Pestic. Inf.*, 1972, No. 10, p. 73) and by K. Wagner & H. Scheinflug (*Pflanzenchutz Nachr. (Engl. Ed.)*, 1975, **28**, 210). Introduced by Kureha Chemical Co. Ltd and manufactured by a process licensed from Bayer AG (JPP 575 584 Kureha; DEAS 1 643 347 Bayer) as code no. 'KF-32' (Kureha),'Bayer 96 610'; trade mark 'Rabcide' (Kureha).

**Properties.**

It is a colourless crystalline solid; m.p. 209-210 °C. Solubility at 25 °C: 2.5 mg/l water; sparingly soluble in most organic solvents and oils. It is incompatible with strongly alkaline compounds.

**Uses.**

It is a fungicide with specific action against *Pyricularia oryzae* on rice. It is applied at 270-750 g a.i. (as w.p.) in 800-1500 l water/ha whenever a blast infection is expected or lesions are observed on the rice plant.

**Toxicology.**

Acute oral $LD_{50}$ for rats and mice >10 000 mg/kg. Acute percutaneous $LD_{50}$ for rats and mice >10 000 mg/kg; non-irritant to the eyes and shaved skin of rabbits. $LC_{50}$ (48-h) for young carp is >320 mg a.i. (as tech. or dust)/l, 135 mg a.i. (as 50% w.p.)/l.

**Formulations.**

These include: w.p. (300 or 500 g a.i./kg); dust (25 g/kg); s.c. (200 g/l, mainly used for aerial spraying).

**Analysis.**

Product analysis is by glc with TCD. Residues may be determined by glc with ECD (H. Nagayoshi *et al., Bull. Agric. Chem. Inspect. Stn.*, 1973, **13**, 27). Detailed analytical methods and other information may be obtained from Kureha Chemical Industry Co., Ltd.

$C_{10}H_9Cl_4O_4P$ (366.0)
GR BG DG EYU1GOPO&O1&O1 &&(Z)
FORM

$(CH_3O)_2P(O)O$, H / C=C / Cl — (E)
$(CH_3O)_2P(O)O$, Cl / C=C / H — (Z)

**Nomenclature and development.**

Common name tetrachlorvinphos (BSI, E-ISO, F-ISO);stirofos (ESA);CVMP (JMAF). Chemical name (IUPAC) (*Z*)-2-chloro-1-(2,4,5-trichlorophenyl)vinyl dimethyl phosphate (I). (*C.A.*) (*Z*)-2-chloro-1-(2,4,5-trichlorophenyl)ethenyl dimethyl phosphate (9CI); (I) (8CI); Reg. No. *[22248-79-9]* (*Z*)-isomer, *[22350-76-1]* (*E*)-isomer, *[961-11-5]* mixed isomers. OMS 595, ENT 25 841. Its insecticidal properties were described by R. R. Whetsone *et al.* (*J. Agric. Food Chem.*, 1966, **14**, 352). Introduced by Shell Chemical Company, USA, (USP 3 102 842) as code no. 'SD 8447'; trade marks 'Gardona' (for crop protection use), 'Rabond' (for veterinary use).

**Properties.**

Produced by the reaction of triethyl phosphite with 2,2,2',4',5'-pentachloroacetophenone, the technical product, typically 98% (*Z*)-isomer, is a colourless crystalline solid; m.p. 95-97 °C. Solubility at 20 °C: 11 mg/l water; <200 g/kg acetone; 400 g/kg chloroform, dichloromethane; <150 g /kg xylene. It is stable <100 °C. Slowly hydrolysed in water, 50% loss occurring at 50 °C in 54 d at pH3, 44 d at pH7, 80 h at pH10.5. The pure (*Z*)-isomer has v.p. 5.6 μPa at 20 °C.

**Uses.**

It is a selective insecticide controlling: lepidopterous and dipterous pests on fruit at 25-75 g a.i./100 l; lepidopterous pests on cotton and maize at 0.75-2.0 kg/ha, on rice at 250-600 g/ha, on tobacco at 0.5-1.5 kg/ha and on vegetables at 0.25-1.0 kg/ha. With certain exceptions, it does not exhibit high activity against hemipterous insects and other sucking pests and, because of rapid breakdown, is not effective against soil dwelling pests. It is used against flies in dairies and in livestock barns, against certain ectoparasites on poultry and against pests of stored products. It is also used against insects that damage forests and pasture. Its metabolism and breakdown have been reviewed (K. I. Beynon *et al., Residue Rev.*, 1973, **47**, 55).

**Toxicology.**

Acute oral $LD_{50}$: for rats 4000-5000 mg/kg; for mice 2500-5000 mg/kg; for mallard duck and Chukar partridge >2000 mg/kg; for various birds 1500-2600 mg/kg. Acute percutaneous $LD_{50}$ for rabbits >2500 mg/kg. In 2-y feeding trials NEL was: for rats 125 mg/kg diet; for dogs 200 mg/kg diet. It is rapidly metabolised in mammals and metabolites are eliminated within a few days. No significant residues of tetrachlorvinphos or its metabolites are found in milk or tissues of exposed animals. Reproduction studies on rats receiving 1000 mg/kg diet showed no adverse effect. $LC_{50}$ (24-h) in the range 0.3-6.0 mg/l for various species of fish.

**Formulations.**

These include: e.c. (240 g a.i./l); w.p. (500 and 750 g/kg); granules (50 g/kg); s.c. (700 g/l).

**Analysis.**

Product analysis is by i.r. spectrometry or glc (details from Shell International Chemical Co.). Residues of tetrachlorvinphos and its major metabolites in plants may be determined by glc (*Anal. Methods Pestic. Plant Growth Regul.*, 1973, **7**, 297; K. I. Beynon & A. N. Wright, *J. Sci. Food Agric.*, 1969, **20**, 250; K. I. Beynon *et al., Pestic. Sci.*, 1970, **1**, 250, 254, 259).

$C_{12}H_6Cl_4O_2S$ (356.0)
GR DSWR BG DG EG

**Nomenclature and development.**

Common name tetradifon (BSI, E-ISO, F-ISO, JMAF); tedion (Turkey, USSR); *exception* (Portugal). Chemical name (IUPAC) 4-chlorophenyl 2,4,5-trichlorophenyl sulphone; 2,4,4',5-tetrachlorodiphenyl sulphone. (*C.A.*) 1,2,4-trichloro-5[(4-chlorophenyl)sulfonyl]benzene (9CI); *p*-chlorophenyl 2,4,5-trichlorophenyl sulfone (8CI); Reg. No. *[116-29-0]*. ENT 23 737. Its acaricidal properties were described by H. O. Huisman *et al.* (*Nature (London)*, 1955, **176**, 515). Introduced by N. V. Philips-Roxane (now Duphar B.V.) (NLP 81 359; USP 2 812 281) as code no. 'V-18'; trade mark 'Tedion V18'.

**Properties.**

Technical grade tetradifon (≥95% pure) is an off-white to slightly yellow crystalline solid; m.p. ≥144 °C (148-149 °C for the pure compound); v.p. 32 nPa at 20 °C. Solubility at 10 °C: 50 µg/l water; at 20 °C: 82 g/l acetone; 148 g/l benzene; 255 g/l chloroform; 200 g/l cyclohexanone; 10 g/l kerosene; 10 g/l methanol; 135 g/l toluene; 115 g/l xylene. No change occurred in 1 h at 100 °C in 25% aqueous sodium hydroxide or concentrated hydrochloric acid. It is resistant to strong oxidising agents and stable to sunlight.

**Uses.**

It is a non-systemic acaricide toxic to eggs and all non-adult stages of phytophagous mites. The w.p. is recommended for citrus, coffee, bush fruit, top fruit, grapes, nursery stock, ornamentals and vegetables at 20 g a.i./100 l; the e.c. on cotton at 240-400 g a.i./100 l, groundnuts at 10-20 g a.i./100 l, tea at 16 g a.i./100 l. At these rates it is non-phytotoxic and harmless to honeybees and the natural enemies of spider mites.

**Toxicology.**

Acute oral $LD_{50}$ for male rats $>$14 700 mg/kg. Acute percutaneous $LD_{50}$ for rabbits $>$10 000 mg/kg. In 2-y feeding trials rats receiving 300 mg a.i./kg diet suffered no ill-effect; offspring of rats receiving 1000 mg/kg diet were normal. $LC_{50}$ (8-d) for bob-white quail, Japanese quail, pheasant and mallard duck was $>$500 mg/kg diet. $LC_{50}$ (3 h) for carp is $>$10 mg/l.

**Formulations.**

'Tedion V18' e.c. (75.2 g a.i./l); 'Tedion V18' w.p. (188 g/kg). 'Childion', e.c. (125 g tetradifon + 400 g dicofol/l).

**Analysis.**

Product analysis is by glc (A. van Rossum *et al., J. Assoc. Off. Anal. Chem.*, 1981, **64**, 829; *CIPAC Handbook*, 1983, **1B**, in press). Residues may be determined by glc (*AOAC Methods*, 1980, 29.029-29.034; *Anal. Methods Pestic. Plant Growth Regul.*, 1972, **6**, 488; A. van Rossum *et al., ibid*, 1978, **10**, 119).

$C_{19}H_{25}NO_4$ (331.4)
T56 BVNV&TJ C1OV- BL3TJ A1 A1
C1UY1&1

(CH3)2C=CH— —CO.OCH2N
CH3 CH3

**Nomenclature and development.**

Common name tetramethrin (BSI, E-ISO, ANSI); tétraméthrine (F-ISO); phthalthrin (JMAF). Chemical name (IUPAC) cyclohex-1-ene-1,2-dicarboximidomethyl (1*RS*,3*RS*; 1*RS*,3*SR*)-2,2-dimethyl-3-(2-methylprop-1-enyl)cyclopropanecarboxylate; cyclohex-1-ene-1,2-dicarboximidomethyl (1*RS*)-*cis-trans*-2,2-dimethyl-3-(2-methylprop-1-enyl)cyclopropanecarboxylate; cyclohex-1-ene-1,2-dicarboximidomethyl (1*RS*)-*cis-trans*-chrysanthemate; 3,4,5,6-tetrahydrophthalimidomethyl (1*RS*)-*cis-trans*-chrysanthemate. (*C.A.*) (1,3,4,5,6,7-hexahydro-1,3-dioxo-2*H*-isoindol-2-yl)methyl 2,2-dimethyl-3-(2-methyl-1-propenyl)cyclopropanecarboxylate (9CI); 2,2-dimethyl-3-(2-methylpropenyl)cyclopropanecarboxylic acid ester with *N*-(hydroxymethyl)-1-cyclohexane-1,2-dicarboximide (8CI); Reg. No. *[7696-12-0]* for racemate, *[51384-90-4]* for (1*R*)-*cis*-cyclopropane moiety, *[1166-46-7]* for (1*R*)-*trans*-cyclopropane moiety. Trivial name *d*-tetramethrin for product rich in (1*R*)-cyclopropane moieties. OMS 1011. Introduced by Sumitomo Chemical Co. and later by FMC Corp. (who no longer manufacture or market it) (JPP 453 929; 462 108; USP 3 268 398 all to Sumitomo for the racemate; JPP 1 023 548; USP 3 634 023 for Sumitomo for mixture rich in (1*R*)-isomers of cyclopropane moiety) as code no. 'FMC 9260' (for racemate); trade marks 'Neo-Pynamin' (for racemate), 'Neo-Pynamin Forte' (for product rich in (1*R*)-isomers cyclopropane moiety) (*cis*:*trans* ratio 20:80 *m*/*m*) both to Sumitomo.

**Properties.**

Technical tetramethrin (racemate) is a colourless to light yellow-brown solid; m.p. 60-80 °C; v.p. 944 µPa at 30 °C; $d_{20}^{20}$ 1.11. Solubility at 30 °C: 4.6 mg/l water; at 25 °C: 20 g/kg hexane; 53 g/kg methanol; 1 kg/kg xylene. It is stable under normal storage conditions.

The technical material rich in (1*R*)-isomers is a yellow or brown viscous liquid, which sometimes tends to crystallise partly at room temperature; $d_{25}^{25}$ 1.11, $n_D^{20}$ 1.510-1.530, v.p. 320 µPa at 20 °C. Solubility at 23 °C: 2-4 mg/l water; >1 kg/kg hexane, methanol, xylene.

**Uses.**

Tetramethrin and the (1*R*)-rich isomeric product are contact insecticides with a strong 'knockdown' action on cockroaches, flies, mosquitoes and other pests of public health. Mixtures with other insecticides and synergists are used to secure a greater kill of the pests than with tetramethrin itself.

**Toxicology.**

Tetramethrin (racemate). Acute oral $LD_{50}$: for rats >5000 mg/kg; for mallard duck and bobwhite quail >1000 mg/kg. Acute percutaneous $LD_{50}$ for rats >5000 mg/kg. In 0.5-y feeding trials NEL for rats was 1500 mg/kg diet. $LC_{50}$ (96-h) for bluegill is 21 µg/l. Its metabolic fate in mammals has been studied (J. Miyamoto *et al.*, *Agric. Biol. Chem.*, 1968, **32**, 628).

(1*R*)-Rich product. Acute oral $LD_{50}$ for rats >5000 mg/kg. Acute percutaneous $LD_{50}$ for rats >5000 mg/kg. $LC_{50}$ (96-h) for bluegill is 69 µg/l.

**Formulations.**

These include: concentrates for the preparation of oil-based and water-based pressurised formulations; space and residual sprays containing tetramethrin or the (1*R*)-rich product. Also dust. Mixtures with other insecticides and/or with piperonyl butoxide. 'Pesguard-A NS' and 'Pesguard-A NS W' for aerosol concentrates. 'Pesguard NS' liquid and e.c. for public health, the former was developed for ULV spraying or thermal fogging and the latter for power sprayer or mist blower.

**Analysis.**

Product analysis is by glc (M. Horiba *et al.*, *Botyu-Kagaku*, 1975, **40**, 123), or by u.v. spectrometry (J. Miyamoto, *Anal. Methods Pestic. Plant Growth Regul.*, 1973, **7**, 345).

# $O,O,O',O'$-Tetrapropyl dithiopyrophosphate

$C_{12}H_{28}O_5P_2S_3$ (378.4)
3OPS&O3&OPS&O3&O3

$(CH_3CH_2CH_2O)_2P(=S)-O-P(=S)(OCH_2CH_2CH_3)_2$

**Nomenclature and development.**

Chemical name (IUPAC) $O,O,O',O'$-tetrapropyl dithiopyrophosphate. (*C.A.*) tetrapropyl thiodiphosphate (9CI); tetrapropyl thiopyrophosphate (8CI); Reg. No. *[3244-90-4]*. ENT 16 894. Insecticide described by A. D. F. Toy (*J. Am. Chem. Soc.*, 1951, **73**, 4670). Introduced by Stauffer Chemical Co. (USP 2 663 722) as code number 'ASP-51'; trade mark 'Aspon'.

**Properties.**

Technical product (93-96% pure) is a straw to amber-coloured liquid, with a faint aromatic odour; b.p. *c.* 170 °C/1 mmHg; v.p. *c.* 13 mPa at 25 °C; $d_{20}^{20}$ 1.119-1.123, $n_D^{21}$ 1.471. Solubility at 20 °C: 30 mg/l water; miscible with acetone, ethanol, kerosene, 4-methylpentan-2-one, xylene. In water buffered at pH 7, 50% loss occurs in 32 d at 40 °C. Thermally stable $<100$ °C. Corrosive to steel.

**Uses.**

A non-systemic insecticide particularly effective at 8.6. kg a.i./ha for the control of *Blissus leucopterus* in turf. It is persistent in soil.

**Toxicology.**

Acute oral $LD_{50}$: for male albino rats 2800 mg/kg, for females 740 mg/kg. Acute percutaneous $LD_{50}$ for rabbits $>2000$ mg/kg; mildly irritating to skin and non-irritating to their eyes. In 90-d feeding trials at sublethal rates, rats showed some depression of red blood cell cholinesterase.

**Formulations.**

'Aspon E', e.c. (240 or 720 g a.i./l). 'Aspon 5GA', 'Aspon 25G', granules (50 or 250 g/kg).

**Analysis.**

Product analysis is by glc. Residues in soil and turf are determined by glc. Analytical methods are available from the Stauffer Chemical Co.

$C_{12}H_6Cl_4S$ (324.1)
GR DSR BG DG EG

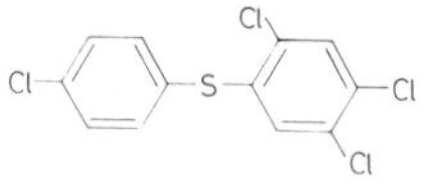

**Nomenclature and development.**

Common name tetrasul (BSI, E-ISO, F-ISO); tetradisul (Canada); diphenylsulphide (JMAF); *exception* (Federal Republic of Germany, Italy). Chemical name (IUPAC) 4-chlorophenyl 2,4,5-trichlorophenyl sulphide; 2,4,4',5-tetrachlorodiphenyl sulphide. (*C.A.*) 1,2,4-trichloro-5-[4-(chlorophenyl)thio]benzene (9CI); *p*-chlorophenyl 2,4,5-trichlorophenyl sulfide (8CI); Reg. No. *[2227-13-6]*. OMS 755, ENT 27 115. Its acaricidal properties were described by J. Meltzer & F. C. Dietvoorts (*Proc. Int. Congr. Crop Prot., 4th*, 1957, **1**, 669). Introduced by N. V. Philips-Roxane (now Duphar B.V.) (NLP 94 329; USP 3 054 719) as code no. 'V-101'; trade mark 'Animert V101'.

**Properties.**

Technical grade tetrasul is a cream to yellow-brown solid ($\geqslant$77.5% pure); m.p. 75-85 °C (pure compound 88.4-88.6 °C); v.p. 100 μPa. Solubility at 20 °C of the pure compound 30 mg/l water; approximate values for the technical grade at 20 °C: 274 g/l 1,2-dichloroethane; 20 g/l ethanol; 400 g/l xylene. Stable at ambient temperature in the dark; will be slowly oxidised via the corresponding sulphoxide to tetradifon (11430) under prolonged exposure to sunlight. No conversion after 30 min at 90 °C in a methanolic potassium hydroxide solution; <2% decomposition after 7 d at pH6 as a 25 mg/l suspension in water.

**Uses.**

It is a non-systemic acaricide, particularly suitable for the control of various phytophagous mites which hibernate in the winter egg form. It must be applied when the winter eggs are about to hatch. It is also active as an ovicide in summer but for this type of application it has no advantage over the related product tetradifon (11430). It is recommended for use on cucumber, bush fruit, top fruit and grapes at 36 g a.i./100 l. At this concentration it is non-phytotoxic and harmless to beneficial insects.

**Toxicology.**

Acute oral $LD_{50}$: for male rats 8250 mg/kg, for females 6810 mg/kg; for male and female mice 5010 mg/kg; for male guinea-pigs 8250 mg/kg, for females 8800 mg/kg. Acute percutaneous $LD_{50}$ for rabbits >2000 mg/kg. In 2-y feeding trials NEL for rats was 10 mg/kg diet. In a 3-generation study on rats, NEL was 20 mg/kg diet. In 90-d feeding trials beagle-dogs tolerated 200 mg/kg diet. $LC_{50}$ (8-d) bobwhite quail was 1225 mg/kg diet. $LC_{50}$ (24-h) for rainbow trout, goldfish, black bullhead, bluegill is >10 mg/l.

**Formulations.**

'Animert V101', w.p. (180 g a.i./kg).

**Analysis.**

Product analysis is by glc (L. R. Mitchell, *J. Assoc. Off. Anal. Chem.*, 1976, **59**, 209). Residues may be determined by glc (*idem, ibid.*; *AOAC Methods*, 1980, 29.029-29.034).

$C_{10}H_7N_3S$ (201.2)
T56 BM DNJ C- ET5N CSJ

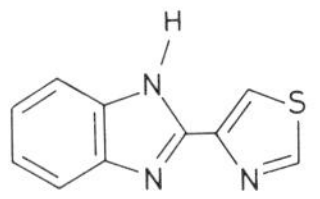

**Nomenclature and development.**

Common name thiabendazole (BSI, E-ISO, F-ISO, BPC, JMAF). Chemical name (IUPAC) 2-(thiazol-4-yl)benzimidazole. (*C.A.*) 2-(4-thiazolyl)-1*H*-benzimidazole (9CI); 2-(4-thiazolyl)benzimidazole (8CI); Reg. No. *[148-79-8]*. Trivial name TBZ. Its fungicidal properties were described by H. J. Robinson *et al.* (*J. Invest. Dermatol.*, 1964, **42**, 479) and T. Staron & C. Allard (*Phytriatr.-Phytopharm.*, 1964, **13**, 163) and its systemic properties in plants by D. C. Erwin *et al.* (*Phytopathology*, 1968, **58**, 860). It was originally introduced by Merck & Co. Inc. as an anthelmintic and later as an agricultural fungicide (USP 3 017 415) as code no. 'MK-360'; trade marks 'Mertect', 'Tecto', 'Storite'.

**Properties.**

Thiabendazole is a colourless powder; m.p. 304-305 °C; non-volatile at room temperature. Solubility at 25 °C; *c.* 10 g/l water at pH2, <50 mg/l water at pH5-12, >50 mg/l water at pH12, 4.2 g/l acetone, 7.9 g/l ethanol, 2.1 g/l ethyl acetate; at room temperature: 230 mg/l benzene, 80 mg/l chloroform, 39 g/l dimethylformamide, 80 g/l dimethyl sulphoxide, 9.3 g/l methanol. Under normal conditions it is stable to hydrolysis, light and heat.

**Uses.**

Thiabendazole is a fungicide controlling diseases of the following crops: asparagus, avocado, banana, cabbage, celery, cherry, citrus, cotton, certain cucurbits, grapes, mushrooms, onions (and garlic), ornamentals (bulbs, corms and flowers), pome fruit, potatoes, rice, soyabean, sugar beet, sweet potato, tobacco, tomatoes, turf and wheat. Pathogenic fungi controlled include species of *Aspergillus, Botrytis, Ceratocystis, Cercopora, Colletotrichum, Diaporthe, Fusarium, Gibberella, Gloesporium, Oospora, Penicillium, Phoma, Rhizoctonia, Sclerotinia, Septoria*, and *Verticillium*. It is also effective at 0.2-5.0 g a.i./l for the post-harvest treatment of fruit and vegetables to control storage diseases. It is also used as an anthelmintic in human and veterinary medicine.

**Toxicology.**

Acute oral $LD_{50}$: for rats 3300 mg a.i./kg; for mice 3810 mg/kg; for rabbits 3850 mg/kg. No observable clinical effect followed chronic inhalation at 70 mg/m$^3$. In 2-y feeding trials rats receiving 40 mg/kg daily suffered no ill-effect.

**Formulations.**

These include w.p. (400, 600 or 900 g a.i./kg); flowable suspension (450 g/l); fumigation tablets (7 g a.i.).

**Analysis.**

Product analysis is by u.v. spectrometry. Residues in food crops are determined fluorimetrically (J. S. Wood, *Anal. Methods Pestic. Plant Growth Regul.*, 1976, **8**, 299; *Pestic. Anal. Man.*, 1979, **I**, 201-I; R. Mestres *et al., Ann. Falsif. Expert. Chim.*, 1974, **67**, 585; 1976, **69**, 369).

# Thiazafluron



$C_6H_7F_3N_4OS$ (240.2)
T5NN DSJ CXFFF EN1&VM1

**Nomenclature and development.**

Common name thiazafluron (BSI, E-ISO, F-ISO); thiazfluron (Canada). Chemical name (IUPAC) 1,3-dimethyl-1-(5-trifluoromethyl-1,3,4-thiadiazol-2-yl)urea (I). (*C.A.*) *N*,*N*'-dimethyl-*N*-[(5-trifluoromethyl)-1,3,4-thiadiazol-2-yl]urea (9CI); (I) (8CI); Reg. No. *[25366-23-8]*. Its insecticidal properties were described by G. Müller *et al.* (*C. R. Journ. Etud. Herbic. Conf. COLUMA, 7th*, 1973, p. 32). Introduced by Ciba-Geigy AG (BEP 725 984; GBP 1 254 468) as code no. 'GS 29 696'; trade mark 'Erbotan'.

**Properties.**

Pure thiazafluron forms colourless crystals; m.p. 136-137 °C; v.p. 267 μPa at 20 °C; ρ 1.60 g/cm³ at 20 °C. Solubility at 20 °C: 2.1 g/l water; 12 g/kg benzene; 146 g/kg dichloromethane; 100 mg/l hexane; 257 g/kg methanol; 60 g/kg octan-1-ol. No significant hydrolysis occurred over a period of 28 d at pH values of 1, 5 or 7.

**Uses.**

It is a non-selective herbicide used pre- and post-em. for industrial weed control, effective against most annual and perennial mono- and di-cotyledonous weeds. It is active mainly through plant roots, foliar activity is poor. The dosage range is 2-8 kg a.i./ha in moist and temperate and 6-12 kg/ha in warm and dry regions.

**Toxicology.**

Acute oral $LD_{50}$ for rats 278 mg tech./kg. Acute percutaneous $LD_{50}$ for rats >2150 mg/kg; slight irritant to eyes and non-irritant to skin of rabbits. In 90-d feeding trials NEL for rats was 160 mg/kg diet (11 mg/kg daily); in 105-d trials on dogs 250 mg/kg diet (8 mg/kg daily). $LC_{50}$ (96-h) is: for rainbow trout 82 mg/l; for bluegill and crucian carp >100 mg/l. It is slightly toxic to birds and practically non-toxic to honeybees.

**Formulations.**

'Erbotan' 50WP and 80WP, w.p. (500 or 800 g a.i./kg); 'Erbotan' 10G, granules (100 g/kg).

**Analysis.**

Product analysis is by hplc. Residues may be determined by glc with FPD. Particulars are available from Ciba-Geigy AG.

$C_9H_8N_4OS$ (220.2)
T5NNSJ DMVMR

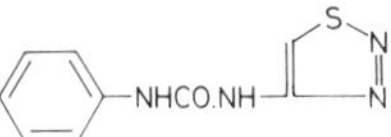


**Nomenclature and development.**

Common name thidiazuron (BSI, E-ISO, F-ISO, ANSI). Chemical name (IUPAC) 1-phenyl-3-(1,2,3-thiadiazol-5-yl)urea. (*C.A.*) *N*-phenyl-*N*'-1,2,3-thiadiazol-5-ylurea (9CI); Reg. No. *[51707-55-2]*. Its plant growth regulating effects were described by F. Arndt *et al.* (*Plant Physiol.*, 1976, **57**, Supplement p.99). Introduced by Schering AG (DEP 2 506 690; 2 214 632) as code no. 'SN 49 537'; trade mark 'Dropp'.

**Properties.**

It is a colourless crystalline solid; m.p. 213 °C (decomp.); v.p. 4 nPa at 25 °C. Solubility at 23 °C: 20 mg/l water; 8 g/l acetone; 21 g/l cyclohexane; >500 g/l dimethylformamide, dimethyl sulphoxide; 0.8 g/l ethyl acetate; 4.5 g/l methanol. It is stable at 23 °C, and between pH5-pH9.

**Uses.**

It is a plant growth regulator used at 50-200 g a.i./ha for the defoliation of cotton, the leaves dropping while still green.

**Toxicology.**

Acute oral $LD_{50}$: for rats >4000 mg a.i./kg; for mice >5000 mg/kg; for ducks >1320 mg a.i. (as w.p.)/kg; for Japanese quail >16 000 mg a.i. (as w.p.)/kg. Acute percutaneous $LD_{50}$ for rats and rabbits >1000 mg a.i./kg. In 2-y feeding trials there was no abnormal response in rats receiving 500 mg/kg diet. $LC_{50}$ (96-d) for trout, bluegill and catfish is >1000 mg/l.

**Formulations.**

A w.p. (500 g a.i./kg).

**Analysis.**

Product analysis is by hplc. Residues may be determined by reversed phase hplc with u.v. detection, or by hydrolysis to the aniline followed by glc of a suitable derivative. Details of methods are available from Schering AG.

$C_{12}H_{16}ClNOS$ (257.8)
GR D1SVN2&2

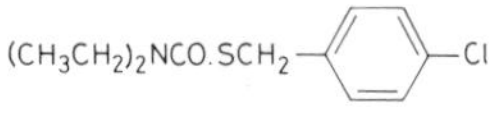

$(CH_3CH_2)_2NCO.SCH_2-C_6H_4-Cl$

**Nomenclature and development.**

Common name thiobencarb (BSI, E-ISO, ANSI, WSSA); thiobencarbe (F-ISO); benthiocarb (JMAF). Chemical name (IUPAC) *S*-4-chlorobenzyl diethyl(thiocarbamate). (*C.A.*) *S*-[(4-chlorophenyl)methyl] diethylcarbamothioate (9CI); *S*-(*p*-chlorobenzyl) diethylthiocarbamate (8CI); Reg. No. *[28249-77-6]*. Its herbicidal properties were described (*Jpn. Pestic. Inf.*, 1970, No. 2, p. 29). Introduced by Kumiai Chemical Industry Co. Ltd. and by Chevron Chemical Co. as code no. 'B-3015' (Kumiai); trade marks 'Saturn' (to Kumiai), 'Bolero' (both firms).

**Properties.**

Pure thiobencarb is a pale yellow liquid; m.p. 3.3 °C; b.p. 126-129 °C/0.008 mmHg; $d^{20}$ 1.145-1.180. Solubility at 20 °C: *c.* 30 mg/l water; soluble in most organic solvents. It is stable under acid and moderately alkaline conditions. The technical grade is *c.* 93% pure.

**Uses.**

It is a herbicide of use in rice. For direct-seeded rice the concentrate is applied at 3-6 kg a.i./ha to the surface of the water 3-5 d before or 5-10 d after sowing. For transplanted rice it is applied at 3-6 kg a.i./ha to the water 3-7 d after transplanting.

**Toxicology.**

Acute oral $LD_{50}$: for rats 1300 mg/kg; for mice 560 mg/kg. Acute percutaneous $LD_{50}$ for rats 2900 mg/kg. $LC_{50}$ (48-h) for carp is 3.6 mg/l.

**Formulations.**

These include: e.c. (500 g/kg); 'Bolero', e.c. (480 or 960 g/l); 'Bolero', granules (50 or 100 g/kg). Mixtures include: 'Saturn' (thiobencarb + simetryne).

**Analysis.**

Product and residue analysis is by glc with FID (K. Kejima *et al.*, *U.S.A.—Japan Seminar Envir. Toxicol. Pestic.*, 1971; K. Ishikawa *et al.*, *Agric. Biol. Chem.*, 1971, **35**, 1161; S. K. De Datta *et al.*, *Weed Res.*, 1971, **11**, 41; Y. Ishii, *Jpn. Pestic. Inf.*, 1974, No. 19, p. 21).

$C_5H_{11}NS_3$ (181.3)
T6SSSTJ EN1&1
*thiocyclam hydrogen oxalate*
$C_7H_{13}NO_4S_3$ (271.4)
T6SSSTJ EN1&1 &&$(COOH)_2$ SALT

$(CH_3)_2N$ S S S

**Nomenclature and development.**

Common name thiocyclam (BSI, E-ISO, JMAF); thiocyclame (F-ISO). Chemical name (IUPAC) *N,N*-dimethyl-1,2,3-trithian-5-ylamine. (*C.A.*) *N,N*-dimethyl-1,2,3-trithian-5-amine (9CI); Reg. No. *[31895-21-3]* (thiocyclam), *[31895-22-4]* (hydrogen oxalate). Its insecticidal properties were described by W. Berg & H. J. Knutti (*Proc. Br. Insectic. Fungic. Conf. 8th*, 1975, **2**, 683). Its hydrogen oxalate was introduced by Sandoz AG (DEOS 2 039 555) as code no. 'SAN 155I'; trade marks 'Evisect', 'Evisekt'.

**Properties.**

Pure thiocyclam hydrogen oxalate is a colourless solid; m.p. 125-128 °C (decomp.); v.p. 545 µPa at 20 °C. Solubility at 23 °C: 84 g/l water; 500 mg/l acetone; 1.2 g/l acetonitrile; 92 g/l dimethyl sulphoxide; 1.9 g/l ethanol; 17 g/l methanol. At 25 °C, 50% loss by hydrolysis occurs in 0.5 y at pH5, 5-7 d at pH7-9. It is stable during storage (shelf life at 20 °C $\geqslant$2 y), but very sensitive to light.

**Uses.**

Thiocyclam is a selective insecticide acting as stomach poison and by contact with a 7-14 d residual activity and translocated acropetally. It is effective against lepidopterous and coleopterous pests, particularly *Leptinotarsa decemlineata* larvae on potatoes at 150 g a.i./ha, on rape, rice and vegetables at 200-500 g/ha.

**Toxicology.**

Acute oral $LD_{50}$: for male rats 310 mg a.i./kg; for male mice 273 mg/kg. In 2-y feeding trials NEL was: for rats 100 mg/kg diet; for dogs 75 mg/kg diet. $LC_{50}$ (96-h) for carp is 1.03 mg/l. It is moderately toxic to honeybees.

**Formulations.**

'Evisect S', water soluble powder (500 g a.i./kg); 'Evisect G', granules (50 g/kg).

**Analysis.**

Product analysis is by hplc with u.v. detection (M. Wisson *et al.*, *Anal. Methods Pestic. Plant Growth Regul.*, 1980, **11**, 185) or by tlc and subsequent u.v. determination of the eluted compound (details available from Sandoz AG). Residues are determined by glc with FPD or ECD (*idem, ibid.*, p. 190).

It is the aim of the editors of *The Pesticide Manual* to present accurate and up-to-date information and its contents have been stored in a computer. This will simplify modification of the text for future editions. The contents and periodic updates of the *Manual* will be available on-line as the Pesticide Databank from 1984.

The easiest way to find a given compound is by its ***Entry Number***. Because the majority of pesticides are known by various synonyms it is recommended that the reader ascertains the required ***Entry Number*** from one or more of the Indexes.

**Index 1** Wiswesser Line-Formula Notation (for description of this see p. 613).

**Index 2** Molecular formulae.

**Index 3** Code numbers used by official bodies, the manufacturers and suppliers. This index is arranged numerically and whether the qualifying letters are, for example, BAY or 'Bayer' is immaterial.

**Index 4** Chemical, common, and trivial names and trade marks.
Each of these give you the ***Entry Number***. A simple number (e.g. 1230) indicates that the entry is in the ***Main Section***, while a preceding asterisk (*2400) shows it is in the ***Superseded Compounds Section*** printed on coloured paper so that it is easy to find.

**Read the Section on How to Use Part I of the Manual,** page xvi.

Remember: an equals sign ( = ) at the end of a line means run on to the next line (omitting = ); chemical names ending in -, ( or [ run straight on. Otherwise the word finishes on a given line.

$C_9H_{18}N_2O_2S$ (218.3)
1X1&1&Y1S1&UNOVM1

$$CH_3SCH_2C(=N-OCO.NHCH_3)-C(CH_3)_3$$

**Nomenclature and development.**

Common name thiofanox (BSI, E-ISO, F-ISO, ANSI); thiofanocarb (Republic of South Africa). Chemical name (IUPAC) 1-(2,2-dimethyl-1-methylthiomethylpropylideneamino-oxy)-*N*-methylformamide; 3,3-dimethyl-1-methylthiobutanone *O*-methylcarbamoyloxime. (*C.A.*) 3,3-dimethyl-1-(methylthio)-2-butanone *O*-[(methylamino)carbonyl]oxime (9CI); Reg. No. *[39196-18-4]*. Its insecticidal properties were described by R. L. Schauer (*Proc. Br. Insect. Fungic. Conf., 7th*, 1973, **2**, 713). Relationship between chemical structure and biological activity of analogues was reported by T. A. Magee & L. E. Limpel (*J. Agric. Food Chem.*, 1977, **25**, 1376). Introduced by Diamond Shamrock Chemical Co. as code no. 'DS 15 647'; trade mark 'Dacamox'.

**Properties.**

It is a colourless solid with a pungent odour; m.p. 56.5-57.5 °C; v.p. 22.6 mPa at 25 °C. Solubility at 22 °C: 5.2 g/l water; sparingly soluble in aliphatic hydrocarbons; very soluble in chlorinated and aromatic hydrocarbons, ketones. Stable at normal storage temperatures and reasonably stable to hydrolysis $<30$ °C at $5<pH<9$.

**Uses.**

It is a systemic acaricide and insecticide, used as a soil or seed treatment and giving protection from foliage-feeding and soil-dwelling pests. In-furrow applications of: 0.4-1.0 kg a.i./ha to sugar beet controlled *Aphis fabae, Atomaria linearis, Chaetocnema concinna, Myzus persicae* and *Pegomya betae*; 1-3 kg/ha to potatoes controlled *Empoasca fabae, Epitrix* spp, *Leptinotarsa decemlineata* and *Macrosiphum* spp.

**Toxicology.**

Acute oral $LD_{50}$: for albino rats 8.5 mg/kg; for mallard duck 109 mg/kg; for bobwhite quail 43 mg/kg. Acute percutaneous $LD_{50}$ for albino rabbits 39 mg/kg. In 90-d feeding trials: NEL for beagle dogs was 1.0 mg/kg daily; the weight gain of rats receiving 100 mg/kg diet was not affected. $LC_{50}$ (96-h) is: for bluegill 0.33 mg/l; for rainbow trout 0.13 mg/l. Its degradation has been studied (W. T. Chin *et al.*, *ibid.*, 1976, **24**, 1001, 1071).

**Formulations.**

These include: granules, 'Dacamox 5G', 'Dacamox 10G', 'Dacamox 15G' (50, 100 and 150 g a.i./kg, respectively); seed treatment, 'Dacamox ST' (429 g/l).

**Analysis.**

Residues, which are likely to include thiofanox, its sulphoxide and sulphone, are oxidised to the sulphone, which is determined by glc giving the total residue (*idem, ibid.*, 1975, **23**, 963; M. B. Szalkowski *et al.*, *J. Chromatogr.*, 1976, **128**, 426). Details are available from Diamond Shamrock Corp.

$C_6H_{15}O_2PS_3$ (246.3)
2S2SPS&O1&O1

$$CH_3CH_2SCH_2CH_2S\overset{S}{\overset{\|}{P}}(OCH_3)_2$$

**Nomenclature and development.**

Common name thiometon (BSI, E-ISO, F-ISO, JMAF);dithiométon (France);M-81 (USSR);exceptions (Federal Republic of Germany, Portugal, Turkey). Chemical name (IUPAC) *S*-2-ethylthioethyl *O*,*O*-dimethyl phosphorodithioate. (*C.A.*) *S*-[2-(ethylthio)ethyl] *O*,*O*-dimethyl phosphorodithioate (8 & 9CI); Reg. No. *[640-15-3]*. Introduced by Bayer AG (who no longer manufacture or market it) and independently by Sandoz AG (DEP 917 668 to Bayer; CHP 319 579 to Sandoz) as code no. 'Bayer 23 129'; trade mark 'Ekatin' (to Sandoz AG).

**Properties.**

Pure thiometon is a colourless oil, with a characteristic odour; b.p. 110 °C/0.1 mmHg; v.p. 23 mPa at 20 °C; $d_4^{20}$ 1.209, $n_D^{20}$ 1.5515. Solubility at 25 °C: 200 mg/l water; readily soluble in most organic solvents; slightly soluble in petroleum spirit. At 25 °C, 50% loss by hydrolysis takes 25 d at pH 3, 27 d at pH 6, 17 d at pH 9. It is unstable in its pure form, but dilute solutions in nonpolar solvents and its formulations are stable (shelf-life at 20 °C *c*. 2 y).

**Uses.**

It is a systemic insecticide and acaricide which controls sucking insects, mainly aphids, and mites on most crops. It is metabolised in plants, animals and soils to the sulphoxide and sulphone, to the corresponding phosphorothioates, and finally to derivatives of *O*,*O*-dimethyl hydrogen phosphate and 2-ethylthioethyl mercaptan.

**Toxicology.**

Acute oral $LD_{50}$ for rats 120-130 mg/kg. Acute percutaneous $LD_{50}$ for rats $>$1000 mg/kg. In 2-y feeding trials NEL was: for dogs 6 mg/kg diet; for rats 2.5 mg/kg diet. $LC_{50}$ (96-h) for carp is 13.2 mg/l.

**Formulations.**

These include: 'Ekatin', e.c. (245 g a.i./l), 'Ekatin ULV', concentrate (125 g/l). 'Ekatin WF', e.c. (250 g thiometon + 100 g parathion/kg); 'Ekatin WF' ULV, concentrate (150 g thiometon + 60 g parathion/l); 'Thiolonan', e.c. (100 g thiometon + 100 g parathion/l). 'Serk' (70 g thiometon + 200 g endosulfan/l). 'Tombel', e.c. (160 g thiometon + 160 g quinalphos/l).

**Analysis.**

Product analysis is by glc (M. Wisson *et al.*, *Anal. Methods Pestic. Plant Growth Regul.*, 1976, **8**, 239), by i.r. spectroscopy, or by paper chromatography, followed by combustion and subsequent determination of phosphate by standard methods (*CIPAC Handbook*, 1980, **1A**, 1355). Residues may be determined by glc (M. Wisson, *loc. cit.*, p.244; *Pestic. Anal. Man.*, 1979, **I**, 201-H, 201-I; D. C. Abbott *et al.*, *Pestic. Sci.*, 1970, **1**, 10).

$C_8H_{13}N_2O_3PS$ (248.2)
T6N DNJ BOPS&O2&O2

$$\text{pyrazin-2-yl-}OP(S)(OCH_2CH_3)_2$$

**Nomenclature and development.**

Common name thionazin (BSI, E-ISO); thionazine (F-ISO). Chemical name (IUPAC) *O,O*-diethyl *O*-pyrazin-2-yl phosphorothioate. (*C.A.*) *O,O*-diethyl *O*-(2-pyrazinyl) phosphorothioate (8 & 9CI); Reg. No. *[297-97-2]*. ENT 25 580. Introduced by American Cyanamid Co. (USP 2 918 468; 2 938 831; 3 091 614) as code no. 'Experimental Nematicide 18 133'; trade marks 'Nemafos', 'Zinofos', in USA only 'Cynem', in UK only 'Neal's Bulb Dip'.

**Properties.**

Pure thionazin is a colourless to pale yellow liquid, m.p. −1.67 °C; v.p. 400 mPa at 30 °C; $d^{25}$ 1.207; $n_D^{25}$ 1.5080. Solubility at 27 °C: 1.14 g/l water; miscible with most organic solvents. It is readily hydrolysed by alkali (pH>9). The technical grade (*c*.90% pure) is a dark brown liquid $d^{25}$ 1.204-1.210.

**Uses.**

Thionazin is a soil insecticide and nematicide effective against a number of plant-parasitic as well as free-living nematodes, including those attacking buds, bulbs, leaves and roots, as well as against soil dwelling pests such as root maggots and symphylids and foliar insects such as aphids and leaf miners. When incorporated in mushroom compost at spawning it is effective against mushroom flies. It is of short persistence.

**Toxicology.**

Acute oral $LD_{50}$ for rats 12 mg tech./kg. Acute percutaneous $LD_{50}$ for rats 11 mg/kg. In 90-d feeding trials, rats receiving 25 mg/kg diet for 30 d and 50 mg/kg diet for the remaining 60 d showed a moderate depression in growth rate, but no abnormal behavioural reactions. $LC_{50}$ (5-d) was: for bobwhite quail 65 mg/kg diet; for pheasants 72 mg/kg diet. $LC_{50}$ (48-h) for harlequin fish is 0.09 mg (as 48% e.c.)/l. Topical $LD_{50}$ for honeybees is 0.042 µg/l.

**Formulations.**

An e.c. (460 g a.i./l); granules (50 or 100 g/kg).

**Analysis.**

Product analysis is by hydrolysis with fluorimetric determination of the sodium salt of pyrazin-2-ol (U. Kiigemagi *et al*., *J. Agric. Food Chem*., 1963, **11**, 293). Residues may be determined by glc (D. R. Coahran, *Bull. Environ. Contam. Toxicol*., 1966, **1**, 208).

$C_{14}H_{18}N_4O_4S_2$ (370.4)
2OVMYUS&MR BMYUS&MVO2

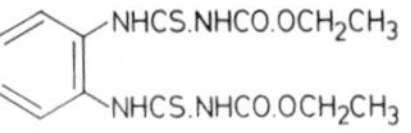


**Nomenclature and development.**

Common name thiophanate (BSI, E-ISO, F-ISO, JMAF, BPC); thiophanate-éthyl (France). Chemical name (IUPAC) diethyl 4,4'-(*o*-phenylene)bis(3-thioallophanate) (I); 1,2-di-(3-ethoxycarbonyl-2-thioureido)benzene. (*C.A.*) diethyl [1,2-phenylenebis(iminocarbonothioyl)]biscarbamate (9CI); (I) (8CI); Reg. No. *[23564-06-9]*. Its fungicidal properties were described by K. Ishii (*Abstr. Int. Congr. Plant Prot., 7th, Paris*, 1970, p. 200); he also reviewed it (*Jpn. Pestic. Inf.*, 1971, No. 7, p. 27). Introduced by Nippon Soda Co. Ltd (DEP 1 930 540) as code no. 'NF 35'; trade marks 'Topsin', 'Cercobin', 'Nemafax' (May & Baker Ltd as an anthelmintic).

**Properties.**

Thiophanate is a colourless crystalline solid; m.p. 195 °C (decomp.). Solubility: almost insoluble in water; sparingly soluble in most organic solvents. It forms unstable solutions of salts with aqueous alkali and forms complexes with divalent transition metal ions, for example copper.

**Uses.**

It is a systemic fungicide with a broad range of action, effective against *Venturia* spp. on apple and pear crops, powdery mildews, *Botrytis* spp. and *Sclerotinia* spp. on various crops. In plants it is converted into ethyl benzimidazol-2-ylcarbamate.

**Toxicology.**

Acute oral $LD_{50}$ for male and female rats >15 000 mg/kg. Acute percutaneous $LD_{50}$ for male and female rats >15 000 mg/kg. In 2-y feeding trials no ill-effect was observed in rats and mice. $LC_{50}$ (48-h) for carp is 20 mg a.i./l.

**Formulations.**

A w.p. (500 g a.i./kg).

**Analysis.**

Product analysis is by u.v. spectrometry.

$C_{12}H_{14}N_4O_4S_2$ (342.4)
1OVMYUS&MR BMYUS&MVO1

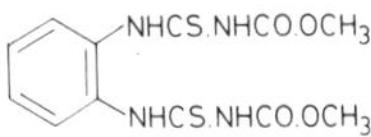


**Nomenclature and development.**

Common name thiophanate-methyl (BSI, E-ISO, F-ISO, ANSI, JMAF). Chemical name (IUPAC) dimethyl 4,4'-(*o*-phenylene)bis(3-thioallophanate) (I); 1,2-di-(3-methoxycarbonyl-2-thioureido)benzene. (*C.A.*) dimethyl [1,2-phenylenebis(iminocarbonothioyl)]biscarbamate (9CI); (I) (8CI); Reg. No. *[23564-05-8]*. Its fungicidal properties were described by K. Ishii (*Abstr. Int. Congr. Plant Prot., 7th, Paris*, 1970, p. 200). Introduced by the Nippon Soda Co. Ltd (DEP 1 930 540) as code no. 'NF 44'; trade marks 'Topsin M', 'Cercobin M' (Nippon Soda), 'Mildothane' (May & Baker Ltd), 'Cycosin' (American Cyanamid Co.).

**Properties.**

Thiophanate-methyl is a colourless crystalline solid; m.p. 172 °C (decomp.). Solubility at 20 °C: sparingly soluble in water; 58 g/kg acetone; 26 g/kg chloroform; 29 g/kg methanol. It forms complexes with copper salts. It is compatible with other agricultural chemicals that are neither highly alkaline nor contain copper. It has been shown to break down to carbendazim (1990) in plant tissues and on prolonged storage of aqueous suspensions. It, therefore, has activities similar to carbendazim and related fungicides, but is a more potent fungicide and has greater systemic properties than thiophanate (11660).

**Uses.**

It is a fungicide used at 30-50 g a.i./100 l and is effective against a wide range of fungal pathogens including: *Venturia* spp. on apples and pears; *Mycosphaerella musicola* on bananas; powdery mildews (*Erysiphe* and *Podosphaera* spp., and *Sphaerotheca fuliginea*) on apples, cucurbits, small grains, pears and vines; *Piricularia oryzae* on rice; *Botrytis, Cercospora* and *Sclerotinia* spp. on various crops.

**Toxicology.**

Acute oral $LD_{50}$: for male rats 7500 mg/kg, for females 6640 mg/kg. Acute percutaneous $LD_{50}$ for male and female rats >10 000 mg/kg; for Japanese quail >5000 mg/kg. In 2-y feeding trials NEL was: for rats and mice 160 mg/kg diet; for dogs 50 mg/kg diet. $LC_{50}$ (48-h) for carp is 11 mg/l.

**Formulations.**

These include: w.p. (500 or 700 g a.i./kg). Mixtures include. 'Labilit', w.p. (200 g thiophanate-methyl + 500 g maneb/kg). 'Homai', w.p. (500 g thiophanate-methyl + 300 g thiram/kg). Thiophanate-methyl + anthraquinone + gamma-HCH + 2-methoxyethylmercury silicate + pyracarbolid.

**Analysis.**

Product analysis is by u.v. spectrometry. Residues may be determined by colorimetry (V. L. Miller *et al., J. Assoc. Off. Anal. Chem.*, 1977, **60**, 1154; *Pestic. Anal. Man.*, 1979, **II**); by u.v. spectrometry (S. Ono *et al., Bunseki Kagaku*, 1975, **24**, 750) or by hplc (*idem, Nihon Noyaku Gakkaishi*, 1982, **7**, 363; N. Shiga *et al., ibid.*, 1977, **2**, 27).

$C_6H_{12}N_2S_4$ (240.4)
1N1&YUS&SSYUS&N1&1

$(CH_3)_2NCS.SS.CS.N(CH_3)_2$

**Nomenclature and development.**

Common name thiram (BSI, E-ISO); thirame (F-ISO); TMTD (USSR); thiuram (JMAF). Chemical name (IUPAC) tetramethylthiuram disulphide; bis(dimethylthiocarbamoyl) disulphide. (*C.A.*) tetramethylthioperoxydicarbonic diamide (9CI); bis(dimethylthiocarbomoyl) disulfide (8CI); Reg. No. *[137-26-8]*. ENT 987. Its fungicidal properties were described by W. H. Tisdale & A. L. Flenner (*Ind. Eng. Chem.*, 1942, **34**, 501). Introduced by E. I. du Pont de Nemours & Co. (Inc.)., Bayer AG and later by other companies. (USP 1 972 961 to du Pont; DRP 642 532 to I. G. Farbenindustrie) as trade marks 'Arasan', 'Tersan' (both to du Pont), 'Pomarsol' (to Bayer AG), 'Ferna-Col', 'Fernasan' (both to ICI Plant Protection Division).

**Properties.**

Pure thiram forms colourless crystals; m.p. 146 °C; $d_4^{20}$ 1.29; v.p. negligible at room temperature. Solubility at room temperature: *c*. 30 mg/l water; 80 g/l acetone; 230 g/l chloroform; $<$10 g/l ethanol. Some deterioration occurs on prolonged exposure to air, heat or moisture.

**Uses.**

It is a protective fungicide suitable for application to foliage to control: *Botrytis* spp. on lettuces, ornamentals, soft fruit and vegetables; rusts on ornamentals; *Venturia pirina* on pears. Also as a seed treatment, sometimes with added insecticide or other fungicide, to control 'damping off' diseases of maize, ornamentals and vegetables. At high doses it is repellent to birds, rodents and deer in fields and orchards.

**Toxicology.**

Acute oral $LD_{50}$: for rats 780-865 mg/kg; for mice 1500-2000 mg/kg; for rabbits 210 mg/kg; for redwing blackbird $>$100 mg/kg. Application of the dry powder to the skin of humans produced very slight erythema in 9% of cases. In chronic feeding trials NEL is: for rats $<$250 mg/kg diet (2.5 mg/kg daily); for dogs 200 mg/kg diet (5.0 mg/kg daily). $LC_{50}$ (48-h) is: for bluegill 0.23 mg/l; for carp 4 mg/l; for trout 0.13 mg/l.

**Formulations.**

These include: w.p. ($\leqslant$ 800 g a.i./kg); colloidal suspension; seed treatment. Mixtures include: thiram + gamma-HCH.

**Analysis.**

Product analysis is by conversion to carbon disulphide and colorimetric estimation using standard methods (*AOAC Methods*, 1980, 6.342-6.345), or by conversion to dimethylamine which is estimated by titration (*CIPAC Handbook*, 1970, **1**, 672; 1980, **1A**, 1360). Residues may be determined by conversion to carbon disulphide estimated by glc or colorimetry of a derivative (*AOAC Methods*, 1980, 29.165-29.171; H. L. Pease, *J. Assoc. Off. Agric. Chem.*, 1957, **40**, 1113; G. E. Keppel *J. Assoc. Off. Anal. Chem.*, 1969, **52**, 162; G. E. Keppel *et al.*, *ibid.*, 1971, **54**, 528; R. Mestres *et al.*, *Trav. Soc. Pharm. Montpellier*, 1973, **33**, 191).

$C_{16}H_{25}NOS$ (279.4)
2Y1&NVS1R&Y2&1

$(CH_3CH_2CH(CH_3))_2NCO.SCH_2-C_6H_5$

**Nomenclature and development.**

Common name tiocarbazil (BSI, draft E-ISO). Chemical name (IUPAC) *S*-benzyl di-*sec*-butyl(thiocarbamate). (*C.A.*) *S*-phenylmethyl bis(1-methylpropyl)carbamothioate (9CI); Reg. No. *[36756-79-3]*. Its herbicidal properties were described by N. Caracalli *et al.* (*Proc. Congr. Risicultura, 8th*, 1973, p. 446). Introduced by Montedison S.p.A. (ITP 907 710; DEP 2 144 700) as trade mark 'Drepamon'.

**Properties.**

Pure tiocarbazil is a colourless liquid; b.p. 130-132 °C 0.1 mmHg; v.p. 93 mPa at 50 °C; $d_4^{20}$ 1.023; $n_D^{20}$ 1.535. Solubility at 30 °C: 2.5 mg/l water; miscible with most organic solvents. It is stable to hydrolysis at $5.6<pH<8.4$ but is slightly decomposed after 30 d at 40 °C in aqueous ethanolic solution at pH 1.5. It suffered no decomposition on storage for 60 d at 40 °C nor in aqueous solutions exposed to sunlight for 100 h.

**Uses.**

It is a pre-em. and post-em. herbicide effective against weeds in rice, in particular *Cyperus* spp., *Echinochloa colonum, E. crusgalli, Leptochloa fascicularis* and *Lolium perenne* at 4.0 kg a.i./ha, provided that the rice field remains submerged for ⩾20 d after application. No ill-effect was seen in the rice varieties tested.

**Toxicology.**

Acute oral $LD_{50}$: for rats, guinea-pigs, rabbits, chickens, quails and pheasants > 10 000 mg tech./kg; for mice 8000 mg/kg. There was no mortality to rats or rabbits treated percutaneously with 1200 mg/kg. In 2-y feeding trials albino rats and beagle dogs receiving 1000 mg tech./kg diet suffered no ill-effect except for a slight weight loss in male dogs. There was no significant difference between controls and albino rats receiving 300 mg/kg diet for 3 generations. The $LC_{50}$ for the fish species tested is ⩾8 mg/l; for the mollusc *Australorbis glabratus* >60 mg/l.

**Formulations.**

These include: w.s.c. (500 or 700 g a.i./l); seed treatment (700 g/l); granules (50 and 75 g/kg).

**Analysis.**

Product analysis is by glc with FID (R. Fabbrini & G. Galluzzi, *Anal. Methods Pestic. Plant Growth Regul.*, 1980, **11**, 307). Residues may be determined by glc with FPD of a derivative (*idem, ibid.*). Details of methods are available from Montedison S.p.A.

$C_9H_{11}Cl_2O_3PS$ (301.1)
GR CG E1 BOPS&O1&O1



**Nomenclature and development.**

Common name tolclofos-methyl (BSI, E-ISO, F-ISO). Chemical name (IUPAC) *O*-2,6-dichloro-*p*-tolyl *O,O*-dimethyl phosphorothioate. (*C.A.*) *O*-(2,6-dichloro-4-methylphenyl) *O,O*-dimethyl phosphorothioate; (9CI); Reg. No. *[57018-04-9]*. Introduced by Sumitomo Chemical Co. (GBP 1 467 561; USP 4 039 635) as code no. 'S-3349'; trade mark 'Rizolex'.

**Properties.**

The technical grade is a colourless to light brown solid, m.p. 78-80 °C; v.p. 57 mPa at 20 °C. Solubility at 23 °C: 0.3-0.4 mg/l water; at 26 °C: 1 kg/kg acetone; 537 g/kg cyclohexanone; 537 g/kg xylene.

**Uses.**

It is a fungicide for controlling soil borne diseases caused by *Rhizoctonia solani, Sclerotium rolfsii*, and *Typhula* spp. on various crops. Typical rates are: 5-10 kg a.i./ha for soil incorporation; 50-200 g/100 m of row for in-furrow spray, 100-200 g/t of seed potatoes; 200-300 g/100 kg of cotton seeds, 0.5-1.0 kg/ha for foliar spray.

**Toxicology.**

Acute oral $LD_{50}$: for rats *c.* 5000 mg/kg; for bobwhite quail and mallard duck >5000 mg/kg. $LC_{50}$ (96-h) for carp is 2.13 mg/l.

**Formulations.**

These include: e.c. (200 g a.i./l); dust (50, 100 or 200 g/kg); w.p. (500 g/kg); s.c. (250 g/kg).

**Analysis.**

Details of methods for product and residue analysis are available from Sumitomo Chemical Co.

$C_{10}H_{13}Cl_2FN_2O_2S_2$ (347.2)
GXGFSNR D1&SWN1&1

$(CH_3)_2N.SO_2.N - SCCl_2F$ (on a benzene ring bearing $CH_3$ in the para position)

**Nomenclature and development.**

Common name tolylfluanid (BSI, E-ISO); tolylfluanide (F-ISO). Chemical name (IUPAC) *N*-dichlorofluoromethylthio-*N'*,*N'*-dimethyl-*N*-*p*-tolysulphamide; *N*-dichlorofluoromethanesulphenyl-*N'*,*N'*-dimethyl-*N*-*p*-tolylsulphamide. (*C.A.*) 1,1-dichloro-*N*-[(dimethylamino)sulfonyl]-1-fluoro-*N*-(4-methylphenyl)methanesulfenamide (9CI); *N*-[(dichlorofluoromethyl)thio]-*N'*,*N'*-dimethyl-*N*-*p*-tolysulfamide (8CI); Reg. No. *[737-27-1]*. Its fungicidal properties were described by H. Kaspers & F. Grewe (*Abstr. Intr. Congr. Crop. Prot., 6th Vienna*, 1967, p. 345). Introduced by Bayer AG (DEP 1 193 498) as code no. 'Bayer 49 854'; trade mark 'Euparen M'.

**Properties.**

It is a colourless to pale yellow powder; m.p. 95-97 °C; v.p. 13 µPa at 45 °C. Solubility at room temperature: 4 g/l water; 570 g/l benzene; 46 g/l methanol; 230 g/l xylene.

**Uses.**

It is a protective fungicide with a broad range of activity. It is mainly used in deciduous fruit crops and, when used regularly for control of *Venturia* spp., there is usually no need for additional steps to control *Phodosphaera leucotricha* on apples; or red spider mites (W. Kolbe, *Pflanzenschutz-Nachr. (Engl. Ed.)*, 1972, **25**, 123; R. Wäckers & C. van der Berge, *ibid.*, p.163).

**Toxicology.**

Acute oral $LD_{50}$: for rats >1000 mg/kg; for male mice >1100 mg/kg; for female guinea-pigs 250-500 mg/kg; for male rabbits 500 mg/kg; for female canaries 1000 mg/kg. Acute percutaneous $LD_{50}$ for male rats >500 mg/kg. In 90-d feeding trials rats receiving 1000 mg/kg showed no effect. $LC_{50}$ (96-h) is: for carp 0.25-0.5 mg/l; for golden orfe 0.07-0.25 mg/l.

**Formulations.**

A w.p. (500 g a.i./kg).

**Analysis.**

Product and residue analysis is by glc, details are available from Bayer AG.

$C_{22}H_{19}Br_4NO_3$ (665.0)
L3TJ A1 A1 BYEXEEE CVOYCN&R
COR &&(1R)-CIS-(RS) FORM

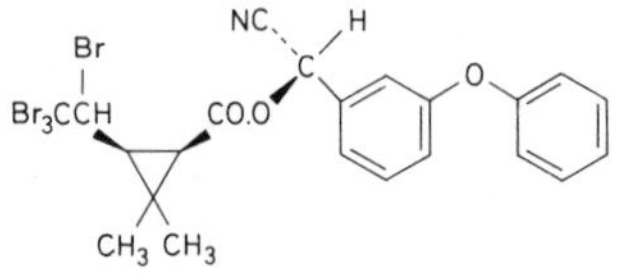

**Nomenclature and development.**

Common name tralomethrin (BSI, draft E-ISO); tralométhrine (draft F-ISO). Chemical name (IUPAC) (*S*)-α-cyano-3-phenoxybenzyl (1*R*,3*S*)-2,2-dimethyl-3-[(*RS*)-1,2,2,2-tetrabromoethyl]cyclpropanecarboxylate; (1*R*)-*cis*-2,2-dimethyl-3-[(*RS*)-1,2,2,2-tetrabromoethyl]cyclopropanecarboxylate. (*C.A.*) cyano(3-phenoxyphenyl)methyl 2,2-dimethyl-3-(1,2,2,2-tetrabromoethyl)cyclopropanecarboxylate (9CI); Reg. No. *[66841-25-6]* (unstated stereochemistry). Discovered by Roussel Uclaf and introduced (FP 2 364 884) as code no. 'RU 25 474','HAG 107' (Roussel Uclaf).

**Properties.**

Technical grade tralomethrin (>93% a.i.) is an orange to yellow resinous solid. The pure material is a 60:40 mixture of two diasteroisomers. It has $d^{20}$ 1.700; v.p. 17 pPa; $[\alpha]_D$ +21-27° (50 g/l toluene). Solubility: 70 mg/kg water; >1 kg/l acetone, dichloromethane, toluene, xylene; >500 g/l dimethyl sulphoxide; >180 g/l ethanol. Acidic media increase stability to hydrolysis and reduce epimerisation.

**Uses.**

It is a broad range insecticide effective in controlling most foliar insect pests of cotton at 16-22 g a.i./ha. It controls coleopterous, lepidopterous and other insects in potatoes (15-20 g/ha), in soyabeans and vegetables (11-15 g/ha).

**Toxicology.**

Acute oral $LD_{50}$: for rats ranges from 99-3000 mg/kg depending on the carrier used; for quail >2510 mg/kg; for dogs >500 mg (in capsules)/kg. Acute percutaneous $LD_{50}$ for rabbits >2000 mg/kg; slightly irritant to skin and eyes. Acute inhalation $LC_{50}$ for rats >286 mg/m³ air. In 90-d feeding trials NEL was: for rats 6 mg/kg daily; for dogs 1 mg/kg daily. $LC_{50}$ (8-d) was: for quail 4301 mg/kg diet; for ducks 7716 mg/kg diet. In laboratory tests $LC_{50}$ (96-h) is: for rainbow trout 1.6 μg/l; for bluegill 4.3 μg/l; $LC_{50}$ (48-h) for *Daphnia* 0.038 μg/l. In normal field use it is non-toxic to fish.

**Formulations.**

Concentrate (300 g tech./kg xylene); e.c. (360 g a.i./l); s.c. (150 g/l).

**Analysis.**

Product analysis is by hplc. Residues may be determined by glc with ECD.

$C_{14}H_{16}ClN_3O_2$ (293.8)
T5NN DNJ AYOR DG&VX1&1&1

**Nomenclature and development.**

Common name triadimefon (BSI, E-ISO); triadiméfone (F-ISO). Chemical name (IUPAC) 1-(4-chlorophenoxy)-3,3-dimethyl-1-(1*H*-1,2,4-triazol-1-yl)butanone. (*C.A.*) 1-(4-chlorophenoxy)-3,3-dimethyl-1-(1*H*-1,2,4-triazol-1-yl)-2-butanone (9CI); Reg. No. *[43121-43-3]*. Its fungicidal properties were described by P. E. Frohberger (*Mitt. Biol. Bundesanst. Land-Forstwirtsch. Berlin-Dahlem*, 1973, **151**, 61) and by F. Grewe & K. H. Büchel (*ibid.*, p. 208). It was introduced by Bayer AG (BEP 793 867; USP 3 912 752) as code no. 'BAY MEB 6447'; trade mark 'Bayleton'.

**Properties.**

It is a colourless solid; m.p. 82.3 °C; v.p. <100 µPa at 20 °C. Solubility at 20 °C: 260 mg/l water; 0.6-1.2 kg/kg cyclohexanone; 1.2 kg/kg dichloromethane; 200-400 g/kg propan-2-ol; 400-600 g/kg toluene. It is stable at pH 1 and pH 13 for 7 d at 20 °C.

**Uses.**

It is a systemic fungicide with protective and curative action, effective against mildews and rusts attacking cereals, coffee, grapes, ornamentals, stone fruit and vegetables at 2.5-6.2 g a.i./100 l (H. Buchenauer, *Pflanzenschutz-Nachr. (Engl. Ed.)*, 1976, **29**, 266; R. Siebert, *ibid.*, p. 303; W. Kolbe, *ibid.*, p. 310). Its mode of action has been investigated (H. Buchenauer, *ibid.*, p. 281, *idem, Pestic. Sci.*, 1978, **9**, 497) and the formation of the corresponding alcohol, the fungicide triadimenol (11830) demonstrated in plants (T. Clark *et al.*, *ibid.*, p. 503; H. Buchenauer, *Pflanzenschutz-Nachr. (Engl. Ed.)*, 1976, **29**, 266).

**Toxicology.**

Acute oral $LD_{50}$: for rats 363-568 mg/kg; for mice 989-1071 mg/kg; for Japanese quail 1750-2500 mg/kg. Acute percutaneous $LD_{50}$ for rats >1000 mg/kg. In 2-y feeding trials NEL was: for male rats 500 mg/kg diet, for females 50 mg/kg diet; for dogs 330 mg/kg diet. $LC_{50}$ (96-h) for goldfish is 10-50 mg/l. It is non-toxic to honeybees.

**Formulations.**

These include: w.p. (50 or 250 g a.i./kg); e.c. (100 g/l); dust (10 g/kg).

**Analysis.**

Product analysis is by i.r. spectrometry. Residues of triadimefon and the corresponding alcohol (triadimenol) may be determined by glc (W. Specht, *ibid.*, 1977, **30**, 55; W. Specht & M. Tillkes, *ibid.*, 1980, **33**, 61).

# Triadimenol 11830

$C_{14}H_{18}ClN_3O_2$ (295.8)
T5NN DNJ AYOR DG&YQX1&1&1

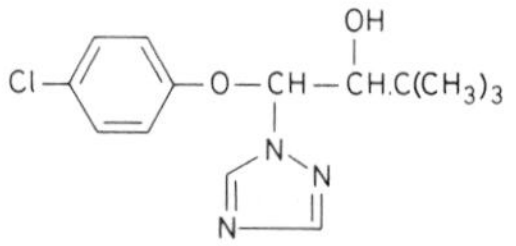

**Nomenclature and development.**

Common name triadimenol (BSI, E-ISO, F-ISO). Chemical name (IUPAC) 1-(4-chlorophenoxy)-3,3-dimethyl-1-(1*H*-1,2,4-triazol-1-yl)butan-2-ol, it is a mixture of two diastereoisomeric forms. (*C.A.*) β-(4-chlorophenoxy)-α-(1,1-dimethylethyl)-1*H*-1,2,4-triazole-1-ethanol (9CI); Reg. No. *[55219-65-3]*. Its fungicidal properties were described by P. E. Frohberger (*Pflanzenschutz-Nachr. (Engl. Ed.)*, 1978, **31**, 11). Introduced by Bayer AG (DEPS 2 324 010) as code no 'BAY KWG 0519'; trade mark 'Baytan'.

**Properties.**

It forms colourless crystals; m.p. *c.* 121-127 °C; v.p. $<1$ mPa at 20 °C. Solubility at 20 °C: 95 mg/l water; 100-200 g/l dichloromethane; $<1$ g/kg light petroleum (b.p. 80-110 °C); 10-100 g/l propan-2-ol.

**Uses.**

It is a systemic fungicide used as a cereal seed treatment to control *Tilletia caries*, and *Ustilago* spp.

**Toxicology.**

Acute oral $LD_{50}$ for rats 700-1200 mg/kg. Acute percutaneous $LD_{50}$ for rats $>5000$ mg/kg. In 90-d feeding trials rats receiving 600 mg/kg diet showed no symptom of poisoning. $LC_{50}$ (96-h) is: for goldfish 10-50 mg/l; for rainbow trout 23.5 mg/l.

**Formulations.**

Powder for dry seed treatment; water dispersible powder for slurry treatment; s.c. for seed treatment. 'Baytan', (Reg. No. *[72767-01-2]*), seed treatment (triadimenol + fuberidazole).

**Analysis.**

Details of product analysis are available from Bayer AG. Residues may be determined by glc (W. Specht & M. Tillkes, *Pflanzenschutz-Nachr. (Engl. Ed.)*, 1980, **33**, 61).

$C_{10}H_{16}Cl_3NOS$ (304.7)
GYGUYG1SVNY1&1&Y1&1

$[(CH_3)_2CH]_2NCO.SCH_2C(Cl){=}CCl_2$

**Nomenclature and development.**

Common name tri-allate (BSI, E-ISO); triallate (F-ISO, WSSA). Chemical name (IUPAC) *S*-2,3,3-trichloroallyl di-isopropyl(thiocarbamate). (*C.A.*) *S*-(2,3,3-trichloro-2-propenyl) bis(1-methylethyl)carbamothioate (9CI); *S*-(2,3,3-trichloroallyl) diisopropylthiocarbamate (8CI); Reg. No. *[2303-17-5]*. Its herbicidal properties were described by G. Friesen (*Res. Proc. Nat. Weed Comm., Can.*, 1960). Introduced by Monsanto Co. (USP 3 330 821; 3 330 642) as code no. 'CP 23 426'; trade marks 'Avadex BW', 'Far-Go'.

**Properties.**

Pure tri-allate is an amber oil; m.p. 29-30 °C; b.p. 117 °C/0.3 mmHg; $d^{25}_{15.6}$ 1.273; v.p. 16 mPa at 25 °C. Solubility at 25 °C: 4 mg/l water; soluble in organic solvents. Stable to light, decomposition temperature >200 °C, non-flammable.

**Uses.**

It is a pre-plant or post-plant soil-incorporated herbicide used at 1.12-1.68 kg a.i./ha to control wild oats in barley, lentils, peas and spring and winter wheat. Absorbed mainly by coleoptile of wild oats, inhibiting cell division and affecting cell elongation. In soil, main loss is microbiological (or by volatilisation, if not incorporated in soil).

**Toxicology.**

Acute oral $LD_{50}$: for rats 1675-2165 mg/kg; for bobwhite quail >2251 mg/kg. Acute percutaneous $LD_{50}$ for rabbits 8200 mg tech./kg; slightly irritating to eye and moderately to skin of rabbits; the e.c. is severely irritating to skin. In 2-y feeding trials no adverse effect was observed in rats receiving 200 mg/kg diet or in dogs at 15 mg/kg daily (highest doses tested). $LC_{50}$ (8-d) for mallard ducklings and bobwhite quail >5000 mg/kg diet. $LC_{50}$ (96-h) is: for bluegill 1.3 mg/l; for rainbow trout 1.2 mg/l.

**Formulations.**

These include: e.c. (480 g a.i./l); granules (100 g/kg).

**Analysis.**

Product analysis is by glc. Residues in soil, grain or straw may be determined by glc with ECD (T. H. Byast *et al.*, *Tech. Rep. ARC Weed Res. Organ.*, No. 15 (2nd Ed.), p.17).

$C_{12}H_{16}N_3O_3PS$ (313.3)
T5NN DNJ AR& COPS&O2&O2

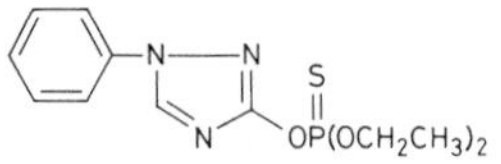

**Nomenclature and development.**

Common name triazophos (BSI, E-ISO, F-ISO). Chemical name (IUPAC) *O,O*-diethyl *O*-1-phenyl-1*H*-1,2,4-triazol-3-yl phosphorothioate. (*C.A.*) *O,O*-diethyl *O*-(1-phenyl-1*H*-1,2,4-triazol-3-yl) phosphorothioate (8 & 9CI); Reg. No. *[24017-47-8]*. Its insecticidal properties were described by M. Vulic *et al.*, (*Abstr. Int. Congr. Plant Prot., 7th, Paris*, 1970, p.123). Introduced by Hoechst AG (DEP 1 670 876; 1 299 924; USP 3 686 200) as code no. 'Hoe 2960'; trade mark 'Hostathion'.

**Properties.**

Pure triazophos is a yellowish oil; m.p. 2-5 °C; v.p. 13 mPa at 55 °C; $d_4^{20}$ 1.247; $n_D^{20}$ 1.5500-1.5503. Solubility at 20 °C: 30-40 mg/kg water; >1 kg/kg acetone, ethyl acetate; >330 g/kg ethanol, toluene; 9 g/kg hexane. The technical grade (≥92% pure) has m.p. 0-5 °C.

**Uses.**

It is a broad spectrum insecticide and acaricide with some nematicidal properties. It controls aphids on cereals at 320-600 g a.i. (40% e.c.)/ha, on fruit at 75-125 g a.i./100 l. It is also used to control Lepidoptera on fruit and vegetables. When incorporated at 1-2 kg a.i. (40% e.c.)/ha in the soil prior to planting, it controls *Agrotis* spp. and other cutworms. It can penetrate plant tissue, but has no systemic activity.

**Toxicology.**

Acute oral $LD_{50}$: for dogs >320 mg/kg; and, depending on carrier and sex: for rats 57-68 mg/kg, for Japanese quail 4.2-27.1 mg/kg. Acute percutaneous $LD_{50}$ for rats 1100 mg/kg. In 2-y feeding trials rats receiving 1 mg/kg diet and dogs 0.3 mg/kg diet, inhibition of blood serum cholinesterase was the only effect noted. $LC_{50}$ (96-h) is: for carp 5.6 mg/l; for golden orfe 11 mg/l.

**Formulations.**

These include: e.c. (245 g a.i./l or 428 g/l); granules (10-100 g/kg); various ULV concentrates.

**Analysis.**

Product analysis is by glc, details are available from Hoechst AG. Residues may be determined by glc with FPD (W. G. Thier *et al.*, *Anal. Methods Pestic. Plant Growth Regul.*, 1978, **10**, 127). Details are available from Hoechst AG.

$C_{12}H_{27}OPS_3$ (314.5)
4SPO&S4&S4

$(CH_3CH_2CH_2CH_2S)_3P{=}O$

**Nomenclature and development.**

Chemical name (IUPAC) *S,S,S*-tributyl phosphorothioate (I). (*C.A.*) (I) (8 & 9CI); Reg. No. *[78-48-8]*. Having been tested as a defoliant by Ethyl Corp. (now Boots-Hercules Agrochemical Co.) it was introduced by Chemagro Corp. (USP 2 943 107; 2 965 467) as code no. 'Chemagro B-1776'; trade mark 'Def Defoliant'.

**Properties.**

It is a pale yellow liquid, with a mercaptan-like odour; b.p. 150 °C/0.3 mmHg; m.p. <−25 °C; $d^{20}$ 1.06; $n_D^{25}$ 1.532. Practically insoluble in water; soluble in most organic solvents, including chlorinated hydrocarbons. It is relatively stable to heat and to acids but is slowly hydrolysed under alkaline conditions.

**Uses.**

It is strongly phytotoxic and is recommended for the defoliation of cotton; for complete defoliation 1.25-2.0 kg a.i./ha, for bottom defoliation 1.0-1.5 kg/ha.

**Toxicology.**

Acute oral $LD_{50}$ for female rats 325 mg/kg (S. D. Murphy & K. P. DuBois, *AMA Arch. Ind. Health*, 1959, **20**, 161). Acute percutaneous $LD_{50}$ for male rats 850 mg/kg. In 84-d feeding trials dogs receiving 25 mg/kg diet showed no ill-effect.

**Formulations.**

Spray concentrate (700 g a.i./l).

**Analysis.**

Product analysis is by i.r. spectroscopy. Residues in cotton seed may be determined by glc (R. F. Thomas & T. H. Harris, *J. Agric. Food Chem.*, 1965, **13**, 505; D. MacDougall, *Anal. Methods Pestic., Plant Growth Regul. Food Addit.*, 1964, **4**, 89; *Anal. Methods Pestic. Plant Growth Regul.*, 1972, **6**, 627). Details of methods are available from Chemagro Corp.

$C_{12}H_{27}PS_3$ (298.5)
4SPS4&S4

$(CH_3CH_2CH_2CH_2S)_3P$

**Nomenclature and development.**

Chemical name (IUPAC) tributyl phosphorotrithioite (I). (*C.A.*) (I) (8 & 9CI); Reg. No. *[150-50-5]*. Trivial name merphos. Introduced by the Mobil Chemical Co (who no longer manufacture or market it) and later by Rhône-Poulenc Phytosanitaire (USP 2 955 803 to Mobil Chemical Co.) as trade mark 'Folex'.

**Properties.**

The pure compound is a colourless to pale yellow liquid; b.p. 115-134 °C/0.08 mmHg; $d_4^{20}$ 0.99-1.01; $n_D^{25}$ 1.55. Solubility: sparingly soluble in water; very soluble in most organic solvents. The technical grade is ⩾95% pure.

**Uses.**

It is used at 1.2-2.5 kg a.i./ha to defoliate cotton and can induce leaf abscission in some other plants, such as roses and hydrangeas.

**Toxicology.**

Acute oral $LD_{50}$ for male albino rats 1272 mg a.i./kg. Acute percutaneous $LD_{50}$ for albino rabbits >4600 mg/kg. In 90-d feeding trials dogs and rats receiving 750 mg/kg diet showed depression of cholinesterase but no other effect on pathology or histology.

**Formulations.**

'Folex Cotton Defoliant' (720 g a.i./l).

**Analysis.**

Product analysis is by glc. Residues may be determined by glc with MCD.

$C_{10}H_7Cl_5O_2$ (336.4)
GXGGYOV1&R CG DG

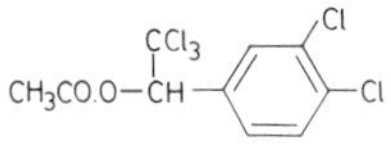

CH3CO.O—CH(CCl3)—C6H3(Cl)2

**Nomenclature and development.**

Chemical name (IUPAC) 2,2,2-trichloro-1-(3,4-dichlorophenyl)ethyl acetate. (*C.A.*) 3,4-dichloro-α-(trichloromethyl)benzenemethyl acetate (9CI); 3,4-dichloro-α-(trichloromethyl)benzyl acetate (8CI); Reg. No. *[51366-25-7]*. Its insecticidal properties were reported by W. Behrenz *et al.* (*Pflanzenschutz-Nachr. (Engl. Ed.)*, 1977, **30**, 237). Introduced by Bayer AG (DEP 2 110 056) as code no. 'BAY MEB 6046'; trade mark 'Baygon MEB'.

**Properties.**

The pure compound is a colourless crystalline solid; m.p. 84.5 °C; v.p. 14 μPa at 20 °C. Solubility at 20 °C: 50 mg/kg water; >600 g/kg cyclohexanone; <10 g/kg propan-2-ol.

**Uses.**

It is an insecticide, recommended for use against such pests as clothes moths, flies and mosquitoes.

**Toxicology.**

Acute oral $LD_{50}$: for rats >10 000 mg/kg; for hens >2500 mg/kg. Acute percutaneous $LD_{50}$ for rats >1000 mg/kg. Inhalation $LC_{50}$ (4-h) for rats >561- >742 mg/m³. In 90-d feeding trials NEL for rats was 1000 mg/kg diet. $LC_{50}$ (96-h) for golden orfe 0.5-1.0 mg/l.

**Formulations.**

Aerosol concentrates (20 g/l), in combination with dichlorvos; oil-based spray (10 g/l) same ingredients.

**Analysis.**

Product and residue analysis are by glc - details are available from Bayer AG.

$C_{10}H_{12}Cl_3O_2PS$ (333.6)
GR BG DG EOPS&2&O2

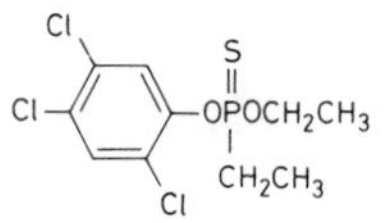

**Nomenclature and development.**

Common name trichloronate (BSI);trichloronat (E-ISO, F-ISO). Chemical name (IUPAC) *O*-ethyl *O*-2,4,5-trichlorophenyl ethylphosphonothioate. (*C.A.*) *O*-ethyl *O*-( 2,4,5-trichlorophenyl) ethylphosphonothioate (8 & 9CI); Reg. No. *[327-98-0]*. OMS 412, OMS 578, ENT 25 712; Its insecticidal properties were described by R. O. Drummond (*J. Econ. Entomol.*, 1963, **56**, 831). Introduced by Bayer AG (DEP 1 099 530) as code no. 'Bayer 37 289','S 4400'; trade marks 'Agrisil', 'Agritox', 'Phytosol'.

**Properties.**

It is an amber-coloured liquid; b.p. 108 °C/0.01 mmHg; $d_4^{20}$ 1.365. Solubility at 20 °C: 50 mg/l water; >1.2 kg/kg dichloromethane, propan-2-ol. It is hydrolysed by alkali.

**Uses.**

It is a non-systemic insecticide recommended for the control of root maggots, other soil-dwelling insects and wireworms.

**Toxicology.**

Acute oral $LD_{50}$: for rats 16-37.5 mg/kg; for rabbits 25-50 mg/kg; for chickens 45 mg/kg. Acute percutaneous $LD_{50}$ for male rats 135-341 mg/kg. In 2-y feeding trials rats receiving 3 mg/kg diet showed no symptom of poisoning. $LC_{50}$ (96-h) for rainbow trout is 0.2 mg/l; $LC_{50}$ (24-h) for goldfish >10 mg/l.

**Formulations.**

Granules, e.c. and seed dressings of various a.i. content.

**Analysis.**

Product analysis is by u.v. spectroscopy after hydrolysis to 2,4,5-trichlorophenol. Residues may be determined by glc (E. Möllhoff, *Pflanzenschutz-Nachr. (Engl. Ed.)*, 1968, **21**, 331). Details of both methods may be obtained from Bayer AG.

$C_4H_8Cl_3O_4P$ (257.4)
GXGGYQPO&O1&O1

$$Cl_3CCH(OH)P(O)(OCH_3)_2$$

**Nomenclature and development.**

Common name trichlorphon (BSI);trichlorfon (E-ISO, F-ISO);chlorophos (USSR);dipterex (Turkey);metriphonate (BPC);DEP (JMAF). Chemical name (IUPAC) dimethyl 2,2,2-trichloro-1-hydroxyethylphosphonate. (*C.A.*) dimethyl (2,2,2-trichloro-1-hydroxyethyl)phosphonate (8 & 9CI); Reg. No. *[52-68-6]*. OMS 800, ENT 19 763. Its insecticidal properties were described by G. Unterstenhöfer (*Anz. Schaedlingskd.*, 1957, **30**, 7). It was first prepared by W. Lorenz, and introduced by Bayer AG (USP 2 701 225) as code no. 'Bayer 15 922','Bayer L13/59'; trade marks 'Dipterex', 'Neguvon', 'Tugon', and in the USA 'Dylox'.

**Properties.**

It is a colourless crystalline powder; m.p. 83-84 °C; v.p. 1.0 mPa at 20 °C; $d_4^{20}$ 1.73. Solubility at 25 °C: 154 g/l water; soluble in benzene, ethanol and most chlorinated hydrocarbons; poorly soluble in carbon tetrachloride, diethyl ether; insoluble in petroleum oils. It is stable at room temperature but is decomposed by hot water and at $pH<5.5$ forming dichlorvos (4490).

**Uses.**

Trichlorphon is a contact and stomach insecticide with penetrant action recommended for agricultural use against fruit flies and lepidopterous larvae at 75-120 g a.i./100 l, for controlling household pests especially flies (with special formulations) and for the control of ectoparasites of domestic animals. Its activity is attributed to its metabolic conversion to dichlorvos (R. L. Metcalf *et al.*, *J. Econ, Entomol.*, 1959, **52**, 44).

**Toxicology.**

Acute oral $LD_{50}$ for rats 560-630 mg/kg. Acute percutaneous $LD_{50}$ for rats >2000 mg/kg. In 2-y feeding trials rats receiving 500 mg/kg diet suffered no ill-effect. $LC_{50}$ (48-h) is: for carp 6.2 mg/l; for goldfish >10 mg/l.

**Formulations.**

These include: w.p. (500 g a.i./kg); s.p. (500, 800 or 950 g/kg); s.c. (500 g/l); ULV concentrate (250, 500 or 750 g/l); dust (50 g/kg); granules (25 or 50 g/kg); 'Tugon Stable Spray' (300 g/l); 'Tugon Fly Bait' (10 g/kg); 'Nevugon' preparations for veterinary use.

**Analysis.**

Product analysis is by polarography (P. A. Giang & R. L. Caswell, *J. Agric. Food Chem.*, 1957, **5**, 753) or by titration of the chloride ion obtained by hydrolysis (*CIPAC Handbook*, 1970, **1**, 684; FAO Specification (CP/51)). Residues may be determined by glc (*Pestic. Anal. Man.*, 1979, **I**, 201-I; D. C. Abbott *et al., Pestic. Sci.*, 1970, **1**, 10; M. A. Luke *et al.*, *J. Assoc. Off. Anal. Chem.*, 1981, **64**, 1187). Details of these and other methods may be obtained from Bayer AG.

$C_7H_4Cl_3NO_3$ (256.5)
T6NJ BO1VQ CG EG FG
*triclopyr-triethylammonium*
$C_{13}H_{19}Cl_3N_2O_3$ (371.7)
T6NJ BO1VQ CG EG FG &&$(CH_3CH_2)_3N$
SALT
*triclopyr-(2-butoxyethyl)*
$C_{13}H_{16}Cl_3NO_4$ (356.6)
T6NJ BO1VO2O4 CG EG FG

**Nomenclature and development.**

Common name triclopyr (BSI, E-ISO, F-ISO, ANSI, WSSA);*exception* (Eire). Chemical name (IUPAC) 3,5,6-trichloro-2-pyridyloxyacetic acid. (*C.A.*) [(3,5,6-trichloro-2-pyridinyl)oxy]acetic acid (8 & 9CI); Reg. No. *[55335-06-3]*. Introduced by Dow Chemical Co. as code no. 'Dowco 233'; trade mark 'Garlon'.

**Properties.**

Triclopyr is a fluffy colourless solid; m.p. 148-150 °C; v.p. 168 μPa at 25 °C. Solubility at 25 °C: 440 mg/l; 989 g/kg acetone; 27.3 g/kg chloroform; 410 mg/kg hexane; 307 g/kg octan-1-ol. It is stable under normal storage conditions and to hydrolysis, but subject to photolysis with 50% loss in $<$12 h.

It is acidic (p$K_a$ 2.68) forming water-soluble salts, and oil-soluble esters, *e.g.* triclopyr-(2-butoxyethyl) (Reg. No. *[64470-88-8]*).

**Uses.**

It is a herbicide which is rapidly absorbed by foliage and roots and translocated throughout the plant. It induces auxin-type responses in susceptible species namely most broad-leaved plants whilst Gramineae are usually unaffected at normal application rates. Used at 90 g a.e./ha in combination with 2,4-D and propanil in rice and wheat (but not in barley). Also in plantation crops (oil palm, rubber) at 125-250 g a.e./ha for covercrop maintenance and at 0.72-1.0 kg a.e/ha to control *Eupatorium odoratum* and other problem plants; in pastures at 1-2 kg/ha to control annual and perennial herbaceous weeds and 2-4 kg/ha against *Rubus* spp. and other woody plants; in forestry with 4-8 kg/ha for site preparation and 1.5-2.0 kg/ha for conifer release; for industrial sites 2-8 kg/ha; in rangeland at 0.24-1.0 kg/ha.

It undergoes relatively rapid microbial degradation in soil, under non-leaching conditions and those favouring microbial activity 50% loss averages 46 d.

**Toxicology.**

Acute oral $LD_{50}$: for rats 713 mg/kg; for rabbits 550 mg/kg; for guinea-pigs 310 mg/kg; for mallard duck 1698 mg/kg. Acute percutaneous $LD_{50}$ for rabbits $>$2000 mg/kg; it is essentially non-irritating to skin and a mild irritant to eyes. In 2-y feeding trials NEL was: for rats 10 mg/kg daily; for mice 240 mg/kg diet. $LC_{50}$ (8-d) was: for mallard duck $>$5000 mg/kg diet; for bobwhite quail 2935 mg/kg; for Japanese quail 3278 mg/kg. $LC_{50}$ (96-h) is: for rainbow trout 117 mg/l; for bluegill 148 mg/l. non-toxic to honeybees at 60.4 μg/bee.

**Formulations.**

These include: aqueous concentrate triclopyr-triethylammonium (360 g/l); e.c. (triclopyr-(2-butoxy)ethyl 100, 200 or 300 g a.e./l). Mixtures include: triclopyr + 2,4-D (as esters); triclopyr + picloram (as amine salts); triclopyr + 3,6-dichloropicolinic acid.

**Analysis.**

Product analysis is by hplc. Residues may be determined by glc. Details of methods are available from Dow Chemical Co.

$C_9H_7N_3S$ (189.2)
T B556 BN DNNGSJ L1

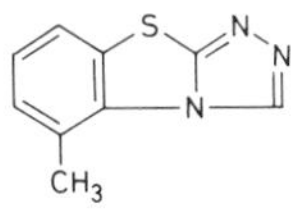

**Nomenclature and development.**

Common name tricyclazole (BSI, E-ISO, F-ISO, ANSI). Chemical name (IUPAC) 5-methyl-1,2,4-triazolo[3,4-*b*]benzothiazole (I). *(C.A.)* (I) (9CI); Reg. No. *[41814-78-2]*. Its fungicidal properties were described by J. D. Froyd *et al*., (*Phytopathology*, 1976, **66**, 1135). Introduced by Eli Lilly & Co. (GBP 1 419 121) as code no. 'EL-291'; trade marks 'Beam', 'Bim', 'Blascide'.

**Properties.**

It is a crystalline solid; m.p. 187-188 °C; v.p. 27 μPa at 25 °C. Solubility at 25 °C: 1.6 g/l water, 10.4 g/l acetone; 25 g/l methanol; 2.1 g/l xylene.

**Uses.**

It is a fungicide for control of *Pyricularia oryzae* in transplanted and direct-seeded rice by flat drench (1.5-2.4 g a.i./flat), root soak (0.5-2.0 g/$m^2$ of nursery bed) or foliar application (150-400 g/ha). One or 2 applications by one or more of these methods give a season-long control of the disease.

**Toxicology.**

Acute oral $LD_{50}$ for mice 245 mg/kg; for rats 314 mg/kg; for mallard duck and bobwhite quail >100 mg/kg; for beagle dogs >50 mg/kg. Acute percutaneous $LD_{50}$ >2000 mg/kg for rabbits; no skin irritation was seen at this dose, slight irritation was observed when 0.1 ml (78 mg) was applied to their eyes. In 2-y feeding trials NEL was: for rats 275 mg/kg; for mice 400 mg/kg. $LC_{50}$ (96-h) is: for goldfish fingerlings 13.5 mg/l; for bluegill 16.0 mg/l; for rainbow trout 7.3 mg/l.

**Formulations.**

These include: w.p. (200 or 750 g a.i./kg); dust (10 g/kg); granules (30 g/kg).

**Analysis.**

Product analysis is by glc with FID or by hplc (E. W. Day *et al. Anal. Methods Pestic. Plant Growth Regul*., 1980, **11**, 263). Residues in plant tissue are determined by glc with FPD, the main metabolite (hydroxymethyl analogue) first being converted to a derivative (*idem, ibid*.). Detailed analytical procedures and other information may be obtained from Eli Lilly & Co.

$C_{19}H_{39}NO$ approx. (297.5) approx.
T6N DOTJ A13 C1 E1 &&$C_{11}$-$C_{14}$
ISOMERS

$CH_3-(C_nH_{2n})-N$ [2,6-dimethylmorpholine ring: $CH_3$, O, $CH_3$]

n = 10, 11, 12 (60-70%) or 13

**Nomenclature and development.**

Common name tridemorph (BSI, E-ISO); tridémorphe (F-ISO) for a reaction mixture (see below). Chemical name (IUPAC) for the original definition 2,6-dimethyl-4-tridecylmorpholine (I); the reaction mixture has now been shown to comprise $C_{11}$ to $C_{14}$ 4-*alkyl*-2,6-dimethylmorpholine homologues containing 60-70% of 4-tridecyl isomers. (*C.A.*) (I) for original structure; Reg. No. *[24602-86-6]* (for original structure). Its fungicidal properties were described by J. Kradel *et al.* (*Proc. Br. Insectic. Fungic. Conf., 5th*, 1969, **1**, 16) and by E. -H. Pommer *et al.* (*ibid.*, 1969, **2**, 347). Introduced by BASF AG as code no. 'BASF 220'; trade marks 'Calixin'; 'Cosmic' (BASF) for a mixture with carbendazim and maneb.

**Properties.**

Tridemorph is a colourless oil, with a slight amine-like odour; b.p. 134 °C/0.5 mmHg; v.p. 40 mPa at 20 °C. Solubility: miscible with water, acetone, benzene, chloroform, cyclohexane, ethanol, olive oil. It is stable in unopened containers $\geqslant$2 y. It is compatible with most herbicides used in cereals.

**Uses.**

It is an eradicant fungicide with systemic action, being absorbed through foliage and roots to give some protectant action. It controls cereal mildews when applied between the 5-leaf and jointing stages of growth at 525 g a.i. in $\geqslant$250 l/ha, often in combination with a hormone-type herbicide or with chlormequat chloride. This treatment gives protection for 21-28 d when a second application may be made. The mixture with carbendazim plus maneb controls a wider range of diseases, including *Rhynchosporium* spp. on barley and eyespot on winter barley and winter wheat, also moderate attacks of brown rust and, on winter wheat, moderate attacks of yellow rust, and light to moderate attacks of mildew.

**Toxicology.**

Acute oral $LD_{50}$: for rats 825 mg a.i. (as e.c.)/kg; for rabbits 562 mg a.i. (as e.c.)/kg. Acute percutaneous $LD_{50}$ for rabbits *c.* 1350 mg a.i. (as e.c.)/kg. Rats survived a single exposure (8-h) to air saturated with the vapour at 20 °C.

**Formulations.**

'Calixin', e.c. (750 g a.i./l). 'Cosmic', w.p. (tridemorph + carbendazim + maneb).

**Analysis.**

Product analysis is by acid-base titration. Residues in cereal straw and soil may be determined by colorimetry of a derivative.

$C_9H_{16}ClN_5$ (229.7)
T6N CN ENJ BN2&2 DM2 FG

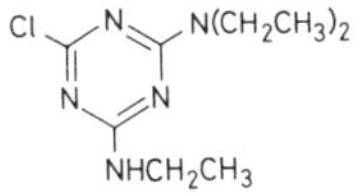

Cl, N, N(CH$_2$CH$_3$)$_2$, N, N, NHCH$_2$CH$_3$

**Nomenclature and development.**

Common name trietazine (BSI, E-ISO, F-ISO, ANSI, JMAF);*exception* (India). Chemical name (IUPAC) 6-chloro-*N*,*N*,*N*'-triethyl-1,3,5-triazine-2,4-diyldiamine; 2-chloro-4-diethylamino-6-ethylamino-1,3,5-triazine. (*C.A.*) 6-chloro-*N*,*N*,*N*'-triethyl-1,3,5-triazine-2,4-diamine (9CI); 2-chloro-4-(diethylamino)-6-(ethylamino)-*s*-triazine (8CI); Reg. No. *[1912-26-1]*. Its herbicidal properties were reported by H. Gysin & E. Knüsli (*Proc. Br. Weed Control Conf., 4th*, 1958, p. 225). It was discovered by J. R. Geigy S.A. (now Ciba-Geigy AG) and introduced commercially by Fisons Ltd (now FBC Limited) in combinations with other herbicides (CHP 329 277; GBP 814 947; USP 2 819 855 to Ciba-Geigy AG) as code no. 'G 27 901' (Ciba-Geigy AG), 'NC 1667' (FBC Limited); 'Bronox', 'Remtal' (both FBC Limited), 'Aventox' (Murphy Chemical Ltd.).

**Properties.**

It is a crystalline solid; m.p. 100-101 °C. Solubility at 25 °C: 20 mg/l water; 170 g/l acetone; 200 g/l benzene; >500 g/l chloroform; 100 g/l dioxane; 30 g/l ethanol. It is stable to air and water.

**Uses.**

Trietazine is a herbicide that is taken up by roots and foliage and inhibits the Hill reaction. It is used with linuron for weed control in potatoes, and with simazine in peas at 1.6-4.5 kg/ha according to crop and soil type.

**Toxicology.**

Acute oral $LD_{50}$: for rats 2830- >4000 mg/kg; for quail 800 mg/kg. Application (24-h) of 1000 mg/kg to the skin of rats produced no ill-effect. In 90-d feeding trials rats receiving 16 mg/kg diet showed no ill-effect. $LC_{50}$ (24-h) for guppies is 5.5 mg/l.

**Formulations.**

'Aventox', 'Remtal', s.c. (trietazine + simazine). 'Bronox', w.p. (trietazine + linuron). '*Furol*' is no longer available.

**Analysis.**

Residues may be determined by glc with FID (T. H. Byast *et al.*, *Tech. Rep. ARC Weed Res. Organ.*, No. 15 (2nd Ed.), pp. 40, 81).

$C_{13}H_{16}F_3N_3O_4$ (335.5)
FXFFR CNW ENW DN3&3

**Nomenclature and development.**

Common name trifluralin (BSI, E-ISO, ANSI, WSSA, JMAF);trifluraline (F-ISO). Chemical name (IUPAC) α,α,α-trifluoro-2,6-dinitro-*N*,*N*-dipropyl-*p*-toluidine (I); 2,6-dinitro-*N*,*N*-dipropyl-4-trifluoromethylaniline. (*C.A.*) 2,6-dinitro-*N*,*N*-dipropyl-4-(trifluoromethyl)benzenamine (9CI); (I) (8CI); Reg. No. *[1582-09-8]*. Its herbicidal properties were described by E. F. Alder *et al.* (*Proc. North Cent. Weed Control Conf.*, 1960, p. 23). Introduced by Eli Lilly & Co. (USP 3 257 190) as code no. 'L-36 352'; trade marks 'Treflan', 'Elancolan'.

**Properties.**

Technical trifluralin (98% pure) is an orange crystalline solid; m.p. 48.5-49 °C; v.p. 13.7 mPa at 25 °C. Solubility at 27 °C: <1 mg/l water; 400 g/l acetone; 580 g/l xylene. The technical grade (⩾95% pure) has m.p. >42 °C. It is stable, though susceptible to decomposition by u.v. radiation.

**Uses.**

It is a pre-em. herbicide with little activity post-em. When incorporated in the soil at 0.5-1.0 kg a.i./ha, it is effective for the control of annual grasses and broad-leaved weeds in beans, brassicas, cotton, groundnuts, forage legumes, orchards, ornamentals, transplanted peppers, soyabeans, sugar beet, sunflowers, tomatoes and vineyards. A combination with linuron controls broad-leaved and grass weeds when applied pre-em. to the soil surface in winter cereals. Trifluralin plus 2,4-D is used as a post-planting herbicide in transplanted rice. Trifluralin shows a loss in soil of 85-90% in 0.5-1.0 y (G. W. Probst *et al., J. Agric. Food Chem.*, 1967, **15**, 592).

**Toxicology.**

Acute oral $LD_{50}$: for rats >10 000 mg/kg; for mice 500 mg/kg; for dogs, rabbits and chickens >2000 mg/kg. Skin applications to rabbits of 2000 mg/kg caused neither toxicity nor irritation. In 2-y feeding trials rats receiving 2000 mg/kg diet and dogs 1000 mg/kg diet suffered no ill-effect. $LC_{50}$ (96-h) for bluegill fingerlings is 0.089 mg/l.

**Formulations.**

These include: e.c. and granules. 'Chandor', 'Mudekan', 'Trinulan' (Reg. No. *[8072-92-6]*) (trifluralin + linuron). 'Neepex' (trifluralin + napropamide).

**Analysis.**

Product analysis is by glc (*CIPAC Handbook*, 1980, **1A**, 1362). Residues are determined by glc with ECD (T. H. Byast *et al., Tech. Rep. ARC Weed Res. Organ.*, No. 15 (2nd Ed.) p. 11; J. B. Tepe & R. E. Scroggs, *Anal. Methods Pestic., Plant Growth Regul. Food Addit.*, 1967, **5**, 527; *Anal. Methods Pestic. Plant Growth Regul.*, 1972, **6**, 703).

$C_{10}H_{14}Cl_6N_4O_2$ (435.0)
T6N DNTJ AYMVHXGGG
DYMVHXGGG

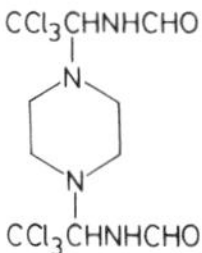


**Nomenclature and development.**

Common name triforine (BSI, E-ISO, F-ISO, ANSI, JMAF). Chemical name (IUPAC) 1,4-bis(2,2,2-trichloro-1-formamidoethyl)piperazine; 1,1'-piperazine-1,4-diyldi-[*N*-(2,2,2-trichloroethyl)formamide]. (*C.A.*) *N,N'*-[1,4-piperazinediylbis(2,2,2-trichloroethylidene)]bisformamide (8 & 9CI); Reg. No. *[26644-46-2]*. Its fungicidal properties were described by P. Schicks & K. H. Veen (*Proc. Br. Insectic. Fungic. Conf., 5th*, 1969, **2**, 569). Introduced by Cela GmbH after being synthesised by C. H. Boehringer Sohn (now Celamerck GmbH Co.) (DEOS 1 901 421) as code no. 'Cela W524'.

**Properties.**

Pure triforine forms colourless crystals; m.p. 155 °C; v.p. 27 μPa at 25 °C. Solubility at room temperature: 6 mg/l water; 11 g/l acetone; 1 g/l benzene; 1 g/l dichloromethane; 330 g/l dimethylformamide; 10 g/l methanol; 476 g/l 1-methylpyrrolid-2-one. It is rapidly decomposed to trichloroacetaldehyde and piperazine salts by concentrated sulphuric or hydrochloric acids; slowly decomposed to chloroform and piperazine by alkali (pH >9). It is non-persistent in soil. The pathways of its photochemical and metabolic breakdown have been established (H. Buchenauer, *Pestic. Sci.*, 1975, **6**, 553; A. Fuchs *et al.*, 1976, **7**, 115, 127; S. Darda *et al.*, *ibid.*, 1977, **8**, 183; S. Darda, *ibid.*, p.173; J. P. Rouchard *et al.*, *ibid.*, p.65; 1978, **9**, 74, 139, 587; J. P. Rouchard, *Bull. Environ. Contam. Toxicol.*, 1977, **18**, 184).

**Uses.**

It is a systemic fungicide effective against: powdery mildew, scab and other diseases of fruit and berries at 20-25 g/100 l; black spot, powdery mildew and rust on ornamentals at 15 g/100 l; powdery mildew and other diseases of vegetables at 25 g/100 l; powdery mildew and other leaf diseases on cereals at 200-250 g/ha, rust on cereals at 300 g/ha. It is active against storage diseases of fruit and suppresses red spider mite activity.

**Toxicology.**

Acute oral $LD_{50}$: for rats >16 000 mg/kg; for mice >6000 mg/kg; for bobwhite quail >5000 mg/kg. Acute percutaneous $LD_{50}$ for rats >10 000 mg/kg. Acute inhalation $LC_{50}$ (1-h) for rats >4.5 mg/l. In 2-y feeding trials NEL was: for rats 625 mg/kg diet; for dogs 100 mg/kg diet. The $LC_{50}$ (96-h) for bluegill and rainbow trout is >1000 mg/l. No toxic effect is observed on honeybees ≤1000 mg/kg diet.

**Formulations.**

'Saprol', e.c. (200 g a.i./l); 'Funginex', e.c. (190 g/l). 'Brolly', e.c. (100 g triforine + 100 g carbendazim/l) (Pan Britannica Industries); 'Nimrod T', e.c. (62.5 g triforine + 62.5 g bupirimate/l) (ICI Plant Protection Division).

**Analysis.**

Product and residue analysis is by polarography or glc (after degradation to chloral hydrate) (R. Darskus & D. Eichler, *Anal. Methods Pestic. Plant Growth Regul.*, 1978, **10**, 243). Particulars are also available from Celamerck GmbH Co.

$C_{20}H_{35}NO_{13}$ (497.5)
T6OTJ B1Q CQ DQ EQ FO- BL6TJ A1Q CQ DQ EM- CL6UTJ A1Q DQ EQ FQ

**Nomenclature and development.**

Common name validamycin (JMAF); validamycin A (Japanese Antiobiotics Research Association). Chemical name (IUPAC) 1L-(1,3,4/2,6)-2,3-dihydroxy-6-hydroxymethyl-4-[(1*S*,4*R*,5*S*,6*S*)-4,5,6-trihydroxy-3-hydroxymethylcyclohex-2-enylamino]cyclohexylβ-D-glucopyranoside. (*C.A.*) [1*S*-(1α,4α,5β,6α)]-1,5,6-trideoxy-3-*O*-β-D-glucopyranosyl-4-( hydroxymethyl)-1-[[4,5,6-trihydroxy-3-(hydroxymethyl)-2-cyclohexen-1-yl]amino]-D-*chiro*-inositol (9CI); Reg. No. *[37248-47-8]*. This antibiotic was described by T. Iwasa *et al.* (*J. Antibiot.*, 1970, **23**, 595). Following earlier chemical studies, a structure was proposed for validamycin A. The results of recent investigations by T. Suami *et al.* (*ibid.*, 1980, **33**, 98) have shown that the compound is isomeric with the structure proposed earlier, and validamycin has the structure given above. Introduced by Takeda Chemical Industries Ltd as trade marks 'Validacin', 'Valimon'.

**Properties.**

Produced by the fermentation of *Streptomyces hygroscopicus* var. *limoneus* nov. var. it is a colourless powder; it decomposes over a wide temperature range *c.* 130 °C; $[\alpha]_D^{24}$ + 110 ° (water). Solubility: very soluble in water; soluble in dimethylformamide, dimethyl sulphoxide, methanol; sparingly soluble in acetone, ethanol. It is stable at room temperature, in neutral and alkaline solution, but slightly unstable under acidic conditions. It is weakly basic $pK_a$ 6.0.

**Uses.**

It is fungistatic against a narrow range of fungi and is used to control *Pellicularia (Corticium) sasakii* of rice at 45-90 g a.i./ha, black scurf of potatoes and damping off of vegetable seedlings caused by *Rhizoctonia solani* (O. Wakae & K. Matsuura, *Rev. Plant Prot. Res.*, 1975, **8**, 81; M. Bakkeren *et al., Proc. Br. Crop Prot. Conf. - Pests Dis.*, 1977, **2**, 541).

**Toxicology.**

Acute oral $LD_{50}$ for rats and mice >20 000 mg/kg. Application of a solution (500 g/l) to the skin of rabbits produced no irritation. In 90-d feeding trials rats receiving 1000 mg/kg daily and mice 2000 mg/kg daily suffered no ill-effect.

**Formulations.**

These include: solution (30 g a.i./l); dust (3 g/kg).

**Analysis.**

Product and residue analysis is by glc of a derivative with FID (K. Nishi & K. Konishi, *Anal. Methods Pestic. Plant Growth Regul.*, 1976, **8**, 309).

$C_8H_{18}NO_4PS_2$ (287.3)
1OPO&O1D2&SY1&VM1

$$CH_3NHCO.\overset{\displaystyle CH_3}{\overset{|}{C}}HSCH_2CH_2S\overset{\displaystyle O}{\overset{\|}{P}}(OCH_3)_2$$

**Nomenclature and development.**

Common name vamidothion (BSI, E-ISO, F-ISO, JMAF). Chemical name (IUPAC) *O,O*-dimethyl *S*-2-(1-methylcarbamoylethylthio)ethyl phosphorothioate. (*C.A.*) *O,O*-dimethyl *S*-[2-[[1-methyl-2-(methylamino)-2-oxoethyl]thio]ethyl] phosphorothioate (9CI); *O,O*-dimethyl phosphorothioate *S*-ester with 2-[(2-mercaptoethyl)thio]-*N*-methylpropionamide (8CI); Reg. No. *[2275-23-2]*. ENT 26 613. Its insecticidal properties were described by J. Desmoras *et al.* (*Phytiatr.-Phytopharm.*, 1962, **11**, 107). Introduced by Rhône-Poulenc Phytosanitaire (GBP 872 823; BEP 575 106) as code no. '10 465 RP',' NPH 83'; trade mark 'Kilval' (Rhône-Poulenc and May & Baker Ltd).

**Properties.**

Pure vamidothion forms colourless needles; m.p. 46-48 °C. The technical grade is an off-white waxy solid; m.p. *c.* 40 °C; v.p. negligible at 20 °C. Solubility: 4 kg/l water; very soluble in most organic solvents; practically insoluble in cyclohexane, light petroleum. The technical grade and, to a lesser extent, the pure compound undergo slight decomposition at room temperature, but this is prevented by certain solvents (anisole, butanone).

**Uses.**

It is a systemic acaricide and insecticide giving persistent control of *Eriosoma lanigerum* and used against insects and mites on apples, cotton, hops, peaches, pears, plums, rice etc., at 37-50 g a.i./100 l. It is metabolised in plants to the corresponding sulphoxide which is of similar activity to vamidothion but of greater persistence.

**Toxicology.**

Acute oral $LD_{50}$: for rats 64-105 mg/kg; for mice 34-57 mg/kg; for pheasants 35 mg/kg. Acute percutaneous $LD_{50}$: for mice 1460 mg/kg; for rabbits 1160 mg/kg. The oral toxicity of the sulphoxide is about half that of vamidothion. In 90-d feeding trials the growth rate of rats receiving 50 mg vamidothion/kg diet or 100 mg of its sulphoxide/kg diet was unaffected. Goldfish survived 14 d in water containing 10 mg vamidothion/l.

**Formulations.**

'Kilval', 'Kilvar' (in Japan), 'Trucidor', solutions (400 g a.i./l).

**Analysis.**

Product analysis is by glc (J. Desmoras *et al., Anal. Methods Pestic. Plant Growth Regul.*, 1973, 7, 479). Residues may be determined by glc (*idem, ibid.*).

$C_{10}H_{21}NOS$ (203.3)
3SVN3&3

$(CH_3CH_2CH_2)_2NCO.SCH_2CH_2CH_3$

**Nomenclature and development.**

Common name vernolate (BSI, E-ISO, F-ISO, WSSA, JMAF). Chemical name (IUPAC) *S*-propyl dipropyl(thiocarbamate). (*C.A.*) *S*-propyl dipropylcarbamothioate (9 CI); *S*-propyl dipropylthiocarbamate (8 CI); Reg. No. *[1929-77-7]*. Introduced as a herbicide by Stauffer Chemical Co. (USP 2 913 327) as code no. 'R-1607'; trademark 'Vernam'.

**Properties.**

Vernolate (99.9% pure) is a clear liquid, with a faint aromatic odour; b.p. 150 °C/30 mmHg; v.p. 1.39 Pa at 25 °C; $d_{20}^{20}$ 0.954; $n_D^{20}$ 1.474. Solubility at 20 °C: 90 mg/l water; miscible with acetone, ethanol, kerosene, 4-methylpentan-2-one, xylene. In water buffered at pH7, 50% loss occurs in 13 d at 40 °C. Stable <200 °C.

**Uses.**

It is toxic to germinating broad-leaved and grassy weeds and is recommended for the pre-em. weeding of groundnuts and soyabeans at 1.5-3.0 kg a.i./ha followed by soil incorporation. Its persistence in soil is brief.

**Toxicology.**

Acute oral $LD_{50}$: for male rats 1500 mg/kg, for females 1550 mg/kg. Acute percutaneous $LD_{50}$ for rabbits >5000 mg/kg; non-irritating to both skin and eyes.

**Formulations.**

'Vernam E', e.c. (720 or 840 a.i./l). 'Vernan G', granules (50 or 100 g/kg). 'Surpass E' (Reg. No. *[68924-90-3]*), e.c. (800 g vernolate + *N,N*-diallyl-2,2-dichloroacetamide/l). 'Vernam' atrazine 10-5G, granules (100 g vernolate + 50 g atrazine/kg).

**Analysis.**

Product analysis is by glc (*CIPAC Handbook*, 1983, **1B**, in press). Residues in crops and soils are determined by glc or by colorimetry after conversion to a derivative (G.G. Patchett & G. H. Batchelder, *Anal. Methods Pestic., Plant Growth Regul. Food Addit.*, 1967, **5**, 337; W. J. Ja, *Anal. Methods Pestic. Plant Growth Regul.*, 1972, **6**, 708). Analytical methods are available from the Stafffer Chemical Co.

**Nomenclature and development.**

Scientific name *Verticillium lecanii* (Zimm.) Viégas;formerly *Cephalosporium lecanii*. Trivial name white halo fungus. An isolate of the Deuteromycete (Moniliales) fungus *V. lecanii* was obtained from the aphid *Macrosiphoniella sanborni* by R. A. Hall (*J. Inverteb. Pathol.*, 1976, **27**, 41) during a natural epidemic. Later he isolated a second isolate, with a different host range, from the glasshouse whitefly *Trialeurodes vaporariorum* (*idem, Ann. Appl. Biol.*, 1979, **101**, 1). These isolates were developed by Tate & Lyle Ltd as trade marks 'Vertalec', 'Mycotal' for the products from the two isolates.

**Properties.**

*V. lecanii* isolates are cultured on a sterile undefined medium and the spores harvested by concentration and drying. Germinating spores infect target insects by enzymic degradation of the cuticle followed by growth within the haemolymph and their tissues. The infection process requires freely available water and takes *c.* 24 h. The insect dies 5-14 d later and, at high r.h. fresh spores are produced on the outside of the cadaver. These subsequently infect healthy insects and produce an epidemic which can give 84-112 d effective control from one application, provided the r.h. is suitable. *V. lecanii* is compatible with a range of pesticides but is susceptible to certain fungicides, notably dithiocarbamates.

**Uses.**

*V. lecanii* as 'Vertalec' is used to control aphids under protected cultivation on crops such as year-round and potted chrysanthemums. As 'Mycotal' it is used to control *T. vaporariorum* on protected crops such as cucumbers and tomatoes. Both products have been employed extensively and with complete safety in integrated control programmes where other biological agents (*Encarsia formosa, Phytoseiulus persimilis*) are used. *V. lecanii* is neither phytotoxic nor phytopathogenic.

**Toxicology.**

There is no evidence of acute or chronic toxicity, infectivity, eye or skin irritation or hypersensitivity to mammals. No allergic responses or health problems have been observed by research workers, manufacturing staff or users.

**Formulations.**

'Vertalec', w.p. containing spores of the 'aphid-active' strain of *V. lecanii* ($2 \times 10^8$ viable colony-forming units/g dry wt). Products should be stored at 4 °C and have a 120-d shelf life.

**Analysis.**

Activity of *V. lecanii* is measured in terms of spore count and efficacy against insects. Spore count measured as viable colony-forming unit count can be assayed by conventional techniques. Efficacy assays are made against *Aphis fabae* for 'Vertalec', and scales of *T. vaporariorum* for 'Mycotal'.

$C_{12}H_9Cl_2NO_3$ (286.1)
T5OVNV EHJ CR CG EG& E1U1 E1

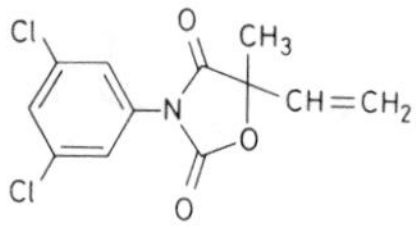

**Nomenclature and development.**

Common name vinclozolin (BSI, E-ISO, JMAF);vinclozoline (F-ISO). Chemical name (IUPAC) (*RS*)-3-(3,5-dichlorophenyl)-5-methyl-5-vinyl-1,3-oxazolidine-2,4-dione. (*C.A.*) (*RS*)-3(3,5-dichlorophenyl)-5-ethenyl-5-methyl-2,4-oxazolidinedione (9CI); Reg. No. *[50471-44-8]*. Its fungicidal properties were reported by E. -H. Pommer & D. Mangold (*Meded. Fac. Landbouwwet. Rijksuniv. Gent*, 1975, **40**, 713). Introduced by BASF AG as code no. 'BAS 352F'; trade mark 'Ronilan'.

**Properties.**

Pure vinclozolin is a colourless crystalline solid; m.p. 108 °C; v.p. <10 mPa at 20 °C. Solubility at 20 °C: 1 g/l water; 435 g/kg acetone; 146 g/kg benzene; 319 g/kg chloroform; 253 g/kg ethyl acetate. It is stable ⩽50 °C; in water at pH1 and pH7, but at pH13 50% hydrolysis occurs in 3.8 h. The technical grade is ⩾93% *m/m* pure.

**Uses.**

It is a selective contact fungicide effective against *Botrytis cinerea, Sclerotinia sclerotiorum* and *Monilia* spp. It is used on fruits, hops, ornamentals, rape, strawberries, vegetables and vines at 0.75-1.0 kg a.i./ha.

**Toxicology.**

Acute oral $LD_{50}$: for rats 10 000 mg tech./kg; for guinea-pigs *c.* 8000 mg/kg. In 90-d feeding trials NEL was: for rats 450 mg/kg diet; for dogs 300 mg/kg diet. $LC_{50}$ (96-h) is: for guppies 32.5 mg a.i. (as w.p.)/l; for trout 52.5 mg a.i./l. It is non-toxic to honeybees and to earthworms.

**Formulations.**

'Ronilan', w.p. (500 g a.i./kg).

**Analysis.**

Product analysis is by glc. Residues may be determined by hydrolysis to 3,5-dichloroaniline, a derivative of which is measured by glc with ECD. Details are available from BASF AG.

$C_{19}H_{16}O_4$ (308.4)
T66 BOVJ DYR&1V1 EQ

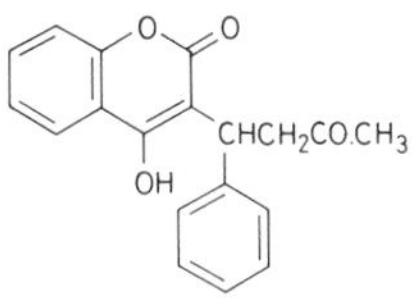

**Nomenclature and development.**

Common name warfarin (BSI, E-ISO, BPC);warfarine (F-ISO);coumaféne (France);zoocoumarin (USSR);coumarins a term also applied to coumatetralyl (JMAF);*exception* The Netherlands. Chemical name (IUPAC) 4-hydroxy-3-(3-oxo-1-phenylbutyl)coumarin; 3-(α-acetonylbenzyl)-4-hydroxycoumarin (I). (*C.A.*) 4-hydroxy-3-(3-oxo-1-phenylbutyl)-2*H*-1-benzopyran-2-one (9CI); (I) (8CI); Reg. No. *[81-81-2]*. Its anticoagulant properties were described by K. P. Link *et al.* (*J. Biol. Chem.*, 1944, **153**, 5) at the Wisconsin Alumni Research Foundation (USP 2 427 578) and known as code no. 'WARF 42'.

**Properties.**

The racemate forms colourless crystals, m.p. 159-161 °C. It is practically insoluble in water; readily soluble in acetone, dioxane; moderately soluble in alcohols. It is acidic; the sodium salt being soluble in water, insoluble in organic solvents.

**Uses.**

It is an anticoagulant rodenticide to which rats do not develop 'bait-shyness'. Rats are killed by 5 daily doses of 1 mg/kg. It is used at 50 mg a.i./kg bait material to control *Rattus norvegicus*, 250 mg/kg for *R. rattus* and mice, baiting is continued for 14 d. The racemate has been resolved, the (*S*)-isomer showing 7-fold greater rodenticidal activity than the (*R*)-isomer (B. D. West *et al.*, *J. Am. Chem. Soc.*, 1961, **83**, 2676). The anticoagulant properties of warfarin (as 'Coumadin' or the sodium salt 'Marevan') are used in medicine to reduce the risk of post-operative thrombosis.

**Toxicology.**

Death followed 5 daily doses of: 3 mg/kg for cats; 1 mg/kg for pigs; poultry are more resistant.

**Formulations.**

Dust (10 g a.i./kg) for use in holes and runs; dust (1 and 5 g/kg) for admixture with suitable protein-rich bait.

**Analysis.**

Product analysis is by u.v. spectrometry (*AOAC Methods*, 1980, 6.141-6.142; *CIPAC Handbook*, 1970, **1**, 696; C. H. Schroeder & J. N. Eble, *Anal. Methods Pestic., Plant Growth Regul. Food Addit.*, 1964, **3**, 197; C. H. Schroeder & J. Sherma, *Anal. Methods Pestic. Plant Growth Regul.*, 1973, **7**, 677).

$C_{10}H_{13}NO_2$ (179.2)
1MVOR C1 D1

**Nomenclature and development.**

Common name xylylcarb (BSI, draft E-ISO); MPMC (JMAF). Chemical name (IUPAC) 3,4-xylyl methylcarbamate (I). (*C.A.*) 3,4-dimethylphenyl methylcarbamate (9CI); (I) (8CI); Reg. No. *[2425-10-7]*. Its insecticidal properties were described by R. L. Metcalf *et al.* (*J. Econ. Entomol.*, 1963, **56**, 862). Introduced by Sumitomo Chemical Co. as code no. 'S-1046'; trade mark 'Meobal'.

**Properties.**

Technical grade xylylcarb is a colourless solid; m.p. 71.5-76.0 °C; v.p. 70 mPa at 20 °C. Solubility at 20 °C: 580 mg/l water; at room temperature: 930 g/kg acetonitrile; 770 g/kg cyclohexanone; 134 g/kg xylene. It is unstable in alkaline media.

**Uses.**

It is an insecticide used for the control of sucking insects on rice, leafhoppers, planthoppers and scales on fruits at *c.* 40 g a.i./100 l.

**Toxicology.**

Acute oral $LD_{50}$: for male rats 375 mg/kg, for females 325 mg/kg. Acute percutaneous $LD_{50}$ for rats $>$1000 mg/kg. $LC_{50}$ (48-h) for carp is 10 mg/l.

**Formulations.**

These include: e.c. (300 g a.i./l); w.p. (500 g/kg); dust (20 g/kg); microgranule (20 g/kg).

**Analysis.**

Product analysis is by hplc (S. Sakaue *et al., Nippon Nogei Kagaku Kaishi*, 1981, **55**, 1237) or by u.v. spectrometry. Residues may be determined by glc of a derivative with ECD (J. Miyamoto *et al., Nihon Hoyaku Gakkaishi,* 1978, **3**, 119).

# 3,5-Xylyl methylcarbamate

$C_{10}H_{13}NO_2$ (179.2)
1MVOR C1 E1

**Nomenclature and development.**

Common name XMC (JMAF). Chemical name (IUPAC) 3,5-xylyl methylcarbamate (I). (*C.A.*) 3,5-dimethylphenyl methylcarbamate (9CI); (I) (8CI); Reg. No. *[2655-14-3]*. Introduced as an insecticide by Hokko Chemical Industry Co. Ltd and Hodogaya Chemical Co. Ltd as code no. 'H-69'; trade mark 'Macbal'.

**Properties.**

The technical grade (97% pure) is a colourless crystalline solid; m.p. 99 °C. Practically insoluble in water; soluble in most organic solvents including benzene, cyclohexanone, 3,5,5-trimethylcyclohex-2-enone. Stable to light and at temperatures 90 °C.

**Uses.**

It is a systemic insecticide recommended for use at 600 g a.i./ha against leafhoppers and plant hoppers on rice.

**Toxicology.**

Acute oral $LD_{50}$: for rats 542 mg/kg; for rabbits 445 mg/kg. In 90-d feeding trials NEL for rats and mice was 230 mg/kg daily. $LC_{50}$ (48-h) for carp is >40 mg/l.

**Formulations.**

These include: e.c. (200 g a.i./l); w.p. (500 g/kg); dust (20 g/kg); microgranule (30 g/kg).

**Analysis.**

Product analysis is by hydrolysis to 3,5-xylenol, which is measured by u.v. spectroscopy. Residues may be determined by hydrolysis to 3,5-xylenol a derivative of which is measured by glc.

$$\begin{bmatrix} [(-S.CS.NH.CH_2CH_2NH.CS.S-)^{--}Zn(NH_3)^{++}]_3 \\ (-S.CS.NH.CH_2CH_2NH.CS.S-) \end{bmatrix}_x$$

**Nomenclature and development.**

Common name metiram (New Zealand, JMAF); métirame zinc (France). No common name was accepted by ISO (and metiram was withdrawn by BSI) because the product appears to be a mixture rather than a complex. Chemical name (IUPAC) zinc ammoniate ethylenebis(dithiocarbamate) - poly[ethylenebis(thiuram disulphide)]. (*C.A.*) indexed as metiram (composition not specified); Reg. No. *[9006-42-2]*. Introduced by BASF AG (GBP 840 211; USP 3 248 400) as code no. 'FMC 9102' (FMC Corp.); trade mark 'Polyram'.

**Properties.**

It is a yellowish powder which begins to decompose *c.* 140 °C; v.p. $<10$ μPa at 20 °C. Solubility: practically insoluble in water, acetone, benzene, ethanol; soluble (with decomposition) in pyridine. It is unstable under strongly acidic or alkaline conditions.

**Uses.**

It is a non-systemic fungicide used for foliage application to control *Bremia lactucae* on lettuce, *Phytophthora infestans* on potatoes and tomatoes, *Plasmopara viticola* on vines, *Pseudoperonospora humuli* on hops; also *Alternaria* spp., *Colletotrichum lindemuthianum, Puccinia* spp. and other phytopathogenic fungi. It is also used in combination with tridemorph to control *Puccinia striiformis* on barley and winter wheat and with nitrothal-isopropyl to control powdery mildew on top fruit.

**Toxicology.**

Acute oral $LD_{50}$: for rats $>10\,000$ mg/kg; for male mice $>5400$ mg/kg; for female guinea-pigs 2400-4800 mg/kg. In feeding trials no ill-effect was observed in dogs receiving 7.5 mg/kg daily for 1.92 y or 45 mg/kg daily for 90 d; rats receiving 10 000 mg/kg diet, but not those 1000 mg/kg diet, for 14 d showed an increase in thyroid weight and decreased uptake of iodide. $LC_{50}$ (48-h) for harlequin fish is 17 mg/l.

**Formulations.**

These include: 'Polyram' and 'Polyram Combi', w.p. (800 g/kg); dusts of various strength. 'Pallinal', w.p. (mixture with nitrothal-isopropyl).

**Analysis.**

Product analysis is by acid hydrolysis to produce carbon disulphide which is estimated by titration. Residues may be determined by hydrolysis to carbon disulphide which is measured by glc or by colorimetry of a derivative (*Analyst (London)*, 1981, **106**, 782; G. E. Keppel *J. Assoc. Off. Anal. Chem.*, 1971, **54**, 528).

$P_2Zn_3$ (258.1)
.ZN3.P2

$Zn_3P_2$

**Nomenclature and development.**

The traditional chemical name zinc phosphide (I) (E-ISO, JMAF);phosphure de zinc (F-ISO) is accepted in lieu of a common name. Chemical name (IUPAC) trizinc diphosphide. (*C.A.*) (I) (8 & 9CI); Reg. No. *[1314-84-7]*. It has long been used as a poison against rodents.

**Properties.**

Zinc phosphide is a grey powder; *d* 4.55; m.p. 420 °C (when heated in the absence of oxygen). Solubility: practically insoluble in water (with which it reacts), ethanol; soluble in benzene, carbon disulphide. It is stable when dry but decomposes slowly in moist air; it is decomposed violently by acids to produce phosphine which is a potent mammalian poison, and impurities which render the gas spontaneously flammable. The technical product is 80-95% pure.

**Uses.**

Zinc phoshide is a rodenticide, mainly restricted to use by trained personnel against rats, field mice and gophers. Used as baits.

**Toxicology.**

Acute oral $LD_{50}$ for rats 45.7 mg/kg. It is extremely poisonous to mammals and birds.

**Formulations.**

Pastes (25 or 50 g a.i./kg) for bait preparation.

**Analysis.**

Product analysis is by reaction with acid, the phosphine produced is estimated by titration (*CIPAC Handbook*, 1970, **1**, 703), or oxidised to phosphoric acid which is estimated by standard methods. (J. W. Elmore & F. R. Roth, *J. Assoc. Off. Agric. Chem.*, 1943, **26**, 559; 1947, **30**, 213; B. L. Griswold *et al., Anal. Chem.*, 1951, **23**, 192).

$(C4_4H_6N_2S_4Zn)_\chi$ (275.8)χ
SUYSHM2MYUS&SH &&Zn COMPLEX

$\left[-S.CS.NHCH_2CH_2NHCS.S.Zn-\right]_x$

**Nomenclature and development.**

Common name zineb (BSI, E-ISO, JMAF);zinèbe (F-ISO);*exception* (Federal Republic of Germany). Chemical name (IUPAC) zinc ethylenebis(dithiocarbamate) (polymeric). (*C.A.*) [[1,2-ethanediylbis[carbamodithioato]](2-)]zinc complex (9CI); [ethylenebis[dithiocarbamato]]zinc (8CI); Reg. No. *[12122-67-7]*. ENT 14 874. J. M. Heuberger & T. F. Manns (*Phytopathology*, 1943, **33**, 113) reported that the addition of zinc sulphate improved the field performance of nabam (8740) as a fungicide. This led to the introduction of zineb by Rohm & Haas Co. and by E. I. du Pont de Nemours & Co. (Inc.) (who no longer manufacture or market it) (USP 2 457 674; 3 050 439) as trade marks 'Dithane Z-78' (Rohm & Haas), formerly '*Parzate*' (Du Pont), 'Tiezene' (Farmoplant S.p.A.).

**Properties.**

Produced by precipitation from aqueous nabam on addition of soluble zinc salts, it is a light-coloured powder; it decomposes without melting; v.p. negligible at room temperature. Solubility at room temperature: *c.* 10 mg/l water; soluble in carbon disulphide, pyridine. It is somewhat unstable to light, heat and moisture. When precipitated from concentrated solution a polymer is formed which is of lower fungicidal activity.

**Uses.**

It is a fungicide used to protect foliage and fruit of a wide range of crops against diseases such as potato and tomato blight, *Botrytis* spp., downy mildews and rusts. It is generally non-phytotoxic except to zinc-sensitive varieties.

**Toxicology.**

Acute oral $LD_{50}$ for rats >5200 mg/kg. Cases of irritation of skin and the mucous membranes have been reported.

**Formulations.**

These include: w.p. (700 or 750 g a.i./kg); dust ('Dithane' dust). Several related products in which various proportions of other metal compounds (copper, manganese, zinc) are also available, their chemical structure being unknown at present.

**Analysis.**

Product analysis is by acid hydrolysis, converting the carbon disulphide evolved into dithiocarbonate which is estimated by titration with iodine (*CIPAC Handbook*, 1970, **1**, 706; *AOAC Methods*, 1980, 6.338-6.341). Residues may be determined by acid hydrolysis converting the carbon disulphide evolved into a copper/amine complex which is estimated by colorimetry (H. L. Pease, *J. Assoc. Off. Agric. Chem.*, 1957, **30**, 1113; G. E. Keppel, *J. Assoc. Off. Anal. Chem.*, 1969, **52**, 162; 1971, **54**, 528; R. Mestres *et al., Trav. Soc. Pharm. Montpellier*, 1973, **33**, 191; W. K. Lowen & H. L. Pease, *Anal. Methods Pestic., Plant Growth Regul. Food Addit.*, 1964, **3**, 69).

$C_6H_{12}N_2S_4Zn$ (305.8)
SUYSHN1&1 &&Zn SALT 2:1

$\left[(CH_3)_2NCS.S^-\right]_2 Zn^{++}$

**Nomenclature and development.**

Common name ziram (BSI, E-ISO, JMAF);zirame (F-ISO). Chemical name (IUPAC) zinc dimethyldithiocarbamate. (*C.A.*) (T-4)-bis(dimethyldithiocarbamato-*S*,*S*')zinc (9CI); bis(dimethyldithiocarbamato)zinc (8CI); Reg. No. *[137-30-4]*. Introduced as a fungicide by E. I. du Pont de Nemours & Co. (Inc). (who no longer manufacture or market it) as trade marks '*Milbam*', '*Zerlate*', 'Fuklasin' (Schering AG), and as a rodent repellent 'Aaprotect' (Aagrunol).

**Properties.**

Pure ziram is a colourless powder; m.p. 240 °C; v.p. negligible at room temperature. Solubility at 25 °C: 65 mg/l water; slightly soluble in diethyl ether, ethanol; moderately soluble in acetone; soluble in dilute alkali, carbon disulphide, chloroform. The technical grade has m.p. 240-244 °C. Decomposed by acids; incompatible with copper and mercury compounds.

**Uses.**

Ziram is a protectant fungicide for use on fruit and vegetable crops against *Alternaria* and *Septoria* spp. It is non-phytotoxic except to zinc-sensitive plants. It has also been used as a repellent to birds and rodents.

**Toxicology.**

Acute oral $LD_{50}$ for rats 1400 mg/kg. It may cause irritation of the skin and mucous membranes. In 1-y feeding trials rats receiving 5 mg a.i./kg daily showed no effect, neither did weanling rats receiving 100 mg/kg diet for 30 d.

**Formulations.**

A w.p. (900 g a.i./kg); 'Aaprotect', repellent paste (370 g/kg + sticker).

**Analysis.**

Product analysis is by acid hydrolysis, the carbon disulphide liberated being converted to potassium *O*-methyl dithiocarbonate which is estimated by titration with iodine (*CIPAC Handbook*, 1970, **1**, 716). Residues may be determined by acid hydrolysis followed by colorimetry of the liberated carbon disulphide (W. K. Lowen & H. L. Pease, *Anal. Methods Pestic., Plant Growth Regul. Food Addit.*, 1964, **3**, 69; G. E. Keppel, *J. Assoc. Off. Anal. Chem.*, 1971, **54**, 528; *Analyst (London)*, 1981, **106**, 782).

# PART II

## Superseded Compounds

Compounds at one time marketed or widely reported, but at present of little commercial interest in relation to the subjects covered by the *Pesticide Manual*

## *SUPERSEDED COMPOUNDS*

The compounds in this section, though at one time marketed or widely reported, are believed to be currently of little commercial interest in the subjects covered by the *Pesticide Manual*. Some are still widely used in other areas. If available, the following information is included.

Entry number.
Common names (BSI, E-ISO, F-ISO, ANSI, WSSA, JMAF and significantly different names, but countries in which the ISO name is unacceptable are *not* listed).
Chemical names (IUPAC and *C.A.*) and *C.A.* Registry No.
Trivial names.
Code numbers (official bodies and manufacturers').
Trade marks and the firm(s) concerned.
Molecula formula.
Wiswesser Line-Formula Notation.
Reference to last edition of the *Pesticide Manual* and page number on which they were described.
Use and early scientific reference.

The entries in this Section are readily identified in the indexes by an asterisk before the entry number (*e.g.* *2400).

Chemical names that end a line with a hyphen (-), opening parenthesis (, or bracket [ run on to the next line without a space and retain the last printed character; lines ending in an equals sign ( = ) run straight on to the next line (but omitting the = ). In the Wiswesser Line-Formula, a hyphen (-) is an integral part of the notation; equals ( = ) is used at the end of a line (as above) to show the next line continues directly without a space after the last normal character.

Sometimes conflicting reports have been received from various countries about the current status of a given pesticide. The Editors will be pleased to receive details of compounds in this list which are still in commercial use and are apparently likely to remain so for the forseeable future.

**20** *S*-2-Acetamidoethyl *O,O*-dimethyl phosphorodithioate (IUPAC). (*C.A.*) *S*-[2-(acetylamino)ethyl] *O,O*-dimethyl phosphorodithioate (9CI); *O,O*-dimethyl phosphorodithioate *S*-ester with *N*-(2-mercaptoethyl)acetamide; Reg. No. *[13265-60-6]*. DAEP (JMAF). 'Amiphos' (Nippon Soda Co. Ltd). $C_6H_{14}NO_3PS_2$; 1VM2SPS&O1&O1; insecticide.

**50** Aclonifen (BSI, draft E-ISO). (IUPAC) 2-chloro-6-nitro-3-phenoxyaniline. (*C.A.*) 2-chloro-6-nitro-3-phenoxybenzenamine (9CI). 'KUB 3359'; 'Bandur' (Celamerck GmbH KG). $C_{12}H_9ClN_2O_3$; WNR BZ CG DOR; herbicide.

**70** Acrylonitrile (I) (IUPAC) (BSI, E-ISO) accepted in lieu of a common name. (*C.A.*) 2-propenenitrile (9CI); (I) (8CI); Reg. No. *[107-13-1]*. Vinyl cyanide. ENT 54. 'Ventox' (American Cyanamid Co. and Degesch AG). $C_3H_3N$; NC1U1; *PM5*, p. 3. Insecticidal fumigant, *The Chemistry of Acrylonitrile*.

**210** Allyxycarb (BSI, E-ISO); allyxycarbe (F-ISO); APC (JMAF). (IUPAC) 4-diallylamino-3,5-xylyl methylcarbamate (I). (*C.A.*) 4-(di-2-propenylamino)-3,5-dimethylphenyl methylcarbamate (9CI); (I) (8CI); Reg. No. *[6392-46-7]*. OMS 773. 'Bayer 50 282', 'A 546'; 'Hydrol' (Bayer AG). $C_{16}H_{22}N_2O_2$; 1U2N2U1&R B1 F1 DOVM1; *PM4*, p. 10. Insecticide

**260** Amidithion (BSI, E-ISO, F-ISO, ex-ANSI); amidiphos (France). (IUPAC) *S*-2-methoxyethylcarbamoylmethyl *O,O*-dimethyl phosphorodithioate. (*C.A.*) *S*-[2-[(2-methoxyethyl)amino]-2-oxoethyl] *O,O*-dimethyl phosphorodithioate (9CI); *O,O*-dimethyl phophorodithioate *S*-ester with 2-mercapto-*N*-(2-methoxyethyl)acetamide (8CI); Reg. No. *[919-76-6]*. ENT 27 160. 'Ciba 2446'; 'Thiocron' (Ciba-Geigy AG). $C_7H_{16}NO_4PS_2$; 1OPS&O1&S1VM2O1; *PM2*, p. 16. Acaricide and insecticide (V.Dittrich & F. Bachman, *Proc. Br. Insectic. Fungic. Conf., 2nd*, 1963, p. 421).

**280** 2-Amino-3-chloro-1,4-naphthoquinone (I) (IUPAC). (*C.A.*) 2-amino-3-chloro-1,4-naphthalenedione (9CI); (I) (8CI); Reg. No. *[2797-51-5]*. '06K' (Uniroyal Inc.). $C_{10}H_6ClNO_2$; L66 BV EVJ CZ DG; fungicide.

**320** Amiton (BSI). (IUPAC) *S*-2-diethylaminoethyl *O,O*-diethyl phosphorothioate. (*C.A.*) *S*-[2-(diethylamino)ethyl] *O,O*-diethyl phosphorothioate (8 & 9CI); Reg. No. *[78-53-5]*. ENT 24 980-X. 'Tetram' (ICI Plant Protection Division). $C_{10}H_{24}NO_3PS$; 2OPO&O2&S2N2&2; Amiton hydrogen oxalate (Reg. No. *[3734-97-2]*; ENT 20 993) was experimental insecticide/acaricide (R. Ghosh & J.F. Newman, *Chem. Ind. (London)*, 1955, p. 118).

**380** Anabasine (JMAF). (IUPAC) (*S*)-3-(piperidin-2-yl)pyridine. (*C.A.*) (*S*)-3-(2-piperidinyl)pyridine (9CI); anabasine (8CI); Reg. No. *[494-52-0]*. 2-(3-Pyridyl)piperidine. 'Neonicotine' (racemate). $C_{10}H_{14}N_2$; T6NJ C- BT6MTJ; insecticide from *Anabasis aphylla*.

**480** Athidathion (BSI, E-ISO, F-ISO). (IUPAC) *O,O*-diethyl *S*-2,3-dihydro-5-methoxy-2-oxo-1,3,4-thiadiazol-3-ylmethyl phosphorodithioate. (*C.A.*) *O,O*-diethyl *S*-[(5-methoxy-2-oxo-1,3,4-thiadiazol-3(2*H*)-yl)methyl] phosphorodithioate (9CI); *O,O*-diethyl phosphorodithioate *S*-ester with 4-(mercaptomethyl)-2-methoxy-$\Delta^2$-1,3,4-thiadiazolin-5-one (8CI); Reg. No. *[19691-80-6]*. 'G 13 006' (Ciba-Geigy AG). $C_8H_{15}N_2O_4PS_3$; T5NNVSJ B1SPS&O2&O2 EO1; insecticide reported by K. Rüfenacht (*Helv. Chim. Acta*, 1968, **51**, 518).

**490** Atraton (BSI, E-ISO, F-ISO); atratone (France). (IUPAC) *N*-ethyl-*N* -isopropyl-6-methoxy-1,3,5-triazine-2,4-diyldiamine; 2-ethylamino-4-isopropylamino-6-methoxy-1,3,5-triazine. (*C.A.*) *N*-ethyl-6-methoxy-*N*'-(methylethyl)-1,3,5-triazine-2,4-diamine (9CI); 2-(ethylamino)-4-(isopropylamino)-6-methoxy-*s*-triazine (8CI); Reg. No. *[1610-17-9]*. 'G 32 293'; 'Gesatamin' (Ciba-Geigy AG). $C_9H_{17}N_5O$; T6N CN ENJ BO1 DMY1&1 FM2; Herbicide (E. Knüsli, *Phytiatr.-Phytopharm.*, 1958, **7**, 81).

**520** 6-Azido-*N-tert*-butyl-*N'*-ethyl-1,3,5-triazine-2,4-diyldiamine (IUPAC); 2-azido-4-*tert*-butylamino-6-ethylamino-1,3,5-triazine. (*C.A.*) 6-azido-*N*-(1,1-dimethylethyl)-*N'*-ethyl-1,3,5-triazine-2,4-diamine (9CI); 2-azido-4-(*tert*-butylamino)-6-(ethylamino)-*s*-triazine (8CI); Reg. No. *[2854-70-8]*. 'WL 9385' (Shell Research Ltd). $C_9H_{16}N_8$; T6N CN ENJ BNNN DMX1&1&1 FM2; Herbicide (R. A. Abbott & G. E. Barnsley, *J. Sci. Food Agric.*, 1968 **19**, 16).

**560** Azithiram (BSI, E-ISO); azithirame (F-ISO). (IUPAC) *N,N'*-bis(dimethylamino)thiuram disulphide; bis(dimethylaminocarbonyl) disulphide. (*C.A.*) 2,2,2',2'-tetramethylthioperoxydicarbonic dihydrazide (9CI); bis(3,3-dimethylthiocarbazoyl) disulfide (8CI); Reg. No. *[5834-94-6]*. 'PP 447' (ICI Plant Protection Division). $C_6H_{14}N_4S_4$; 1N1&MYUS&SSYUS&MN1&1; fungicide.

**570** Azobenzene (I) (IUPAC, accepted by BSI, E-ISO, F-ISO). *(C.A.)* Diphenyldiazene (9CI); (I) (8CI); Reg. No. *[103-33-3]*. ENT 14 611. $C_{12}H_{10}N_2$; RNUNR; *PM4*, p. 27. Acaricide (W. E. Blauvelt, *N. Y. State Flower Grow. Bull.*, 1945, No. 2).

**590** Azothoate (BSI, E-ISO, F-ISO). (IUPAC) *O*-4-(4-chlorophenylazo)phenyl *O,O*-dimethyl phosphorothioate. (*C.A.*) *O*-[4-[(4-chlorophenyl)azo]phenyl] *O,O*-dimethyl phosphorothioate (9CI); *O*-[*p*-[(*p*-chlorophenyl)azo]phenyl] *O,O*-dimethyl phosphorothioate (8CI); Reg. No. *[5834-96-8]*. OMS 1089. 'L 1058'; 'Slam C' (Montecatini S.p.A.). $C_{14}H_{14}ClN_2O_3PS$; GR DNUNR DOPS&O1&O1; acaracide and insecticide.

**630** Barium carbonate [IUPAC, *C.A.* (8 & 9CI)] Reg. No. *[513-77-9]*. $CBaO_3$; .BA..C-O3; *PM5*, p. 28. Rodenticide.

**650** Barium polysulphide (accepted by E-ISO) (IUPAC); polysulfure de baryum (F-ISO). (*C.A.*) barium polysulfide Reg. No. *[50864-67-0]*. 'Solbar' (Bayer AG). $BaS_x$; .BA..S-X; insecticide and fungicide.

**700** Benfuresate (BSI, draft E-ISO). (IUPAC) 2,3-dihydro-3,3-dimethylbenzofuran-5-yl ethanesulphonate. (*C.A.*) 2,3-dihydro-3,3-dimethyl-5-benzofuranyl ethanesulfonate (9CI); Reg. No. *[68505-69-1]*. $C_{12}H_{16}O_4S$; T56 BOT&J D1 D1 GOSW2; herbicide (K. W. Chisholm *et al., Proc. South. Weed Sci. Soc.*, 1980, **33**, 326).

**730** Benquinox (BSI, E-ISO, F-ISO); tserenox (USSR). (IUPAC) 1,4-benzoquinone 1-benzoylhydrazone 4-oxime. (*C.A.)* benzoic acid [4-(hydroxyimino)-2,5-cyclohexadien-1-ylidene]hydrazide (9CI); benzoic acid (4-oxo-2,5-cyclohexadien-1-ylidene)hydrazide 4-oxime (8CI); Reg. No. *[495-73-8]*. COBH. 'Bayer 15 080'; 'Ceredon' (Bayer AG). 'Ceredon special', Cereline', 'Tillantox' (with phenylmercury chloride), 'Ceredon T' (with thiram), 'Rhizoctol slurry', 'Rhizoctol combi' (Reg. No. [*8066-69-1*]) (with methylarsenic sulphide). $C_{13}H_{11}N_3O_2$; L6Y DYJ AUNQ DUNMVR; *PM4*, p. 35. Fungicide (P. E. Frohberger, *Phytopath. Z.*, 1956, **37**, 427).

**780** Benzadox (BSI, ANSI, WSSA). (IUPAC) benzamido-oxyacetic acid. *(C.A.)* [(benzoylamino)oxy]acetic acid (9CI); [(benzamido)oxy]acetic acid (8CI); Reg. No. *[5251-93-4]*. 'MC 0035' (Murphy Chemical Ltd), 'S 6173'; 'Topcide' (Gulf Oil Corp.). $C_9H_9NO_4$; QV1OMVR; *PM4*, p. 38. Herbicide *(Farm Chem.*, 1967, **130**, 86).

**800** Benzamorf (BSI, E-ISO); benzamorphe (F-ISO). (IUPAC) morpholinium 4-dodecylbenzenesulphonate. (*C.A.*) 4-dodecylbenzenesulfonic acid compound with morpholine (1:1) (8 & 9CI); Reg. No. *[12068-08-5]*. 'BAS 276F' (BASF AG). $C_{22}H_{39}NO_4S$; T6N DOTJ &&4-$C_{12}H_{25}C_6H_4SO_3H$ SALT; WSQR D 12 && $O[(CH_2)_2]_2NH$ SALT; fungicide.

**810** Benzipram (BSI, E-ISO, ANSI, WSSA); benziprame (F-ISO). (IUPAC) *N*-benzyl-*N*-isopropyl-3,5-dimethylbenzamide (I). *(C.A.)* 3,5-dimethyl-*N*-(1-methylethyl)-*N*-( phenylmethyl)benzamide (9CI); (I) (8CI); Reg. No. *[35256-86-1]*. 'S-18 510' (Gulf Oil Corp.). $C_{19}H_{23}NO$; 1Y1&N1R&VR; herbicide.

**890** *S*-Benzyl *O,O*-diethyl phosphorothioate (I) (IUPAC). (*C.A.*) *O,O*-diethyl *S*-phenylmethyl phosphorothioate (9CI); (I) (8CI); Reg. No. *[13286-32-3]*. EBP (JMAF). 'Kitazin' (Kumiai Chemical Industry Ltd). $C_{11}H_{17}O_3PS$; R1SPO&O2&O2; fungicide.

**910** *S*-Benzyl *O*-ethyl phenylphosphonothioate (I) (IUPAC). *(C.A.) O*-ethyl *S*-(phenylmethyl) phenylphosphonothioate (9CI); (I) (8CI); Reg. No. *[21722-85-0]*. ESBP (JMAF). 'Inezin' (Nissan Chemical Industries Ltd). $C_{15}H_{17}O_2PS$; 2OPO&R&S1R; fungicide.

**1020** 5-[Bis[2-(2-butoxyethoxy)ethoxy]methyl]-1,3-benzodioxole (I) (IUPAC); 1-bis[2-(2-butoxyethoxy)ethoxy]methyl-3,4-methylenedioxybenzene. *(C.A.)* (I) (9CI); piperonal bis[2-(2-butoxyethoxy)ethyl]acetal (8CI); Reg. No. *[5281-13-0]*. Piprotal. 'Tropital' (McLaughlin Gormley King Co.). $C_{24}H_{40}O_8$; T56 BO DO CHJ GYO2O2O4&O2O2O4; *PM4*, p. 166. Pyrethrum synergist (L. O. Hopkins & D. R. Maciver, *Pyrethrum Post*, 1965, **8**(2), 3).

**1030** Bis(2-chloroethyl) ether (I) (IUPAC). *(C.A.)* 1,1'-oxybis[2-chloroethane] (9CI); (I) (8CI); Reg. No. *[111-44-4]*. Di-(2-chloroethyl) ether. ENT 4505. $C_4H_8Cl_2O$; G2O2G; *PM4*, p. 177. Fumigant insecticide (R. C. Roark & R. T. Cotton, *USDA Tech. Bull.*, 1929, No. 162).

**1050** Bis(4-chlorophenoxy)methane (IUPAC); di-(4-chlorophenoxy)methane. *(C.A.)* 1,1'-[methylenebis(oxy)]bis(4-chlorobenzene) (9CI); bis(*p*-chlorophenoxy)methane (8CI); Reg. No. *[555-89-5]*. DCPM (JMAF), oxythane. 'K-1875'; 'Neotran' (Dow Chemical Co.). $C_{13}H_{10}Cl_2O_2$; GR DO1OR DG; *PM1*, p. 39. Acaricide (L. R. Jeppson, *J. Econ. Entomol.*, 1946, **39**, 813).

**1070** 1,1-Bis(4-chlorophenyl)-2-ethoxyethanol (IUPAC). *(C.A.)* 4-chloro-α-(4-chlorophenyl)-α-(ethoxymethyl)benzenemethanol (9CI); 4,4'-dichloro-α-(ethoxymethyl)benzhydrol (8CI); Reg. No. *[6012-83-5]*. 'Geigy 337', 'G 23 645'; 'Etoxinol' (Ciba-Geigy AG). $C_{16}H_{16}Cl_2O_2$; GR DXQR DG&1O2; acaricide.

**1080** 1,1-Bis(4-chlorophenyl)-2-nitrobutane (i) + 1,1-bis(4-chlorophenyl)-2-nitropropane (ii) (IUPAC) mixture. (*C.A.*) 1,1'-(2-nitrobutylidene)bis(4-chlorobenzene) (i) + 1,1'-(2-nitropropylidene)bis(4-chlorobenzene) (ii) (9CI); 1,1'-bis(*p*-chlorophenyl)-2-nitrobutane (i) + 1,1'-bis(*p*-chlorophenyl)-2-nitropropane (ii) (8CI); Reg. No. *[117-26-0]* (i), *[117-27-1]* (ii), *[8002-82-2]* (i + ii). ENT 18 066 [2:1 mixture of (i) + (ii)]; 'CS 674A' (i), 'CS 645A' (ii), 'CS 708' (mixture); 'Bulan' (i), 'Prolan' (ii); 'Dilan' (mixture) (Commercial Solvents Corp.). (i) $C_{16}H_{15}Cl_2NO_2$; WNY2&YR DG&R DG; (ii) $C_{15}H_{13}Cl_2NO_2$; WNY1&YR DG&R DG; *PM5*, p. 176. Insecticide.

**1130** Bis(diethoxyphosphinothioyl)disulphide (i) + bis(di-isopropoxyphosphinothioyl)disulphide (ii) 75:25 mixture (IUPAC). (*C.A.*) bis(diethoxyphosphinothioyl) disulfide (i); bis[di=(1-methylethoxy)phosphinothioyl] disulfide (ii) (9CI); tetraethyl thioperoxydiphosphate (i); tetraisopropyl thioperoxydiphosphate (ii) (8CI); Reg. No. *[2901-90-8]* (i), *[3031-21-8]* (ii). ENT 23 584 (i + ii). 'FMC 1137'; 'Phostex' (FMC Corp.) (i) $C_8H_{20}O_4P_2S_4$; 2OPS&O2&SSPS&O2&O2; (ii) $C_{12}H_{28}O_4P_2S_4$; 1Y1&OPS&OY1&1&SSPS&OY1&1&OY1&1; acaricide and insecticide.

**1160** *N,N'*-Bis(3-methoxypropyl)-6-methylthio-1,3,5-triazine-2,4-diyldiamine (IUPAC); 2,4-bis(3-methoxypropylamino)-6-(methylthio)-1,3,5-triazine. (*C.A.*) *N,N'*-bis(3-methoxypropyl)-6-methylthio-1,3,5-triazine-2,4-diamine (9CI); 2,4-bis(3-methoxypropyl)-6-methylthio-*s*-triazine (8CI); Reg. No. *[845-52-3]*. MPMT. 'CP 17 029'; 'Lambast' (Monsanto Co.). $C_{12}H_{23}N_5O_2S$; T6N CN ENJ BS1 DM3O1 FM3O1; herbicide.

**1170** Bis(methylmercury) sulphate (IUPAC). (*C.A.*) dimethyl-μ-sulfatodimercury (8 & 9CI); bis(methylmercury) sulfate; Reg. No. *[3810-81-9]*. 'Cerewet', Aretan-nieuw' (Bayer AG). $C_2H_6Hg_2O_4S$; 1-HG-OSWO-HG-1; *PM2*, p. 47. Fungicide.

**1215** (1*R*,2*R*,4*R*)-Bornyl thiocyanatoacetate (IUPAC). (*C.A.*) *exo*-1,7,7-trimethylbicyclo[2.2.1]hept-2-yl thiocyanatoacetate (9CI); isobornyl thiocyanatoacetate (8CI); Reg. No. *[115-31-1]*. ENT 92. 'Thanite' (Hercules Inc. now Boots-Hercules Agrochemical Co.). $C_{13}H_{19}NO_2S$; L55 ATJ A1 A1 B1 COV1SCN; *PM5*, p. 309. Insecticide (R. L. Pierpont, *Del. Agric. Exp. Sta. Bull.*, 1945, No. 253).

**1310** 1-Bromo-2-chloroethane (I) (IUPAC). *(C.A.)* (I) (8 & 9CI); Reg. No. *[107-04-0]*. Ethylene chlorobromide, *sym*-chlorobromoethane. $C_2H_4BrCl$; G2E; *PM4*, p. 255. Insecticidal fumigant for soils (M. W. Stone & F. B. Foley, *J. Econ. Entomol.*, 1951, **44**, 711).

**1320** 3-Bromo-1-chloroprop-1-ene (IUPAC). *(C.A.)* 3-bromo-1-chloro-1-propene (8 & 9CI); Reg. No. *[3737-00-6]*. 'CBP-55' (Shell Development Co.). $C_3H_4BrCl$; G1U2E; Soil fumigant.

**1430** Brompyrazone (BSI, F-ISO); brompyrazon (E-ISO). (IUPAC) 5-amino-4-bromo-2-phenylpyridazin-3(2*H*)-one. *(C.A.)* 5-amino-4-bromo-2-phenyl-3(2*H*)-pyridazinone (8 & 9CI); Reg. No. *[3042-84-0]*. 'Basanor' (Reg. No. *[66402-41-3]*) (BASF AG) for mixture with isonoruron. $C_{10}H_8BrN_3O$; T6NNVJ BR& DE EZ; *PM5*, p. 55. Herbicide (A. Fischer, *Weed Res.*, 1962, **2**, 177).

**1500** Butacarb (BSI, E-ISO); butacarbe (F-ISO). (IUPAC) 3,5-di-*tert*-butylphenyl methylcarbamate (I). (*C.A.*) 3,5-bis(1,1-dimethylethyl)phenyl methylcarbamate (9CI); (I) (8CI); Reg. No. *[2655-19-8]*. 'RD 14 639', 'BTS 14 639' (Boots Co. Ltd). $C_{16}H_{25}NO_2$; 1X1&1&R COVM1 EX1&1&1; *PM5*, p. 59. Insecticide (J. Fraser *et al.*, *J. Sci. Food Agric.*, 1967, **18**, 372) used as mixture with gamma-HCH against animal ectoparasites.

**1570** Butonate (BSI, E-ISO, F-ISO); butilchlorofos (USSR). (IUPAC) 2,2,2-trichloro-1-(dimethoxyphosphinoyl)ethyl butyrate; dimethyl 1-butyryloxy-2,2,2-trichloroethylphosphonate. *(C.A.)* 2,2,2-trichloro-1-(dimethoxyphosphinyl)ethyl butanoate (9CI); butyrate ester of dimethyl (2,2,2-trichloro-1-hydroxyethyl)phosphonate (8CI); Reg. No. [*126-22-7]*. ENT 20 852. $C_8H_{14}Cl_3O_5P$; GXGGYOV3&PO&O1&O1; *PM4*, p. 61. Insecticide (B. W. Arthur & J. E. Casida, *J. Agric. Food Chem.*, 1958, **6**, 360) introduced by Prentiss Drug & Chemical Co.

**1580** Butopyronoxyl (US Pharmacopoeia). (IUPAC) butyl 5,6,-dihydro-6,6-dimethyl-4-oxo-4*H*-pyran-2-carboxylate; butyl dihydro-6,6-dimethyl-4-oxopyran-2-carboxylate. (*C.A.*) butyl 3,4-dihydro-6,6-dimethyl-4-oxo-2*H*-pyran-6-carboxylate (8 & 9CI); Reg. No. *[532-34-3]*. Dihydropyrone, butyl mesityloxide oxalate. ENT 9. 'Indalone' (FMC Corp.). $C_{12}H_{18}O_4$; T6O DV BUTJ BVO4 F1 F1; 4OVYQU1V1UY1&1; *PM6*, p. 66. Insect repellent, existing mainly as the dihydropyran structure, introduced by Kilgore Chemicals.

**1600** 2-(2-Butoxyethoxy)ethyl piperonylate (I) (IUPAC). (*C.A.*) 2-(2-butoxyethoxy)ethyl 1,3-benzodioxole-5-carboxylate (9CI); (I) (8CI); Reg. No. *[136-63-0]*. BCP. 'Bucarpolate' (Bush Boake & Allen). $C_{16}H_{22}O_6$; T56 BO DO CHJ GVO2O2O4; synergist for pyrethrins.

**1610** 2-(2-Butoxyethoxy)ethyl thiocyanate (I) (IUPAC). (*C.A.*) (I) (8 & 9CI); Reg. No. *[112-56-1]*. Butyl 'Carbitol' thiocyanate; butyl 'Carbitol' rhodanate. ENT 6. 'Lethane 384' (Rohm & Haas Co.). $C_9H_{17}NO_2S$; NCS2O2O4; *PM5*, p. 62. Insecticide.

**1620** Butoxy(polypropylene glycol) (ESA). (IUPAC) α-hydroxypropyl-ω-butoxypoly[oxy(1-methylethylene)]. (*C.A.*) α-butyl-ω-hydroxypoly[oxy(methyl-1,2-ethanediyl)] (9CI); polypropylene glycol monobutyl ether (8CI); Reg. No. *[9003-13-8]*. ENT 8286. 'Crag Fly Repellent' (Union Carbide Corp.). / ★ Y1&1O ★ /; fly repellent.

**1700** 6-*tert*-Butyl-4,5-dihydro-3-isopropylisothiazolo[3,4-*d*]pyrimidin-4-one (IUPAC). (*C.A.*) 6-(1,1-dimethylethyl)-3-(1-methylethyl)isothiazolo[3,4-*d*]pyrimidin-4(5*H*)-one (9CI); Reg. No. *[40915-86-4]*. 'FMC 19 873' (FMC Corp.). $C_{12}H_{17}N_3OS$; T56 BNS FVM INJ DY1&1 HX1&1&1; *PM4*, p. 66. Herbicide.

**1710** 6-*tert*-Butyl-4,5-dihydro-3-isopropylisoxazolo[3,4-*d*]pyrimidin-4-one (IUPAC). (*C.A.*) 6-(1,1-dimethylethyl)-3-(1-methylethyl)isoxazolo[3,4-*d*]pyrimidin-4(5*H*)-one (9CI); Reg. No. *[38897-15-3]*. 'FMC 23 486' (FMC Corp.). $C_{12}H_{17}N_3O_2$; T56 BNO FVM INJ DY1&1 HX1&1&1; *PM5*, p. 66. Herbicide (W. M. Dest *et al., Proc. Northeast. Weed Sci. Soc.*, 1973, **27**, 31).

**1720** 6-*tert*-Butyl-4,5-dihydro-3-isopropylisoxazolo[5,4-*d*]pyrimidin-4-one (IUPAC). (*C.A.*) 6-(1,1-dimethylethyl)-3-(1-methylethyl)isoxazolo[5,4-*d*]pyrimidin-4(5*H*)-one (9CI); Reg. No. *[35258-87-8]*. 'FMC 21 844' (FMC Corp.). $C_{12}H_{17}N_3O_2$; T56 BON FVM INJ DY1&1 HX1&1&1; *PM4*, p. 68. Herbicide.

**1730** 6-*tert*-Butyl-4,5-dihydro-3-propylisoxazolo[5,4-*d*]pyrimidin-4-one (IUPAC). (*C.A.*) 6-(1,1-dimethylethyl)-3-propylisoxazolo[5,4-*d*]pyrimidin-4(5*H*)-one (9CI); Reg. No. *[35260-91-4]*. 'FMC 21 861' (FMC Corp.). $C_{12}H_{17}N_3O_2$; T56 BON FVM INJ D3 HX1&1&1; *PM3*, p. 75. Herbicide.

**1760** 2-(4-*tert*-Butylphenoxy)-1-methylethyl 2-chloroethyl sulphite (IUPAC). (*C.A.*) 2-chloroethyl 2-[4-(1,1-dimethylethyl)phenoxy]-1-methylethyl sulfite (9CI); 2-(*p*-*tert*-butylphenoxy)-1-methylethyl 2-chloroethyl sulfite (8CI); Reg. No. *[140-57-8]*. Aramite (JMAF). ENT 16 519. '88-R'; 'Aramite' (Uniroyal Inc.). $C_{15}H_{23}ClO_4S$; G2OSO&OY1&1OR DX1&1&1; *PM3*, p. 73. Acaricide (W. D. Harris & J. W. Zukel, *J. Agric. Food Chem.*, 1954, **2**, 140).

**1820** Cadmium calcium copper zinc chromate sulphate (IUPAC). (*C.A.*) Cadmium calcium copper zinc chromate sulfate (8 & 9CI); Reg. No. *[12001-20-6]*. 'Fungicide 531'; 'Crab Turf Fungicide' (Union Carbide Corp.). Fungicide.

**1860** Calcium cyanamide (I) (IUPAC). (*C.A.*) (I) (8 & 9CI); Reg. No. *[156-62-7]*. 'Cyanamid' (American Cyanamid Co.). $CCaN_2$; NCN &-CA-; *PM5*, p. 72. Herbicide.

**1910** Cambendichlor (BSI, E-ISO); cambendichlore (F-ISO). (IUPAC) 2,2'-(phenylimino)diethylene bis(3,6-dichloro-*o*-anisate). (*C.A.*) (phenylimino)di-2,1-ethanediyl bis(3,6-dichloro-2-methoxybenzoate) (9CI); Reg. No. *[56141-00-5]*. 'Vel 4207' (Velsicol Chemical Corp.). $C_{26}H_{23}Cl_4NO_6$; 1OR BG EG FVO2NR&2OVR BG EG FO1; herbicide.

**1950** Carbamorph (BSI, E-ISO); carbamorphe (F-ISO). (IUPAC) morpholinomethyl dimethyldithiocarbamate (I). (*C.A.*) *S*-morpholinomethyl dimethylcarbamodithioate (9CI); (I) (8CI); Reg. No. *[31848-11-0]*. 'MC 833' (Murphy Chemical Co.). $C_8H_{16}N_2OS_2$; T6N DOTJ A1SYUS&N1&1; fungicide.

**1960** Carbanolate (BSI, E-ISO, F-ISO, JMAF). (IUPAC) 6-chloro-3,4-xylyl methylcarbamate (I). (*C.A.*) 2-chloro-4,5-dimethylphenyl methylcarbamate (9CI); (I) (8CI); Reg. No. *[671-04-5]*. OMS 174. 'U 12 927'; 'Banol' (Upjohn Co.). $C_{10}H_{12}ClNO_2$; GR C1 D1 FOVM1; insecticide/ixodicide used against animal ectoparasites.

**2160** Chloraniformethan (BSI, E-ISO); chloraniforméthane (F-ISO). (IUPAC) *N*-[2,2,2-trichloro-1-(3,4-dichloroanilino)ethyl]formamide (I). (*C.A.*) *N*-[[2,2,2-trichloro-1-(3,4-dichlorophenyl)amino]ethyl]formamide (9CI); (I) (8CI); Reg. No. *[20856-57-9]*. 'Bayer 79 770'; 'Imugan', 'Milfaron' (Bayer AG). $C_9H_7Cl_5N_2O$; GXGGYMVHMR CG DG; *PM4*, p. 91. Fungicide (A. O. Paulus *et al., Calif. Agric.*, 1968, **22**(3), 10).

**2170** Chloranil (acceptable to BSI, E-ISO); chloranile (F-ISO). (IUPAC) tetrachloro-*p*-benzoquinone; 2,3,5,6-tetrachloro-1,4-benzoquinone. (*C.A.*) 2,3,5,6-tetrachloro-2,5-cyclohexadiene-1,4-dione (9CI); 2,3,5,6-tetrachloro-*p*-benzoquinone (8CI); Reg. No. *[118-75-2]*. ENT 3797. 'Spergon' (Uniroyal Inc.). $C_6Cl_4O_2$; L6V DVJ BG CG EG FG; *PM5*, p. 90. Fungicide (H. S. Cunningham & E. G. Shavelle, *Phytopathology*, 1940, **30**, 4).

**2180** Chloranocryl (BSI, E-ISO, F-ISO); dicryl (Canada, WSSA, ex ANSI). (IUPAC) 3',4'-dichloro-2-methylacrylanilide (I). (*C.A.*) *N*-(3,4-dichlorophenyl)-2-methyl-2-propenamide (9CI); (I) (8CI); Reg. No. *[2164-09-2]*. 'FMC 4556'; 'Dicryl' (FMC Corp.). $C_{10}H_9Cl_2NO$; GR BG DMVY1&U1; *PM2*, p. 85. Herbicide.

**2200** Chlorazine (BSI, ANSI). (IUPAC) 6-chloro-*N*,*N*,*N'*,*N'*-tetraethyl-1,3,5-triazine-2,4-diyldiamine; 2-chloro-4,6-bis(diethylamino)-1,3,5-triazine. (*C.A.*) 6-chloro-*N*,*N*,*N'*,*N'*-tetraethyl-1,3,5-triazine-2,4-diamine (9CI); 2-chloro-4,6-bis(diethylamino)-*s*-triazine (8CI); Reg. No. *[580-48-3]*. 'G 30 031' (Ciba-Geigy AG). $C_{11}H_{20}ClN_5$; T6N CN ENJ BN2&2 FG; Herbicide.

**2210** Chlorbenside (BSI, E-ISO, F-ISO, ex ANSI). (IUPAC) 4-chlorobenzyl 4-chlorophenyl sulphide. (*C.A.*) 1-chloro-4-[[(4-chlorophenyl)methyl]thio]benzene (9CI); *p*-chlorobenzyl *p*-chlorophenyl sulfide (8CI); Reg. No. *[103-17-3]*. ENT 20 696. 'HRS 860', 'RD 2195'; 'Chlorparacide', 'Chlorsulphacide' (FBC Limited formerly Boots Co. Ltd.). Mixtures with malathion or parathion. $C_{13}H_{10}Cl_2S$; GR DS1R DG; *PM3*, p. 97. Acaricide (J. E. Cranham *et al., Chem. Ind. (London)*, 1953, p. 1206).

**2220** Chlorbicyclen (BSI, E-ISO); chlorbicyclène (F-ISO). (IUPAC) 1,2,3,4,7,7-hexachloro-5,6-bis(chloromethyl)-8,9,10-trinorborn-2-ene; 1,2,3,4,7,7-hexachloro-5,6-bis(chloromethyl)bicyclo[2.2.1]hept-2-ene (I). (*C.A.*) (I) (9CI); 1,2,3,4,7,7-hexachloro-5,6-bis(chloromethyl)-2-norbornene (8CI); Reg. No. *[2550-75-6]*. ENT 211; ENT 785. 'Hercules 426'; 'Alodan' (Boots-Hercules Agrochemical Co.). $C_9H_6Cl_8$: L55 A CUTJ AG AG BG CG DG EG F1G G1G; insecticide.

**2260** Chlordecone (BSI, E-ISO, F-ISO). (IUPAC) decachloropentacyclo[5.2.1.$0^{2,6}$.$0^{3,9}$.$0^{5,8}$]decan-4-one. (*C.A.*) 1,1a,3,3a,4,5,5,5a,5b,6-decachlorooctahydro-1,3,4-metheno-2*H*-cyclobuta[*cd*]pentalen-2-one (8 & 9CI); Reg. No. *[143-50-0]*. ENT 16 391. 'GC-1189'; 'Kepone' (Allied Chemical Corp.). $C_{10}Cl_{10}O$; L545 B4 C5 D 4ABCE J DVTJ AG BG CG EG FG GG HG HG IG JG; *PM5*, p. 94. Insecticide.

**2310** Chlorfenazole (BSI, draft E-ISO). (IUPAC) 2-(2-chlorophenyl)benzimidazole. (*C.A.*) 2-(2-chlorophenyl)-1*H*-benzimidazole (9CI); Reg. No. *[3574-96-7]*. 'CUR 616' (Celamerck). $C_{13}H_9ClN_2$; T56 BM DNJ CR BG; Fungicide.

**2350** Chlorfensulphide (BSI, E-ISO); chlorfensulfide (F-ISO); CPAS (JMAF). (IUPAC) 4-chlorophenyl 2,4,5-trichlorobenzenediazosulphide; 4-chlorophenyl 2,4,5-trichlorophenylazosulphide. (*C.A.*) [(4-chlorophenyl)thio](2,4,5-trichlorophenyl)diazene (9CI); [(*p*-chlorophenyl)thio](2,4,5-trichlorophenyl)diimide (8CI); Reg. No. *[2274-74-0]*. 'Milbex' (Reg. No. *[8072-20-6]*) (Nippon Soda Co. Ltd) as mixture with chlorfenethol. $C_{12}H_6Cl_4N_2S$; GR DSNUNR BG DG EG; *PM4*, p. 101. Acaricide.

**2370** Chlorflurazole (BSI, E-ISO, F-ISO). (IUPAC) 4,5-dichloro-2-trifluoromethylbenzimidazole. (*C.A.*) 4,5-dichloro-2-(trifluoromethyl)-1*H*-benzimidazole (9CI); 4,5-dichloro-2-(trifluoromethyl)benzimidazole (8CI); Reg. No. *[3615-21-2]*. 'NC 3363' (FBC Limited). $C_8H_3Cl_2F_3N_2$; T56 BM DNJ CXFFF FG GG; experimental herbicide.

**2390** Chlorfluren (BSI, E-ISO); Chlorflurène (F-ISO). (IUPAC) 2-chlorofluorene-9-carboxylic acid (I). (*C.A.*) 2-chloro-9*H*-fluorene-9-carboxylic acid (9CI); (I) (8CI); Reg. No. *[24539-66-0]* (acid), *[22909-50-8]* (chlorfluren-methyl). 'IT-5732' for chlorfluren-methyl (E. Merck). Chlorfluren $C_{14}H_9ClO_2$; L B656 HHJ EG HVQ; chlorfluren-methyl $C_{15}H_{11}ClO_2$; L B656 HHJ EG HVO1; chlorfluren was plant growth regulator.

**2530** 2-Chloro-*N*-(2-cyanoethyl)acetamide (IUPAC) (I). (*C.A.*) (I) (8 & 9CI); Reg. No. *[17756-81-9]*. CECA (JMAF). 'NF 21'; 'Udonkor' (Nippon Soda Co. Ltd). $C_5H_7ClN_2O$; NC2MV1G; fungicide.

**2540** 1-Chloro-2,4-dinitronaphthalene (IUPAC) (I). (*C.A.*) (I) (8 & 9CI); Reg. No. *[2401-85-6]*. $C_{10}H_5ClN_2O_4$; L66J BG CNW ENW; fungicide.

**2620** Chloromebuform (BSI, E-ISO); chloromébuforme (F-ISO). (IUPAC) $N^1$-butyl-$N^2$-(4-chloro-*o*-tolyl)-$N^1$-methylformamidine. (*C.A.*) *N*-butyl-*N'*-(4-chloro-2-methylphenyl)-*N*-methylmethanimidamide (9CI); *N*-butyl-*N'*-(4-chloro-*o*-tolyl)-*N'*-methylformamidine; Reg. No.*[37407-77-5]*. 'C 22 598'; 'Ektomin' (Ciba-Geigy AG). $C_{13}H_{19}ClN_2$; GR C1 DNU1N4&1; acaricide.

**2670** *O*-2-Chloro-4-(methylthio)phenyl *O*-methyl ethylphosphoramidothioate (IUPAC). (*C.A.*) *O*-[2-chloro-4-(methylthio)phenyl] *O*-methyl ethylphosphoramidothioate (9CI); Reg. No. *[54381-26-9]*. Amidothioate (JMAF). 'Mitemate' (Nippon Kayaku Co. Ltd). $C_{10}H_{15}C1NO_2PS_2$; 2MPS&O1&OR BG DS1; acaricide.

**2710** *O*-3-Chloro-4-nitrophenyl *O,O*-dimethyl phosphorothioate (IUPAC). (*C.A.*) *O*-(3-chloro-4-nitrophenyl) *O,O*-dimethyl phosphorothioate (8 & 9CI); Reg. No. *[500-28-7]*. OMS 217. 'Bayer 22/190'; 'Chlorthion' (Bayer AG). $C_8H_9ClNO_5PS$; WNR BG DOPS&O1&O1; insecticide.

**2740** 1-Chloro-2-nitropropane (IUPAC) (I). (*C.A.*) (I) (8 & 9CI); Reg. No. *[2425-66-3]*. 'FMC 5916'; 'Lanstan' (FMC Corp.). $C_3H_6ClNO_2$; WNY1&1G; *PM2*, p. 104. Soil fungicide.

**2790** 1-(4-Chlorophenyl)-3-(2,6-dichlorobenzoyl)urea (I) (IUPAC). (*C.A.*) 2,6-dichloro-*N*-[[(4-chlorophenyl)amino]carbonyl]benzamide (9CI); Reg. No. *[35409-97-3]*. 'PH 60-38' (Philips Duphar now Duphar B.V.). $C_{14}H_9Cl_3N_2O_2$; GR DMVMVR BG FG; experimental insecticide.

**2820** 5-Chloro-4-phenyl-1,2-dithiol-3-one (IUPAC). (*C.A.*) 5-chloro-4-phenyl-3*H*-1,2-dithiol-3-one (8 & 9CI); Reg. No. *[2425-05-0]*. 'Hercules 3944' (Boots-Hercules Agrochemical Co.). $C_9H_5ClOS_2$; T5SSVJ DR& EG; *PM3*, p. 115. Fungicide.

**2840** 3-(4-Chlorophenyl)-5-methylrhodanine (I) (IUPAC); 3-(4-chlorophenyl)-5-methyl-2-thioxo-1,3-thiazolidin-4-one. (*C.A.*) 3-(4-chlorophenyl)-5-methyl-2-thioxo-4-thiazolidinone (9CI); (I) (8CI); Reg. No. *[6012-92-6]*. 'N 244' (Stauffer Chemical Co.). $C_{10}H_8ClNOS_2$; T5SYNV EHJ BUS CR DG& E1; Fungicide and nematicide.

**2850** 4-Chlorophenyl phenyl sulphone (IUPAC). (*C.A.*) 1-chloro-4-(phenylsulfonyl)benzene (9CI); *p*-chlorophenyl phenyl sulfone (8CI); Reg. No. *[80-00-2]*. 'R-242'; 'Sulphenone' (Stauffer Chemical Co.). $C_{12}H_9ClO_2S$; WSR&R DG; acaricide.

**2870** *S*-4-Chlorophenylthiomethyl *O,O*-dimethyl phosphorodithioate (IUPAC). (*C.A.*) *S*-[[(4-chlorophenyl)thio]methyl] *O,O*-dimethyl phosphorodithioate (9CI); *S*-[[(*p*-chlorophenyl)thio]methyl] *O,O*-dimethyl phosphorodithioate (8CI); Reg. No. *[953-17-3]*. ENT 25 599. 'R-1492'; 'Methyl Trithion', 'Tri-Me' (Stauffer Chemical Co.). $C_9H_{12}ClO_2PS_3$; GR DS1SPS&O1&O1; *PM6*, p. 111. Insecticide and acaricide (J. A. Harding, *J. Econ. Entomol.*, 1959, **52**, 1219).

**2960** 2-Chlorovinyl diethyl phosphate (I) (IUPAC). (*C.A.*) 2-chloroethenyl diethyl phosphate (9CI); (I) (8CI); Reg. No. *[311-47-7]*. 'SD 1836', 'OS 1836' (Shell Chemical Co.). $C_6H_{12}ClO_4P$; G1U1OPO&O2&O2; insecticide.

**2980** 2-(4-Chloro-3,5-xylyloxy)ethanol (IUPAC). (*C.A.*) 2-(4-chloro-3,5-dimethylphenoxy)ethanol (9CI); 2-[(4-chloro-3,5-xylyl)oxy]ethanol (8CI); Reg. No. *[5825-79-6]*. 'Experimental Chemotherapeutant 1182' (Union Carbide Corp.). $C_{10}H_{13}ClO_2$; Q2OR DG C1 E1; insecticide.

**3030** Chlorprocarb (BSI, E-ISO); chlorprocarbe (F-ISO). (IUPAC) 3-methoxycarbonylaminophenyl 1-chloromethylpropylcarbamate. (*C.A.*) 3-[(methoxycarbonyl)amino]phenyl [1-(chloromethyl)propyl]carbamate (9CI); methyl *m*-hydroxycarbanilate [1-(chloromethyl)propyl]carbamate ester(8CI); Reg. No. *[23121-99-5]*. 'BAS 379H' (BASF AG). $C_{13}H_{17}ClN_2O_4$; G1Y2&MVOR CMVO1; herbicide.

**3070** Chlorquinox (BSI, E-ISO, F-ISO). (IUPAC) 5,6,7,8-tetrachloroquinoxaline (I). (*C.A.*) (I) (8 & 9CI); Reg. No. *[3495-42-9]*. 'Lucel' (FBC Limited). $C_8H_2Cl_4N_2$; T66 BN ENJ GG HG IG JG; *PM5*, p. 119. Fungicide.

**3180** Cliodinate (BSI, draft E-ISO). (IUPAC) 2-chloro-3,5-di-iodo-4-pyridyl acetate. (*C.A.*) 2-chloro-3,5-diiodo-4-pyridinyl acetate (9CI); Reg. No. *[69148-12-5]*. 'ASD 2288' (Celamerck). $C_7H_4ClI_2NO_2$; T6NJ BG CI DOV1 EI; herbicide.

**3200** Clofop (BSI, E-ISO, F-ISO, WSSA). (IUPAC) (±)-2-[4-(4-chlorophenoxy)phenoxy]propionic acid. (*C.A.*) (±)-2-[4-(4-chlorophenoxy)phenoxy]propanoic acid (9CI); Reg. No. *[51337-71-4]* (isobutyl ester); 'Hoe 22 870' (Hoechst AG). (Acid) $C_{15}H_{13}ClO_4$; QVY1&OR DOR DG; (isobutyl ester) $C_{19}H_{21}ClO_4$; GR DOR DOY1&VO1Y1&1; *PM6*, p. 125. The isobutyl ester was used as a herbicide (F. Schwerdtle *et al., Mitt. Biol, Bundesanst. Land-Forstwirtsch. Berlin-Dahlem*, 1975, **165**, 171).

**3240** Copper bis(3-phenylsalicylate) (IUPAC). (*C.A.*) bis(2-hydroxy[1,1'-biphenyl]-3-carboxylato-$O^2$,$O^0$)copper (9CI); bis(3-phenylsalicylato)copper(II) (8CI); Reg. No. *[5328-04-1]*. $C_{26}H_{18}CuO_6$; QVR BQ CR &&Cu COMPLEX 2:1; fungicide.

**3310** Copper zinc chromate (indefinite composition). 'Crag Fungicide 658', 'Experimental Fungicide 658' (Union Carbide Corp). Fungicide.

**3330** Coumaphos (BSI, E-ISO, F-ISO, BPC). *O*-3-chloro-4-methyl-2-oxo-2*H*-chromen-7-yl *O,O*-diethyl phosphorothioate (IUPAC); *O*-3-chloro-4-methylcoumarin-7-yl *O,O*-diethyl phosphorothioate. (*C.A.*) *O*-(3-chloro-4-methyl-2-oxo-2*H*-benzopyran-7-yl) *O,O*-diethyl phosphorothioate (9CI); 3-chloro-7-hydroxy-4-methylcoumarin *O*-ester with *O,O*-diethyl phosphorothioate (8CI); Reg. No. *[56-72-4]*. OMS 485; ENT 17 957. 'Bayer 21/199'; 'Asuntol', 'Muscatox', 'Resitox' (Bayer AG), 'Co-Ral' (Baychem Corp.); 'Negasunt'. $C_{14}H_{16}ClO_5PS$; T66 BOVJ DG E1 IOPS&O2&O2; *PM2*, p. 120. Veterinary insecticide.

**3400** Crufomate (BSI, E-ISO, F-ISO, ANSI). (IUPAC) 4-*tert*-butyl-2-chlorophenyl methyl methylphosphoramidate (I). (*C.A.*) 2-chloro-4-(1,1-dimethylethyl)phenyl methyl methylphosphoramidate (9CI); (I) (8CI); Reg. No. *[299-86-5]*. ENT 26 602-X. 'Dowco 132'; 'Ruelene' (Dow Chemical Co.). $C_{12}H_{19}ClNO_3P$; 1X1&1&R CG DOPO&O1&M1 *PM6*, p. 131. Insecticide and anthelmintic used against pests on cattle (J. F. Landram & R. J. Shaver, *J. Parasitol.*, 1959, **45**, 55).

**3425** *m*-Cumenyl methylcarbamate (I) (IUPAC); 3-isopropylphenyl methylcarbamate. (*C.A.*) 3-(methylethyl)phenyl methylcarbamate (9CI); (I) (8CI); Reg. No. *[64-00-6]*. ENT 25 500. 'AC 5727' (Hercules Inc.), 'UC 10 854' (Union Carbide Corp.). $C_{11}H_{15}NO_2$; 1Y1&R COVM1; *PM1*, p. 263. Insecticide.

**3430** Cupric hydrazinium sulphate; copper(II) dihydrazinium disulphate (IUPAC). (*C.A.*) bis(hydrazine)bis(hydrogen sulfato)copper (9CI); copper dihydrazine disulfate; Reg. No. *[33271-65-7]*. 'Mathieson 466'; 'Omazene' (Olin Mathieson Chemical Corp.). $CuH_{10}N_4O_8S_2$; .Cu..ZZ.2.SO4*2; fungicide.

**3460** Cyanatryn (BSI, E-ISO); cyanatryne (F-ISO). (IUPAC) 2-(4-ethylamino-6-methylthio-1,3,5-triazin-2-ylamino)-2-methylpropionitrile; 2-(1-cyano-1-methylethylamino)-4-ethylamino-6-methylthio-1,3,5-triazine. (*C.A.*) 2-[[4-(ethylamino)-6-(methylthio)-1,3,5-triazin-2-yl]amino]-2-methylpropanenitrile (9CI); 2-[[4-(ethylamino)-6-(methylthio)-1,3,5-triazin-2-yl]amino]-2-methylpropionitrile (8CI); Reg. No. *[21689-84-9]*. 'WL 63 611' (Shell Research Ltd). $C_{10}H_{16}N_6S$; T6N CN ENJ BS1 DMX1&1&CN FM2; experimental herbicide (V. V. Dovlatyan & F. V. Avetisyan, *Arm. Khim. Zh.* 1972, **25**, 880).

**3480** 2-Cyano-3-(2,4-dichlorophenyl)acrylic acid (I) (IUPAC). (*C.A.*) 2-cyano-3-(2,4-dichlorophenyl)-2-propenoic acid (9CI); (I) (8CI); Reg. No. *[6013-05-4]*. 'Ethyl 214' (Ethyl Corp.). $C_{10}H_5Cl_2NO_2$; QVYCN&U1R BG DG; plant growth regulator.

**3560** Cyanthoate (BSI, E-ISO, F-ISO). (IUPAC) *S*-[*N*-(1-cyano-1-methylethyl)carbamoylmethyl] *O,O*-diethyl phosphorothioate. (*C.A.*) *S*-[2-[(1-cyano-1-methylethyl)amino]-2-oxoethyl] *O,O*-diethyl phosphorothioate (9CI); *O,O*-diethyl phosphorothioate *S*-ester with *N*-(1-cyano-1-methylethyl)-2-mercaptoacetamide (8CI); Reg. No. *[3734-95-0]*. 'M 1568'; 'Tartan' (Montedison S.p.A.). $C_{10}H_{19}N_2O_4PS$; NCX1&1&MV1SPO&O2&O2; *PM5*, p. 138. Insecticide and acaricide (F. Galbaiti, *Proc. Br. Insectic. Fungic. Conf., 1st*, 1961, **2**, 507).

**3570** Cyclafuramid (BSI, E-ISO); cyclafuramide (F-ISO). (IUPAC) *N*-cyclohexyl-2,5-dimethyl-3-furamide (I). (*C.A.*) *N*-cyclohexyl-2,5-dimethyl-3-furancarboxamide (9CI); (I) (8CI); Reg. No. *[34849-42-8]*. 'BAS 327F' (BASF AG). $C_{13}H_{19}NO_2$; T5OJ B1 E1 CVM-AL6TJ; fungicide.

**3620** Cycluron (BSI, E-ISO, F-ISO, WSSA); COMU (JMAF). (IUPAC) 3-cyclo-octyl-1,1-dimethylurea (I). *(C.A.)* *N*'-cyclooctyl-*N,N*-dimethylurea (9CI); (I) (8CI); Reg. No. *[2163-69-1]*. OMU. 'Alipur' (Reg. No. *[8015-55-2]*) (BASF AG) for mixture with chlorbufam. $C_{11}H_{22}N_2O$; L8TJ AMVN1&1; *PM6*, p. 141. Herbicide (A. Fischer, *Z. Pflanzenkr. Pflanzenpath. Pflanzenschutz*, 1960, **67**, 577).

**3680** Cypendazole (BSI, E-ISO, F-ISO, JMAF). (IUPAC) methyl 1-(5-cyanopentylcarbamoyl)benzimidazol-2-ylcarbamate. (*C.A.*) methyl [1-[(5-cyanopentyl)amino]carbonyl]-1*H*-benzimidazol-2-ylcarbamate (9CI); methyl 1-(5-cyanopentylcarbamoyl)benzimidazole-2-carbamate (8CI); Reg. No. *[28559-00-4]*. 'DAM 18 654'; 'Folcidin' (Bayer AG). $C_{16}H_{19}N_5O_3$; T56 BN DNJ BVM5CN CMVO1; *PM4*, p. 145. Fungicide.

**3700** Cyperquat (BSI, E-ISO, F-ISO, ANSI, WSSA). (IUPAC) 1-methyl-4-phenylpyridinium ion. (*C.A.*) 1-methyl-4-phenylpyridinium (9CI); Reg. No. *[48134-75-4]* (ion), *[39794-99-5]* (chloride). 'S 21 634' (Gulf Oil Chemicals). (Ion) $C_{12}H_{12}N$; T6KJ A1 DR; (chloride) $C_{12}H_{12}ClN$; T6KJ A1 DR &&Cl SALT; chloride is herbicide.

**3720** Cyprazine (BSI, E-ISO, F-ISO, ANSI, WSSA) (IUPAC) 6-chloro-*N*-cyclopropyl-*N*'-isopropyl-1,3,5-triazine-2,4-diyldiamine; 2-chloro-4-cyclopropylamino-6-isopropylamino-1,3,5-triazine. (*C.A.*) 6-chloro-*N*-cyclopropyl-*N*'-(1-methylethyl)-1,3,5-triazine-2,4-diamine (9CI); 2-chloro-4-(cyclopropylamino)-6-(isopropylamino)-*s*-triazine (8CI); Reg. No. *[22936-86-3]*. 'S 6115'; 'Outfox' (Gulf Oil Corp.). $C_9H_{14}ClN_5$; T6N CN ENJ BMY1&1 DG FM- AL3TJ; *PM4*, p. 146. Herbicide (O. C. Burnside *et al., Proc. North Cent. Weed Control Conf.*, 1969, p. 21).

**3730** Cyprazole (BSI, E-ISO, F-ISO, ANSI, WSSA). (IUPAC) *N*-[5-(2-chloro-1,1-dimethylethyl)-1,3,4-thiadiazol-2-yl]cyclopropanecarboxamide (I). (*C.A.*) (I) (8 & 9CI); Reg. No. *[42089-03-2]*. 'S 19 073' (Gulf Oil Chemicals). $C_{10}H_{14}ClN_3OS$; T5NN DSJ CX1&1&1G EMV- AL3TJ; herbicide.

**3750** Cypromid (BSI, E-ISO, ANSI, WSSA, JMAF). (IUPAC) 3',4'-dichlorocyclopropanecarboxanilide (I). (*C.A.*) *N*-(3,4-dichlorophenyl)cyclopropanecarbox = amide (9CI); (I) (8CI); Reg. No. *[2759-71-9]*. 'S 6000'; 'Clobber' (Gulf Oil Corp.). $C_{10}H_9Cl_2NO$; L3TJ AVMR CG DG; *PM3*, p.148. Herbicide (T. R. Hopkins *et al., Proc. Symp. New Herbic., 2nd*, Paris, 1965, p. 187).

**3840** Decafentin (BSI, E-ISO, F-ISO). (IUPAC) decyltriphenylphosphonium bromochlorotriphenylstannate(IV) (I). (*C.A.*) (TB-5-12)-decyltriphenylphosphonium bromochlorotriphenylstannate(1-) (9CI); (I) (8CI); Reg. No. *[15652-38-7]*. 'A 36'; 'Stanoram' (Celamerck). $C_{46}H_{51}BrClPSn$; G-SN-ER&R&R &&$CH_3[CH_2]_9P(C_6H_5)_3$ SALT; 1ØPR&R&R &&$(C_6H_5)_3SnBrCl$ SALT; fungicide.

**3870** Dehydroacetic acid (trivial name). (IUPAC) 2-acetyl-5-methyl-3-oxopent-4-en-5-olide (i); 3-acetyl-6-methylpyran-2,4-dione (i). (*C.A.*) 3-acetyl-6-methyl-2*H*-pyran-2,4(3*H*)-dione (i); 3-acetyl-4-hydroxy-6-methyl-2*H*-pyran-2-one (ii); 3-acetyl-2-hydroxy-6-methyl-4*H*-pyran-4-one (iii) (8 & 9CI); Reg. No. *[520-45-6]* (i), *[771-03-9]* (ii), *[16807-48-0]* (iii). DHA. $C_8H_8O_4$; T6OV DV CHJ CV1 F1; *PM6*, p. 152. Fungicide, formerly produced by Dow Chemical Co. to protect fruit and vegetables from post-harvest attack by microorganisms (P. A. Wolf, *Food Technol. (Chicago)*, 1950, **4**, 294).

**3880** Delachlor (BSI, E-ISO, ANSI); délachlore (F-ISO). (IUPAC) 2-chloro-*N*-(isobutoxymethyl)acet-2',6'-xylidide. (*C.A.*) 2-chloro-*N*-(2,6-dimethylphenyl)-*N*-[(2-methylpropoxy)methyl]acetamide (9CI); α-chloro-*N*-isobutoxymethylacet-2',6'-xylidide (8CI); Reg. No. *[24353-58-0]*. 'CP 52 223' (Monsanto Co.). $C_{15}H_{22}ClNO_2$; 1Y1&1O1NV1GR B1 F1; *PM2*, p. 100. Herbicide (R. F. Husted *et al., Proc. North Cent. Weed Control Conf.*, 1967)

**3900** Demephion (BSI) for mixture of demephion-O (i) and demephion-S (ii) (BSI, E-ISO, F-ISO for both). (IUPAC) *O,O*-dimethyl *O*-2-methylthioethyl phosphorothioate (i); *O,O*-dimethyl *S*-2-methylthioethyl phosphorothioate (ii). (*C.A.*) *O,O*-dimethyl *O*-[2-(methylthio)ethyl] phosphorothioate (i); *O,O*-dimethyl *S*-[2-(methylthio)ethyl] phosphorothioate (ii) (8 & 9CI for both); Reg. No. *[682-80-4]* (i), *[2587-90-8]* (ii), *[8065-62-1]* (i + ii). '*Cymetox*' (Cyanamid of Great Britain), 'Pyracide' (BASF UK), 'Atlasetox' (Atlas Products), 'Tinox' in Eastern Europe. (i or ii) $C_5H_{13}O_3PS_2$; (i) 1S2OPS&O1&O1; (ii) 1S2SPO&O1&O1; *PM6*, p. 153. Acaricide and insecticide (H. Rueppold, *Wiss. Z. Martin-Luther Univ. Halle-Wittenberg Math.-Naturwiss. Reihe*, 1955, **5**, 219).

**3940** 2,4-DEP (WSSA). (IUPAC) mixture of tris[2-(2,4-dichlorophenoxy)ethyl] phosphite (I) (i) + bis[2-(2,4-dichlorophenoxy)ethyl] phosphonate (II) (ii). (*C.A.*) (I) + (II) (8 & 9CI); Reg. No. *[39420-34-3]* (i + ii), *[94-84-8]* (i). '3Y9'; 'Falone' (Uniroyal Inc.). (i) $C_{24}H_{21}Cl_6O_6P$; GR CG DO2OPO2OR BG DG&O2OR BG DG; (ii) $C_{16}H_{15}Cl_4O_5P$; GR CG DO2OPHO&O2OR BG DG; *PM2*, p. 143. Herbicide.

**4010** Diamidafos (BSI, E-ISO, ANSI); diamidaphos (F-ISO). (IUPAC) phenyl *N,N'*-dimethylphosphorodiamidate (I). (*C.A.*) (I) (8 & 9CI); Reg. No. *[1754-58-1]*. 'Dowco 169'; 'Nellite' (Dow Chemical Co.). $C_8H_{13}N_2O_2P$; 1MPO&OR&M1; *PM5*, p. 413. Nematicide (C. R. Youngson & C. A. I. Goring, *Down Earth*, 1963, **18**(4), 3).

**4060** 1,2-Dibromo-3-chloropropane (I) (IUPAC). (*C.A.*) (I) (8 & 9CI); Reg. No. *[96-12-8]*. DBCP (JMAF). 'OS1897'; 'Nemagon' (Shell Development Co.), 'Fumazone' (Dow Chemical Co.). $C_3H_5Br_2Cl$; G1YE1E; *PM6*, p. 164. Nematicide (C. W. McBeth & G. B. Bergeson, *Plant Dis. Rep.*, 1955, **39**, 223).

**4070** Dibutyl adipate (I) (IUPAC). (*C.A.*) dibutyl hexanedioate (9CI); (I) (8CI); Reg. No. *[105-99-7]*. 'Experimental Tick Repellent 3' (Union Carbide Corp.). $C_{14}H_{26}O_4$; 4OV4VO4; insect repellent.

**4080** Dibutyl phthalate (I) (IUPAC, accepted BSI, E-ISO). (*C.A.*) dibutyl 1,2-benzenedicarboxylate (9CI); (I) (8CI); Reg. No. *[84-74-2]*. DBP. $C_{16}H_{22}O_4$; 4OVR BVO4; *PM5*, p. 163. Insect repellent (F. M. Snyder & F. A. Morton, *J. Econ. Entomol.*, 1947, **40**, 586).

**4090** Dibutyl succinate (I) (IUPAC). (*C.A.*) Dibutyl butanedioate (9CI); (I) (8CI); Reg. No. *[141-03-7]*. 'Tabatrex' (Glenn Chemical Co.), formerly '*Tabutrex*'. $C_{12}H_{22}O_4$; 4OV2VO4; *PM5*, p. 164. Insect repellent.

**4110** Dicamba-methyl (BSI, E-ISO, F-ISO); disugran (ANSI). (IUPAC) methyl 3,6-dichloro-*o*-anisate (I). (*C.A.*) methyl 3,6-dichloro-2-methoxybenzoate (9CI); (I) (8CI); Reg. No. *[6597-78-0]*. 'Racuza' (Velsicol Chemical Co.). $C_9H_8Cl_2O_3$; 1OVR BG EG FO1; plant growth regulator.

**4120** Dicapthon (ex ANSI). (IUPAC) *O*-2-chloro-4-nitrophenyl *O,O*-dimethyl phosphorothioate. (*C.A.*) *O*-(2-chloro-4-nitrophenyl) *O,O*-dimethyl phosphorothioate (8 & 9CI); Reg. No. *[2463-84-5]*. OMS 214; ENT 17 035. 'Experimental Insecticide 4124'; 'Dicaptan' (American Cyanamid Co.). $C_8H_9ClNO_5PS$; WNR CG DOPS&O1&O1; *PM3*, p. 169. Insecticide (T. B. Davich & J. W. Apple, *J. Econ. Entomol.*, 1951, **44**, 528).

**4160** Dichlone (BSI, E-ISO, F-ISO, JMAF). (IUPAC) 2,3-dichloro-1,4-naphthoquinone (I). (*C.A.*) 2,3-dichloro-1,4-naphthalenedione (9CI); (I) (8CI); Reg. No. *[117-80-6]*. ENT 3776. 'USR 604'; 'Phygon' (Uniroyal Inc.). $C_{10}H_4Cl_2O_2$; L66 BV EVJ CG DG; *PM5*, p. 169. Fungicide (W. P. ter Horst & E. L. Felix, *Ind. Eng. Chem.*, 1943, **35**, 1255).

**4170** Dichloralurea (accepted BSI, E-ISO); dichloralurée (F-ISO);DCU (WSSA). (IUPAC) 1,3-bis(2,2,2-trichloro-1-hydroxyethyl)urea (I). (*C.A.*) *N,N'*-bis(2,2,2-trichloro-1-hydroxyethyl)urea (9CI); (I) (8CI); Reg. No. *[116-52-9]*. 'Crag Herbicide 2'; 'Crag Herbicide 2' (Union Carbide Corp.). $C_5H_6Cl_6N_2O_3$; GXGGYQMVMYQXGGG; *PM1*, p. 139. Herbicide (L. J. King, *Proc. Northeast. Weed Control Conf.*, 1950, p. 302).

**4180** Dichlorflurecol (BSI, Canada); dichlorflurenol (E-ISO); dichloroflurénol (F-ISO). (IUPAC) 2,7-dichloro-9-hydroxyfluorene-9-carboxylic acid (I). (*C.A.*) 2,7-dichloro-9-hydroxy-9*H*-fluorene-9-carboxylic acid (9CI); (I) (8CI); Reg. No. *[21634-96-8]* (dichlorflurecol-methyl). IT -5733 for dichlorflurecol-methyl (E. Merck). Dichlor = flurecol $C_{14}H_8C1_2O_3$; L B656 HHJ EG HVQ HQ KG; dichlorflurecol-methyl $C_{15}H_{10}C1_2O_3$; L B656 HHJ EG HVO1 HQ KG; dichlorflurecol-methyl was plant growth regulator.

**4190** Dichlormate (BSI, E-ISO, F-ISO, ANSI). (IUPAC) 3,4-dichlorobenzyl methylcarbamate (I). (*C.A.*) 3,4-dichlorobenzenemethyl methylcarbamate (9CI); (I) (8CI); Reg. No. *[1966-58-1]*. 'UC 22 463A' 'Rowmate' (Union Carbide Corp.). $C_9H_9Cl_2NO_2$; GR BG D1OVM1; *PM1*, p. 141. Herbicide (R. A. Herrett & R. V. Berthold, *Science*, 1965, **149**, 191).

**4230** 1,1-Dichloro-2,2-bis(4-ethylphenyl)ethane (IUPAC); 1,1-dichloro-2,2-di-(4-ethylphenyl)ethane. (*C.A.*) 1,1'-(2,2-dichloroethylidene)bis[4-ethylbenzene] (9CI); 1,1-dichloro-2,2-bis[*p*-ethylphenyl]ethane (8CI); Reg. No. *[72-56-0]*. 'Q-137'; 'Perthane' (Rohm & Haas Co.). $C_{18}H_{20}Cl_2$; GYGYR D2&R D2; *PM6*, p. 170. Insecticide.

**4260** *O*-2,5-Dichloro-4-iodophenyl *O*-ethyl ethylphosphonothioate (IUPAC). (*C.A.*) *O*-(2,5-dichloro-4-iodophenyl) *O*-ethyl ethylphosphonothioate (8 & 9CI); Reg. No. *[25177-27-9]*. 'C 18 244' (Ciba-Geigy AG). $C_{10}H_{12}Cl_2IO_2PS$; IR BG EG DOPS&2&O2; *PM2*, p. 231. Insecticide.

**4290** 1,1-Dichloro-1-nitroethane (I) (IUPAC). (*C.A.*) (I) (8 & 9CI); Reg. No. *[594-72-9]*. 'Ethide' (Commercial Solvents Corp.). $C_2H_3Cl_2NO_2$; WNXGG1; *PM3*, p. 177. Fumigant insecticide (W. C. O'Kane & H. W. Smith, *J. Econ. Entomol.*, 1941, **34**, 438).

**4310** 2,4-Dichlorophenyl benzenesulphonate (IUPAC). (*C.A.*) 2,4-dichlorophenyl benzenesulfonate (8 & 9CI); Reg. No. *[97-16-5]*. 'EM 293' 'Genite', 'Genitol' (Allied Chemical Corp.). $C_{12}H_8Cl_2O_3S$; WSR&OR BG DG; *PM4*, p. 179. Acaricide.

**4360** *N*-3,5-Dichlorophenylsuccinimide (IUPAC). (*C.A.*) 1-(3,5-dichlorophenyl)-2,5-pyrrolidinedione (9CI); *N*-(3,5-dichlorophenyl)succinimide (8CI); Reg. No. *[24096-53-5]*. 'S-47127'; 'Ohric' (Sumitomo Chemical Co. Ltd). $C_{10}H_7Cl_2NO_2$; T5VNVJ BR CG EG; fungicide.

**4430** 1,3-Dichloro-1,1,3,3-tetrafluoropropane-2,2-diol (i); 1,3-dichloro-1,1,3,3,-tetrafluoroacetone hydrate (ii) (IUPAC). (*C.A.*) 1,3-dichloro-1,1,3,3-tetrafluoro-2,2-propanediol (i) (8 & 9CI); 1,3-dichloro-1,1,3,3-tetrafluoro-2-propanone hydrate (ii) (8 & 9CI); Reg. No. *[993-57-8]* (i), *[121-21-9]* (ii). 'GC 9832' (Allied Chemical Corp.). $C_3H_2Cl_2F_4O_2$; (i) GXFFXQQXGFF; (ii) GXFFVXGFF &&$H_2O$; *PM4*, p. 185. Fungicide.

**4440** 3,4-Dichlorotetrahydrothiophene 1,1-dioxide (I) (IUPAC). (*C.A.*) (I) (8 & 9CI); Reg. No. *[3001-57-8]*. Dichlorothiolane dioxide. 'PRD Experimental Nematicide' (Diamond Shamrock Chemical Co.). $C_4H_6Cl_2O_2S$; T5SWTJ CG DG; nematicide.

**4470** 2,2-Dichlorovinyl 2-ethylsulphinylethyl methyl phosphate (IUPAC). (*C.A.*) 2,2-dichloroethenyl 2-(ethylsulfinyl)ethyl methyl phosphate (9CI); 2,2-dichlorovinyl 2-(ethylsulfinyl)ethyl methyl phosphate (8CI); Reg. No. *[7076-53-1]*. 'Nexion 1378' (Celamerck). $C_7H_{13}Cl_2O_5PS$; OS2&2OPO&O1&O1UYGG; insecticide.

**4500** Dichlozoline (BSI, E-ISO, F-ISO, JMAF, ex ANSI). (IUPAC) 3-(3,5-dichlorophenyl)-5,5-dimethyl-1,3-oxazolidine-2,4-dione. (*C.A.*) 3-(3,5-dichlorophenyl)-5,5-dimethyl-2,4-oxazolidinedione (8 & 9CI); Reg. No. *[24201-58-9]*. DDOD. 'CS 8890','Ortho 8890'; 'Sclex' (Chevron Chemical Co.). $C_{11}H_9Cl_2NO_3$; T5OVNV EHJ CR CG EG& E1 E1; fungicide.

**4600** Diethamquat (BSI, E-ISO, F-ISO). (IUPAC) 1,1'-bis(diethylcarbamoylmethyl)-4,4'-bipyridinium ion; 1,1'-di(diethylcarbamoylmethyl)-4,4'-bipyridyl-1,1'-diylium ion. (*C.A.*) 1,1'-bis[2-(diethylamino)-2-oxoethyl]-4,4'-bipyridinium (9CI). 'PP 831' for dichloride (ICI Plant Protection Division). $C_{22}H_{32}N_4O_2$; T6KJ A1VN2&2 D- DT6KJ A1VN2&2; dichloride $C_{22}H_{32}Cl_2N_4O_2$; T6KJ A1VN2&2 D- DT6KJ A1VN2&2 &&2Cl SALT; herbicide.

**4620** *O,O*-Diethyl *O*-4-methyl-2-oxo-2*H*-chromen-7-yl phosphorothioate (IUPAC); *O,O*-diethyl *O*-4-methylcoumarin-7-yl phosporothioate (*C.A.*) *O,O*-diethyl *O*-(4-methyl-2-oxo-2*H*-1-benzopyran-7-yl) phosphorothioate (9CI); *O,O*-diethyl phosphorothioate *O*-ester with 7-hydroxy-4-methylcoumarin (8CI); Reg. No. *[299-45-6]*. 'E 838'; 'Potasan' (Bayer AG). $C_{14}H_{17}O_5PS$; T66 BOVJ E1 IOPS&O2&O2; *PM1*, p. 158. Insecticide.

**4630** *O,O*-Diethyl *O*-6-methyl-2-propylpyrimidin-4-yl phosphorothioate (IUPAC). (*C.A.*) *O,O*-diethyl *O*-[6-methyl-2-propyl-4-pyrimidinyl] phosphorothioate (8 & 9CI); Reg. No. *[5826-91-5]*. 'G 24 622'; 'Pyrazinon' (Ciba-Geigy AG). $C_{12}H_{21}N_2O_3PS$; T6N CNJ B3 DOPS&O2&O2 F1; insecticide.

**4640** Diethyl 5-methylpyrazol-3-yl phosphate (IUPAC). (*C.A.*) diethyl 5-methyl-1*H*-pyrazol-3-yl phosphate (9CI); diethyl 5-methyl-3-pyrazolyl phosphate (8CI); Reg. No. *[108-34-9]*. ENT 24 723. 'G 24 483'; 'Pyrazoxon'. $C_8H_{15}N_2O_4P$; T5MNJ COPO&O2&O2 E1; insecticide.

**4650** *O,O*-Diethyl naphthalimido-oxyphosphonothioate (IUPAC). (*C.A.*) 2-[(diethoxyphosphinothioyl)oxy]-1*H*-benz[*de*]isoquinoline-1,3(2*H*)-dione (9CI); *N*-hydroxynaphthalimide *O*-(*O,O*-diethyl phosphorothioate); Reg. No. *[2668-92-0]*. ENT 24 970. 'Bayer 22 408','S 125' (Bayer AG). $C_{16}H_{16}NO_5PS$; T666 1A M CVNVJ DOPS&O2&O2; *PM1*, p. 159. Insecticide.

**4760** 2,3-Dihydro-5,6-diphenyl-1,4-oxathi-ine (IUPAC); 2,3-dihydro-5,6-diphenyl-1,4-oxathi-in (I). (*C.A.*) (I) (9CI); Reg. No. *[58041-19-8]*. 'P 293' (Uniroyal Inc.). $C_{16}H_{14}OS$; T6O DS BUTJ BR& CR; plant growth regulator.

**4790** 2,3-Dihydro-5-phenyl-1,4-dithi-ine 1,1,4,4-tetraoxide (IUPAC); 2,3-dihydro-5-phenyl-1,4-dithi-in 1,1,4,4-tetraoxide (I). (*C.A.*) (I) (9CI); 'P 368' (Uniroyal Inc.). $C_{10}H_{10}O_4S_2$; T6SW DSW BUTJ BR; fungicide.

**4820** Dimefox (BSI, E-ISO, F-ISO). (IUPAC) tetramethylphosphorodiamidic fluoride (I); bis(dimethylamino)fluorophosphine oxide. (*C.A.*) (I) (8 & 9CI); Reg. No. *[115-26-4]*. ENT 19 109. '*Pestox XIV*' (Murphy Chemical Ltd), 'Terra Sytam' (Wacker-Chemie GmbH). $C_4H_{12}FN_2OP$; 1N1&PO&FN1&1; *PM6*, p. 196. Acaricide and insecticide (H. Kükenthal & G. Schrader, *B.I.O.S. Final Report*, 1946, 1095).

**4940** 2-(4,5-Dimethyl-1,3-dioxolan-2-yl)phenyl methylcarbamate (I) (IUPAC). (*C.A.*) (I) (9CI); *O*-(dimethyl-1,3-dioxolan-2-yl)phenyl methylcarbamate (8CI); Reg. No. *[7122-04-5]*. ENT 27 410. 'C 10 015'; 'Fondaren' (Ciba-Geigy AG). $C_{13}H_{17}NO_4$; T5O COTJ BR BOVM1& D1 E1; *PM3*, p. 198. Insecticide.

**4960** 5,5-Dimethyl-3-oxocyclohex-1-enyl dimethylcarbamate (IUPAC). (*C.A.*) 5,5-dimethyl-3-oxo-1-cyclohexen-1-yl dimethylcarbamate (8 & 9CI); Reg. No. *[122-15-6]*. ENT 24 728. 'G 19 258'; 'Dimetan' (Ciba-Geigy AG). $C_{11}H_{17}NO_3$; L6V BUTJ COVN1&1 E1 E1; insecticide.

**5010** *O*-4-Dimethylsulphamoylphenyl *O,O*-diethyl phosphorothioate (IUPAC). (*C.A.*) *O*-[4-[(dimethylamino)sulfonyl]phenyl] *O,O*-diethyl phosphorothioate (9CI); *O,O*-diethyl phosphorothioate *O*-ester with *N,N*-dimethyl-*p*-hydroxybenzenesulfonamide (8CI); Reg. No. *[3078-97-5]*. DSP (JMAF). 'NK 0795'; 'Kaya-ace' (Nippon Soda Co. Ltd). $C_{12}H_{20}NO_5PS_2$; 2OPS&O2&OR DSWN1&1; insecticide.

**5030** Dimetilan (BSI, ESA). (IUPAC) 1-dimethylcarbamoyl-5-methylpyrazol-3-yl dimethylcarbamate. (*C.A.*) 1-[(dimethylamino)carbonyl]-5-methyl-1*H*-pyrazol-3-yl dimethylcarbamate (9CI): dimethylcarbamate ester of 3-hydroxy-*N,N*,5-trimethylpyrazole-1-carboxamide (8CI); Reg. No. *[644-64-4]*. OMS 479; ENT 25 922. 'G 22 870', 'GS 13 332'; 'Snip' (Ciba-Geigy AG). $C_{10}H_{16}N_4O_3$; T5NNJ AVN1&1 COVN1&1 E1; *PM6*, p. 205. Insecticide (H. Gysin, *Chimia*, 1954, **8**, 205, 221).

**5040** Dimexan (BSI); dimexano (E-ISO, F-ISO). (IUPAC) *O,O*-dimethyl dithiobis(thioformate) (I). (*C.A.*) dimethyl thioperoxydicarbonate (9CI); (I) (8CI); Reg. No. *[1468-37-7]*. 'Tri-PE' (Vondelingenplaat N.V.). $C_4H_6O_2S_4$; SUYO1&SSYUS&O1; *PM5*, p. 206. Herbicide.

**5050** Dimidazon (BSI, E-ISO); dimidazone (F-ISO); *dimethazone* (withdrawn BSI name). (IUPAC) 4,5-dimethoxy-2-phenylpyridazin-3(2*H*)-one. (*C.A.*) 4,5-dimethoxy-2-phenyl-3(2*H*)-pyridazinone (8 & 9CI); Reg. No. *[3295-78-1]*. 'BAS 255H' (BASF AG). $C_{12}H_{12}N_2O_3$; T6NNVJ BR& DO1 EO1; herbicide.

**5060** Dinex (BSI, E-ISO, F-ISO); pedinex (France); DN (JMAF). (IUPAC) 2-cyclohexyl-4,6-dinitrophenol (I). (*C.A.*) (I) (8 & 9CI); Reg. No. *[131-89-5]*. DNOCHP. ENT 157. 'DN1' (Dow Chemical Co.). $C_{12}H_{14}N_2O_5$; L6TJ AR BQ CNW ENW; acaricide and insecticide.

**5065** Dinex-dicyclohexylammonium (BSI, E-ISO, F-ISO); pedinex-dicyclohexylammonium (France). Dicyclohexylammonium 2-cyclohexyl-4,6-dinitrophenoxide (IUPAC). (*C.A.*) 2-cyclohexyl-4,6-dinitrophenol compound with *N*-cyclohexylcyclohexanamine (1:1) (8CI); Reg. No. *[317-83-9]*. ENT 30 828. 'DN 111'; (Dow Chemical Co.). '*Dynone II*' (Fisons Ltd). $C_{24}H_{37}N_3O_5$; L6TJ AR BQ CNW ENW &&$[(CH_2)_5CH]_2NH$ SALT; insecticide, acaricide.

**5120** Dinocton (BSI, E-ISO, F-ISO), a reaction mixture comprising isomeric dinitro(octyl)phenyl methyl carbonates (see below). (IUPAC) major components include: 2,4-dinitro-6-(1-propylpentyl)phenyl methyl carbonate (I) (i); 2-(1-ethylhexyl)-4,6-dinitrophenyl methyl carbonate (II) (ii); 2,6-dinitro-4-(1-propylpentyl)phenyl methyl carbonate (III) (iii); 4-(1-ethylhexyl)-2,6-dinitrophenyl methyl carbonate (IV) (iv). (*C.A.*) (I), (II), (III), (IV) (8 & 9CI); Reg. No. *[32534-96-6]* (i + ii), *[19000-58-9]* (i), *[19000-52-3]* (ii), *[32535-08-3]* (iii + iv), *[6465-51-6]* (iii), *[6465-60-7]* (iv). Dinocton-6 (i + ii), dinocton-4 (iii + iv). 'MC 1945' (i + ii), 'MC 1947' (iii + iv) (Murphy Chemical Ltd). $C_{16}H_{22}N_2O_7$ (all isomers); (i) 4Y3&R CNW ENW BOVO1; (ii) 5Y2&R CNW ENW BOVO1; (iii) 4Y3&R CNW ENW DOVO1; *PM4*, p. 210 (i + ii), p. 209 (iii + iv). Acaricide and fungicide (i + ii), acaricide (iii + iv) (M. Pianka, *Crop Prot. Symp., 2nd, Magdeburg*, 1966).

**5230** Dinoterb acetate (BSI, E-ISO); dinoterbe-acétate (F-ISO). (IUPAC) 2-*tert*-butyl-4,6-dinitrophenyl acetate (I). (*C.A.*) 2-(1,1-dimethylethyl)-4,6-dinitrophenyl acetate (9CI); (I) (8CI); Reg. No. *[3204-27-1]*. 'P 1108' (Murphy Chemical Ltd). $C_{12}H_{14}N_2O_6$; 1X1&1&R CNW ENW BOV1; *PM4*, p. 214. Herbicide and acaricide (G. A. Emery, *et al., Proc. Conf. EWRC/COLUMA, 2nd*, 1965, p. 141; F. Colliot & B. Henrion, *Proc. 6th Br. Insectic. Fungic. Conf.*, 1971, **2**, 529).

**5300** Diphenyl sulphone (IUPAC) (accepted BSI, E-ISO); diphénylsulfone (accepted F-ISO). (*C.A.*) 1,1'-sulfonylbis[benzene] (9CI); phenyl sulfone (8CI); Reg. No. *[127-63-9]*. DPS. $C_{12}H_{10}O_2S$; WSR&R; *PM4*, p. 220. Acaricide.

**5360** Ditalimfos (BSI, E-ISO, F-ISO ANSI). (IUPAC) *O,O*-diethyl phthalimidophosphonothioate (I). (*C.A.*) *O,O*-diethyl (1,3-dihydro-1,3-dioxo-2*H*-isoindol-2-yl)phosphonothioate (9CI); (I) (8CI); Reg. No. *[5131-24-8]*. 'Dowco 199'; 'Plondrel' (Dow Chemical Co.). $C_{12}H_{14}NO_4PS$; T56 BVNVJ CPS&O2&O2; *PM6*, p. 222. Fungicide (H. Tolkmith, *Nature (London)*, 1966, **211**, 522).

**5390** 2-(1,3-Dithiolan-2-yl)phenyl dimethylcarbamate (I) (IUPAC). (*C.A.*) (I) (9CI); *o*-(1,3-dithiolan-2-yl)phenyl dimethylcarbamate (8CI); Reg. No. *[21709-44-4]*. 'C 13 963' (Ciba-Geigy AG). $C_{12}H_{15}NO_2S_2$; T5S CSTJ BR BOVN1&1; *PM2*, p. 209. Insecticide (F. Bachmann & J. B. Legge, *J. Sci. Food Agric.*, Suppl., 1968, p. 39).

**5420** DMPA (WSSA). (IUPAC) *O*-2,4-dichlorophenyl *O*-methyl isopropylphosphoramidothioate. (*C.A.*) *O*-(2,4-dichlorophenyl) *O*-methyl (1-methylethyl)phosphoramidothioate (9CI); *O*-(2,4-dichlorophenyl) *O*-methyl isopropylphosphoramidothioate (8CI); Reg. No. *[299-85-4]*. OMS 115, ENT 25 647. 'K 22 023', 'Dowco 118'; 'Zytron' (Dow Chemical Co.). $C_{10}H_{14}Cl_2NO_2PS$; GR CG DOPS&O1&MY1&1; *PM2*, p. 211. Herbicide.

**5550** Endothion (BSI, E-ISO, F-ISO, ex ANSI). (IUPAC) *S*-5-methoxy-4-oxo-4*H*-pyran-2-ylmethyl *O,O*-dimethyl phosphorothioate.(*C.A.*) *S*-[(5-methoxy-4-oxo-4*H*-pyran-2-yl)methyl] *O,O*-dimethyl phosphorothioate (9CI); *O,O*-dimethyl phosphorothioate *S*-ester with 2-mercaptomethyl-5-methoxy-4*H*-pyran-4-one (8CI); Reg. No. *[2778-04-3]*. ENT 24 653. '7175 RP' (Rhône Poulenc S.A.), 'AC 18 737' (American Cyanamid Co.), 'FMC 5767' (FMC Corp.); 'Endocide' (Rhône Poulenc S.A.). $C_9H_{13}O_6PS$; T6O DVJ B1SPO&O1&O1 EO1; *PM5*, p. 234. Insecticide and acaricide (F. Chaboussou & P. Ramadier, *Rev. Zool. Agric. Appl.*, 1957, **55**, 116).

**5580** Epofenonane (BSI, E-ISO, F-ISO). (IUPAC) 6,7-epoxy-3-ethyl-7-methylnonyl 4-ethylphenyl ether. (*C.A.*) 2-ethyl-3-[3-ethyl-5-(4-ethylphenoxy)pentyl]-2-methyloxirane (9CI); Reg. No. *[57342-02-6]*. 'Ro 10-3108' (Hoffman-La Roche & Co.). $C_{20}H_{32}O_2$; T3OTJ B2 B1 C2Y2&2OR D2; insect growth regulator.

**5600** Epronaz (BSI, E-ISO, F-ISO). (IUPAC) *N*-ethyl-*N*-propyl-3-propylsulphonyl-1*H*-1,2,4-triazole-1-carboxamide; 1-(*N*-ethyl-*N*-propylcarbamoyl)-3-propylsulphonyl-1*H*-1,2,4-triazole. (*C.A.*) *N*-ethyl-*N*-propyl-3-(propylsulfonyl)-1*H*-1,2,4-triazole-1-carboxamide (9CI); Reg. No. *[59026-08-3]*. 'BTS 30 843' (Boots Co. Ltd now FBC Limited). $C_{11}H_{20}N_4O_3S$; T5NN DNJ AVN3&2 CSW3; herbicide (L. G. Copping & R. F. Brookes, *Proc. Br. Weed Control Conf., 12th*, 1974, **2**, p. 809).

**5620** Erbon (E-ISO, F-ISO, ANSI, WSSA); *exception* (BSI). (IUPAC) 2-(2,4,5-trichlorophenoxy)ethyl 2,2-dichloropropionate (I). (*C.A.*) 2-(2,4,5-trichlorophenoxy)ethyl 2,2-dichloropropanoate (9CI); (I) (8CI); Reg. No. *[136-25-4]*. 'Baron' (Dow Chemical Co.). $C_{11}H_9Cl_5O_3$; GXG1&VO2OR BG DG EG; *PM5*, p. 239. Herbicide.

**5660** Etem (BSI). (IUPAC) 5,6-dihydroimidazo[2,1-*c*]-1,2,4-dithiazole-3-thione. (*C.A.*) 5,6-dihydro-3*H*-imidazo[2,1-*c*]-1,2,4-dithiazole-3-thione (9CI); Reg. No. *[33813-20-6]*. 'UCP/21' (Universal Crop Protection); 'Vegita' (Kumiai Chemical Industry Co. Ltd and Hokko Chemical Industry Co. Ltd). (i) $C_4H_4N_2S_3$; T55 ANYSS FN EU&TJ BUS; (ii) $C_4H_6N_2S_3$; T7MYSYMTJ BUS DUS; *PM6*, p. 239. Fungicide originally believed to be ETM (JMAF); 1,3,6-thiadiazepane-2,7-dithione (IUPAC); ethylenethiuram monosulphide; (*C.A.*) hexahydro-1,3,6,-thiadiazepine-2,7-dithione (8 & 9CI); Reg. No. *[5782-83-3]*.

**5710** Ethiolate (BSI, E-ISO, F-ISO, ANSI, WSSA). (IUPAC) *S*-ethyl diethyl(thiocarbamate). (*C.A.*) *S*-ethyl diethylcarbamothioate (9CI); *S*-ethyl diethylthiocarbamate (8CI); Reg. No. *[2941-55-1]*. 'Prefox' (Reg. No. *[8071-40-7]*) mixture with cyprazine (Gulf Oil Chemicals Co.). $C_7H_{15}NOS$; 2SVN2&2; *PM4*, p. 246. Herbicide.

**5740** Ethoate-methyl (BSI, E-ISO); éthoate-méthyle (F-ISO).(IUPAC) *S*-ethylcarbamoylmethyl *O,O*-dimethyl phosphorodithioate. (*C.A.*) *S*-[2-(ethylamino)-2-oxoethyl] *O,O*-dimethyl phosphorodithioate (9CI); *O,O*-dimethyl phosphorodithioate *S*-ester with *N*-ethyl-2-mercaptoacetamide (8CI); Reg. No. *[116-01-8]*. OMS 252; ENT 25 506. 'B/77'; 'Fitios' (Snia Viscosa). $C_6H_{14}NO_3PS_2$; 2MV1SPS&O1&O1; *PM5*, p. 245. Insecticide and acaricide (G. Lemetre *et al.*, *Not. Mal. Piante*, 1963, No. 64).

**5820** Ethylene bis(trichloroacetate) (IUPAC). (*C.A.*) 1,2-ethanediyl bis(trichloroacetate) (9CI); 1,2-bis(trichloroacetoxy)ethane (8CI); Reg. No. *[2514-53-6]*. Ethylene glycol bis(trichloroacetate). 'Glytac' (Hooker Chemical Corp.). $C_6H_4Cl_6O_4$; GXGGVO2OVXGGG; *PM5*, p. 252. Herbicide.

**5830** Ethyl hexanediol (ESA); ethohexadiol (US Pharmacopoeia). (IUPAC) 2-ethylhexane-1,3-diol. (*C.A.*) 2-ethyl-1,3-hexanediol (8 & 9CI); Reg. No. *[94-96-2]*. Ethyl hexylene glycol. ENT 375. 'Rutgers 612'. $C_8H_{18}O_2$; QY3&Y2&1Q; *PM6*, p. 247. Insect repellent (P. Granett & H. L. Haynes, *J. Econ. Entomol.*, 1945, **38**, 671; W. V. King, *ibid.*, 1951, **44**, 339; B. V. Travis & C. N. Smith, *ibid.*, p. 428).

**5850** *N*-(Ethylmercurio)-4-toluenesulphonanilide (IUPAC); *N*-(ethylmercury)-*p*-toluenesulphonanilide. (*C.A.*) ethyl(4-methyl-*N*-phenylbenzenesulfonamidato-*N*)mercury (9CI); ethyl(*p*-toluenesulfonanilidato)mercury (8CI); Reg. No. *[517-16-8]*. 'Ceresan M', 'Granosan M' (E. I. du Pont de Nemours & Co. (Inc.)). $C_{15}H_{17}HgNO_2S$; 2-HG-NR&SWR D1; *PM3*, p. 252. Fungicide.

**5890** 2-Ethyl-5-methyl-1,3-dioxan-2-yl 2-methylbenzyl ether (IUPAC); 2-ethyl-5-methyl-5-(2-methylbenzyloxy)-1,3-dioxane. (*C.A.*) 2-ethyl-5-methyl-5-(2-methylphenylmethoxy)-1,3-dioxane (9CI); Reg. No. *[41129-10-6]* (*cis*-isomer). 'FMC 25 213'. $C_{15}H_{22}O_3$; T6O COTJ B2 EO1R B1& E1; *PM5*, p. 254. Herbicide.

**5940** 4-(Ethylthio)phenyl methylcarbamate (I) (IUPAC). (*C.A.*) (I) (9CI); *p*-(ethylthio)phenyl methylcarbamate (8CI); Reg. No. *[18809-57-9]*. EMPC (JMAF). 'NK-1'; 'Toxamate' (Nippon Kayaku Co. Ltd). $C_{10}H_{13}NO_2S$; 2SR DOVM1; insecticide.

**5950** Etinofen (BSI, E-ISO); étinofène (F-ISO). (IUPAC) α-ethoxy-4,6-dinitro-*o*-cresol (I). (*C.A.*) 2-(ethoxymethyl)-4,6-dinitrophenol (9CI); (I) (8CI); Reg. No. *[2544-94-7]*. 'Dinethon' (C. H. Boehringer Sohn). $C_9H_{10}N_2O_6$; WNR BQ ENW C1O2; herbicide.

**5990** EXD (WSSA). (IUPAC) *O,O*-diethyl dithiobis(thioformate) (I). (*C.A.*) diethyl thioperoxydicarbonate (9CI); (I) (8CI); Reg. No. *[502-55-6]*. 'Herbisan' (Roberts Chemicals Inc.), 'Sulfasan' (Monsanto Chemical Co.). $C_6H_{10}O_2S_4$; SUYO2&SSYUS&O2; *PM5*, p. 260. Herbicide (E. K. Alban & L. McCombs, *Proc. North Cent. Weed Control Conf.*, 1949, p. 81).

**6020** Fenapanil (BSI, ANSI, draft E-ISO). (IUPAC) (±)-2-(imidazol-1-ylmethyl)-2-phenylhexanenitrile. (*C.A.*) α-butyl-α-phenyl-1*H*-imidazole-1-propanenitrile (9CI); Reg. No. *[61019-78-1]*. 'RH-2161'; 'Sisthane' (Rohm & Haas Co.). $C_{16}H_{19}N_3$; T5N CNJ A1X4&R&CN; fungicide.

**6050** Fenazaflor (BSI, E-ISO, F-ISO). (IUPAC) phenyl 5,6-dichloro-2-trifluoromethylbenzimidazole-1-carboxylate. (*C.A.*) phenyl 5,6-dichloro-2-(trifluoromethyl)-1*H*-benzimidazole-1-carboxylate (9CI); phenyl 5,6-dichloro-2-(trifluoromethyl)-1-benzimidazolecarboxylate (8CI); Reg. No. *[14255-88-0]*. OMS 1243. 'NC 5016'; 'Lovozal' (FBC Limited). $C_{15}H_7Cl_2F_3N_2O_2$; T56 BN DNJ BVOR& CXFFF GG HG; *PM4*, p. 267. Acaricide (D. T. Saggers & M. L. Clark, *Nature (London)*, 1967, **215**, 275).

**6070** Fenchlorphos (BSI, E-ISO, F-ISO, BPC); ronnel (ANSI, Canada). (IUPAC) *O,O*-dimethyl *O*-2,4,5-trichlorophenyl phosphorothioate. (*C.A.*) *O,O*-dimethyl *O*-(2,4,5-trichlorophenyl) phosphorothioate (8 & 9CI); Reg. No. *[299-84-3]*. OMS 123; ENT 23 284. 'Dow ET-14', 'Dow ET-57'; 'Nankor', 'Trolene', 'Korlan' (Dow Chemical Co.). $C_8H_8Cl_3O_3PS$; GR BG DG EOPS&O1&O1; *PM6*, p. 260. Insecticide.

**6080** Fenethacarb (BSI, E-ISO); phénétacarbe (F-ISO). (IUPAC) 3,5-diethylphenyl methylcarbamate (I). (*C.A.*) (I) (8 & 9CI); Reg. No. *[30087-47-9]*. 'BAS 235I' (BASF AG). $C_{12}H_{17}NO_2$; 2R C2 EOVM1; insecticide.

**6310** Fentrifanil (BSI, E-ISO, F-ISO); hexafluoramin (Republic of South Africa). (IUPAC) *N*-(6-chloro-α,α,α-trifluoro-*m*-tolyl)-α,α,α-trifluoro-4,6-dinitro-*o*-toluidine; 2'-chloro-2,4-dinitro-5',6-bis(trifluoromethyl)diphenylamine. (*C.A.*) *N*-[2-chloro-5-(trifluoromethyl)phenyl]-2,4-dinitro-6-(trifluoromethyl)benzenamine (9CI); Reg. No. *[62441-54-7]*. 'PP 199' (ICI Plant Protection Division). $C_{14}H_6ClF_6N_3O_4$; FXFFR CNW ENW BMR BG EXFFF; acaricide (N. Morton, *Proc. 1977 Br. Crop Prot. Conf. - Pests Dis.*, 1977, **2**, 349).

**6450** Fluenetil (BSI, E-ISO, F-ISO); fluénéthyl (France). (IUPAC) 2-fluoroethyl biphenyl-4-ylacetate. (*C.A.*) 2-fluoroethyl [1,1'-biphenyl]-4-acetate (9CI); 2-fluoroethyl 4-biphenylacetate (8CI); Reg. No. *[4301-50-2]*. 'M 2060'; 'Lambrol' (Montecatini-Edison S.p.A.). $C_{16}H_{15}FO_2$; F2OV1R DR; *PM2*, p. 254. Insecticide and acaricide (P. de Pietri-Tonelli *et al., Proc. Br. Insectic. Fungic. Conf., 3rd*, 1965, p. 478).

**6490** Flumezin (BSI, E-ISO); flumézine (F-ISO). (IUPAC) 2-methyl-4-(α,α,α-trifluoro-*m*-tolyl)-1,2,4-oxadazinane-3,5-dione; 2-methyl-4-(α,α,α-trifluoro-*m*-tolyl)-2*H*-1,2,4-oxadiazine-3,5(4*H*,6*H*)-dione (I). (*C.A.*) 2-methyl-4-[3-(trifluoromethyl)phenyl]-2*H*-1,2,4-oxadiazine-3,5(4*H*,6*H*)-dione (9CI); (I) (8CI); Reg. No. *[25475-73-4]*. 'BAS 348H' (BASF AG). $C_{11}H_9F_3N_2O_3$; T6ONZNV FHJ B1 DR CXFFF; herbicide.

**6520** Fluorbenside (BSI, E-ISO, F-ISO). (IUPAC) 4-chlorobenzyl 4-fluorophenyl sulphide. (*C.A.*) 1-chloro-4-[[(4-fluorophenyl)thio]methyl]benzene (9CI); *p*-chlorobenzyl *p*-fluorophenyl sulfide (8CI); Reg. No. *[405-30-1]*. 'HRS 924', 'RD 2454'; 'Fluorparacide', 'Fluorsulphacide' (FBC Limited). $C_{13}H_{10}ClFS$; GR D1SR DF; *PM1*, p. 235. Acaricide (N. G. Clark *et al., J. Sci. Food Agric.*, 1957, **8**, 566).

**6570** 2-Fluoro-*N*-methyl-*N*-1-naphthylacetamide (I) (IUPAC). (*C.A.*) 2-fluoro-*N*-methyl-*N*-(1-naphthalenyl)acetamide (9CI); (I) (8CI); Reg. No. *[5903-13-9]*. MNFA (JMAF). ENT 27 403. 'NA-26'; 'Nissol' (Nippon Soda Co. Ltd). $C_{13}H_{12}FNO$; L66J BN1&V1F; acaricide.

**6590** Fluothiuron (BSI, E-ISO, F-ISO). (IUPAC) 3-[3-chloro-4-(chlorodifluoromethylthio)phenyl]-1,1-dimethylurea. (*C.A.*) *N'*-[3-chloro-4-[(chlorodifluoromethyl)thio]phenyl]-*N*,*N*-dimethylurea (9CI); Reg. No. *[33439-45-1]*. 'BAY KUE 2079A'; 'Clearcide' (Bayer AG). $C_{10}H_{10}Cl_2F_2N_2OS$; GXFFSR BG DMVN1&1; *PM5*, p. 107. Herbicide.

**6650** Fluromidine (BSI, F-ISO); fluoromidine (E-ISO). (IUPAC) 6-chloro-2-trifluoromethyl-3*H*-imidazo[4,5-*b*]pyridine. (*C.A.*) 6-chloro-2-(trifluoromethyl)-1*H*-imidazo[4,5-*b*]pyridine (8 & 9CI); Reg. No. *[13577-71-4]*. 'NC 4780' (Fisons Ltd now FBC Limited). $C_7H_3ClF_3N_3$; T56 BM DN FNJ CXFFF HG; herbicide.

**6800** Fospirate (BSI, E-ISO, F-ISO, ANSI). (IUPAC) dimethyl 3,5,6-trichloro-2-pyridyl phosphate (I). (*C.A.*) dimethyl 3,4,5-trichloro-2-pyridinyl phosphate (9CI); Reg. No. *[5598-52-7]*. OMS 1168; ENT 27 521. 'Dowco 217' (Dow Chemical Co.). $C_7H_7Cl_3NO_4P$; T6NJ BOPO&O1&O1 CG EG FG; insecticide.

**6860** Furcarbanil (BSI, E-ISO, F-ISO). (IUPAC) 3,5-dimethyl-3-furanilide (I). (*C.A.*) 3,5-dimethyl-*N*-phenylfurancarboxamide (9CI); (I) (8CI); Reg. No. *[28562-70-1]*. 'BAS 319F' (BASF AG). $C_{13}H_{13}NO_2$; T5OJ B1 CVMR& E1; fungicide.

**6870** 3-Furfuryl-2-methyl-4-oxocyclopent-2-enyl (1*RS*,3*RS*; 1*RS*,3*SR*)-2,2-dimethyl-3-(2-methylprop-1-enyl)cyclopropanecarboxylate (IUPAC); 3-furfuryl-2-methyl-4-oxocyclopent-2-enyl (1*RS*)-*cis-trans*-2,2-dimethyl-3-(2-methylprop-1-enyl)cyclopropanecarboxylate; 3-furfuryl-2-methyl-4-oxocyclopent-2-enyl (±)-*cis-trans*-chrysanthemate. (*C.A.*) 3-(2-furanylmethyl)-2-methyl-4-oxo-2-cyclopenten-1-yl (±)-2,2-dimethyl-3-(2-methyl-1-propenyl)cyclopropanecarboxylate (9CI); 3-(2-furylmethyl)-2-methyl-4-oxo-2-cyclopenten-1-yl (±)-2,2-dimethyl-3-(2-methylpropenyl)cyclopropanecarboxylate (8CI); Reg. No. *[17080-02-3]* (formerly *[7076-49-5]*). Furethrin. $C_{21}H_{26}O_4$; T5OJ B1- BL5V BUTJ C1 DOV- BL3TJ A1 A1 C1UY1&1; *PM2*, p. 264. Insecticide (M. Matsui *et al., J. Am. Chem. Soc.*, 1952, **74**, 2181). The esters from the (*RS*)-alcohol and (1*R*)-*trans* acid are more potent insecticides than the complete mixture of stereoisomers (W. A. Gersdorff & N. Mitlin, *J. Econ. Entomol.*, 1952, **45**, 849).

**6890** Furophanate (BSI, E-ISO, F-ISO, ANSI). (IUPAC) Methyl 4-(2-furfurylideneaminophenyl)-3-thioallophanate. (*C.A.*) methyl [[[2-[(2-furanylmethylene)amino]phenyl]amino]thioxomethyl]carbamate (9CI); Reg. No. *[53878-17-4]*. 'RH-3928' (Rohm & Haas Co.). $C_{14}H_{13}N_3O_3S$; T5OJ B1UNR BMYUS&MVO1; fungicide.

**6920** Gliotoxin. (IUPAC) (3*R*,5a*S*,6*S*,10a*R*)-2,3,5a,6,10,10-hexahydro-6-hydroxy-3-hydroxymethyl-2-methyl-3,10a-epidithiopyrazino[1,2-*a*]indole-1,4-dione. (*C.A.*) [3*R*-3α,5aβ,6β,10aα]-2,3,5a,6-tetrahydro-6-hydroxy-3-hydroxymethyl-2-methyl-10*H*-3,10a-epidithiopyrazino[1,2-*a*]indole-1,4-dione; Reg. No.*[67-99-2]*. $C_{13}H_{14}N_2O_4S_2$; T C6 B566/JO A 2BJ O AVN JXSS NNV EU GU MHTT&&J DQ M1Q N1; *PM4*, p. 287. Fungicide.

**6940** Glyodin (E-ISO, F-ISO). (IUPAC) 2-heptadecyl-2-imidazoline acetate. (*C.A.*) 2-heptadecyl-3,4-dihydro-1*H*-imidazolyl acetate (1:1) (9CI); 2-heptadecyl-2-imadazoline monoacetate (8CI); Reg. No. *[556-22-9]*. 'Crag Fungicide 341' (Union Carbide Corp.). $C_{22}H_{44}N_2O_2$; T5M CN BUTJ B17 &&CH$_3$COOH SALT *PM1*, p. 245, Fungicide (R. H. Wellman & S. E. A. McCallan, *Contrib. Boyce Thompson Inst.*, 1946, **14**, 151).

**6970** Griseofulvin (BSI, E-ISO, BPC, JMAF); griséofulvine (F-ISO). (IUPAC) 7-chloro-2',4,6-trimethoxy-6'-methylspiro[benzofuran-2(3*H*),1'cyclohex-2-ene]-3,4'-dione; 7-chloro-4,6-dimethoxycoumaran-3-one-2-spiro-1'-(2'-methoxy-6'-methylcyclohex-2'-en-4'-one). (*C.A.*) (1'*S*)-*trans*-7-chloro-2',4,6-trimethoxy-6'-methylspiro[benzofuran-2(3*H*),1'-[2]cyclohexene]-3,4'-dione (9CI); 7-chloro-2',4,6-trimethoxy-6'β-methylspiro[benzofuran-2(3*H*),1-[2]cyclohexene]-3,4'-dione (8CI); Reg. No. *[126-07-8]*. $C_{17}H_{17}ClO_6$; T56 BOXVJ FO1 HO1 IG C-& DL6V DX BUTJ CO1 E1; *PM5*, p. 290. Fungicide (A. E. Oxford *et al. Biochem. J.*, 1939, **33**, 240; J. F. Grove, *Q. Rev. Chem. Soc.*, 1963, **17**, 1), no longer used against plant diseases.

**6990** Halacrinate (BSI, E-ISO, F-ISO). (IUPAC) 7-bromo-5-chloro-8-quinolyl acrylate. (*C.A.*) 7-bromo-5-chloro-8-quinolinyl 2-propenoate (9CI); Reg. No. *[34462-96-9]*. 'CGA 30 599'; '*Tilt*' (Reg. No. *[61144-41-0]*) mixture with captafol (Ciba-Geigy AG). $C_{12}H_7BrClNO_2$; T66 BNJ GG IE JOV1U1; *PM5*, p. 292. Fungicide (J. M. Smith *et al. Proc. Br. Insectic. Fungic. Conf., 8th*, 1975, **2**, 421).

**7000** Haloxydine (BSI, E-ISO, F-ISO). (IUPAC) 3,5-dichloro-2,6-difluoropyridin-4-ol. (*C.A.*) 3,5-dichloro-2,6-difluoro-4-pyridinol (8 & 9CI); Reg. No. *[2693-61-0]*. 'PP 493' (ICI Plant Protection Division). $C_5HCl_2F_2NO$; T6NJ BG CF DQ EF FG; herbicide.

**7030** 2-(2-Heptadecyl-2-imidazolin-1-yl)ethanol (IUPAC); 2-heptadecyl-1-(2-hydroxyethyl)imidazoline. (*C.A.*) 2-heptadecyl-4,5-dihydro-1*H*-imidazole-1-ethanol (9CI); 2-heptadecyl-2-imidazoline-1-ethanol (8CI); Reg. No. *[95-19-2]*. 'Fungicide 337' (Union Carbide Corp.). $C_{22}H_{44}N_2O$; T5N CN AUTJ B17 C2Q; fungicide.

**7060** Hexachloroacetone (IUPAC, accepted by E-ISO); hexachloracetone (F-ISO). (*C.A.*) 1,1,1,3,3,3-hexachloropropan-2-one (8 & 9CI); Reg. No. *[116-16-5]*. HCA (former WSSA common name). 'GC 1106'; 'HCA Weedkiller', 'Urox' (Allied Chemical Co.). 'Urox 379' (mixture with bromacil). $C_3Cl_6O$; GXGGVXGGG; *PM5*, p. 296. Herbicide.

**7090** Hexadecyl cyclopropanecarboxylate (I) (IUPAC). (*C.A.*) (I) (8 & 9CI). Reg. No. *[54460-46-7]*. *Cycloprate* (rejected proposed common name). 'ZR 856'; 'Zardex' (Zoecon Corp.). $C_{20}H_{38}O_2$; L3TJ AVO16; acaricide.

**7100** Hexafluoroacetone trihydrate (I) (i) (IUPAC); 1,1,1,3,3,3-hexafluoropropane-2,2-diol dihydrate (II) (ii) (IUPAC). (*C.A.*) (I) (i) (8 & 9CI); 1,1,1,3,3,3-hexafluoro-2,2-propanediol (ii) (8 & 9CI); Reg. No. *[993-58-8]*. 'GC 7887' (Allied Chemical Corp.). $C_3H_6F_6O_4$; (i) FXFFVXFFF &&3$H_2O$; (ii) FXFFXQQXFFF &&2$H_2O$; *PM4*, p. 295. Herbicide.

**7110** 1,1,1,7,7,7-Hexafluoro-4-methyl-2,6-bis(trifluoromethyl)hept-3-ene-2,6-diol (i) + 1,1,1,7,7,7-hexafluoro-4-methylene-2,6-bis(trifluoromethyl)heptane-2,6-diol (ii) (IUPAC). 1,1,1,7,7,7-hexafluoro-4-methyl-2,6-bis(trifluoromethyl)-3-heptene-2,6-diol (i) + 1,1,1,7,7,7-hexafluoro-4-methylene-2,6-bis(trifluoromethyl)-2,6-heptanediol (ii) (8 & 9CI); Reg. No. *[756-91-2]* (i), *[16202-91-8]* (ii), *[19493-94-8]* (monosodium salts). 'ACD 10 614' (i + ii), 'ACD 10 435' (monosodium salts) (Allied Chemical Co.). (i or ii) $C_{10}H_8F_{12}O_2$; (i) FXFFXQXFFF1YU1&1XQXFFFXFFF; (ii) FXFFXQXFFF1Y1&U1XQXFFFXFFF; *PM4*, p. 296. Experimental herbicides and as monosodium 1,1,1,7,7,7-4-methyl-2,6-bis(trifluoromethyl)hept-3-ene-2,6-diolate, and monosodium 1,1,1,7,7,7-hexafluoro-4-methylene-2,6-bis(trifluoromethyl)heptane-2,6-diolate.

**7120** Hexaflurate (WSSA). (IUPAC) potassium hexafluoroarsenate (I). (*C.A.*) potassium hexafluoroarsenate(1-) (9CI); (I) (8CI); Reg. No. *[17029-22-0]*. 'Nopalmate' (Pennwalt Corp.). $AsF_6K$; .KA..AS-F6; *PM5*, p. 298. Herbicide.

**7200** 1-Hydroxy-1*H*-pyridine-2-thione (i); tautomeric with pyridine-2-thiol 1-oxide (ii) (IUPAC). (*C.A.*) 1-hydroxy-2(1*H*)-pyridinethione (i); 2-pyridinethiol 1-oxide (ii) (8 & 9CI); Reg. No. *[1121-30-8]* (i), *[1121-31-9]* (ii). 'Omadine' (Olin Mathieson Chemical Corp.). $C_5H_5NOS$; (i) T6NYJ AQ BUS; (ii) T6NJ AO BSH; *PM2*, p. 395. Bactericide and fungicide (M. Szkolnik & J. M. Hamilton, *Plant Dis. Rep.*, 1957, **41**, 289, 293, 301), usually as salts: iron derivative,'Omadine OM 1565'; manganese(II) derivative, 'Omadine OM 1564', (Reg. No. *[32255-90-6]*); zinc derivative, 'Omadine OM 1563', *[13463-41-7]*.

**7240** 2-Imidazolidone (IUPAC); imidazolidin-2-one. (*C.A.*) 2-imidazolidinone (8 & 9CI); Reg. No. *[120-93-4]*. Ethyleneurea. $C_3H_6N_2O$; T5MVMTJ; *PM1*, p. 214. Insecticide (H. G. Simkover, *J. Econ. Entomol.*, 1964, **57**, 574).

**7320** Ipazine (BSI, E-ISO, F-ISO, WSSA). (IUPAC) 6-chloro-*N*,*N*-diethyl-*N*'-isopropyl-1,3,5-triazine-2,4-diyldiamine; 2-chloro-4-diethylamino-6-isopropylamino-1,3,5-triazine. (*C.A.*) 6-chloro-*N*,*N*-diethyl-*N*'-(1-methylethyl)-1,3,5-triazine-2,4-diamine (9CI); 2-chloro-4-diethylamino-6-isopropylamino-*s*-triazine (8CI); Reg. No. *[1912-25-0]*. 'G 30 031'; 'Gesabal' (Ciba-Geigy AG). $C_{10}H_{18}ClN_5$; T6N CN ENJ BN2&2 DMY1&1 FG; herbicide.

**7340** Iprymidam (BSI, E-ISO, F-ISO). (IUPAC) 6-chloro-$N^4$-isopropylpyrimidinediyl-2,4-diamine; 2-amino-4-chloro-6-isopropylaminopyrimidine. (*C.A.*) 6-chloro-$N^4$-(1-methylethyl)-2,4-pyrimidinediamine (9CI); 'SAN 52 123H' (Sandoz AG). $C_7H_{11}ClN_4$; T6N CNJ BZ DMY1&1 FG; herbicide.

**7360** Isobenzan (BSI, E-ISO, F-ISO); telodrin (JMAF). (IUPAC) 1,3,4,5,6,7,8,8-octachloro-1,3,3a,4,7,7a-hexahydro-4,7-methanoisobenzofuran(I). (*C.A.*) (I) (8 & 9CI); Reg. No. *[297-78-9]*. OMS 206. ENT 25 545. 'SD 4402'; 'Telodrin' (Shell International Chemical Co.). $C_9H_4Cl_8O$; T C555 A EOTJ AG AG BG DG FG HG IG JG; insecticide.

**7390** Isocil (BSI, E-ISO, F-ISO, ANSI, WSSA); isoprocil (France, Republic of South Africa). (IUPAC) 5-bromo-3-isopropyl-6-methyluracil (I). (*C.A.*) 5-bromo-6-methyl-3-(1-methylethyl)-2,4(1*H*,3*H*)-pyrimidinedione (9CI); (I) (8CI); Reg. No. *[314-42-1]*. 'Du Pont Herbicide 82'; 'Hyvar' (E. I. du Pont de Nemours & Co. Inc.). $C_8H_{11}BrN_2O_2$; T6MVNVJ CY1&1 EE F1; *PM1*, p. 260. Herbicide.

**7400** Isodrin (BSI). (IUPAC) (1*R*,4*S*,5*R*,8*S*)-1,2,3,4,10,10-hexachloro-1,4,4a,5,8,8a-hexahydro-1,4:5,8-dimethanonaphthalene. (*C.A.*) (1α,4α,4aβ,5β,8β,8aβ)-1,2,3,4,10,10-hexachloro-1,4,4a,5,8,8a-hexahydro-1,4:5,8-dimethanonaphthalene (9CI); *endo,endo*-1,2,3,4,10,10-hexachloro-1,4,4a,5,8,8a-hexahydro-1,4:5,8-dimethanonaphthalene (8CI); Reg. No. *[465-73-6]*. 'Compound 711' (Shell International Co.). $C_{12}H_8Cl_6$; L D5 C555 A D-EU JUTJ AG AG BG IG JG KG &&(1R,4S,5R,8S) FORM; insecticide.

**7420** Isomethiozin (BSI, E-ISO); isométhiozine (F-ISO). (IUPAC) 6-*tert*-butyl-4-isobutylideneamino-3-methylthio-1,2,4-triazin-5(4*H*)-one. (*C.A.*) 6-(1,1-dimethylethyl)-4-[(2-methylpropylidene)amino]-3-(methylthio)-1,2,4-triazin-5(4*H*)-one (9CI); Reg. No. *[57052-04-7]*. 'BAY DIC 1577'; 'Tantizon' (Bayer AG); 'Tantizon Combi', 'Tantizon DP', (Reg. No. *[61641-75-6]*), mixtures with dichlorprop. $C_{12}H_{20}N_4OS$; T6NN DNVJ CS1 DNU1Y1&1 FX1&1&1; *PM6*, p. 310. Herbicide (H. Hack, *Mitt. Biol. Bundesanst. Land- Fortwirtsch. Berlin-Dahlem*, 1975, **165**, 179).

**7430** Isonoruron (BSI, E-ISO, F-ISO). (IUPAC) mixture 1,1-dimethyl-3-(perhydro-4,7-methoanoinden-1-yl)urea (i) + 1,1-dimethyl-3-(perhydro-4,7-methanoinden-2-yl)urea (ii). (*C.A.*) *N,N*-dimethyl-*N'*-[2,3,3a,4,5,6,7,7a-octahydro-4,7-methano-1*H*-inden-1-yl]urea (i) + *N,N*-dimethyl-*N'*-[2,3,3a,4,5,6,7,7a-octahydro-4,7-methano-1*H*-inden-2-yl]urea (ii) (9CI); 3-(hexahydro-4,7-methanoindan-1-yl)-1,1-dimethylurea (i) + 3-( hexahydro-4,7-methanoindan-2-yl)-1,1-dimethylurea (ii) (8CI); Reg. No. *[28346-65-8]* (i + ii). 'Basanor' (Reg. No. *[66402-41-3]*) (mixture with brompyrazone), 'Basfitox' (mixture with buturon) (BASF AG). $C_{13}H_{22}N_2O$; (i) L C555 ATJ DMVN1&1;(ii) L C555 ATJ EMVN1&1; *PM5*, p. 313. Herbicide (A. Fischer, *Abstr. Int. Congr. Plant Prot., 6th*, 1967, p. 446).

**7480** 1-Isopropyl-3-methylpyrazol-5-yl dimethylcarbamate (I) (IUPAC). (*C.A.*) 3-methyl-1-(1-methylethyl)-1*H*-pyrazol-5-yl dimethylcarbamate (9CI); (I) 9(CI); Reg. No. *[119-38-0]*. OMS 62, ENT 19 060. 'G 23 611'; 'Isolan' (Ciba-Geigy AG). $C_{10}H_{17}N_3O_2$; T5NNJ AY1&1 C1 EOVN1&1; *PM1*, p. 262. Insecticide

**7530** Isopyrimol (BSI, E-ISO, F-ISO, ANSI). (IUPAC) 1-(4-chlorophenyl)-2-methyl-1-pyrimidin-5-ylpropan-1-ol. (*C.A.*) α-(4-chlorophenyl)-α-(1-methylethyl)-5-pyrimidinemethanol (9CI); Reg. No. *[55283-69-7]*. $C_{14}H_{15}ClN_2O$; T6N CNJ EXQR DG&Y1&1; plant growth regulator.

**7580** 2-Isovalerylindan-1,3-dione (IUPAC). (*C.A.*) 2-(3-methyl-1-oxobutyl)-1*H*-indene-1,3(2*H*)-dione (9CI); 2-isovaleryl-1,3-indandione (8CI); Reg. No. *[83-28-3]*. 'Valone' (Kilgore Chemical Co.). $C_{14}H_{14}O_3$; L56 BV DV CHJ CV1Y1&1; *PM5*, p. 318. Insecticide (L. B. Kilgore *et al., Ind. Eng. Chem.*, 1942, **34**, 494).

**7620** Kelevan (BSI, E-ISO); kélévane (F-ISO). (IUPAC) ethyl 5-(1,2,4,5,6,7,8,8,9,10-decachloro-3-hydroxypentacyclo[5.3.0$^{2,6}$. 0$^{4,10}$. 0$^{5,9}$]dec-3-yl)-4-oxovalerate. (*C.A.*) ethyl 1,1a,3,3a,4,5,5,5a,5b,6-decachlorooctahydro-α-2-hydroxy-γ-oxo-1,3,4-metheno-1*H*-cyclobuta[*cd*]pentalene-2-pentanoate (9CI); ethyl 1,1a,3,3a,4,5,5a,5b,6-decachlorooctahydro-2-hydroxy-1,3,4-metheno-1*H*-cyclobuta[*cd*]pentalene-2-levulinate (8CI); Reg. No. *[4234-79-1]*. 'GC-9160' (Allied Chemical Corp.); 'Despirol'; 'Despirol Plus' (mixture with mancozeb) (C. F. Spiess & Sohn). $C_{17}H_{12}Cl_{10}O_4$; L545 B4 C5 D 4ABCE JTJ AG BG CG DQ D1V1VO2 EG FG GG HG HG IG JG; insecticide (E. E. Gilbert *et al., J. Agric. Food Chem.*, 1966, **14**, 111; H. Maier-Bode, *Residue Rev.*, 1976, **63**, 45).

**7670** Leptophos (BSI, E-ISO, F-ISO, ANSI); MBCP (JMAF). (IUPAC) *O*-4-bromo-2,5-dichlorophenyl *O*-methyl phenylphosphonothioate. (*C.A.*) *O*-(4-bromo-2,5-dichlorophenyl) *O*-methyl phenylphosphonothioate (8 & 9CI); Reg. No. *[21609-90-5]*. OMS 1438. 'VCS 506'; 'Phosvel', 'Abar' (Velsicol Chemical Corp.). $C_{13}H_{10}BrCl_2O_2PS$; GR DG BE EOPS&R&O1; *PM6*, p. 318. Insecticide (A. K. Azab, *Proc. Br. Insectic. Fungic. Conf.*, 5th, 1969, **2**, 550).

**7700** Lirimfos (BSI, E-ISO, F-ISO). (IUPAC) *O*-6-ethoxy-2-isopropylpyrimidin-4-yl *O,O*-dimethyl phosphorothioate. (*C.A.*) *O*-[6-ethoxy-2-(1-methylethyl)-4-pyrimidinyl] *O,O*-dimethyl phosphorothioate (9CI); Reg. No. *[38260-63-8]*. 'SAN 201I' (Sandoz AG). $C_{11}H_{19}N_2O_4PS$; T6N CNJ BY1&1 DOPS&O1&O1 FO2; insecticide.

**7710** Lythidathion (BSI, E-ISO, F-ISO). (IUPAC) *S*-5-ethoxy-2,3-dihydro-2-oxo-1,3,4-thiadiazol-3-ylmethyl *O,O*-dimethyl phosphorodithioate. (*C.A.*) *S*-[5-ethoxy-2-oxo-1,3,4-thiadiazol-3(2*H*)-ylmethyl] *O,O*-dimethyl phosphorodithioate (9CI); *O,O*-dimethyl phosphorodithioate *S*-ester with 2-ethoxy-4-(mercaptomethyl)-$\Delta^2$-1,3,4-thiadiazolin-5-one (8CI); Reg. No. *[2669-32-1]*. 'GS 12 968' (Ciba-Geigy AG), 'NC 2962' (FBC Limited). $C_7H_{13}N_2O_4PS_3$; T5NNVSJ B1SPS&O1&O1 EO2; insecticide

**7820** Mebenil (BSI, E-ISO, F-ISO). (IUPAC) *o*-toluanilide (I). (*C.A.*) 2-methyl-*N*-phenylbenzamide (9CI); (I) (8CI); Reg. No. *[7055-03-0]*. 'BAS 305F', 'BAS 3050F', 'BAS 3053F' (BASF AG). $C_{14}H_{13}NO$; 1R BVMR; *PM5*, p. 332. Fungicide (E. -H. Pommer & J. Kradel, *Proc. Br. Insectic. Fungic. Conf., 5th*, 1969, **2**, 563).

**7840** Mecarbinzide (BSI, F-ISO);mecarbinzid (E-ISO). (IUPAC) methyl 1-(2-methylthioethylcarbamoyl)benzimidazol-2-ylcarbamate. (*C.A.*) methyl 1-[[(methylthioethyl)amino]carbonyl]-1*H*-benzimidazol-2-ylcarbamate (9CI); methyl 1-(2-methylthioethylcarbamoyl)benzimidazole-2-carbamate (8CI); Reg. No. *[27386-64-7]*. $C_{13}H_{16}N_4O_3S$; T56 BN DNJ BVM2S1 CMVO1; fungicide (BASF AG).

**7850** Mecarphon (BSI, E-ISO, F-ISO). (IUPAC) methyl *N*-[methoxy(methyl)thiophosphorylthio = acetyl]-*N*-methylcarbamate; *S*-(*N*-methoxycarbonyl-*N*-methylcarbamoylmethyl) *O*-methyl methylphosphonodithioate. (*C.A.*) methyl 3,7-dimethyl-6-oxo-2-oxa-4-thia-7-aza-3-phosphaoctan-8-oate 3-sulfide (9CI); methyl [[methoxy(methylphosphinothioyl)thio = o]acetyl]methylcarbamate (former 9CI); methyl (mercaptoacetyl)methylcarbamate *S*-ester with *O*-methyl methylphosphonodithioate (8CI); Reg. No. *[29173-31-7]*. OMS 1478. 'MC 2420' (Murphy Chemical Ltd). $C_7H_{14}NO_4PS_2$; 1OVN1&V1SPS&1&O1; *PM4*, p. 329. Insecticide (M. Pianka & W. S. Catling, *Int. Congr. Entomol., 13th*, Moscow, 1968).

**7870** Medinoterb acetate (BSI, E-ISO); médinoterbe acétate (F-ISO). (IUPAC) 6-*tert*-butyl-2,4-dinitro-*m*-tolyl acetate (I); 6-*tert*-butyl-3-methyl-2,4-dinitrophenyl acetate. (*C.A.*) 6-(1,1-dimethylethyl)-3-methyl-2,4-dinitrophenyl acetate (9CI); (I) (8CI); Reg. No. *[2487-01-6]*. 'P 1488', 'MC 1488' (Murphy Chemical Ltd). $C_{13}H_{16}N_2O_6$; 1X1&1&R D1 CNW ENW BOV1; *PM5*, p. 335. Herbicide (G. A. Emery *et al., Proc. EWRC Symp. New Herbic., 2nd*, 1965, p. 141).

**7900** Menazon (BSI, E-ISO, F-ISO, JMAF, ESA); azidithion (France). (IUPAC) *S*-4,6-diamino-1,3,5,-triazin-2-ylmethyl *O,O*-dimethyl phosphorodithioate. (*C.A.*) *S*-(4,6-diamino-1,3,5-triazin-2-yl)methyl *O,O*-dimethyl phosphorodithioate (9CI); *S*-(4,6-diamino-*s*-triazin-2-yl)methyl *O,O*-dimethyl phosphorodithioate (8CI); Reg. No. *[78-57-9]*. OMS 503; ENT 25 760. 'PP 175'. 'Sayfos', 'Saphizon', 'Saphicol', 'SAIsan; 'Abol X' (mixture with gamma-HCH) (ICI Plant Protection Division). $C_6H_{12}N_5O_2PS_2$; T6N CN ENJ BZ DZ F1SPS&O1&O1; *PM6*, p. 331. Aphicide (A. Calderbank *et al., Chem. Ind. (London)*, 1961, p. 630).

**7970** Mesoprazine (BSI, E-ISO, F-ISO). (IUPAC) 6-chloro-*N*-isopropyl-*N'*-3-methoxypropyl-1,3,5-triazine-2,4-diyldiamine; 2-chloro-4-isopropylamino-6-(3-methoxypropylamino)-1,3,5-triazine. (*C.A.*) 6-chloro-*N*-(3-methoxypropyl)-*N'*-(1-methylethyl)-1,3,5-triazine-2,4-diamine (9CI); 2-chloro-4-(isopropylamino)-6-(3-methoxypropylamino)-*s*-triazine (8CI); Reg. No. *[1824-09-5]*. 'G 34 698', 'CGA 4999' (Ciba-Geigy AG). $C_{10}H_{18}ClN_5O$; T6N CN ENJ BMY1&1 DM3O1 FG; herbicide.

**8030** Metazoxolon (BSI, E-ISO, F-ISO). (IUPAC) 4-(3-chlorophenylhydrazono)-3-methylisoxazol-5(4*H*)-one. (*C.A.*) 3-methyl-4,5-isoxazoledione 4-[(3-chlorophenyl)hydrazone] (9CI); 3-methyl-4,5-isoxazoledione 4-[(*m*-chlorophenyl)hydrazone] (8CI); Reg. No. *[5707-73-3]*. 'PP 395' (ICI Plant Protection Division). $C_{10}H_8ClN_3O_2$; T5NOVYJ DUNMR CG& E1; fungicide (T. J. Purnell, *Proc. Br. Insectic. Fungic. Conf., 7th*, 1973, **2**, 603).

**8040** Metflurazone (BSI, F-ISO); metflurazon (E-ISO). (IUPAC) 4-chloro-5-dimethylamino-2-(α,α,α-trifluoro-*m*-tolyl)pyridazin-3(2*H*)-one. (*C.A.*) 4-chloro-5-(dimethylamino)-2-[(3-trifluoromethyl)phenyl]-3(2*H*)-pyridazinone (9CI); Reg. No. *[23576-23-0]*. 'SAN 6706H' (Sandoz AG). $C_{13}H_{11}ClF_3N_3O$; T6NNVJ BR CXFFF& DG EN1&1; herbicide.

**8100** Methanesulphonyl fluoride (IUPAC). (*C.A.*) methanesulfonyl fluoride (8 & 9CI); Reg. No. *[558-25-8]*. MSF. 'Fumette' (Bayer AG). $CH_3FO_2S$; WSF1; insecticide.

**8160** Methiuron (BSI, E-ISO, F-ISO). (IUPAC) 1,1-dimethyl-3-*m*-tolyl-2-thiourea (I). (*C.A.*) *N*,*N*-dimethyl-*N*'-(3-methylphenyl)thiourea (9CI); (I) (8CI); Reg. No. *[21540-35-2]*. 'MH 090' (Yorkshire Tar Distillers Ltd). $C_{10}H_{14}N_2S$; 1N1&YUS&MR C1; *PM2*, p. 306. Herbicide

**8180** Methometon (BSI, E-ISO, F-ISO). (IUPAC) 2-methoxy-*N*,*N*'-bis(3-methoxypropyl)-1,3,5-triazine-2,4-diyldiamine; 2-methoxy-4,6-bis(3-methoxypropylamino)-1,3,5-triazine. (*C.A.*) 6-methoxy-*N*,*N*'-bis(3-methoxypropyl)-1,3,5-triazine-2,4-diamine (9CI); 2-methoxy-4,6-bis(3-methoxypropylamino)-*s*-triazine (8CI); Reg. No. *[1771-07-9]*. 'G 34 690' (Ciba-Geigy AG). $C_{12}H_{23}N_5O_3$; T6N CN ENJ BO1 DM3O1 FM 3O1; herbicide.

**8310** Methylarsenic sulphide (IUPAC). (*C.A.*) methylthioxoarsine (8 & 9CI); Reg. No. *[2533-82-6]*. MAS. 'Rhizoctol', 'Urbasulf'; 'Rhizoctol combi', 'Rhizoctol slurry' (both mixtures with benquinox) (Bayer AG). $CH_3AsS$; S-AS-1; *PM4*, p. 349. Fungicide.

**8320** Methylarsinediyl bis(dimethyldithiocarbamate) (IUPAC). (*C.A.*) dimethylcarbamodithioic acid bis(anhydrosulfide) with methylarsonodithious acid (9CI); dimethyldithiocarbamic acid bis(anhydrosulfide) with dithiomethanearsonous acid (8CI); Reg. No. *[2445-07-0]*. Urbacid (JMAF). 'Urbacid', 'Monzet'; 'Tuzet' (Reg. No. *[8066-44-2]*) (mixture with thiram + ziram) (Bayer AG). $C_7H_{15}AsN_2S_4$; 1N1&YUS&S-AS-1&SYUS&N1&1; *PM5*, p. 356. Fungicide.

**8360** *S*-Methyl *N*-(carbamoyloxy)thioacetimidate (IUPAC). (*C.A.*) methyl *N*-[(aminocarbonyl)oxy]ethanimidothioate (9CI); methyl *N*-(carbamoyloxy)thioacetimidate (8CI); Reg. No. *[16960-39-7]*. ENT 27 411. 'EI 1642' (E. I. du Pont de Nemours & Co. (Inc.).). $C_4H_8N_2O_2S$; ZVONUY1&S1; *PM3*, p. 341. Insecticide.

**8405** Methylmercury dicyandiamide (accepted BSI). (IUPAC) 1-cyano-3-(methylmercurio)guanidine; (methylmercurio)guanidinocarbonitrile. (*C.A.*) (cyanoguanidinato-*N*')methylmercury (9CI); (cyanoguanidinato)methylmercury (8CI); Reg. No. *[502-39-6]*. 'Panogen' (Panogen Inc.). $C_3H_6HgN_4$; NCMYUM&M-HG-1; *PM5*, p. 361. Fungicide.

**8440** 3-Methyl-1-phenylpyrazol-5-yl dimethylcarbamate (I) (IUPAC). (*C.A.*) 3-methyl-1-phenyl-1*H*-pyrazol-5-yl dimethylcarbamate (9CI); (I) (8CI); Reg. No. *[87-47-8]*. OMS 20, ENT 17 588. 'G 22 008'; 'Pyrolan' (Ciba-Geigy AG). $C_{13}H_{15}N_3O_2$; T5NNJ AR& C1 EOVN1&1; insecticide.

**8450** 2-(*N*-Methyl-*N*-prop-2-ynylamino)phenyl methylcarbamate (IUPAC). (*C.A.*) 2-(methyl-2-propynylamino)phenyl methylcarbamate (9CI); *o*-(methyl-2-propynylamino)phenyl methylcarbamate (8CI); Reg. No. *[23504-07-6]*. 'C 17 018' (Ciba-Geigy AG). $C_{12}H_{14}N_2O_2$; 1UU2N1&R BOVM1; *PM3*, p. 339. Insecticide.

**8460** 4-(*N*-Methyl-*N*-prop-2-ynylamino)-3,5-xylyl methylcarbamate (IUPAC). (*C.A.*) 3,5-dimethyl-4-(methyl-2-propynylamino)phenyl methylcarbamate (9CI); 4-(methyl-2-propynylamino)-3,5-xylyl methylcarbamate (8CI); Reg. No. *[23623-49-6]*. 'C 20 132' (Ciba-Geigy AG). $C_{14}H_{18}N_2O_2$; 1UU2N1&R B1 F1 DOVM1; *PM2*, p. 320. Insecticide.

**8490** Methyl 2,3,5,6-tetrachloro-*N*-methoxy-*N*-methylterephthalamate (I) (IUPAC). (*C.A.*) methyl 2,3,5,6-tetrachloro-4-[methoxy(methylamino)carbonyl]benzoate (9CI); (I) (8CI); Reg. No. *[14419-01-3]*. 'OCS 21 693' (Velsicol Corp.). $C_{11}H_9Cl_4NO_4$; 1OVR BG CG EG FG DVN1&O1; *PM4*, p. 354. Herbicide.

**8510** 5-Methyl-6-thioxo-1,3,5-thiadiazinan-3-ylcetic acid (IUPAC); tetrahydro-5-methyl-6-thioxo-2*H*-1,3,5-thiadiazin-3-ylacetic acid. (*C.A.*) dihydro-5-methyl-6-thioxo-2*H*-1,3,5-thiadiazine-3(4*H*)-acetic acid (8 & 9CI); Reg. No. *[3655-88-7]*. 'Terracur' (Bayer AG). $C_6H_{10}N_2O_2S_2$; T6NYS ENTJ A1 BUS E1VQ; *PM1*, p. 71. Nematicide.

**8600** Mexacarbate (BSI, E-ISO, F-ISO, ANSI). (IUPAC) 4-dimethylamino-3,5-xylyl methylcarbamate. (*C.A.*) 3,5-dimethyl-4-(dimethylamino)phenyl methylcarbamate (9CI); 4-(dimethylamino)-3,5-xylyl methylcarbamate (8CI); Reg. No. *[315-18-4]*. OMS 47, OMS 639, ENT 25 766. 'Dowco 139'; 'Zectran' ( Dow Chemical Co.). $C_{12}H_{18}N_2O_2$; 1N1&R B1 F1 DOVM1; *PM4*, p. 359. Insecticide.

**8610** Milneb (BSI, ANSI); thiadiazine (JMAF). (IUPAC) 4,4',6,6'-tetramethyl-3,3'-ethylenedi-1,3,5-thiadiazinane-2-thione; 4,4',6,6'-tetramethyl-3,3'-ethylenebis(tetrahydro-1*H*-1,3,5-thiadiazine-2-thione) (I). (*C.A.*) 3,3'-(1,2-ethanediyl)bis[tetrahydro-4,6-dimethyl-2*H*-1,3,5-thiadiazine-2-thione] (9CI); 3,3-ethylenebis[tetrahydro-4,6-dimethyl-2*H*-1,3,5-thiadiazine-2-thione] (8CI); Reg. No. *[3773-49-7]*. 'Experimental Fungicide 328'; 'Banlate' (E. I. du Pont de Nemours & Co. (Inc.)). $C_{12}H_{22}N_4S_4$; T6NYS EMTJ BUS D1 F1 A2- AT6NYS EMTJ BUS D1 F1; fungicide.

**8620** Mipafox (BSI, E-ISO, F-ISO). (IUPAC) *N,N'*-di-isopropylphosphorodiamidic fluoride. (*C.A.*) *N,N'*-bis(1-methylethyl)phosphorodiamidic fluoride (9CI); *N,N'*-diisopropylphosphorodiamidic fluoride (8CI); Reg. No. *[371-86-8]*. 'Isopestox', 'Pestox 15' (FBC Limited). $C_6H_{16}FN_2OP$; 1Y1&MPO&FMY1&1; acaricide and insecticide.

**8630** Mirex (ESA). (IUPAC) dodecachloropentacyclo[5.2.1.0$^{2,6}$.0$^{3,9}$.0$^{5,8}$]decane. (*C.A.*) 1,1a,2,2,3,3a,4,5,5,5a,5b,6-dodecachlorooctahydro-1,3,4-metheno-1*H*-cyclobuta[*cd*]pentalene (8 & 9CI); Reg. No. *[2385-85-5]*. Perchlordécone. ENT 25 719. 'GC 1283' (Allied Chemical Corp.); 'Mirex' (Seppic). $C_{10}Cl_{12}$; L545 B4 C5 D 4ABCE JTJ AG BG CG DG DG EG FG GG HG HG IG JG; *PM5*, p. 368. Insecticide.

**8700** Morfamquat dichloride (BSI, E-ISO, F-ISO). (IUPAC) 1,1'-bis(3,5-dimethylmorpholinocarbonylmethyl)-4,4'-bipyridinium dichloride; 1,1'-bis(3,5-dimethylmorpholinocarbonylmethyl)-4,4'-bipyridyldiylium dichloride. (*C.A.*) 1,1'-bis[2-(3,5-dimethyl-4-morpholinyl)-2-oxoethyl]-4,4'-bipyridinium dichloride (9CI); 1,1'-bis[[(3,5-dimethylmorpholino)carbonyl]methyl]-4,4'-bipyridinium dichloride (8CI); Reg. No. *[4638-83-3]*. 'PP 745'; 'Morfoxone' (ICI Plant Protection Division). $C_{26}H_{36}Cl_2N_4O_4$; T6N DOTJ B1 F1 AV1- AT6KJ D- DT6KJ A1V- AT6N DOTJ B1 F1 &&2Cl SALT; *PM3*, p. 355. Herbicide (H. M. Fox, *Proc. Br. Weed Control Conf., 7th*, 1964, 29; H. M. Fox & C. R. Beech, *ibid.*, p. 108).

**8710** Morphothion (BSI, E-ISO, F-ISO). (IUPAC) *O,O*-dimethyl *S*-morpholinocarbonylmethyl phosphorodithioate. (*C.A.*) *O,O*-dimethyl *S*-[2-(4-morpholinyl)-2-oxoethyl] phosphorodithioate (9CI); *O,O*-dimethyl phosphorodithioate *S*-ester with 4-(mercaptoacetyl)morpholine (8CI); Reg. No. *[144-41-2]*. 'Ekatin M' (Sandoz AG). $C_8H_{16}NO_4PS_2$; T6N DOTJ AV1SPS&O1&O1; insecticide.

**8940** Nitrilacarb (BSI, E-ISO, ANSI); nitrilacarbe (F-ISO). (IUPAC) 4,4-dimethyl-5-(methylcarbamoyloxyimino)pentanenitrile. (*C.A.*) 4,4-dimethyl-5-[[[(methylamino)carbonyl]oxy]imino]pentanenitrile (9CI); Reg. No. *[29672-19-3]*. 'CL 72 613', 'AC 82 258' 1:1 complex with zinc dichloride; 'Accotril' (American Cyanamid Co.). $C_9H_{15}N_3O_2$; NC2X1&1&1UNOVM1; (zinc dichloride complex) $C_9H_{15}Cl_2N_3O_2Zn$;. NC2X1&1&1UNOVM1 &&$ZnCl_2$ COMPLEX; *PM6*, p. 385 Zinc dichloride 1:1 complex (Reg. No. *[58270-08-9]* was *[61332-32-9]*) reported as insecticide (W. K. Whitney & J. L. Aston, *Proc. Br. Insectic. Fungic. Conf., 8th*, 1975, **2**, 633).

**8960** Nitrofluorfen (BSI, E-ISO, ANSI, WSSA); nitrofluorfène (F-ISO). (IUPAC) 2-chloro-$\alpha,\alpha,\alpha$-trifluoro-*p*-tolyl 4-nitrophenyl ether. (*C.A.*) 2-chloro-1-(4-nitrophenoxy)-4-(trifluoromethyl)benzene (9CI); Reg. No. *[42874-01-1]*. 'RH-2512' (Rohm & Haas Co.). $C_{13}H_7ClF_3NO_3$; WNR DOR BG DXFFF; herbicide.

**8980** *N*-3-Nitrophenylitaconimide (IUPAC). (*C.A.*) 3-methylene-*N*-(3-nitrophenyl)pyrrolidine-2,5-dione (9CI); 2-methylene-3-(*m*-nitrophenyl)succinimide (8CI); Reg. No. *[4137-12-6]*. 'B 720' (Uniroyal Inc.). $C_{11}H_8N_2O_4$; T5VNVY EHJ BR CNW& DU1; fungicide (B. von Schmeling, *Phytopathology*, 1962, **52**, 819).

**9000** 4-(2-Nitroprop-1-enyl)phenyl thiocyanate (IUPAC). (*C.A.*) 4-(2-nitro-1-propenyl)phenylthiocyanate (9CI); Reg. No. *[950-00-5]*. Nitrostyrene (JMAF). 'Styrocide' (Nippon Kayaku Co. Ltd). $C_{10}H_8N_2O_2S$; WNY1&U1R DSCN; fungicide.

**9050** Nornicotine. (IUPAC) 3-(pyrrolidin-2-yl)pyridine. (*C.A.*) (*S*)-3-(2-pyrrolidinyl)pyridine (9CI); 1'-demethylnicotine (8CI); Reg. No. *[494-97-3]*. 2-(3-Pyridyl)pyrrolidine. $C_9H_{12}N_2$; T6NJ C- BT5MTJ; insecticide.

**9060** Noruron (BSI, E-ISO, F-ISO); norea (ANSI, WSSA). (IUPAC) 1,1-dimethyl-3-(perhydro-4,7-methanoinden-5-yl)urea; 3-(hexahydro-4,7-methanoindan-5-yl)-1,1-dimethylurea (I). (*C.A.*) *N,N*-dimethyl-*N'*-(octahydro-4,7-methano-1*H*-inden-5-yl)-urea 3aα,4α,5α,7α,7aα-isomer (9CI); (I) (8CI); Reg. No. *[18530-56-8]* for 3aα,4α,5α,7α,7aα-isomer, *[2163-79-3]* unstated stereochemistry. 'Hercules 7531'; 'Herban' (Hercules Inc. now Boots-Hercules Agrochemical Co.). $C_{13}H_{22}N_2O$; L C555 ATJ IMVN1&1; *PM3*, p. 368. Herbicide (G. A. Buntin *et al., Abstr. Am. Chem. Soc. Meeting*, Los Angeles, 1963).

**9090** Octachlorocyclohex-2-en-1-one (IUPAC). (*C.A.*) 2,3,4,4,5,5,6,6-octachloro-2-cyclohexen-1-one (8 & 9CI); Reg. No. *[4024-81-1]*. OCH. 'Oktone' (P. F. Goodrich Chemical Co.). $C_6Cl_8O$; L6V BUTJ BG CG DG DG EG EG FG FG; fungicide and herbicide.

**9190** Oxapyrazone (BSI, F-ISO);oxapyrazon (E-ISO). (IUPAC) 2-hydroxyethyldimethylammonium 5-bromo-1,6-dihydro-6-oxo-1-phenylpyridazin-4-yloxamate. (*C.A.*) [(5-bromo-1,6-dihydro-6-oxo-1-phenyl-4-pyridazinyl)amino]oxoacetic acid compound with 2-(dimethylamino)ethanol (1:1) (9CI); (5-bromo-1,6-dihydro-6-oxo-1-phenyl-4-pyridazinyl)oxamic acid compound with 2-(dimethylamino)ethanol (1:1) (8CI); Reg. No. *[25316-57-8]*. 'BAS 350H' (BASF AG). $C_{16}H_{19}BrN_4O_5$; T6NNVJ BR& DE EMVVQ &&$(CH_3)_2NCH_2CH_2OH$ SALT; herbicide.

**9200** Oxapyrazone-sodium (BSI, F-ISO); oxapyrazon-sodium (E-ISO). (IUPAC) sodium 5-bromo-1,6-dihydro-6-oxo-1-phenylpyridazin-4-yloxamate; sodium *N*-(5-bromo-6-oxo-1-phenylpyridazin-4-yl)oxamate. (*C.A.*) sodium *N*-(5-bromo-6-oxo-1-phenyl-4(1*H*)-pyridazinyl)oxamate (9CI);sodium *N*-(5-bromo-1,6-dihydro-6-oxo-1-phenyl-4-pyridazinyl)oxamate (8CI); Reg. No. *[25316-56-7]*. 'BAS 3380H' (BASF AG). $C_{12}H_7BrN_3NaO_4$; T6NNVJ BR& DE FMVVO &&Na SALT; herbicide.

**9260** Oxydisulfoton (BSI, E-ISO, F-ISO). (IUPAC) *O,O*-diethyl *S*-2-ethylsulphinylethyl phorodithioate. (*C.A.*) *O,O*-diethyl *S*-[2-(ethylsulfinyl)ethyl] phosphorodithioate (8 & 9CI); Reg. No. *[2497-07-6]*. 'Bayer 23 323', 'S 309', 'L 16/184'; 'Disyston-S' (Bayer AG). $C_8H_{19}O_3PS_3$; OS2&2SPS&O2&O2; *PM4*, p. 389. Acaricide and insecticide.

**9310** Parafluron (BSI, E-ISO, F-ISO). (IUPAC) 1,1-dimethyl-3-(α,α,α-trifluoro-*p*-tolyl)urea (I). (*C.A.*) *N,N*-dimethyl-*N'*-[4-(trifluoromethyl)phenyl]urea (9CI); (I) (8CI); Reg. No. *[7159-99-1]*. 'C 15 935' (Ciba-Geigy AG). $C_{10}H_{11}F_3N_2O$; FXFFR DMVN1&1; herbicide.

**9490** Phenkapton (BSI, E-ISO, F-ISO); CMP (JMAF). (IUPAC) *S*-2,5-dichlorophenylthiomethyl *O,O*-diethyl phosphorodithioate. (*C.A.*) *S*-[(2,5-dichlorophenyl)thio]methyl *O,O*-diethyl phosphorodithioate (8 & 9CI); Reg. No. *[2275-14-1]*. ENT 25 585. 'G 28 029' (Ciba-Geigy AG). $C_{11}H_{15}Cl_2O_2PS_3$; GR DG BS1SPS&O2&O2; acaricide and insecticide.

**9510** Phenmedipham-ethyl (BSI, E-ISO); phenmédiphame-éthyl (F-ISO). (IUPAC) ethyl 3-(3-methylcarbaniloyloxy)carbanilate; 3-ethoxycarbonylaminophenyl 3-methylcarbanilate. (*C.A.*) 3-[(ethoxycarbonyl)amino]phenyl (3-methylphenyl)carbamate (9CI); ethyl *m*-hydroxycarbanilate *m*-methylcarbanilate (ester) (8CI). 'SN38 574' (Schering AG). $C_{17}H_{18}N_2O_4$; 2OVMR DOVMR C1; herbicide.

**9520** Phenobenzuron (BSI, E-ISO, F-ISO). (IUPAC) 1-benzoyl-1-(3,4-dichlorophenyl)-3,3-dimethylurea (I). (*C.A.*) *N*-(3,4-dichlorophenyl)-*N*-[(dimethylamino)carbonyl]benzamide (9CI); (I) (8CI); Reg. No. *[3134-12-1]*. 'PP 65-25' 'Benzomarc' (Rhône-Poulenc Phytosanitaire). $C_{16}H_{14}Cl_2N_2O_2$; GR BG DNVR&VN1&1; *PM5*, p. 409. Herbicide (P. Poignant *et al., Symp. New Herbic., 2nd,* 1965, p. 1).

**9560** 2-Phenyl-4*H*-3,1-benzoxazin-4-one (I) (IUPAC). (*C.A.*) (I) (8 & 9CI); Reg. No. *[1022-46-4]*. 'H-170'; 'Linurotox' (BASF AG). $C_{14}H_9NO_2$; T66 BVO ENJ DR; *PM5*, p. 412. Herbicide (A. Fischer, *Meded. Landbouwhogesch. Opzoekingsstn. Staat Gent*, 1965, **30**, 163).

**9590** Phenylmercury dimethyldithiocarbamate (IUPAC). (*C.A.*) (dimethylcarbamodithioato-*S,S'*)phenylmercury (9CI); (dimethyldithiocarbamato)phenylmercury (8CI); Reg. No. *[32407-99-1]*. 'Phelam' (Steetley Chemicals Ltd.). $C_9H_{11}HgNS_2$; 1N1&YUS&S-HG-R; *PM6*, p. 417. Fungicide.

**9600** Phenylmercury nitrate (BSI, draft E-ISO); nitrate de phénylmercure (draft F-ISO) in lieu of common name. (IUPAC) phenylmercury hydroxide + phenylmercury nitrate; phenylmercury nitrate (basic). (*C.A.*) hydroxy(nitrato)diphenyldimercury (8 & 9CI); Reg. No. *[8003-05-2]* was *[6059-33-1]*. 'Harvesan Plus' (with gamma-HCH); 'Murcocide' (with captan). $C_{12}H_{11}Hg_2NO_4$; WNO&-HG-R &Q-HG-R; *PM6*, p. 418. Fungicide.

**9650** Phosacetim (BSI, E-ISO); phosacétime (F-ISO). (IUPAC) *O,O*-bis(4-chlorophenyl) *N*-acetimidoylphosphoramidothioate. (*C.A.*) *O,O*-bis(4-chlorophenyl) (1-iminoethyl)phosphoramidothioate (9CI); *O,O*-bis(*p*-chlorophenyl) acetimidoylphosphoramidothioate (8CI); Reg. No. *[4104-14-7]*. 'Bayer 38 819'; 'Gophacide' (Bayer AG). $C_{14}H_{13}Cl_2N_2O_2PS$; MUY1&MPS&OR DG&OR DG; *PM3*, p. 394. Rodenticide.

**9800** Piperonyl cyclonene.(IUPAC) 5-( ,3-benzodioxol-5-yl)-3-hexylcyclohex-2-enone. (*C.A.*) 5-( 5-benzo-1,3-dioxolyl)-3-hexyl-2-cyclohexen-1-one (9CI); 3-hexyl-5-(3,4-methylenedioxyphenyl)-2-cyclohexen-1-one (8CI); Reg. No. *[8022-12-4]*. $C_{19}H_{24}O_3$; T56 BO DO CHJ G- EL6V BUTJ C6; synergist for insecticides.

**9870** Polychlorodicyclpentadiene isomers (IUPAC). (*C.A.*) 'Bandane' (8 CI); Reg. No. *[8029-29-6]*. 'Bandane' (Velsicol Chemical Corp.). Herbicide and insecticide.

**9880** Polychloroterpenes; heptachloro-2,2-dimethyl-3-methylenenorbornane (IUPAC). (*C.A.*) chlorinated mixed terpenes (8 & 9CI); Reg. No. *[8001-50-1]*. Terpene polychlorinates. ENT 19 442. 'Compound 3961'; 'Strobane' (B. F. Goodrich & Co.). *PM1*, p. 410. Acaricide and insecticide (D. L. Kent *et al., Soap Chem. Spec.*, 1953, **19**(6), 157).

**9920** Potassium cyanate (I) (IUPAC). (*C.A.*) (I) (8 & 9CI); Reg. No. *[590-28-3]*. 'Aero' cyanate (American Cyanamid Co.). CKNO; .KA..OCN; *PM5*, p. 432. Herbicide.

**9980** Primidophos (BSI, E-ISO, F-ISO). (IUPAC) *O,O*-diethyl *O*-(2-*N*-ethylacetamido-6-methylpyrimidin-4-yl) phosphorothioate. (*C.A.*) *O*-[2-(acetylethylamino)-6-methyl-4-pyrimidinyl] *O,O*-diethyl phosphorothioate (8 & 9CI); Reg. No. *[39247-96-6]*. 'PP 484' (ICI Plant Protection Division). $C_{13}H_{22}N_3O_4PS$; T6N CNJ BN2&V1 DOPS&O2&O2 F1; insecticide.

**10000** Proclonol (BPC). (IUPAC) 4,4'-dichloro-α-cyclopropylbenzhydrol. (*C.A.*) 4-chloro-α-(4-chlorophenyl)-α-cyclopropylbenzenemethanol (9CI); bis(*p*-chlorophenyl)cyclopropylmethanol (8CI); Reg. No. *[14088-71-2]*. $C_{16}H_{14}Cl_2O$; L3TJ A1R DG &R DG &Q; acaricide (Janssen Pharmaceutica).

**10210** Propyl isome (ESA). (IUPAC) dipropyl 5,6,7,8-tetrahydro-7-methylnaphtho[2,3-*d*]-1,3-dioxole-5,6-dicarboxylate (I); dipropyl 1,2,3,4-tetrahydro-3-methyl-6,7-methylenedioxynaphthalene-1,2-dicarboxylate. (*C.A.*) (I) (8 & 9CI); Reg. No. *[83-59-0]*. Dipropyl maleate isosafrole condensate. $C_{20}H_{26}O_6$; T C566 DO FO EH&&TJ JVO3 KVO3 L1; *PM5*, p. 446. Synergist for insecticides (S. B. Penwick & Co.).

**10300** Proxan (BSI, Canada, New Zealand) for acid; proxan-sodium (E-ISO) for salt; proxane-sodium (F-ISO) for salt. (IUPAC) *O*-isopropyl hydrogen dithiocarbonate (i); sodium *O*-isopropyl dithiocarbonate (ii); sodium *O*-isopropyl xanthate (ii). (*C.A.*) *O*-(1-methylethyl) carbonodithioic acid (i); sodium *O*-(1-methylethyl) carbonodithioate (ii) (9CI); *O*-isopropyl dithiocarbonic acid (i); *O*-isopropyl sodium dithiocarbonate (ii) (8CI); Reg. No. *[140-93-2]* (ii). 'Good-rite n.i.x.' (B. F. Goodrich Chemical Co.). (i) $C_4H_8OS_2$; SUYSHOY1&1; (ii) $C_4H_7NaOS_2$; SUYSHOY1&1 &&Na SALT; *PM1*, p. 360. Herbicide.

**10310** Prynachlor (BSI, E-ISO, ANSI, WSSA); prynachlore (F-ISO). (IUPAC) 2-chloro-*N*-(1-methylprop-2-ynyl)acetanilide. (*C.A.*) 2-chloro-*N*-(1-methyl-2-propynyl)-*N*-phenylacetamide (9CI); 2-chloro-*N*-(1-methyl-2-propynyl)acetanilide (8CI); Reg. No. *[21267-72-1]*. 'BAS 290H'; 'Basamaize' (BASF AG). $C_{12}H_{12}ClNO$; G1VNR&Y1&1UU1; herbicide.

**10320** Pydanon (BSI, E-ISO, F-ISO). (IUPAC) (±)-hexahydro-4-hydroxy-3,6-dioxopyridazin-4-ylacetic acid. (*C.A.*) hexahydro-4-hydroxy-3,6-dioxo-4-pyridazineacetic acid (8 & 9CI); Reg. No. *[22571-07-9]*. 'H 1244' (C. F. Spiess & Sohn). $C_6H_8N_2O_5$; T6VMMVTJ EQ E1VQ; plant growth regulator.

**10380** Pyrichlor (WSSA). (IUPAC) 2,3,5-trichloropyridin-4-ol. (*C.A.*) 2,3,5-trichloro-4-pyridinol (8 & 9CI); Reg. No. *[1970-40-7]*. '*Daxtron*' (Dow Chemical Co.). $C_5H_2Cl_3NO$; T6NJ BG CG DQ EG; *PM1*, p. 363. Herbicide (M. J. Huraux & H. M. Lawson, *Symp. New Herbic., 2nd*, 1965, p. 269).

**10395** Pyridazin-3-yl *o*-tolyl ether (IUPAC). (*C.A.*) 3-(2-methylphenoxy)pyridazine (9CI); 3-*o*-tolyloxypyridazine (8CI); Reg. No. *[14491-59-9]*. Credazine (JMAF). 'H 722', 'SW-6701', 'SW-6721'; 'Kusakira' (Sankyo Chemical Ltd). $C_{11}H_{10}N_2O$; T6NNJ COR B1; *PM6*, p. 522. Herbicide (T. Jojima *et al., Agric. Biol. Chem.*, 1968, **32**, 1376; 1969, **33**, 96).

**10400** Pyridinitril (BSI, E-ISO); pyridinitrile (F-ISO); DDPP (JMAF). (IUPAC) 2,6-dichloro-4-phenylpyridine-3,5-dicarbonitrile. (*C.A.*) 2,6-dichloro-4-phenyl-3,5-pyridinedicarbonitrile (9CI); 2,6-dichloro-3,5-dicyano-4-phenylpyridine (8CI); Reg. No. *[1086-02-8]*. 'IT 3296'; 'Ciluan' (E. Merck). $C_{13}H_5Cl_2N_3$; T6NJ BG CCN DR& ECN FG; *PM5*, p. 454. Fungicide (G. Mohr *et al., Meded. Rijksfac. Landbouwwet. Gent*, 1968, **33**, 1293).

**10410** 2-Pyridyl 1-(2,5-xylyl)ethyl sulphone 1-oxide (IUPAC). (*C.A.*) 2-[[1-(2,5-dimethylphenyl)ethyl]sulfonyl]pyridine 1-oxide; Reg. No. *[60263-88-9]*. 'UBI-S734' (Uniroyal Chemical Co.). $C_{15}H_{17}NO_3S$; T6NJ AO BSWY1&R B1 E1; herbicide.

**10420** Pyrimithate (BSI, BPC); pyrimitate (E-ISO, F-ISO). (IUPAC) *O*-2-dimethylamino-6-methylpyrimidin-4-yl *O,O*-diethyl phosphorothioate. (*C.A.*) *O*-[2-(dimethylamino)-6-methyl-4-pyrimidinyl] *O,O*-diethyl phosphorothioate (8 & 9CI); Reg. No.*[5221-49-8]*. 'ICI 29 661'; 'Diothyl' (ICI Pharmaceuticals Division). $C_{11}H_{20}N_3O_3PS$; T6N CNJ BN1&1 DOPS&O2&O2 F1; veterinary acaricide and insecticide.

**10430** Pyrinuron (BSI, E-ISO, F-ISO, ANSI). (IUPAC) 1-(4-nitrophenyl)-3-(3-pyridylmethyl)urea. (*C.A.*) *N*-(4-nitrophenyl)-*N'*-(3-pyridinylmethyl)urea (9CI); Reg. No. *[53558-25-1]*. 'RH-787'; 'Vacor' (Rohm & Haas Co.). $C_{13}H_{12}N_4O_3$; T6NJ C1MVMR DNW; rodenticide.

**10450** Pyroxychlor (BSI, E-ISO, ANSI); pyroxychlore (F-ISO). (IUPAC) 2-chloro-6-methoxy-4-trichloromethylpyridine. (*C.A.*) 2-chloro-6-methoxy-4-(trichloromethyl)pyridine (8 & 9CI); Reg. No. *[7159-34-4]*. 'Dowco 269'; 'Nurelle', 'Lorvek' (Dow Chemical Co.). $C_7H_5Cl_4NO$; T6NJ BO1 DXGGG FG; fungicide.

**10470** *N*-Pyrrolidinosuccinamic acid (IUPAC). (*C.A.*) 4-oxo-4-(1-pyrrolidinylamino)butanoic acid (9CI); *N*-1-pyrrolidinylsuccinamic acid (8CI); Reg. No. *[23744-05-0]*. 'F 529'; 'F-Five' (Uniroyal Inc.). $C_8H_{14}N_2O_3$; T5NTJ AMV2VQ; plant growth regulator.

**10480** Quinacetol sulphate (BSI, E-ISO); quinacétol sulfate (F-ISO). (IUPAC) bis(5-acetyl-8-hydroxyquinolinium) sulphate. (*C.A.*) 1-(8-hydroxy-5-quinolinyl)ethanone sulfate (2:1) (9CI); Reg. No. *[57130-91-3]*. 'G 20 072'; 'Fongoren', 'Risoter' (Ciba-Geigy AG). $C_{22}H_{20}N_2O_8S$; T66 BNJ GV1 JQ &&$H_2SO_4$ SALT 2:1; *PM5*, p. 455. Fungicide (S. D. Hocombe *et al., Proc. Br. Insectic. Fungic. Conf., 7th*, 1973, **1**, 365).

**10500** Quinalphos-methyl (BSI, E-ISO, F-ISO). (IUPAC) *O,O*-dimethyl *O*-quinoxalin-2-yl phosphorothioate. (*C.A.*) *O,O*-dimethyl *O*-(2-quinoxalinyl) phosphorothioate (8 & 9CI); Reg. No. *[13593-08-3]*. 'SAN 52 056I' (Sandoz AG). $C_{10}H_{11}N_2O_3PS$; T66 BN ENJ COPS&O1&O1; insecticide.

**10510** Quinazamid (BSI, E-ISO); quinazamide (F-ISO). (IUPAC) *p*-benzoquinone monosemicarbazone (I). (*C.A.*) 2-(4-oxo-2,5-cyclohexadien-1-ylidene)hydrazinecarboxamide (9CI); (I) (8CI); 'RD 8684','BTS 8684' (Boots Co. Ltd now FBC Limited) $C_7H_7N_3O_2$; L6V DYJ DUNMVZ; fungicide.

**10570** Rabenazole (BSI, draft E-ISO). (IUPAC) 2-(3,5-dimethylpyrazol-1-yl)benzimidazole. (*C.A.*) 2-(3,5-dimethyl-1*H*-pyrazol-1-yl)-1*H*-benzimidazole (9CI); Reg. No. *[40341-04-6]*. 'Ciriom'; 'Ciriom F' (Bayer AG) (Reg. No. *[74725-93-2]*) for combination with fuberidazole. $C_{12}H_{12}N_4$; T56 BM DNJ C- AT5NNJ C1 E1; fungicide (W. Specht & M. Tillkes, *Pflanzenschutz-Nachr. (Engl. Ed.)* 1980, **33**, 61).

**10620** Ryanodine (I) for the main toxicant of ryania, the ground stemwood of the shrub *Ryania speciosa*. (IUPAC) (2*S*,3*S*,4*R*,4a*S*,5*S*,5a*S*,8*S*,9*R*,9a*R*,9b*R*)-2,3,4a,5a,9,9b-hexahydro-3-isopropyl-2a,5,8-trimethylperhydro-2,5-methanobenzo[1,2]pentaleno[1,6-*bc*]furan-4-yl pyrrole-2-carboxylate. (*C.A.*) [3*S*-(3α,4β,4a*S*★,6a,6aα,7a,8β,8aα,8bβ,9β,9aα)]-octahydro-3,6a,9-trimethyl-7-(1-methylethyl)-6,9-methanobenzo[1,2]pentaleno[1,6-*bc*]furan-4,6,7,8,8a,8b,9a-heptol 3-(1*H*)-pyrrole-2-carboxylate (9CI); (I) (8CI); Reg. No. *[15662-33-6]* formerly *[15800-60-9]* (ryanodine), *[8047-13-0]* (ryania).$C_{25}H_{35}NO_9$; T5 G6556/BN/FO 5AABEL O LXOTJ AQ B1 CY1&1 CQ EQ F1 GQ J1 KQ NQ DOV- BT5MJ; *PM6*, p. 469. Ryania was formerly marketed as an insecticide by S. B. Penick & Co. (B. E. Pepper & L. A. Carruth, *J. Econ. Entomol.*, 1945, **38**, 59).

**10630** Sabadilla; cevadilla. *PM5*, p. 464. Insecticidal extract from *Schoenocaulon officinale*.

**10650** Salicylanilide (I) (IUPAC) (accepted by BSI, E-ISO, F-ISO, JMAF). (*C.A.*) 2-hydroxy-*N*-phenylbenzamide (9CI); (I) (8CI); Reg. No. *[87-17-2]*. 'Shirlan' (ICI). $C_{13}H_{11}NO_2$; QR BVMR; *PM5*, p. 465. Fungicide (R. G. Fargher *et al., Mem. Shirley Inst.*, 1930, **9**, 37).

**10660** Schradan (BSI, E-ISO, JMAF); schradane (F-ISO). (IUPAC) octamethylpyrophosphoric tetra-amide. (*C.A.*) octamethyldiphosphoramide (9CI); octamethylpyrophosphoramide (8CI); Reg. No. *[151-26-9]*. OMPA. '*Pestox 3*' (Fisons Ltd now FBC Limited), 'Sytam' (Murphy Chemical Ltd), Wacker-Chemie GmbH. $C_8H_{24}N_4O_3P_2$; 1N1&PO&N1&1&OPO&N1&1&N1&1; *PM6*, p. 470. Systemic acaricide and insecticide (G. Schrader, *Die Entwicklung neuer insektizider Phosphorsaüre-Ester*, 3rd Ed. p. 88).

**10690** Sesamex (ESA). (IUPAC) 5-1-[2-(2-ethoxyethoxy)ethoxy]ethoxy-1,3-benzodioxole (I); 2-(1,3-benzodioxol-5-yloxy)-3,6,9-trioxaundecane. (*C.A.*) (I) (9CI); 4-[1-[2-(2-ethoxyethoxy)ethoxy]ethoxy]-1,2-methylenedioxybenzene; acetaldehyde 2-(2-ethoxyethoxy)ethyl 3,4-methylenedioxyphenyl acetal (8CI); Reg. No. *[51-14-9]*. ENT 20 871. 'Sesoxane' (Shulton Inc.). $C_{15}H_{22}O_6$; T56 BO DO CHJ GOY1&O2O2O2; *PM6*, p. 472. Synergist for pyrethrins (M. Beroza, *J. Agric. Food Chem.*, 1956, **4**, 49).

**10700** Sesasmolin (ESA). (IUPAC) 1,3-benzodioxol-5-yl (1*R*,3a*R*,4*S*,6a*R*)-4-(1,3-benzodioxol-5-yl)perhydrofuro[3,4-*c*]furan-1-yl ether; 1-(1,3-benzodioxol-5-yl)-4-(1,3-benzodioxol-5-yloxy)tetrahydrofuro[3,4-*c*]furan; 2-(1,3-benzodioxol-5-yl)-6-(1,3-benzodioxol-5-yloxy)-3,7-dioxabicyclo[3.3.0]octane. (*C.A.*) [1*S*-(1α,3aα,4α,6aα)]-5-[4-(1,3-benzodioxol-5-yloxy)tetrahydro-1*H*,3*H*-furo[3,4-*c*]furan-1-yl]-1,3-benzodioxole (9CI); (1*R*,3a*R*,4*S*,6a*R*)-tetrahydro-1-[3,4-(methylenedioxy)phenoxy]-4-[3,4-(methylenedioxy)phenyl]-1*H*,3*H*-furo[3,4-*c*]furan (8CI); Reg. No. *[526-07-8]*. $C_{20}H_{18}O_7$; T56 BO DO CHJ GO- BT55 CO GOTJ F- GT56 BO DO CHJ; the related sesamin, **not** sesasmolin, is described in *PM5*, p. 469. Synergist for insecticides which is derived from sesame oil from *Sesamum indicum*.

**10750** Simeton (BSI). (IUPAC) *N*,*N*'-diethyl-6-methoxy-1,3,5-triazine-2,4-diyldiamine; 2,4-bis(ethylamino)-6-methoxy-1,3,5-triazine. (*C.A.*) *N*,*N*'-diethyl-6-methoxy-1,3,5-triazine-2,4-diamine (9CI); 2,4-bis(ethylamino)-6-methoxy-*s*-triazine (8CI); Reg. No. *[673-04-1]*. 'G 30 044' (Ciba-Geigy AG). $C_8H_{15}N_5O$; T6N CN ENJ BO1 DM2 FM2; herbicide.

**10800** Sodium (*Z*)-3-chloroacrylate (I) (IUPAC). (*C.A.*) sodium (*Z*)-3-chloro-2-propenoate (9CI); (I) (8CI); Reg. No. *[4312-97-4]*. Sodium *cis*-3-chloroacrylate. 'UC 20 299'; 'Prep' (Union Carbide Corp.). $C_3H_2ClNaO_2$; QV1U1G &&Na SALT; *PM1*, p. 386. Defoliant (R. A. Herrett & A. N. Kurtz, *Science*, 1963, **141**, 1192).

**10920** Sodium selenate (IUPAC). (*C.A.*) disodium selenate (8 & 9CI); Reg. No. *[13410-01-0]*. $Na_2O_4Se$; .NA2.SE-04; insecticide.

**11000** Sulfallate (BSI, E-ISO, F-ISO); CDEC (WSSA). (IUPAC) 2-chloroallyl diethyldithiocarbamate (I). (*C.A.*) 2-chloro-2-propenyl diethylcarbamodithioate (9CI); (I) (8CI); Reg. No. *[95-06-7]*. 'CP 4742'; 'Vegadex' (Monsanto Co.). $C_8H_{14}ClNS_2$; 2N2&YUS&S1YGU1; *PM6*, p. 483. Herbicide.

**11040** Sulfoxide (ESA). (IUPAC) 2-(1,3-benzodioxol-5-yl)ethyl octyl sulphoxide; 1-methyl-2-(3,4-methylenedioxyphenyl)ethyl octyl sulphoxide. (*C.A.*) 5-[2-(octylsulfinyl)propyl]-1,3-benzodioxole (9CI); 1,2-methylenedioxy-4-[2-(octylsulfinyl)propyl]benzene (8CI); Reg. No. *[120-62-7]*. ENT 16 634. 'Sulfocide' (S. B. Penick & Co.). $C_{18}H_{28}O_3S$; T56 BO DO CHJ G1Y1&SO&8; *PM6*, p. 485. Insecticide synergist (M.E. Synerholm *et al., Contrib. Boyce Thompson Inst.*, 1947, **15**, 35).

**11050** Sulglycapin (BSI, E-ISO, F-ISO). (IUPAC) azepan-1-ylcarbonylmethyl methylsulphamate; perhydroazepin-1-ylcarbonylmethyl methylsulphamate. (*C.A.*) 2-(hexahydro-1*H*-azepin-1-yl)-2-oxoethyl methylsulfamate (9CI); Reg. No. *[51068-60-1]*. 'BAS 461H' (BASF AG). $C_9H_{18}N_2O_4S$; T7NTJ AV1OSWM1; herbicide (E. Eysell *et al., Proc. Br. Crop Prot. Conf. - Weeds*, 1976, **2**, 709).

**11070** Sulphuric acid (I) (E-ISO); acide sulfurique (F-ISO) accepted in lieu of a common name. (IUPAC) (I). (*C.A.*) sulfuric acid (8 & 9CI); Reg. No. *[7664-93-9]*. *Brown oil of vitriol*, BOV for commercial grade. $H_2O_4S$; ..H2.S-O4; *PM6*, p. 487. Herbicide (C. D. Woods & J. M. Bartlett, *Bull. Me. Agric. Exp. Stn.*, 1909, No. 167).

**11100** Sultropen (BSI, E-ISO); sultropène (F-ISO). (IUPAC) 2,4-dinitrophenyl pentyl sulphone. (*C.A.*) 2,4-dinitro-1-(pentylsulfonyl)benzene (9CI); 2,4-dinitrophenyl pentyl sulfone (8CI); Reg. No. *[963-22-4]*. 'RD 7901', 'BTS 7901' (Boots Co. Ltd now FBC Limited). $C_{11}H_{14}N_2O_6S$; WS5&R BNW DNW; fungicide.

**11110** Swep (ANSI, WSSA); MCC (JMAF). (IUPAC) methyl 3,4-dichlorocarbanilate (I). (*C.A.*) methyl 3,4-dichlorophenylcarbamate (9CI); (I) (8CI); Reg. No. *[1918-18-9]*. 'FMC 2995' (FMC Corp.). $C_8H_7Cl_2NO_2$; GR BG DMVO1; *PM5*, p. 486. Herbicide (H. R. Hudgins, *Proc. South. Weed Control Conf.*, 1963, **16**, 118).

**11140** Tazimcarb (BSI, E-ISO); tazimcarbe (F-ISO). (IUPAC) *N*-methyl-1-(3,5,5-trimethyl-4-oxo-1,3-thiazolidin-2-ylideneamino-oxy)formamide; 3,5,5-trimethyl-2-methylcarbamoyloxyimino-1,3-thiazolidin-4-one. (*C.A.*) 2-[[[(aminocarbonyl)oxy]methyl]imino]-3,5,5-trimethyl-4-thiazolidinone (9CI); formerly 3,5,5-trimethyl-2,4-thiazolidinedione 2-[*O*-[(methylamino)carbonyl]oxime] (9CI); Reg. No. *[62113-03-5]*. 'PP 505' (ICI Plant Protection Division). $C_8H_{13}N_3O_3S$; T5SYNV EHJ BUNOVM1 C1 E1 E1; insecticide and molluscicide.

**11180** TDE (BSI, E-ISO, F-ISO). (IUPAC) 1,1-dichloro-2,2-bis(4-chlorophenyl)ethane. (*C.A.*) 1,1'-(2,2-dichloroethylidene)bis[4-chlorobenzene] (9CI); 1,1-dichloro-2,2-bis(*p*-chlorophenyl)ethane (8CI); Reg. No. *[72-54-8]*. DDD.OMS 1078, ENT 4225. 'Rhothane' (Rohm & Haas Co.). $C_{14}H_{10}Cl_4$; GYGYR DG&R DG; *PM3*, p. 457. Insecticide (P. Läuger *et al., Helv. Chim. Acta*, 1944, **27**, 892). Though no longer manufactured commercially, is found in nature as a degradation product of DDT (3830).

**11280** Terbucarb (BSI, E-ISO); terbucarbe (F-ISO); MBPMC (JMAF). (IUPAC) 2,6-di-*tert*-butyl-*p*-tolyl methylcarbamate (I). (*C.A.*) 2,6-bis(1,1-dimethylethyl)-4-methylphenyl methylcarbamate (9CI); (I) (8CI); Reg. No. *[1918-11-2]*. 'Hercules 9573'; 'Azak' (Hercules Inc. now Boots-Hercules Agrochemical Co.). $C_{17}H_{27}NO_2$; 1X1&1&R C1 FOVM1 EX1&1&1; *PM4*, p. 476. Herbicide (A. H. Haubein & J. R. Hansen, *J. Agric. Food Chem.*, 1965, **13**, 555).

**11290** Terbuchlor (BSI, E-ISO, ANSI); terbuchlore (F-ISO). (IUPAC) *N*-butoxymethyl-6'-*tert*-butyl-2-chloroacet-*o*-toluidide. (*C.A.*) *N*-(butoxymethyl)-2-chloro-*N*-[2-(1,1-dimethylethyl)-6-methylphenyl]acetamide (9CI); *N*-(butoxymethyl)-6'-*tert*-butyl-2-chloroacet-*o*-toluidide (8CI); Reg. No. *[4212-93-5]*. 'MON 0358', 'CP 46 358' (Monsanto Co.). $C_{18}H_{28}ClNO_2$; G1VN1O4&R B1 FX1&1&1; herbicide.

**11385** 2,2',3,3'-Tetrachloro-4,4'-oxydibut-2-en-4-olide (IUPAC); 5,5'-oxybis(3,4-dichloro-5*H*-furan-2-one). (*C.A.*) 5,5'-oxybis[3,4-dichloro-2(5*H*)-furanone] (9CI); bis(3,4-dichloro-2,5-dihydro-5-oxo-2-furanyl) ether (8CI); Reg. No. *[4412-09-3]*. Mucochloric anhydride. 'GC 2466' (Allied Chemical Corp.). $C_8H_2Cl_4O_5$; T5OV EHJ CG DG EO- ET5OV EHJ CG DG; *PM2*, p. 334. Fungicide (A. E. Rich, *Plant Dis. Rep.*, 1960, **44**, 306).

**11400** Tetrachlorothiophene (I) (IUPAC). (*C.A.*) (I) (8 & 9CI); Reg. No. *[6012-97-1]*. TCTP. ENT 25 764. 'Penphene' (Pennwalt). $C_4Cl_4S$; T5SJ BG CG DG EG; *PM3*, p. 464. Nematicide.

**11560** Thiocarboxime (BSI, E-ISO, F-ISO). (IUPAC) 3-[1-(methylcarbamoyloxyimino)ethylthio=]propionitrile. (*C.A.*) 2-cyanoethyl *N*-[[(methylamino)carbonyl]oxy]ethanimidothioate (9CI); *N*-[(methylcarbamoyl)oxy]thioacetimidic acid ester with 3-mercaptopropionitrile (8CI); Reg. No. *[25171-63-5]*. 'WL 21 959'; 'Talcord' (Shell Research Ltd). $C_7H_{11}N_3O_2S$; NC2SYUNOVM1; experimental acaricide and insecticide.

**11590** 2-Thiocyanatoethyl laurate (I) (IUPAC). (*C.A.*) 2-thiocyanatoethyl dodecanoate (9CI); (I) (8CI); Reg. No. *[301-11-1]*. 'Lethane 60' (Rohm & Haas Co.). $C_{15}H_{27}NO_2S$; SCN2OV11; insecticide.

**11680** Thioquinox (BSI, E-ISO, F-ISO). (IUPAC) 1,3-dithiolo[4,5-*b*]quinoxaline-2-thione (I); quinoxaline-2,3-diyl trithiocarbonate; 2-thioxo-1,3-dithiolo[4,5-*b*]quinoxaline. (*C.A.*) (I) (9CI); cyclic quinoxaline-2,3-diyl trithiocarbonate (8CI); Reg. No. *[93-75-4]*. ENT 25 579. 'Bayer 30 686', 'Ss 1451'; 'Eradex' (Bayer AG). $C_9H_4N_2S_3$; T C566 BN DSYS HNJ EUS; *PM4*, p. 491. Acaricide and fungicide (G. Unterstenhöfer, *Hoefchen-Briefe, (Engl. Ed.)*, 1960, **13**, 207).

**11850** Triamiphos (BSI, E-ISO, F-ISO). (IUPAC) 5-amino-3-phenyl-1*H*-1,2,4-triazol-1-yl-*N,N,N',N'*-tetramethylphosphonic diamide; *P*-amino-3-phenyl-1*H*-1,2,4-triazol-1-yl-*N,N,N',N'*-tetramethylphosphonic diamide; 5-amino-1-(bisdimethylaminophosphinyl)-3-phenyl-1*H*-1,2,4-triazole. (*C.A.*) *P*-(5-amino-3-phenyl-1*H*-1,2,4-triazol-1-yl)-*N,N,N',N'*-tetramethylphosphonic diamide (8 & 9CI); Reg. No. *[1031-47-6]*. ENT 27 223. 'WP 155'; 'Wepsyn 155' (Philips-Duphar B.V.). $C_{12}H_{19}N_6OP$; T5NN DNJ APO&N1&1&N1&1 CR& EZ; *PM6*, p. 525. Acaricide, fungicide & insecticide (B. G. van den Bos *et al., Rec. Trav. Chim. Pays-Bas*, 1960, **79**, 807).

**11860** Triarimol (BSI, E-ISO, F-ISO, ANSI). (IUPAC) 2,4-dichloro-α-(pyrimidin-5-yl)benzhydryl alcohol. (*C.A.*) α-(2,4-dichlorophenyl)-α-phenyl-5-pyrimidinemethanol (8 & 9CI); Reg. No. *[26766-27-8]*. 'EL 273'; '*Trimidal*' (Eli Lilly & Co.). $C_{17}H_{12}Cl_2N_2O$; T6N CNJ EXQR&R BG DG; *PM3*, p. 481. Fungicide (J. V. Gramlich *et al., Proc. Br. Insectic. Fungic. Conf., 5th*, 1969, **2**, 576).

**11870** Triazbutil (BSI, E-ISO, F-ISO, ANSI). (IUPAC) 4-butyl-4*H*-1,2,4-triazole (I). (*C.A.*) (I) (8 & 9CI); Reg. No. *[16227-10-4]*. 'RH-124'; 'Indar' (Rohm & Haas Co.). $C_6H_{11}N_3$; T5NN DNJ D4; fungicide.

**11920** Tricamba (BSI, E-ISO, F-ISO, ANSI). (IUPAC) 3,5,6-trichloro-*o*-anisic acid (I). (*C.A.*) 2,3,5-trichloro-6-methoxybenzoic acid (9CI); (I) (8CI); Reg. No. *[2307-49-5]*. 'Banvel T' (Velsicol Corp.). $C_8H_5Cl_3O_3$; QVR BG CG EG FO1; herbicide.

**11930** 4,5,7-Trichloro-2,1,3-benzothiadiazole (I) (IUPAC). (*C.A.*) (I) (8 & 9CI); Reg. No. *[1982-55-4]*. 'PH 40-21' (Philips Duphar N.V.). $C_6HCl_3N_2S$; T56 BNSNJ FG GG IG; *PM3*, p. 483. Herbicide (J. Dams *et al., Proc. Br. Weed Control Conf., 7th*, 1964, p. 1091).

**11940** Trichlorobenzyl chloride (IUPAC). (*C.A.*) trichloro(chloromethyl)benzene (9CI); α,ar,ar,ar-tetrachlorotoluene (8CI); Reg. No. *[1344-32-7]*, previously *[25429-36-1]*. TCBC. 'Randon-T' (Monsanto Co.). $C_7H_4Cl_4$; G1R XG XG XG; *PM4*, p. 500. Herbicide.

**11950** 1-(2,3,6-Trichlorobenzyloxy)propan-2-ol (IUPAC). (*C.A.*) 1-[(2,3,6-trichlorophenyl)methoxy]-2-propanol (9CI); 1-(2,3,6-trichlorobenzyloxy)-2-propanol (8CI); Reg. No. *[1861-44-5]*. 'Tritac' (Hooker Chemical Corp.). $C_{10}H_{11}Cl_3O_2$; QY1&1O1R BG CG FG; herbicide.

**12070** Trifenmorph (BSI, E-ISO); triphenmorphe (F-ISO). (IUPAC) 4-tritylmorpholine (I); 4-(triphenylmethyl)morpholine (II). (*C.A.*) (I) (9CI); (II) (8CI); Reg. No. *[1420-06-0]*. 'WL 8008'; 'Frescon' (Shell Research Ltd). $C_{23}H_{23}NO$; T6N DOTJ AXR&R&R; *PM6*, p. 536. Molluscicide (C. B. C. Boyce *et al., Nature (London)*, 1966, **210**, 1140).

**12080** Trifenofos (BSI, E-ISO, F-ISO, ANSI). (IUPAC) *O*-ethyl *S*-propyl *O*-2,4,6-trichlorophenyl phosphorothioate. (*C.A.*) *O*-ethyl *S*-propyl *O*-(2,4,6-trichlorophenyl) phosphorothioate; Reg. No. *[38524-82-2]*. 'RH-218' (Rohm & Haas Co.). $C_{11}H_{14}Cl_3O_3PS$; GR CG EG BOPO&S3&O2; insecticide and acaricide.

**12190** Triprene (BSI, E-ISO, F-ISO, ANSI). (IUPAC) *S*-ethyl (±)-(*E,E*)-11-methoxy-3,7,11-trimethyldodeca-2,4-dienethioate. (*C.A.*) *S*-ethyl (*E,E*)-11-methoxy-3,7,11-trimethyl-2,4-dodecadienethioate (9CI); Reg. No. *[40596-80-3]*. 'ZR 519'; 'Altorick' (Zoecon Corp.). $C_{18}H_{32}O_2S$; 2SV1UY1&1U2Y1&3X1&1&O1 &&(E,E) FORM; insect growth regulator.

**12210** Tris[1-dodecyl-3-methyl-2-phenylbenzimidazolium] hexacyanoferrate (IUPAC). (*C.A.*) tris[1-dodecyl-3-methyl-2-phenyl-1*H*-benzimidazolium] hexakis(cyano-*C*)ferrate (9CI); tris[1-dodecyl-3-methyl-2-phenylbenzimidazolium] ferricyanide (8CI); 'Bayer 32 394', 'B 169 Ferricyanide'. 'Fungilon' (Bayer AG). $C_{84}H_{111}FeN_{12}$; T56 BK DNJ B12 CR& D1 &&Fe(CN)$_6$ SALT 3:1; *PM2*, p. 213. Fungicide.

**12330** Xylachlor (WSSA). (IUPAC) 2-chloro-*N*-isopropylacet-2',3'-xylidide. (*C.A.*) 2-chloro-*N*-(2,3-dimethylphenyl)-*N*-(1-methylethyl)acetamide (9CI); Reg. No. *[36114-77-2]*. 'AC 206 784'; 'Combat' (American Cyanamid Co.). $C_{13}H_{18}ClNO$; 1Y1&NV1GR B1 C1; herbicide.

**12430** Zolaprofos (BSI, draft E-ISO, draft F-ISO). (IUPAC) *O*-ethyl *S*-3-methylisoxazol-5-ylmethyl *S*-propyl phosphorodithioate. (*C.A.*) *O*-ethyl *S*-[(3-methyl-5-isoxazolyl)methyl] *S*-propyl phosphorodithioate (9CI); Reg. No. *[63771-69-7]*. 'BAS 268I' (BASF AG). $C_{10}H_{18}NO_3PS_2$; T5NOJ CSPO&S3&O2 E1; insecticide.

## ADDITIONAL INFORMATION FOR PART I AND PART II

The following data were received or noted after the pages on the respective entries had been set from the computer records.

**Folpet.** $C_9H_4Cl_3NO_2S$ (296.6); T56 BVNVJ CSXGGG; full details are given in *PM6* p. 281. Common name folpet (BSI, ANSI, Canada, JMAF); folpel (France). Chemical name (IUPAC) *N*-(trichloromethylthio)phthalimide (I); *N*-(trichloro= methanesulphenyl)phthalimide. (*C.A.*) 2-[(trichloromethyl)thio]-1*H*-isoindole-1,3(2*H*)-dione (9CI); (I) (8CI); Reg. No. *[133-07-3]*. Its fungicidal properties were described by A. R. Kittleson (*Science*, 1952, **115**, 84). Introduced by the Standard Oil Development Co. and later by Chevron Chemical Co. (USP 2 553 770; 2 553 771; 2 553 776) as trade mark 'Phaltan'.
It is a protectant fungicide used alone ('Phaltan') or in mixtures with various systemic fungicides including: benalaxyl ('Galben F'); cymoxanil ('Fytospore'); cyprofuram; metalaxyl.

**Fomesafen.** $C_{15}H_{10}ClF_3N_2O_6F$ (438.5); WS1&MVR BNW EOR BG DXFFF; further details will be available in the on-line version of the database. Common name fomesafen (BSI, ANSI, draft E-ISO). Chemical name (IUPAC) 5-(2-chloro-α,α,α-trifluoro-*p*-tolyloxy)-*N*-methylsulphonyl-2-nitrobenzamide. (*C.A.*) 5-[2-chloro-4-(trifluoromethyl)pehnoxy]-*N*-(methylsulfonyl)-2-nitrobenzamide (9CI); Reg. No. *[72178-02-0]*. Developed as a herbicide by ICI Plant Protection Division (EP 3416) as code no.'PP 021'; trade mark 'Flex'.
It is used as a post-em. spray to control broad-leaved weeds in soyabeans. Formulated as an aqueous solution of fomesafen-sodium.

**Paclobutrazol.** $C_{15}H_{20}ClN_3O$ (293.5); T5NN DNJ AY1R DG&YQX1&1&1 &&(2RS,3RS FORM; further details will be available in the on-line version of the database. Chemical name (IUPAC) (2*RS*,3*RS*)-1-(4-chlorophenyl)-4,4-dimethyl-2-(1*H*-1,2,4-triazol-1-yl)pentan-3-ol. (*C.A.*) β-[(4-chlorophenyl)methyl]-α-(1,1-dimethylethyl)-1*H*-1,2,4-triazole-1-ethanol (9CI); Reg. No. *[76738-62-0]*.
Synthesised and developed as a plant growth regulator by ICI Plant Protection Division (GBP 1 595 696, 1 595 697) as code no.'PP 333'.
It is a plant growth regulator for use on a wide range of crops. Formulations include w.p. and s.c.

300 **Aminotriazole.** BSI has adopted the common name amitrole.

1090 **Clofentezine.** Also draft F-ISO name.

6200 **Fenpropathrin.** Acute oral $LD_{50}$ for male rats 70.6 mg (in corn oil)/kg, for females 66.7 mg/kg.

6380 **Flamprop ((*R*)-ISOMER).** Agreement obtained within ISO means that this acid will have the common name flamprop-M, and its isopropyl ester flamprop-M-isopropyl.

*6490 **Flumezin.** The preferred IUPAC name should read 2-methyl-4-(α,α,α,-trifluoro-*m*-tolyl)-1,2,4-oxadiazinane-3,5-dione. The Wiswesser Line-Formula Notation should read T6ONVNV FHJ B1 DR CXFFF.

6910 **Gibberellic acid.** The preferred IUPAC name should read (3*S*,3a*R*,4*S*,4a*S*,7*S*,= 9a*R*,9b*R*,12*S*)-7,12-dihydroxy-3-methyl-6-methylene-2-oxoperhydro-4a,7-methano-9b,3-propenoazuleno[1,2-*b*]furan-4-carboxylic acid.

7690 **Linuron.** Formulations (mixtures) include: 'Neminfest' (120 g linuron + 240 g trifluralin/l) (Montedison S.p.A.).

*continued overleaf*

7760 **Mancozeb.** Formulations of mancozeb itself should include: 'Manzate 200' (800 g a.i./hg), (E. I. du Pont de Nemours & Co., Inc.) ($\geqslant$ 800 g.a.i./kg); 'Nemispor' (Montedison S.p.A.).

12120 **Trifluralin.** Formulations include: 'Digermin' (480 g/l) (Montedison S.p.A.); mixtures include: 'Neminfest' (240 g trifluralin + 120 g linuron/l) (Montedison S.p.A.).

# APPENDICES

# APPENDIX I

## Scientific and common names of organisms quoted in the Toxicology Section of the main entries (PART I)

*Agelaius phoeniceus* ... Blackbird, red-winged
*Alectoris chukar* .......... Partridge, chukar
*Anas platyrhynchos* .......... Duck, mallard
*Anas platyrhynchos domesticus* .. Duck, Pekin
*Apis mellifera* .................. Bee, honey
*Australorbis glabratus*....................

*Bacillus subtilis*..........................
Barbel .................... *Baubus barbus*
Barbel, southern ........ *Barbus meridionalis*
*Barbus barbus*..................... Barbel
*Barbus meridionalis* ........ Barbel, southern
Bass, largemouth ..... *Micropterus salmoides*
Bass, largemouth, black......... *Micropterus* .............................. *salmoides*
Bee, honey ................. *Apis mellifera*
Blackbird.................. *Turdus merula*
Blackbird, red-winged ... *Agelaius phoeniceus*
Bluegill (Sunfish, bluegill) ................. ................... *Lepomis marcrochirus*
*Bos taurus*..................... Cattle; cow
Bullhead, black............. *Ictaluras melas*

Canary .................... *Serinus canaria*
*Canis familiaris*...................... Dog
*Carassius auratus* ................ Goldfish
Carp..................... *Cyprinus carpio*
Carp, crucian ........... *Cyprinus carassius*
Carp, Japanese..........................
Carp, mirror .............. *Cyprinus carpio*
Cat..................... *Felis domesticus*
Catfish ................................ .......... Siluroids, covering several families
Catfish, channel ........ *Ictalurus punctatus*
Cattle........................ *Bos taurus*
*Cavia cobaya*.................. Guinea-pig
Chicken...................... *Gallus gallus*
*Colinus virginianus* ......... Quail, bobwhite
*Columba* spp...................... Pigeon
*Coturnix coturnix*........... Quail, common
*Coturnix coturnix japonica*... Quail, Japanese
Cow........................ *Bos taurus*
Crab, fiddler...........................
*Crassotrea virginica* ................ Oyster
Crayfish.............. *Procambarus acutus*
*Cynomys ludovicianus* .................... .................... Prairie dog (a rodent)
*Cyprinodon variegatus* ................... ..................... Minnow, sheepshead
*Cyprinus carassius* ........... Carp, crucian
*Cyprinus carpio* ............. Carp (mirror)

*Daphnia* spp.................. Flea, water
Dog..................... *Canis familiaris*
Dog, beagle ................ *Canis familiaris*
Dove, mourning ........ *Zenaidura macroura*
Dove, mallard .......... *Anas platyrhynchos*
Dove, Pekin ............ *Anas platyrhynchos* .............................. *domesticus*

*Felis domesticus*..................... Cat
Finfish ............... *Fundulus heteroclitus*
Flea, water.................... *Daphnia* spp.
Fowl, domestic................ *Gallus gallus*
*Fundulus heteroclitus* ............... Finfish

*Gallus gallus* ............................ ........... Chicken; hen; hen, Rhode Island
*Gambusia affinis* ............. Mosquito fish
*Gammarus* spp..................... Shrimp
Gerbil.................................
Goldfish ................. *Carassius auratus*
Guinea-pig................... *Cavia cobaya*
Guppy................ *Libistes reticulatus* / *Poecilia reticulata*

Hamster, golden ....... *Mesocricetus auratus*
Harlequin fish ....... *Rasbora heteromorpha*
Hen ........................ *Gallus gallus*
Hen, Rhode Island............ *Gallus gallus*
Honeybee ................... *Apis mellifera*

*Ictaluras melas*.............. Bullhead, black
*Ictaluras punctatus* ......... Catfish, channel
Ide..................... *Idus melanotus* / *Leuciscus idus*
*Idus melanotus* ............ Ide; Golden orfe

Killifish, Japanese........... *Oryzias laticeps*

*Lepomis cyanellus*............ Sunfish, green
*Lepomis macrochirus*....... Bluegill (sunfish)
*Lepomis microlophus* ........ Sunfish, redear
*Libistes reticulatus*................. Guppy

*Macaca mulatta* ............ Monkey, rhesus
*Meleagris galloparvo*............... Turkey
*Mesocricetus auratus* ....... Hamster, golden
*Micropterus salmoides* ..... Bass, largemouth
Minnow ................ *Phoxinus phoxinus*
Minnow, fathead........ *Pimephals promelas*
Minnow, sheepshead .. *Cyprinodon variegatus*
Monkey, rhesus ............ *Macaca mulatta*
Mosquito fish ............. *Gambusia affinis*
Mouse, albino ............... *Mus musculus*
Mouse, house................ *Mus musculus*
*Mugil cephalus* ............... Mullet, black

*Mugil curema*................. Mullet, silver
Mullet, black ............... *Mugil cephalus*
Mullet, silver................. *Mugil curema*
*Mus musculus* .. Mouse, albino; Mouse, house

*Oncorhynchus kisutch*......... Salmon, coho
*Oncorhynchus tschawytscha*. Salmon, chinook
Orge, golden ............... *Idus melanotus*
*Orizias laticeps* .........................
........ Ricefish, Japanese; Killifish, Japanese
*Oryctolagus cuniculus* ..... Rabbit, European
*Ovis aries* .......................... Sheep
Oyster ................. *Crassotrea virginica*

Partridge, chukar .......... *Alectoris chukar*
Partridge, European, grey...... *Perdix perdix*
*Passer domesticus* .......... Sparrow (house)
*Perca flavescens* .............. Perch, yellow
Perch, yellow .............. *Perca flavescens*
*Perdix perdix*...... Partridge, European, grey
*Phasianus colchicus* .. Pheasant, ring-neck; . Pheasant (ring necked)
Pheasant .................. *Phasianus* spp.
Pheasant, ring-neck; Pheasant, ring-necked . *Phasianus colchicus*
*Phoxinus phoxinus* ................ Minnow
Pig ....................... *Sus domesticus*
Pigeon...................... *Columba* spp.
*Pimephals promelas*........ Minnow, fathead
*Poecilia reticulata*.................. Guppy
Prairie dog (a rodent) .. *Cynomys ludovicianus*
*Procambarus acutus*.............. Crayfish

Quail, bobwhite ......... *Colinus virginianus*
Quail, common (European) ...............
....................... *Coturnix coturnix*
Quail Japanese ...........................
................ *Coturnix coturnix japonica*

Rabbit ............... *Oryctolagus cuniculus*
*Rasbora heteromorpha* ....... Harlequin fish
Rat, brown............... *Rattus norvegicus*
Rat (black) ................... *Rattus rattus*
Rat, Wistair ............................
*Rattus hawaiiensis* .......................
*Rattus norvegicus* .. Rat, brown; Rat, Norway
*Rattus rattus* .................. Rat (black)
Rhesus monkey............. *Macaca mulatta*
Ricefish, Japanese ........... *Oryzias latipes*

*Salmo* spp. ......................... Trout
*Salmo gairdneri*............. Trout, rainbow
*Salmo trutta* ................. Trout, brown
Salmon, chinook .........................
................ *Oncorhynchus ischawytscha*
Salmon, coho; Salmon, silver ...... *Oncorhynchus kisutch*
*Salmonella typhimurium*
*Salvelinus fontinalis* ........... Trout, brook
*Serinus canaria* ................... Canary
Sheep ......................... *Ovis aries*
Shrimp .................... *Gammarus* spp.
Sparrow (house) .......... *Passer domesticus*
Starling, common........... *Sturnus vulgaris*
*Sturnus vulgaris*........... Starling, common
Sunfish .................................
....... A general term covering several classes
Sunfish, bluegill ....... *Lepomis macrochirus*
Sunfish, green............ *Lepomis cyanellus*
Sunfish, pumpkinseed....................
Sunfish, redear ........ *Lepomis microlophus*
*Sus domesticus* ...................... Pig

Tilapia....................... *Tilapia* spp.
*Tilapia* spp........................ Tilapia
Trout ....................... *Salmo* spp.
Trout, brook ........... *Salvelinus fontinalis*
Trout, brown ................ *Salmo trutta*
Trout, rainbow ............. *Salmo gairdneri*
*Turdus merula*.................. Blackbird
Turkey................ *Meleagris galloparvo*

*Zenaidura macroura* ........ Dove, mourning

## APPENDIX II

### BOOKS MENTIONED IN THE TEXT OR USED IN PREPARATION OF THE MANUAL

*Note:* in several cases other volumes in a given series deal with compounds other than those described in detail in this edition of the *Manual* or with general methods or principles and are not listed below.


*Advances in Chemistry Series*, American Chemical Society.

*Advances in Pest Control Research,* ed. R. L. Metcalf. New York: Interscience. Volume **8,** 1968.

*Analytical Methods for Pesticides, Plant Growth Regulators and Food Additives.* Series editor, G. Zweig, New York: Academic Press. Volumes: **2,** 1964; **3,** 1964; **4,** 1964. (see also next entry).

*Analytical Methods for Pesticides and Plant Growth Regulators.* (*cf.* previous entry) Series editor, G. Zweig, New York; Academic Press. Volumes: **5,** ed. G. Zweig & J. Sherma, 1967; **6,** ed. G. Zweig & J. Sherma, 1972; **7,** ed. J. Sherma & G. Zweig, 1973; **8,** ed. G. Zweig & J. Sherma, 1976; **10,** ed. G. Zweig & J. Sherma, 1978; **11,** ed. G. Zweig & J. Sherma, 1980.

*Approved Names* (and supplements). British Pharmocopoeia Commission. London: HMSO (1981).

*Chemistry and Action of Herbicide Antidotes*, ed. F. C. Pallos & J. E. Casida. New York: Academic Press (1978).

*Chemistry of Acrylonitrile,* New York: American Cyanamid Co. (1951).

*CIPAC Handbook*, Volume **1**, ed. G. R. Raw. Harpenden: Collaborative International Pesticides Council Ltd (1970). Volume **1A**, Cambridge: Heffer (1980); Volume **1B** is due to be published in 1983 (Cambridge: Heffer). Material has been assembled for later volumes and is quoted in the text as 'in press'.

*Common Names for Pesticide and Other Agrochemicals,* ISO 1750 (1981); Addendum 1 (1983). Geneva: International Organization for Standardization. (Other addenda are in press).

*Degradation of Herbicides*, ed. P. C. Kearney & D. D. Kaufmann. New York: Dekker (1969).

*Deltamethrin Monograph*, ed. J. Lhoste (English translation). Avignon: Roussel Uclaf (1982).

*Die Entwicklung neuer insektizider Phosphorsäure-Ester*, (3rd Ed.), G. Schrader. Weinheim: Verlag Chemie GmbH (1963).

*Guidelines on Good Analytical Practice in Residue Analysis and Recommendations for Methods of Analysis for Pesticide Residues*, GIFAP *Technical Monograph*, No. 8 (1983).

*Herbicides: Chemistry, Degradation and Mode of Action*, (2nd revised Ed.), ed. P. C. Kearney & D. D. Kaufmann. New York: Dekker. Volume **1,** 1975; Volume **2,** 1976.

*The Hormone Weedkillers*, C. Kirby. Croydon: BCPC Publications (1980).

*Insecticide and Fungicide Handbook for Crop Protection*, (5th Ed.), ed. H. Martin & C. R. Worthing. Oxford: Blackwell Scientific Publications (1976).

*Lindane,* ed. E. Ulmann. Freiburg: Verlag Karl Schillinger (1972). Supplement I (1974), Supplement II (1976).

*Official Methods of Analysis of the Association of Official Analytical Chemists*, (13th Ed.), ed. W. Horwitz. Washington: Association of Official Analytical Chemists (1980). Also amendments to methods, as described in *Journal of the Association of Official Analytical Chemists.*

*Organic Insecticides: their Chemistry and Mode of Action*, R. L. Metcalf. New York: Interscience (1955).

[*Pathology of the Silkworm*], K. Ishikawa. Tokyo: Meinbundo (1902). (In Japanese).

*Pentachlorophenol*, ed. K. R. Rao. Proceedings of a Symposium, Pensacola, Florida, 1977. (*Environmental Science Research*, Volume **12**). Plenum Press (1978).

*Pest and Disease Control Handbook*, ed. N. Scopes and M. Ledieu. Croydon: BCPC Publications (1979). (New edition in press).

*Pesticide Analytical Manual: Methods for Individual Pesticide Residues.* Vols I & II. Washington: Food & Drug Administration, USA (1979).

*Pesticide Manufacturing and Toxic Materials Control Encyclopedia*, ed. Marshall Sittig. Park Ridge, New Jersey: Noyes Data Corporation (1980).

*Pesticides Considered Not to Require Common Names.* ISO 765 (1976). Geneva: International Organization for Standardization. (New edition in preparation).

*Pesticides: Synonyms and Chemical Names* (1st Ed.). Tokyo: The Society of Agricultural Chemical Industry (Japan) (1981).

*Recommended Common Names for Pesticides.* BS 1831: 1969. Amendment Slip No. 1 (1969); Supplement No. 1 (1970); Supplement No. 2 (1970); Supplement No. 3 (1974). Amendment Slips No. 1 (1975), No. 2 (1983). London: British Standards Institution.

*Specifications for Pesticides*
*Insecticides.* (1st Rev.), Geneva: WHO.
*Organochlorine Insecticides 1973.* Rome: FAO, (1973); MAFF (1973).
*Pesticides Used in Public Health*, Geneva: WHO (1973).
*Herbicides 1977*, Rome: FAO (1977); MAFF (1977).

*The Wiswesser Line-Formula Chemical Notation,* (3rd Ed.), E. G. Smith & P. A. Baker. Cherry Hill, New Jersey: Chemical Information Management Inc. (1976).

# APPENDIX III

## Addresses of firms mentioned in text

List of the main firms mentioned in the text as discoverers, producers or suppliers of the compounds concerned. In general, the address of the firm's headquarters is given; the regional offices may be able to help with additional information but there is no guarantee of this. In several cases firms have changed ownership after the main text of the Manual went to press.

**Aagrunol B.V.,** Groningen, Netherlands.

**Abbott Laboratories,** Agriculture and Veterinary Division, North Chicago, Illinois 60064, USA.

**ACF Chemiefarma NV,** P.O. Box 5, Straatweg 2, Maarssen, Netherlands.

**Allied Chemical Corporation** (For former Agricultural Division, *see* Hopkins Agricultrual Chemical Company).

**Amchem Products Incorporated** (Now part of Union Carbide Agricultural Products Company, Inc., *which see*).

**American Cyanamid Company,** Agricultural Division, P.O. Box 400, Princeton, New Jersey 08540, USA.

**Applied Horticulture Limited,** Horticultural Chemicals, Marringdean Road, Billingshurst, West Sussex, England.

**Atlas Agrochemicals Limited,** Fraser Road, Erith, Kent DA8 1PN, England. (Incorporates former Cropsafe Limited).

**BASF AG,** Landwirtschaftliche Versuchsstation, Postfach 220, 6703 Limburgerhof, Birkenweg 2, Federal Republic of Germany.

**BASF (UK) Limited,** Agrochemical Division, Lady Lane, Hadleigh, Ipswich, Suffolk IP7 6BQ, England.

**Bayer AG,** PF-AT Beratung, 5090 Leverkusen Bayerwerk, Federal Republic of Germany.

**The Boots Company PLC,** 1 Thane Road West, Nottingham NG2 3AA, England.

**Boots-Hercules Agrochemical Co.** (*see* FBC Limited; formerly Agricultural Divisions of Boots Limited and Hercules Incorporated).

**Borax** (*see* United States Borax and Chemical) Corporation).

**BP Oil Limited,** CMCL Division, BP House, Victoria Street, London SW1E 5NJ, England.

**J. D. Campbell & Sons Limited,** 18 Liverpool Road, Great Sankey, Warrington, Lancs WA5 1QR, England.

**Celamerck GmbH & Co. KG,** 6507 Ingelheim am Rhein, Federal Republic of Germany.

**Cequisia,** Muntaner, 322, Barcelona-21, Spain.

**Chemie Linz AG,** Postfach 296, A-4021 Linz, Austria.

**Chevron Chemical Company,** 940 Hensley Street, Richmond, California 94804, USA.

**Ciba AG** (now amalgamated as Ciba-Geigy AG).

**Ciba-Geigy AG,** CH-4002, Basle, Switzerland.

**Cropsafe Limited,** (*see* Atlas Agrochemicals).

**Crown Chemicals,** Agricultural Division (*see* Hopkins Agricultural Chemical Company).

**Crystal Chemical Co.,** 1525N Post Oak Road, Houston, Texas 77055, USA.

**Diamond Alkali Company** (*see* Diamond Shamrock Corporation).

**Diamond Shamrock Corporation,** Agricultural Chemicals Division, 1100 Superior Avenue, Cleveland, Ohio 44114, USA.

**Dow Chemical Company,** Ag-Organics Department, Midland, Michigan 48640, USA.

**Duphar B.V.,** Crop Protection Division, Goolilust Zuidereinde 49, 1243 kl 'S-graveland, The Netherlands.

**E.I. Du Pont de Nemours and Company Incorporated,** Biochemicals Department, 14348 Brandywine Building, Wilmington, Delaware 19898, USA.

**Du Pont Seppic Phyto,** 19 rue de Passy, 75016 Paris, France.

**EGYT Pharmaceutica Works,** Budapest, POB 100, Hungary.

**Elanco Products** (*see* (Eli) Lilly and Company).

**Eli Lilly and Company** (*see* (Eli) Lilly and Company).

**Farmoplant S.p.A., Centro Ricerche Antiparassitari,** Via Bonfadini, 148, 20138 Milan, Italy.

**FBC Limited,** Hauxton, Cambridge CB2 5HU, England. (Since the preparation of the Manual, it has been announced that Schering AG is to acquire FBC Limited).

**Fisons PLC** (Horticultural Division), Paper Mill Lane, Bramford, Ipswich, Suffolk IP8 4BZ, England.

**FMC Corporation,** Agricultural Chemical Division, 100 Niagara Street, Middleport, New York 14105, USA.

**J. R. Giegy S.A.** (now amalgamated as Ciba-Geigy AG).

**Gist-Brocades N.V.,** P.O. Box 1, Delft, Netherlands.

**Hercules Incorporated** (the Agricultural Division was combined with Boots Agrochemicals to form Boots-Hercules Agrochemical Co. Then this became part of FBC Limited (which see) — in USA, BFC Limited. Its name may change because Schering have since bought FBC Limited).

**Hoechst AG,** Verkauf Landwirtschaft, Beratung, Postfach 80 03 20, Frankfurt am Main 80, Federal Republic of Germany.

**Hokko Chemical Industry Company Limited,** Mitsui Bldg. No. 2, 24, Nihonbashi Hongokucho, Chuo-ku, Tokyo, Japan.

**Hooker Chemical Corporation,** Industrial Chemical Division, P.O. Box 344, Niagara Falls, New York 14302, USA.

**Hopkins Agricultural Chemical Company,** P.O. Box 7532, Madison, Wisconsin 53707, USA. (Formerly Agricultural Division of Allied Chemical Corporation, bought by Crown Chemicals).

**ICI** (*see* Imperial Chemical Industries PLC).

**I. G. Farbenindustrie** — disbanded after 1945, forming what became BASF AG, Bayer AG and Hoechst AG.

**Imperial Chemical Industries PLC,** Pharmaceuticals Division, Alderley House, Alderly Park, Macclesfield, Cheshire SK10 4TF, England.

**Imperial Chemical Industries PLC,** Plant Protection Division, Fernhurst, Haslemere, Surrey GU27 3JE, England.

**Janssen Pharmaceutica N.V.,** Turnhoutseweg 30, B-2340 Beerse, Belgium.

**Kenogard A.B.,** Box 11033, S-100 61 Stockholm, Sweden.

**Kumiai Chemical Industry Company Limited,** 4-26, Ikenohata 1-Chome, Taitoh-ku, Tokyo 110, Japan.

**Kureha Chemical Industry Company Limited,** 9-11, 1-Chome, Nihonbashi, Huridome-Cho, Chuo-Ku, Tokyo 103, Japan.

**Eli Lilly and Company,** Lilly Research Laboratories, P.O. Box 708, Greenfield, Indianapolis, Indiana 46206, USA.

**Lipha S.A.,** B.P. No. 106, 69212 Lyon, Cedex 1, France.

**3M Company,** Agrochemical Project, 3M Centre, St. Paul, Minnesota 55101, USA.

**Dr R. Maag Limited,** Chemical Works, CH-8157 Dielsdorf, Switzerland.

**Mallinckrodt Incorporated,** P.O. Box 5439, St. Louis, Missouri 63147, USA.

**A. H. Marks and Company Limited,** Wyke Lane, Wyke, Bradford, West Yorkshire BD12 9EJ, England.

**May & Baker Limited,** Ongar Research Station, Fyfield Road, Ongar, Essex CM5 0HW, England.

**McLaughlin, Gormley King and Company,** 8810 Tenth Avenue N, Minneapolis, Minnesota 55427, USA.

**Merck Animal Health Division,** Merck and Company Incorporated, Lincoln Avenue, Rahway, New Jersey 07065, USA.

**MGK** (*see* McLaughlin, Gormley King and Company).

**Mitsubishi Chemical Industries Limited,** Mitsubishi Bldg., 5-2, Marunouchi 2-Chome, Chiyoda-ku, Tokyo 100, Japan.

**Monsanto Commercial Products Company,** Agricultural Division, 800 N. Lindbergh Boulevard, St. Louis, Missouri 63166, USA.

**Montecatini S.p.A.** (became Montedison S.p.A., for Agrochemical Division *see* Farmoplant S.p.A).

**Montedison S.p.A.,** (for Agrochemical Division *see* Farmoplant S.p.A.)

**Murphy Chemical Limited,** Latchmore Court, Brand Street, Hitchin, Hertfordshire SG5 1HZ. (now owned by Dow Chemical Company).

**Nihon Nohyaku Company Limited,** Eitaro Bldg., 1-2-5, Nihonbashi, Chuo-ku, Tokyo Japan.

**Nippon Kayaku Company Limited,** Kaijo Bldg., 1-2-1, Marunouchi, Chiyoda-ku, Tokyo, Japan.

**Nippon Soda Company Limited,** New-Ohtemachi Bldg., Chiyoda-ku, Tokyo, Japan.

**Nissan Chemical Industries Limited,** Kowa-Hitosubashi Bldg., 7-1, 3-Chrome, Kanda-Nishiki-Cho, Chiyoda-ku, Tokyo 101, Japan.

**Nitrokémia Ipartelepek,** 8184 Fúzfógyártelep, Hungary.

**Olin Corporation,** Agricultural Division, P.O. Box 991, Little Rock, Arkansas 72203, USA.

**Pan Britannica Industries Limited,** Britannica House, Waltham Cross, Hertfordshire EN8 7DY, England.

**Penick Corporation,** 1050 Wall Street West, Lyndhurst, New Jersey 07071, USA.

**Pennwalt Corporation,** Agchem. Div., Three Parkway, Philadelphia, Pennsylvania 19102, USA.

**Perifleur Products Limited,** Hangleton Lane, Ferring, Worthing, West Sussex BN12 6PP, England.

**Philips-Duphar B.V.** (*see* Duphar B.V.).

**Plant Protection Division**<br>**Plant Protection Limited** } (*see* Imperial Chemical Industries PLC, Plant Protection Division).

**Rentokil Limited,** Felcourt, East Grinstead, Sussex RH19 2JY, England.

**Rhône-Poulenc-Phytosanitaire,** BP9163, Lyon 09-69263, Lyon, Cedex 1, France.

**Rohm and Haas Company,** Independence Mall West, Philadelphia, Pennsylvania 19105, USA.

**Roussel Uclaf,** Division Agro-Vétérinaire, 163 Avenue Gambetta, 75020 Paris, France.

**Sandoz AG,** Agrochemicals Division, Research, CH-4002 Basle, Switzerland.

**Sankyo Company Limited,** No. 7-12 Ginza 2-chome, Chuo-ku, Tokyo 104, Japan.

**Schering AG,** Agrochemical Division, Postfach 650311, D 1000 Berlin 65, Federal Republic of Germany.

**Seppic,** (*see* Du Pont Seppic Phyto).

**Shell International Chemical Company Limited** (Ref. CAMK/343), Shell Centre, London SE1 7PG, England.

**Sorex (London) Limited,** Agricultural Research Department, Halebank Factory, Lower Road, Widnes, Cheshire WA8 8NS, England.

**G. F. Spiess and Sohn GmbH & Company,** Chemische Fabrik, 6719 Kleinkarlbach Über Grünstadt/Pfalz, Federal Republic of Germany.

**Stanhope Chemical Products Limited,** 37 Broadwater Road, Welwyn Garden City, Herts AL7 1JW, England

**Stauffer Chemical Company,** Westport, Connecticut 06880, USA.

**Sumitomo Chemical Company Limited,** 15, 5-chome, Kitahama, Higashi-ku, Osaka, Japan.

**Synchemicals Limited,** Agrochemicals and Garden Products, 44 Grange Walk, London SE1 3EN, England.

**Takeda Chemical Industries Limited,** 12-10 Nihonbashi 2-chome, Chuo-ku, Tokyo 103, Japan.

**Tate & Lyle,** P.O. Box 68, Reading RG6 2BX, England.

**Union Carbide Agricultural Products Company Inc.,** P.O. Box 12014, T. W. Alexander Drive, Research Triangle Park, NC 27709, USA.

**Uniroyal Chemical Division,** Agricultural Chemicals, Research and Development, Amity Road, Connecticut 06525, USA.

**The Upjohn Company,** Agricultural Division, Plant Health Research and Development, 301 Henrietta Street, Kalamazoo, Michigan 49001, USA.

**United States Borax and Chemical Corporation,** U.S. Borax Research Corporation, 412 Crescent Way, Anaheim, California 92801, USA.

**Velsicol Chemical Corporation,** Commercial Development Department, 341 East Ohio Street, Chicago, Illinois 60611, USA.

**Wacker-Chemie GmbH,** Prinzregentenstrasse 22, D-800 München 22, Federal Republic of Germany.

**Wellcome Research Laboratories (Berkhamsted),** Berkhamsted Hill, Berkhamsted, Hertfordshire HP4 2QE, England.

**Zoecon Corporation,** 975 California Avenue, Palo Alto, California 93404, USA. (now owned by Sandoz AG).

# INDEXES

# INDEX 1

## Wisswesser Line-Formula Notation

These entries follow the hierarchial order:

'nothing'

.

&

-

other punctuation including the /

letters in alphabetical order

digits Ø 1 2 3 4 5 6 7 8 9

Note: Because the comparison is made in each sign (or space) in turn the correct numeric order is 1, 1Ø, 11 etc preceding 2 etc.

When possible WLNs are broken in this index at spaces between fragments. When this is not possible an equals sign (=) is placed at the end of the line, signifying that the following line runs immediately after it. A hyphen (-) has its usual significance in WLN (see p. xiv).

References are to **Entry Numbers**, for example 1450, in PART I. References preceded by an asterisk, for example *1310, are in PART II (Superseded Compounds) printed on yellow paper.

# INDEX 2

## Molecular Formulae

These entries include those in both PART I and PART II. Individual salts and esters mentioned only under the sections on formulations are not included. Ions (*e.g.* diquat) and salts (*e.g.* diquat dibromide, sodium fluoroacetate) are listed as such so as to agree with the formulae in the text. Compounds (*e.g.* zineb) believed to be polymeric are indexed as monomers.

References are to **Entry Numbers**, for example 1450, in PART I. References preceded by an asterisk, for example *1310, are in PART II (Superseded Compounds) printed on yellow paper.

# INDEX 3

## Code Numbers

This index includes codes used to identify pesticides. All the codes are made up of an identifier and a serial number. It contains codes used by official bodies (qualifier AI3-AN4-, ENT, OMS etc.) and those used by discoverers, manufacturers and suppliers (indicated by single quotation marks'' ). Some less well known common names based on letters and numbers (*e.g.* M-81) are also included.

The index is arranged in order of the *numerical* component of the code. When more than one entry has the same numeral, separation is by the alphabetical qualifier. Numbers of the type 12AB34 or 12/34 or 12-34 are indexed in all possible ways, *e.g.* under 12, 34 and 1234.

References are to the **Entry Numbers**, for example 1450, in PART I. References preceded by an asterisk, for example *1310, are in PART II (Superseded Compounds) printed on yellow paper.

# INDEX 4

## Chemical, Common and Trivial Names, and Trade Marks

Numerical and alphabetical prefixes and infixes signifying positions in chemical names or stereochemical descriptors have been omitted. Brackets [ ] and parentheses ( ) have been introduced for convenience in distinguishing between, *e.g.* chloro(methyl) (two groups) and (chloromethyl) (one group). An equals sign ( = ) at the end of a line means the chemical name runs straight on into the next line without a break ( = being omitted). Lines ending in a hyphen, opening parentheses (, or brackets [ run straight on, too. Otherwise there is a break (two words), for example—

(dichloro =
methyl) 2,4-D-
butyl chloro(
methyl)naphthalene are one word;

(ethylthio)
diethyl is two words.

Trade marks have, as far as possible, been placed within single quotation marks ' ' and their use, even without this indication, is not to be considered to imply that they are not protected by law. No attempt has been made to distinguish between registered and unregistered trade marks. Qualifying numbers or letters are included only when these indicate additional active ingredient(s) in the product — not the concentration of a.i.

All references are to **Entry Numbers**; unqualified numbers (1230) are in PART I those preceded by an asterisk (*2400) are in PART II (Superseded Compounds) printed on yellow paper.

Index terms in roman print refer to entries in which the details of the parent compound are given. Terms in *italic print* mainly refer to entries where the compound or product is mentioned as a mixture. *Italic print* is also used for index terms that are scientific names or organisms and these are indicated by a single dagger, *e.g.* †*Bacillus thuringiensis. Italic print* is also used for common names and for trade marks the original meaning of which has been discontinued; they are indicated by a double dagger, *e.g.* ‡*Arprocarb,* ‡ *'Ambush'*.

Data in Additional Information are not included in the index.

This computer-generated index has provided a different ordering of entries from that in previous editions of the *Pesticide Manual.* Index terms comprising more than one word have been sorted into alphabetical order on each one in turn, the space between words being given priority. Thus all entries (there are about fifty of them) that begin with **dimethyl** as a separate word come before any beginning with **dimethylamino** (however long the word). It is important to understand the ordering rules when searching for frequently occurring terms such as diethyl, dimethyl, ethyl and methyl. Salts and esters of acids referred to by their common name can occur much later in the order than does the parent acid. A typical order is: 2,4-D, dalapon, dimethyl tetrachloroterephthalate, (dimethylamino)succinamic acid, 2,4-D-isooctyl, dodemorph, 2,4-D-sodium. All accents are ignored when deciding order.

Readers should bear in mind national variations in spelling, especially 'f', and 'ph', 't' and 'th', and the presence or absence of a final 'e' in syllables such as 'ol(e)', 'on(e)' and 'yn(e)'.

# The British Crop Protection Council

| | |
|---|---|
| President | Sir James Scott-Hopkins, MEP |
| Chairman | Mr. E. Lester |
| Vice Chairmen | Mr. B. W. Cox    Mr. S. A. Evans |
| Hon. Treasurer and Secretary | Dr. Q. A. Geering |

**Corporate Members:**

Agricultural Engineers Association
Agricultural Research Council
Association of Applied Biologists
British Agrochemicals Association Limited
British Society for Plant Pathology
Department of Agriculture and Fisheries for Scotland
Department of Agriculture, Northern Ireland
Department of the Environment
Ministry of Agriculture, Fisheries and Food
National Association of Agricultural Contractors
National Farmers' Union
Natural Environment Research Council
Overseas Development Administration
Society of Chemical Industry, Pesticides Group
United Kingdom Agricultural Supply Trade Association Ltd

**Individual Members:**

Dr. J. T. Braunholtz
Dr. D. H. S. Drennan
Professor J. D. Fryer
Dr. H. C. Gough
Mr. A. G. Harris
Mr. D. J. Higgons
Dr. D. Rudd-Jones

**Hon. Vice-Presidents**

Mr. A. W. Billitt
Mr. M. S. Bradford
Professor L. Broadbent
Mr. J. S. W. Simonds

## The British Crop Protection Council

The British Crop Protection Council exists to promote the knowledge and understanding of crop protection. It was founded in 1968 when the British Weed Control Council, set up in 1953, and the British Insecticide and Fungicide Council, set up in 1962, merged to form a single body concernced with all aspects of crop protection. The BCPC is essentially a British organisation but its work is rapidly becoming international in outlook.

The Council is composed of corporate members including Government bodies, research and advisory services, the farming and agrochemical industries, distribution and contracting services, environmental bodies and other organisations, as well as individual members with special qualifications and experience in the field of crop protection. This blend is probably unique.

## Objectives

Members of The BCPC have a common objective—to promote and encourage the science and practice of pest, disease and weed control, and allied subjects both in the UK and overseas. To achieve this the Council aims:

to compile and arrange the publication of information and recommendations on crop protection for specialists;

to help the public to understand the nature of pests, diseases and weeds, and their control, and the part their control plays in food production;

to provide a forum for discussion at conferences and other meetings on matters relating to crop protection and to publish and distribute the proceedings of these meetings;

to identify short- and long-term requirements for research and development in the field of crop protection;

to act a a liaison agency and to collaborate with other organisations with similar objectives.

Further information about The BCPC, its organisation and its work can be obtained from:

The Administrative Secretary,
The British Crop Protection Council
144-150 London Road
Croydon CR0 2TD

BCPC publications are available from:

'Shirley'
Westfields
Cradley
Malvern
Worcs. WR13 5LP